DICTIONARY OF FOOD
SCIENCE AND TECHNOLOGY

DICTIONARY OF FOOD SCIENCE AND TECHNOLOGY

This *Dictionary* was compiled and edited by the following past or present members of the Editorial & Production team at IFIS Publishing:

L. M. Merryweather (BSc, PhD)
S. H. A. Hill (BSc, PhD)
A. L. Robinson (BA)
H. A. Spencer (BSc)
K. A. Rudge (BSc, PhD)
G. R. J. Taylor (BA, BM, BCh)
A. J. D. Wheatcroft (BSc)
J. J. Weeks (BSc, MSc)
S. S. Sahibdeen (BSc)
P. J. Flynn (BSc, PhD)
M. Blowers (BSc, PhD)
L. I. Murphy (BSc, PhD)
J. T. Murray (BSc, MSc)
B. A. Waites (BSc, PhD)
M. E. Goodchild (BSc)
C. Morgan (BSc)
S. L. Hogarth (BSc, MSc)
D. S. Buckley (BSc, PhD)
F. C. Rogers (BSc)

DICTIONARY OF FOOD SCIENCE AND TECHNOLOGY

Compiled and edited by the
International Food Information Service

Blackwell
Publishing

Blackwell Publishing
Editorial Offices:
Blackwell Publishing Ltd, 9600 Garsington Road, Oxford OX4 2DQ, UK
  Tel: +44 (0)1865 776868
Blackwell Publishing Professional, 2121 State Avenue, Ames, Iowa 50014-8300, USA
  Tel: +1 515 292 0140
Blackwell Publishing Asia, 550 Swanston Street, Carlton, Victoria 3053, Australia
  Tel: +61 (0)3 8359 1011

IFIS Publishing, Lane End House, Shinfield Road, Shinfield, Reading RG2 9BB, UK
Telephone +44 118 988 3895, email ifis@ifis.org, or visit www.foodsciencecentral.com

First published 2005

ISBN 10: 1-4051-2505-5
ISBN 13: 978-14051-2505-5

A catalogue record for this title is available from the British Library

Typeset by IFIS Publishing
Printed and bound in India
By Gopsons Papers Ltd, Noida

The publisher's policy is to use permanent paper from mills that operate a sustainable forestry policy, and which has been manufactured from pulp processed using acid-free and elementary chlorine-free practices. Furthermore, the publisher ensures that the text paper and cover board used have met acceptable environmental accreditation standards.

For further information on Blackwell Publishing, visit our website:
www.blackwellpublishing.com

CONTENTS

FOREWORD

Work on the *Dictionary of Food Science and Technology* started at IFIS Publishing back in the summer of 2000. The original idea was to create the *Dictionary* from the companion *Thesaurus* to the bibliographic database *FSTA – Food Science and Technology Abstracts®* which is also produced by IFIS. Since the *Thesaurus* is compiled on the basis of frequency of current use of terms in *FSTA*, it was felt that it would be an excellent starting point for identifying terms which should also be in a dictionary of food science and technology. All the terms in the *Thesaurus* were reviewed as candidate terms for the *Dictionary* and over the last four years the *Thesaurus* and *Dictionary* have been updated concomitantly.

All the terms in this first edition of the *Dictionary* have been defined by specialist scientific staff at IFIS with the aim of satisfying the needs of academia and the food industry worldwide. It is also hoped that the *Dictionary* will be of value to people working in a number of related fields who may need to learn the meaning of specific terms in food science and technology.

It is probably inevitable that some deserving terms will be missing from the *Dictionary*. We intend to continue to identify candidate terms and would also be pleased to hear from users of the *Dictionary* who may wish to suggest entries for future editions.

Professor Jeremy D. Selman
Managing Director, IFIS Publishing
Shinfield, Reading
January 2005

PREFACE AND GUIDE TO DICTIONARY USAGE

The *Dictionary of Food Science and Technology* has been largely compiled from terms contained in the most recent edition (2004) of the companion *Thesaurus* (ISBN 0 86014 188 8) to the bibliographic database *FSTA – Food Science and Technology Abstracts®*. It, therefore, contains a large number of definitions of terms which are specific to food science and technology (covering sensory analysis, consumer research, food composition, nutrition [food related not clinical aspects], catering, and food safety) and is augmented with definitions of terms from cognate disciplines (including chemistry, biochemistry, physics, microbiology, public health, economics, engineering and packaging). The *Dictionary* also contains a large number of definitions covering food commodities of every description including processed and prepared foods of all types together with: alcoholic and non-alcoholic beverages; fruits; vegetables; nuts; cocoa, chocolate and sugar confectionery products; sugars; syrups; starches; cereals and bakery products; fats; oils; margarine; milk and dairy products (including cream, butter, cheese, cultured milk products etc); eggs and egg products; fish and marine products; meat; poultry; game; additives; spices; and condiments. Whenever appropriate, local names, synonyms and Latin names also appear.

Probably the biggest impact on food science and technology in the last ten years or so has been made by the application of biotechnology in its most modern incarnation; this is reflected in the *Dictionary* by the inclusion of definitions of a large number of related terms such as Gene cloning, Genetic engineering, Gene transfer, Immobilization, Protein engineering, PCR (polymerase chain reaction), and Bioremediation. There are 7852 defined terms in this edition of the *Dictionary*.

Alphabetical order in the *Dictionary* is determined on a letter by letter basis (not word by word) as follows:

Acetates
Acetic acid
Acetic acid bacteria
Acetic fermentation
Acetobacter.

Characters such as numbers, hyphens, primes, subscripts and superscripts are ignored when ordering terms; neither are small capitals, hyphenated modifiers and alphabetic Greek characters used to determine primary alphabetic order. For example, *N*-Acetylglucosamine, D-Amino acids, and 2-Aminobutane all appear under the letter A. Similarly, α-Carotene and β-Carotene both appear under the letter C. The Greek alphabet is given at Appendix A.

Terms in the *Dictionary* are shown in bold type face. Cross references within definitions to other terms appearing in the Dictionary are also shown in bold. For example,

> **Ale** Historically, a **beer** type made without **hops**; in modern usage, a range of British-style beers, commonly brewed with top-fermenting **brewers yeasts.**

Thus, the entry for ale given above shows that the *Dictionary* also contains definitions for the terms beer, hops and brewers yeasts. Similarly, the entry for bacteriocins which follows:

> **Bacteriocins Peptides** produced by specific bacteria that possess **antibacterial activity**. Both purified bacteriocins and bacteriocin-producing bacteria are used in the food industry, applications including inhibition of the growth of **pathogens** and **spoilage** organisms.

indicates that the *Dictionary* also has definitions for the terms peptides, antibacterial activity, pathogens and spoilage.

The definitions in the *Dictionary* have been compiled and edited by specialist scientific staff at IFIS Publishing who also produce *FSTA* and the companion *Thesaurus*. IFIS is an acronym for the International Food Information Service which was founded in 1968. IFIS Publishing is a not for profit organisation (Charity No. 1068176) and a company limited by guarantee (Company No. 3507902).

The *Dictionary* has been compiled to appeal to a wide range of users – not just students of food science and technology and their teachers or researchers in this field or food processors. It is hoped that the *Dictionary* will also become a valuable tool for people working in related fields or anyone who has a general interest in the issues facing the international food sector. The *Dictionary* has been designed to be comprehensive, clear and easy to use. Some deserving terms may be missing and future editions are planned to make good any omissions. We would be pleased to hear from users who may wish to comment on this edition or suggest candidate terms for future editions. Correspondence concerning the *Dictionary* should be addressed to the Head of Publishing, IFIS Publishing, Lane End House, Shinfield Road, Shinfield, Reading RG2 9BB, UK; e-mail: ifis@ifis.org.

A

AAS Abbreviation for **atomic absorption spectroscopy**.

Abalones Marine gastropod **molluscs** belonging to the family Haliotidae, which contains around 70 species; widely distributed, but found mainly in the Western Pacific (Japan and Australia), and also off California and Southern Africa. Only the adductor muscle is edible, having a mild sweet **flavour**; this muscle is normally tenderized to soften the naturally tough, rubbery texture. Marketed in a variety of forms, including powdered, brined and canned products.

Abate Alternative term for the pesticide **temephos**.

Abattoirs Types of **slaughterhouses** where animals are slaughtered for **meat** and **offal**. Abattoirs usually include lairage (a holding area for live animals), a slaughtering line and cold stores. Facilities for processing of by-products (blood, intestines, skins, fat, bristle, unusable waste products), and treatment of waste water and air are often included.

Abondance cheese French semi-hard mountain **cheese** made from **milk** of cows of the breeds Abondance, Montbeliard and Tarine. Characterized by a strong **aroma** and a complex **flavour**. The crust and a grey layer beneath are removed before consumption.

Abreh Alternative term for **abrey**.

Abrey Sudanese non-alcoholic **fermented beverages** made from **sorghum**.

Abscisic acid Plant growth regulator, important in ripening of **fruits** and **cereals**.

Absidia Genus of **fungi** of the class Zygomycetes. Occur as saprotrophs on decaying vegetable matter, grains, soil or dung, and meat, or as **parasites** or **pathogens** of plants or animals. Some species may be used in the production of **chitosan** (e.g. *Absidia coerulea*, *A. glauca* and *A. atrospora*).

Absinthe Liqueur flavoured with **aniseed** and **wormwood**. Widely believed to be neurotoxic as a result of **thujone** derived from wormwood; sale is prohibited in many countries for this reason.

Absorbents Materials or substances that are capable of **absorption**. Uses of absorbents include incorporation within food **packaging** (to absorb oxygen as a preservation technique, to control humidity, and to manage aroma and flavour problems in packaged foods) and for **purification** of foods and beverages, such as drinking water and liquid foods.

Absorption Process involving molecules of one substance being taken directly into another substance. Absorption may be either a physical or a chemical process, physical absorption involving such factors as solubility and vapour-pressure relationships and chemical absorption involving chemical reactions between the absorbed substance and the absorbing medium. Absorption includes such processes as the passage of **nutrients** and other substances from the **gastrointestinal tract** into the blood and lymph, and also the uptake of **water**, **fats** and other substances into foods.

Acaricides **Pesticides** used to control **mites** and ticks (family Acaridae), many of which are responsible for animal diseases and spoilage of stored crops. Commonly used examples are **amitraz**, bromopropylate, **coumaphos** and **fluvalinate**. Residues in foods may represent a health hazard to consumers.

ACC Abbreviation for the plant growth regulator, **1-aminocyclopropane-1-carboxylic acid**.

Acceptability The degree to which the quality of a food is regarded as satisfactory.

Acceptable daily intake A safety level for substances used as **food additives**. The acceptable daily intake (ADI) is usually calculated as $1/100^{th}$ of the maximum dose of the substance that causes no adverse effects in humans.

Acceptance The willingness to regard the quality of a food as satisfactory.

Acephate Systemic insecticide used to control a wide range of chewing and sucking **insects** (e.g. **aphids**, sawflies and leafhoppers) in fruits and vegetables. Classified by WHO as slightly toxic (WHO III).

Acerola Alternative term for **Barbados cherries**.

Acesulfame K Non-nutritive artificial sweetener (trade name Sunett), approximately 200 times as sweet as **sucrose**. Potassium salt derived from acetoacetic acid, with good heat stability and a synergistic effect in sweetener blends. Used in a variety of food applications, including **yoghurt**, table-top **sweeteners**, **soft drinks**, **candy** and other **confectionery**.

Acetaldehyde Aldehyde, synonym ethanal. One of the common **flavour compounds** in many foods and **beverages**. May cause **taints** in some foods. Toxic at excessive concentrations.

Acetals Group of diethers which occur as natural **flavour compounds** in foods such as **fruits** and **herbs**, and **alcoholic beverages**. May be used in **flavourings**.

Acetan Anionic, **xanthan**-like **exopolysaccharides** formed by *Acetobacter* xylinum. Of potential use in **thickeners** or **gelling agents**.

Acetates Salts or esters of **acetic acid**. **Flavour compounds** in many foods and **beverages**. May be used as **preservatives**.

Acetic acid Member of the short chain **fatty acids** group, which occurs in a range of foods and **beverages**. May be one of the **flavour compounds**, or cause **taints**, depending on food or beverage type and the concentration at which it is present. Acetic acid is the main constituent of **vinegar**. It may be used for **preservation** or flavouring of foods.

Acetic acid bacteria Any Gram negative, aerobic, rod-shaped **bacteria**, e.g. *Acetobacter* species and *Gluconobacter* species, capable of oxidizing **ethanol** to **acetic acid**. Occur on the surface of fruits, vegetables and flowers, and in soil. Used industrially in the manufacture of **vinegar**. May cause **spoilage** of **beer** and **wines**.

Acetic fermentation The process by which certain **microorganisms** (e.g. *Acetobacter* and *Acetomonas* spp., and *Gluconobacter* oxydans) metabolize an alcoholic substrate to form **acetic acid**, the main constituent of **vinegar**. Alcoholic substrates can be obtained from a variety of sources, such as **fruits**, **vegetables** and **grain**.

Acetobacter Genus of Gram negative, strictly anaerobic, rod-shaped **acetic acid bacteria** of the family Acetobacteraceae, that are capable of oxidizing **ethanol** to **acetic acid**. Occur on **fruits** and flowers. May be responsible for **spoilage** of **beer** and **wines**. *Acetobacter aceti, A. aceti* subsp. *xylinum, A. acetigenum* and *A. schuezenbachii* are used in commercial production of **vinegar**.

Acetoin Flavour compound found commonly in **dairy products** and **wines**. Synonyms include **3-hydroxy-2-butanone** and **acetylmethylcarbinol**.

α-Acetolactate Precursor of the flavour compound **diacetyl**.

Acetolactate decarboxylases EC 4.1.1.5. These enzymes can be used to reduce the maturation time in **winemaking** by converting acetolactate to **acetoin**, and to analyse **diacetyl** and acetoin concentrations in **beer** during **brewing**. Expression of these enzymes in **brewers yeasts** has also been used to reduce the levels of diacetyl, thus reducing the time required for lagering.

Acetomonas Former name for the genus *Gluconobacter*.

Acetone Smallest of the **ketones**, synonym **propanone**. Widely used as a solvent in food analyses, particularly for lipids and related compounds. Produced along with **butanol** and **ethanol** as a microbial **fermentation** product from unconventional feedstocks including food processing wastes.

Acetophenone Aromatic ketone that is among the **flavour compounds** in foods such as **herbs**, **honeys** and **katsuobushi**.

Acetylacetone Ketone, synonym pentanedione, which occurs in the **flavour compounds** of foods and **beverages**, including **beer**, **coffee** and **fermented dairy products**.

Acetylation Introduction of acetyl groups into a compound or substance. Usually achieved by reaction with acetic anhydride, **acetic acid** or an acetate such as vinyl acetate. Sometimes used to protect hydroxyl groups during organic syntheses. Such modification is also used to alter the **physicochemical properties**, **functional properties** or nutritional quality of substances such as **starch**, **proteins** and **carbohydrates**.

Acetylcholinesterases EC 3.1.1.7. Convert acetylcholine to **choline** and **acetates**. Act on a variety of acetic esters and also possess transacetylase activity. A number of inhibitors of these enzymes are used as **pesticides**. As such, the enzymes can be used to detect the presence of **residues** of these inhibitors in foods and **beverages**.

Acetylene Hydrocarbon which acts as a plant growth regulator and can be used to control **ripening** of **fruits**.

N-Acetylglucosamine Derivative of the amino sugar **glucosamine** in which the amino group is acetylated. Component of cell walls and **chitin**.

Acetylglucosaminidases Comprise α-N-acetylglucosaminidases (EC 3.2.1.50), which hydrolyse terminal non-reducing N-acetyl-D-glucosamine residues in N-acetyl-α-D-glucosaminides, and mannosyl-glycoprotein endo-β-N-acetylglucosamidases (EC 3.2.1.96), which catalyse endo-hydrolysis of the N,N'-diacetylchitobiosyl unit in high-mannose glycopeptides and glycoproteins containing the [Man(GlcNAc)$_2$]Asn structure; one N-acetyl-D-glucosamine residue remains attached to the protein, while the rest of the oligosaccharide is released intact.

N-Acetyllactosamine synthases Alternative term for **lactose synthases**.

Acetylmethylcarbinol Flavour compound found commonly in **dairy products** and **wines**. Synonym of **acetoin**.

N-Acetylneuraminic acid One of the **organic acids**, synonym **sialic acid**. A nitrogen-containing sugar derivative with a carbonyl functional group found ubiquitously in complex carbohydrates.

Acha Species of cereal crop, *Digitaria exilis*, indigenous to West and North Africa and grown for its grain.

Achromobacter Obsolete genus of Gram negative, rod-shaped **bacteria** of the family Achromobactereaceae, in which several parasitic and pathogenic species were formerly grouped. Occur in soil, and fresh and salt water. Some representatives have now been designated to the genera *Acinetobacter* and *Alcaligenes*.

Acid casein **Casein** produced by acid precipitation from **milk** at its isoelectric point, pH 4.7. Acidification can be achieved by direct addition of an acid or through the action of **lactic acid bacteria**.

Acid curd cheese A **cheese** produced by microbial ripening of **quarg**, ripening proceeding from the outside of the cheese. Cultures used include bacteria, fungi and yeasts, the selection depending on the type of cheese being made.

Acidification Process by which the **pH** of a substance is decreased to below 7 making it acidic.

Acidity The degree to which a substance or solution is acidic, being dependent upon the concentration of hydrogen ions. Level of acidity is expressed using **pH**.

Acidocins **Bacteriocins** produced by *Lactobacillus acidophilus*.

Acidophilin **Fermented milk** prepared by fermentation of milk with a mixture of **lactic acid bacteria**, including *Lactobacillus* acidophilus, and **kefir grains**.

Acidophilus milk **Fermented milk** produced by fermentation of milk with *Lactobacillus acidophilus*. Consumption of acidophilus milk has beneficial effects on the intestine.

Acidophilus pastes **Cultured milk products** made using **curd** resulting from milk acidification with *Lactobacillus* acidophilus.

Acid phosphatases EC 3.1.3.2. **Enzymes** with wide specificity which catalyse hydrolysis of orthophosphoric monoesters into an alcohol and orthphosphate. Also catalyse transphosphorylation. Important for **flavour** development, e.g. in **cheese** and also used as a marker of thermal processing in **meat**.

Acid rain Rain which has low **pH** caused by formation of **acids** due to interaction of industrial gas emissions with water. Studies with simulated acid rain have shown adverse effects on yield and quality of exposed crops, especially **fruits** such as **apples**, **pears** and **peaches**. Fruit marketability and composition were affected.

Acids Chemical compounds which release hydrogen ions when dissolved in water, or whose H can be replaced by metal atoms or basic radicals, or which react with bases to form salts and water. Include both **organic acids** and **inorganic acids**. Inorganic acids may be used in food processing or cleaning of equipment. Organic acids of many types are constituents of a wide range of foods, both as natural constituents and as processing aids. Important types of organic acids in foods include **fatty acids**, **amino acids** and **carboxylic acids**.

α-Acids Humulones. The main **bitter compounds** of **hops** resins, used to impart a bitter taste to **beer**. Converted to the more soluble and more bitter **iso-α-acids** during boiling of **worts**.

β-Acids Lupulones. Low-solubility resin constituents in **hops** which have little bittering capacity in **beer**.

Acids resistance Ability of organisms to withstand acidic conditions. Important for survival of **microorganisms** in acid environments such as the gastrointestinal tract and during **fermentation** of foods.

Acidulants **Organic acids** used in foods to control **pH** and fulfil a variety of functions. Applications include **preservation** of **meat** products, flavour enhancement, prevention of discoloration in sliced **fruits**, and prevention of development of **rancidity** in **oils** and **fats**. Commonly used acidulants in the food industry include **citric acid**, **acetic acid**, **propionic acid** and **lactic acid**.

Acid values The level of free **fatty acids** present in **lipids**. The acid value, also known as the acid number, is determined by measuring the amount of KOH in milligrammes that neutralizes 1 g of the lipid. Acid values of fresh edible **fats** tend to be low and increase with storage as the **glycerides** present in the lipids break down to generate free fatty acids.

Acid whey **Whey** produced by acid **coagulation** of **milk** during **cheesemaking**.

Acinetobacter Genus of Gram negative, aerobic, rod-shaped psychrotrophic **bacteria** of the family Neisseriaceae. Occur in soil, water and **raw milk**, and on the surfaces of chilled **meat** and **fish**. Some species may be used in production of **lipases** (e.g. *Acinetobacter radioresistens* and *A. calcoaceticus*).

Ackee Common name for *Blighia sapida*, also known as akee. This fruit was introduced to the West Indies from West Africa and is particularly popular in Jamaica. Fruits are pear shaped and can be consumed raw, cooked, or in canned or frozen forms. Unripe ackee contains hypoglycine A, a toxic amino acid, which can cause the potentially fatal Jamaican vomiting sickness. Levels of hypoglycine A rapidly diminish at maturity, but damaged or fallen fruit should not be consumed.

Aconitic acid One of the **organic acids** found in **sugar cane**. Used in **flavourings** and **acidulants** for the food industry and also in the manufacture of **emulsifying agents**, plastics and detergents.

Acorns **Nuts** obtained from the oak tree (*Quercus* spp.). Widely available, and used as a source of food by some populations, particularly in times of need. Can be pounded into **meal** for use in baked goods or used as **coffee substitutes**. Acorns are high in **starch** and are used in Korea to produce an edible starch gel known as **mook**. They also represent a source of edible **oils**.

Acoustics Study of the physical properties of sound; also refers to techniques based on transmission, generation or reception of sound. Acoustic devices have been used to detect **insects** infestation of grain. Acoustics has also been used in examining the structure of materials, including foods such as **pasta**, and as the basis of a non-destructive method to determine the **firmness** of **fruits**, especially **peaches**.

Acquired immunodeficiency syndrome Epidemic disease commonly abbreviated to **AIDS**.

Acremonium Genus of **fungi** of the class Plectomycetes. May be used in production of **cellulases** (e.g. *Acremonium cellulolyticus* and *A. alcalophilum*).

Acrocomia Genus of **palms**, including *Acrocomia mexicana*, which has edible fruits, and is used as a source of **palm oils** (oil of coyal) and in manufacture of **palm wines**. Also includes *A. sclerocarpa*, which has edible fruits that are used as the source of coconut-like **oils**.

Acrolein Aldehyde, synonym propenal. Member of the **aroma compounds** group, which may cause **off odour** or **off flavour** in overheated **fats** or **spirits**.

Acrylonitrile Monomer used in manufacture of a range of **plastics** used in **packaging materials** or other food contact applications. Acrylonitrile residues may migrate out of plastics items and cause contamination of foods.

F-actin Filamentous **actins**, formed by longitudinal **polymerization** of G-actin (globular actin) monomers. Two strands of F-actin coil spirally around one another to form the superhelix, which is characteristic of actin myofilaments within myofibrils.

Actinidains EC 3.4.22.14. Cysteine **proteinases** found in **kiwifruit** (Chinese gooseberries) with specificity similar to that of **papain**. Also known as actinidins.

Actinidins Alternative term for **actinidains**.

Actinomucor Genus of **fungi** of the class Zygomycetes. Occur as saprotrophs on decaying vegetable matter, soil or dung, or as **parasites** or **pathogens** of plants or animals. *Actinomucor elegans* and *A. taiwanensis* are used in production of East Asian speciality foods, such as **sufu** and **meitauza**.

Actinomyces Genus of Gram positive, facultatively anaerobic **bacteria** of the family Actinomycetaceae. Occur as the normal flora of the mouth and throat or as **pathogens** in humans and cattle. *Actinomyces pyogenes* is the cause of summer **mastitis** in cattle, and can therefore contaminate their **milk**.

Actinomycetales Order of Gram positive, aerobic **bacteria**. Occur in soil, composts and aquatic habitats. Most species are free-living and saprophytic, but some form symbiotic associations and others are pathogenic to man, other animals, and plants.

Actinomycetes Gram positive, aerobic **bacteria** of the order Actinomycetales. Occur in soil and water. Some are plant or animal **pathogens**. May be used in the production of **enzymes** (e.g. **lipases** and **cellulolytic enzymes**).

Actinoplanes Genus of Gram positive, aerobic **bacteria** of the family Actinoplanaceae. Occur in soil, plant litter and aquatic habitats. *Actinoplanes missouriensis* may be used in production of **enzymes** (e.g. **glucose isomerases** and **xylose isomerases**).

Actinospectacin Alternative term for the antibiotic **spectinomycin**.

Actins A family of multifunctional intracellular **proteins**, best known as a myofibrillar component of striated muscle fibres. They constitute about 13% of muscle proteins and are the major components of the I-band or thin filament of the sarcomere. Actins contain high levels of the amino acid **proline**. Imino-groups within proline contribute to the folding of actin molecules and result in formation of G-actin (globular actin). G-actin, a spherical molecule approximately 5.5 nm in diameter, constitutes the monomeric form of actin. In the presence of potassium chloride and **ATP**, G-actin polymerizes into long fibres of F-actin. Most vertebrate genomes contain numerous actin genes with high sequence homology in protein coding regions, but considerable variability in intron size and number. This genetic diversity can be utilized for livestock speciation and meat **authenticity** tests. Determination of actin content has been proposed as a means of calculating the **meat** content of meat products.

Activated C Alternative term for **activated carbon**.

Activated carbon Amorphous forms of elemental carbon, particularly **charcoal**, which have been treated, e.g. by acid or heat, to improve their powers of **absorption**. Used for a variety of food and industrial applications, including drinking water purification, decoloration of sugar solutions and sorption of residues of **pesticides** from **wines**.

Activation energy Minimum energy required for a chemical reaction to proceed; the difference in energy between that of the reactants and that at the transition state of the reaction. Activation energy determines the way in which the rate of a reaction varies with temperature.

Active packaging **Packaging materials** which have functions additional to their basic barrier action. Used for packaging a wide range of foods and **beverages**. Types of active packaging include: packs which adsorb ethylene to control **ripening** of **fruits**; **packs** which regulate moisture levels; packs which contain **oxygen scavengers**; packs which contain CO_2 scavengers or generators; packs which release or absorb flavours or aromas; antimicrobial packaging (e.g. packs which release ethanol to control the growth of fungi); packs with special microwave heating properties; and packaging with monitoring systems (time/temp. exposure indicators or temp. control).

Active sites Locations on the surface of **catalysts** at which reactions occur. On **enzymes**, substrates are bound at the active sites, the shape of the site being important for strong and specific binding to occur.

Actomyosin A complex of the two major muscle **proteins**, **actins** and **myosin**. Actomyosin is formed during muscle contraction with simultaneous hydrolysis of **ATP** to **ADP**. Within myofibrils during contraction, each myosin head region on a thick myofilament attaches to a G-actin molecule within a thin myofilament. This interaction leads to formation of crossbridges between actin and myosin and to formation of the actomyosin complex. Formation of actomyosin results in rigidity and lack of extensibility in muscles. In the presence of ATP, as in living animals, the actomyosin complex dissociates rapidly; however, *post mortem*, actomyosin is the dominant form of myofibrillar protein and it plays a major role in the development of *rigor mortis*. During *post mortem* storage, **tenderness** of **meat** is affected by modification of the actin-myosin interaction. Thermal denaturation of actomyosin occurs at temp. between 30 and 50°C.

Acylamidases Alternative term for **amidases**.

Acylases Alternative term for **amidases** and **aminoacylases**.

Acylation Introduction of acyl groups into a compound or substance. Usually achieved by reaction with an acyl halide or carboxylic acid anhydride. Such modification is used to alter the **physicochemical** properties, **functional properties** or nutritional quality of substances such as **starch**, **proteins** and **sugars**.

Acylglycerols Systematic name for **fatty acid esters** of glycerol, such as **monoacylglycerols**, **diacylglycerols** and **triacylglycerols**. Major components of natural **fats** and **oils** (particularly as triacylglycerols); also used as **emulsifiers**. Synonym for **glycerides**.

Additives Ingredients added in low quantities to foods during processing for one or more specific purposes. These include prevention of chemical and microbial **spoilage**, enhancement of **flavour** or **colour**, improvement of **nutritional values** or as an aid to processing. The most common types of additives include **preservatives**, **colorants**, **sweeteners**, **flavourings**, emulsifiers, **thickeners** and **stabilizers**.

Adenine Purine, synonym 6-aminopurine. Component base of **nucleic acids**, **nucleosides** and **nucleotides**.

Adenosine Nucleoside of adenine and ribose, synonym adenine riboside. Constituent of **nucleotides** and **nucleic acids**.

Adenosine diphosphate Phosphorylated adenoside derivative, and breakdown product of the nucleotide **adenosine triphosphate** (ATP). Level may be used as an indicator of **freshness** in foods such as meat and fish. Usually abbreviated to ADP.

Adenosine monophosphate Nucleotide formed by breakdown of **nucleic acids**, **adenosine triphosphate** (ATP) or **adenosine diphosphate** (ADP). Level may be used as an indicator of **freshness** in foods such as meat and fish. Commonly abbreviated to AMP.

Adenosinetriphosphatases Alternative term for **ATPases**.

Adenosine triphosphate Nucleotide which is important in energy metabolism. Ratios of adenosine triphosphate to its decomposition products may be used as indicators of **freshness** in foods such as **meat** and **fish**. Levels may also be used as an indicator of microbial counts in foods. Commonly abbreviated to ATP.

Adenoviruses Genus of double stranded DNA-containing **viruses** of the family Adenoviridae which can infect mammals and birds. Infection, which can occur via **water** or **shellfish**, may be asymptomatic or result in disease.

Adherence Binding of **microorganisms** specifically or non-specifically to a substratum or to other cells. May be mediated by specialized microbial components or structures (e.g. **adhesins** and prostheca).

Adherence to a particular host tissue is a preliminary stage in **pathogenesis** for many **pathogens**.

Adhesins Bacterial cell surface appendages or extra-cellular macromolecular components that facilitate **adherence** of a cell to a surface or to other cells. Important in the colonization of mucous membranes, e.g. the intestinal mucous membranes by enteropathogenic *Escherichia coli*. Also facilitate adherence of **bacteria** to surfaces such as glass, ceramics and synthetics.

Adhesion Attachment and sticking together of one or more substance. **Adhesives** may be used to promote adhesion, e.g. in **packaging materials**. Sometimes used to refer to **adherence** of **microorganisms** to a substratum or other cells. This may be mediated by specialized microbial components or structures such as **adhesins** or prostheca. This type of adhesion is important for the action of the microorganism, e.g. a preliminary step in **pathogenesis** of **pathogens**.

Adhesives Substances used to stick items together. Most adhesives form a bond by filling in the minute pits and fissures normally present even in very smooth surfaces. Effectiveness of an adhesive depends on several factors, including resistance to slippage and shrinkage, malleability, cohesive strength, and surface tension, which determines how far the adhesive penetrates the tiny depressions in the bonding surfaces.

Adhumulone α-**Acids** fraction of the **bitter compounds** of **hops**.

ADI Abbreviation for **acceptable daily intake**.

Adipic acid Synonym for hexanedioic acid. Used in **acidulants**, antimicrobial **preservatives** or starch-modifying agents. Adipic acid **esters** are used as **plasticizers** in **plastics**.

Adipose tissues **Connective tissues** which function as an energy reserve and insulation layer composed of cells (adipocytes) which synthesize and store large lipid globules.

Adjunct cultures Non-starter cultures used in addition to **starters**, mainly in **cheesemaking**, to produce a specific benefit, e.g. smoother **texture**, improved **flavour** or accelerated **ripening** of **cheese**. In production of **yoghurt**, adjunct cultures have been used to manufacture products with increased levels of nutrients such as **folates**.

Adjuvants Ingredients added to a mixture to improve the effectiveness of the primary ingredient. For example colour adjuvants are used to enhance food **colour**.

Adlay Alternative term for **Jobs tears**.

ADP Abbreviation for **adenosine diphosphate**.

ADPglucose pyrophosphorylases Alternative term for **glucose-1-phosphate adenylyltransferases**.

β-Adrenergic agonists Group of non-hormonal **growth promoters**. Used to enhance growth rates and improve feed efficiency and lean meat content of animals; also used in veterinary medicine as bronchodilatory and tocolytic agents. In general, rapidly excreted from the body; non-authorized use during withdrawal period has resulted in cases of human **food poisoning**. Banned for use as growth-promoting agents in farm animals in many countries, including European Union member states and the USA. Commonly used examples are **cimaterol**, **clenbuterol**, rimiterol and terbutaline.

Adsorbents Substance that are capable of **adsorption**. Used widely in the food and biotechnology industries. Uses include removal of unwanted materials in foods and beverages that affect either food safety or food quality. Examples include removal of **proteins** from **white wines**, **pathogens** from **drinking water** sources, **radioelements** from foods, oxidation products from **frying oils** allowing oil recovery and reuse, and **bitter compounds** from **fruit juices**. Amongst other applications are: isolation of compounds with potential for use in foods; immobilization of enzymes; as agents in analytical techniques such as gas analysis and **chromatography**; and removal of unwanted **aroma** and **flavour** in packaged foods.

Adsorption **Adhesion** of the molecules of liquids, gases and dissolved substances to the surfaces of solids, in contrast to **absorption**, in which the molecules actually enter the medium. Adsorption is employed in **hydrogenation** of **oils**, in gas analysis, and in **chromatography**.

Adulteration Addition of substances to foods, or substitution of food ingredients with inferior substances, with the intent of lowering the quality of and costs of producing the food and defrauding the purchaser, e.g. addition of starch to **spices**, and of water to **milk** or **beer**.

Adzuki beans Common name for seeds produced by *Vigna angularis,* also known as azuki beans. Small red beans with a mild, sweet **flavour**, which are widely cultivated in Japan and China. Traditionally consumed boiled, ground into meal or used to make sweet bean pastes known as **ann** or an. Seeds may also be germinated to produce **bean sprouts**.

Aerated confectionery **Confectionery** produced with incorporation of air as an ingredient. Use of air adds bulk to the product without increasing its weight, improving product **texture** and **flavour**. Aeration of confectionery results in a range of products with densities ranging from 0.2 to 1.0 g/cm^3. Such products include chews, mallows, honeycomb, **mousses** and **meringues**.

Aeration Introduction of air into a product to enhance **texture**, **mouthfeel**, **rheology** and visual appeal. The following methods are used to aerate foods: **fermentation**; **whipping** or shaking of low-medium viscosity liquids; **mixing** of doughs or high viscosity pastes, in which air bubbles are entrapped as surfaces come together; steam generation during slow to moderate **cooking**, **baking** or **frying**; entrapment of air between sheeted layers, as in **pastries** and **croissants**, or between pulled strands, as in pulled taffy and **candy**; frying in very hot oils, such that internal steam rapidly forms, causing the product to puff; use of chemical **raising agents** such as **baking powders** or **sodium bicarbonate**; rapid dry heating of small or thin products to induce blistering or slight puffing; gas injection (e.g. air, **carbon dioxide**, nitrogen and nitrous oxide); expansion **extrusion**; pressure beating (dissolution of air or gas under pressure in a syrup, fat mixture or chocolate); **puffing**, in which products such as **breakfast cereals** containing superheated moisture are subjected to a sudden release of pressure; and vacuum expansion, followed by rapid cooling to set the expanded products.

Aerobacter Obsolete genus of Gram negative, rod-shaped **bacteria** of the family **Enterobacteriaceae**, the species of which have now been reclassified into the genera **Enterobacter** and **Klebsiella**.

Aerobes Organisms that require atmospheric oxygen to live. Often refers to aerobic **bacteria** or other **microorganisms**. Facultative anaerobes are aerobes that can also grow under anaerobic conditions.

Aerococcus Genus of Gram positive, coccoid **bacteria** of the family Streptococcaceae. *Aerococcus viridans* may be used for the production of **lactate 2-monooxygenases**.

Aerolysins Cytolytic **toxins** secreted by *Aeromonas hydrophila*. Form channels in cell lipid bilayers, leading to destruction of the membrane permeability barrier and osmotic lysis.

Aeromonas Genus of Gram negative, facultatively anaerobic rod-shaped **bacteria**. Occur in salt and fresh water, sewage and soil. *Aeromonas hydrophila*, frequently found in **fish** and **shellfish** and occasionally in red **meat** and **poultry meat**, may cause septicaemia, meningitis and **gastroenteritis** in humans.

Aerosol packs Containers for pressurized liquids, which are released in the form of a spray or foam when a valve is pressed. Aerosol propellants, usually liquefied gases, are used in the packs. Used as **dispensers** for a variety of foods.

Aerosols Substances, including foods, stored under pressure in a container (for example in aerosol cans) containing a propellant and released as a fine spray or froth. Also, in a chemical sense, suspensions of submicroscopic particles dispersed in air or gas.

Afalon Alternative term for the herbicide **linuron**.

Affination The first stage in processing of raw **sugar**, in which the layer of mother liquor surrounding the crystals is softened and removed. Raw sugar is mixed with a warm, concentrated syrup of slightly higher purity than the syrup layer so that it will not dissolve the crystals. The resulting magma is centrifuged to separate the crystals from the syrup, thus removing the greater part of the impurities from the input sugar and leaving the crystals ready for dissolving before further treatment. The liquor which results from dissolving the washed crystals still contains some colour, fine particles, gums and resins and other non-sugars.

Affinity chromatography **Chromatography** technique in which an immobilized ligand is used to retain an analyte that is later eluted under conditions where the binding affinity is reduced. The ligand, which may be a substance such as an enzyme, hormone or antigen, is bound to a matrix such as silica.

Aflatoxicosis **Mycotoxicosis** caused by ingestion of **aflatoxins** in contaminated foods or feeds.

Aflatoxin B_1 A potent hepatocarcinogen. Toxic to many species, including humans, birds, fish and rodents.

Aflatoxin B_2 Moderate hepatocarcinogen compared with **aflatoxin B_1**.

Aflatoxin B_3 Hepatocarcinogen with a **toxicity** similar to that of **aflatoxin B_1**.

Aflatoxin D_1 Carboxylated product of **aflatoxin B_1**. Possesses less **toxicity** than aflatoxin B_1.

Aflatoxin G_1 Potent hepatocarcinogen with a **toxicity** similar to that of **aflatoxin B_1**.

Aflatoxin G_2 Dihydroxylated product of **aflatoxin G_1**. Possesses less **toxicity** than aflatoxin G_1.

Aflatoxin M_1 Metabolic product of **aflatoxin B_1** in animals. Usually excreted in the **milk** of cattle and other mammalian species that have consumed aflatoxin B_1-contaminated foods or feeds. Possesses less **toxicity** than aflatoxin B_1.

Aflatoxin M_2 Metabolic product of **aflatoxin B_2** in animals. Usually excreted in the **milk** of cattle and other mammalian species that have consumed aflatoxin B_2-contaminated foods or feeds. Possesses less **toxicity** than aflatoxin B_2.

Aflatoxin P_1 Demethylated and hydroxylated product of **aflatoxin B_1** found in animals. Very weak toxin compared to aflatoxin B_1.

Aflatoxin Q_1 Main metabolite of **aflatoxin B_1** found in humans and primates.

Aflatoxins **Mycotoxins** produced by certain strains of the fungi *Aspergillus* *flavus* and *A. parasiticus*. Formed during the growth of these **fungi** on commodities such as **corn**, **peanuts**, **cottonseeds** and **soybeans**. Hepatotoxic and hepatocarcinogenic in humans and animals.

African breadfruit seeds Kernels of **fruits** produced by the tree *Treculia africana*. Eaten roasted as **nuts** or ground into **meal** which is used to fortify foods or to prepare a type of **porridge**.

African locust beans Seeds produced by *Parkia filicoidea* or *P. biglobosa*. Not eaten raw, but fermented to produce food **flavourings** or protein-rich **iru** or **dawadawa**. The yellowish pulp surrounding the seeds can also be eaten, either raw or as an ingredient in **soups**, stews and **beverages**.

African mangoes Common name for the African tree species, *Irvingia gabonensis*. Also known as bush mango or wild mango. **Fruits** resemble cultivated **mangoes**, but they are botanically unrelated. Pulp of the fruit is eaten fresh or used for the preparation of products such as juices and jams. **African mango seeds**, also known as **dika nuts**, have a variety of food uses.

African mango seeds Seeds from the tropical African tree *Irvingia gabonensis* which are rich in **fats** and are used in Africa to make **bread** as well as a type of **butter**.

African nutmeg Seeds of the African tree, *Monodora myristica*. Used as **spices** in Nigeria and other parts of Africa.

African oil beans Edible **oilseeds** of the leguminous tree *Pentaclethra macrophylla*, native to tropical Africa. Cooked seeds are fermented to produce **ugba**.

African spider herb Common name for *Cleome gynandra*, also known as cat's whiskers. The plant grows wild in most tropical countries, and is mainly consumed as a leafy vegetable. Leaves are a rich source **vitamin A**, **vitamin C** and **minerals** such as **calcium** and **iron**. Leaves also contain **glucosinolates** and **phenols**, which can impart **astringency**.

African yam beans **Beans** produced by *Sphenostylis stenocarpa*. Popular grain legume of West Africa and other areas of tropical Africa. Beans have a distinctive **flavour** and are high in **starch** and moderately high in **proteins**. Prolonged cooking time is recommended to inactivate **antinutritional factors** present in the beans. The plant also produces edible **tubers**.

Aftertaste A **flavour**, often unpleasant, that lingers in the mouth after a food has been swallowed.

Afuega'l Pitu cheese Unpasteurized Spanish cheese usually made mainly from **cow milk**. Fresh red **chil-**lies are added to the cheese and more are rubbed into the rind as the cheese is allowed to mature, giving the rind a buff to deep orange colour. The rind also has a dusting of white mould.

Agar Extract obtained from various species of red **seaweeds** belonging to *Eucheuma*, *Gelidium* and *Graciliria* genera. Contains **agarose** and agaropectin **polysaccharides**. Sets following dissolution in warm water to form **agar gels**, which are widely used as **thickeners** and **stabilizers** in the food industry. Additionally used in **gelling agents** to prepare culture media for bacteriological plate counts. Also known as agar-agar.

Agar-agar Alternative term for **agar**.

Agarases EC 3.2.1.81. **Enzymes**, produced often by marine **bacteria**, which catalyse hydrolysis of 1,3-β-D-galactosidic linkages in agarose, forming the tetramer as the predominant product. **Oligosaccharides** formed have potential for use in foods. Also act on porphyran.

Agar gels Gels formed by dissolving **agar** in water. Widely used as **thickeners** and **stabilizers**, e.g. in **ice cream**, **soups**, **jellies**, **sauces**, **glazes** and **meat** products.

Agaricus Genus which includes some **edible fungi**, such as the widely cultivated common mushroom, *Agaricus bisporus,* which is sold commercially in flat, cup or button forms. Other edible species include the wild **mushrooms** *A. campestris* (field mushroom) and *A. arvensis* (horse mushroom).

Agaritine Genotoxic substance present in raw **mushrooms**.

Agarose Neutral gelling fraction of **agar**, a complex polysaccharide produced by **algae** of the class Rhodophyceae. Composed of alternating β-D-galactose and 3,6-anhydro-β-L-galactose **sugars**. Uses in foods include as **thickeners** and **gelling agents**, e.g. in canned **meat**, **sugar confectionery**, **bakers confectionery** and **dairy products**. Also used as a matrix for electrophoretic separation of large molecules, most frequently **DNA**.

Agave Plants of the genus *Agave*, the flowers, leaves, stalks and sap of which are used as a source of food or beverages. **Starch** in buds is converted into sugar causing a sweet nectar to be exuded from the flowers. Sap is used to make a refreshing beverage or can be boiled to make **sugar syrups**. **Fermentation** of the sap produces **vinegar** or the alcoholic beverage **pulque**. Fermented sap from *A. tequilana* is distilled to make **tequila**.

Ageing Process in which properties change over time. Ageing includes the intentional storage of foods and beverages to induce desirable changes in sensory properties, such as for wines and cheeses (also **ripening**). The term is also used to denote the artificial hastening of this process, such as treatment of flour with ammonium persulfate to produce a more resilient dough.

Agglomerates   Masses or collections of particles or items.

Agglomeration   The process by which particles or items are collected together and formed into a mass.

Agglutination   The clumping together of cells, such as **bacteria**, due to cross-linking by proteins such as **antibodies**. Agglutination is utilized in **immunological techniques** for detecting bacteria in foods. In food processing, however, agglutination of **starters**, such as those used in the manufacture of **dairy products** including certain **cheese** varieties, can have detrimental consequences for the process outcome.

Agglutination tests   **Immunological techniques** in which **antigens** on the surface of particulate material, such as **bacteria**, or inorganic particles, such as latex, are precipitated with **antibodies**. Antibodies react with the antigens causing the cells to clump together and form visible aggregates or agglutinates. Applications include detection of *Escherichia coli* O157:H7.

Agglutinins   Substances, such as **antibodies** and **lectins** found in plant seeds, which cause **agglutination** of cells to form clumps.

Aggregation   The process for forming a whole by combining several different elements or items.

Agitation   The process of **stirring**, shaking or disturbing briskly, particularly applied to a liquid.

Aglycones   The part of a **glycosides** molecule which is not a sugar residue, e.g. the **anthocyanidins** component of **anthocyanins**.

Agmatine   One of the **biogenic amines**, which occurs in a wide range of foods, including **fish**, **cheese** and **alcoholic beverages**. Concentrations in foods may increase with increasing storage time.

Agricultural produce   Collective name for crops and other commodities obtained as a result of agriculture and used for provision of food, fibre or other materials. Examples include fruits, cereals, cotton and livestock. Used in a similar way to the term **agricultural products**.

Agricultural products   Term used in a similar way to **agricultural produce**. Collective name for crops and other commodities obtained as a result of agriculture and used for provision of food, fibre or other materials. Examples include fruits, cereals, cotton and livestock.

Agrobacterium   Genus of Gram negative, aerobic, rod-shaped **bacteria** of the family Rhizobiaceae. Occur in soil. Typically plant pathogens that form galls or tumours on roots or stems. *Agrobacterium rhizogenes* causes hairy root, *A. rubi* causes cane gall and *A. tumefaciens* causes crown gall.

Agrocybe   Genus including **edible fungi** such as *Agrocybe cylindracea,* a mushroom with similar characteristics to matsutake (*Tricholoma matsutake*), *A. aegerita* and *A. parasitica.*

AIDS   Common abbreviation for acquired immunodeficiency syndrome, an epidemic disease caused by infection with **human immunodeficiency viruses** (HIV) and spread through direct contact with body fluids. The HIV retroviruses cause immune system failure. There is concern over their possible transmission to infants of infected mothers through breast feeding.

Aiele fruits   Olive-like **fruits** produced by the aiele tree (*Canarium schweinfurthii*) which are widely consumed in West African countries. Pulp and kernel are rich in **oleic acid** and **palmitic acid**. **Oils** produced from the fruits show similarities to **olive oils**. Also known as African black olives, mbeu or black fruit.

Air drying   Removal of moisture or liquid from a substance using air, or to preserve an item by evaporation.

Airflow properties   Characteristics of the **flow** of air through, or across the surface of, a substance or piece of equipment. Airflow properties are utilized in designing **ovens** and **driers** and in determining the most appropriate ways of storing large quantities of foods such as fruits, vegetables, cereals and carcasses in order to minimize **spoilage**.

Airline meals   **Meals** provided for consumption during aircraft travel, designed to be served and consumed in a limited amount of space. Menu items are prepared and packaged at a central location either by the **catering** branch of the airline company or a contracted **foods service** operator. Chilled or frozen items are then reheated in special ovens during the flight.

Air quality   Measure of the condition of the air, especially with respect to the requirements for specific environments. In food processing and packaging facilities, air quality is important for food safety and **shelf life**, and health of personnel. Special filtration systems are used to remove airborne hazards such as **microorganisms**, **insects** and dust from the atmosphere.

Ajowan   Common name for the umbelliferous plant, *Trachyspermum ammi* (syn. *Carum copticum*). Cultivated in parts of Egypt and Asia for its pungent, aromatic seeds, typically used in **flavourings** for Indian foods. Related to **caraway** and **cumin**, but with a strong flavour of **thyme**. Also used as a source of **thymol**.

Akamu **Cereal products** produced by boiling the starchy extract from fermented **corn**, **millet** or **sorghum** until complete **gelatinization** occurs.

Akara Deep fried **pastes** made from **cowpeas**, seasoned and flavoured with chopped **capsicums**, **onions** and salt. Popular foods in West Africa, where they are consumed as **snack foods**, side dishes or **fast foods**. Steamed cowpea paste is known as **moinmoin**.

Alachlor Selective systemic herbicide used preemergence to control annual grasses and many broadleaved weeds among many types of vegetables and nuts. Classified by WHO as slightly toxic (WHO III).

Alanine One of the non-essential **amino acids**, occurring in most food proteins.

Alar Alternative term for the plant growth regulator **daminozide**.

Alaska pollack Commercially important **marine fish** species (*Theregra chalcogramma*) belonging to the cod family (Gadidae); widely distributed in the Pacific Ocean. Flesh has a moderate to low fat content and a mild, slightly sweet **flavour**. Normally marketed in frozen form and processed into fillets, blocks and **surimi**, but is also sold fresh or as a cured product. Also known as **walleye pollack**.

Albacore **Marine fish** species (*Thunnus alalunga*) belonging to the **tuna** family which is widely distributed in tropical and temperate waters. Flesh is lighter in **colour** and has a milder **flavour** than that from other tuna species. Widely considered to be the best tuna species for canning, but is also marketed fresh, smoked and frozen.

Albendazole Anthelmintic widely used in sheep and cattle for treating roundworms and flukes. Along with its various metabolites, is normally depleted rapidly from edible tissues and milk.

Albumen Alternative term for **egg whites**.

Albumins Proteins which are soluble in water or dilute salt solutions and coagulable by heat. Albumins occurring in foods include **conalbumin**, **lactalbumins** and **ovalbumins**.

Alcaligenes Genus of Gram negative, aerobic, rodshaped, **bacteria** of the family Achromobacteraceae. Occur in the intestinal tracts of vertebrates, soil, water, **milk**, and as part of the normal skin flora. May be involved in food **spoilage**. *Alcaligenes viscolactis* may cause **ropiness** in milk.

Alcohol Common name for **ethanol**, especially in the context of **alcoholic beverages**.

Alcohol dehydrogenases Group of **enzymes** catalysing the oxidation of **alcohols**. Alcohol dehydrogenases (EC 1.1.1.1) catalyse the oxidation of alcohols to **aldehydes** or **ketones** with concomitant re-

duction of NAD^+. Also known as aldehyde reductases, these enzymes act on primary and secondary alcohols, and also on hemi-acetals. Catalyse the final step of **alcoholic fermentation**. Alcohol dehydrogenases ($NADP^+$), EC 1.1.1.2, catalyse the oxidation of alcohols to aldehydes with concomitant reduction of $NADP^+$. Some members act only on primary alcohols, while others also act on secondary alcohols. Alcohol dehydrogenases ($NAD(P)^+$), EC 1.1.1.71, catalyse the oxidation of alcohols to aldehydes with concomitant reduction of $NAD(P)^+$. Reduce aliphatic aldehydes of carbon chain length 2-14, with greatest activity on C_4, C_6 and C_8 aldehydes. Also known as retinal reductases, since they can reduce retinal to retinol. Alcohol dehydrogenases (acceptor), EC 1.1.99.8, catalyse the oxidation of primary alcohols to aldehydes in the presence of an acceptor.

Alcohol free beverages **Beverages** of types normally containing **ethanol**, which have been formulated or processed to be free from ethanol.

Alcoholic beverages **Beverages** containing a significant concentration of **ethanol**. Major types include **beer**, **wines**, **spirits**, **liqueurs** and **rice wines**.

Alcoholic fermentation Process by which certain **microorganisms** (mainly **yeasts**) metabolize sugars anaerobically to produce **alcohols**. In this process, **glucose** is converted to **pyruvic acid**, which is decarboxylated to **acetaldehyde**. The acetaldehyde is subsequently reduced to **ethanol**. A wide variety of substrates can be used to produce **alcoholic beverages**, e.g. **grain** for production of **beer**, and **grapes** and other **fruits** for production of **wines**. However, the constituent **sugars** must be released from these substrates prior to **fermentation**. Fermentation can be carried out by endogenous yeasts or by addition of **starters**. The most common yeasts used in the manufacture of alcoholic beverages are *Saccharomyces cerevisiae* and *S. carlsbergensis*. Synonymous with **ethanolic fermentation**.

Alcoholic soft drinks **Beverages** with **flavour** and other properties typical of **soft drinks** (e.g. **fruits** flavoured beverages), but with addition of a significant concentration of **alcohol**.

Alcohol reduced beer **Beer** in which the **ethanol** content has been reduced.

Alcohol reduced beverages Beverages in which the **ethanol** content has been reduced.

Alcohol reduced wines Wines in which the **ethanol** content has been reduced.

Alcohols Alkyl or **aromatic compounds** containing a hydroxyl (OH) group. Classes of alcohols important in the context of foods include aliphatic alcohols, e.g. methanol, **ethanol** and higher alcohols, polyols, glycols, aromatic alcohols, terpene alcohols and sterols.

Al compounds Alternative term for **aluminium compounds**.

Aldehyde dehydrogenases Include members of subclass EC 1.2. **Dehydrogenases** which catalyse oxidation of **aldehydes** to the corresponding acids. In most cases, the acceptor is NAD^+ or $NADP^+$. Used in techniques to determine aldehyde levels in foods and **beverages**.

Aldehyde reductases EC 1.1.1.21. **Enzymes** with wide specificity, catalysing the conversion of **alditols** and $NAD(P)^+$ to the corresponding aldoses and $NAD(P)H$. Can be used to convert **xylose** to **xylitol**, useful as a food sweetener.

Aldehydes **Carbonyl compounds** containing the CHO radical. Many are important for **flavour** or **off flavour** in foods and **beverages**. Aldehydes formed by oxidation of **fatty acids** are important sources of flavour deterioration of **lipids** rich foods.

Aldicarb Systemic insecticide, acaricide and nematocide used for control of chewing and sucking **insects** (especially **aphids**, whitefly, leaf miners and soil-dwelling insects) in a wide range of fruit and vegetable crops. Classified by WHO as extremely hazardous (WHO Ia).

Alditols General term for **polyols, sugar alcohols** produced by reduction of **sugars** on an aldehyde group. Examples of alditols include D-**sorbitol**, D-**mannitol** and **xylitol**.

Aldolases Alternative term for **fructose-bisphosphate aldolases**.

Aldose 1-epimerases EC 5.1.3.3. Convert α-D-glucose to β-D-glucose but also act on L-arabinose, D-xylose, D-galactose, maltose and lactose. Have been used extensively as components of **biosensors** for analysis of **sugars**. Also known as mutarotases and aldose mutarotases.

Aldrin Cyclodiene insecticide used to control root worms, **beetles** and termites in soil around fruit and vegetable crops. A potent neurotoxin, which has been banned for use on crops in most countries; very stable, persisting in soil for many years.

Ale Historically, a **beer** type made without **hops**; in modern usage, a range of British-style beers, commonly brewed with top-fermenting **brewers yeasts**.

Aleurone Layer of cells found under the **bran** coat and outside the endosperm of cereal grains. Rich in **cereal proteins** and minerals as well as containing non-digestible carbohydrates and **phytic acid**.

Alewife Marine **fish** species (*Alosa pseudoharengus*) belonging to the **herring** family (Clupeidae); occurs in marine and estuarine waters along the Atlantic coast of North America. Marketed in fresh, dried/salted, smoked and frozen form; popularly consumed as a fried product.

Alexandrium Genus of **dinoflagellates** responsible for outbreaks of **paralytic shellfish poisoning**. Common species include *Alexandrium catenella, A. minutum* and *A. tamarense*.

Alfalfa Common name for the leguminous plant, *Medicago sativa*, also known as **lucerne**, generally grown as a fodder plant, although young leaves and **alfalfa sprouts** can be used as a vegetable, e.g. in Chinese cooking.

Alfalfa seeds **Seeds** produced by **alfalfa** (*Medicago sativa*) which are germinated to make **alfalfa sprouts** for human consumption. Sprouts are generally eaten raw in **sandwiches** and **salads**.

Alfalfa sprouts Crisp sprouts obtained by **germination** of **alfalfa** seeds. Popular in **salads** and **sandwiches**.

Al foils Abbreviation for **aluminium foils**.

Algae A heterogeneous group of unicellular and multicellular eukaryotic photosynthetic organisms which most occur in aquatic habitats. Certain algae are harvested for commercial production of **thickeners** (e.g. **agar, alginates, carrageenans**) or proteins (e.g. **single cell proteins**). A few algae produce **toxins** that may cause **food poisoning** in humans via consumption of **fish** and **shellfish**.

Algicides Chemicals used to control growth of algae in water bodies or water containers. Examples include bethoxazin, dichlone, quinoclamine and **simazine**.

Alginate gels **Gels** derived from **alginates**. Calcium alginate gels are commonly used for immobilization of **biocatalysts**.

Alginate lyases Alternative term for **poly(β-D-mannuronate) lyases**.

Alginates Any of several derivatives of **alginic acid** (e.g. sodium, calcium or potassium salts or propylene glycol alginate). Used as **stabilizers, thickeners** and **gelling agents** in foods.

Alginic acid Polysaccharide (polymer of D-mannuronic acid) obtained from brown algae such as *Macrocystis pyrifera* or *Laminaria*. Possesses significant hydrocolloidal properties making it suitable for thickening, emulsifying and stabilizing applications. Authorized for use in foods in various forms, including as sodium, calcium and potassium **alginates**.

Alicyclobacillus Genus of Gram positive, aerobic or facultatively anaerobic, rod-shaped, spore-forming **bacteria**. *Alicyclobacillus acidoterrestris* and *A. acidocaldarius* may cause **spoilage** of **fruit juices**.

Alimentary pastes Alternative term for **pasta**.

Alitame High intensity dipeptide sweetener formed from L-aspartic acid, D-alanine and a novel amine. It has good water solubility, and sweetness is approximately 2000 times that of **sucrose** at typical usage levels. Alitame has the potential for use in a broad range of foods and **beverages**, such as **dairy products, frozen desserts** and **chewing gums**, and in tabletop **sweeteners**.

Alkalies Bases which are soluble in water and include the strongly basic hydroxides of sodium, potassium or ammonium. Neutralize, or are neutralized by, **acids**. Solutions have a pH higher than 7. Alkalies are used in the food industry during **processing** (e.g. **peeling** of **potatoes**) or in **cleaning** applications. Alternative spelling is alkalis.

Alkaline phosphatases EC 3.1.3.1. Catalyse formation of orthophosphate and an alcohol from an orthophosphoric monoester, and also catalyse transphosphorylation. Enzymes with wide specificity. Uses include analysis of **tannins** in **grapes** and **red wines**, detection of the adequacy of **pasteurization** of **milk** and **dairy products**, and detection of **phosphates** in **drinking water**.

Alkalinity The degree to which a substance is alkaline. Level of alkalinity is expressed using **pH**.

Alkalization Process by which the **pH** of a substance is increased to above 7 making it alkaline.

Alkaloids Organic nitrogenous bases. Many have pharmacological activity. Some foods contain toxic **alkaloids**, e.g. **solanine** in **potatoes**. Some alkaloids are desirable food constituents, e.g. the purine alkaloids **caffeine** and **theobromine** in **tea, coffee, chocolate** and **cocoa**.

Alkanes Saturated **hydrocarbons** of the methane series, including methane, ethane, propane and butane.

Alkylresorcinols **Phenols** with antifungal activity found in **rye** and other **cereals**, cashew nut shells and some **bacteria** and **algae**. Similar in structure to commercially used **antioxidants** such as **BHA** and **BHT**. Like other resorcinolic lipids, display biological properties and have been reported also to have **antitumour activity, antimicrobial activity** and antiparasitic activity.

Alleles Alternative forms of **genes** or **DNA** sequences that occupy the same position (locus) on either of two homologous **chromosomes** in a diploid organism. If both chromosomes have the same allele, then the organism is homozygous for this allele. If the allele is different, the organism is heterozygous for this particular allele.

Allergenicity The ability of a substances to act as **allergens**.

Allergens **Antigens** that are capable of inducing an allergic reaction when they come in contact with specific tissues of susceptible individuals. Allergens may induce formation of reaginic **antibodies**. Common food allergens include **proteins** from **shellfish, nuts, eggs, fish** and **milk**.

Allergies Hypersensitivity states induced by the body in reaction to foreign **antigens** that are harmless to other individuals in similar doses. Allergic reactions are of four basic types and can be immediate or delayed in their onset. Type I reactions, which involve release of histamine from mast cells by immunoglobulin E, can be induced by many food allergens often resulting in respiratory and dermatological symptoms. Severe type I reactions include anaphylaxis. Most foods have been demonstrated to produce allergic reactions in certain individuals, however, common causes of food allergy in adults include **shellfish, nuts** and **eggs**. In children, the pattern of food allergy differs from that in adults, with allergies to eggs, **milk, peanuts** and **fruits** being common. In contrast to adults, children can outgrow allergies, especially to milk and **soy infant formulas**.

Allicin One of the **organic sulfur compounds** occurring in **onions** and other *Allium* spp. **vegetables**. Important **flavour compounds** fraction with antibacterial properties.

Alligator meat **Meat** from **alligators**. Most of the **meat** from alligator carcasses is in the tail; however, jaw meat is favoured because it has a very low content of **fats** and the muscle fibres are shorter. Usually, alligator meat is trimmed heavily of fat because the fat has an unpleasant flavour. Each carcass includes both light and dark meat. In comparison with free-range alligator farming, indoor farming may be associated with an increased prevalence of salmonellae. Due to biomagnification, **alligators** living in polluted areas can accumulate substantial concentrations of **heavy metals**.

Alligator pears Alternative term for **avocados**.

Alligators Large semi-aquatic predatory reptiles in the genus *Alligator* of the family Alligatoridae. There are two species, namely the American alligator (*A. mississippiensis*) and the Chinese alligator (*A. sinensis*). They are hunted or farmed (free-range or indoor production systems) for **alligator meat** and skins.

Alliin One of the **organic sulfur compounds** contributing to the **flavour compounds** fraction in **garlic** and *Allium* spp. **vegetables**.

Alliinases Alternative term for **alliin lyases**.

Alliin lyases EC 4.4.1.4. Also known as alliinases, these enzymes are found in **onions** and **garlic**, where they are responsible for formation of the characteristic **flavour**. They also catalyse formation of **allicin**, thought to have a number of health benefits. Have been used to determine **alliin** contents in garlic extracts.

Allium Genus of low-growing perennial plants, that includes cultivated vegetables such as **onions**, **leeks**, **shallots** and **garlic**, and many wild edible species. Noted for their distinctive **flavour** and **pungency**, due to the presence of **organic sulfur compounds** such as **alliin**. These compounds are also associated with the therapeutic properties noted for garlic and other *Allium.*

Allspice Spice obtained from the dried **fruits** of the tropical tree, *Pimenta officinalis* (syn. *P. dioica*). **Flavour** resembles a blend of **cinnamon**, **cloves**, **nutmeg**, **ginger** and **pepper**. Used in **flavourings** for **meat** products and **bakery products**. Also known as **pimento** or Jamaican pepper.

Allura Red General-purpose, water-soluble artificial colorant. Also known as **FDC red** 40. Used to impart a reddish-yellow colour to foods such as **desserts**, **confectionery** and **cereal products**.

***S*-Allylcysteine** Sulfur containing amino acid which is one of the major **organic sulfur compounds** in **garlic**. Responsible in part for some of the health benefits of garlic, including **hypolipaemic activity**, **anticarcinogenicity** and **radical scavenging activity**.

Allyl isothiocyanate Naturally occurring volatile **organic sulfur compounds** found in *Brassica* vegetables and some other plants, such as **cassava**. Largely responsible for the **pungency** of foods such as **mustard** and **horseradish**. Possess antimicrobial properties and are used in food **preservatives** and as antifermentative agents in **winemaking**. Like other **isothiocyanates**, display goitrogenic properties.

Almond oils Oils rich in **oleic acid** and low in cholesterol derived mainly from the seeds of bitter **almonds** (*Prunus dulcis*). Used in **cooking** and in foods as well as in the cosmetics industry.

Almonds One of the most widely grown type of **nuts**. Produced on the tree *Prunus dulcis* (syn. *P. amygdalus, Amygdalus communis*). **Sweet almonds** (*P. dulcis* var. *dulcis*) are grown for their edible nuts which are important ingredients in many **confectionery** products, such as **marzipan**, **macaroons** and **sugar almonds**. Bitter varieties (*P. dulcis.* var. *amara*) are cultivated for the **almond oils**, which are used as **flavourings**.

Aloe Plants of the genus *Aloe* (family Lilaceae), such as aloe vera. Used in the manufacture of foods, **beverages**, and pharmaceutical and cosmetic products due to their characteristic **flavour**, **aroma** and **biological activity** (attributed mainly to the presence of aloins).

Aloin Bitter tasting compound which is a major component of **aloe** leaves. An anthroquinone which on its own is used as a laxative but which also displays **antifungal activity** and analgaesic effects.

Alternan **Glucans** fraction derived from **fungi** of the genus ***Alternaria***. Has potential for use in **thickeners** or **stabilizers** for foods.

Alternansucrases EC 2.4.1.140. **Glycosyltransferases** that transfer α-D-glucosyl residues to the non-reducing terminal residues of α-D-glucans, producing glucans with alternating α-1,6- and α-1,3- linkages. Enzyme from ***Leuconostoc*** *mesenteroides* produces **alternan**, a glucan with potential applications in **food additives**.

Alternaria Genus of **fungi** belonging to the class Hyphomycetes. Occur in soil and vegetable matter. Many species are pathogenic to plants. *Alternaria solani* may cause early **blights** of **potatoes** and **tomatoes**. Some species may produce **mycotoxins**, e.g. alternariol and alternariol-monomethyl-ether, on foods such as **rice**, **fruits** and **vegetables**.

Alteromonas Genus of Gram negative, aerobic, rod-shaped **bacteria**. Occur in coastal and marine habitats. *Alteromonas nigrifaciens* may cause **spoilage** of **fish** and **meat**.

Alum Double salts of aluminium sulfate and the sulfate of a monovalent metal. Alums are used as coagulating agents in **purification** of **drinking water**. Other food applications include use in **colorants**, manufacture of cured **jellyfish** products, modification of properties of **starch**, and use in processed fruit products and vegetable products.

Aluminium Light metal, chemical symbol Al, which may be used in food packs or food processing equipment. Occurs in the **trace elements** fraction in the diet; there is no known nutritional requirement. There is concern that excessive intake may be toxic, and dietary aluminium has been implicated as a causative factor in Alzheimer's disease.

Aluminium compounds Chemical compounds of **aluminium**. May be food constituents, **additives** or **contaminants**. There is concern about possible adverse health effects of high intakes of aluminium compounds via foods or **beverages**.

Aluminium foils Aluminium **packaging materials** which are used to decorate, protect and preserve foods, providing a barrier to external factors, such as light, oxygen and water vapour. Food applications include: foil containers and lids; metallized films; and wrappings. Also used in laminated packaging to enhance the barrier properties and rigidity of other packaging materials such as **plastics** and **paper**. There is very little migration of aluminium from aluminium foil containers into food. Environmental considerations include the importance of **recycling** and the use of aluminium foil laminates to fuel incineration processes.

Aluminium phosphide Synonym for **phostoxin**. Used in **fumigants** for stored grain, as it releases the toxic gas **phosphine**.

Alveograms Records of air pressure inside bubbles formed by inflating pieces of **dough** until rupture, a test performed on **alveographs**.

Alveographs Apparatus used to analyse the **physical properties** of **dough** and the **baking properties** of **wheat**. A piece of dough is inflated using air until it forms a bubble and bursts. Traces of the pressure inside the bubble (**alveograms**) are used to indicate dough strength, stability and distensibility.

Amadori compounds Intermediates of the **Maillard reaction** occurring between amino groups and **reducing sugars**. Amadori compounds are produced by rearrangement of nitrogen-containing carbohydrate ring structures and their fate is dependent on the conditions present in the reaction medium. Acid hydrolysis of these compounds can result in unsaturated ring systems that have a characteristic **flavour** and **aroma**, which under less acidic conditions may polymerize to form an insoluble dark-coloured material.

Amala Traditional Nigerian paste-like product made by reconstituting yam meal in boiling water. Sometimes fortified with legume meal, e.g. **cowpea meal** or **soy meal**, to improve the protein content and nutritional quality. Typically, amala is dark brown in **colour** and is eaten with **soups**.

Amanita Genus of soft, fleshy fungi, which includes both edible and highly poisonous species. Edible species include *Amanita rubescens*, which should not be eaten raw, and *A. caesarea*. Care should be taken in the identification of these mushrooms as many cases of poisoning have occurred due to unintentional ingestion of related, lethal species, such as *A. phalloides* (death cap mushroom).

Amanitins Class of **amatoxins**. Also known as amanitoxins or amantines.

Amaranth Red food **colorants** stable to light, made from small, pigmented flowers of plants of the genus *Amaranthus*.

Amaranth flour **Amaranth grain** that is milled for food use.

Amaranth grain Seeds from plants of the genus *Amaranthus*, which are high in **starch**, protein, **lysine** and minerals. Also known as grain amaranth.

Amaranth starch **Starch** extracted from **amaranth grain**. Most commonly utilized in parts of South America, Africa and Asia where amaranth is cultivated as a food crop.

Amaranthus Genus of dicotyledenous plants of the family Amaranthaceae. Certain species of *Amaranthus* are grown for **amaranth grain** or **grain amaranth**, which is high in **starch**, **proteins**, **lysine** and **minerals**. Other species are grown for their **spinach**-like leaves, which are good sources of protein, **vitamin C**, minerals and **β-carotene**.

Amasi Traditional Zimbabwean **fermented milk** resembling thick **curd**. **Fermentation** is performed at ambient temperature and naturally fermented cream may be added to improve **viscosity**. Often eaten with stiff corn **porridge**.

Amatoxins Powerful **toxins** produced by several species of **mushrooms** of the genus *Amanita* (such as *Amanita phalloides* (Death Cap), *A. virosa* (Destroying Angel) and *A. verna* (Fool's Mushroom). Ingestion results in abdominal pain, persistent vomiting and watery diarrhoea, usually followed by death due to organ failure.

Ambaritsa Raw **dry sausages**, traditionally made in Bulgaria. They are prepared primarily from **pork**, but include smaller amounts of **beef**. Moisture content should be <33% (by wt.).

Amberjack Alternative term for **yellowtail**.

American groundnuts Common name for seeds produced by *Apios Americana*, a legume native to North America, which also produces small edible **tubers**. The tubers can be dried and ground into a powder which is added to **flour** or used in **sweeteners** and **thickeners**.

American lobsters **Lobsters** of the species *Homarus americanus*. Found in the north Atlantic Ocean. Also known as Atlantic lobsters or true lobsters.

Ames test Technique used to assess the **mutagenicity** of chemicals. Samples are incubated in medium containing liver homogenate and derivatives formed are mixed with a mutant strain of *Salmonella* Typhimurium that lacks autotrophic properties towards **histidine**. These properties are restored by metabolic derivatives formed in the sample during incubation in the presence of liver enzymes.

Amidases EC 3.5.1.4. Convert monocarboxylic acid amides to monocarboxylates and **ammonia**. Have been used for production of D-alanine from DL-alaninamide.

Amides **Organic nitrogen compounds** containing the $CO.NH_2$ radical which are common constituents of foods. Include **capsaicin** and **urea**.

Amine oxidases Two enzymes: EC 1.4.3.4 (flavin-containing), also known as monoamine oxidases and tyramine oxidases; and EC 1.4.3.6 (copper-containing), also known as diamine oxidases. The former act on primary, and usually secondary and tertiary, **amines**

to form **aldehydes**, while the latter act on primary monoamines, diamines and **histamine**. Several **bacteria** are able to degrade **biogenic amines** through production of diamine oxidases and these enzymes have been used in **biosensors** for determination of biogenic amines in foods.

Amines **Organic nitrogen compounds** derived from NH_3 by substitution of organic radicals for the H atoms. Depending on whether 1, 2 or 3 H atoms are replaced, they are classed as primary, secondary or tertiary amines. Include a wide range of compounds important for **flavour** and **aroma** of foods. Amines are formed during breakdown of proteins and contribute to the characteristic odour of spoiled foods such as fish. **Biogenic amines** such as **histamine** may be toxic.

Amino acids **Organic acids** characterized by possession of one or more COOH and NH_2 groups. Amino acids are the main constituents of proteins. 10 amino acids (arginine, histidine, isoleucine, leucine, lysine, methionine, phenylalanine, threonine, tryptophan and valine) are essential nutrients in the human diet.

D-**Amino acids** Amino acid **enantiomers** with a specific configuration around a chosen chiral element, usually the α-carbon atom. These **amino acids** have the opposite configuration to L-amino acids. Many D-amino acids are naturally occurring in **microorganisms**, plants and animals, and some are of especial interest for the synthesis of novel **sweeteners**.

Aminoacylases EC 3.5.1.14. Hydrolyse *N*-acyl-L-amino acids, releasing the corresponding L-amino acids. Can be used for purification of L-amino acids from racemic mixtures of the corresponding *N*-acyl-DL-amino acids. Can also be used for acylation of **amino acids** in organic solvents.

Aminobenzoic acid Aromatic acid used in antimicrobial **preservatives** for use in foods.

2-Aminobutane Alternative term for (*RS*)-sec-butylamine, a fungicide used for fumigation of seed and ware potatoes for control of **skin spot** and gangrene. Also used as a fungicidal dip or spray on harvested fruit (e.g. control of blue and green moulds on **citrus fruits**, and stem rot of **oranges**).

1-Aminocyclopropane-1-carboxylate oxidases Catalyse the final step in **ethylene** biosynthesis in higher plants, converting **1-aminocyclopropane-1-carboxylic acid** (ACC) to ethylene, and are involved in **ripening** of **fruits**.

1-Aminocyclopropane-1-carboxylate synthases EC 4.4.1.14. Catalyse the rate-limiting step in **ethylene** biosynthesis in higher plants which leads to **ripening** of **fruits**.

1-Aminocyclopropane-1-carboxylic acid Plant growth regulator important in **ripening** of **fruits**. Often abbreviated to ACC.

Aminoethanol Synonym for ethanolamine. Amine which in pure form exists as a colourless, combustible, hygroscopic liquid with an **aroma** of ammonia. A member of the **biogenic amines** group, which occurs in various foods, including **wines** and **cheese**.

Aminoethoxyvinylglycine Plant growth regulator which acts by blocking **ethylene** synthesis through inhibition of **1-aminocyclopropane-1-carboxylate synthases**.

Amino N Nitrogen which is present in foods and other substances in the form of amino (NH_2) groups.

α-Amino N Index of the amino acid N content of foods, **beverages** or their raw materials and intermediate materials, e.g. in **brewing**.

Aminopeptidases EC 3.4.11. Exoproteinases that hydrolyse peptide bonds and remove **amino acids** one at a time from the chains of proteins, working from the amino terminus. Used for reducing the **bitterness** of proteolytic hydrolysates, and important in **flavour** development in **dairy products** and **meat**.

Amino sugars General term for **sugars** substituted with an amino group at the carbon-2 position. Examples of amino sugars include **galactosamine**, **glucosamine** and **furosine**, an important indicator of **Maillard reaction** in **dairy products**.

Amitraz Non-systemic acaricide and insecticide used for control of **mites**, scale insects, whitefly and **aphids** on pome fruits, citrus fruits and vegetables such as capiscums and tomatoes. Classified by WHO as slightly toxic (WHO III).

Amla **Fruits** of the sub-tropical deciduous tree *Emblica officinalis* Gaertn. (syn. *Phyllanthus emblica*), also known as aonla or Indian gooseberry. Fruit are usually processed into products such as **pickles**, **fruit juices** and **syrups**, as the raw fruit is highly acidic and astringent. Amla is a rich source of **vitamin C** and also contains **tannins**, **alkaloids**, **auxins** and **minerals**. Reported to have hypocholesterolaemic and **antioxidative activity** and is widely used in traditional Indian medicine.

Ammonia Gas, chemical formula NH_3, which is formed on breakdown of nitrogen-containing compounds such as **proteins**, **peptides** and **amino acids**. Has a characteristic pungent odour and is toxic at high concentrations in air. May be used as in **refrigerants** for **freezing** or **cooling** systems.

Ammonium compounds Group of compounds containing the NH_4 radical. In the context of foods, important members include **betaine**, inorganic ammonium salts (e.g. ammonium bicarbonate used as a leavening agent, and ammonium salts used as nutrients for **yeasts**) and **quaternary ammonium compounds** used as **disinfectants**.

Amnesic shellfish poisoning Disease resulting from ingestion of **shellfish** (commonly **mussels**) containing the neurotoxin **domoic acid** (produced by certain toxigenic marine diatoms). Symptoms include abdominal cramps, vomiting, disorientation and memory loss.

Amoebae Common name for a number of species of unicellular, usually microscopic, organisms of the order Amoebida and the class Sarcodina. Occur in fresh and salt water, moist soil, and as **parasites** in humans and animals. Characterized by ability to alter their shape, generally by the extrusion of one or more pseudopodia.

Amoebiasis Specifically refers to an infection of the intestine, liver, or other sites with *Entamoeba histolytica*, a pathogenic amoeba, acquired by ingesting contaminated water or foods. In general, may be any infection caused by any amoebic parasite. Characterized by severe bloody diarrhoea, abdominal pain, fever, vomiting and ulceration of the colon. Also known as amoebic dysentery.

Amoxicillin Penicillin antibiotic used against a wide variety of bacterial infections in farm animals. Becomes widely distributed in animal tissues following administration, but is rapidly eliminated; typically undetectable in **livers** and **kidneys** of animals 5 days after withdrawal.

Amoxycillin Alternative spelling for **amoxicillin**.

AMP Abbreviation for **adenosine monophosphate**.

Amperometry Technique based on measurement of current resulting from oxidation or reduction of an electroactive species. A constant potential is maintained at a working electrode or on an array of electrodes with respect to a reference electrode. The current is correlated with the content of the electroactive species.

Ampicillin Broad-spectrum semisynthetic penicillin antibiotic used in the treatment of several diseases in cattle, swine, sheep and poultry. Rapidly excreted, primarily in unchanged form in the urine; relatively small amounts are excreted in milk.

Amycolatopsis Genus of Gram positive aerobic **bacteria**, type species *Amycolatopsis orientalis*, of the order **Actinomycetales**. Isolated from soil, vegetable matter and clinical specimens. Some species produce **antibiotics** or biotechnologically significant **enzymes**. One strain has been used in **biotransformations** to produce **vanillin** from **ferulic acid**.

Amygdalin **Glycosides** fraction present in **bitter almonds** which is hydrolysed by water to yield **hydrocyanic acid** and **benzaldehyde**.

Amyl alcohol Synonym for **pentanol**. One of the higher alcohols, comprising five carbon atoms and a single alcohol group. Of importance in the **flavour compounds** fraction of **alcoholic beverages**. Forms part of the toxic **fusel oils** fraction of **spirits**. Used as a solvent and as a substrate for production of the flavouring amyl acetate.

Amylases **Enzymes** that hydrolyse the α-1,4 glycosidic linkages in both **amyloses** and **amylopectins**. Act on **starch**, **glycogen**, and related **polysaccharides** and **oligosaccharides**. Specific types are **α-amylases** and **β-amylases**.

α-Amylases EC 3.2.1.1. Catalyse endohydrolysis of 1,4-α-D-glucosidic linkages in **polysaccharides** containing three or more 1,4-α-linked D-glucose units. Act on **starch**, **glycogen**, and related polysaccharides and **oligosaccharides** in a random manner; reducing groups are liberated in the α configuration. Used to convert starch to **dextrins** in the production of **corn syrups**, as a flour supplement to aid **yeasts** growth and gas production in **dough** making, and for solubilization of **brewing adjuncts**.

β-Amylases EC 3.2.1.2. Hydrolyse 1,4-α-D-glucosidic linkages in **polysaccharides**, removing successive **maltose** units from the non-reducing ends of the chains. Act on **starch**, **glycogen**, and related polysaccharides and **oligosaccharides**, producing β-maltose by an inversion reaction. Used for production of high maltose **syrups**.

Amylases inhibitors Substances that inhibit the activity of **amylases** (including **α-amylases** and **β-amylases**) which catalyse the breakdown of **starch** into **sugars**. **α-Amylases inhibitors** present in foods can act as **antinutritional factors** by inhibiting the breakdown of starch into sugars by amylases present in the saliva and pancreatic secretions.

α-Amylases inhibitors Components of foods that inhibit **α-amylases**, enzymes present in saliva and pancreatic secretions which are involved in digestion of dietary **starch**. Presence of α-amylase inhibitors in starch-rich foods can reduce the rate of starch digestion and release of glucose into the bloodstream. Types of α-amylase inhibitor include proteins of higher plants (such as cereals and legumes), and polypeptides and nitrogen-containing carbohydrates produced by *Streptomyces* spp.

Amyloglucosidases Alternative term for **glucan 1,4-α-glucosidases**.

Amylograms   Records of results obtained using **amylographs** to investigate **flour** or **starch viscosity** as a function of temperature.

Amylographs   Instruments used to measure the **viscosity** of cereal flours or other **starch**-based products during variations in temperature. Samples are mixed at a constant speed and viscosity is recorded on charts (**amylograms**).

Amylolytic enzymes   Term that encompasses α-**amylases**, glucan **1,4-α-glucosidases**, β-**amylases**, **α-dextrin endo-1,6-α-glucosidases** and **α-glucosidases**.

Amylopectins   High molecular weight polymers that, together with **amyloses**, form **starch**. Composed of α-1,4-linked glucopyranose chains connected by α-1,6-linkages. 3-6% of glucose residues are α-1,6-linked, giving rise to a highly branched polymer. Starch that is almost exclusively composed of amylopectin is termed waxy, e.g. waxy corn (>99% amylopectin and <1% amylose); in starch of this type, **retrogradation** is slow or absent, thus pastes of gelatinized waxy starch are non-gelling but gum-like.

Amyloses   **Polysaccharides** composed of chains of α-1,4-linked glucopyranose residues that, together with **amylopectins** are constituents of **starch**. Amyloses have much lower molecular weights than amylopectins (at least 100-fold less) and are non-branched. In contrast to amylopectins, **retrogradation** of cooked amyloses is rapid, and thus gel formation occurs.

Amylovorins   Small, heat-stable and strongly hydrophobic **bacteriocins** synthesized by *Lactobacillus amylovorus*. Show a relatively narrow inhibitory spectrum, mainly against related *Lactobacillus* species, although some species of *Clostridium* and *Listeria* are also sensitive.

Amyrin   Triterpene **alcohols** fraction which occurs in the unsaponifiable fraction of some **fats**, and may be used as a marker of origin or **authenticity** of fats (e.g. for detection of **cocoa butter substitutes** in **chocolate**).

An   Alternative term for **ann**.

Anabaena   Genus of filamentous **cyanobacteria**. Some species, such as *Anabaena flos-aquae* and *A. circinalis*, can form algal blooms in fresh water, producing **anatoxins** which are **neurotoxins**.

Anabolic agents   Natural and synthetic hormonal-type growth promoting substances. Most are derivatives of reproductive steroid **hormones** (**oestrogens, progesterone** and **testosterone**); non-steroidal compounds (naturally or non-naturally occurring) such as **zeranol** and stilbene oestrogens are also available. Widely used in many countries to promote weight gain and feed efficiency in farm animals (prin-cipally in cattle); their use is not permitted in the EU, although many types may be used illegally. Also known as **anabolic drugs**.

Anabolic drugs   Chemical substances based on natural or synthetic growth promoting hormones. Most are derived from reproductive **steroids** (**oestrogens, progesterone** and **testosterone**) while a few are based on polypeptide hormones (e.g. recombinant **bovine somatotropin**). Used to promote weight gain and feed efficiency in farm animals; use is not permitted in the EU, although illegal use has been reported.

Anabolic steroids   **Anabolic agents** derived from or similar in structure to reproductive steroid **hormones**. Examples of naturally produced steroids used in animal production include oestradiol-17β and **progesterone** (female steroids), and **testosterone** (male steroid); synthetic examples include melengestrol acetate and **trenbolone acetate**. Used to promote growth and feed conversion efficiency in a range of farm animals. Anabolic steroids are banned for use in animal production in the EU, although many are used illegally.

Anacystis   Genus of **cyanobacteria** of the Cyanophyceae family. *Anacystis nidulans* may be used in the production of **single cell proteins**.

Anaerobes   Organisms that do not require atmospheric oxygen to live, or cannot survive in the presence of oxygen. Often refers to anaerobic **bacteria** or other **microorganisms**. Facultative **aerobes** refer to anaerobes that can also grow under aerobic conditions.

Analogues   In relation to foods, products that are made to resemble and act as substitutes for specific commodities. Similar to **simulated foods**. Reasons for producing analogues include to provide alternatives to **meat** for vegetarians, for consumption by those with special dietary requirements or to reduce costs.

Analysers   Instruments used in analysis.

Analytical techniques   Methods used in analysis.

Anaphylaxis   A severe type I allergic reaction occurring rapidly in sensitized individuals following exposure to small amounts of **allergens**. Symptoms can range from itching and angioedema to widespread tissue oedema, airway constriction, respiratory distress and circulatory collapse. Foods that can induce anaphylaxis include **peanuts, eggs** and **sea foods**.

Anardana   Dried seeds of wild **pomegranates** (*Punica granatum*). Added in **condiments** or **acidulants** to a number of Indian foods including **chutneys** and **curries**.

Anasazi   Ancient variety of *Phaseolus vulgaris*, reintroduced onto the market following the successful cultivation of samples discovered in a New Mexico cave. The purple and white **beans** have a delicate **flavour**, similar to that of **pinto beans**, and a relatively low content of indigestible sugars compared with other beans.

Anatoxins **Neurotoxins** produced in fresh water by some species of filamentous **cyanobacteria** of the genus *Anabaena*, especially *A. flos-aquae*. Include the **alkaloids** anatoxin-a and anatoxin-a(s). Extremely poisonous, sometimes killing animals drinking contaminated water within a few minutes. May represent a hazard for **drinking water** safety.

Anchoveta Small **herring**-like fish which occurs abundantly in Pacific waters off the western coast of South America. Anchoveta (*Engraulis ringens*) are a commercially important source of **fish meal** and **fish oils**.

Anchovy Group of **herring**-like **marine fish** species belonging to the family Engraulidae. Commercially important species include European anchovy (*Engraulis encrasicolus*), northern anchovy (*E. mordax*) and Japanese anchovy (*E. japonica*). Anchovy are marketed in fresh, dried, smoked, canned and frozen forms and are also used to make **anchovy pastes**.

Anchovy oils **Oils** derived from the muscle of *Engraulis* spp. which are rich in **eicosapentaenoic acid** and **docosahexaenoic acid**.

Anchovy pastes Processed **fish** products comprising ground **anchovy** (*Engraulis* and *Anchoa* spp.) mixed with ingredients such as **vegetable oils** and **seasonings**. Often used in **toppings** for **pizzas** and as a component of pasta **sauces** and **salad dressings**.

Androlla Dry cured **pork sausages** traditionally made in Galicia, Spain.

Androstenone Steroid hormone with a characteristic odour; implicated in **boar taint** occurring in **pork** produced from non-castrated male swine.

Anencephaly A lethal neural tube defect characterized by the absence of the cranial vault and the majority or all of the cerebral and cerebellar hemispheres. Anencephaly results from failure of the neural tube to close during embryogenesis. The risk for developing anencephaly, as with other **neural tube defects**, is reduced by increasing the level of **folic acid** in the maternal diet during pregnancy.

Anethole Synonym for *p*-allylphenyl methyl ether. One of the **flavour compounds** which occurs in **herbs** and **spices**, especially **anise** and **fennel**.

Aneurin Alternative term for **thiamin** (**vitamin B₁**), used commonly in Europe. Also spelt **aneurine**.

Aneurine Alternative spelling for **aneurin**.

Angel cakes Very light, airy **sponge cakes** made with stiffly beaten **egg whites** and no **egg yolks** or **fats**. Also known as angel food cakes.

Angelica Herb obtained from umbelliferous plants of the genus *Angelica*, particularly *A. archangelica*, which is grown extensively in southern Europe. The young **celery**-like stalks are crystallized and used for decorating **cakes** and **confectionery** products. Leaves are occasionally used for flavouring stews, while roots and seeds are used as **flavourings** for some types of **gin** and **liqueurs**, respectively.

Angiotensin I-converting enzymes Alternative term for **peptidyl-dipeptidase A**.

Angkak Red pigment produced by fermentation of rice with *Monascus* spp. Used in natural food **colorants** in the Far East.

Anhydrous milk fats **Milk fats** with a very high fat content and negligible moisture content. Sometimes called water free milk fats.

Aniline Synonym for aminobenzene or phenylamine. Toxic **amines** fraction which is used in chemical syntheses, e.g. for dyes. Aniline may occur as a contaminant in foods.

Animal carcasses Dead bodies of animals, particularly those used for **meat** production. The term is used by butchers to describe animal bodies after removal of the heads, limbs, hides and offal; these processed carcasses are also called dressed carcasses. Major animal carcass meats in Europe and the USA are produced from **cattle**, **sheep** and **swine**, whilst in the Middle East, Africa and Asia, water **buffaloes**, **camels** and **goats** are more important. Conditioning or ageing of carcasses results in break down of muscle **glycogen** into **lactic acid**, which tends to improve **tenderness** and **shelf life** of **meat**.

Animal diseases Pathological conditions that occur in animals that are used as sources of foods and may affect the quality or safety of the foods. Examples that affect food quality or safety include **mastitis** and **malignant hyperthermia**.

Animal fats Lipid products derived from animal sources. Include **butter**, **lard**, **tallow**, **suet** and **fish oils**.

Animal foods Foods derived from sources in the animal kingdom. Examples include **aquatic foods** (**sea foods** and **aquaculture products**), **dairy products**, **eggs** and egg products, **animal fats**, **insect foods**, **meat** and meat products and other animals such as worms.

Animal models Animals used to simulate human physiological and pathological processes. Animal models allow investigations that would not be ethical or practical in humans.

Animal proteins Proteins that are derived from animal sources, such as **meat**, **fish**, **eggs** and **dairy products**.

Animal rennets Proteinases present in the abomasum of young ruminants, e.g. calves, and used for clotting of milk during **cheesemaking**. Comprise a mixture of the main enzyme, **chymosin**, and pepsin, the ratio of the two enzymes affecting the final properties of the cheese. Due to shortages of animal rennets and the increasing popularity of vegetarian cheeses, **microbial rennets**, genetically-engineered enzyme preparations synthesized by various microorganisms and **milk clotting enzymes** of plant origin (**vegetable rennets**) have been developed.

Animal science Discipline relating to the science and technology of the production, management and distribution of animals, including those intended for food use.

Animal tissues Alternative term for **meat**.

Animal welfare Protection of the rights of animals, whether in the wild or in captivity. For animals used in agriculture as food sources, conditions (and possibly food quality) can be improved by high quality care and humane use. Implementation of high standards of care for animals used in research is believed to improve the quality of the resultant scientific data.

Anion exchange Type of **ion exchange** in which hydrogen ions and anions may be displaced from the ion exchange resin.

Anions Negatively charged particles that have gained one or more electrons. Anions migrate towards positively charged electrodes (anodes).

Anisakiasis Infection in humans caused by the third larval stage of the parasitic nematode *Anisakis simplex*, usually as a result of eating contaminated raw or undercooked **sea foods**. *Pseudoterranova* larvae have also been implicated as causative organisms. Also known as **anisakidosis**.

Anisakidosis Alternative term for **anisakiasis**.

Anisakis Genus of parasitic nematodes of the family Anisakidae. *Anisakis simplex* has been implicated in **anisakiasis**, an infection caused by consumption of contaminated raw or undercooked **sea foods**.

Anisaldehyde Common name for *p*-methoxybenzaldehyde. One of the **flavour compounds** occurring in a wide range of foods.

Anise Alternative term for **aniseed**.

Aniseed Liquorice-flavoured, fragrant seeds of *Pimpinella anisum*. Used as **spices** and **flavourings** for many foods and **beverages**, including **confectionery** and **alcoholic beverages** such as **anisette**.

Anisette Aniseed-flavoured **liqueurs** manufactured in France.

Anisole Phenolic compounds which occur naturally in a range of foods. Chlorinated anisoles derivatives may cause **taints**, e.g. in **corks** and **wines**.

Ann Traditional Japanese **bean jams** used as the base for many **confectionery** products. Usually made from **adzuki beans**, although other beans may be used. Typically prepared by boiling and pounding the beans and adding **syrups** to form a paste.

Annatto Yellowish red natural colorant obtained from seeds of the tropical tree *Bixa orellana*. Contains a fat-soluble component (**bixin**) and a water-soluble component (**norbixin**). Used to add **colour** to **cheese**, **sausage casings** and **bakery products**.

Annealing To heat an item and allow it to cool slowly, so as to remove internal stresses.

Anserine Synonym for N-β-alanyl-1-methylhistidine. Peptide which occurs in **fish** and **meat**, and may contribute to their **sensory properties**.

Antelope meat Meat from **antelopes**, sometimes referred to as **venison**. Antelope meat has a lower fat content than lean beef, but has a similar content of essential amino acids. It may be cooked by roasting, but requires basting to prevent the meat from becoming too dry.

Antelopes Various species of swift running, deer-like, hollow-horned, hoofed ruminant mammals of the subfamily Antilopinae. The major well-known species include elands, gnus, **gazelles** and **impala**. Many are hunted for their **meat** and some species, for example the blackbuck antelope (*Antilope cervicapra*), have been farmed successfully to produce **antelope meat** of a high quality.

Anthelmintics Drugs used to treat internal infections of animals caused by parasitic worms (**nematodes** and **cestodes**). Most frequently used in younger farm animals which are more susceptible to parasitic infections. Residues are most likely to be found in milk when withdrawal periods have not been strictly observed; livers may also contain residues. Examples include **albendazole**, **dichlorvos**, **ivermectin** and **thiabendazole**.

Anthocyanidins Flavylium salts which are the aglycone component of **pigments** of the **anthocyanins** group.

Anthocyanins Class of organic **pigments** (**glycosides** of **malvidin**, **pelargonidin**, **peonidin**, **cyanidin**, **delphinidin** and **petunidin**) giving pink, red, blue and purple **colour** to many foods and **beverages** of plant origin (including **fruits** and **red wines**). Extracted anthocyanins may be used as food **colorants**. Colour is pH-sensitive, and stability differs from that of **artificial colorants**.

Anthocyanogens Alternative term for **leucoanthocyanins**, **anthocyanins** found in a range of plant foods, and also in **wines**. In a polymerized form, constituents of **polyphenols** and condensed **tannins**.

Anthracene One of the **polycyclic aromatic hydrocarbons** (PAH). Occurs as an environmental contaminant in a wide range of foods, water and **packaging materials**. May also be formed during **smoking** or **cooking** of foods.

Anthracnose Any of several **plant diseases** caused by **fungi** (*Colletotrichum* spp.). Characterized by dark spots that appear on leaves, stems or **fruits**. One of the main postharvest diseases, affecting the quality of stored produce such as **bananas**, **citrus fruits** and **mangoes**.

Anthraquinones **Pigments** of the **quinones** group which occur in a range of plants and plant products.

Antiatherogenic activity Ability of foods or food components to slow, inhibit or reverse the process of **atherosclerosis**, the pathological process underlying cardiovascular disease. Consumption of foods possessing antiatherogenic activity is potentially beneficial for health as a result of the consequent decreased risk for **cardiovascular diseases**.

Antibacterial activity Ability to kill or inhibit the growth of **bacteria**.

Antibacterial compounds Compounds that possess **antibacterial activity**, e.g. certain **antibiotics**, **antiseptics** and **disinfectants**.

Antibiotics Substances produced by **microorganisms** that can kill or inhibit other microorganisms; used to treat bacterial and fungal infections in humans and animals. Grouped into several different classes, the most widely used being **β-lactam antibiotics** (including **penicillins** and **cephalosporins**). Other classes include aminocyclitols, aminoglycosides, amphenicols, macrolides, nitrofurans and **quinolones**. Residues may occur in animal foods; toxic effects are unlikely, but potential hazards include allergic responses in consumers and development of resistant strains of **bacteria**.

Antibiotics resistance Ability of **microorganisms** to be unaffected by treatment with specific **antibiotics**. Resistance can result from a range of mechanisms, including decreased permeability of the organism to the drug, modification of drug or receptor, and production of a modified protein that is unaffected by the antibiotic. Organisms can become resistant either by undergoing spontaneous mutations or by acquiring resistance genes from other resistant organisms through the processes of conjugation and transduction. **Plasmids** containing multiple resistance genes can be transferred not only amongst similar but also quite different **bacteria**.

Antibodies **Proteins**, also known as **immunoglobulins**, that are produced by the body in response to foreign substances (**antigens**) and are capable of forming complexes with the antigens. Mechanisms by which antibodies protect the body include **agglutination** or precipitation of foreign antigens, lysis of foreign cells, and neutralization of **toxins**.

Anticaking agents Anhydrous compounds that are added in small amounts to dry foods (e.g. **salt**, **baking powders**, **pudding mixes**) to prevent the particles **caking** together and thus ensure the product remains dry and free-flowing. Typical anticaking agents for the food industry include magnesium and calcium carbonates, magnesium stearate, calcium silicate and calcium stearate.

Anticarcinogenicity Ability of a food or food component to slow, inhibit or reverse the process of **carcinogenesis**, in particular, the ability to attenuate carcinoma formation in response to application of known **carcinogens**. Anticarcinogenicity of a substance can be determined *in vitro* using cell culture or *in vivo* using animals treated with carcinogens or a carcinoma cell line.

Anticarcinogens Substances that inhibit the formation of carcinomas induced by application of **carcinogens**. Potential dietary anticarcinogens include **phytoestrogens** (**isoflavonoids**, **lignans**), **flavonoids**, **lycopene**, **glucosinolates**, **terpenes**, allyl sulfides and simple **phenols**.

Antifoaming agents Used in a similar manner to **defoaming agents** to control **foams** formation during food processing. Examples include **dimethylpolysiloxane**.

Antifreeze proteins **Proteins** occurring naturally in a range of organisms (especially cold water **fish**), which prevent or minimize freezing of tissues on exposure to low temperatures. Of potential use in the food industry for lowering the **freezing point** of foods and inhibiting recrystallization of ice. Possible applications include in **ice cream**, **frozen foods** or chilled meat products.

Antifungal activity Ability to kill or inhibit the growth of **fungi**.

Antifungal agents Substances that possess **antifungal activity**.

Antifungal compounds Compounds that possess **antifungal activity**.

Antigenicity Ability of substances to act as **antigens** by eliciting an antibody-mediated or cellular **immune response**.

Antigenotoxicity Measure of the ability to prevent damage to **DNA** caused by genotoxins.

Antigens Substances that induce an **immune response**, either by stimulating formation of **antibodies** or eliciting a cellular response.

Antihypertensive activity Ability of a substance to alleviate or reduce high blood pressure (**hypertension**). Food components that demonstrate antihypertensive activity often act as inhibitors of angiotensin I-converting enzyme (ACE; **peptidyl-dipeptidase A**). Potential dietary antihypertensive agents include **bioactive peptides** in **dairy products**, **adenosine** in plant foods and **garlic** constituents.

Anti-inflammatory activity Ability to inhibit or counteract the inflammatory response, which is an innate **immune response** to tissue injury by stimuli such as chemicals, trauma, extremes of temperature or microbial attack. Many foods and food components possess anti-inflammatory activity. These include some **fatty acids**, **tocotrienol**, **lactoferrin**, **colostrum**, **wines** and **honeys**.

Antimicrobial activity Ability to kill or inhibit the growth of **microorganisms**.

Antimicrobial compounds Compounds that possess **antimicrobial activity**.

Antimicrobial packaging films **Packaging films**, e.g. **polyethylene films**, that contain an antimicrobial substance such as **chlorine dioxide**. The aim of using such films for packaging foods is to inhibit microbial growth on the foods and thus extend their **shelf life**.

Antimony Toxic member of the **trace elements** group, chemical symbol Sb, which may occur in foods.

Antimutagenicity Ability of substance to reduce either spontaneous mutation rates or mutation rates induced by known **mutagens**. Antimutagenicity of a substance against a mutagen can be determined using the **Ames test**.

Antimutagens Substances capable of reducing background spontaneous mutation rates or reducing the ability of known **mutagens** to cause DNA damage. There is a wide range of antimutagens in foods and beverages, such as **fruits**, **vegetables**, **spices** and **green tea**, including **catechols**, **flavonoids**, **Maillard reaction products** and other **polyphenols**. Antimutagens are also produced by certain **probiotic bacteria** and **bacteria** used to produce **fermented foods**.

Antimycotics Alternative term for **antifungal agents**.

Antinutritional factors Substances that reduce the nutritional value of a food by reducing its **nutrients** content, **bioavailability**, **digestibility** or utilization. Antinutritional factors include **enzyme inhibitors** (**proteinases inhibitors** and **amylases inhibitors** present in a wide range of foods and microorganisms), **inositol** and its derivatives (including **phytates** and **phytic acid** present in legumes and cereals) and **an-**

tivitamins such as thiaminase, dicoumarol, **theophylline**.

Antioxidant compounds Natural compounds present in foods that exhibit **antioxidative activity**.

Antioxidants Substances used in **preservation** of foods by retarding deterioration, **rancidity** or **discoloration** due to **oxidation**. The most commonly used food antioxidants include **BHA** (butylated hydroxyanisole), **BHT** (butylated hydroxytoluene) and **propyl gallate**. Naturally occurring **antioxidant compounds** include **tocopherols** and **ascorbic acid**.

Antioxidative activity Ability of a substance to inhibit **oxidation**. Substances possessing antioxidative activity can be utilized in foods, such as oils, to inhibit oxidation, thus improving shelf life and quality. Foods possessing a high antioxidative activity have also been investigated as potentially health promoting foods, as lipid oxidation has been associated with a range of pathological processes, including **atherosclerosis**.

Antioxidative properties Alternative term for **antioxidative activity**.

Antisense technology Use of **DNA** or **RNA** sequences to bind *in vivo* to complementary DNA or **mRNA** strands, respectively, preventing correct **gene expression**. Can be used to turn off selectively production of certain **proteins**. Has been used to delay **ripening** in **fruits**, modify the composition of **fatty acids** in **oilseeds** and modify the **starch** contents of potato tubers.

Antiseptics **Antimicrobial compounds** used to treat human and animal body surfaces (particularly skin).

Antisera Sera which contain **antibodies** that are either specific to **antigens** (monovalent antisera) or reactive against more than one antigen (polyvalent antisera). Antisera can be produced by immunization of an animal either by injection of antigen(s) or infection with microorganisms that contain the antigen(s).

Antisprouting agents **Plant growth regulators** used to prevent **sprouting** of crops (especially root or bulb crops, e.g. **potatoes**) during storage.

Antithrombotic activity Ability to prevent or regulate the formation of blood clots or thrombi, and thus protect against **coronary heart diseases** and **cardiovascular diseases** such as **stroke**. Foods and beverages displaying antithrombotic activity include plant derived products, **fish oils** and **dairy products** containing **bioactive peptides**.

Antithyroid agents **Drugs** that inhibit the production of **hormones** produced in the thyroid gland; used to increase meat yield in animals by reducing their basal metabolism, lowering gastrointestinal motility and stimulating extracellular water retention. May cause excess accumulation of water in muscle tissues, resulting in poorer quality meat; residues may be a potential risk to consumer health. Examples include **thiouracil** and methimazole.

Antitranspirants **Plant growth regulators** which reduce the intensity of transpiration of food crops; used to improve yield, and product quality or **shelf life**.

Antitumorigenicity Ability of a substance to slow, inhibit or reverse the process of tumorigenesis, in particular, the ability to attenuate tumour formation in the presence of tumour promoters or **carcinogens**. Antitumorigenicity of foods and food components can be determined either *in vitro* using **cell culture** or *in vivo* using **animal models**.

Antitumour activity Ability of a substance to inhibit or reverse the progression of established tumours.

Antiviral activity Ability to kill or inhibit the growth of **viruses**. Many food components possess antiviral activity. These include **lactoferrin** and other constituents of **milk** and **dairy products**, **polyphenols**, **tannins** and **polysaccharides** from some **mushrooms**.

Antivitamins **Antinutritional factors** that destroy or inhibit the metabolic effects of **vitamins**. Examples of antivitamins in foods include thiaminase (antivitamin B_1, present in raw **fish** and other **animal foods**), **caramel colorants** (antivitamin B_6) and dicoumarol (antivitamin K).

Ants Common name for narrow-waisted, generally wingless **insects** of the family Formicidae. May be consumed as **insect foods**. Can also act as insect **pests**.

Anu Common name for *Tropaeolum tuberosum,* also known as mashua or ulluco. An important tuber crop of the Andes, which is closely related to the garden nasturtium. Consumption of the raw tuber is limited due to the bitter taste, associated with **isothiocyanates**, but **flavour** generally becomes milder when the tubers are boiled. Nutrient contents compare well with those of other **tubers**. Anu flowers are also edible.

Aonla Alternative term for **amla**.

Aperitifs **Alcoholic beverages** intended to be consumed before meals to promote appetite. Proprietary aperitifs include products based on flavoured **wines** or **spirits**.

Aphids Common name for plant parasites of the family Aphididae. Includes insects that suck plant sap and exude sugary secretions favoured by **ants**. Some species are important vectors of plant **viruses**.

Apigenin Yellow **pigments** of the **flavonoids** group which occur in a wide range of plants and plant-derived foods.

Apocarotenal Member of the **carotenoids** group of natural **pigments** which occurs in **oranges** and other plant foods. May be used in **natural colorants**.

Apoproteins Term describing the protein component of conjugated **proteins**, e.g. the globin component of **haemoglobin**.

Apoptosis Controlled destruction of cells which occurs as a natural process during tissue growth and development. Also referred to as programmed cell death. Failure of apoptosis is thought to be involved in uncontrolled cell growth in some types of **cancer**, and also autoimmune diseases.

Apparent density Weight of a porous material per unit volume. Apparent density of a porous substance is always lower than the theoretical density of its constituents.

Appearance Perception of the outward form of a substance. The appearance of a food contributes to its overall **sensory properties**.

Appenzeller cheese Swiss semi-hard cheese made from **cow milk**. It can be made with **skim milk** and brine cured for 12 months, or with **whole milk** and cured with brine, pepper and sediment from white winemaking.

Appetite A natural longing to satisfy bodily needs, particularly, but not exclusively, the recurring desire for food. Appetite is increased in the state of hunger and decreased during **satiety**. Appetite for foods, in general, and for particular foods, may become modified over time. A particularly intense appetite for certain foods occurs during **cravings**.

Apple brandy **Spirits** manufactured by distillation of fermented **mashes** based on **apples**. Well known apple brandy types include **calvados**.

Apple cider Alternative term for **cider**.

Apple juice concentrates **Apple juices** which have been concentrated. May be diluted to produce normal strength apple juices or used in manufacture of other **beverages** or foods.

Apple juices Juices extracted from **apples** (*Pyrus malus*). Commonly consumed as **beverages**, but may be fermented to **cider** or used in manufacture of **apple brandy**.

Apple musts Alternative term for **apple juices**, especially those to be fermented in manufacture of **cider**.

Apple pectins **Pectins** obtained from **apples**. **Apple pomaces** are one of the main commercial sources of pectins.

Apple peel Outer skins of **apples**; used as a source of **apple pectins**.

Apple pomaces The solids residue remaining after extraction of **apple juices** or **apple musts**.

Apple pulps Soft mass prepared from the flesh of **apples** by processes such as **slicing**, **chopping** and mashing. Typically available in dried, frozen or canned forms and used in products such as **sauces**, **infant foods** and **desserts**.

Apple purees Thick, smooth preparations made from cooked, strained apples. Used in products such as **infant foods** and **apple sauces**.

Apples One of most widely grown and economically important **fruits** of temperate regions. Most commercial varieties are derived from *Malus pumila* and are grown for production of dessert, cooking, ornamental or **cider apples**. Fruits are large round pomes that range in **flavour** from sweet to sharp, and in **colour** from green and yellow to red and brown. Useful source of **vitamin C**, **potassium** and **dietary fibre**. Cooking apples are usually green and larger and more acidic than dessert apples. Crab apples grow wild in many regions; these are barely edible, but can be used to make **jelly**.

Apple sauces Sauces made by stewing chopped **apples** with **sugar** to form a pulp. Available in canned or bottled form. Used in **desserts** and as an accompaniment to **meat** dishes, especially **pork**.

Apple vinegar Vinegar made using **apples** as the starting material. Similar to **cider vinegar**.

Apple wines Alternative term for **cider**.

Apricot jams Jams made from fresh or dried **apricots**. Used as **spreads**, as **glazes** for **pies** and **cakes**, or as **confectionery** ingredients.

Apricot juices Juices extracted from **apricots** (*Prunus armeniaca*).

Apricot kernels Constituents of **apricot seeds**, rich in **oils** and **proteins**, but limited in use by the presence of **amygdalin** (yielding toxic hydrogen cyanide (HCN)). Detoxified apricot kernels are used in the manufacture of **bitter almond oils**, **persipan** and **marzipan** substitutes. Also consumed as roasted, salted or dried products in some countries.

Apricot nectars Fruit nectars prepared by addition of water and/or **sugar** to **apricot juices**.

Apricot pulps Soft, succulent flesh from **apricots**, which is used in a range of processed foods, such as **fruit juices**, **ice cream** and **infant foods**. Sheets of apricot pulp are dried to make apricot leathers.

Apricot purees Flesh of **apricots** that has been mashed to a thick, paste-like consistency by various means, such as **sieving**, mashing or processing in a blender. Used in a range of products including **infant foods**, **cakes** and **fruit juices**.

Apricots Stone fruits from *Prunus armeniaca* (syn. *Armeniaca vulgaris*), a tree which originated in ancient China and is now widely cultivated in warm temperate zones. The orange/yellow coloured fruits are utilized in a similar manner to **peaches** and are eaten fresh, canned or dried. The distinctive **aroma** makes the fruit suitable for manufacture into **apricot jams** and **apricot juices** or for incorporation into **flavourings** for products such as **ice cream**, **desserts** and **infant foods**. Compared with other fruits, apricots have a high nutritional value, including high amounts of **vitamin A**, **carotenes**, **proteins**, **potassium** and **iron**.

Apricot seeds Hard seeds found in the centre of the flesh of **apricots**. The kernels within the outer casing are utilized as a source of **oils** and in making a form of **marzipan** substitute as well as being eaten roasted, salted or dried. Also called stones.

Apricot wines Fruit wines manufactured by **alcoholic fermentation** of **mashes** prepared from **apricots** (*Prunus armeniaca*).

Aquaculture Production of aquatic organisms under controlled or semi-controlled conditions; mainly for food purposes. A wide range of **aquaculture products**, including **farmed fish**, **farmed shellfish**, aquatic plants and **algae** are produced commercially across the world.

Aquaculture products Aquatic organisms (such as **fish**, **shellfish** and aquatic plants) produced by **aquaculture** for food or industrial purposes.

Aqualysins Thermostable alkaline serine **proteinases** secreted by various **bacteria**, especially *Thermus aquaticus*.

Aquatic foods Foods derived from aquatic organisms, including **fish**, **shellfish**, aquatic plants and **algae**.

Aquavit Scandinavian **spirits**, distilled from fermented **mashes** based on grain or **potatoes**, and commonly flavoured with aromatic seeds and **spices**. Also known as akvavit.

Arabans Polysaccharides in which the main constituent sugar is arabinose. Present in **fruits** and **fruit juices**, and may be used as additives such as **bulking agents** in foods.

Arabic bread Flat round bread composed of **yeasts** leavened **dough** which, when baked, is easily split to make **sandwiches**. Also known as **pita bread**.

Arabidopsis Non-commercial genus of the mustard family of plants. *Arabidopsis thaliana* is commonly used as a model for plant research studies.

α-N-Arabinofuranosidases EC 3.2.1.55. Hydrolyse terminal, non-reducing α-L-arabinofuranoside residues in α-L-arabinosides. Release arabinose from α-L-arabinofuranosides, α-L-arabinans containing (1,3)- and/or (1,5)-linkages, arabinoxylans and arabinogalactans. Useful in the **winemaking** industry where they can be used to release monoterpenols in **grape juices**, and for improving the pore structure and distribution in **bread**. Also known as arabinosidases.

Arabinogalactans **Polysaccharides** in which the main constituent sugars are **arabinose** and **galactose**. Occur in the **pectic substances** fractions of a wide range of plant foods, including **fruits**, **vegetables** and **cereals**. May be of importance for the processing properties of plant foods.

Arabinose Monosaccharide of five carbon atoms (**pentoses**) found predominantly in plants as a component of complex **polysaccharides**, such as **gums** and **pectins**.

Arabinosidases Alternative term for **α-N-arabinofuranosidases**.

Arabinoxylans **Polysaccharides** in which the main constituent sugars are **arabinose** and **xylose**. Form part of the **pentosans** fraction in **cereals** and **cereal products**, and may be of importance for technological properties in processes such as **baking** and **brewing**.

Arabitol Polyol synthesized by reduction of **arabinose** or produced by microbial **fermentation** of plant hydrolysates.

Arachidic acid One of the **saturated fatty acids** with 20 carbon atoms. Occurs at low concentrations in a wide range of **fats**, **oils** and tissue lipids.

Arachidonic acid One of the ω-6 **polyunsaturated fatty acids** with 20 carbon atoms. Widely distributed in foods and essential in the human diet.

Arachin One of the two major **globulins** present in **peanuts**, the other being **conarachin**. As well as having good nutritional quality, both globulins play an important role in **flavour** development during peanut processing.

Arachis oils Alternative term for **groundnut oils**.

Arak Asian **spirits** which may be manufactured from a range of raw materials, including palm juices, **sugar juices**, **dates** or **rice**. Also know as arrack.

Arare Alternative term for **rice cakes**.

Arbutus berries **Fruits** of the Mediterranean shrub *Arbutus unedo*, also known as **strawberry tree fruits** or madrona fruits. The bitter-tasting red **berries** are rarely eaten fresh, but are used in a range of fruit products, including **jellies**, **jams** and **wines**. Also used to make **liqueurs** in France and Portugal.

Arcobacter Genus of Gram negative, microaerophilic rod-shaped **bacteria**. Occur in the reproductive and intestinal tracts of animals and humans. Some species are pathogenic, e.g. *Arcobacter butzleri*, which frequently contaminates raw **chicken meat**. **Raw milk** is also a source of infections.

Arctic char A salmonid fish (*Salvelinus alpinus*) from northern Europe and North America which occurs in fresh and marine water; some forms are landlocked, spending their whole lives in freshwater, while highly migratory forms spend most of their lives at sea. Flesh **flavour** is highly regarded. Marketed fresh, smoked, canned, and frozen.

Areca Any of various Asiatic palm trees of the genus *Areca*, including *A. catechu*, the source of **betel nuts**.

Areca nuts Alternative term for **betel nuts**.

Arecastrum Genus of **palms** which includes *Arecastrum romanzoffianum* (syn. *Syagrus romanzoffianum*), also known as queen palm or pindo palm. Stems are utilized for **starch** (**sago**), while young buds are consumed as a vegetable. Seed kernels have been reported to have potential as a source of **vegetable fats**.

Arenga Genus of **palms**, some of which are used as a source of edible fruits, palm sugar and **palm wines**.

Arepas Alternative name used in Colombia for **tortillas**, round, thin unleavened **pancakes** originating from Mexico which are traditionally made with **corn flour** and baked on a hot surface.

Argemone oils **Oils** derived from any species of the genus *Argemone* (prickly poppies) which are found in North America and the West Indies.

Arginine One of the basic **amino acids**, present in most food proteins and essential in the human diet.

Arkshells A group of bivalve **molluscs** similar to **cockles**. Edible species include *Scapharca subcrenata*, *Arca noae* and *Anadara broughtoni*.

Armagnac A high-quality **brandy** manufactured in a specified region of the Gers district in southwest France.

Armillaria Genus of mainly lignicolous **fungi** belonging to the family Agaricales and once called *Armillariella*. Species include the **edible fungi** *Armillaria mellea*, also known as the honey fungus, bootlace fungus and shoestring fungus.

Armillariella Former name for the genus of **fungi** *Armillaria* which includes edible species.

Arochlor Commercial name for a range of **polychlorinated biphenyls** (PCB), which occur as **contaminants** in foods.

Aroma Physiological sensation, also known as smell, that results from stimulation of olfactory receptors in the nasal mucosae and the interpretation of this information by a specialized area of the cerebral cortex. Food aroma, which is generated by release of volatile **aroma compounds** from the food, makes a marked contribution to overall **flavour**.

Aroma compounds Volatile compounds that are present in foods and contribute towards **aroma**.

Aroma concentrates Concentrates typically obtained by extracting and/or concentrating **volatile compounds** from a source material, e.g. **fruit juices**, **coffee** or **butter**. Can be used as **flavourings** in various foods or to restore **aroma** lost during processing. Other methods of producing aroma concentrates include **fermentation** and enzymic modification (e.g. for cheese flavour concentrates).

Aromatic compounds Organic compounds characterized by a cyclic, conjugated structure, such as occurs in **benzene**. Some aromatic compounds, such as **polycyclic aromatic hydrocarbons** (PAH), may occur as toxic or carcinogenic **contaminants** in foods. Also refers, more generally, to **flavour compounds** or **aroma compounds** present in foods and beverages.

Aromatization Procedure for increasing the **aroma** of a food or beverage. Strategies include the addition of **aroma compounds** to the product or container, and the facilitation of aroma compound release through chemical or mechanical means. Also refers to the chemical conversion of non-aromatic compounds into **aromatic compounds**.

Aromatized wines Wines, often **fortified wines**, which have been flavoured with **herbs**, **spices** or other plant-derived ingredients.

Aromatizing agents Alternative term for **flavourings**.

Aroma volatiles Alternative term for **aroma compounds**.

Aronia Violet-black **berries** produced by *Aronia melanocarpa*, also known as black chokeberries. Contain high amounts of **anthocyanins**, **folic acid** and **minerals**, and are believed to possess health giving properties. Used as the source of juices and in the production of natural food **colorants**. May also be used to impart **colour** and **flavour** to other **beverages**, **dairy products**, **confectionery** and **snack foods**.

Arracacha Common name for *Arracacia xanthorrhiza*, a member of the umbellifer family, which is grown in South and Central America, primarily for its large, starchy, edible roots, which resemble **carrots** or **parsnips** in appearance. Roots, which are also known as Peruvian carrots or Peruvian parsnips, are cooked and consumed as a vegetable, or processed into a variety of products including **infant foods**, **soups**, **bakery products** and **alcoholic beverages** such as **chicha**. The young stems can be used in **salads** and have similar characteristics to **celery**.

Arrack Alternative term for **arak**.

Arrowhead Common name for *Sagittaria sagittifolia*, a perennial herb with arrowhead-shaped leaves which grows in ponds, rice fields and swamps in parts of South-East Asia. The starchy roots (corms) are peeled, sliced and cooked in stews or fried. Widely cultivated in China and Japan.

Arrowroot Starch obtained from rhizomes of *Maranta arundinacea*, a West Indian plant. Neutral in **flavour** and easily digestible, it is used as a thickener in invalid diets, and also in fruit **sauces**, **pie fillings** and **desserts**, where it imparts a clear finish. Can also refer to starch obtained from roots or rhizomes of several other tropical plants.

Arrowtooth flounder A relatively under-exploited **flatfish** species (*Atheresthes stomias*) occuring in north to mid-Pacific waters. Flesh **texture** is less firm than that of most other flatfish, due to presence of a cysteine proteinase in flesh; this species therefore has a low market value compared with other flatfish. Marketed in fillet form; also frozen into blocks and processed into portions.

Arsenates Toxic salts of arsenic acid, which may occur as **contaminants**, especially in **drinking water**.

Arsenic Toxic element which may occur as a contaminant in a range of substances, including water and **sea foods**. Chemical symbol As.

Artemisia Genus of plants used as the source of **spices**. Includes **davana** (*Artemisia pallensis*), **tarragon** (*A. dracunculus*), **wormwood** (*A. absinthium*) and **mugwort** (*A. vulgaris*).

Arthritis Inflammation of one or more joints resulting in swelling, redness and pain. A range of conditions that includes rheumatoid, autoimmune, infectious and osteoarthritides. Increased risk for certain arthritides has been linked with dietary and nutritional factors, including poor nutrition and consumption of **meat**, **fried foods** and **fats**. Oils rich in *n*-3 fatty acids, such as **fish oils**, **borage oils** and **evening primrose oils**, vegetarian diets and nutrients with **antioxidative activity** have been associated with symptomatic relief.

Arthrobacter Genus of Gram positive, obligately aerobic **bacteria** of irregular cell form. Occur in soil. *Arthrobacter nicotianae* may be used as **starters** in the production of smear-ripened **cheese**. Other species may be used in the production of industrial **enzymes**.

Artichokes Term generally applied to the edible buds from *Cynara scolymus* (**globe artichokes**). May also refer to the edible tubers from *Helianthus tuberosus* (**Jerusalem artichokes**), *Stachys sieboldii* (Japanese artichokes) and *S. affinis* (Chinese artichokes).

Artificial colorants Colorants which have been manufactured synthetically, as opposed to those extracted from natural sources (**natural colorants**).

Tend to be less expensive and have better **colour** intensity, uniformity and stability than natural colorants. Examples include **azo dyes** and **FDC colours**.

Artificial flavourings **Flavourings** which contain one or more artificial components not yet identified in a natural material. Synthetic flavourings containing the same chemicals as those found in a natural product are known as nature-identical. Synthetic flavourings are usually less expensive than **natural flavourings**, and less likely to vary in quality, availability and processing stability.

Artificial foods Alternative term for **simulated foods**.

Artificial neural networks Systems of computer programs and data structures which are modelled on the human nervous system and brain. Incorporate large numbers of processors operating in parallel, each with an individual sphere of knowledge which has been fed into it along with rules about relationships. Networks can use this information to recognize patterns in large amounts of data. Used in the food industry in modelling of processes and predicting the behaviour of foods under specific conditions. Also known as **neural networks**.

Artificial sweeteners Synthetic non-nutritive **sweeteners**, usually many times sweeter than **sucrose**. Examples include **aspartame**, **saccharin** and **alitame**. Widespread applications include **low calorie foods**, **soft drinks** and sugar free foods.

Aryl-alcohol oxidases EC 1.1.3.7. Catalyse oxidation of primary **alcohols** with aromatic rings to form aromatic **aldehydes**, including some **aroma compounds** and **flavour compounds** such as **benzaldehyde**, and hydrogen peroxide. Involved in **lignin** degradation by white rot **fungi**.

Arzua cheese Spanish semi-soft **cheese** made from pasteurized **cow milk**. Elastic fine rind and creamy body. Eaten as a dessert with honey, as a sandwich filling or in cooking.

As Chemical symbol for **arsenic**.

Asafoetida Bitter, strong smelling resin extracted from the roots of the umbelliferous plant *Ferula foetida*. The pungent **garlic**-like **aroma** and **flavour** are due to the presence of sulfur compounds. Used in **spices** for Asian foods, **pickles** and **Worcestershire sauces**.

Asbestos Fibrous magnesium calcium silicates, which may be used for thermal insulation or in **filter aids**. Some types are carcinogenic. Asbestos fibres may occur as contaminants in substances such as water.

Ascidians Small marine filter feeding organisms which are primitive chordates (not invertebrates) and are widely distributed around the seas of the world.

Benthic non-motile organisms, often attached to outer surfaces of boats, jetties, and oil rigs. Some species are utilized as sea foods, particularly *Halocynthia roretzi*, *Styela clava* and *S. plicata*.

Ascochyta Genus of **fungi** of the order Dothideales. Species may cause pulse crop diseases. *Ascochyta pisi* causes leaf spot disease in pea plants.

Ascomycetes Former term for a large class of **fungi** containing approximately 2000 genera. Still commonly used to describe members of the subdivision Ascomycotina. Typically terrestrial saprotrophs or **parasites**. Includes most **yeasts**, the edible morels (***Morchella***) and truffles (***Tuber***), the cup fungi, the powdery mildews, black mildews and sooty moulds.

Ascorbases Alternative term for **L-ascorbate oxidases**.

Ascorbate oxidases Alternative term for **L-ascorbate oxidases**.

L-Ascorbate oxidases EC 1.10.3.3. Oxidize **ascorbic acid** to **dehydroascorbic acid**. Can be used to determine the levels of ascorbic acid (**vitamin C**) in foods and **beverages**, and as part of an antioxidant protection system for food **preservation**.

Ascorbates Salts of **ascorbic acid**, including sodium ascorbate and calcium ascorbate, which can be used as **food additives**. Food uses include as **antioxidants** in products such as meat products, as **browning inhibitors** for fruits and vegetables, and as **bakery additives**.

Ascorbic acid Synonym for **vitamin C**, an antioxidant nutrient present in a wide range of foods. Necessary for growth of bones and teeth, for maintenance of blood vessel walls and subcutaneous tissues, and for wound healing; dietary deficiency results in scurvy. Used in **food additives**, with applications in **food antioxidants** and **bakery additives**.

Ascorbyl palmitate One of the **fatty acid esters** that are used as **food antioxidants**. Formed by **esterification** of **ascorbic acid** and **palmitic acid**. Particular applications include in **oils** and **meat** products.

Aseptic packaging Packaging technique in which an aseptic product is placed into an aseptic container in an aseptic environment. The sealed container is designed to maintain aseptic conditions until the seal is broken. Used to enhance **shelf life** of foods, e.g. **fruit juices**. Advantages over conventional **sterilization** techniques include high product quality, optimization of sterilization, minimum energy consumption and low production costs. Aseptic packaging is not suitable for use with products containing large particles, and shelf life stability is shorter than for sterilized foods.

Aseptic processing High-temperature, short-time process which results in products with improved **texture, colour, flavour** and **nutritional values** compared with conventional **canning**. This technology involves filling of pre-sterilized **containers** with a commercially sterile cooled product, followed by aseptic hermetic sealing with a pre-sterilized closure in an atmosphere free of **microorganisms**.

Ash Mineral content of foods, determined by combustion of the sample under defined conditions and weighing of the residue.

Asiago cheese Unpasteurized Italian **hard cheese** originally made from **ewe milk**, but now made entirely from **cow milk**. Two types of Asiago are made, i.e. a lightly pressed cheese made from **whole milk** and matured for 20-30 days (**Asiago Pressato cheese**) and a mature cheese made with **skim milk** (Asiago d'Allevo). An intense **flavour** develops in cheese matured for 2 years.

Asiago Pressato cheese Type of **Asiago cheese** that is mild and delicately flavoured. Interior of this **fresh cheese** is white with a hint of straw colouring.

Asian pears **Pears** produced by *Pyrus pyrifolia* and *P. serotina*. Grown extensively in Asia, particularly Japan, China and Korea, and currently gaining popularity in the West, partly because of their distinctive crisp **texture**, which remains unchanged after picking and long-term **cold storage**. Frequently called apple pears due to their crisp, juicy qualities, they are also known as Oriental pears or **Japanese pears**.

Asparagine One of the non-essential **amino acids**, occurring in most food **proteins**.

Asparagus Liliaceous plants of the genus *Asparagus*, particularly *A. officinalis*, which is widely cultivated in Europe and the USA for its edible young shoots (spears). Lightly cooked asparagus spears are regarded as a luxury vegetable and can be eaten hot or cold. They are also widely used in **soups**.

Asparagus beans Common name for seeds produced by *Vigna sesquipedalis*. Long thin **legumes** that resemble **string beans**, but which are actually related to **cowpeas**. **Flavour** is similar to that of string beans and has also been likened to that of **asparagus**. Also known as sitao, Chinese long beans or yard-long beans due to their ability to grow up to 3 feet in length. Asparagus beans are picked before reaching this stage and used in **salads** or stir-fries. Young leaves and stems are steamed and consumed as vegetables.

Asparagus peas Common name for *Lotus tetragonolobus*, a southern European plant, occasionally grown for its edible pods, which are harvested before maturity and consumed as a vegetable (usually steamed).

Aspartame Low calorie artificial sweetener (chemical name aspartyl phenylalanine methyl ester, also known as NutraSweet). A dipeptide (**aspartic acid** and **phenylalanine**) ester, approximately 160-200 times as sweet as **sucrose**. Non-cariogenic and without an **aftertaste**. Loses sweetness on prolonged storage and exposure to heat (unsuitable for **baking**). Safe for diabetics but not for individuals with **phenylketonuria** as phenylalanine is released during metabolism of aspartame.

Aspartate aminotransferases EC 2.6.1.1. Catalyse the conversion of L-aspartic acid and 2-oxoglutaric acid to L-glutamic acid and **oxaloacetic acid**. Also act on L-tyrosine, L-phenylalanine and L-tryptophan.

Aspartic acid One of the non-essential **amino acids**, occurring in most food **proteins**.

Aspartyl phenylalanine methyl ester Systematic name for **aspartame**.

Aspergillic acid Antifungal compound produced by *Aspergillus flavus*.

Aspergillus Genus of **fungi** of the class Hyphomycetes. Some species can cause food **spoilage** (e.g. *Aspergillus flavus*, *A. parasiticus* and *A. niger*). Many species produce **mycotoxins** (e.g. **aflatoxins, cyclopiazonic acid, ochratoxins, patulin**). Certain species are used in production of industrial **enzymes** (e.g. synthesis of **amylases, catalases, proteinases** and **lipases** by *A. niger*). Also involved in production of **fermented foods** (e.g. manufacture of **koji, miso, sake** and **soy sauces** by *A. oryzae*) and other agents for the food industry (e.g. **citric acid** and **gluconic acid** production by *A. niger*).

Aspic Savoury clear jelly made from meat or fish **stocks**, often made with **gelatin**. Used as a setting gel or for glazes on foods such as meat and vegetables. Also available in powdered form.

Aspirators Instruments or equipment for drawing fluids by suction from vessels or cavities.

Ass milk Milk obtained from asses. Close in composition to **human milk**.

Astacene **Pigments** fraction of the **carotenoids** group, occurring in **crustacea**.

Astaxanthin **Pigments** fraction of the **carotenoids** group, occurring in **crustacea**.

Asthma A breathing disorder that results from spasm of the muscles surrounding the airways of the lungs (bronchospasm) that is generally reversible. Narrowed airways causes shortness of breath, wheezing, coughing and congestion. Atopic (allergic) asthma is most common and can be associated with other food **allergies**. A wide range of asthma triggers have been identified, including environmental pollutants, drugs, cold air and exercise. Asthma triggered by foods is rare;

food triggers include **sulfites** and sulfiting agents found in a range of foods and known food **allergens**.

Astringency A sensation of dryness in the mouth combined with roughening of the oral epithelium and puckering of the muscles of the face and cheeks. It is induced by foods containing chemicals such as **tannins** and other **polyphenols**, **acids** and **aluminium** salts. Sensory perception of astringency has been attributed to binding of tannins to salivary proteins.

Astrocaryum Genus of **palms** including *Astrocaryum vulgare*, which, along with other palms, is a source of tucuma **oils** and edible fruits.

Atherosclerosis A pathological process resulting in thickening and hardening of the walls of medium and large arteries due to formation of atherosclerotic plaques. **Cardiovascular diseases** produced by occlusion of the affected arteries can be of gradual onset (angina, peripheral vascular disease) or sudden onset (**stroke**, myocardial infarction). Rate of development of atherosclerosis is affected by many factors including lifestyle and diet.

Atlantic mackerel Commercially important pelagic **marine fish** species (*Scomber scombrus*) which occurs abundantly in cold and temperate coastal waters, often forming large shoals near the surface. Flesh is firm and fatty with a distinctive savoury **flavour**. The species is marketed in fresh, frozen, smoked and canned forms and is popularly consumed fried, grilled or baked.

Atlantic salmon A well known **freshwater fish/marine fish** species (*Salmo salar*) of high commercial importance; indigenous to geographical areas linked to the Atlantic ocean but is also cultured in other areas. World Atlantic salmon production is more than half a million tonnes per year. Flesh has a highly valued **flavour**. It is marketed and consumed in a wide range of forms, including fresh, frozen, smoked and canned products.

Atomic absorption spectrophotometry Alternative term for **atomic absorption spectroscopy**.

Atomic absorption spectroscopy Technique in which the mineral composition of a sample is determined from the absorption of light by atoms. A monochromatic source of light at a specific absorption wavelength is passed through the sample following atomization by various means. Often abbreviated to AAS.

Atomic emission spectroscopy Technique in which the mineral composition of a sample is determined from the emission of light from excited atoms at wavelengths characteristic of the atoms.

Atomic force microscopy **Imaging** technique in which the surface of the sample is scanned using a small tip to construct a 3-dimensional image. The tip may be in contact with or just above the surface. Molecular forces exerted against the tip by the surface are used by image processing software to give information about the surface.

Atomizers Devices that convert a substance into very fine particles or droplets.

ATP Abbreviation for **adenosine triphosphate**.

ATPases Include EC 3.6.1.3 and EC 3.6.1.32-39. Catalyse hydrolysis of **ATP** to **ADP**. Can be used as an index of protein **denaturation** in **meat** during storage. Functional ATPases are required for growth of **yeasts** and **lactic acid bacteria** in the presence of acids, e.g. during **fermentation** of **wines** and **dairy products**.

Atrazine Selective systemic herbicide used for pre- and post-emergent control of annual grasses and broad-leaved weeds in a range of cereals (particularly corn and sorghum), fruits, vegetables, coffee, oil palms and sugar cane. Often used in combination with other **herbicides**. Classified by Environmental Protection Agency as slightly toxic (Environmental Protection Agency III).

Atta Indian wholemeal **wheat flour** used in preparation of Indian **bread**.

Attalea Genus of **palms**, including *Attalea colenda* and *A. cohune*, used as a source of **palm oils**.

Attieke A traditional product of the Ivory Coast made by **fermentation** and steam-cooking of **cassava** roots.

Aubergines Egg-shaped fruits of *Solanum melongena*, a native plant of tropical Asia, but now cultivated widely in tropical and warm temperate regions. Fruits are usually black or dark purple in **colour**, although green, creamy white or yellow varieties are also available. Consumed as a vegetable, typically fried or stuffed, or used as an ingredient in ratatouille, moussaka and **curries**. Also known as egg plants in North America and brinjal in India and Africa.

Aureobasidium Genus of yeast-like **fungi** of the class Hyphomycetes. Occur in **sea foods**, **fruits** and **vegetables**. *Aureobasidium pullulans* may cause **spoilage** of stored **beef**.

Auricularia Genus of **fungi** which includes edible species, e.g. *Auricularia auricula* and *A. polytricha*.

Austamide Mycotoxin produced by *Aspergillus ustus*.

Australian chestnuts Seeds produced by the tree, *Castanospermum australe*. Also known as Moreton Bay chestnuts or blackbeans. Poisonous when fresh, but can be consumed after roasting to remove toxins. Common to some parts of Australia, where they are consumed by aborigines. Contain castanospermine, an antiviral of potential use in AIDS therapy.

Authenticity The genuineness of foods and **beverages**; can be with respect to various factors, such as ingredient content, processing methods and geographical origin. For certain foods and beverages, labelling schemes have been implemented to indicate authenticity. A range of methods is used to test authenticity depending on the potential method of **adulteration**.

Autoclaves Strong containers employed in processes using high pressures and temperatures, e.g. steam **sterilization**.

Autolysins Endogeneous **enzymes** found in cell walls which can hydrolyse certain structural cell components (e.g. peptidoglycans in **bacteria**) to bring about **autolysis**.

Autolysis Process by which the structural components of cells are degraded by their **autolysins**. Usually occurs after the cells have experienced a traumatic event such as injury or death. May result in the release of intracellular enzymes from cells, which may play an important role in **cheese ripening**. Can be responsible for inactive cultures or for sensory defects (by autolytic products) in **wines** and **beer**.

Autoxidation An autocatalytic oxidation reaction that occurs spontaneously in the atmosphere. Initiators include heat and **light**. **Unsaturated fatty acids** present in foods are susceptible to autoxidation when exposed to the air, with the reaction proceeding by a free-radical mechanism. The reaction may result in production of stable nonpropagating products that contribute to **off flavour**. In addition, **radicals** produced by autoxidation may cause **bleaching** of food **colour** and destruction of **vitamin A**, **vitamin C** and **vitamin E**. This type of deterioration is prevalent in fried **snack foods**, **nuts**, **oils**, and **margarines**.

Auxins **Plant growth regulators** important for **ripening** and quality of **fruits**.

Availability The extent to which dietary **nutrients** are present in a form that can be absorbed and utilized. Similarly, **bioavailability**.

Avenasterol **Sterols** fraction which occurs in the unsaponifiable fraction of many **vegetable oils**. In combination with other sterols, avenasterol concentration may be used as an index for identification and for monitoring the **authenticity** of vegetable oils samples.

Avenins **Glutelins** present in **oats**; the major **storage proteins** of this cereal.

Avermectins **Insecticides** and **acaricides** which may be used for control of **pests** on plants and **parasites** on animals. May occur as contaminant residues in foods.

Avidin **Glycoproteins** fraction which occurs in **egg whites** and binds **biotin**.

Avocado oils Unsaturated **oils** rich in **oleic acid** derived from the pulp of **avocados** (*Persea americana*).

Avocados Common name for *Persea americana*, also known as alligator pears. A pear-shaped fruit with a leathery green or black skin, enclosing yellow to orange flesh and a single pit. Compared with other fruits, avocados have high protein and oil contents. Traditionally marketed fresh and used like a vegetable, they can also be processed into **guacamole** or used as a source of **avocado oils**. Fruits do not ripen if left on the tree and are usually treated with **ethylene** in ripening rooms to ensure uniform maturation.

Avoparcin Narrow-spectrum glycopeptide antibiotic active against **Gram positive bacteria**; used mainly for growth-promoting purposes (improves absorption of nutrients from the gastrointestinal tract) in **chickens**, **turkeys**, **swine** and **calves**. Remains virtually unabsorbed within the gastrointestinal tract and is rapidly eliminated in the form of the parent compound; no withdrawal period is required.

a$_w$ The symbol for **water activity**, which is a measure of the **water vapour** generated by the moisture present in a hygroscopic product. It is defined as the ratio of the partial pressure of water vapour to the partial pressure of water vapour above pure water at the same temperature. In foods, it represents water not bound to food molecules. Level of unbound water has marked effects on the chemical, microbiological and enzymic stability of foods.

Awamori **Rice**-derived **spirits** originating in the Okinawa region of Japan.

Ayu A fish species (*Plecoglossus altivelis*) distributed in western North Pacific waters that regularly migrates between the sea and freshwater; some forms remain in lakes and rivers for long periods. Ayu are cultured in several Asian countries and their flesh **flavour** is highly regarded. Usually marketed fresh and consumed fresh, fried and grilled.

Azaperone Sedative drug used primarily to reduce **stress** in **swine** prior to and during transportation. Frequently administered to animals a few hours prior to **slaughter**; high levels of active residues in edible tissues is a potential hazard to consumers. Also known as Stresnil.

Azaspiracids Group of **toxins** produced by marine **algae**. Cause **food poisoning** in people eating contaminated **shellfish**, especially **mussels**.

Azinphos-ethyl Non-systemic insecticide and acaricide used for control of chewing and sucking **insects** and spider **mites** on fruit trees, vegetables, cereals and coffee plants. Classified by WHO as highly hazardous (WHO Ib).

Azinphos-methyl Non-systemic insecticide used for control of chewing and sucking **insects** on fruit trees, vegetables, cereals, nuts, sugar cane and coffee plants. Classified by WHO as highly hazardous (WHO Ib). Also known as guthion.

Azodicarbonamide Oxidizing **bakery additives** used to age and bleach cereal **flour**, and to condition **dough** for **breadmaking**.

Azodrin Alternative term for the insecticide **monocrotophos**.

Azo dyes Series of **artificial colorants** containing at least 1 chromophoric azo group. Examples include **amaranth**, **tartrazine**, **Sunset Yellow** and **Carmoisine**.

Azotobacter Genus of Gram negative, aerobic, rod-shaped **bacteria** of the family Azotobacteraceae. Occur in soil and water. Capable of nitrogen fixation, thereby converting atmospheric nitrogen into a chemical form which is usable by plants.

Azoxystrobin Member of the strobilurins class of **fungicides**. Active against a broad spectrum of **fungi** and used on a wide range of crops.

B

Babaco Common name for *Carica pentagona*. A seedless pentagonal-shaped fruit, which is related to **pawpaws** and believed to have originated in Ecuador. The ripe fruit is golden yellow in **colour** and has a delicate strawberry-like aroma. Flesh is very juicy, slightly acidic, low in **sugar** and rich in **vitamin C**. Immature green fruit can be used as a vegetable.

Babassu oils Edible **oils** derived from the babassu (Brazilian palm nut), which have similar **fatty acids** composition and **physical properties** to **coconut oils**. Used as a cooking oil, as well as in the manufacture of soaps and cosmetics.

Babassu palm kernels Softer, central parts of the babassu nut (Brazilian palm nut) which form the source of **babassu oils**.

Baby corn Small ears of immature **corn**, generally harvested between 2 days before and 3 days after silking. Baby (dwarf) corn is sold fresh or canned and generally measures around 4-9 cm in length and 1-1.5 cm in diameter. Popular in Oriental cuisine.

Baby foods Alternative term for **infant foods**.

Bacilli Generally refers to any rod-shaped bacterial cells. May be used specifically to refer to a member of the genus *Bacillus*.

Bacillus Genus of Gram positive, aerobic or facultatively anaerobic, rod-shaped, spore-forming **bacteria** of the family Bacillaceae. Occur in soil and water. *Bacillus cereus* can cause **spoilage** of **pasteurized milk** and **cream**, while other species can cause spoilage of low acid canned foods. Some species are used commercially as sources of **enzymes** (e.g. **glucose isomerases, subtilisins**). *B. cereus* can cause a diarrhoeal type of **food poisoning** (associated with consumption of meat, milk, vegetables and fish) or an emetic type of food poisoning (associated with consumption of rice, potatoes, pasta, sauces, puddings and soups). *B. thuringiensis* is an important insect pathogen used as an agent for **biocontrol**. Some species have now been transferred to the new genus *Geobacillus*, including *G. stearothermophilus* and *G. thermoleovorans*.

Bacitracin Peptide antibiotic produced by the bacteria *Bacillus* subtilis and *B. licheniformis*; active against Gram positive microorganisms and used (in the form of zinc bacitracin) to promote growth in calves, lambs, swine and turkeys. Also used to enhance egg production in poultry and for treatment of **mastitis** in cows. Remains virtually unabsorbed in the gastrointestinal tract of animals; distribution in edible tissues is considered negligible.

Backfat Fatty tissue covering the back area on **animal carcasses**. In **swine**, a particularly thick fat layer is present in the back region, which is thick enough to be separated and used independently. Swine backfat is fairly soft at room temperature. Backfat thickness is thought to affect attributes of **pork** such as **flavour** and **eating quality**.

Bacon Meat from the sides, backs and bellies of **swine**, preserved by **curing**; it may be smoked or unsmoked. When bacon is sold after curing but before smoking, it is called green bacon, pancetta or raw kaiserfleisch. **Smoking** produces a strong flavour in bacon. In order to decrease the retail price per kilogram, some bacon manufacturers increase the weight of their product using water, phosphates and other ingredients. Most bacon is sliced into rashers before retail; middle rashers have a round eye of lean meat, whilst streaky bacon is the tail end of the loin. A rasher of bacon can contain up to 40% fat.

Baconburgers Round, flat cakes of chopped or minced **bacon**, cooked by grilling or frying. Baconburgers are usually eaten in **bread rolls**, and can be served with lettuce, **tomatoes, onions, pickles, mustard** and **tomato ketchups**.

Bacteria Heterogeneous group of usually unicellular prokaryotic **microorganisms**, generally possessing a characteristic cell wall, and found in virtually all environments. Some cause diseases in humans and animals, while others are used in the manufacture of foods (e.g. **dairy products**).

Bacterial biomass Quantitative estimate of the total **bacteria** present in a given habitat, in terms of mass, volume, or energy.

Bacterial counts Estimations of numbers of **bacteria** in a sample.

Bacterial spoilage Spoilage caused by the action of **bacteria**.

Bacterial spores Spores (either endospores or exospores) formed by **bacteria** (e.g. *Bacillus* and *Clostridium* species) under conditions of nutrient limitation. Endospores are resistant and may be disseminative, rather than reproductive, while bacterial exospores are characteristically reproductive and disseminative. They are generally more resistant than vegetative cells to heat, desiccation, antimicrobial compounds and radiation, and can remain dormant for long periods.

Bactericides Biological, chemical or physical agents that kill **bacteria**, but not necessarily their endospores. Include formaldehyde, peracetic acid, hydrogen peroxide or activated carbon.

Bacteriocins **Peptides** produced by specific **bacteria** that possess **antibacterial activity**. Both purified bacteriocins and bacteriocin-producing bacteria are used in the food industry, applications including inhibition of the growth of **pathogens** and **spoilage** organisms.

Bacteriological quality Extent to which a substance (e.g. a food) is contaminated with **bacteria**.

Bacteriology Scientific study of **bacteria**.

Bacteriophages **Viruses** that infect **bacteria**. In the case of lytic **phages**, bacterial synthesis of **DNA**, **RNA** and **proteins** ceases following infection, and new phage constituents are synthesized using the host's transcription and translation apparatus. Following self-assembly of phages, host cells rupture, releasing several hundred new phage particles. Many phages, however, are lysogenic and integrate into the host cell DNA as prophages. These remain dormant and only undergo the lytic cycle under appropriate environmental conditions. Bacteriophage infection of **starters** causes significant losses in the manufacture of **cheese** and other **fermented dairy products**. Altered forms of bacteriophages are often used as DNA **cloning vectors**.

Bacteriophages resistance Resistance of **bacteria** to infection by **bacteriophages**. Resistance may be mediated by alteration of the cell wall or by various intracellular mechanisms, such as restriction modification systems. Several resistance mechanisms have been found to be **plasmids**-based and, potentially, can be introduced into bacteria in order to increase their resistance to infection.

Bacteriostats Chemical agents that inhibit the growth and multiplication of **bacteria**. Includes several **disinfectants**, **spices** and **antibiotics**.

Bacteroides Genus of Gram negative, obligately anaerobic, rod-shaped **bacteria** of the family Bacteroidaceae. Occur in the oral cavity and intestinal tract of humans and other animals. Some species are opportunistic **pathogens**.

Bactofugation High speed **centrifugation** process used to remove most bacterial endospores, **yeasts** and **fungi** from **milk**, thereby extending its **shelf life**. Used to produce milk with a low spore count for **cheese** production to prevent late **blowing** of **hard cheese**.

Bactris Genus of **palms** which includes *Bactris gasipaes*, also known as pupunha or peach palm, a species utilized for its edible fruits and **palm hearts**.

Bacuri **Fruits** similar to **mangosteens** produced by *Platonia insignis* or *P. esculenta*, trees growing in the Amazonian forests of South America. Yellow, with a leathery shell enclosing creamy white flesh. Flesh is eaten fresh or canned, or used in manufacture of products such as **purees**, **jams**, **ice cream**, **fruit juices** and **liqueurs**.

Bael fruit Thick-shelled fruits of *Aegle marmelos*, a rutaceous tree native to India. The citrus-like fruits are rich in **vitamin C**, with slight astringency, and are consumed fresh or processed into products such as juices and jams. Fresh fruits have a yellow pulp, which turns reddish brown when dried. Particularly prized for their medicinal properties, especially as a treatment for dysentery. Also known as Bengal quinces or Indian quinces.

Bagasse **Cane sugar** processing waste that is composed of unextracted **sugar** and the remains of the **sugar cane** after milling. Used as a fuel source, in feeds, as a substrate for microbial **fermentation** and for paper and board manufacture. Also called **sugar cane bagasse** and megass.

Bagels **Yeasts**-leavened **rolls** with a hole in the middle, characterized by a glazed crust and a tough chewy **texture**. Made by dropping into boiling water briefly before **baking**.

Bagging Packing of substances, such as foods, into **bags**.

Bag in box packaging Packaging consisting of a flexible inner bag, which closely fits inside a box. The product is contained in the inner bag, which acts to keep out atmospheric oxygen. The rigid outer box protects the contents. Used widely for **breakfast cereals** and also for storing and dispensing **wines**.

Bagoong Fermented salted fish paste originating from the Philippines; usually made from an **anchovy**-like fish called dilis (*Stolephorus indicus*) or from young **herring**.

Bags **Containers** with a single opening that are used for storing or carrying items. Made from a variety of flexible materials. Bags for food use are usually made of **paper** or **plastics**. The term is also used for small perforated paper sacks in which **tea leaves** or **coffee grounds** are placed and which are used to make small quantities of **tea** or **coffee beverages**.

Baguettes Small narrow loaves of crusty **bread** containing little or no shortening. Often used to make **sandwiches**.

Bajra Indian **millet**, *Pennisetum typhoideum*.

Baked beans **Haricot beans** (usually **navy beans**) that have been baked and canned in **tomato sauces**. Other ingredients include **modified starches**, water, **sugar**, **salt** and **spices**. A good source of **proteins** and **dietary fibre**.

Bakeries Facilities in which **bakery products** are manufactured. Also refers to retail outlets in which bakery products are sold.

Bakers confectionery Alternative term for **bakery products**, especially those of a sweet nature, e.g. **cakes**.

Bakers yeasts Leavening agents (e.g. *Saccharomyces cerevisiae*) used in making **bread** and other **bakery products**. Available in compressed, liquid and dried forms.

Bakery additives Ingredients used in making **bakery products** with the aim of prolonging **shelf life** or improving the quality of the finished products. Include **humectants**, **antifoaming agents**, anti-staling agents, crumb softeners and texture improvers.

Bakery fillings Fillings used in **bakery products**, e.g. **cakes** and **biscuits**.

Bakery product mixes Pre-mixed dry formulations which usually require the addition of liquid ingredients to make **batters** or **dough**.

Bakery products Products in which **flour** based components are major ingredients, and which are cooked by **baking**. Include **biscuits** or **cookies**, **bread**, **cakes**, **doughnuts**, **scones** and **tortillas**.

Baking Cooking of foods in **ovens** by surrounding with dry heat. The temperature of the oven is varied depending on the type of food that is to be cooked.

Baking ovens Enclosed chambers or compartments in which foods are cooked or heated by application of dry heat (**baking**).

Baking powders **Bakery additives** comprising mixtures of **sodium bicarbonate**, **starch** and one or more acidic substance (e.g. cream of tartar). When moistened and heated, they act as **raising agents** by generating **carbon dioxide**, bubbles of which have a leavening effect.

Baking properties Characteristics of **cereals**, **bakery additives**, **flour** or **dough** associated with their suitability for use in **baking**.

Baking quality Extent to which a **flour** is able to produce a well leavened **bread** which has optimal **texture** and an even distribution of air pockets formed during **fermentation**, or good quality **bakery products**.

Balady Middle Eastern unleavened **sourdough** flat **bread**, especially popular in Egypt.

Balsamic vinegar Richly-flavoured dark **vinegar** produced in Modena, Northern Italy, by slow ageing of **grape juices** in wooden casks. Frequently used in **salad dressings** and **marinades**.

Balsam pears Alternative term for **bitter gourds**.

Bambara beans Alternative term for **bambara groundnuts**.

Bambara groundnuts **Fruits** of *Voandzeia subterranea* (syn. *Vigna subterranea*), also known as bambara beans. Grown extensively in the tropics, particularly Africa. Seeds are rich in **starch** and **proteins**, but low in **minerals** and contain only about half the oil content of true groundnuts (**peanuts**). Can be eaten fresh, boiled or roasted, or ground into **meal** to make **porridge** or bean cakes. Immature seeds are sweeter and easier to cook than mature, hard seeds.

Bamboo Tall tree-like plants belonging to the grass family and characterized by hollow woody stems and edible young **bamboo shoots**. Of great economic importance in many areas. Species utilized for bamboo shoots include those belonging to the *Bambussa*, *Phyllostachys* and *Dendrocalamus* genera.

Bamboo shoots Emerging ivory-coloured shoots of several species of **bamboo**. These include *Bambusa oldhamii*, *Dendrocalamus latiflorus* and *Phyllostachys edulis*. An important component of Oriental cuisine, bamboo shoots are available fresh or canned and have a crispy texture. Bitter-tasting shoots require precooking due to the presence of cyanogenic **glucosides**.

Banaba Common name for the plant *Lagerstroemia speciosa*, the leaves of which are extracted to make **banaba tea** which is drunk as a **herb tea**, principally in the Philippines and Japan. Banaba leaf extracts also have blood sugar lowering activity, making them useful in treating diabetes mellitus and as major components in weight reduction products.

Banaba tea Aqueous extract prepared from the leaves of the banaba tree (*Lagerstroemia speciosa*) which is drunk as a **herb tea**, principally in the Philippines and Japan. Claimed to have many beneficial properties for health, including insulin-like activity.

Banana juices **Fruit juices** extracted from **bananas** (*Musa* spp.).

Banana peel Thick outer skin of **bananas**, which helps protect the fruit and whose **colour** provides a good indication of ripeness. Occasionally incorporated into **jams**.

Banana pulps　Banana flesh or a preparation made from it by mashing. Used as the starting material for manufacture of various products, including banana **milkshakes**, **fruit juices** and **infant foods**.

Bananas　**Fruits** produced by large tropical plants of the genus *Musa*. Wild fruits contain seeds and are inedible, whereas edible cultivars are seedless (sterile) hybrids, and a good source of **carbohydrates** and **vitamin A**. Yellow dessert bananas are relatively high in **sugar** and are consumed fresh, whereas starchier **plantains** (green bananas) are used like a vegetable in cooking. Bananas are also widely used in foods such as **fruit purees**, **fruit juices** and **bakery products**.

Bannocks　Traditional Scottish flat **bread** or **cakes** made usually from **barley flour** or **oatmeal**. Cooked on a griddle and eaten plain or flavoured, with breakfast or evening meals.

Banvel　Alternative term for the herbicide **dicamba**.

Baobab　Common name for *Adansonia digitata*, a giant tree of the Bombacaceae family, used as a source of foods in many parts of Africa. Baobab **fruits** are similar in appearance to **gourds** and yield an edible pulp known as monkey bread, which is used in foods and beverages. Leaves are also edible and can be made into **soups** or stews, while seeds are ground to produce a **meal** (frequently mixed with millet) or used for production of **baobab oils**. Mixtures of **milk** and baobab **fruit juices** are popular **beverages** in some areas.

Baobab oils　Oils produced from the gourd-like fruit of large trees of tropical Africa which belong to the genus *Adansonia*.

Barbados cherries　**Fruits** from *Malpighia glabra* (syn. *Malpighia emarginata*), a large shrub native to the West Indies and South America. Also known as acerola or West Indian cherry. The bright red **stone fruits** are about the size of **cherries** and can be eaten fresh or processed into products such as **jams** and **preserves**. Fruits are the richest known natural source of **vitamin C** and represent an important commercial source of the vitamin.

Barbados cherry juices　**Fruit juices** extracted from **Barbados cherries** (*Malphigia punicifolia*). A rich source of **vitamin C**.

Barbecued foods　Meat and other foods cooked out of doors on a barbecue (originally a revolving spit over an open fire, nowadays more likely to be a wire grid placed over hot charcoals or a gas fire source). Popular barbecued foods include **sausages**, **burgers** and **fish** or **meat steaks**.

Barberry figs　Alternative term for **prickly pears**.

Barbiturates　General name used for **drugs** that act on the central nervous system to produce a sedative effect or induce anaesthesia; used to reduce **stress** in farm animals, particularly prior to and during transportation. High levels of active residues in edible tissues form a potential hazard to consumers. Examples include **azaperone**, nembutal and propiopromazine.

Bar codes　Machine-readable codes which contain product specific information. Bar codes are formed by patterns of parallel lines of varying thickness with spaces of varying length between them. Information is usually read from bar codes using light pens or laser scanners. Standard international codes are used. Benefits of using bar codes include: rapid and efficient data capture; improved product traceability; the possibility of automated product storage; improved control of product storage and distribution; time and costs savings; and improved customer service. Consumer unit bar codes, which encode fixed information, are used on primary packaging of products intended for sale directly to consumers at retail outlets. Traded unit bar codes, which include fixed as well as supplementary product information (e.g. product weight, batch number and time of production), are often compulsory within product supply chains. Transport unit bar codes used to label pallets and encode shipping containers are used to track pallets through supply chains.

Barley　Edible grain from *Hordeum vulgare* used as a cereal and livestock feed and in **malt** production. Contains little **gluten**, and so is unsuitable for **breadmaking**. Most popular form is **pearl barley** in which the outer husk and part of the **bran** layer are removed by **polishing**. Provides a source of **vitamins** (e.g. niacin, folates) and **minerals** (e.g. zinc, copper, iron).

Barley flour　Ground hulled **barley** used to make unleavened **bread** and **porridge**.

Barley malt　**Malt** prepared from special **malting barley** cv; mainly used in **brewing**. Barley malt is the main malt type used in brewing worldwide.

Barley starch　**Starch** isolated from **barley**.

Barracuda　Pelagic predatory **marine fish** species (*Sphyraena* spp.); widely distributed in warmer regions of the Atlantic and Pacific Oceans. Flesh is firm in **texture** with moderate fat content. Marketed fresh and as a salted or dried product.

Barramundi　**Fish** species (*Lates calcarifer*) of considerable economic importance; found in coastal waters, estuaries and lagoons in the southwest Pacific region. Sold in fresh and frozen form and consumed steamed, pan-fried, grilled and baked. Cultured in Thailand, Indonesia and Australia and can reach 1500-3000 g in one year in ponds under optimum conditions.

Barrels Cylindrical **containers** for liquids and dry materials. Traditionally made of wooden staves held together by metal hoops, but may also be made of cheaper and/or more durable materials, such as metal or plastics. Oak barrels are used for the ageing of **wines** and **spirits**; constituents of the wood (e.g. **tannins**, **lignin** and fragments, **carbohydrates**, **acids** and **esters**, volatile **phenols**, oak **lactones**, **pyrazines**, **furfural** and norisoprenoids) have major effects on **flavour** of wines and spirits. Barrels are also used as measures for liquids, e.g. **beer** and **oils**, based on the capacity of standard barrels. Also known as **casks** or **kegs**.

Bartail flatheads Bottom dwelling **fish** (*Platycephalus indicus*) found in coastal waters and estuaries in South Pacific and Indian Ocean regions; also occurs in the eastern Mediterranean, where it was recently introduced. A valued food fish that is normally marketed fresh and is cultured commercially in Japan.

Basidiomycetes Biological classification for a large group of **fungi**, in which spores are produced in the basidia. Contains many types of **edible fungi**, including *Agaricus* mushrooms, but not truffles, which belong to the Ascomycetes group.

Basil Herb obtained from the genus *Ocimum*. The main varieties used in **cooking** are **sweet basil** (*O. basilicum*) and bush basil (*O. minimum*). **Flavour** of the fresh leaves has been likened to a blend of **liquorice** and **cloves**, while dried leaves are more lemony and less pungent. Much used in Italian cuisine (particularly tomato-based dishes) and a key ingredient of **pesto**.

Baskets Perforated containers used to hold or carry food. Made from interwoven strips of wood (e.g. bamboo), twigs, wire, or other lightweight flexible materials. The open structure of baskets allows ventilation of the product. Compared with solid containers, the increased flow of air allows greater **cooling** rates.

Basmati rice Aromatic long grain variety of **rice** from India.

Bass Name given to a variety of **marine fish** and **freshwater fish**. In Europe, the name particularly refers to a marine fish species (*Dicentrarchus labrax*) widely distributed in eastern Atlantic regions from North Africa up to Norway. Enters coastal waters and river mouths in summer, but migrates offshore in colder weather and occurs in deep water during winter. A highly valued food fish; usually marketed fresh or smoked. Also known as European sea bass.

Bastard halibut Marine **flatfish** species (*Paralichthys olivaceus*) from the flounder family (Paralicthyidae), which occurs in the western Pacific Ocean. Highly prized as a food fish in Japan. Usually marketed fresh. Also known as hirame and **Japanese flounders**.

Batters Thin liquid mixtures of pouring consistency made from **flour**, **milk** and **eggs**. May be used as **coatings** for foods such as **fish** prior to **frying**, or cooked on their own to make products such as **pancakes**, **waffles** and Yorkshire puddings.

Baumkuchen Moist almond **sponge cakes**, often baked in the shape of a pyramid.

Bavaricins **Bacteriocins** produced by *Lactobacillus* spp.

Bavarois Cold **desserts** made with **eggs**, **gelatin** and **whipped cream**. Also known as Bavarian cream.

Bavistin Alternative term for the fungicide **carbendazim**.

Bay Alternative term for **laurel** (*Laurus nobilis*), a small, evergreen tree. May also refer to **bay leaves**, the **herbs** obtained from this tree.

Bay leaves Aromatic leaves obtained from the **laurel** tree, *Laurus nobilis*. Used as a herb to flavour to stews, sauces and many other foods. Generally added whole and removed before serving.

Bayrusil Alternative term for the insecticide **quinalphos**.

Baytex Alternative term for the insecticide **fenthion**.

Bdellovibrio Genus of Gram negative, aerobic **bacteria**. Occur in soil, sewage and in both fresh and marine waters. Characteristically intracellular **parasites** of other Gram negative bacteria, reproducing between the cell wall and plasma membrane of the bacterium and ultimately killing it.

Beach peas **Seeds** produced by *Lathyrus maritimus* or *L. japonicus*, leguminous plants growing particularly along the shores of Arctic and sub-Arctic regions, but also in coastal areas of Europe and Asia. New stalks may be cooked by **stir frying**, **steaming** or **boiling**. After the plant has flowered, young pods are cooked and eaten like **snow peas**. These young pods are rich in vitamin B complex, **β-carotene** and **proteins**. Also known as sea peas and seaside peas.

Beakers Tall, wide-mouthed **plastics** or **glass containers**, often with a pointed lip for pouring. Also used to describe simple drinking vessels without handles commonly made from **clays** or plastics.

Bean curd Coagulated product obtained from **beans**. Used particularly with reference to **soy curd** (**tofu**).

Bean jams Sweet **bean pastes**, such as **ann**, which form the basis of many Japanese **confectionery** products.

Bean pastes **Pastes** prepared from beans such as **soybeans**, e.g. **miso** or **ann**.

Beans **Seeds** which grow in pods produced by plants such as *Phaseolus* species. Some beans are eaten fresh, frozen or canned, but most are dried to form a long-life staple food in many parts of the world. Beans are typically kidney-shaped and a good, inexpensive source of **proteins**, **fibre** and **folates**. The term is also commonly applied to seeds which resemble beans, such as **coffee beans** and **cocoa beans**.

Bean sprouts Young shoots of germinated **beans**, particularly **mung beans**. Rich in **vitamins** and **minerals** and a common ingredient in **salads** and Oriental dishes.

Bearberries Berries produced by the bush *Arctostaphylos uva-ursi*, which grows wild in northern and Arctic areas of Europe, Asia and North America. Similar in size to **currants**, with a tough skin and mealy white pulp containing hard seeds. Eaten raw as an emergency food or used as an extender with other berries in **bakery products** such as **fruit pies**. Extracts of leaves from the bush have **antioxidative activity**, making them of interest in production of natural **antioxidants** for use in foods.

Bear meat Meat from **bears**. In comparison with **beef**, it has high protein and low fat contents. Bear steaks can be cooked like beef, but the meat may be tough so it is often marinated for a couple of days in oil and wine or vinegar. In some countries, such as Thailand, wild bear meat may be consumed raw or partially cooked, and is consequently a source of **trichinosis**.

Bears Members of the widespread mammalian family Ursidae; there are several species including Asiatic black bears (*Selenarctos thibetanus*), polar bears (*Thalarctos maritimus*) and grizzly bears (*Ursus arctos*). Bears are hunted for their skins and for **bear meat**.

Beating Vigorous **stirring** of cooking ingredients, usually in a circular motion with the intention of incorporating air.

Beauvericin Insecticidal and antimicrobial hexadepsipeptide mycotoxin produced by several *Fusarium* strains. May be produced in *Fusarium*-infected **cereals**.

Beche de mer Name commonly given to edible **sea cucumbers** (Holothuroidae; *Stichopus* spp. and *Cucumaria* spp.); a popular delicacy in Japan, China and the Philippines. Marketed in gutted, boiled and dried forms.

Beech nut oils Yellow **oils** derived from the kernels of *Fagus sylvatica*, which are rich in **olein** and contain **stearin** and palmitin. Used as a cooking oil and salad oil.

Beef Meat from **cattle**, including bulls, calves, cows, steers and oxen. Quality is determined largely by breed, age and gender of the animal; it is also influenced by animal feeding, slaughtering technique and treatment of the meat post-slaughter. **Tenderness** and **flavour** are increased by hanging cattle **carcasses** (**ageing**/conditioning). Raw fresh beef is usually bright red in colour with creamy coloured **marbling**; however, meat from older cattle, particularly bulls, tends to be darker in colour. Composition varies with fat content and between different cuts, e.g. brisket, forerib, rump and silverside. Cuts which contain little **connective tissues** can be cooked by **roasting**, **frying** or **grilling**; however, tougher cuts should be cooked by **stewing** or **braising**, in order to soften the connective tissue. During the 1980s and 1990s, markets for beef have been affected negatively by consumer health concerns relating to high levels of saturated fats in red meat and to **prion diseases**, particularly bovine spongiform encephalopathy (**BSE**). Alternative term for beef muscles, bovine muscles, bull muscles, calf meat, calf muscles, cattle muscles and cattle tissues.

Beefburgers Round, flat cakes of **beef mince**, cooked by **grilling** or **frying**. Beefburgers are usually prepared from beef mince with a high content of fat. They are commonly eaten in **bread rolls**, served with lettuce, slices of onion and **tomato ketchups**.

Beef extracts Water-soluble extracts prepared from **beef**, used widely as **flavourings**. Preparation involves immersion of **beef mince** in boiling water to leach out the water-soluble extractives, and concentration. Direct extract can be produced by exhaustive extraction of beef; it contains a high concentration of **gelatin**. Beef extracts are rich nutritional sources of the **vitamin B group**; they can be formulated for use as **spreads** for bread, as flavourings, and, when mixed with water, as **beverages**. Beef extracts can also be used in preparation of beef tea, an extract of stewing beef that may be used as a food for invalids.

Beef loaf Meat products prepared primarily from **beef mince**, but also containing **pork mince** or pork **sausagemeat**. Other ingredients may include **onions**, **tomato purees**, **garlic**, white **bread**, **milk**, **herbs**, **eggs** and **seasonings**. The ingredients are mixed before baking in a loaf tin. Once cold, beef loaf can be cut into firm slices. Generally, it is served cold.

Beef mince Meat mince prepared from **beef** which is available in several grades; these primarily relate to the percentage of fat in the mince. For example, beef mince may be graded as: extra lean; lean, which has good flavour but does not shrink excessively on cooking; or regular, which is usually made from lower cost cuts of beef. Also known as ground beef or minced beef.

Beef muscles Alternative term for **beef**.

Beef patties Meat patties prepared from **beef mince**. They include **hamburgers**.

Beef roasts Joints of **beef** which are intended for cooking or have been cooked by **roasting**.

Beef sausages Sausages made primarily from **beef**. They may include **pork**, but the proportion of this is less than that of beef.

Beef steaks Thick slices of high-quality beef taken from the hindquarters of cattle **carcasses**, including sirloin, porterhouse, T-bone, fillet and rump steaks. They are usually cooked by **grilling** or **frying**.

Beer Alcoholic beverages manufactured by **alcoholic fermentation** of **worts** using either top or bottom fermenting **brewers yeasts**. The **malt** is commonly **barley malt**, but other malt types, including **wheat malt** or **sorghum malt** may be used. Non-malted cereals or other **brewing adjuncts** may be used in combination with the malt. Beer is commonly, but not always, flavoured with **hops**.

Beermaking Alternative term for **brewing**.

Beer manufacture Alternative term for **brewing**.

Bees Insects of the order Hymenoptera that are of commercial importance due to the ability of some species to produce **beeswax**, **honeys** and **royal jelly**. Some bee species of Halictidae or Apidae families have evolved to living in social groups or colonies. One of these species, the honeybee (*Apis mellifera*), produces a bee colony or comb, constructed of hexagonal cells composed of beeswax, in which to store food (honeys), and house insect eggs and larvae and the reproducing female bee or queen. Colonies of domestic honeybees also produce honeys which are collected as nectar and stored within the combs. Bees also have an important role in pollination of plants, including fruit trees.

Beeswax Yellow-coloured substance secreted by bees to make honeycombs. Solid, but easily moulded when warm. Consists of esters, cerotic acid and hydrocarbons. Used to make edible **wax coatings** for foods and edible films. Aqueous extracts may be used as **flavourings**.

Beetles Members of the large insect order Coleoptera, characterized by thickened shell-like forewings and membranous hind wings. *Tribolium* castaneum and *Sitophilus* oryzae are common insect **pests** of stored grain. Larvae of some species may be consumed as **insect foods**.

Beet molasses Molasses produced as a by-product of **beet sugar** refining. Beet molasses commonly contain approximately 60% **sucrose**. Also called beet sugar molasses.

Beetroot juices Juices extracted from **beetroots** (bulbous roots of *Beta vulgaris*). Consumed on their own or mixed with other **vegetable juices**, e.g. **carrot juices**, or **fruit juices**. Also drunk after **fermentation**. Useful as **natural colorants** due to the presence of the red pigment **betanin**. High contents of **nitrates** and **nitrites**, which might limit this application, can be removed by incubation with denitrifying **microorganisms**.

Beetroots Bulbous crimson red roots of *Beta vulgaris*, grown widely in Europe and America. Consumed as a boiled vegetable, pickled, or used as the basis for **borshch**. The red pigmentation of the root is due to the presence to **betanin**.

Beets Fleshy roots produced by plants of the genus *Beta*, such as **sugar beets**, used as a source of sugar, and **beetroots**, which are eaten as a vegetable.

Beet sugar Sucrose purified from roots of **sugar beets** (*Beta vulgaris*). Stages of beet sugar manufacture include: cleaning and cutting of roots; hot water extraction of **sugars**; purification of **beet sugar juices** by precipitation of impurities with lime-phosphoric acid or lime-CO_2 treatments; filtration to remove solids; concentration of the purified beet sugar juices; and crystallization of the pure beet sugar. Commercially available beet sugar comprises ≥99.80% sucrose and <0.05% moisture content.

Beet sugar factories Factories that contain processing lines equipped for **refining** of **sugar** from **sugar beets** (*Beta vulgaris*). Factories also usually contain sugar storage and packaging facilities.

Beet sugar juices Aqueous solutions of **beet sugar** produced during processing of roots of **sugar beets**. Raw juices are solutions produced by direct hot water extraction of the roots and contain beet sugar and impurities. Thin juices are purified beet sugar solutions and thick juices are formed by concentration of the thin juices.

Beet sugar molasses Alternative term for **beet molasses**.

Beet sugar syrups Highly concentrated aqueous solutions of **beet sugar** produced by **evaporation** of purified **beet sugar juices** (thin beet sugar juices).

Behavioural effects Alterations in human behaviour that can result from dietary constituents. Examples include modulation of **mood**, **cravings** and **cognitive performance**.

Behenic acid Synonym for docosanoic acid. One of the constituent **fatty acids** of the lipids fraction in various food plants. A low uptake from the digestive tract makes it potentially useful in preparation of low-calorie natural fat products.

Bell peppers Large, sweet-tasting **fruits** of *Capsicum annuum* with bell shaped pods that can vary in **colour** from green and white through to shades of red, orange, yellow and purple. One of the most popular types of **sweet peppers**; many different cultivars are available, most of which are non pungent. Can be eaten raw in **salads** or added to a variety of cooked dishes.

Belona Commercial cereal-based product composed of **wheat**, wheat protein concentrate, defatted **soy meal**, refined **soybean oils**, **vitamins** and **minerals**. Used in **weaning foods** in Nigeria.

Beluga Freshwater fish species (*Huso huso*); the largest member of the sturgeon family (Acipenseridae); also known as great sturgeon. Found in the basins of the Black and Caspian seas in Europe. Highly valued and sought after, mainly for its roe (**caviar**); flesh is also sold fresh, smoked and frozen. Bester, a hybrid of female beluga and male sterlet (*Acipenser ruthenus)*, has been successfully cultured for production of high quality caviar.

Beluga whales Species of **whales** (*Delphinapterus leucas*) widely distributed in the Arctic Ocean, which is still hunted on a subsistence level by indigenous people of Canada and Alaska for **whale meat**, **blubber** and other raw materials. Also known as **white whales**.

Bengal gram Indian name for **chick peas**.

Bengal quinces Alternative term for **bael fruit**.

Benlate Alternative term for the fungicide **benomyl**.

Benomyl Systemic fungicide used for control of a wide range of fungal diseases of fruits, vegetables and cereals. Example uses include: control of *Botrytis* in vegetables; control of *Verticillium* wilt in tomatoes; and control of snow mould in **wheat**. Often used in conjunction with other pesticides. Degradation in plants and animals is relatively slow. Classified by WHO as slightly toxic (WHO III).

Bentazone Selective contact herbicide used for control of certain weeds in cereal and vegetable crops. Rapidly metabolized to various derivatives in plants and animals and degrades rapidly in soils. Classified by WHO as slightly toxic (WHO III).

Bentonite Type of absorbent clay (a colloidal hydrated aluminium silicate) formed by the breakdown of volcanic ash that has the ability to absorb water with an increase in volume. Bentonite uses in the food industry include **fining agents** for **winemaking**, **clarifiers** for **fruit juices** and **vegetable oils**, **bakery additives** to reduce **staling**, **stabilizers** and **filter aids**.

Benzaldehyde Aromatic aldehyde which is one of the **flavour compounds** in a wide range of foods.

Benzene Aromatic hydrocarbon which exists as a colourless liquid with a sweet odour and which can evaporate into the air and dissolve in water. Widely used in industry in the manufacture of chemicals and a range of substances including **plastics**, **rubber**, dyes, detergents, **drugs** and **pesticides**. Carcinogenic in humans at high doses. Present as a pollutant of air from a variety of sources, and has also been found as a contaminant in **drinking water**, **mineral waters** and **soft drinks**. Contamination of carbon dioxide used in processing can lead to the presence of benzene in **carbonated beverages**.

Benzidine Toxic and carcinogenic aromatic amine which may occur as a contaminant in foods, and especially some **colorants**. Benzidine and its derivatives are also used as reagents in food analyses.

Benzimidazole One of a group of **pesticides** and **anthelmintics**, residues of which may occur as **contaminants** in foods.

Benzoates Salts of **benzoic acid**, used as antimicrobial **preservatives** in foods.

Benzoic acid Organic acid which, along with its salts, is used in antimicrobial **preservatives** for a wide range of foods.

Benzopyrene Carcinogenic and mutagenic **polycyclic aromatic hydrocarbons** (PAH) fraction which occurs as a contaminant in foods.

Benzothiazole Member of the **heterocyclic compounds** class of **flavour compounds**, occurring in a range of foods. May cause **taints** in some foods.

Benzyladenine One of the **plant growth regulators** which may be used to improve **ripening** and quality of **fruits**. May also be used as a thinning agent in cultivation of fruits.

Benzyl alcohol Aromatic alcohol which is a constituent of the **flavour compounds** and **aroma compounds** in various **fruits** and **spices**, and in plant-derived products such as **alcoholic beverages**.

6-Benzylaminopurine Plant growth regulator used to control processes such as **ripening** and **senescence**, and composition of **fruits**, **vegetables** and **cereals**.

Benzyl isothiocyanate One of the typical **flavour compounds** in **vegetables** and **spices** of the family **Cruciferae**; formed by hydrolysis of **glucosinolates**. May display **cytotoxicity** and **anticarcinogenicity**.

Benzylpenicillin Alternative term for the antibiotic **penicillin G**.

Berberries **Berries** produced by *Berberis vulgaris*. Ripe fruits are edible, but unripe berries contain toxic **alkaloids**. Bright orange red when ripe with a tart **flavour**. Can be made into **jellies**, pickled, used as a garnish or made into **spirits** and **liqueurs**. Their juice is rich in **vitamin C**. Also known as barberries.

Ber fruits Alternative term for **jujubes**.

Bergamot essential oils **Essential oils** obtained from the bergamot orange. Main use is in **flavourings** for Earl Grey tea. Also used in citrus flavourings for **soft drinks** and in some natural fruit flavourings, such as apricot. Contains bergapten, a skin sensitizer. Alternative term for **bergamot oils**.

Bergamot oils Alternative term for **bergamot essential oils**.

Bergapten Furocoumarin of the **psoralens** group of **flavour compounds**, characteristic of **bergamot essential oils**. Also occurs in **celery** and **parsley**.

Bergkaese cheese **Hard cheese** made from unpasteurized **cow milk** in Switzerland, Austria and Germany. Traditionally made from milk of cows grazing mountain pastures. Similar to **Emmental cheese**. Alternative spelling is bergkase cheese.

Berries Name commonly applied to various small, juicy, stone-less fruits. Include **strawberries**, **bilberries** and **loganberries**. In a botanical sense, the term relates to fruits having a pulpy edible part containing one or more seeds, such as **cranberries**, **grapes** and **bananas**.

Berry juices **Fruit juices** extracted from any of a range of **berries**, including: **bilberry juices**, **blackcurrant juices**, **cranberry juices**, **elderberry juices**, **hawthorn juices**, **raspberry juices**, **redcurrant juices** and **strawberry juices**.

Betacoccus Former name for the genus *Leuconostoc*.

Betacyanins Red/violet **pigments** of the **betalains** group, which occur naturally in **red beets** and other plant foods. Used as **natural colorants** in foods.

Betaine Soluble **nitrogen compounds** occurring in a range of foods, especially **sugar beets**, **molasses** and **beet sugar** factory wastes. May be included in **flavour compounds**, and have **antioxidative activity**.

Betalaines Alternative term for **betalains**.

Betalains Class of **pigments** naturally occurring in **fruits** and **vegetables**, especially those derived from plants of the Caryophyllales family. Include red/violet **betacyanins** and yellow **betaxanthin**. May be used as food **colorants**.

Betanin Member of the **betacyanins** group of **pigments**, characteristic of **red beets**. May be used as **natural colorants**.

Betaxanthin Yellow pigment of the **betalains** group.

Betel leaves Aromatic leaves of the Asian climbing plant, betel vine (*Piper betle*). Used to wrap **betel nuts** for the ritual chewing of betel quid. Also used as an edible wrapping for food in some Asian countries.

Betel nuts Acorn-shaped seeds of the betel palm, *Areca catechu*, also known as areca nuts. Seeds are used medicinally as an antihelminthic, but are most commonly used for the ritual chewing of betel quid, a popular masticatory, comprising betel nuts, slaked lime and **spices** wrapped in **betel leaves** (*Piper betle*). Chewing of this preparation is widespread throughout Asia, and causes mild stimulation due to the presence of **alkaloids** such as arecoline. Chewing of betel quid is associated with an increased risk of oral cancer.

Beutelwurst Types of **blood sausages** derived from **pork** and swine **offal** (including intestine and brain), and encased in swine intestines. A regional speciality in Germany.

Beverage concentrates Concentrated solutions or **syrups** which may be diluted to prepare **beverages**, e.g. **soft drinks**.

Beverage mixes Mixtures of ingredients which may be dissolved to prepare **beverages**, e.g. **soft drinks**.

Beverage powders **Beverage mixes** in the form of powders, to be dissolved in water or other liquids prior to dispensing or consumption.

Beverages Liquids intended for drinking. Types include **alcoholic beverages**, **soft drinks**, **teas**, **coffee**, **cocoa beverages**, **dairy beverages**, **health beverages**, **fruit beverages**, **soy beverages** and **drinking water**.

Beverages factories Factories in which **beverages** are manufactured or processed.

Beyaz cheese Turkish semi-soft **cheese** made from raw **ewe milk**. Usually made with **vegetable rennets** and stored in **brines** for at least 6 months before consumption. Used in **salads**, **pastries** and many local dishes. Similar to **feta cheese**.

BHA Abbreviation for **butylated hydroxyanisole**.

BHC Abbreviation for benzene hexachloride. Alternative term for the insecticide **HCH**.

BHT Abbreviation for **butylated hydroxytoluene**.

Bierschinken Ham **sausages** containing coarsely cut pieces of **meat**, originally made in Germany. Top quality bierschinken contains >60% coarsely cut, cured, tendon-free meat, with good cohesion in 1-mm thick slices. Medium quality bierschinken contains ≥50% coarsely cut meat, including pieces of meat which vary in size from 2-cm-sided cubes to egg-sized pieces.

Bierwurst Chunky, tubular, dark red coloured, cooked German **sausages**. They are prepared from **beef** and **pork**; the meat is chopped and blended, and **seasonings**, such as garlic, are added. The sausages are cooked at high temperature and smoked. They are usually sliced and served cold in sandwiches. Alternative term for beerwurst or beer **salami**.

Bifidobacterium Genus of Gram positive, anaerobic, rod-shaped **bacteria** of the family Actinomycetacea.

Occur among the normal flora of the urogenital and gastrointestinal tracts. *Bifidobacterium bifidum* may be incorporated in some **starters** used for the manufacture of **fermented dairy products**.

Bifidus factors Dietary constituents, particularly a component of **human milk**, that promote growth of *Bifidobacterium* in the **gastrointestinal tract**. This activity is demonstrated by certain prebiotic **oligosaccharides**, **lactulose** and derivatives of **glycoproteins**.

Bifidus milk **Fermented milk** containing *Bifidobacterium* species that make the product beneficial for intestinal health.

Bigeye snapper **Marine fish** species (*Lutjanus lutjanus* or *L. lineolatus*) belonging to the family Lutjanidae (snappers) and of high commercial importance. Widely distributed across the Pacific ocean. Snapper are normally marketed fresh.

Bigeye tuna **Marine fish** species (*Thunnus obesus*) from the tuna family. Found in the Atlantic, Indian and Pacific Oceans but absent in the Mediterranean. Flesh from this tuna species is highly prized; used for **sashimi** production in Japan. Marketed mainly canned or frozen but also sold fresh.

Bighead carp **Freshwater fish** species (*Aristichthys nobilis*) belonging to the carp family (Cyprinidae) and of high commercial importance. Widely distributed throughout the world. Marketed fresh and frozen.

Bilberries Dark blue berries produced by the European shrub *Vaccinium myrtillus*. Also known as whortleberries and similar in flavour to American **blueberries**. Rich in **vitamin C**, they can be eaten raw or used in products such as **pies**, **jams**, **jellies** and **fruit wines**.

Bilberry juices **Fruit juices** extracted from **bilberries** (*Vaccinium myrtilis*).

Bile acids Steroidal acids present in bile, which play an important role in **digestion** and **absorption** of **fats**. Cholic acid and chenodeoxycholic acids (primary bile acids) are produced by the liver from **cholesterol** and are secreted as glyco- and tauroconjugates into bile. On secretion of bile into the lumen of the **gastrointestinal tract**, bile salts bind colipase, allowing **lipolysis** of **triglycerides**, and also participate in formation of micelles facilitating absorption of lipids. Dehydroxylation of primary bile acids by intestinal bacteria generates secondary bile acids (deoxycholic and lithocholic acids). Bile acids can be reabsorbed as part of the enterohepatic circulation.

Bile salt hydrolases Alternative term for **choloylglycine hydrolases**.

Bile salts Alkaline salts present in bile involved in **emulsification** of fats in the intestine. Include sodium glycocholate and sodium taurocholate.

Biltong Traditional South African intermediate moisture meat product prepared from **meat** of domestic animals or **game**, but mainly from **beef**. Meat is cut into strips, trimmed and dipped in a solution of **salt** and sometimes **preservatives** and **spices** prior to drying to the desired moisture content. The dried product may also be smoked. Consumed by chewing the strips or by grating to a powder which can be spread on **bread**.

Binders Alternative term for **binding agents**.

Binding agents Substances used to hold ingredient mixtures together, providing **adhesion**, **solidification** and correct **consistency**. For example, **carrageenans** and **soy proteins** can be used to bind free moisture and improve **texture** and **juiciness** in reduced fat or restructured **meat** products. Other applications for binding agents include **chewing gums**, extruded foods and **confectionery**.

Binding capacity Ability of one substance to attach to another.

Bins Large containers used for storing specified substances or containers used for depositing rubbish. Also used to describe partitioned stands for storing bottles of **wines**.

Bioactive compounds Substances which display biological activity, e.g. **immunomodulation**, opioid activity, **antihypertensive activity** or **hypolipaemic activity**, upon ingestion. Found in a range of foods, and are of interest to the **functional foods** sector. Include **bioactive peptides** (occur widely in **dairy products**), many **vitamins** and **fatty acids**, **flavonoids** and **phytosterols**.

Bioactive peptides **Peptides** produced from plant or animal proteins, which display biological activity (e.g. opioid, immunostimulatory or **antihypertensive activity**), and are of interest to the **functional foods** sector. **Milk proteins** are a particularly rich source of bioactive peptides, such as casein phosphopeptides and β-casomorphins. Peptides that inhibit activity of **peptidyl-dipeptidase A** are found in a number of food sources and have potential use as antihypertensive functional foods ingredients.

Bioassay Technique for measuring the **biological activity** of a substance by testing its effects in living material such as a **cell culture**.

Bioavailability Extent to which a dietary component can be absorbed and utilized by the target tissue of the body. **Nutrients** with low bioavailability may be in a form that is poorly absorbed from the **gastrointestinal tract** (e.g. **lysine** combined with **reducing**

sugars as a results of the **Maillard reaction, minerals** in the presence of **antinutritional factors** such as **phytates**) or may be biologically inactive once absorbed.

Biocatalysts Substances that catalyse biochemical processes in living organisms. The most well known examples are **enzymes**, although **RNA** may also fulfil this function.

Biochemical oxygen demand Alternative term for **biological oxygen demand**.

Biochemistry Science of the chemistry of living organisms.

Biocides Chemical agents, such as **pesticides, herbicides** and **fungicides**, that are toxic or lethal to living organisms.

Biocontrol Deliberate exploitation by humans of one species of organism to eliminate or control another. Commonly involves introduction into the environment of **parasites, insects** or **pathogens** which can infect and kill or disable particular insect **pests** or weeds of crop plants. Also known as biological control.

Bioconversions Utilize the catalytic activity of living organisms to convert a defined substrate to a defined product in a process involving several reactions/steps. The term is often used interchangeably with **biotransformations**. Advantages include the ability to operate under mild conditions, the ability to produce specific **enantiomers** and the ability to carry out reactions not possible using conventional chemical synthesis. Bioconversions differ considerably from **fermentation**, since in the latter, the products often bear no structural resemblance to the pool of compounds given to the **microorganisms**.

Biodegradability Ability of a substance to undergo **biodegradation**.

Biodegradation Degradation of a substance as a result of biological (usually microbial) activity, rendering it less noxious to the environment.

Biodeterioration Deterioration (**spoilage**) of an object or material as a result of biological (usually microbial) activity. Biodeterioration of foods causes them to become less palatable and sometimes toxic, and can involve alterations in **flavour, aroma**, appearance or **texture**. The organisms involved are typically **bacteria** and **fungi**, and their activity is dependent on factors such as nutrients present, a_w, **pH, temperature** and degree of aeration.

Biofilms Films of **microorganisms**, usually embedded in extracellular polymers, which adhere to surfaces submerged in or subjected to aquatic environments. Frequently cause fouling of the surfaces of water pipes.

Bioflavonoids Flavonoids present in a wide range of plant foods, which have beneficial effects on health.

Bio foods Term used to describe **biotechnologically derived foods** or **functional foods**.

Biogarde German yoghurt-like **acidophilus milk** usually made with **starters** containing *Streptococcus thermophilus, Lactobacillus acidophilus* and *Bifidobacterium bifidum*.

Biogas A mixture of gases produced by anaerobic digestion of organic wastes, comprising mainly methane and **carbon dioxide** with traces of **hydrogen, nitrogen** and **water vapour**. Used as a fuel. Product of **bioremediation** of many types of food processing wastes.

Biogenic amines Amines (e.g. **histamine, tyramine, tryptamine, putrescine**) synthesized by decarboxylation and hydroxylation of **amino acids** by microbial **enzymes**. Can cause allergic reactions. May be formed in **cheese, wines, chocolate** and **fermented foods**.

Bioghurt German yoghurt-like **acidophilus milk** usually made with **starters** containing *Streptococcus thermophilus* and *Lactobacillus acidophilus*.

Biological activity Activity of compounds, generally organic in origin, within living organisms. For food-derived chemicals, this is generally a non-nutritional property, such as **antimicrobial activity, antioxidative activity, immunomodulation**, or other **physiological effects**.

Biological membranes Selectively permeable **membranes** containing mainly **lipids** and **proteins** that surround the cytoplasm in eukaryotic and prokaryotic cells. Can also contain **carbohydrates** and **sterols**. The precise composition depends on the species and, in some cases, on growth conditions and age of the cells. The lipids (**phospholipids** and **glycolipids**) usually form a bilayer within which proteins are partly or wholly embedded, some spanning the entire width of the bilayer. Artificial biological membranes (liposomes) are often used to transport biological molecules.

Biological oxygen demand Amount of dissolved oxygen required for microbial oxidation of biodegradable matter in an aquatic environment containing organic matter, such as sewage, water or milk. Gives an indication of contamination by **microorganisms** which take up oxygen for their metabolism. Also known as biochemical oxygen demand or by the abbreviation BOD.

Biological values Indication of the nutritional value of food **proteins**. Relative measure of the amount of absorbed proteins retained by the body, assuming no loss of protein nitrogen during digestion. Values are highest for **egg proteins** (0.9-1.0) and **milk proteins** (0.85), with meat proteins and **fish proteins** (0.7-0.8), **cereal proteins** (0.5-0.7) and **gelatin** (0) having lower values.

Biology Science of the properties of living organisms and the interactions of these organisms with their environment.

Bioluminescence Production of light as a product of biochemical reactions by organisms including **bacteria**, **fungi**, some **fish** and fireflies.

Biomarkers Biological indicators of exposure and response to a chemical, physical or biological agent. May be used to indicate the presence of that agent or the progress of a specific biological state, such as a disease.

Biomass Quantitative estimate of the total population of living organisms present in a given habitat, in terms of mass, volume, or energy.

Biomycin Alternative term for **chlortetracycline**.

Biopolymers Polymers which occur in living organisms. Included in this group of macromolecules are **polysaccharides**, **proteins** and **nucleic acids**.

Bioreactors Reactors used for generating products using the synthetic or chemical conversion capacity of a biological system, such as enzyme reactions and **fermentation**. They can be **fermenters**, **cell culture** perfusion systems or enzyme bioreactors. For production of **enzymes** or **proteins**, **recombinant microorganisms**, such as **bacteria**, **yeasts**, **fungi** and plant cells, are usually chosen. Also used for **bioremediation** of industrial **effluents**, such as food industry **waste water**.

Bioremediation Use of **microorganisms** and/or **enzymes** to reduce the pollution potential of industrial **effluents**, such as food industry **waste water**, converting them to less hazardous forms. Can also be used to generate **biomass** and **biogas**.

Biosensors Biomolecular probes that can be used to measure a variety of parameters in biological systems by translating a biochemical interaction at the probe surface into a quantifiable physical signal. **Immobilization** of **enzymes**, **antibodies**, receptors, **DNA**, cells or organelles on the surface of a transducer forms the basis of various biosensors. Used widely in the food industry for measuring levels of various components in foods and **beverages**, detection of **contamination** and **adulteration**, and for monitoring and **process control** of **fermentation** processes, **bioconversions** and **biotransformations**.

Biosurfactants Potent **surface active agents** produced by a variety of microorganisms, including *Pseudomonas*, *Rhodococcus*, *Candida*, *Corynebacterium*, *Mycobacterium*, *Acinetobacter* spp., *Bacillus* subtilis, *Serratia* and *Thiobacillus* spp. Low molecular weight biosurfactants are often **glycolipids**, and high molecular weight biosurfactants are generally either polyanionic heteropolysaccharides containing covalently-linked hydrophobic side chains or complexes containing both polysaccharides and proteins. Biosurfactants have a number of advantages over their chemical counterparts, such as biodegradability, effectiveness at extremes of temperature and pH, and lower toxicity. Biosurfactants are used in the food industry as **emulsifiers** and **stabilizers**.

Biotechnologically derived foods Foods produced by means of **biotechnology**.

Biotechnology In its broadest sense, any industrial process in which **microorganisms** are used. More commonly used for those industrial processes in which **genetic engineering** techniques have been used to construct novel strains to improve their properties and produce new products.

Biotechnology products Products produced by **microorganisms** in biotechnological processes.

Biotin A water-soluble nutrient that is involved in the physiological biosynthesis of fats and the utilization of carbon dioxide. Rich dietary sources include **egg yolks**, **cattle livers** and **yeasts**. Avidin, a protein present in raw **egg whites**, can act as a biotin antivitamin by binding biotin and reducing its **bioavailability**. Also known as **vitamin H** and coenzyme R.

Biotransformations Specific modification of a defined compound to a defined product with structural similarity through the use of biological catalysts (**enzymes**, or whole dead or resting **microorganisms**). Advantages are the same as those for **bioconversions**.

Biphenyl Fungicides which inhibit fungal sporulation; used primarily to control fungal growth on the surface of stored **citrus fruits**. Residues on fruits sometimes persist throughout the storage period. Classified by Environmental Protection Agency as slightly toxic (Environmental Protection Agency III). Also known as diphenyl.

Bird rape **Oilseeds** produced by *Brassica* rapa or *B. campestris*.

Birds Warm-blooded vertebrates in the class Aves that have wings and feathers and lay eggs.

Birds nests Edible birds nests are nests made by swifts and swallows, especially species of the genus *Collocalia*, in which minor feathers are mixed with gelatinous strands of saliva. Used in traditional Chinese medicine and Chinese cuisine. Most commonly eaten in birds nest **soups**, but also used in other dishes. Nests are often relatively inaccessible, making them expensive and prone to fraud. **Authenticity** can be established by analysis of either the **amino acids** or the **oligosaccharides** of the **glycoproteins**.

Birefringence The optical property of a substance, usually a crystal, in which a ray of light passing through the substance is separated into two plane-polarized rays (double refraction). The effect can occur when the velocity of light in the material is not equivalent in all directions, resulting in different refractive indices for light polarized in different planes.

Biscuit dough Dough used to make **biscuits**.

Biscuit factories Factories in which **biscuits** are manufactured.

Biscuits Flour-based products that vary greatly in size, shape and texture, but are generally small, thin and short or crisp. Usually made with flour, **butter** or vegetable **shortenings**, sugar and sometimes a leavening agent; other ingredients, e.g. cocoa, chocolate chips, dried fruits, nuts, cheese or flavourings, are added according to the type of biscuits to be made. Usually eaten as **snack foods**, often with **beverages**. Can be eaten as part of a meal along with cheese. Called **cookies** in the USA, where the term biscuits refers to soft, scone-like products.

Bison Humpbacked, shaggy coated members of the family Bovidae. There are two species: the North American bison (*Bison bison*); and the European bison (*B. bonasus*). Bison are reared on game farms for **bison meat** production, particularly in the USA and Canada.

Bison meat Meat from **bison**. Bison meat is very lean and tender, and has a similar **flavour** to lean **beef**; it has no pronounced gamey flavour.

Bisphenol A Common name for 4,4'-isopropylidenediphenol, an intermediate used in production of epoxy, polycarboate and phenolic resins. **Polycarbonates**, **plastics** used in a wide range of products including microwave cookware and food **containers**, are formed by reaction of bisphenol A with phosgene. Bisphenol A is also used in coatings for **cans**. There are concerns over the possibility of **migration** of bisphenol A monomers from cans or containers into foods as intake might have endocrine disrupting effects.

Bisphenol A diglycidyl ether Constituent of epoxy resin coatings used in food **cans** or food storage **containers**. Residues may migrate into the foods in the cans or containers. Often abbreviated to BADGE.

Bisulfites Hydrogen sulfite salts used in antimicrobial **preservatives** and **antioxidants** in foods and **beverages**.

Bitter acids Bitter compounds in **hops**, specifically α-acids (**humulones**) and β-acids (**lupulones**).

Bitter almond oils Oils rich in **oleic acid** derived from seeds of **bitter almonds**. Contain **benzaldehyde** and **hydrocyanic acid**; the latter compound, which is toxic, is removed during extraction. Used in **flavourings**.

Bitter almonds Common name for nuts produced by *Prunus dulcis* (syn. *P. amygdalus*). Too bitter for fresh consumption and also contain highly toxic **hydrocyanic acid**, or hydrogen cyanide. Cultivated mainly for manufacture of **bitter almond oils** (principal component **benzaldehyde**), which are used as **flavourings** following removal of the hydrocyanic acid.

Bitter compounds Compounds with a bitter taste; these may be used as **flavourings** in foods or **beverages**, e.g. **hops bitter acids** in **beer**, or **quinine** in **soft drinks**.

Bitter gourds Ovoid orange-yellow **fruits** from the tropical climbing plant *Momordica charantia*. Fruits are also known as balsam pears and have a characteristic bitter taste, which can be minimized by salt-water treatment, and by selecting young fruits. Bitter gourds can be eaten raw in **salads**, cooked as a vegetable or used in **pickles** and **curries**. Contents of **vitamins**, **minerals** and essential **amino acids** are similar or superior to those of other Cucurbitaceae. Young shoots can also be eaten as a substitute for **spinach**.

Bitterness Flavour produced by **bitter compounds** such as **caffeine** and other **alkaloids**, often at low thresholds.

Bitter peptides Peptides, formed during enzymic hydrolysis of **proteins**, which have a bitter taste and may impair the sensory quality of the food. Bitter peptides derived from **casein** may be a particular problem in **cheesemaking**. Bitter peptides may also cause problems in soy products and **protein hydrolysates**. Treatment with **peptidases** may eliminate quality problems attributable to bitter peptides.

Bitter pit Physiological disorder of **apples** associated with low **calcium** concentrations in the fruit.

Bitter principles Alternative term for **bitter compounds**.

Bitto cheese Italian **cheese** made on an artisanal scale from **cow milk** or cow milk mixed with not more than 10% **goat milk**. Granted Denomination of Origin status. Rind has a characteristic straw-yellow **colour** which intensifies with **ripening**. **Texture** and **flavour** vary with stage of **ageing**, young cheese being soft with a sweet and slightly aromatic flavour, while older cheese acquires a piquant flavour and is firmer. Used as an ingredient in local cooked dishes and **salads**.

Bivalves Molluscs from class Bivalvia having paired shells (valves) usually connected by a hinge that per-

uses; for example, they are used as **binders** in **sausages** and as ingredients, particularly as **emulsifiers**, in other foods.

Blood sausages Cooked **sausages** produced using cattle and/or swine blood, but also including ingredients such as diced, cooked **pork** fat, cooked **meat mince**, **gelatin**-producing materials, **oatmeal**, **bread**, **apples**, **chestnuts**, **onions**, **cream** and a wide range of **spices**. **Flavourings** used vary from region to region, and national preference for **texture** also varies. For example, French, Italian and Spanish-style blood sausages are moist and have a loose texture, whilst German and English versions are more compact. Blood sausages are usually sold precooked, but tend to be grilled or fried whole, or cut into slices and grilled before serving. Varieties produced in different countries include **blutwurst** in Germany, morcilla in Spain, **black puddings** in the UK, boudin noir in France and biroldo in Italy.

Bloom **Spoilage** of **chocolate** by fat or **sugar** deposition. Fat bloom is the appearance of white spots (composed of crystals) on the surface of chocolate, caused either by **cocoa butter triacylglycerols** undergoing polymorphic transition at the surface, or by cocoa fats rising to the surface of the chocolate as a result of high or fluctuating temperatures. Sugar bloom is deposition of sugar crystals on the surface of chocolate, caused by excessive moisture.

Blossom end rot Physiological disorder of plants associated with **calcium** deficiency. Affects **tomatoes** and some other crops such as **watermelons** and **peppers**.

Blotting Transfer of **nucleic acids** and/or **proteins** either directly or from a gel to a chemically reactive matrix (e.g. nitrocellulose), on which the nucleic acids/proteins bind covalently in a pattern identical to that on the original gel. After blotting, target molecules are detected through the use of complementary labelled nucleic acids or **antibodies**. Includes **northern blotting**, **Southern blotting** and **western blotting**.

Blowing Defect of **cheese**, also known as late blowing, that leads to gas formation and abnormal flavour development during **ripening**. Caused by **butyric acid** fermentation by *Clostridium* *tyrobutyricum*, a pasteurization-resistant contaminant of milk that can occur when animals have been fed silage.

Blubber Thick, subdermal lipid layer found in **cetacea** and other large marine animals, often forming up to 25% of the animal's total weight and acting as an insulator. May often become contaminated by organochlorine compounds such as **polychlorinated biphenyls** (PCB). Frequently consumed by Arctic inhabitants.

Blueberries Edible, smooth-skinned dark blue **berries** of several species of *Vaccinium*, grown predominantly in North America. Fruit of *V. corynbosum* are known as highbush blueberries, while those of *V. angustifolium* are known as lowbush blueberries. Berries are similar in flavour to European **bilberries** and contain a wide range of **phytochemicals** and antioxidant **vitamins**. Eaten raw or consumed in baked or other processed foods.

Blue cheese Hard white cheese with blue veins. It usually has a tangy or spicy flavour. In addition to being eaten as a dessert cheese, it is also used in **salad dressings**, **dips** and **sauces**.

Blue crabs Marine **crabs** (*Callinectus sapidus*) found on the Atlantic coast and Gulf coast of the USA. The most commercially valuable crab species consumed in the USA. Usually marketed fresh, in soft and hard shell stages.

Bluefish **Marine fish** species (*Pomatmus saltatrix*) of high commercial value belonging to the family Pomatomidae. Widely distributed in all world oceans with the exception of the eastern and north-western Pacific Ocean. Mainly marketed fresh, but also sold frozen, dried or salted. A popular game fish, produced commercially by **aquaculture**.

Blue green algae Older term for **cyanobacteria**, a large group of prokaryotic, photosynthetic, unicellular or filamentous organisms, which differ from other bacteria in that they possess chlorophyll a and carry out photosynthesis. Occur in fresh, brackish, marine and hypersaline waters. Some freshwater bloom-forming strains (e.g. *Anabaena*, *Microcystis* and *Nodularia*) produce potent cyanotoxins (**saxitoxin**, **microcystins** and nodularin, respectively) which may contaminate reservoirs. Tainting of **drinking water** supplies by such cyanobacterial blooms may cause illness or death in humans and animals which drink the water.

Blue mussels Common name for **mussels** of the species *Mytilus edulis* or *M. galloprovincialis* which have high commercial value. Distributed worldwide. Due to high demand, wild populations are supplemented with mussels produced by **aquaculture**.

Blue whiting **Marine fish** species (*Micromesistius poutassou*) from the **cod** family (Gadidae) which is widely distributed across the northern Atlantic Ocean and also occurs in the west Mediterranean Sea. Has tender white fine-textured flesh with a flaky delicate **flavour**. Marketed fresh and frozen (mainly headed and gutted), and also processed for **fish oils** and **fish meal**. In France, this fish is utilized for the production of **fish balls**.

Blutwurst Salty, spicy **blood sausages**, originally made in Germany, that are prepared from **pork**, **beef** and cattle **blood**. Commonly, they are eaten as a snack or mixed with **sauerkraut**. Sold precooked, but usually heated before serving.

Board Long, flat, usually rectangular, piece of rigid or semi-rigid composite material used to make containers, e.g. **boxes** or **cartons**. Available in various types that differ in composition and thickness, including **cardboard**, **paperboard** and **fibreboard**.

Boar meat **Pork** produced from entire male swine **carcasses**. In comparison with pork produced from castrated male swine, boar meat comes from leaner carcasses which have larger eye muscles, and the meat has a better flavour. However, consumers may avoid boar meat because it is perceived to be associated with **boar taint**. Other problems may include poor slicing properties, poor rind finish and soft fat, skin blemishes from fighting, poorer yields of matured bacon, high carcass pH values, lower keeping quality, and presence of the **PSE defect**.

Boars Mature entire male **swine**. Usually those used for breeding, but can also relate to those used for meat production. Production of **boar meat** has certain economic advantages to the producer, as boars grow faster and use feeds more efficiently than castrated male swine; they are thus a source of cheaper lean meat. However, boars are more excitable than castrated male swine and fighting may occur during transport and lairage, causing stress, skin damage, high carcass pH values, rapid spoilage and the **PSE defect** in the meat produced. In broader use, the term is used to describe mature males of certain other mammalian species, such as guinea pigs or hedgehogs.

Boar taint An unpleasant **off odour** and **off flavour**, which arises when **meat** from certain **swine** is heated. Also known as boar odour. Boar taint occurs mainly in **boar meat** but may also occur, to a lesser extent, in **pork** from female and castrated male swine. Boar taint sensitivity differs markedly between consumers; notably, more women than men are able to detect it. The major compounds responsible are **androstenone** and **skatole** in swine fat. Taint associated with skatole is characterized by the descriptors mothball and musty, and that associated with androstenone by parsnips, silage, sweaty and dirty. An integrated approach to management of boar taint has been proposed, involving techniques such as immunocastration, genetic selection and processing of tainted meat.

Bockwurst Mildly flavoured, fresh or parboiled German **sausages**, a type of **bruehwurst**, made from **veal** and **pork**, usually with a higher proportion of veal. Recipes often include **chives**, chopped **parsley**, **eggs** and **milk**. The sausages have a short shelf life and require thorough cooking before they are eaten. Traditionally, they are served with bock **beer**, especially during Bavarian bock beer festivals in Germany.

BOD Abbreviation for **biological oxygen demand**.

Body Texture term relating to the fullness of a product; especially applied to **wines**.

Bodying agents **Additives** used to impart desirable **body**, **viscosity** and **consistency** to foods. Often used to improve **texture** of **low calorie foods**.

Body mass index Index of human obesity which is calculated as the weight in kilograms divided by the square of the height in metres. The normal range is usually quoted as 20-25 kg/m^2.

Bogue **Marine fish** species (*Boops boops*) of high commercial importance belonging to the porgies family (Sparidae). Found in the eastern Atlantic Ocean, Mediterranean Sea and Black Sea. Marketed fresh and frozen and eaten pan-fried, broiled or baked.

Boiled ham Boneless ham, which is cured, shaped and fully cooked by **steaming** or **boiling**. It is sold whole or is sliced and packaged before retail.

Boiled sausages **Sausages** that are heat processed by boiling during manufacture. Examples include Pariska, Posebna and Hrenovke sausages.

Boiled sweets **Sugar confectionery** products formed by boiling **sugar** and **glucose syrups** with **flavourings** and other ingredients as required, to form a glassy mass upon cooling.

Boilers Fuel-burning devices for heating water in which foods can be immersed and cooked.

Boiling Process of raising the temperature of a liquid, by application of heat, to the point where it bubbles and turns to vapour. The term also means **cooking** food in a boiling liquid.

Boiling point Temperature at which a liquid boils. This occurs when the vapour pressure of the liquid is equal to the atmospheric pressure of the surrounding environment. Usually abbreviated to b.p.

Boletus A large genus of wild **fungi**, many of which are edible. *B. edulis* (also known as cep) is one of the best known edible species. It is found throughout Europe and can be fried, baked in casseroles, used in **salads** or **soups**, or dried. *B. badius* is another edible species with a similar **flavour** to *B. edulis*.

Bologna Smoked, cooked **sausages** prepared from finely minced, cured **pork** and/or **beef**. Their name originates from the city of Bologna in Italy, but true Italian bologna is known as **mortadella**. Bologna of various diameters can be purchased in rings, rolls or slices; they are retailed fully cooked and ready to serve. Types of bologna include chub bologna, beef bologna and ham-style bologna.

Bolti Common name, used especially in Egypt, for the **freshwater fish** *Oreochromis niloticus* (formerly *Tilapia nilotica*), also known as **Nile tilapia**. Of high commercial importance. Widely distributed in lakes and rivers in Africa and also produced by **aquaculture**. Marketed fresh and frozen.

Bombay duck **Marine fish** species (*Harpodon nehereus*) from the Indo-West Pacific; primarily caught along the coast of the Maharashtra region of India. Regarded as an excellent food fish with jelly-like flesh having high moisture content. Marketed fresh and as a dried/salted product.

Bonbons Generic term for **sugar confectionery**. Applied to a variety of types of sweets, often with a chewy centre.

Bone marrow Soft, gelatinous, highly vascular connective tissues that occur in certain long **bones**. Responsible for producing red blood cells as well as many white blood cells. When bones of high marrow content are used during mechanical recovery of **meat**, the lipid and haem concentrations of the recovered product are increased, and the tendency of the meat to undergo oxidation is increased.

Bone mineral density Level of mineralization of bone. Measurements can be taken by dual energy X-ray absorptiometry or ultrasound, and are used in the clinical assessment of **osteoporosis** risk. Many dietary and lifestyle factors have been proposed to modulate bone mineral density, including positive effects being reported for dietary **calcium**, **tea**, soy **isoflavones** and **vegetable proteins**.

Bones Components of the skeleton made from hard, rigid structural material. Bones are composed of an organic matrix of **collagen**, osseoalbumoid and osseomucoid; this is impregnated with mineral **salts**, particularly **calcium** phosphate, calcium carbonate and magnesium phosphate. **Fluorides** and **sulfates** are also present. The central cavities of most bones contain red **bone marrow**; however, the cavities of the long bones contain yellow marrow. Bones are processed to produce **fats** and **gelatin**. Chopped bones may be boiled to prepare bone broth. Degreased animal bones are used to prepare bone meal, which is used as a supplementary source of calcium and phosphates in foods and feeds, and as a source of phosphates in plant fertilizers. Bone charcoal is used in sugar refining and in bleaching. Specialized bone powders are used to remove fluorine from drinking water.

Bongkrek Traditional Indonesian type of **tempeh** made by **fermentation** of presscake of **coconuts** or **coconut milk** residue with *Rhizopus oligosporus*. Consumption can lead to fatal **food poisoning** due to contamination of the product with *Pseudomonas cocovenenans* (*Burkholderia* cocovenenans), strains of which produce the toxins toxoflavin and bongkrekic acid; favourable conditions for optimum production coincide with conditions under which bongkrek is manufactured.

Boning Removal of the **bones** from **meat** or **fish**, usually before **cooking**.

Bonito Any of several species of medium sized **tuna**, especially those from the genus *Sarda*, including *S. sarda* (Atlantic bonito), *S. chilliensis* (Pacific bonito) and *S. orientalis* (Oriental bonito). Fat content of flesh ranges from moderate to high; the most strongly flavoured of the tunas. Marketed mainly fresh; also dried-salted, canned and frozen.

Boondi Deep fried **fritters** made with **Bengal gram meal** and eaten as **snack foods** in India.

Borage Common name for the Mediterranean herb, *Borago officinalis*. Leaves have a **flavour** reminiscent of **cucumbers** and are consumed as a vegetable or used in **flavourings** for **beverages**, **soups** and **salads**. The purple star-shaped flowers are used as a garnish, often in crystallized form. Seeds are used for production of **borage oils**.

Borage oils **Oils** derived from the European herb, *Borago officinalis*, which contain γ-linolenic acid, **palmitic acid**, **oleic acid** and **linoleic acid**.

Borassus Genus of **palms** which includes *Borassus flabellifer*, also known as **palmyra** palm, a species which yields edible fruit and whose inflorescence is a source of **palm wines**, **sugar** and **vinegar**. *B. aethiopum* (black rum palm) is also utilized as a source of ingredients for foods and **beverages**.

Bordeaux mixture Broad-spectrum fungicide originally developed in France to control disease in **grapes**. Made by mixing copper sulfate and hydrated lime in water. Used to control disease in a wide range of tree **fruits**, **nuts** and vine fruits.

Boric acid Acid used in **preservatives** for **sea foods** and **wines**.

Borneol Member of the terpene alcohols class of **flavour compounds** present in many **fruits**, **herbs** and **spices**.

Borneo tallow Seed lipid from *Shorea stenoptera* which is rich in **palmitic acid**, **stearic acid** and **oleic acid**. Shows a sharp melting profile due to its high content of a single triacylglycerol (stearic acid-oleic acid-stearic acid, SOS) and may be used in **cocoa butter equivalents**.

Boronia Genus of woody flowering **plants**. Flowers of some species, especially *Boronia megastima*, yield **essential oils** which are used in fruit **flavourings** for foods and **beverages**.

Borshch **Soups** made from meat **stocks**, **beetroots** and **cabbages**. Popular in Russia, the Ukraine and Poland. Served hot or cold, usually with a dash of **sour cream**.

Botrytis   Genus of **fungi** of the class Hyphomycetes. Includes many species which are plant **pathogens**, e.g. *Botrytis aclada*, *B. cinerea* and *B. fabae*.

Bottled mineral waters   **Mineral waters** which have been packaged in bottles for distribution and retail sale.

Bottled water   Any type of **potable water** which has been packaged in bottles for distribution and retail sale.

Bottles   Portable **containers** often made from **glass** or plastics, used to hold or store liquids. Typically with narrow necks, which can be closed with **caps**, **corks** or **stoppers**. The term also refers to metal containers which are used to transport and store liquefied **gases**.

Bottling   Process of putting substances into **bottles** for storage and **preservation**. Most commonly applied to **beverages**, such as **wines**, **fruit juices** and **beer**, but also used to describe a method of preserving **fruits** in **syrups** or in the form of **jams**. After the products have been placed in the bottles, the containers are sealed with **corks** or other **closures** to prevent air or **microorganisms** from entering and causing **spoilage**.

Bottom fermenting yeasts   **Brewers yeasts** (*Saccharomyces* spp.) which flocculate and collect at the bottom of the **fermentation** tank during fermentation of **beer**. Used in brewing of a range of beer types, including **lager**.

Botulism   Foodborne disease caused by ingestion of food contaminated with **botulotoxins** produced by *Clostridium botulinum*. Symptoms include vomiting, abdominal pain, visual disturbances and difficulty in speaking and swallowing. Foods commonly implicated are low-acid, low-salt foods (e.g. improperly canned **vegetables** and **soups**, and **fish** and **meat** products).

Botulotoxins   Extremely potent **neurotoxins** produced by *Clostridium botulinum*, which cause **botulism**. Also known as botulinus toxins, botulinum toxins, botulins and botulismotoxins.

Bouillon   Thin, unclarified **broths** or **soups** typically made by boiling beef or chicken in water. Similar to **stocks**.

Bouquet   **Aroma** of foods or **beverages**, in particular, that of **wines**.

Bourbon whiskey   A type of American corn **whiskey**, originally made in Bourbon County, Kentucky, USA.

Bovine   Relating to or belonging to the **cattle** family.

Bovine immunodeficiency viruses   Lentiviruses which cause lymphadenopathy, lymphocytosis, central nervous system lesions, progressive weakness and emaciation in **cattle**.

Bovine muscles   Alternative term for **beef**.

Bovine serum albumin   Protein fraction present in **cattle blood** and frequently used as a model in studies on factors influencing properties and behaviour of food **proteins**.

Bovine somatotropin   Alternative term for bovine growth hormone. Recombinant bovine somatotropin may be administered to cattle to modify **milk** production, growth rate, or composition of cattle **carcasses** or **beef**. This application is permitted in some countries but prohibited in others due to concerns about the safety of food products.

Bovine spongiform encephalopathy   Commonly abbreviated to BSE, one of a group of **prion diseases**, this one affecting **cattle**. BSE can be transmitted to humans and other animal species via contaminated cattle-derived foods and **feeds**. Individual cattle in the UK were probably first infected with BSE in the 1970s, but BSE was not formally identified until November 1986. It is believed to have developed because of intensive farming practices, particularly the inclusion of meat and bone meal from animal carcasses in cattle feeds. Currently, it is believed that BSE may have originated in an individual cow as a consequence of a gene mutation and spread to other cattle because of recycling of cattle remains in cattle feeds. In cattle, BSE has a 5-yr incubation period. Despite various warnings, including the realization in 1990 that BSE could be transmitted to cats, BSE was not widely recognized as a threat to human health until variant **Creutzfeldt-Jakob disease** (CJD) was identified in 1996. This discovery was followed rapidly by a worldwide ban on all UK beef exports by the European Union and the banning of cattle of >30 months of age from the UK food chain. Health and safety concerns relating to BSE have severely damaged the UK farming industry and threaten to have similar effects in continental Europe.

Bovista   Genus of **fungi** of the class Basidiomycetes. *Bovista plumbea* (lead puffball) and *B. nigrescens* (dark puffball) are edible **mushrooms**.

Bowels   Common name for the large and/or small **intestines**. They form a part of edible **offal**.

Bowls   Round, concave **containers**, usually hemispherical and open at the top which can be used for holding foods. Made from various materials, including **glass**, **wood** and **plastics**. Also a specific type of drinking goblets.

Boxes   **Containers** with the four sides perpendicular to the base and a cover or lid. Made from various materials, including **wood**, **cardboard** and **plastics**. Can be used to store or package foods.

Boxthorn Common name for the solanaceous plant *Lycium chinense*, also known as matrimony vine. **Fruits** and leaves are used as foods (vegetables) and/or tea infusions in the Orient.

Boysenberries *Rubus* hybrid **berries** obtained by crossing **loganberries**, **raspberries** and **blackberries**. Purple red in **colour**, and a rich source of **anthocyanins**.

Boza **Fermented beverages** consumed traditionally in Turkey, Bulgaria and other Balkan countries. Made from various types of cereal, most commonly **bulgur wheat** or **millet**, but also **barley**, **oats** or **corn**, which is cooked in water and then crushed before fermenting with **yeasts**. The thick beverage has an unusual sweet and sour flavour. It is served chilled, sprinkled with **cinnamon** and garnished with roasted **chick peas**.

b.p. Abbreviation for **boiling point**.

Bracken Edible **ferns** from *Pteridium aquilinum* and related species. Consumed as a vegetable in some parts of the world, including Japan, New Zealand and Canada. Also used as a source of **starch**. Bracken contains a number of toxins such as ptaquiloside, a potent carcinogen, which can also be transmitted via **milk** from cattle feeding on bracken. Curled, undeveloped bracken fronds (**fiddleheads**), which are consumed as a delicacy in some areas, are particularly hazardous to health, as they contain **carcinogens** that must be destroyed by **roasting** before consumption. Bracken fiddleheads can be consumed accidentally as they resemble those of the **ostrich fern**, which are not poisonous.

Brains The main organs of the central nervous system (CNS), located within skulls. Brains of slaughtered animals are a part of edible **offal**; however, in many European countries, recent concerns relating to **prion diseases** have led to the exclusion of brains and other **central nervous system tissues** from the food chain. Nevertheless, the protein component of brains has a well-balanced amino acid composition. Brains have a very high content of fat, a large proportion of which is made up of complex phospholipids and glycolipids. Brains are a rich source of **minerals** (especially Fe, P, Ca and Mn) and vitamins B_2 and B_{12}.

Braising Light **frying** of foods (usually **meat** or **vegetables**) followed by **stewing** in a small amount of liquid at low heat for a lengthy period of time in a closed (tightly covered) container. The long, slow cooking develops **flavour** and causes **tenderization** of the food by gently breaking down fibres. Braising can be undertaken on stoves or in **ovens**.

Bran Protective outer layer of the seeds of cereals that is separated from the kernels during **milling**. Often used in **breakfast cereals** and other products as a source of **dietary fibre**.

Branching enzymes Alternative term for **1,4-α-glucan branching enzymes**.

Brandies **Spirits** manufactured by **distillation** of fermented fruit-based **mashes** or **wines** (including grape wines or **fruit wines**) or wine by-products.

Branding Process of applying a trademark or distinctive logo to a food or its packaging to identify its manufacturer or retailer.

Brandy **Spirits** manufactured by **distillation** of **wines**; unless further qualified, the term brandy generally refers to spirits distilled from grape wines.

Brassica Genus of plants belonging to the Cruciferae family. Native to the Mediterranean region and cultivated widely in Europe. Important *Brassica* crops include **cabbages**, **cauliflowers**, **broccoli**, **swedes**, **turnips**, rape and mustard. Also of interest due to their contents of compounds believed to protect against **cancer** (e.g. **indoles** and **glucosinolates**).

Brassica seeds **Oilseeds** produced by plants of the genus *Brassica*, including **rapeseeds** and **mustard seeds**.

Brassicasterol **Phytosterols** fraction which is characteristic of **vegetable oils** derived from *Brassica* oilseeds (including **rapeseed oils**) and may be used as an indicator of **adulteration** of other oils with rapeseed oils.

Brassinosteroids Hydroxysteroids which act as **plant growth regulators** in a wide range of plants, including food crops.

Bratwurst Fresh **sausages**, usually made from a mixture of highly seasoned **pork**, **veal** and **onions**. **Seasonings** may include **ginger**, **nutmeg**, **coriander** or **caraway**. Numerous different types of bratwurst are produced in Germany; many districts produce their own special varieties. Traditional bratwurst must not contain **nitrites** as **curing** salts. Although some pre-cooked bratwurst is sold, most requires cooking before it is eaten. Other product names may also include the term bratwurst (e.g. bauernbratwurst and smoked bratwurst), but these sausages are produced by hot **smoking** and **fermentation** and are classified as raw dry sausages, **rohwurst**.

Brawn Meat products prepared from **pork**, swine ears and swine **tongues**. Ingredients are boiled with herbs and peppercorns before mincing and pressing into a mould. Mock brawn differs from brawn as it is prepared from different types of **offal**.

Brazil nut oils **Oils** derived from **Brazil nuts**, large, edible seeds of the South American tree *Bertholletia excelsa*. Contain **olein**, palmitin and **stearin**.

Brazil nuts **Nuts** produced by the South American tree, *Bertholletia excelsa*. Eaten raw, salted or roasted, or added to other foods such as **ice cream** and **confectionery**. Good source of several B **vitamins** and minerals, with particularly high amounts of bioavailable selenium. Used as a source of **brazil nut oils**.

Brazzein High potency, sweet-tasting thermostable protein from the west African plant *Pentadiplandra brazzeana*. Has potential as a low-calorie sweetener.

Bread Baked **dough** product made from cereal grains (mostly commonly **wheat**) ground into **flour**, moistened and kneaded into a dough and then baked. Often leavened by the action of **bakers yeasts** or by addition of **sodium bicarbonate**.

Bread crumb The soft inner part of **bread**, which is surrounded by the **bread crust**.

Breadcrumbs Small fragments prepared by grinding **bread**. Used in **coatings**, usually for foods to be fried, **stuffings** and in some **desserts**.

Bread crust Crisp, outer part of **bread**, which is dehydrated and browned during **baking**.

Bread dough Unbaked thick, plastic mixture of **flour** and liquid (e.g. water) that is kneaded, shaped and rolled to make **bread**. The elasticity of bread dough is dependent upon the amount of **gluten** contained in the flour.

Breadfruit Green, starchy **fruits** from the breadfruit tree (*Artocarpus communis* syn. *A. altilis*). An important subsistence crop of the Pacific islands, where it provides a significant source of energy due to its high **starch** content. Also popular in the Caribbean and throughout the tropics. Fruit is typically roasted, boiled or fried and consumed as a vegetable; it can also be fermented or processed into **meal**. Breadfruit seeds are also eaten in some areas and are a good source of **proteins** (the seeded form of breadfruit is known as breadnut).

Breading Coating of foods with **breadcrumbs** or other crumbs usually before **frying** or **baking**. The food is dipped first into a liquid (e.g. beaten **eggs**, **milk** or **beer**), then into the crumbs, which may be seasoned with **herbs** or **spices**. The breaded product is then fried or baked. Breading serves to retain the moisture content of the food and forms a crisp crust after **cooking**.

Breadings **Breadcrumbs** and other types of crumb used in **breading** foods, usually before **frying** or **baking**.

Breadmaking Process by which **bread** is prepared from **flour** and other ingredients that vary according to the type of bread to be made. Steps involved include **fermentation**, **kneading** of the **dough**, **proofing** and **baking**.

Breadmaking properties Characteristics of **cereals**, **flour** or **dough** that determine their suitability for **bread** manufacture.

Bread manufacture Alternative term for **breadmaking**.

Bread rolls **Bread** products formed from pieces of **dough** shaped as required before **baking**. May have a soft or crisp **crust**. Also commonly referred to as **rolls**.

Breakfast First meal of the day. Typical breakfast foods include **breakfast cereals**, toast, **bread rolls** and other **morning goods**.

Breakfast cereals Cereal products commonly consumed with milk or cream as part of **breakfast**. Precooked or **ready to eat foods** prepared from cereal grains by processes such as **flaking**, **puffing**, **toasting** or **shredding**. Often sweetened with **sugar**, syrup or **fruits**. May be enriched with **bran** as a source of extra **dietary fibre**.

Breakfast foods General term applied to foods, especially **cereal products** and **bakery products**, which are commonly eaten as part of **breakfast**. Include **breakfast cereals**, **bread**, **muffins**, **bagels**, **pancakes** and **waffles**.

Bream **Freshwater fish** species (*Abramis brama*) distributed across Europe and parts of Asia; a member of the carp family (Cyprinidae). The flesh is bony, insipid and soft. Marketed fresh or frozen and eaten steamed, grilled, fried and baked. In Greece, roe from bream is used to make a type of **caviar**.

Breast cancer Malignant tumour originating from breast tissue. Dietary factors, including fat intake, have been suggested to increase cancer risk. As the tumours are often hormone-dependent, **phytoestrogens**, such as **isoflavones** and **lignans** present in plant foods, have been studied extensively as potential chemopreventive agents.

Breast feeding Feeding of a suckling infant with **human milk** from the mother's breast.

Breast milk Alternative term for **human milk**.

Bregott Swedish **butter** substitute made by mixing approximately 80% **butterfat** with 20% vegetable oil, usually soybean oil, and processing using the same principles as in **buttermaking**. The product has good **spreadability** direct from the refrigerator and a flavour similar to that of butter. It has a relatively high content of **linoleic acid**.

Bresaola Moist, very lean, cured, air-dried meat products, commonly made from **beef**, and sometimes from **horse meat**. A speciality of northern Italy. Usually, bresaola is cut into paper-thin slices before serving with bread and fruits or pickled vegetables. A similar Swiss product is called bundnerfleisch.

Brettanomyces Genus of yeast **fungi** of the class Hyphomycetes. Can cause **spoilage** of **beer**, **grape musts**, **wines**, **soft drinks** and **pickles**.

Brevetoxins Potent **neurotoxins** produced by the unicellular dinoflagellate protozoan *Ptychodiscus brevis*. Also known as brevitoxins. Responsible for **shellfish** poisoning in humans.

Brevibacterium Genus of Gram positive, aerobic, coryneform **bacteria** of the order **Actinomycetales**. Occur in soil, water, dairy products and decomposing matter. *Brevibacterium linens* may be used as a starter in the production of smear-ripened **cheese**.

Breweries Industrial premises used for **brewing** of **beer**.

Breweries effluents **Waste water** from **breweries**.

Brewers grains Alternative term for **brewers spent grains**.

Brewers spent grains By-products from the **clarification** stage of the **brewing** process, comprising the solid residue of substances such as **malt** grist, from which the soluble material has been extracted in the mashing process.

Brewers yeasts **Yeasts** of the genus *Saccharomyces*, now generally classified as *S. cerevisiae* but some formerly classified as *S. carlsbergensis* or *S. uvarum*, used for **brewing** of **beer**. Brewers yeasts may be divided into **top fermenting yeasts** (used for **ale** and similar beer types) and **bottom fermenting yeasts** (used for **lager** and similar beer types) on the basis of their characteristics during the **fermentation** process.

Brewing The process of manufacture of **beer**; important stages of the process include **mashing** to extract soluble material from **malt** and other ingredients, **hopping**, **clarification**, boiling the **worts**, **alcoholic fermentation**, ageing and **filtration**.

Brewing adjuncts Fermentable material other than **malt** used in the **brewing** process. Brewing adjuncts may include unmalted **cereals**, **syrups** or **sugars**.

Brewing by-products By-products of the **brewing** process, including **brewers spent grains**, surplus **brewers yeasts** and **carbon dioxide** formed during **fermentation**.

Brick tea **Tea leaves** which have been compressed to form a solid block. Widely traded and used in Tibet and parts of China and Central Asia.

Brie cheese Soft French cheese made from **cow milk**. Produced as a 1 or 2 kg wheel packed in wooden boxes.

Brightness The intensity of a **colour** (in contrast to its hue or saturation). In addition, the term is used to indicate the level of light appearing to emanate from an object.

Brill A marine **flatfish** species (*Scophthalmus rhombus*) distributed across the northeast Atlantic, Mediterranean and eastern Atlantic along European coasts. Marketed fresh and frozen, whole or gutted and as steaks and fillets.

Brilliance The luminance of an object, which includes both **brightness** and saturation.

Brilliant Blue Artificial colorant also known as FD&C Blue no 1. Imparts a greenish blue tinge to foods. Used in a range of foods, including **sugar confectionery**, **bakery products**, **desserts** and **beverages**.

Brined cucumbers Alternative term for **cucumber pickles**.

Brined meat Meat that is preserved by **brining**. Meat cuts may be immersed in **brines**, which contain **salt** or salt and other **curing agents** dissolved in water. However, it takes a long time for brines to diffuse the whole meat product; thus, in general, only speciality meat products are brined by immersion. A more rapid, uniform distribution of brines throughout meat products can be achieved by direct injection of brines into the meat cuts, methods for which include artery pumping, multiple injection and stitch pumping.

Brines Water saturated or strongly impregnated with **salt** used for **pickling** or **preservation** of foods. Sweeteners such as **sugar** or **molasses** can be added to the brines to make sweet brines.

Brining The process of treating foods with **brines** for **preservation** or **pickling**.

Brinjal Alternative term for **aubergines**.

Brioches Rich buns made with **flour**, **butter** and **eggs** and raised with **yeasts**. Sometimes flavoured with **currants**, **candied fruits**, **chocolate** or **cheese**; sweet or savoury **fillings** such as **custards** or **sausages** may also be added.

Briquetting Use of presses for compacting and processing organic waste or compacting plants to reduce volume for valuable material recycling and disposal. In the food industry, **ice cream** may be formed into bricks, blocks or slabs and sold as briquettes.

Brix values Properties that give an indication of the **density** of **sugar** in a solution at a specific temperature. Brix values are frequently used to express the sugar levels in **fruits**, **beverages** and **sugar juices**. Named after the German inventor A. F. W. Brix.

Broad beans Type of **faba beans** (*Vicia faba*).

Broccoli Common name for certain varieties of *Brassica oleracea*, particularly *B. oleracea* var. *italica*. Green sprouting broccoli (also known as calabrese) has a tight cluster of deep emerald green florets on a thick, edible stem and is prepared and used in a similar manner to **cauliflowers**. Sprouting broccoli has a looser cluster of smaller flower heads. Broccoli is rich in **vitamin A** and **vitamin C** and is attracting interest as a source of **glucosinolates** that may protect against some forms of **cancer**.

Brochothrix Genus of Gram positive, facultatively anaerobic, rod-shaped **bacteria** of the family Listeriaceae. *Brochothrix thermosphacta* may cause **spoilage** of **meat** and meat products.

Broiler meat Meat from **broilers**, specific types of chickens, also used as an alternative term for **chicken meat**.

Broiler muscles Meat from **broilers**, specific types of **chickens**, also used as an alternative term for **chicken meat**.

Broilers Fast-growing strains of **chickens** reared under intensive conditions before slaughter, for meat production at about nine to twelve weeks of age. Development of intensive systems to produce broilers has led to large-scale production at low prices.

Broiling Cooking of foods (usually **meat** or **fish**) by exposure to direct heat. Food can be broiled in **ovens**, directly under a gas or electric heat source, or on a barbecue, directly over charcoal or other heat source.

Bromates Salts of bromic acid, including **potassium bromate**, which is used in **flour improvers**. Bromates can also be formed as **disinfection by-products** during **ozonation** of **drinking water** containing bromide ions.

Bromelains EC 3.4.22.32 (stem bromelain) and EC 3.4.22.33 (fruit bromelain). Cysteine endopeptidases with broad specificity found in the stem and fruit, respectively, of **pineapples**. Have been used in **tenderization** of **meat**, for production of meat and legume **protein hydrolysates**, and for **stabilization** of **beer**.

Bromides Salts of hydrobromic acid. Foods may become contaminated with bromides as a result of **fumigation** with **methyl bromide** to control **infestation** with **insects**.

Bromine Member of the **halogens** group of elements, chemical symbol Br. Occurs naturally in many foods. May also be present in the form of residues of **bromides** following **fumigation**.

Bromocyclen Insecticide and acaricide used for control of ectoparasites (principally **mites** and ticks) in farm animals.

Bromomethane Chemical name for the fumigant **methyl bromide**.

Bromoxynil Herbicide with some systemic activity; used for post-emergence control of annual broad-leaved weeds in cereals and vegetables (particularly *Allium* spp.). Often used in combination with other **herbicides** to extend the spectrum of control. Normally hydrolyses rapidly to less toxic substances in plants and soil. Classified by WHO as moderately toxic (WHO II).

Broths Soups based on **stocks** to which ingredients such as **cereals** and **vegetables** may be added. In microbiology, the term broths refers to liquid culture media.

Brown heart Physiological disorder affecting **apples** and **pears** thought to result from injury caused by **carbon dioxide**. Characterized by internal browning of the fruit flesh, originating in or near to the core. Extent varies from a small spot of brown to almost the entire flesh being affected. The defect is not detectable externally. Symptoms may increase as storage time is extended.

Brownies Moist, fudge-like **cakes** originating from the USA. Made with **chocolate** and frequently nuts, especially **walnuts** or **peanuts**.

Browning Process by which foods become brown, resulting from either enzymic or nonenzymic reactions. **Enzymic browning** occurs in freshly cut **fruits** and **vegetables** due to the oxidation of **phenols** by **catechol oxidases**. However, **nonenzymic browning** is generally a consequence of the **Maillard reaction**. Browning can be either a favourable process and encouraged using **browning agents**, or unfavourable and reduced using **browning inhibitors**.

Browning agents Ingredients or **additives** that promote **browning** of foods during processing, thereby imparting a darker **colour** to the finished product. Examples include **caramel**, **milk** and certain **sugars**. Commercial browning agents are often applied to foods cooked in **microwave ovens** in order to produce surface browning.

Browning inhibitors Substances used to prevent **browning** in foods, also known as anti-browning agents. Use of **sulfur dioxide** or **sulfites** is one of the most widespread chemical means of controlling both **enzymic browning** and **nonenzymic browning**. Other browning inhibitors include **kojic acid**, which inhibits tyrosinase activity, **citric acid** and **cysteine**.

Browning reactions Alternative term for **browning**.

Browning susceptors Alternative term for **microwave susceptors**.

Brown rice Rice from which the **husks** have been removed, leaving the **germ** and outer layers containing the **bran** intact.

Brown rot Any rot resulting in browning and decay of plant tissue, particularly common in fruit trees. Brown rot in **pome fruits** and **stone fruits** is frequently caused by **fungi** of the *Monilinia* genus.

Brown sugar Granulated **white sugar** that has been covered with a layer of **cane sugar syrups** to give it a brown appearance and a caramel-like **flavour**. Brown sugar has a higher moisture content and a higher ash content than white sugar.

Brown trout Freshwater form of the species *Salmo trutta*. Also has a silver bodied migratory form (**sea trout**) which swims out to sea for a period before returning to freshwater to spawn. Marketed fresh, frozen or smoked.

Brucella Genus of Gram negative, aerobic, rod-shaped **bacteria** of the family Brucellaceae. Occur as intracellular **parasites** or **pathogens** in humans and animals. *Brucella abortus* causes undulant fever in man.

Brucellosis Any human or animal disease caused by infection with ***Brucella*** species. In humans, *Brucella abortus*, *B. melitensis* and *B. suis* can cause an acute or chronic systemic disease (undulant fever or Malta fever, respectively) characterized by fever, weakness and general malaise. Transmitted to humans by direct or indirect contact with infected animals or **raw milk**.

Bruehwurst Frankfurter-type **sausages** which are heat treated during preparation. The various types of bruehwurst include **beutelwurst** and **bockwurst**. Skinless varieties are produced by heating the sausage emulsion in a mould; the outer layer of emulsion sets to form a firm skin. For optimum quality, bruehwurst sausages are made from slaughter-warm meat. Major defects include weakness of **flavour** and incorrect use of **seasonings**. In fat-reduced bruehwurst, **texture** may be very firm and rubbery.

Bruising Damage to the surface of foods, particularly **fruits** and **vegetables**, resulting from mechanical impacts.

Brussels sprouts Common name for *Brassica oleracea* var. *gemmifera*. A relatively recent variety of cabbage characterized by a stout stem yielding numerous compact heads (sprouts) resembling miniature **cabbages**. Consumed fresh as a winter vegetable and also available frozen. A good source of **vitamin C, vitamin A, folic acid** and **potassium**. Like other Cruciferae, contain **phytochemicals** such as indole **glucosinolates**, which may help protect against **cancer**.

Bryndza cheese Soft Slovak cheese made from **ewe milk** that is popular throughout Eastern Europe. It is matured for at least 4 weeks and has a fat content of about 45%. Similar cheeses are the Romanian Brinza, Hungarian Brynza, Sirene from Bulgaria and Greek Feta.

BSE Abbreviation for **bovine spongiform encephalopathy**.

Bubble gums Sweetened products made from chicle (gum-like exudate consisting of coagulated milky juice from the bark of the evergreen sapodilla tree, *Achras zapota*) or similar resilient substances (e.g. plasticized rubber or polymers), and chewed for its **flavour**. Bubble gums differ from **chewing gums** in the user's ability to blow bubbles from them during chewing.

Buchu oils Aromatic **oils** which are extracted from leaves of the African shrubs *Agathosma betulina* and *A. crenulata*. Used in **flavourings**.

Buckwheat Grains of *Fagopyrum esculentum* used as a cereal. Unsatisfactory for the manufacture of **bread**. Whole grains are cooked like rice and made into baked **puddings**. Also made into other products such as **porridge**, noodles or griddle cakes. Good source of protein, **niacin** and **vitamin B₁**.

Buckwheat flour **Flour** made from **buckwheat** grains. Most commonly used in making **pancakes**, such as blini.

Buckwheat oils **Oils** extracted from the grains of **buckwheat**.

Budu A type of fermented fish sauce produced from salted **anchovy**. Product has an olive-brown **colour**. Popular in Malaysia and Thailand.

Buffalo butter **Butter** made from **buffalo milk**.

Buffalo cheese **Cheese** made from **buffalo milk**.

Buffaloes The general name for several species of large ruminant mammals in the family Bovidae. Buffaloes are native to sub-Saharan Africa (*Synceros caffer*) and to India and south-east Asia (genus *Bubalus*, in which there are four species); in these regions, domesticated buffaloes are used to provide draft power, **buffalo meat** and **buffalo milk**. Buffaloes are also farmed in other countries, such as Italy, as sources of buffalo meat and buffalo milk. In popular use, the term buffaloes is also used to describe North American **bison**.

Buffalo gourds Common name for *Cucurbita foetidissima*. Potential starch crop or source of **oilseeds**. The root starch has **physicochemical properties** intermediate between those of **cassava** and **corn**.

Buffalo meat **Meat** from **buffaloes**, that has a similar **flavour** to **beef**, but a lower content of fat. During **cooking**, care must be taken to prevent the meat from drying out.

Buffalo milk **Milk** obtained from **buffaloes**. Compared with **cow milk**, buffalo milk has a higher fat content (approximately 8%), higher contents of protein, **calcium** and some other **minerals, vitamin A**

and **biotin**, and lower contents of potassium, and **vitamin B$_2$** and **vitamin B$_6$**. Contains no **carotenes**. Used in making a range of dairy products, including **Mozzarella cheese** and **Domiati cheese**.

Buffalo milk cheese **Cheese** made from **buffalo milk**.

Buffalo Mozzarella cheese Soft Italian plastic, spun-curd cheese made from pasteurized **buffalo milk**. The **curd** is treated with extremely hot water and kneaded into a shiny lump.

Buffalo yoghurt **Yoghurt** made by fermenting **buffalo milk**.

Buffering capacity Ability of a substance or solution to resist change in **acidity** or **alkalinity**.

Bulgur Cracked **wheat** grains prepared from **soaking**, **cooking**, **drying** and light **milling**. May be ground into **flour**; also used in many Middle Eastern dishes.

Bulk density Weight per overall unit volume of a substance. Bulk density is used in particular for porous substances where density is affected by pore volume and can be increased by the presence of pore fluid.

Bulking agents Originally used to describe inert products added to contribute bulk and act as inexpensive fillers/extenders for more expensive ingredients. Now also used in **low calorie foods** and **low fat foods** to produce a feeling of **satiety** and to replace **functional properties**, flavouring characteristics and other qualities of the sugar and/or fat which has been removed. Substances used as bulking agents include **methylcellulose**, **fibre**, **polyols** and **polydextrose**.

Bull muscles **Meat** from specific types of mature, uncastrated, male bovine animals, usually male **cattle**, and also an alternative term for **beef**. Beef from older **bulls** tends to be tough, but toughness is not a problem in beef from younger bulls. Bull beef has a good water binding capacity and a good water holding capacity. Consequently, meat from the forequarters of bull **carcasses** is often used as an ingredient in **sausages**. Bull beef tends to have a lower intramuscular fat content than beef from steers. In comparison with beef from steers or heifers, bull beef may be discriminated against as it tends to be darker in colour and coarser in texture, and may lack finish; it is also associated with an increased incidence of the **DFD defect**.

Bulls Mature, uncastrated, male bovine animals, usually **cattle**. Production of **beef** from bulls has several advantages over production of that from steers; in particular, bulls grow faster, convert feed more efficiently and achieve greater carcass weight than steers.

Buns Small yeast-raised rounded pieces of **bread** which are sometimes sweetened or flavoured, and may contain **dried fruits**. Choux buns are made from choux pastry and usually filled with **whipped cream**.

Burbot **Freshwater fish** species (*Lota lota*) related to cod which is the only freshwater member of the hake family (Lotidae). Widely distributed in Europe and the USA, but rare in Great Britain. Has lean white flesh with a delicate **flavour**. Sold mainly as a salted product, but also marketed fresh in the USA. Liver is sold smoked or canned in Europe. Also utilized as a source of **fish oils** and for **fish meal** production.

Burdock Common name for *Arctium lappa*. Long slender **root vegetables** with a reddish brown skin and greyish flesh. The plant grows wild in many areas of the USA and Europe, and is cultivated in Japan where it is also known as gobo. The root has a crispy **texture** and sweet pungent **flavour**, much prized in Japanese dishes. Also used in the manufacture of **pickles** and **soft drinks**.

Burfee Concentrated **dried dairy products** prepared from **khoa**. Popular as **sweets** in India.

Burfi Alternative term for **burfee**.

Burgers Round, flat cakes of **meat mince**, cooked by **grilling** or **frying**. Specific types of burger include **baconburgers**, **beefburgers**, **cheeseburgers** and **hamburgers**. They are commonly eaten in **bread rolls**, served with lettuce, slices of onion and **tomato ketchups**.

Burgos cheese Spanish **fresh cheese** made from raw **cow milk** or **ewe milk**. Pure white with a slightly acid and salty **flavour**. Used in baking or eaten as a dessert with **sugar** and **honeys**.

Burkholderia Genus of aerobic **Gram negative bacteria**, type species *Burkholderia cepacia*, species of which were formerly classified as belonging to the genus ***Pseudomonas***. Some species, including *B. cepacia*, are of potential biotechnological use as producers of enzymes such as lipases, for bioremediation of sites contaminated with polychlorinated biphenyls or as biocontrol agents. However, some strains are plant and human **pathogens**. The species *B. cocovenenans* produces the toxin bongkrekic acid, responsible for outbreaks of **food poisoning** associated with the fermented coconut product **bongkrek**.

Bush butter African fruit produced by the bush butter or African plum tree (*Dacryodes edulis*). Tough purple peel encases a layer of bitter greenish flesh surrounding a large seed which is fed to livestock. The fruit must be boiled for approximately 1 minute to make it tender enough to eat.

Bush okra Alternative term for **ewedu**.

2,3-Butanediol Alternative term for **2,3-butylene glycol**.

2,3-Butanedione Synonym for the flavour compound **diacetyl**. Yellow, flammable liquid with a strong **aroma** and buttery **flavour** derived from **fermentation** of **glucose**. Soluble in water and alcohol. Used as an aroma carrier in foods and beverages.

Butanoic acid Synonym for **butyric acid**. A member of the short chain **saturated fatty acids** which occur as **flavour compounds** in a wide range of foods and **beverages**. Especially characteristic of **milk fats** and **dairy products**. At high concentrations, it may be responsible for development of **off flavour**.

Butanol Synonym for **butyl alcohol**. Member of the **alcohols** class of **flavour compounds** which occurs in a wide range of foods.

Butanone Member of the ketones class of **flavour compounds**. Occurs in a wide range of foods, especially **dairy products** and **meat** products.

Butcheries **Slaughterhouses** where **animal carcasses** are divided into primal cuts before distribution to the retail **meat** trade. Some **trimming** to remove carcass fat may take place at this stage. The pattern of butchering and the naming of the primal cuts generated vary between regions and countries.

Butter Spreadable water-in-oil emulsion product made from **milk fats** or **cream** by the **buttermaking** process. Usually contains at least 80% fat, with the remainder being water. Salt is sometimes added as a flavouring and **colour** may be adjusted using **annatto** or **β-carotene**. Rich in **vitamin A**; also contains **vitamin E** and **vitamin D**. Main types of butter marketed are cultured cream butter (also known as lactic butter or sour cream butter) and sweet cream butter, the latter having a higher pH value.

Butter beans Alternative term for **lima beans**.

Butterbur Common name for *Petasites japonicus*, a plant with very large, soft leaves formerly used to wrap **butter**. Leaves are eaten as a vegetable in Korea and Japan.

Butter clams Marine **bivalves** (*Saxidomus giganteus*) found in middle to lower intertidal sediments along the Atlantic and Pacific coasts of North America. Smaller specimens are often eaten raw, while larger adults are usually steamed or fried; highly prized for production of clam chowders.

Butter factories Factories in which **butter** is made.

Butterfat Used as an alternative term for **milk fats**, **butter oils** or **ghee**.

Butterfish **Marine fish** species (*Brama brama*; also known as pomfret) with a deep laterally compressed body which is widely distributed in the Atlantic, Pacific and Indian Oceans. Flesh has high fat content with a tender **texture** and a rich, slightly sweet **flavour**. Sold fresh and frozen and cooked in a variety of ways. In the USA, the marine fish *Peprilus triacanthus* is also commonly known as butterfish; this species is marketed in similar forms to *B. brama*, but is also popular as a smoked product.

Butterine Fat product that was developed originally a substitute for **butter**. Composed of approximately 80% vegetable, animal or marine **fats** and 20% water, together with **additives** such as **emulsifiers**, **colorants** and **preservatives**.

Buttermaking Process by which **butter** is made from **milk** or **cream**. Consists of cream ripening, **churning**, washing and working. Cream ripening involves a series of temperature treatments, with or without incubation with **butter starters**, that affect the **consistency** and **flavour** of the final product. Churning breaks up the **milk fat globule membranes**, allowing formation of butter grains which eventually separate from the **buttermilk**. The butter grains are washed with water to remove proteins, sugar and microorganisms, and worked into a homogeneous mass by kneading.

Butter manufacture **Buttermaking** process.

Buttermilk Tangy flavoured residue remaining after separation of butter grains during **buttermaking**. Low in fat but rich in phospholipids and proteins, resulting from breakdown of the **milk fat globule membranes** during **churning** of **cream**. Buttermilk remaining after manufacture of sweet cream butter differs slightly in composition from that resulting from cultured cream buttermaking. A commercial product called **cultured buttermilk** is made by adding **lactic acid bacteria** to **skim milk**, the **lactic acid** produced during fermentation giving the product a tart flavour similar to that of churned buttermilk. Buttermilk is used as a beverage and as an ingredient in baking.

Butternuts **Nuts** produced by *Juglans cinerea*, trees of the walnut family. Shells are hard, but not difficult to open. Characteristics and food applications of the kernels are similar to those of other **walnuts**. Also known as white walnuts.

Butter oils Milk fat products with a very high fat content, usually not less than 90%, and a very low water content (no greater than 0.5%). In anhydrous butter oils, moisture content is no greater than 0.1%.

Butterscotch Hard **confectionery** made by boiling **brown sugar** and **butter** or **corn syrups** together in water. Generally distinguished from **caramels** by the absence of **milk** or milk substitutes among the ingredients.

Butter spreads **Spreads**, often low in fat, based on **milk fats**.

Butter starters Bacterial mixtures, usually *Streptococcus*, used in ripening of **cream** in manufacture of cultured cream butter. Responsible for formation of flavour/aroma compounds such as **diacetyl** and **acetoin**. Affect stability of **milk fat globule membranes**, facilitating their breakdown during **churning**, the next step in the **buttermaking** process.

Butter substitutes Products intended to be similar to **butter** in appearance and properties, but which differ in composition.

Butyl alcohol Synonym for **butanol**. Member of the **alcohols** class of **flavour compounds** which occurs in a wide range of foods.

Butylamine Synonym for 2-aminobutane. Fungicide used to prevent **spoilage** of fresh **fruits** and **vegetables**.

Butylated hydroxyanisole Commonly abbreviated to **BHA**. A synthetic fat-soluble phenolic antioxidant. It has good heat stability and is widely used in the food industry to add stability to **fats** and **oils**. Applications include **cereals**, **confectionery** products, **bakery products** and **packaging materials**. Often used synergistically with **BHT** and other **antioxidants**.

Butylated hydroxytoluene Commonly abbreviated to BHT. A widely used synthetic antioxidant with similar properties to **BHA**, but less stable at high temperatures. Applications include **bakery products**, **breakfast cereals**, and food **packaging materials**. Displays good synergy when used in combination with BHA.

2,3-Butylene glycol Precursor of the flavour compound **diacetyl** (2,3-butanedione), formed by microbial **fermentation** in various foods and **beverages**, especially **wines**. Formation of 2,3-butanediol is a useful parameter for identification and differentiation of **wine yeasts**.

Butyltins Organotin compounds including **tributyltin** and its degradation products, dibutyltin and monobutyltin. **Fish**, **shellfish** and marine mammals may be contaminated by tributyltin, as a result of its use as an antifoulant paint additive on ship and boat hulls, docks, fishing nets and buoys to discourage the growth of marine organisms (e.g. barnacles, **bacteria**, tubeworms, **mussels** and **algae**).

Butyric acid Synonym for **butanoic acid**. One of the short chain **saturated fatty acids** which occurs as one of the **flavour compounds** in a wide range of foods and **beverages**. Especially characteristic of **milk fats** and **dairy products**. At high concentrations, it may be responsible for development of **off flavour**.

Butyric fermentation Process by which certain **bacteria** (mainly *Clostridium* spp.) produce **butyric acid**. Although a useful industrial process, it can cause **cheese spoilage** (**blowing**). Production of butyric acid in the colon by **fermentation** of **dietary fibre** may reduce the risk of certain cancers.

Butyrometers Apparatus used to measure the fats content of **milk**. Samples are mixed with sulfuric acid in special graduated tubes which are then centrifuged. Fat separates as an upper layer, the size of which is measured from the markings on the tube.

Butyrophilin Acidic glycoprotein associated with **milk fat globule membranes**. Has potential roles in lactation and autoimmune disease.

Byssochlamic acid Mycotoxin produced by the heat resistant fungus *Byssochlamys fulva*, responsible for **spoilage** of canned fruit products and **fruit juices**.

Byssochlamys Genus of **fungi** of the order Eurotiales. Occurs in soil. *Byssochlamys fulva* can cause **spoilage** of **canned foods**.

C

C Chemical symbol for **carbon**.

Ca Chemical symbol for **calcium**.

Cabbage juices **Vegetable juices** extracted from cabbages (*Brassica oleracea*). May be blended with other vegetable juices or **fruit juices**, and may be used in the manufacture of lactic acid **fermented beverages**.

Cabbages Any of various cultivated var. of *Brassica oleracea*. Typically have a thick stalk with a large, compact head formed from green or reddish purple edible leaves (e.g. **savoy cabbages**, **white cabbages**). Cabbages that do not form a head are known as **kale**, winter greens or **collards**. Consumed as a vegetable, used as **coleslaw** ingredient or fermented to produce **sauerkraut**. **Red cabbages** are used for **pickling**. **Chinese cabbages** are *Brassica pekinensis*.

Cabrales cheese Spanish hard **blue cheese** made from cow, ewe and goat milks. Matured in natural limestone caverns. It has a creamy **texture**, complex **flavour** and powerful **bouquet**.

Cacao Alternative term for **cocoa**.

Cacao beans Alternative term for **cocoa beans**.

Caciocavallo Palermitano cheese Italian **soft cheese** made from **cow milk**, but said to have been made originally from **mare milk**. A traditional, stretched curd cheese that is gourd-shaped and hung from the thin end to mature. Eaten as a table cheese after 3 months and used for grating after 2 years.

Cacioricotta cheese Italian **cheese** produced from **cow milk**, **goat milk**, **ewe milk** or water buffalo milk. Apulian Cacioricotta is a cheese produced on an artisanal scale from pasteurized goat milk in a specific region of Italy. It is eaten fresh as a soft dessert cheese or ripened and used mainly for grating over local dishes.

Caciotta cheese Italian soft, mild **cheese** made from cow or ewe milk.

CaCl₂ Chemical formula for **calcium chloride**.

Cacodylic acid Alternative term for **dimethylarsinic acid**.

Cacti Large family of spiny, succulent plants, fruits from some of which are edible. The most common edible parts are the fleshy fruits of various species of **prickly pears**. Other types include Barbados gooseberries and **pitayos** (pitaya). The sweet fruits of various cacti can also be fermented to produce **alcoholic beverages**. Garambullo cactus (*Myrtillocactus geometrizans*) produces purple fruits which are a potential source of betalain type **pigments**.

Cactus fruits Alternative term for **cactus pears**.

Cactus pears Spiny **fruits** produced by several varieties of **cacti**, especially *Opuntia ficus-indica*. The soft flesh is similar in **texture** to that of **watermelons**. Usually eaten fresh, but also used as an ingredient for **desserts** and **beverages**. Also known as **prickly pears**, Indian figs and barberry figs.

Cadaverine Toxic, foul-smelling biogenic amine produced by the decarboxylation of **lysine** by various **microorganisms** in decaying **meat** and fish.

Cadmium Toxic heavy metal, chemical symbol Cd. May occur as a contaminant in a wide range of foods and **beverages**.

Caesium Radioelement, chemical symbol Cs, which may occur as the radioactive isotopes ^{137}Cs or ^{134}Cs in foods as **contaminants** from radioactive fallout.

Cafestol Diterpene found in **coffee** which increases plasma triacylglycerol and cholesterol concentrations.

Cafeterias Self service **restaurants**. Often located within larger establishments, such as department stores, schools or universities.

Caffeic acid Member of the **hydroxycinnamic acid** class which occurs in many plants and plant derived foods. Has **antioxidative activity** in foods.

Caffeine One of the xanthine **alkaloids** naturally present in several plant foods, including **tea**, **coffee** and **cola nuts**. Acts as a stimulant. Used as an ingredient in some **soft drinks**, including **cola beverages** and **energy drinks**.

Caffeoylquinic acid Synonym for **chlorogenic acid**. Phenol present in many foods of plant origin. Plays an important role in **enzymic browning** of **fruits** and **vegetables**. Has **antioxidative activity**, and may contribute to possible health-promoting or protective actions of dietary phenolic compounds.

Caja Common name for *Spondias lutea* (syn. *S. mombin*), also known as yellow mombin. A South American fruit, the pulp and skin of which are used locally in the preparation of **fruit juices**, **ice cream** and **liqueurs**.

Cake batters Batters usually prepared from **flour**, **eggs**, **butter** or **margarines**, and **sugar** that are used to make **cakes**. Other ingredients are added according to the type of cakes to be made.

Cake mixes Powdered formulations containing all the ingredients required to make **cakes**.

Cakes Soft **bakery products** produced by baking a batter containing **flour**, **sugar**, **baking powders** and beaten **eggs**, with or without **shortenings**. According to the final product, other ingredients are also included, such as **flavourings**, **nuts**, **chocolate** and **dried fruits**.

Caking **Solidification** of powders or granules into a mass. Caking can be a problem during the storage of **dried foods** and **sugar**.

Calamintha Genus of **herbs** with a **mint** like **aroma**. Includes *Calamintha nepeta*, which is used in **soups** and **sauces**.

Calamus Medicinal herb (*Acorus calumus*) also known as sweet flag. Dried rhizomes are used in the formulation of **vermouths**, **liqueurs** and bitters, and also for medicinal and veterinary purposes.

Calciferol Synonym for **ergocalciferol** and **vitamin D$_2$**; one of the group of **sterols** which constitute **vitamin D**. Synthesized by **irradiation** of the plant provitamin **ergosterol**.

Calcium Mineral with the chemical symbol Ca. Constituent of most foods and an essential nutrient in the human diet, particularly important for strong bones and teeth of which it is a major component. Rich sources include **milk** and **dairy products**, oily **fish** and **spinach**; staple foods are sometimes enriched with calcium. Also important in the setting of **pectins gels**, and the **firmness** of processed fruit and vegetable products.

Calcium chloride Salt, chemical formula $CaCl_2$, widely used as a general purpose food additive. Applications include **flavour** preservation in **pickles**, as a firming agent in **fruits** and **vegetables**, and as a source of calcium for calcium alginate gels.

Calcium tartrate The calcium salt of **tartaric acid**. Calcium tartrate may precipitate in **wines**, forming an undesirable **haze** or sediment. Haze stabilization treatments may be required to prevent this problem.

Calf meat Meat from specific types of young, sexually immature bovine animals, usually milk-fed **cattle**, and also an alternative term for **beef**.

Calf muscles Meat from specific types of young, sexually immature bovine animals, usually milk-fed **cattle**, and also an alternative term for **beef**.

Calf rennets Substance extracted from the abomasum of calves that is used in **coagulation** of **milk** for **cheesemaking**. The active enzyme is **chymosin**; pepsin is also present.

Caliciviruses Genus of RNA-containing **viruses** of the family Caliciviridae. Include **Norwalk viruses** and Norwalk-like viruses, which are responsible for acute **gastroenteritis** in humans and are transmitted by the faecal-oral route via contaminated water and foods (e.g. **shellfish** and **salads**).

Callipyge phenotype In **sheep**, the callipyge locus is involved in muscling. In lambs expressing this gene, weight of some muscles is increased. However, **tenderness** of the **meat** from affected muscles is not as good as in normal **lamb**. Various techniques for **tenderization** of meat from callipyge lambs have been investigated, including **freezing**, **electrical stimulation** and calcium chloride injection of carcasses.

Callus culture Mass of cells with no regular form resulting from the growth of undifferentiated tissue on semisolid agar. Used in tissue culture as the starting material for the propagation of plant clones.

Calmodulin Calcium ion binding protein which can moderate the activity of various metabolic **enzymes** in plants, animals and **microorganisms**.

Calocybe Genus that includes some **edible fungi**, such as the edible milk-white mushroom *Calocybe indica*.

Calories Metric units of energy used widely to indicate the level of energy in foods and nutrients. One normal calorie (also known as the 15° calorie) is the amount of energy required to heat pure water from 14.5 to 15.5°C at atmospheric pressure (equivalent to 4.185 J). The small calorie or therm is equivalent to 4.204 J and is the energy required to heat pure water from 3.5 to 4.5°C.

Calorific values Amount of **calories** in foods or nutrients, indicating the levels of utilizable energy. Also known as energy values.

Calorimetry Technique for measuring the energy content of foods from the number of **calories** formed during combustion of a known amount of sample.

Calpains EC 3.4.22.17. **Enzymes** involved in **meat tenderization** during *post mortem* storage and deterioration of **fish proteins gels**. There are two main types of calpains, one with high Ca^{2+} sensitivity in the micromolar range (μ-calpain) and one with low Ca^{2+} sensitivity in the millimolar range (m-calpain).

Calpastatins **Proteinases inhibitors** present in **meat** which act on **calpains** and play a role in modulating the **tenderness** of meat during storage.

Calvados **Apple brandy** manufactured in a defined district in the Normandy region of France.

Calves Specific types of young, sexually immature bovine animals, usually **cattle** which are <8 months of age, that produce **beef**. Male calves are called bull calves and females are called heifer calves, quey calves or cow calves.

Camel meat Meat from **camels** that has a similar appearance, **colour**, **texture** and palatability to **beef**. Mature camels produce rather tough meat; consequently, meat from young animals is often preferred.

Camel milk Milk obtained from **camels**. Similar in composition to **cow milk**, with approximately 4.2% fat, 3.5% protein, 4.5% lactose and 0.8% ash.

Camels The common name for two species of large, herbivorous, long necked, mainly domesticated, ungulate mammals that are well adapted to living in arid conditions. Camels belong to the genus *Camelus* of the Camelidae family. The one-humped camel is known as the Arabian camel (*C. dromedarius*) whilst the two-humped camel is known as the bactrian camel (*C. ferus*). Camels are reared as a source of **camel milk** and **camel meat**. They are major meat animals in many Arab and sub-Saharan African countries.

Camembert cheese Soft French **cheese** made from **cow milk**. Crumbly and soft at the beginning of **ripening**, it gets creamier over time (usually 2-3 weeks). A genuine Camembert has a delicate salty **flavour**.

Cameros cheese Spanish soft **fresh cheese** made from raw or pasteurized **goat milk**. Characterized by a short shelf life.

Camomile Herb obtained from *Anthemis nobilis* (syn. *Chamaemelum nobile*). The plant is a source of **essential oils** used to flavour **liqueurs**, other **beverages** and **confectionery**. Flowers are used to make **herb tea**. Wild camomile (*Matricaria recutita* syn. *M. chamomilla*) has similar uses. Also known as chamomile.

cAMP Abbreviated name for cyclic adenosine 3',5'-monophosphate, one of the **nucleotides**. A universally distributed metabolite formed by the action of adenylate cyclase on ATP. cAMP is an important mediator in signal transduction pathways, and an activator of several **kinases** and physiological processes, including expression of some virulence-related **genes** in **microorganisms**.

Campesterol Sterol which occurs in many **vegetable oils** and **vegetable fats**. The relative concentrations of campesterol and other sterol fractions may be used as parameters for identification and **authenticity** testing of oils.

Camphechlor Non-systemic contact and stomach insecticide with some acaricidal action. Used for control of a wide range of insect **pests** in crops and soil, often in combination with other **pesticides**. Usage on crops has largely been displaced by less persistent **insecticides**. Also known as toxaphene.

Camphene Monoterpenoid which is one of the **flavour compounds** present in a wide range of **herbs** and **spices**.

Camphor Monoterpene ketone which is one of the **flavour compounds** in a wide range of **herbs** and **spices**.

Campylobacter Genus of Gram negative, microaerophilic rod-shaped **bacteria**. Occur in the reproductive and intestinal tracts of animals and humans. Some species are pathogenic, e.g. *Campylobacter jejuni*, which frequently contaminates raw **chicken meat**. **Raw milk** is also a source of infection.

Campylobacteriosis Any human or animal disease caused by infection with *Campylobacter* species. *C. jejuni* causes **food poisoning** in man characterized by diarrhoea, fever, abdominal pain, nausea, headache and muscle pain.

Camu-camu **Fruits** produced by *Myrciana dubia*, an Amazonian shrub. The round, light orange to purple fruits are the richest source of **vitamin C** discovered so far. Compared with **oranges**, they contain 30 times the vitamin C content, 10 times the content of **iron**, 3 times more **niacin**, twice as much **riboflavin** and fifty percent more **phosphorus**. Fruits are eaten out of hand and the fruit pulp is used to prepare a range of products, including **fruit juices** and **fruit nectars**, **marmalades**, **sherbet**, **vinegar** and **ice cream**. Also known as rumberries.

Canapes Small pieces of **bread**, **toast** or **crackers** spread with savoury **toppings**, such as **cheese** or **pates**. Served as appetizers or cocktail snacks.

Canary grass Annual grass *(Phalaris canariensis)* from the Mediterranean. Its grains are commonly used as food for caged birds, but also consumed by humans.

Canavanine Non-protein amino acid, which is a potentially toxic arginine antimetabolite. Found in **alfalfa** and certain other legumes such as **jack beans**.

Canbra oils **Rapeseed oils** derived from a Canadian variety of **rapeseeds** which contain very low (<2%) amounts of **erucic acid**. Also low in **glucosinolates**.

Cancer A range of malignant diseases characterized by uncontrolled cell proliferation that results in tissue invasion and destruction. Dietary factors have been linked with increased risk for certain cancers (e.g. high intakes of dietary **fats**) and with reduced risk (e.g. increased intakes of **fruits** and **vegetables**).

Candida Genus of yeast **fungi** of the class Saccharomycetes. Occur in soil and on plants. May be used in the production of **fermented foods** (e.g. *Candida kefir* in the production of **kefir** and **koumiss,** and *C. famata* in the production of fermented **sausages**). *C. lipolytica* and *C. zeylanoides* cause **meat spoilage,** while *C. valida* causes spoilage in **wines.** *C. utilis* and *C. lipolytica* may be used for production of **single cell proteins**.

Candied fruits Fruits, usually whole, preserved by softening in water and then soaking in **syrups** of progressively increasing **sucrose** concentrations. After drying, the fruits are coated in **sugar** to make crystallized fruits or dipped in concentrated **sugar syrups** to make glace products, such as glace **cherries**. Often regarded as luxury products, although glace cherries are frequently used as ingredients in **bakery products**.

Candling Technique for determining the quality of **eggs** wherein the egg is held before a light which penetrates the egg and makes it possible to inspect the contents and shell.

Candy Sweet crystallized product formed by boiling of **sugar**. Also a US term for **sugar confectionery** products in general.

Candy floss A fluffy mass of spun **sugar** that is formed from thin threads. Often served on a stick. Also known as **cotton candy,** particularly in the USA and Canada.

Cane molasses Molasses produced as a by-product of **refining** of **sugar** from **sugar cane** (*Saccharum officinarium*). Cane molasses are composed of approximately 40% **sucrose**. Also known as blackstrap molasses and sugar cane molasses.

Canestrato Pugliese cheese Italian **hard cheese** made from unpasteurized **ewe milk**. During manufacture, **peppercorns** are added after the curd has been cut, scalded and salted. **Flavour** and **consistency** vary according to the **ripening** period selected.

Cane sugar Sucrose extracted from stalks of **sugar cane** (*Saccharum officinarium*). Processing of sugar cane to produce cane sugar involves: washing and cutting the cane stalks; extraction of **cane sugar juices** by crushing the stalks using a series of heavy rollers; purification of the raw cane sugar juices by precipitation of impurities (**liming** and **clarification**); **filtration** to remove the precipitates; evaporation of the purified juices which results in concentration of the cane sugar juices and **crystallization** of sucrose. Dried purified cane sugar is composed of ≥99.80% sucrose and has <0.05% moisture content.

Cane sugar factories Factories containing processing lines equipped for extracting **cane sugar** from **sugar cane** (*Saccharum officinarium*). Sugar cane factories located close to where the sugar cane is cultivated (plantation factories) are involved with manufacture from sugar cane of pure **white sugar** or raw cane sugar. Sugar refineries are normally situated nearer to the markets for sugar and are involved in purification of raw or salvaged sugar to produce white sugar.

Cane sugar juices Aqueous solutions of **cane sugar** produced during processing of **sugar cane**. Raw juices are produced by compression of the sugar cane stalks and contain cane sugar and impurities, thin juices are the purified raw juices and thick juices are concentrates of the thin juices.

Cane sugar syrups Highly concentrated aqueous solutions of **cane sugar** produced by evaporation of purified **cane sugar juices** (thin juices).

Canned foods Foods preserved by **canning**. One of the main advantages of canned foods is their ease of storage at ambient temperatures. **Shelf life** is typically around 2 years for canned **fruits** and **vegetables** and longer for canned **meat**.

Canneloni Pasta tubes which may be stuffed with **meat**, **vegetables** or **cheese** and often baked in a tomato or cream **sauces**.

Canneries Factories where foods are canned.

Canning A **sterilization** process in which spoilage and pathogenic **microorganisms** are eliminated in foods prior to hermetic sealing in **containers** (**cans**). Most commercial canning operations are based on the principle that bacterial destruction increases tenfold for each 10°C increase in temperature. Food exposed to high temperatures for short periods of time is known to retain more of its natural **flavour**.

Canning equipment Machinery for **preservation** of foods in sealed **containers** (**cans**).

Canning quality Canning quality scores represent the sum of scores for **colour** (chroma, uniformity, and attractiveness), wholeness, **smoothness**, **firmness**, moistness, lack of fibre, **mouthfeel** and **flavour** of **canned foods**.

Canola Alternative term for **rapeseeds**.

Canola oils Alternative term for **rapeseed oils**.

Canopy Uppermost level of plant vegetation in a forest or area under **cultivation,** such as a vineyard, orchard or vegetable plot. Canopy density and structure affect intensity of light reaching the plant, which may impinge on crop quality.

Cans Rigid cylindrical metal containers made of steel sheet or plate, aluminium, copper or other metals. Used as packaging for foods and **beverages**; most are sealed hermetically for storage and retail over long periods of time.

Cantaloupes One of the main cultivated types of **melons** (*Cucumis melo*). Grown commercially in Europe, they have orange (occasionally green), aromatic flesh and a yellowy-orange ribbed, warty rind.

Canteen meals **Meals** served in **canteens**, i.e. **restaurants** catering for workers in establishments such as schools or factories. Food is usually prepared in large amounts and served from a central point

Canteens **Restaurants** located in establishments such as schools and factories. Usually self service and designed to cater for large numbers of people. Also refers to vessels with caps or other closures used for carrying water or other beverages, especially while travelling.

Cantharellus Genus of **fungi**, which includes chanterelles. True chanterelle (*C. cibarius*) is a much-prized species in France and continental Europe, characterized by a funnel-shaped, apricot-yellow cap and a faint fruity **aroma**. Other edible species include *C. tubiformis* and *C. infundibuliformis*.

Canthaxanthin Red pigment of the **carotenoids** group. Occurs naturally in **crustacea** and salmonid **fish** and has **antioxidative activity**. Used as a feed additive to improve the **colour** of **egg yolks**, skin colour of broilers and flesh colour of aquacultured **salmon** or **trout**.

CAP Abbreviation for **Common Agricultural Policy**.

Capacitance Ability to store energy in the form of electric charge. One of the **electrical properties** used in a wide range of food industry analyses, examples of which include monitoring of **yeasts** in **brewing**, food composition, quality deterioration in **frying oils** and **bottling** efficiency.

Cape gooseberries Small, white or yellow **fruits** produced by *Physalis peruviana* (syn. *P. edulis*). Eaten fresh or used in **jams** and jelly products. Similar in appearance and utilization to ground cherries (*P. pruinosa*), but slightly larger in size and less sweet. Also known as goldenberries.

Capelin **Marine fish** species (*Mallotus villosus*) belonging to the smelt family (Osmeridae) which occurs extensively in the north Atlantic, north Pacific and adjoining regions of the Arctic. Marketed in fresh, frozen, lightly smoked, salted and dried forms. Also utilized as a source of **fish oils** and for **fish meal** production.

Capers Unopened flowers of the shrub, *Capparis spinosa*, pickled in **vinegar** and used as a spice. Commonly used in **pickles**, **sauces** and **toppings** for **pizzas**.

Capillaria Genus of parasitic **nematodes** of the family Trichuridae. *Capillaria philippinensis* and *C. hepatica*, found in **freshwater fish**, are the causative agents of **capillariasis**.

Capillariasis Severe and potentially fatal disease in humans caused by eating raw **fish** contaminated with the larvae of *Capillaria philippinensis* and *C. hepatica*. Symptoms include abdominal pain, nausea, vomiting, diarrhoea and anorexia.

Capillary electrophoresis **Electrophoresis** technique in which separation is performed in buffer filled capillaries across which high voltages are applied. Advantages over conventional electrophoretic techniques include faster analysis and the possibility of incorporating on-line detection of separated species.

Capocollo Italian cured **pork sausages** which are a speciality of the Parma region. **Pork** shoulder is cured, flavoured with **spices** and **seasonings** such as sweet **red peppers**, packed into natural **casings** and air dried. Eaten raw, especially in antipasti platters.

Capons Castrated male **chickens**, which are fattened for eating. Compared with cockerels, capons show slightly increased growth rates, less crowing and fighting behaviour, and greater **meat tenderness**.

Capping devices Alternative term for **caps**.

Cappuccino coffee Type of coffee beverage which is topped with **whipped cream** or frothed **milk**. Often served sprinkled with cocoa powder or cinnamon.

Caprenin Semi-synthetic triacylglycerol used in low calorie **fat substitutes**. Composed of **octanoic acid**, **decanoic acid** and **behenic acid**, the last being only partially absorbed in the gut. Melting profile is similar to that of **cocoa butter**, meaning that it may be used in the manufacture of **confectionery**.

Capretto Lean **meat** from goat kids fed on milk up to 5 months of age. Meat is pale pink in **colour** and finely textured. Low in fat, but rich in protein.

Capric acid Synonym for **decanoic acid**. Medium chain fatty acid which occurs in various **fats**, including **milk fats**. One of the **flavour compounds** found in various foods.

Caprine Relating to or resembling **goats**.

Caproic acid Synonym for **hexanoic acid**. Medium chain fatty acid which occurs in various **fats**, including **milk fats**. One of the **flavour compounds** found in various foods.

Caprylic acid Synonym for **octanoic acid**. Medium chain fatty acid which occurs in various **fats**, including **milk fats**. One of the **flavour compounds** found in various foods.

Caps Protective covers or lids, particularly for **bottles**. May include a thread and be used to reseal containers after use.

Capsaicin One of the **flavour compounds** of **chillies** and other **Capsicums**, in part responsible for their pungent characteristics.

Capsaicinoids Flavour compounds of **chillies** and other **Capsicums** related to **capsaicin** and partly responsible for the pungent characteristics.

Capsanthin Pigment of the **xanthophylls** group which occur in **peppers** (**Capsicums**).

Capsicum annuum Domesticated *Capsicum* species that includes many of the most economically important **capsicums**, including **bell peppers, paprika, pimiento peppers**, and many kinds of **chillies**. Fruits tend to be less pungent than those of *C. frutescens*.

Capsicums Fruits of the *Capsicum* genus, also known as **peppers**. The genus contains several domesticated species, such as the economically-important ***Capsicum annuum*** and *C. frutescens*, and many hundreds of varieties. Capsicums are grown worldwide and vary in pod size, **colour**, shape, **flavour** and **pungency**. Some types are used primarily as a vegetable, while others are used as **spices** or for production of **oleoresins**. Common types of capsicum include **bell peppers, paprika** and **chillies**. Good source of many **nutrients** including the antioxidant vitamins A, C and E. Pungency is due to the presence of **capsaicinoids**.

Captafol Non-systemic fungicide used for control of a wide range of fungal diseases in fruits, vegetables and cereals. Example uses include: control of scab in **apples** and **pears**; control of early and late **blights** in **potatoes**; and control of **celery** leaf spot. Normally degrades rapidly in animals and plants; **phthalic acid** and **ammonia** are commonly formed as hydrolysis products. Classified by WHO as extremely hazardous (WHO Ia).

Captan Fungicide used for control of a wide range of fungal diseases in fruits, vegetables and cereals. Examples uses include: control of brown rot in **stone fruits**; control of leaf spot in **carrots**; and control of downy mildew in grape vines. Classified by Environmental Protection Agency as not acutely toxic (Environmental Protection Agency IV). Also known as orthocide.

Carambolas Common name for *Averrhoa carambola*. Tropical fruits native to Indonesia, and now grown in many hot countries. Rich in **vitamin C**, with a waxy, golden yellow skin and translucent, juicy yellow flesh with large brown seeds. Can be eaten raw or cooked, or processed into **tarts, jams** and juice products. Also known as **five fingers** or **star fruit**, due to their five prominent spokes and star-shaped cross section.

Caramel Complex mixture of brown flavouring/colouring substances produced when **sugars** are heated above their melting point during caramelization. Thermal degradation of the sugars results in a similar bitter-sweet **flavour** profile to that of **molasses** and **maple syrups**. Caramel is used in **flavourings** and **flavour enhancers** for a wide range of foods, including **caramels, cakes** and **biscuits**. Colouring properties are employed in **caramel colorants**.

Caramel colorants Colorants resulting from the carefully controlled heating of carbohydrates (e.g. **sugars** or malt **syrups**) in the presence of small amounts of food-grade acids, alkalis or salts. Widely used to impart a yellow or brown **colour** to numerous foods and **beverages**, including **cola beverages** and other **soft drinks, beer, soy sauces, bakery products, browning agents** and **sausage casings**. Both positively and negatively charged caramel colorants are available (particles of the caramel colorant must have the same charge as the colloidal particles of the product to be coloured, in order to avoid precipitation). Also reported to act as **vitamin antagonists** to **vitamin B$_6$**. **Caramel** is also used in flavourings.

Caramels Sugar confectionery products similar to **toffees** made from sweetened, condensed or evaporated **milk, butter** or **vegetable oils**, and **sugar**. Boiled at lower temperatures than toffees, and may be soft or hard.

Caraway Seeds of the umbelliferous plant *Carum carvi*. Used as a spice in a wide range of products including **bakery products, cheese, meat** and **schnapps**. Caraway **essential oils** are also widely used for flavouring purposes.

Carbadox One of the antibacterial **drugs** which are used as growth promoters in animals. Residues may persist in **meat** from treated animals.

Carbamate pesticides Group of **pesticides** which inhibit activity of **cholinesterases** in **insects**. Used for control of chewing and sucking insects (especially **aphids**, whitefly, leaf miners and soil-dwelling insects) in a wide range of fruit, vegetable and cereal crops. Examples include **aldicarb, carbaryl** and **carbofuran**.

Carbamide Synonym for **urea**. The excretory product of nitrogen metabolism produced in the liver of mammals following the breakdown of **amino acids**. Its formation during the **fermentation** of **wines** is significant, since it is a precursor of **ethyl carbamate**, a well known carcinogen. Used as a fertilizer and as a feed supplement for ruminants, and is found in **milk**.

Carbaryl Insecticide with slight systemic properties; also acts as a plant growth regulator. Used for control of chewing and sucking **insects** in a wide range of fruits, vegetables and cereals. Classified by WHO as moderately toxic (WHO II). Also known as naphthyl-methylcarbamate, sevin and vioxan.

Carbendazim Systemic fungicide used for control of a wide range of fungal diseases in cereals, fruits, vegetables, coffee and sugar cane. Degrades relatively slowly in plants. Classified by WHO as slightly toxic (WHO III). Also known as carbendazole.

Carbendazole Alternative term for the fungicide **carbendazim**.

Carbofos Alternative term for the insecticide **malathion**.

Carbofuran Systemic insecticide and nematicide used for control of soil-dwelling and foliar-feeding **insects** and **nematodes** in vegetables and cereals. Rapidly metabolizes to less harmful compounds in plants and soil. Classified by WHO as hazardous (WHO Ib).

Carbohydrases General name for **enzymes** that hydrolyse **polysaccharides** such as **starch**, **celluloses** and **pectins**. Examples of starch-hydrolysing enzymes include **α-amylases**, **β-amylases**, **α-dextrin endo-1,6-α-glucosidases** and **glucan 1,4-α-glucosidases**. Other carbohydrases include **xylan endo-1,3-β-xylosidases**, **endo-1,3(4)-β-glucanases** and **pectic enzymes**.

Carbohydrates One of the main classes of compounds present in foods, which includes **monosaccharides**, their derivatives such as **glucosides**, **polyols**, **nucleotides** and **nucleosides**, and their oligomers and polymers (**oligosaccharides** and **polysaccharides**). Important carbohydrates in foods include **sugars**, **starch**, **pectins**, **fibre** fractions, **celluloses** and their derivatives, and polysaccharides used as additives such as **gelling agents** and **thickeners**.

Carbolines Pyridoindole compounds which may be formed in foods (e.g. **fish** and **meat**) during **cooking** or **processing**. Tetrahydro-β-carbolines and β-carbolines, generated during the **Maillard reaction**, are potential **carcinogens**.

Carbon Element, chemical symbol C, which is a constituent of all **organic compounds**. A specially modified form, activated carbon, is used in various processing aids, such as **adsorbents**, **filter aids** and deodorizing agents, for foods and **beverages**.

Carbonatation Process used in the manufacture of **white sugar** for purification (**clarification**) of **sugar juices**. Various carbonatation methods have been developed for specific purposes, but the basic principle is the same. The process involves addition of lime (CaO) to sugar juice followed by bubbling of **carbon dioxide** through this mixture. A precipitate of $CaCO_3$ forms that entraps suspended impurities within its crystalline structure and adsorbs soluble impurities. Soluble impurities may also react with the lime to form insoluble Ca salts.

Carbonated beverages **Beverages**, especially **soft drinks**, which have been impregnated with sufficient **carbon dioxide** to cause effervescence.

Carbonation Conversion of a compound into a carbonate, or the impregnation of a liquid with **carbon dioxide** (CO_2) under pressure. CO_2 is added to **beverages** to make them effervescent. Examples of **carbonated beverages** include **lemonade** and sparkling **mineral waters**.

Carbon dioxide A colourless, odourless gas (chemical formula CO_2) produced by the burning of carbon and organic compounds and by respiration, and absorbed by plants in **respiration**. Used in **modified atmosphere packaging** of foods.

Carbon disulfide A colourless, extremely volatile and flammable compound, with chemical formula CS_2, with a disagreeable, fetid odour, used in **insecticides**. Exposure to carbon disulfide can occur by breathing it in from the air and by drinking water or eating foods that contain it.

Carbonic acid Acid formed when CO_2 is dissolved in water. Forms various salts (carbonates and bicarbonates), some of which are important in food processing.

Carbonic maceration A **winemaking** process in which whole **grapes** are macerated under a **carbon dioxide** atmosphere before **alcoholic fermentation**; it is used in manufacture of Beaujolais and similar **wines**. Carbonic maceration enhances the fruity character of the wine **aroma**.

Carbon monoxide Toxic colourless, odourless gas, with the chemical formula, which may be formed by incomplete combustion of carbon-containing materials. May be used in **modified atmosphere packaging** of **meat** or other foods.

Carbon tetrachloride Synonym for **tetrachloromethane**. Organic halogen compound and versatile organic solvent whose use has diminished since the discovery that it is carcinogenic. May be used in **fumigants**. Can occur as a contaminant of treated **drinking water**.

Carbonyl compounds **Organic compounds** which contain the CO radical, including **aldehydes** and **ketones**. Many are important **flavour compounds** and **aroma compounds** in foods.

Carboxylesterases EC 3.1.1.1. Hydrolyse carboxylic esters to **alcohols** and carboxylates. Useful for removing acetyl groups from **hemicelluloses** to form easily fermentable carbohydrate substrates, and for modifying the **gelation** properties and other **rheological properties** of heteropolysaccharides.

Also involved in changes in the **aroma** and **flavour** of **wines** and other **alcoholic beverages**.

Carboxylic acids **Organic acids** characterized by presence of the COOH group.

Carboxymethylcellulose Water-soluble cellulose ether obtained by chemical modification. Widely used in food **stabilizers**, **thickeners** or **binding agents** in a variety of foods including **ice cream**, **puddings**, **batters** and **icings**. Also known by the abbreviation CMC.

Carboxypeptidases EC 3.4.16-3.4.18. Exoproteinases that hydrolyse peptide bonds and remove **amino acids** one at a time from protein chains, working from the carboxyl terminus. Useful for production of **protein hydrolysates** and for modifying the **flavour** of foods, e.g. **dairy products**.

Carboxypeptidase Y Alternative term for **carboxypeptidases**.

Carcass by-products Alternative term for **offal**.

Carcasses Dead bodies of animals and **birds**, especially those prepared for cutting up as **meat**. The term is used by butchers to describe animals' and birds' bodies after removal of the heads, limbs, hides (or feathers in birds) and offal; these types of carcasses are also called dressed carcasses. Bird carcasses are usually chilled whole, whilst animal carcasses are usually split longitudinally into sides before chilling. Many countries operate carcass classification schemes, which are designed to categorize carcasses with common characteristics such as carcass weight, fatness (fat class) and conformation. Usually, carcass classification schemes discriminate against very fat and very lean carcasses.

Carcinogenesis Processes leading to the formation of cancer (tumours).

Carcinogenicity A measure of the relative activity of **carcinogens**.

Carcinogenicity testing Analyses, including the **Ames test**, to determine the **carcinogenicity** of suspected **carcinogens**. Also applied to other chemical compounds as part of routine safety evaluation studies. Tests can include the use of **animal models**, cell cultures or **microorganisms**.

Carcinogens Substances that are able to induce **carcinogenesis**, encompassing direct-acting agents that possess **genotoxicity** and indirect-acting procarcinogens that require activation by cell metabolic pathways. Food sources of carcinogens are widespread, and include **heterocyclic amines** formed in **meat** during cooking, **nitrosamines** in nitrite-treated meat products, urethane in **fermented foods** and **alcoholic beverages**, and **agaritine** in **mushrooms**.

Cardamom Green spice pods containing numerous aromatic seeds produced by *Elettaria cardamomom*, a shrub belonging to the ginger family. Pods and seeds are used extensively in **flavourings** for both sweet and savoury dishes, particularly in Indian and Middle Eastern cuisine. White (bleached) pods are also available.

Cardboard Rigid, moderately thick material made from paper pulp but heavier than **paper**. Used widely to make containers, e.g. **boxes**, for packaging foods.

Cardiovascular diseases Congenital and acquired diseases of the heart or blood vessels including **coronary heart diseases** and **stroke**. Many risk factors for cardiovascular diseases have been identified, including lifestyle (smoking, lack of physical exercise), diseases (obesity, hyperlipaemia) and diet. Cardiovascular risk may be modified by lowering intake of **fats**, modulating dietary **fatty acids** composition and increasing consumption of whole grains, **dietary fibre** and **fruits** and **vegetables**.

Cardoons Common name for *Cynara cardunculus*. The plant is of Mediterranean origin and has many similarities to **globe artichokes**, to which it is related. Cultivated mainly for the fleshy leaf stalks, which can be blanched like **celery**, or used in dishes such as **salads** and stews. Roots can also be cooked and used as a vegetable, while extracts from the dried flowers are used as **vegetable rennets** in **cheese-making**.

Caribou The common name for any of the four North American species of large deer in the genus *Rangifer* within the Cervidae family. Caribou are hunted for their **meat**. Caribou meat is a traditional food for some ethnic groups, e.g. the Baffin Inuit in the Canadian Arctic. Caribou meat is referred to as **venison**.

Caries Alternative term for **dental caries**.

Carmine Water-insoluble aluminium lake of **carminic acid** (the red pigment obtained from **cochineal**). Soluble in alkaline media and widely used in natural red colorants for foods and **beverages**.

Carminic acid Water-soluble red pigment obtained from dried bodies of cochineal insects (*Coccus cacti*). Colour is orange to red, depending on pH. **Carmine** is the insoluble aluminium lake of carminic acid.

Carmoisine Bluish-red artificial **azo dyes** used in **confectionery**, **soft drinks**, **ice cream** and canned **fruits**. Also known as azorubine.

Carnauba wax Yellowish wax exuded by the leaves of the north-eastern Brazilian fan palm. Primarily composed of carnaubic acid, which is also found in many plant oils and resins. Used to prepare **coatings** for foods e.g. **fruits** or **sugar confectionery**, decreasing moisture loss and giving an attractive, shiny appearance. Also used to improve the barrier properties of **packaging films**.

Carnitine Amino acid found in muscle, liver and other tissues. Also known as vitamin B_7 or vitamin Bt. Required for the transport of **fatty acids** into mitochondria for oxidation. Rich dietary sources include **meat** and **dairy products**.

Carnobacterium Genus of Gram positive, aerobic, rod-shaped **lactic acid bacteria** of the family Lactobacillaceae. Species may be responsible for **spoilage** of vacuum packaged **meat** (*Carnobacterium divergens*), **fish** (*C. piscicola*) and **chicken meat** (*C. mobile*).

Carnosine Dipeptide (β-alanylhistidine) which occurs in **meat** and **fish** and displays **antioxidative activity**.

Carob beans Seeds from the leguminous Mediterranean tree *Ceritonia siliqua*. Seeds are encased in a sweetish pulp within the **carob pods**. They are used as the source of **carob gums** or can be ground and used as baking flour. Also known as locust beans.

Carob gums Alternative term for **locust bean gums**, obtained from **carob beans**.

Carob pods Pods from the carob tree (*Ceritonia siliqua*), containing seeds (**carob beans**) encased in a soft, sticky pulp. The pulp is high in **sugar** and has a taste similar to **chocolate**. Powdered pulp is marketed as a chocolate substitute and is also used in the manufacture of **beverages** and **syrups**.

α-Carotene One of the **carotenes** with antioxidant and provitamin A activities found in green and yellow plant foods in association with **chlorophylls**. Has approximately half the vitamin A activity of **β-carotene**. Rich dietary sources include **carrots**, **green beans**, **Swiss chard** and **tomatoes**. As with other **carotenoids**, intake of α-carotene is maximized if foods are eaten raw or lightly cooked.

β-Carotene One of the **carotenoids** with antioxidant and provitamin A activities found in yellow and green plant foods in association with **chlorophylls**. Rich dietary sources include carrots, sweet potatoes, green leafy vegetables and yellow fruits. In general, plant foods with more intense green or yellow colour have greater concentrations of β-carotene.

Carotenes Long chain unsaturated hydrocarbons with provitamin A activity found in green and yellow plant foods such as **carrots**, **sweet potatoes**, green **leafy vegetables** and yellow **fruits**. Carotenes (which include α-carotene and **β-carotene**) are the simplest of the **carotenoids** and are cleaved *in vivo*, generating two molecules of **vitamin A**.

Carotenoids Pigments of the polyenoic **terpenoids** class, which are present in a wide range of plant foods and animal foods. Impart a yellow, orange, red or purple **colour** to foods, and may be used as food **colorants**. Many have **antioxidative activity**; some have **vitamin A** activity.

Carp A group of omnivorous **freshwater fish** from the family Cyprinidae which are widely distributed across Europe and Asia. Several species of carp are valued as food fish; the major commercially important species are common carp (*Cyprinus carpio*), **crucian carp** (*Carassius carassius*), **grass carp** (*Ctenopharyngodon idella*), silver carp (*Hypothalmicthys molitrix*) and big head carp (*H. nobilis*). Commonly cultured (especially *C. carpio*), and marketed and processed in a variety of ways.

Carpet shells Any of several species of edible bivalve **molluscs** in the genera *Tapes* and *Venerupis*, most of which occur along the Atlantic coasts of Europe and North America. Commonly consumed species include *T. decussatus*, *T. virginea*, *T. aureus* and *T. japonica*. Also known as **clovis**.

Carrageenan gels Thermoreversible **gels** formed from κ- and ι-**carrageenans**. κ-Carrageenan gels are strong and brittle, whereas those from ι-carrageenans are softer and more cohesive. Applications include as ingredients in **dairy products**, flans, **puddings** and low calorie **jams** and **jellies**.

Carrageenans **Gums** extracted from red **seaweeds** (mainly *Chondrus* crispus and *Gigartina* stellata). Used as **stabilizers**, **thickeners** and **emulsifiers** in a wide range of foods including **milk beverages**, **processed cheese**, **ice cream**, other **dairy products**, **desserts** and ready to feed **infant formulas**. Can be classified into κ-, ι- and λ-carrageenans on the basis of their solubility and gelation properties. Form thermoreversible **carrageenan gels**, which are also used widely in the food industry.

Carrot chips Deep fried carrot slices, typically consumed as **snack foods**. A **lactic acid fermentation** stage may be incorporated into the manufacture process in order to decrease level of **reducing sugars**.

Carrot juices Juices extracted from **carrots** (*Daucus carota*). Rich in **vitamins**, especially **vitamin A**, and **minerals**.

Carrot pulps **Pulps** prepared from **carrots**. Used in the manufacture of a range of products, including **infant foods**, **confectionery** and pulpy **fruit juices**. Carrot pulp wastes remaining after juice extraction can be utilized as a source of **carotenoids**.

Carrots **Root vegetables** from the umbelliferous plant *Daucus carota*. The most important and well known vegetable umbellifer cultivated worldwide. Wild forms of the species are also abundant. Cultivated roots are typically orange in colour and the best-known plant source of **provitamin A carotenoids**. Widely consumed as **salad vegetables** or cooked **vegetables**. In addition, a large proportion of the crop is further processed by **canning**, **drying** or **freezing**. Also used to make products such as **carrot chips**, carrot cakes and **carrot juices**.

Carthamin A natural red pigment obtained from **safflowers** (*Carthamus tinctorius*). Can be used in natural food **colorants**, but stability is a problem due to susceptibility to discoloration in aqueous solutions.

Cartonboard Thin (usually about 0.25-1.00 mm thick), rigid or semi-rigid material made from one or more layers of fibrous **celluloses**. Used widely to make **cartons**.

Cartoning Process of packaging items such as foods or **beverages** in **cartons**.

Cartons Lightweight containers made from **cartonboard**. Usually delivered to the user in the form of flattened, pre-cut and pre-creased carton blanks.

Carvacrol Phenolic monoterpenoid which is one of the **flavour compounds** in many **herbs** and **spices**, especially **thyme** and **oregano**. Has **antioxidative activity** and **antimicrobial activity**.

Carveol Monterpene alcohol which is one of the **flavour compounds** found in **essential oils** of **herbs** and **spices**, including **mint**, **caraway** and **dill**, and **citrus peel**. Formed by conversion of **limonene**.

Carvone Monocyclic terpenoid ketone which is one of the **flavour compounds** in many **herbs** and **spices**, especially **caraway** and **dill**. Used in **antisprouting agents** for stored **potatoes**.

Carya American tree species. Source of **hickory nuts**. **Pecan nuts** come from *C. illinoensis*.

Caryophyllene Sesquiterpene hydrocarbon which is one of the **flavour compounds** present in a wide range of **herbs**, **spices** and **fruits**.

Casein The main protein of **milk**, representing approximately 80% of the total **milk proteins**. Composed of several fractions, including α_s-casein, α_{s1}-casein, α_{s2}-casein, **β-casein**, **γ-casein** and **κ-casein**. A phosphorus-containing protein that is heat stable, but precipitated by **alcohol**, **rennets** and **acids**. Individual fractions are combined into larger units called **casein micelles**, structure and stability of which are related to **calcium** content.

α_s-Casein The main **casein** fraction in **milk**, accounting for approximately 50% of total casein in **cow milk**. Subdivided into fractions α_{s1}-casein and α_{s2}-casein, each of which exists in several genetic variants that differ in **amino acids** composition. Contains relatively high proportions of **lysine** and **tryptophan**.

α_{s1}-Casein A subfraction of α_s-casein. Found in several genetic variants in **cow milk**. These variants differ in **amino acids** composition and have a bearing on the properties and yield of **milk**.

α_{s2}-Casein A subfraction of α_s-casein. Found in several genetic variants in **cow milk**. These variants differ in **amino acids** composition and have a bearing on the properties and yield of **milk**.

β-Casein One of the main **casein** fractions in **milk**, representing approximately 33% of total casein in **cow milk**. Contains relatively high proportions of essential **amino acids**. Found in several genetic variants that differ in amino acids composition and have a bearing on the properties and yield of milk.

γ-Casein One of the **casein** fractions in **milk**, originating from **β-casein**.

κ-Casein One of the **casein** fractions in **milk**, representing approximately 10% of total casein in **cow milk**. Contains relatively high proportions of **isoleucine** and **threonine**. Located on the surface of **casein micelles**. Found in several genetic variants in cow milk. These variants differ in **amino acids** composition and have a bearing on the properties and yield of **milk**.

Caseinates Salts formed by acid precipitation of **casein** from **milk** followed by **neutralization** and **drying**. Some caseinates, including potassium, sodium and calcium caseinate are widely used as food ingredients due to their nutritional and **functional properties**. Uses include **binding agents**, **emulsifiers**, **whipping** agents and protein supplements in foods.

Casein curd Gel formed by **coagulation** of **milk** by **acids** or **rennets**, e.g. during **cheesemaking**.

Casein micelles Conglomerate of individual **casein** fractions found in **milk**. **κ-Casein** is located on the surface of the micelles. Structure and stability of micelles are related to their **calcium** content.

Caseinomacropeptides Large **peptides** constituting the C-terminal fragment of **κ-casein**, formed by **hydrolysis** with **proteinases**.

Casein whey Liquid remaining after precipitation of **casein** by the action of **acids** or **rennets**. Also called **whey**.

Cashew apples Edible fleshy **fruits** of the cashew tree (*Anacardium occidentale*). Although this tropical tree is grown primarily for its crop of **cashew nuts**, the cashew apple is also of commercial interest. The acidic-tasting apple-like fruits are rich in **vitamin C** and can be eaten raw or processed into **jams**, **jellies** and **ices**. They are also fermented to produce juices and **liqueurs**.

Cashew nuts   Kidney-shaped edible **nuts** from the cashew tree (*Anacardium occidentale*). The nuts protrude from the end of edible fleshy receptacles known as **cashew apples** and are a highly prized commodity on the world market. They are usually consumed roasted or used in **confectionery** products.

Casings   Items used to give processed meat products a uniform or characteristic shape, to hold comminuted products together during further processing and to protect meat products. Casings are most commonly used as forms and containers for **sausages**; these types of casings are specifically known as **sausage casings**. There are two major types of casings: natural and manufactured. Natural casings are derived almost exclusively from the gastrointestinal tract of cattle, sheep and swine. Natural casings are highly permeable to moisture and smoke; moreover, they shrink and thereby remain in close contact with the surface of a meat product as it loses water. Most natural casings are digestible and can be eaten. There are four major classes of manufactured casings, namely cellulose, inedible **collagen**, edible collagen and plastic. Strength, shrinkage and permeability characteristics differ between the different types of casings, providing a range of products suitable for the preparation of many different types of meat products.

Casks   Large **barrels** for the transport and storage of liquids, especially **alcoholic beverages**, such as **draught beer**. Traditionally made from wood, but may also be made from **plastics** or metals.

β-Casomorphins   Pharmacologically active fragments of **β-casein** which exhibit biological effects in mammals.

Cassava   Starchy **tubers** produced by the tropical plant *Cassava esculenta* (syn. *ultissima*), also known as **manioc**. An important staple food in many tropical regions, cassava tubers are a good source of carbohydrate and **vitamin C**, but low in protein, minerals and other vitamins. Tubers are the source of **tapioca** starch, while the leaves can be eaten as a vegetable in **soups** and stews. Fresh cassava roots and leaves (particularly those from bitter cultivars) contain the **cyanogenic glycosides**, **linamarin** and lotaustralin, and must therefore be detoxified prior to consumption in order to prevent cyanide poisoning. Detoxification is achieved by conventional grating, washing and cooking methods, or by fermentation into a variety of products including **gari**, **fufu**, **attieke** and **tape ketela**.

Cassava chips   Product made, mainly in tropical countries, by **peeling cassava** tubers soon after harvesting, **slicing** and **drying** the slices by **solar drying**. This drying process is effective in reducing total cyanide levels in cassava, which contains the **cyano-genic glycosides linamarin** and lotaustralin, thus decreasing the risks of poisoning.

Cassava starch   **Starch** isolated from the **cassava** tuber. Also called **tapioca**.

Casseroles   Meals that are slow cooked, usually in **ovens**, in lidded containers. Casseroles are made with **meat** and/or **vegetables** cooked in **stocks** or **sauces**.

Cassia   **Spices** obtained from the evergreen laurel tree, *Cinnamonum cassia,* and some other *Cinnamonum* species. Related to **cinnamon**, but less delicately flavoured. Cassia bark is often used as a substitute for cinnamon, while leaves can be used in **flavourings** similar to bay leaves, and buds are used in a similar manner to **cloves**. Cassia oil is used in **cola beverages**.

Cassia gums   Galactomannan gums extracted from *Cassia* **seeds**. Swell in water and form high viscosity colloids on boiling. Structure and chemical properties have been likened those of **carob gums** and **guar gums**. Although used mainly in pet foods, cassia gums have potential for use as **thickeners** in a wide range of foods, either alone or in combination with other colloids.

Cassia **seeds**   Seeds produced by leguminous plants of the genus *Cassia*, particularly *C. tora* and *C. obtusifolia*. Source of **cassia gums**.

Cassis   Sweet **blackcurrants** flavoured **liqueurs** manufactured in France.

Castor beans   High-protein **oilseeds** from the castor plant, *Ricinus communis*, from which **castor oils** are extracted. Seeds also contain a toxic albumin (**ricin**) and a highly allergenic protein fraction, which limit its food use after oil extraction. Fermented castor bean **meal** is used in a number of Nigerian foods as a spice and can also serve as the basis of a condiment, known as **ogiri**.

Castor oils   Yellow-brown viscous **oils** derived from **castor beans** (*Ricinius communis*). Rich in **ricinoleic acid**, which is released by hydrolysis in the small intestine when the oils are ingested, giving them a purgative action. Also used industrially in the manufacture of chemicals and resins.

Catalases   EC 1.11.1.6. Break down H_2O_2 to water and O_2. Used for removing the H_2O_2 added to cold-sterilized **milk**, improving the **baking properties** of **dough** and improving the **flavour** of fermented **whey**. Exhibit **antioxidative activity** and play an important role in preventing oxidation of **lipids** in **meat**. In conjunction with D-amino-acid oxidases, catalases can be used for production of α-ketoacids, which are gaining importance as nutraceuticals. The enzymes also protect **microorganisms**, including several foodborne **pathogens**, against various environmental stresses.

Catalysts Substances that promote a chemical reaction by lowering the activation energy, but which are not consumed or altered during the reaction.

Catechin Catechol which occurs in **tea** and many other foods and **beverages**. Catechins are thought to have beneficial effects on health, because of their apparent **antimicrobial activity, antioxidative activity** and anticancer properties.

Catecholamines Phenolic **biogenic amines** which occur in tissues of plants and animals. Some, e.g. adrenaline and noradrenaline, act as **hormones** and high preslaughter levels of these compounds (as a result of stress) may be associated with poor **meat** quality. Aerobic oxidation of catecholamines in the presence of **catechol oxidases** results in formation of **melanins**, and hence **browning** of plant foods.

Catechol oxidases EC 1.10.3.1. A group of copper proteins that act on catechol and a variety of substituted **catechols**. Also known as diphenol oxidases, phenolases, polyphenol oxidases and tyrosinases, these enzymes also catalyse the reaction of **monophenol monooxygenases** (EC 1.14.18.1) under certain conditions. Involved in **enzymic browning** in **fruits, vegetables** and cereal grains.

Catechols Flavan-3-ols which are present in a wide range of foods of plant origin. May be polymerized to form **tannins** by the action of polyphenol oxidases (**catechol oxidases**). Catechols may contribute to the **antioxidative activity** and health benefits of plant-derived **phenols**.

Catering Provision of foods and **beverages** in a commercial or institutional setting, or at a function. Occasionally accommodation is provided also. Includes services provided by hotels, **restaurants, canteens** and hospital kitchens. Also encompasses **foods service**.

Catfish Any of a group of 31 families of scaleless **fish**, often with whisker-like projections around the mouth (barbels) and posterior spines in dorsal and pectoral fins. Most catfish occur in freshwater, and many species around the world are valued as food fish. Flesh tends to be firm with a mild **flavour**. Commonly consumed catfish include **channel catfish** (*Ictalurus punctatus*), which are cultured in large numbers in the USA, *Clarius* spp., which are important food fish in African countries, and *Silurus* spp., found in Asian countries.

Cathepsins Enzymes important in **meat tenderization** during ageing and deterioration of **fish proteins gels**, with subsequent effects on **sensory proper-**

ties. Also exhibit proteolytic activity in **dairy products**.

Cations Positively charged particles that have lost one or more electrons. Cations migrate towards negatively charged electrodes (cathodes).

Catmint Common name for *Nepeta cataria* and related species. Used for flavouring **herb tea** and other **beverages**.

Catsups Synonym for **ketchups**. Originally a spicy pickled fish condiment, nowadays the term refers to various thick piquant **sauces** containing **sugar, spices, vinegar**, and other ingredients such as **tomatoes, mushrooms, nuts** or **fruits**. Tomato **ketchups** are one of the most well known types of catsup and are a popular accompaniment for **French fries, burgers** and many other foods.

Cattle Large ruminant mammals with cloven hooves and often with horns, from the family Bovidae. Worldwide, there are over 1000 cattle breeds, of which 250 are major breeds. Cattle fall into two groups, those developed from *Bos indicus* (Indian cattle or zebus) and those, mainly European breeds, developed from *Bos taurus*. Cattle are mainly domesticated for meat (**beef**) and **milk** production. Different gender and age groups of cattle are known as bulls (adult entire males), steers (adult castrated males), cows (adult females), heifers (in general, young sexually mature females to the end of their first lactation) and calves (in general, sexually immature animals which are less than 8 months old).

Cattle kidneys Kidneys from cattle, part of edible **offal**. They are reddish brown in **colour** and composed of 15-25 lobes, which are partially fused together. Left cattle kidneys have a three-sided shape, whilst right kidneys are elliptical in shape. Kidneys from mature cattle tend to have a stronger **flavour** and are tougher than calf kidneys; they need to be cooked slowly using moist heat and are often used in steak and kidney mixtures. In contrast, calf kidneys are tender, have a delicate flavour, and can be cooked by grilling or sauteing.

Cattle livers Livers from cattle, part of edible **offal**. In particular, calf livers are valued for their smooth **texture** and delicate **flavour**; they are often considered a delicacy. Livers from milk-fed calves are very pale in **colour**. Calf livers are usually cooked by **grilling** or **sauteing**, but may also be braised slowly or roasted whole.

Cattle muscles Alternative term for **beef**.

Cattle tissues Alternative term for **beef**.

Caucas Alternative term for **wild garlic**.

Caulerpa Genus of **seaweeds** commonly found in tropical and subtropical waters around Japan, Indonesia, China, the Philippines and Taiwan. Some *Caulerpa* species are edible; traditionally utilized as a fresh salad accompaniment to Asian dishes. *C. lentillifera* is one of the most favoured species due to its soft and succulent **texture**, while in Thailand, *C. racemosa* is commonly sold for use in spicy **sauces**; both these species are cultured.

Cauliflowers Common name for *Brassica oleracea* var. *botrytis*. A vegetable characterized by large edible flowerheads (curds), composed of a compact mass of tiny, underdeveloped florets, which are usually cream or white in colour, but may also be shades of green or purple. Can be eaten raw in **salads**, cooked in a number of ways or used in **pickles**. A good source of **vitamin C**. Closely related to **broccoli**.

Caviar Salted **roes** (eggs) from various species of sturgeon; prepared by a special process involving washing, salting and ripening. Consumed as a table delicacy, with a highly esteemed **flavour** and **texture**. Black caviar from the **beluga** sturgeon is one of the most highly prized and sought after types of caviar. Marketed in small containers or in barrels. Grainy caviar (where roe are easily separated) and pressed caviar (where roe is pressed to remove excess liquid) are common forms of caviar. Sometimes spelt caviare.

Caviare Alternative spelling for **caviar**.

Caviar substitutes **Roes** (eggs) from fish other than **sturgeon**, which are prepared and packaged in a similar way to **caviar**. Principal fish species used are **bream**, **carp**, **coalfish**, **cod**, **herring**, **mullet**, **pike** and **tuna**. The designation is usually preceded by the name of the fish (e.g. cod caviar) and the name of the country of origin is often included.

Cayenne pepper Pungent powder made from the dried pods of **chillies** (*Capiscum frutescens*) including the seeds. Usually deep orange in **colour**. Used in small quantities as a spice, traditionally in Mexican and Italian cooking, but also in dishes from other regions.

CCC Alternative term for **chlormequat**.

cDNA Abbreviation for complementary **DNA**. Single stranded DNA formed from an mRNA template by reverse transcriptases. Radiolabelled cDNA can be used as a probe in **genetic techniques**.

Cebreiro cheese Spanish soft **fresh cheese** made from **cow milk**. Acidic, slightly bitter **flavour**, similar to that of **yoghurt**.

Cedar nuts Name used for some types of **pine nuts**, particularly those obtained from the Siberian pine.

Ceftazidime Cephalosporin antibiotic active against most Gram negative enteric **bacteria**, particularly *Pseudomonas* *aeruginosa*. Used to treat **mastitis** in cattle and bacterial infections of the respiratory and gastrointestinal tracts in cattle and swine. Rapidly depletes in animal tissues following administration.

Ceftiofur Cephalosporin antibiotic active against both **Gram positive bacteria** and **Gram negative bacteria**. Used to treat bacterial infections in cattle and swine. Rapidly depletes in animal tissues following administration. Use at the approved dosage and route is unlikely to result in residues exceeding the maximum residue limit in milk and edible tissues; no milk withdrawal periods are required and residues are not hazardous to industrial cheese and yoghurt starters.

Celeriac Common name for *Apium graveolens* var. *rapaceum*. A variety of **celery** grown for its globose, edible root rather than the stalk and leaves. The white fleshed root is usually consumed cooked and has a similar flavour to celery. Also known as turnip rooted celery.

Celery Common name for *Apium graveolens* var. *dulce*. A major leafy vegetable of the umbellifer family with many food uses. Celery petioles (leaf stalks) can be eaten raw or cooked and used to impart **flavour** and **texture** to dishes such as stews and **soups**. Their distinctive flavour is due to the presence to **terpenes** and phthalides, which are also found in **celeriac**. **Celery seeds** and leaves are used as **flavourings**.

Celery seeds Small brown aromatic **seeds** of *Apium graveolens*, with a similar **flavour** to **celery** petioles. Both seeds and seed oils can be used to flavour stews and **salads**. Ground seeds can also be mixed with salt to form celery **seasonings**.

Cell counts Numbers of cells present in a given sample quantity.

Cell culture *In vitro* growth or maintenance of cells in or on a medium.

Cell lines Established collections of cells which can be cultured indefinitely and which usually have specific properties which can be exploited in scientific research studies.

Cellobiases Alternative term for **β-glucosidases**.

Cellobiohydrolases Alternative term for **cellulose 1,4-β-cellobiosidases**.

Cellobiose Reducing sugar composed of two molecules of **glucose** linked via a β-1,4-glycosidic bond. Although free cellobiose is not found in nature, it is the monomer unit for **celluloses**, one of the most abundant substances in nature. Cellobiose may be prepared from celluloses by hydrolysis with **cellulases**.

Cellophane Thin, transparent material made from **celluloses**. Used as a wrapping for foods to protect against **contamination** and to preserve **freshness**.

Cellulases EC 3.2.1.4. Catalyse the endohydrolysis of 1,4-β-D-glucosidic linkages in **celluloses**, lichenin and cereal β-D-glucans. Produced commercially from a number of **fungi** and **bacteria**. These enzymes have many applications in the food industry, e.g. processing of **fruits** and **vegetables** and their juices, **brewing**, **winemaking**, improving the **shelf life** of **bakery products**, enhancing the quality of soy protein hydrolysates and hydrolysis of celluloses prior to **ethanolic fermentation**.

Cellulolytic enzymes **Enzymes** that act synergistically to hydrolyse **celluloses** or chemically modified cellulose polymers. These enzymes are traditionally classified into three groups, **cellulose 1,4-β-cellobiosidases**, **cellulases** and **β-glucosidases**. True cellulase systems, produced by a number of **fungi**, are able to hydrolyse crystalline cellulose completely, while low-value cellulase systems can only hydrolyse amorphous cellulose. Cellulolytic enzymes can hydrolyse cellulose waste materials prior to **ethanolic fermentation** and, in conjunction with **pectic enzymes**, represent an alternative to chemical **peeling** of **fruits** and **vegetables**.

Cellulomonas Genus of Gram variable, aerobic or facultatively anaerobic **bacteria** of the family Corynebacteriaceae. Occur in soil. Capable of hydrolysing **celluloses**.

Cellulose acetate Tough polymer made by **acetylation** of **celluloses**. Used as the basis of artificial fibres and **plastics**, and in photographic film and magnetic tapes.

Cellulose 1,4-β-cellobiosidases EC 3.2.1.91. Hydrolyse 1,4-β-D-glucosidic linkages in **celluloses** and cellotetraose, releasing **cellobiose** from the non-reducing ends of the chains. In general, these **enzymes** can hydrolyse amorphous celluloses by themselves but only hydrolyse crystalline celluloses in the presence of **cellulases**.

Cellulose ether Derivatives in which some or all of the hydroxyl groups of **celluloses** are involved in ether linkages. Ethylcelluose, **methylcellulose** and **carboxymethylcellulose** are examples which are used as **food additives**.

Cellulose films Transparent plastic **packaging films** made from **celluloses**. Include **cellulose acetate** films and **cellophane** (regenerated cellulose).

Celluloses Class of β-D-(1→4) glucans which are indigestible **polysaccharides** comprising the majority of plant cell wall material. Occur in large quantities in foods, and comprise much of the **dietary fibre** in plant foods. Derivatives such as modified celluloses and microcrystalline celluloses are used as **food additives**.

Cellulose sausage casings **Sausage casings** made of **celluloses**, which must be removed before sausages are eaten. Various sources of cellulose are used, including cotton linters, which are first dissolved and then regenerated to produce casings. Benefits of use include: ease of use; the variety of available sizes; uniformity of size; stretch and shrinkage properties which mimic those of natural casings; and greater strength and lower microbial levels than natural sausage casings. To add artificial **colour** to sausage surfaces, the inner surface of the casings may be coated with an edible, water soluble dye, which transfers to the sausage surface. Very strong casings can be produced by extruding cellulose onto a paper base material; these casings are used to prepare large sausages, such as bologna. Cellulose casings, removed before retail, are also used to prepare skinless sausages.

Cellulosomes High molecular weight multienzyme cellulolytic complexes produced by *Clostridium thermocellum* and other **bacteria**. They consist of a number of enzymes attached to a scaffolding protein, which contains a cellulose binding domain and several cohesin domains which interact with complementary dockerin domains of the catalytic subunits, integrating them into the complex.

Cellvibrio Genus of **Gram negative bacteria** found in soil. Produce **cellulolytic enzymes** which can be used in production of **saccharides** with potential for food use.

Cell walls Structures that are external to the cytoplasmic membranes of plant, fungal, algal and bacterial cells. Maintain cell shape and rigidity and may protect cells from mechanical damage, osmotic lysis and antibiotics.

Central nervous system tissues Tissues associated with that part of the nervous system in vertebrates which includes the brain, cranial nerves and spinal cord. Due to concerns about a possible link between variant **Creutzfeldt-Jakob disease** (CJD) in humans and **bovine spongiform encephalopathy** (BSE) in cattle, controls are in place in **abattoirs** and **slaughterhouses** to exclude BSE risk materials, such as central nervous system tissues, from the human food chain. The risk materials are considered a source of BSE **prions**, consumption of which could potentially result in the development of CJD. In addition, techniques have been developed to screen **meat** and meat products for the presence of central nervous system material.

Centrifugal separators Machines with rapidly rotating containers used to separate two liquids, solids from a liquid, or a liquid from a gas. In the food industry, these separators are used for **clarification** of **beer** and fermentation broths, during **sugar** processing to separate sugar crystals from **syrups**, and during food hygiene practices (e.g. **cleaning in place**).

Centrifugation Process in which liquids are separated from solids, or heterogeneous liquids are separated, on the basis of differences in **density** using machines (**centrifuges**) with rapidly rotating drums.

Centrifuges Machines with rapidly rotating drums used to separate liquids from solids or heterogeneous liquids on the basis of differences in **density**.

Cephalins Mixtures of glycerophospholipids which can be fractionated into **phosphatidylethanolamine, phosphatidylserine** and **phosphatidylinositol**.

Cephalopods Common name for an advanced group of **molluscs** (class Cephalopoda) characterized by absent or reduced internal shells and heads surrounded by tentacles. Includes **cuttlefish, octopus** and **squid**; many species are commercially important food species.

Cephalosporins Group of semisynthetic **β-lactam antibiotics** derived from the natural antibiotic cephalosporin C. Have a similar mode of action to **penicillins**, but tend to have a broader spectrum of action and wider safety margin. Examples commonly used in treatment of farm animals include **cephapirin**, cephradine and **ceftiofur**.

Cephalosporium Obselete genus of **fungi** of the class Hyphomycetes, some species of which are now classified in the genus **Acremonium**.

Cephapirin Cephalosporin antibiotic, commonly used in the form of benzathine or sodium salts for treatment of **mastitis** in cows; also used for treatment of endometritis in cattle, sheep, goats and swine. Rapidly metabolizes in animals following intramuscular administration.

Ceramic membranes Employed in **ultrafiltration** and **microfiltration** systems, ceramic membranes may be of the following types: flat, hollow fibre or open tubular. These membranes possess a high degree of resistance to chemical and abrasion degradation, and tolerate a wide range of pH and temperature ranges. A wide variety of applications includes those relating to biotechnology and pharmaceuticals, isolation and concentration of **enzymes**, **standardization** of the protein content of **milk**, extraction of **proteins** from **whey**, preparation of **quarg** and fresh **cream cheese** by ultrafiltration, **clarification** of **fruit juices**, microfiltration of **alcoholic beverages**, and **concentration** of whole **eggs** and **egg whites**.

Ceramics Articles made of clay that is permanently hardened by heat. Ceramic materials are non-metallic, inorganic compounds - primarily compounds of oxygen, but also compounds of carbon, nitrogen, boron or silicon. Problems have been found relating to **migration** of **heavy metals**, particularly cadmium and lead, from ceramic containers or containers with ceramic glazes into foods with which they are in contact.

Ceramides Generic term for a class of **sphingolipids**; N-acyl derivatives of a long chain base, e.g. sphingosine. Ceramides are present in a wide range of foods, and may be of importance for human health.

Ceratocystis Genus of **fungi** of the class Plectomycetes. Includes several plant pathogens, e.g. *Ceratocystis fimbriata* and *C. paradoxa* which cause black rot of **sweet potatoes** and **pineapples**, respectively.

Cereal bars Processed cereal grains which are formed into bars and often contain other ingredients such as **dried fruits** and **nuts**.

Cereal bran Protective outer layer of the seeds of edible members of the grass family which is separated from the kernel during milling. Often added to foods as a source of **dietary fibre**.

Cereal by-products Secondary products of cereal processing, e.g. **bran** and **germ** removed during **milling** of cereals to produce refined **flour**.

Cereal flours Flour produced by **milling** of cereals.

Cereal products Generic term for foods which have been formulated using cereals as their main ingredient.

Cereal proteins Proteins found in cereal grains, which may be classed as biologically active **enzymes** or biologically inactive **storage proteins**. Storage proteins make up approximately 80% of total cereal proteins and are often used for varietal classification.

Cereals Plants and seeds from monocotyledonous plants of the grass family. The edible, starchy seeds are suitable for food use and are processed to make a wide range of products.

Cereal wines Non-distilled **alcoholic beverages** made by **fermentation** of saccharified **mashes** made from **cereals**. Examples of cereal wines include **sake** and other **rice wines**.

Cerebrosides Lipids of the **galactosides** class which occur in brain and nervous tissues.

Cervelat Smoked, uncooked, mildly seasoned **sausages** made from chopped pork or a mixture of **pork** and **beef**. There are two kinds, namely: soft cervelat, a semi-dry sausage; and dry cervelat, which is dried slowly to a hard texture. Many countries make cervelat. Varieties manufactured include: Goteborg cervelat from Sweden; Gothaer cervelat from Germany; and Landjaeger cervelat from Switzerland. Cervelat may also be known as **summer sausages**.

Cestodes Parasitic tapeworms of the class Cestoda. Includes species of the genera **Diphyllobothrium**, **Echinococcus** and **Taenia**.

Cetacea Order of mammals including **whales, dolphins** and **porpoises**.

Cetavlon Trade name for the cationic detergent disinfectant **cetyltrimethylammonium bromide** (cetrimide).

Cetylpyridinium chloride Antimicrobial agent used in **disinfectants** for cleaning areas such as food processing equipment.

Cetyltrimethylammonium bromide Cationic detergent disinfectant (cetrimide) with the trade name Cetavlon.

Cevapcici Highly spiced meat products, traditionally produced in the former Yugoslavia. They are sometimes considered to be fresh **sausages** without casings. They are made from **beef mince** and/or **pork mince** mixed with fresh **herbs**; the mixture is formed into logs. Cevapcici are usually cooked by grilling and served with chutney or hot relish and toast.

Ceviche Product prepared by marinating raw **fish fillets** or raw **fish mince** in **lime juices** or **lemon juices** with **olive oils**, **spices**, and sometimes **onions**, **green peppers** or **tomatoes**. **Citric acid** in the juices causes **denaturation** of the **fish proteins**, increasing flesh firmness. Eaten usually as an appetizer particularly in Central and South America. Consumption has been associated with outbreaks of **food poisoning** or **anisakiasis** where infected fish or unhygienic food preparation practices have been used. Alternative spellings include seviche and cebiche.

Ceylon spinach Common name for *Basella rubra* (syn. *B. alba*). Leaves and stems contain high levels of **carotenoids** and **ascorbic acid** and are used as vegetables in a similar manner to **spinach**. Can also be used in **thickeners**, while **fruits** are a source of **natural colorants**. Also known as Malabar nightshade.

Chaconine One of the major toxic **glycoalkaloids** found in **potatoes**.

Chaetomium Genus of **fungi** of the class Sordariomycetes. Occur in soil, paper and textiles. Most species are strongly cellulolytic. Some species (e.g. *Chaetomium globosum*) are used in the industrial production of **enzymes** (e.g. **cellulases**).

Chai Spiced milky **tea** drink which originated in India but is becoming a popular beverage worldwide. Made from **black tea** to which is added **milk**, a mixture of **spices** such as **cardamom**, **cinnamon**, **ginger**, **cloves** and **pepper**, and a sweetener such as **sugar**. Also available are spice mixes for use when preparing chai, and chai mixes to which hot water is added for making the beverage.

Chakka Curd formed during preparation of the Indian dessert, **shrikhand**, made by straining **dahi** through a cloth to remove **whey**.

Chalcones Class of minor **flavonoids**, biochemically related to **flavanones**, and **dihydrochalcones**. Native chalcone glycosides are easily transformed to flavanone glycosides, and are rarely extracted from foods in the chalcone form *per se*. Dietary sources of chalcone compounds include **tomato skins**, **hops** and **liquorice**.

Chalva Alternative term for **halva**.

Chamomile Herb obtained from *Anthemis nobilis* (syn. *Chamaemelum nobile*). The plant is a source of **essential oils** used to flavour **liqueurs**, other **beverages** and **confectionery**. Flowers are used to make **herb tea**. Synonym for **camomile**.

Champagne **Sparkling wines** made by the Methode Champenoise in-bottle secondary **fermentation** process, in a defined area of northeast France

Champagnization The specific **winemaking** process used for manufacture of **champagne**, involving in-bottle secondary **fermentation** under defined conditions.

Champignons French word for **edible fungi**. Typically used to refer to cultivated button **mushrooms** (*Agaricus bisporus*).

Channel catfish A freshwater **catfish** species (*Ictalurus punctatus*) which occurs in rivers and streams in North America. Popular in the USA where it is farmed and marketed fresh, smoked and frozen.

Chantarelles Alternative term for *Cantharellus*.

Chapattis Flat, unleavened disc-shaped breads originating from northern India made with **wheat flour**, water and **salt**, and baked on a griddle.

Chaperones **Proteins** which assist in the correct processing, particularly non-covalent assembly, of other proteins. As well as their role in microbial **pathogenicity**, chaperones and their subclass chaperonins are of interest in biotechnology for the production of correctly folded **recombinant proteins**.

Chaptalization Addition of **sugar** to **grape musts** to increase **alcohol** content in the resulting **wines**. Legal in some **winemaking** countries, prohibited in others.

Char Any of several **trout**-like fish species belonging to the genus *Salvelinus* within the family Salmonidae. Char species include *S. alpinus* (**Arctic char**) *S. fontinalis* (brook trout) and *S. namaycush* (lake trout). Flesh of most species is highly regarded. Usually marketed fresh or frozen.

Charcoal Amorphous, usually impure, form of carbon produced by heating wood or other organic material in the absence of air. Can be used in absorbents (**activated carbon**), as a cooking fuel which produces a distinctive **flavour**, e.g. in **barbecued foods**, or in **fermentation technology**.

Charcuterie products Varieties of cold cooked meats, especially **pork** products, which are cured, smoked or processed. They include **ham**, **pates** and **sausages**. Shops in which these products are produced or sold are known as charcuteries.

Charlock Early flowering annual weed (*Brassica kaber* or *Sinapis arvensis*) native to Europe and North America, seeds of which are used to make a poor quality **mustard**.

Charqui Intermediate moisture (water activity = 0.5-0.7), dried meat products, mainly produced in South America. In Brazil, most charqui is prepared from **beef**, but it is also made from **mutton** and llama meat. In Peru, it is also made from alpaca meat. Strips of meat are cut length-wise, salted and then pressed before **air drying**. In its finished form, charqui is in flat, slightly flaky, thin sheets. Traditional charqui is made without addition of nitrites or nitrates; nevertheless, microbial counts decrease during processing and storage. When good quality raw materials and appropriate handling conditions are used for charqui production, the final product has low microbial counts. Charqui-type products include **jerky**.

Chayote Squashes obtained from the tropical plant *Sechium edule*, also known as mirliton. Similar in shape to a large pear, usually furrowed, and containing a single seed. Chayote fruit are used in a variety of savoury and dessert dishes throughout South America and in Creole cooking. They are low in **calories** and **sodium** and a good source of **trace elements**. **Tubers**, shoots and leaves are also edible.

Cheddar cheese Semi-hard **cow milk cheese** originally made in England but now made all over the world. Natural **colour** ranges from white to pale yellow, but some cheeses have **colorants** added to form a more orange colour. Generally matured for 9-24 months, the **flavour** getting sharper with time.

Cheddaring Process used in manufacture of scalded **cheese**. Pressed **curd** is cut into pieces which are covered and left for 6-10 hours at 15-20°C during which the curd becomes elastic and develops a yellow colour and characteristic **flavour**.

Cheese Dairy products made by separating the solid component of **milk** (**curd**) from the liquid part (**whey**). Made mainly from **cow milk**, but also from milk of many other mammals, commonly **goats**, **ewes** and **buffaloes**. An important part of the diet worldwide due to its nutritional properties and ease of preparation.

Cheese analogues Alternative term for **cheese substitutes**.

Cheeseburgers Beefburgers served in **bread rolls** with a slice of **cheese**.

Cheesecakes Rich **desserts**, typically made from **curd cheese** or **cream cheese**, additional ingredients including **cream**, **eggs**, **sugar** or **flavourings**. Sometimes require to be baked. Usually served cold on a biscuit or pastry base and may be topped with **fruits**.

Cheese curd Protein (**casein**) gel formed by **coagulation** of **milk**, e.g. during **cheesemaking**. Other **milk proteins** are retained in the liquid portion (**whey**).

Cheesemaking Process by which **cheese** is made from **milk**. Depending on the type of cheese being made, steps include preparation of the **cheese milk**, **coagulation** of milk with addition of **cheese starters** and rennets, draining of **whey**, pressing, shaping of curd, **salting** and **ripening**.

Cheesemaking milk Alternative term for **cheese milk**.

Cheese manufacture Alternative term for cheesemaking.

Cheese milk Milk used as the starting material in cheesemaking. Also called **cheesemaking milk**.

Cheese sauces Cheese flavoured white **sauces** used mainly for coating foods, e.g. **macaroni**, **cauliflowers** or **fish**. Can be made at home, or purchased in ready to use format or as **sauce mixes**. Dishes that incorporate a cheese sauce are often known as mornay, e.g. eggs mornay or salmon mornay.

Cheese slices Presliced **cheese** of various types and thicknesses packaged for retail sale.

Cheese spreads Spreadable product made from **cheese** to which other milk products and possibly **emulsifiers** have been added.

Cheese starters Microbial cultures inoculated into **milk** to produce **acidity** by **fermentation** during manufacture of **cheese**. Commercial starter preparations are available in liquid form, or as freeze-dried or deep-frozen powders or granules. Composition of the culture is varied according to the type of cheese being made.

Cheese substitutes Artificial alternative to natural **cheese**.

Cheese varieties Specific types of **cheese**.

Cheese whey By-product of **cheesemaking** formed along with **curd** during **coagulation** of **milk**. Rich in **milk proteins** including **α-lactalbumin** and **β-lactoglobulin**. Whey is produced in large amounts, leading to disposal problems. As well as being utilized as a food ingredient, whey is used as a **fermentation** substrate and in animal feeds. Also known as lactoserum or serum.

Chelating agents Substances which form a stable chelate ring with free metal ions and can therefore be used in foods to help control the reaction of trace metals with other food components. They act as **sequestrants** to prevent metal-catalysed oxidation, unwanted crystal formation and loss of nutritional quality in a variety of foods, and can also be used for the controlled release of metal ions for nutritional purposes or for controlled **gelation** in **thickeners**. Examples of chelating agents include **EDTA** (ethylenediaminetetraaceticacid) and **glucono-δ-lactone**.

Chemesthesis Complex sensation obtained from foods, regarded as a component of the **sensory properties flavour** and **mouthfeel**. Examples include the burn of **capsaicin** in **chillies**, the cooling sensation from **menthol** and the tingle associated with **carbonated beverages**.

Chemical oxygen demand Measure of the quantity of chemically oxidizable components present in water. Often abbreviated to COD.

Chemiluminescence Emission of light during a chemical reaction; may be used to measure that reaction.

Chemisorption Adsorption of a gas by a solid in which the molecules of the adsorbed gas are held on the surface of the adsorbing solid by the formation of chemical bonds.

Chemistry The science of the properties, structure and composition of elements and their compounds, including the transformations which they can undergo and the energy transfer during these reactions.

Chemostats Apparatus for maintaining a microbial population in the exponential phase of growth by regulating the input of a rate-limiting nutrient, and removal of medium and cells. The concentration of **biomass** in the culture vessel remains constant and the culture is normally grown at a sub-maximal growth rate. Under steady-state conditions, the relationship between growth rate and concentration of growth-limiting substrate can often be predicted using the Monod equation, while specific growth rate is numerically equal to the dilution rate.

Cherimoya Common name for *Annona cherimola*, a member of the **custard apples** family. Native to South America, the edible **fruits** have a green, scaly surface and soft, yellowish white flesh containing a number of seeds. Fruits have a flavour similar to **pineapples** and are believed to be one of the finest tasting of the custard apples. They can be eaten raw or used in **flavourings** for **beverages** and foods such as **ice cream**.

Cherries Reddish coloured **stone fruits** from trees of the *Prunus* genus. Can be classified into two main groups, **sweet cherries** (*P. avium*) and **sour cherries** (*P. cerasus*). Available fresh, dried, canned, frozen or brined (e.g. Maraschino cherries). Used as in-

gredients in many food products including **cakes**, **pies**, **cherry brandy**, **cherry juices** and **confectionery**.

Cherry brandy Liqueurs made from **cherries**, which may be made with addition of crushed cherry stones to impart a characteristic bitter almonds **flavour**.

Cherry juices Juices extracted from **cherries** (*Prunus cerasus*).

Cherry laurel Common name for *Prunus laurocerasus* (syn. *Laurocerasus officinalis*). Similar in appearance (but unrelated to) **bay**. Leaves yield **essential oils**, which are used as **flavourings** in various types of foods, including **desserts** and **confectionery**, and **beverages**. Leaves contain **hydrocyanic acid**, which has to be removed from the oils prior to food use.

Cherry salmon A **Pacific salmon** species (*Oncorhynchus masou masou*) from the northwest Pacific region; also known as masu salmon or Japanese char. Some forms remain in fresh water throughout their lives. A valued food fish in Japan, where its market price tends to be considerably higher than that of other salmon. Normally marketed fresh or frozen; also sold as a fermented **sushi**-like product.

Cherry tomatoes Popular small-sized tomatoes characterized by an appealing bright **colour** and good **flavour** characteristics.

Chervil Common name for *Anthriscus cerefolium*. A delicately flavoured herb which is used in a similar manner to **parsley** as a garnish or to flavour **salads**, **sauces**, and **meat** and **fish** dishes.

Chestnuts Edible **nuts** from trees of the genus *Castanea*, particularly, *C. sativa* (Spanish or **sweet chestnuts**), *C. mollissima* (Chinese chestnuts) and *C. crenata* (Japanese chestnuts). Consumed as dessert nuts and also available in canned, pureed or ground forms. Used as an ingredient in **confectionery** and as an accompaniment to savoury dishes. May also refer to water chestnuts (*Trapa natans*) and **Chinese water chestnuts** (*Eleocharis dulcis*).

Chevon Alternative term for **goat meat**; the term is commonly used in India.

Chewiness Texture term relating to the extent to which a product needs chewing, or a measure of the effort needed to chew, i.e. its **toughness**, rubberiness or leatheriness in the mouth.

Chewing gums Sweetened products made from chicle (gum-like exudate consisting of coagulated milky juice from the bark of the evergreen sapodilla tree, *Achras zapota*) or similar resilient substances (e.g. plasticized rubber or polymers), **sugar** or similar **sweeteners**. May also be made using a gum base, softeners and **flavourings**. Some chewing gums are specially formulated to promote dental health. Also known as chicle gums or gum balls.

Chhana Indian style soft **Cottage cheese** analogue prepared by heating **milk** (usually **cow milk**) to nearly boiling, adding acid **coagulants** while the milk is hot and removing **whey** by filtration. Used as a base for various Indian sweets, such as **rasogolla** and **sandesh**. Also known as channa.

Chicha Corn based **alcoholic beverages**, which may be made by a combined **alcoholic fermentation/lactic fermentation** process, originating in Central and South America.

Chicken bones Bones from chicken **carcasses**. During **cooking**, they darken in colour; this **colour** change is increased by **freezing** and **thawing** prior to cooking. Chicken bones are commonly used to prepare chicken **soups** or are processed into animal feeds. Hot-water extracts prepared from chicken bones are used in many types of products, especially in **flavourings**. Exposure of **chicken meat** containing bone to a dose of ionizing radiation results in the formation of long-lived free radicals which give rise to characteristic electron spin resonance (ESR) signals. The presence of these signals provides clear evidence that chicken meat has been irradiated. **Mechanical boning** of chicken meat remains a problem to the meat industry, as bone fragments often remain in chicken fillets, escaping manual or X-ray machine detection.

Chicken drumsticks Lower portions of the legs of **chickens**; they consist of the tibiotarsus and fibula bones with the surrounding **chicken meat**, cartilage and skin. Colour of meat from chicken drumsticks is darker than that of breast meat, primarily because chicken leg meat contains higher concentration of **myoglobin** and **haemoglobin** than breast meat.

Chicken gizzard pickles Pickles made from chicken gizzards. They are usually prepared from sliced, cooked chicken gizzards, salt and water, and are often mustard oil- or vinegar-based. Other ingredients may include **garlic**, **ginger**, **cumin**, red chilli, **aniseed**, **caraway**, **turmeric**, **black pepper**, **cinnamon** or **cloves**.

Chicken livers Livers from **chickens**, part of edible **offal**. They are commonly cooked by **sauteing**, **frying** or **grilling**, or are used to prepare **pates** or mousses.

Chicken meat Meat from **chickens**. Different proportions of red and white myofibrils produce light and dark meat in different parts of chicken **carcasses**. Chicken leg meat is darker than chicken breast meat. Composition of feeds influences **flavour** and **colour** of chicken meat. Compared with chicken meat produced in intensive systems, free-range chicken meat tends to have more flavour; however, it is tougher and, in developed countries, more expensive. Chicken meat can be roasted, grilled, poached or casserolled. Chickens are sold whole, or portioned into joints, including chicken breasts, wings, drumsticks and thighs.

Chicken mince **Meat mince** prepared from **chicken meat**. It may be prepared specifically from light or dark chicken meat. Mince prepared from light coloured chicken meat has a lower content of saturated fats than mince prepared from dark chicken meat. Also known as ground chicken.

Chicken nuggets Breaded, coarsely comminuted chicken products, usually reconstituted from deboned **chicken meat**. Formulations often include spent hen meat and offal. Quality of the product (often prime, choice or economy grades) differs with the proportion of lean meat to **offal**. Economy-type products tend to include higher proportions of offal and show higher cooking losses than the other types.

Chicken patties Meat patties prepared from **chicken mince**.

Chickens Birds of the genus *Galus* belonging to the order Galliformes. These common domestic fowl are kept virtually worldwide for the production of **chicken meat** and **eggs**. Most commercial chicken farms use intensive systems; however, consumer concerns relating to animal welfare have led to an increase in the use of less intensive systems and free-range systems. Different gender and age groups of chickens are known as cocks (adult entire males), capons (adult castrated males), hens (adult females), cockerels and pullets (usually sexually mature young males and females, respectively) and chicks (sexually immature birds with down rather than feathers).

Chicken sausages Sausages prepared from **chicken meat**, often spent hen meat. Commonly they are made from **mechanically recovered meat** or chicken meat trimmings. They also tend to include **chicken skin** and the less preferred components of chicken **offal**, such as gizzards and hearts. Other ingredients may include water, salt, nitrites, pork fat, blood and **phosphates**.

Chicken skin Skin from **chickens**. Antimicrobial treatment of chicken skin is commonly used to decrease bacterial contamination (and cross contamination) of chicken **carcasses** during processing. Most of the **fats** in **chicken meat** are associated with the skin; thus, fat content can be lowered by removing the skin. Chicken skin is used as an ingredient in **sausages**, including **chicken sausages**. Connective tissue proteins recovered from chicken skin are used to manage the added water in comminuted meat products.

After removal of fat and water soluble proteins by aqueous washing, chicken skin is potentially useful as a low-fat ingredient in emulsified meat products. Colour of chicken skin is either white or yellow; density of the yellow pigment is correlated with the amount of **xanthophylls** in chicken feeds.

Chick peas Mild-flavoured **beans** of *Cicer arietinum*. An important pulse in many regions including the Middle East, Mediterranean and Latin America . Chick peas can be divided into two major types: Desi, which are relatively small and dark in **colour** and the larger Kabuli which are of Mediterranean and Middle Eastern origin. Contain high amounts of good-quality protein and are also a good source of **folates** and other B vitamins. They are used in many foods including **salads**, **pasta** and **dips**, and are the basis of **humous** and **falafel**. Also known as **garbanzo beans** and **Bengal gram**.

Chicle gums Alternative term for **chewing gums**.

Chicory Common name for *Cichorium intybus*. Utilized in a number of ways, some cultivars being grown for the root, a powder or extract from which is used as an additive in **coffee**, making a more bitter beverage. Other cultivars are grown for the leaves, which are used in **salads** or cooked as a vegetable. Some cultivars, such as **witloof**, are used to produce blanched leafy growths called chicons, which are eaten raw or cooked. Similar nutritionally to **lettuces** and **endives**.

Chihuahua cheese Mexican semi **hard cheese** made from pasteurized **cow milk**. The interior is pale yellow and the **flavour** varies from mild to a sharp and Cheddar-like. Chihuahua is a stringy **cheese** which melts well, making it suitable for use in **toppings** and **fillings**.

Childrens foods Alternative term for **infant foods**.

Chilean hazelnuts Nuts of the tree *Gevuina avellana*, native to Chile and Argentina but grown also in other parts of the world. Closely related to and similar in quality and size to **macadamia nuts**, but enclosed in a thinner and softer shell. Eaten roasted, but also used as a source of **edible oils**. Also known by several other names, including Chilean nuts, Chile nuts, gevuina nuts, guevin nuts and neufen nuts.

Chilled foods Perishable **foods** that can be stored at chilled (refrigerator) temperature for a specified amount of time. Examples include chilled **ready meals**, **pizzas**, **sandwiches** and many **dairy products**.

Chillers Cold cabinets or **refrigerators** that are capable of rapid **cooling/chilling** of foods to a few degrees above their **freezing point** in order to extend **shelf life**.

Chilli Spices obtained from ground chillies. **Flavour**, **capsaicin** content and **pungency** vary according to type of pepper used. May also refer to chilli-based spice mixtures used for making Mexican dishes such as chilli con carne.

Chillies Very hot, finely tapering **red peppers** of any of several cultivated varieties of **capsicums** (hot peppers). Rich in **vitamin A** and **vitamin C**; good sources of **vitamin E**, potassium and **folic acid**. Used mainly as a flavouring in **cooking**. Also known as chilies, chili peppers and chiles.

Chilling Process of making foods colder to extend their **shelf life**, usually undertaken by application of **refrigeration**.

Chilling injury Disorder of **fruits** and **vegetables** induced by low temperatures. May occur in the field, during transit or in retail or domestic refrigerators. Symptoms include surface lesions, water soaking of tissues, water loss, internal discoloration, failure to ripen, and decay. Critical temperature for chilling injury varies with type of crop. Storage life of produce susceptible to chilling injury is short, as refrigeration cannot be used to preserve quality.

Chinese cabbages Cabbages of the species *Brassica pekinensis* or *B. chinensis*. The crinkly, thickly veined leaves are thin and crisp, cream in colour with green tips, and have a mild flavour. Rich in **vitamin A**, **folic acid** and potassium. Eaten raw or cooked as a vegetable. Many alternative names, including napa cabbage, celery cabbage, Peking cabbage, wong bok, bok choi, pak choi and Chinese white cabbage.

Chinese chives Common name for *Allium tuberosum*. Young leaves and flower stalks, with their **garlic**-like flavour, are used in **seasonings**. Also known as garlic chives and oriental garlic.

Chinese dates Alternative term for **jujubes**.

Chinese gooseberries Alternative term for **kiwifruit**.

Chinese pears Fruits produced by *Pyrus chinensis*, *P. ussuriensis*, *P. bretschneideri* or, more generally, *P. pyrifolia*. Originally cultivated in China. *P. pyrifolia* is the oriental pear, also referred to as Asian pears, Japanese pears and sand pears.

Chinese sausages Fairly hard, **dry sausages** usually made from **pork** meat and pork fat. They are similar in **texture** to **pepperoni**. Chinese sausages are smoked, slightly sweet and highly seasoned. Varieties include the lop chong. Chinese sausages are often added to stir-fry dishes.

Chinese water chestnuts Corms produced at the ends of horizontal rhizomes of *Eleocharis dulcis*, a plant cultivated in marshy areas or lakes in Asia. Skin is brown-black and similar to that of **chestnuts**. The white flesh is crunchy and juicy, with a bland flavour. Used widely in Asian dishes, raw or cooked. Contain moderate amounts of **starch**, **sugar**, B vitamins, vitamin C and **vitamin E**, and relatively high amounts of potassium and phosphorus. Available fresh or canned; a powdered form is used as a thickener, similar to **corn starch**. Also known as matai.

Chinook salmon The largest **Pacific salmon** species (*Oncorhynchus tshawytscha*) found in coastal water and rivers along the Pacific coast of North America, Japan and in the western Arctic; also known as king salmon. High fat, soft-textured flesh is usually red, but some forms are white; the red meat commands a higher price. Marketed fresh, smoked, frozen, and canned in whole (gutted) form, fillets and steaks.

Chipping properties **Functional properties** relating to the ability of different cultivars or varieties of **potatoes** to be processed into good quality **chips**. The most important processing quality parameters for chips are **colour**, **flavour** and **texture**.

Chips Small pieces of food prepared by chopping or cutting, which are then usually fried. Include **potato chips (French fries)**, **corn chips** and **tortilla chips**. The term is frequently used to refer specifically to potato chips in the UK and to **potato crisps** in the USA and continental Europe.

Chistorra Semi-cured sausages that are a speciality of the Basque region of Spain. They are long, thin, flavourful **pork sausages** produced in links. Ingredients include garlic. Chistorra are lightly cured and dried for only a few days. In the Basque region, they are usually cooked lightly before eating with **eggs** or with local **bread**; however, they are also popular as flavourings for cooked dishes such as bean, lentil or rice **casseroles**.

Chitin Homopolysaccharide, consisting of $\beta(1{\rightarrow}4)$-linked D-*N*-acetylglucosamine. Occurs in **shells** of **crustacea** and cell walls of **fungi**, and may be recovered from crustacea shell wastes. Has potential use in **functional foods**.

Chitinases EC 3.2.1.14. Randomly hydrolyse *N*-acetyl-β-D-glucosaminide 1,4-β-linkages in **chitin** and chitodextrins. Produced by plants, **fungi**, **yeasts** and **bacteria**, these **enzymes** exhibit **antifungal activity** and can be used for processing **shellfish wastes**. Also responsible for **haze** formation in **wines** and are major **allergens** of **fruits** such as **avocados**, **bananas**, **chestnuts** and **kiwifruit**, causing latex-fruit syndrome.

Chitin deacetylases EC 3.5.1.41. Catalyse hydrolysis of **chitin** into **chitosan** and acetate, via splitting of the *N*-acetamido groups of *N*-acetyl-D-glucosamine

residues. Chitosan formed is of potential use in the **functional foods** sector.

Chitosan Polysaccharide derived from chitin by partial deacetylation with a strong base. Of potential use in **functional foods**.

Chitosanases EC 3.2.1.132. Hydrolyse β-1,4-linkages between *N*-acetyl-D-glucosamine and D-glucosamine residues in partly acetylated **chitosan**. Act only on polymers with 30-60% acetylation. These enzymes can degrade the cell walls of **microorganisms** that contain glucosamine polymers and can be used for production of chitooligosaccharides, which have a number of potential uses in the food industry.

Chitterlings Term applied to the small **intestines**, usually from swine, when prepared for use as food. May be used as an ingredient of **sausages** or **pies**, or may be eaten raw. Consumption of raw chitterlings has been associated with **food poisoning** where preparation conditions have not been hygienic. Also called chitlings.

Chives Common name for *Allium schoenoprasum*. Fresh leaves have a mild onion-like **flavour** and are chopped and used as a garnishes in **soups** and **salads**. Also available as a dried herb. **Chinese chives** are *A. tuberosum*.

Chlamydomonas Genus of unicellular green **algae** of the family Chlamydomonadaceae. Occur in freshwater habitats and on damp soils. Used as a model for cell and molecular biology research studies.

Chloramine **Antimicrobial compounds** that decompose slowly to release **chlorine**. May be used in the treatment of water supplies.

Chloramine T Organic **chloramine** used as an antiseptic and/or disinfectant. Widely used in the food industry to disinfect equipment before processing.

Chloramphenicol Highly active antibiotic used both in treatment and prophylactically in a range of animals, including poultry, calves, swine and goats. Also used in salmon and trout for the treatment of furunculosis. Potentially genotoxic; use is restricted in many countries and banned in food-producing animals within the EU and USA. Also known as chlormycetin.

Chlorates Salts of chloric acid commonly used for **disinfection** purposes. May be formed in **drinking water** as a result of **chlorination**. Considered to pose a health risk to humans.

Chlordane Non-systemic organochlorine insecticide formerly used for control of a wide range of insect **pests** in crops, soil, industrial and domestic environments, but currently used only rarely. Classified by WHO as moderately toxic (WHO II).

Chlorella Genus of unicellular green **algae** of the family Oocystaceae. Occur in fresh water and soils.

Species (e.g. *Chlorella pyrenoidosa*) may be used in the production of **single cell proteins**, or as **food additives** owing to their nutritional composition (high protein, vitamin B_{12} and iron contents) and beneficial **physiological effects**. Some species are added to foods (e.g. **cakes, cheese, mayonnaise, ice cream** and **rice**) to improve their **flavour**. Due to their high contents of **carotenoids**, they are used as feed additives for the enhancement of the **colour** of **rainbow trout** flesh. *C. protothecoides* produces **lutein**, which is used in food **colorants** for foods such as **pasta**.

Chlorfenvinphos Organophosphorus insecticide and acaricide used for control of soil-based and flying **insects** in citrus fruits, vegetables, cereals and sugar cane; also used to control ectoparasites on animals. Classified by WHO as highly hazardous (WHO Ia).

Chlorides Salts of **hydrochloric acid**. Occur widely in foods and **beverages**, the most important being common **salt**, NaCl, which is used in **food additives** such as **flavourings, preservatives** and **bulking agents**.

Chlorinated hydrocarbons Organic compounds which contain one or more chlorine atoms. Include **pesticides** such as **HCH, heptachlor, aldrin, endrin, dieldrin, PCB, DDE** and **DDT**. Suspected of being carcinogenic, and characterized by accumulation in the food chain and very slow biodegradation. May contaminate **fish** and **shellfish** when discharged into the sea along with industrial effluents.

Chlorination Insertion of a chlorine atom into a compound, or treatment of an item with **chlorine** gas (Cl_2). For example, chlorine gas can be used in **sterilization** of water.

Chlorine Member of the **halogens** group, chemical symbol Cl. Chlorine and it compounds have strong microbicidal activity and are used in the food industry as **disinfectants** and sterilizing agents. Chlorine gas is toxic.

Chlorine dioxide Gaseous **chlorine** compound which is used in **oxidizing agents**-type **disinfectants**, used for **sterilization** of foods and water.

Chlorites Salts of chlorous acid, used as **disinfectants** in the food industry.

Chlormequat Plant growth regulator used for treatment of **fruits, vegetables** and **cereals** to improve **ripening** and quality. Can also be used as a herbicide. Also known as CCC, chlorocholine chloride and cycocel.

Chlorocholine chloride Alternative term for **chlormequat**.

Chlorococcum Genus of unicellular green **microalgae** occurring in damp conditions, e.g. in soil. Produce the pigment **astaxanthin** and other **carotenoids** which can be used as **colorants** for foods.

Chloroethylphosphonic acid Alternative term for **ethephon**.

(2-Chloroethyl)phosphonic acid Chemical name for the plant growth regulator **ethephon**.

Chlorofluorocarbons Abbreviated to CFC. Any class of synthetic compound of carbon, hydrogen, chlorine and fluorine used as **refrigerants** and aerosol propellants. Commercial CFC are nonflammable, non-corrosive, nontoxic and odourless, but are known to be harmful to the ozone layer. The most common commercial CFC, marketed as **Freons**, are trichlorofluoromethane (CFC-11) and dichlorodifluoromethane (CFC-12).

Chloroform Colourless, heavy, volatile, toxic liquid. Used as a solvent, fumigant and insecticide. Also known as **trichloromethane**.

Chlorogenic acid Synonym for **caffeoylquinic acid**. Phenol present in many foods of plant origin. Plays an important role in **enzymic browning** of **fruits** and **vegetables**. Has **antioxidative activity**, and may contribute to possible health-promoting or protective actions of dietary phenolic compounds.

Chloromycetin Alternative term for the antibiotic **chloramphenicol**.

Chlorophenol Organic halogen compound used in **pesticides** and wood preservatives. Formed in water and waste water as a result of **chlorination**. Chlorophenol contamination may cause **taints** in foods, **beverages** or water.

Chlorophos Alternative term for the insecticide **trichlorfon**.

Chlorophyllases EC 3.1.1.14. Involved in **chlorophylls** degradation, catalysing the conversion of chlorophyll to phytol and chlorophyllide. Responsible for production of pheophorbide in **green tea**, which may cause **dermatitis**.

Chlorophylls Green photosynthetic **pigments** of the **porphyrins** class which occur in leaves and other plant tissues. May be used as food **colorants**, but stability is poor.

Chloropicrin Soil fumigant which may occur as residues in foods. Also one of the **disinfection by-products** which may be formed during **chlorination** of **drinking water**.

Chlorothalonil Non-systemic foliar fungicide used for control of many fungal diseases in a wide range of crops, including pome fruits, stone fruits, citrus fruits, vegetables, soybeans, cereals, coffee and tea. In plants, the majority of residue remains as parent compound. Classified by Environmental Protection Agency as not acutely toxic (Environmental Protection Agency IV). Also known as daconil.

Chlorpropham Selective systemic herbicide and plant growth regulator. Used for pre-emergence control of many annual grasses and some broad-leaved weeds in a wide range of vegetable crops; also used as a **sprouting** inhibitor in **potatoes** and **tomatoes**. Classified by Environmental Protection Agency as slightly toxic (Environmental Protection Agency III). Also known as CIPC.

Chlorpyrifos Non-systemic organophosphorus insecticide used for control of biting and chewing **insects** in a wide range of fruits, vegetables and cereals; also used for stored cereals and in animal rearing facilities. Rapidly metabolized by plants and is not absorbed via roots. Classified by WHO as moderately toxic (WHO II). Also known as dursban.

Chlorpyrifos-methyl Non-systemic organophosphorus insecticide and acaricide used for control of biting and chewing **insects** in a wide range of fruits, vegetables and cereals; also used for stored cereals. Classified by Environmental Protection Agency as slightly toxic (Environmental Protection Agency III).

Chlortetracycline Broad-spectrum tetracycline antibiotic used for treatment and control of a wide variety of bacterial infections in farm animals. Readily disperses throughout tissues; rapidly depletes following withdrawal in most cases.

Chocolate A **confectionery** product made from hulled, fermented and roasted **cocoa beans** (nibs), blended with **sugar**, **fats** (**cocoa butter** or **cocoa butter substitutes**) and **lecithins**. **Milk** solids may be added to produce **milk chocolate**. Fat is an important component since its particular melting profile contributes to the **mouthfeel** of the product. Chocolate contains **theobromine**, an alkaloid with effects similar to those of caffeine.

Chocolate bars **Chocolate** products that may or may not contain added ingredients or **fillings**, such as **nuts**, toffee, **biscuits** and **dried fruits**, formed into bars.

Chocolate beverages Hot or cold **beverages** in which **chocolate** is a main ingredient.

Chocolate chips Small pieces of **chocolate** used as ingredients in **confectionery** and **bakery products**.

Chocolate coatings **Chocolate** preparations used to coat various products such as **sugar confectionery**, **bakery products**, fruit or **ice cream**. Formed by pre-crystallization of chocolate, **coating** of the food and cooling. Pre-crystallization and cooling affect the gloss, degree of solidification and coat thickness of the coatings produced.

Chocolate confectionery Collective term for **chocolate** and chocolate products.

Chocolate couverture **Chocolate** which contains maximal levels of **cocoa butter**, used as **coatings** for high quality chocolate products.

Chocolate crumb Intermediate material produced during manufacture of **milk chocolate**, composed of **dried milk**, **sugar** and **cocoa mass**.

Chocolate desserts **Desserts** containing **chocolate** as a main ingredient, e.g. chocolate flavoured **milk puddings** and chocolate **mousses**.

Chocolate dragees **Confectionery** products composed of hard centres coated with **chocolate**.

Chocolate fillings **Chocolate** products used as **fillings** for various products, including **sugar confectionery**, **bakery products** and **snack foods**. May also refer to fillings (e.g. creme fillings) used for **chocolates**.

Chocolate liquor Fermented and roasted **cocoa beans**, which are ground finely to form a paste used in the manufacture of **chocolate** and **cocoa powders**. Grinding releases fats (**cocoa butter**) from the cells of the cocoa beans which helps the chocolate to flow. Also called chocolate mass, cocoa mass and cocoa liquor.

Chocolate mass Alternative term for **chocolate liquor**, produced by grinding dehusked **cocoa beans**, or nibs, to a paste from which **chocolate** and chocolate products are made. Also called cocoa mass and cocoa liquor.

Chocolate milk **Chocolate** flavoured **milk**-based beverage.

Chocolate powders Manufactured from **cocoa powders** which are agglomerated to form larger particle sizes. Used in the manufacture of **chocolate**-based **beverages**.

Chocolates **Sweets** made or coated with chocolate.

Chocolate truffles Small, round **chocolates** with a soft and creamy centre, which may be flavoured, often with fruit **flavourings** or **liqueurs**.

Chokeberries Pea-sized fruits produced in red and black varieties by plants of the genus **Aronia**. Black chokeberries, produced by *A. melanocarpa*, are violet-black in **colour** with a strong sour **flavour**. They are rich in **vitamins** and minerals and have a high content of **flavonoids**. Fruits are eaten fresh or preserved by **canning** or by **drying** whole or as a pulp. Juices may be extracted to make jellies. Also used commercially as a source of **natural colorants**.

Cholecalciferol Synonym for **vitamin D$_3$**; one of the group of **sterols** which constitute **vitamin D**. Fat soluble vitamin necessary for formation of the skeleton and for mineral homeostasis. Produced on exposure to

UV light from the sun from the provitamin 7-dehydrocholesterol, which is found in human skin. Alternative recommended name is calciol.

Cholera Acute infectious human disease characterized by profuse diarrhoea leading to extreme dehydration that can result in shock, renal failure and death. Caused by **cholera toxin** produced by *Vibrio* cholerae. Spread by the faecal-oral route, usually via faeces-contaminated water and food.

Cholera toxin Toxin produced by *Vibrio* cholerae that is responsible for **cholera**.

Cholesterol One of the **sterols**, and the major sterol found in vertebrate mammals. Present in all plasma membranes, but found especially in blood, liver, nerve tissue, brain tissue and **animal fats**. A precursor of many **steroids**, including the **bile acids** and steroid **hormones**. Not an essential dietary requirement; consumption of high levels have been associated with **atherosclerosis** and **coronary heart diseases**. Several **health foods** are claimed to reduce serum cholesterol levels; production of cholesterol-reduced products, especially **dairy products** and **eggs**, is increasing.

Cholesterol oxidases EC 1.1.3.6. Bifunctional flavoenzymes that catalyse the oxidation and **isomerization** of 3-β-hydroxy-5-ene-steroids to 3-keto-4-ene-steroids. Useful for determination of **cholesterol** levels in foods and, potentially, for controlling cholesterol levels in **fermented dairy products** by expression in *Lactobacillus* spp.

Cholesterol oxidation products Oxidized **cholesterol** derivatives, also known as oxysterols, which have been linked to a range of adverse health effects including **cytotoxicity**, atherogenicity and **carcinogenicity**. Cholesterol oxidation products have been identified in a range of foods, including **eggs**, **meat**, **dairy products** and **sea foods**. Their formation can be influenced by food processing and storage conditions.

Cholesterol oxides Type of **cholesterol oxidation products**.

Choline An amino alcohol and biogenic amine precursor with activity similar to that of **vitamin B group** members. Occurs widely in living organisms as a constituent of certain types of **phospholipids (lecithins** and **sphingomyelin)** and in the neurotransmitter acetylcholine. Choline is synthesized in the body, is a ubiquitous component of cell membranes and therefore occurs in all foods. Rich sources include **egg yolks**, **meat**, **livers** and **cereals**.

Cholinesterases EC 3.1.1.7 (**acetylcholinesterases**, also known as true cholinesterases) and EC 3.1.1.8 (also known as pseudocholinesterases). The former have been used in **biosensors** for detection of **insecticides** and **drugs** residues in water and foods, while the latter act on a variety of acetic esters and have been used for determination of **choline** esters.

Choloylglycine hydrolases EC 3.5.1.24. Catalyse hydrolysis of trihydroxycholanoylglycine and dihydroxy derivative into trihydroxycholanate and glycine. Also act on choloyltaurine. Have potential therapeutic uses due to their action in reducing serum cholesterol levels. Activity is common in **lactic acid bacteria**, especially strains of **Bifidobacterium** and **Lactobacillus**. Alternative names include bile salt hydrolases.

Chondrus Genus of **seaweeds** containing the edible species *Chondrus crispus* (Irish moss), which has cartilaginous, dark purplish-red fronds. This species provides a source of **carrageenans** (sulfated polysaccharides) which are used as food **emulsifiers**.

Chopi Common name for the Asian plant, *Zanthoxylum piperitum*. Peel from the dried fruits of this plant is used as a spice and the leaves are also used in **flavourings** for foods. The dried fruits have an aromatic lemon-like **aroma**, while the leaves have a flavour with tones of mint and lime. Extracts of peel and leaves have **antimicrobial activity**. Also known by a variety of other names, including Sichuan pepper and Chinese pepper.

Chopping **Cutting** of foods into bite-sized (or smaller) pieces with repeated, sharp blows with **knives** or cleavers, usually on **chopping boards**. A food processor may also be used to chop foods.

Chopping boards Boards made of **wood** or **plastics** on which food is placed while being cut with a **knives** or cleavers (**chopping**). For safety reasons, it is best to use a separate board for **vegetables** and another (preferably wood) for raw **meat**. Hot water and **detergents** should always be used in conjunction with scrubbing to wash a chopping board after each use. Plastics boards may be cleaned in **dishwashers**.

Chorizo Highly spiced, fermented **pork sausages**, made from coarsely comminuted meat, which is flavoured with garlic, chilli powder and other spices. Three major types are produced, namely fresh, semi-dried and dried. Air dried chorizo is sliced and eaten raw. Other types of chorizo are cooked by grilling or frying, or are added to other meat in spicy casseroles, soups and stews. Smoked versions of chorizo are also produced. Chorizo are used widely in Spanish and Mexican cookery. Spanish chorizo are made from smoked pork and are sold ready-to-eat, whilst Mexican chorizo are made from fresh pork and require **cooking** before eating.

Christstollen Rich bread/cake originally from Germany that contains **dried fruits** and **nuts** and is traditionally eaten at Christmas. Alternative term for **stollen**.

Chromatography Techniques in which components of a gaseous or liquid mixture are separated on the basis of differences in the rate at which they migrate through a liquid or solid stationary phase under the influence of a gas or liquid mobile phase. Once separated, individual components can be measured or identified by various methods. Types of chromatographic techniques include **gas chromatography**, **thin layer chromatography**, **affinity chromatography** and **ion exchange chromatography**, classified according to characteristics of the method.

Chromium Mineral, chemical symbol Cr, which is widespread in foods. An essential nutrient at low concentrations, but toxic in excess.

Chromobacterium Genus of Gram negative, facultatively anaerobic, rod-shaped **bacteria**. Occur in soil and water. *Chromobacterium violaceum* is a pathogen found in water.

Chromoplasts Plastids found in plant cells which contain **pigments** such as **carotenoids** and **xanthophylls**. Present especially in **flowers** and ripe **fruits**.

Chromosomes Self-replicating structures consisting of or containing **DNA** that carries genetic information essential to the cell. Bacterial chromosomes are usually circular and present as a single type within a cell, although many copies may be present in each cell. Eukaryotic chromosomes are complexed with proteins (chromatin) and located in the nucleus. They are present as pairs and each cell may contain a single type or many different types, depending on the organism.

Chrysanthemum Genus of flowering plants, the flowers and leaves from some species of which are consumed as **vegetables**. Commonly used species include the garland chrysanthemum (*Chrysanthemum coronarium*).

Chrysene Member of the carcinogenic **polycyclic aromatic hydrocarbons** (PAH) group which can occur as a contaminant in foods.

Chrysosporium Genus of **fungi** of the class Eurotiomycetes. Species (e.g. *Chrysosporium pannorum*) may cause **spoilage** of fresh **meat**, especially **beef**.

Chub **Freshwater fish** species (*Leuciscus cephalus*) of minor commercial importance belonging to the family Cyprinidae (minnows and carps). Found in rivers and lakes, and sometimes brackish water, in Europe and Asia Minor. Popular as a game fish. Eaten fresh or smoked.

Chub mackerel **Marine fish** species (*Scomber japonicus*) from the mackerel family; widely distributed in the Indian and Pacific Oceans. Commercially

cultured in Japan. Flesh is fatty with a strong **flavour**. Marketed fresh, frozen, smoked, salted and occasionally canned. Also known as **Pacific mackerel** and **Spanish mackerel**.

Chufa nuts Stem tubers of *Cyperus esculentus*, cultivated in West Africa. Eaten raw or roasted, or used to make non-alcoholic beverages. Also known as **tiger-nuts**.

Chukars **Partridges** which are similar to the red-legged partridge, but belonging to the genus *Alectoris*. There are two species. Chukars are hunted as game birds, but are also farmed successfully. Battery-farmed chukars slaughtered at 14-20 weeks of age have ready-to-cook yields (from live weight) of approximately 75%. A large proportion of the boneless cooked meat yield is breast meat.

Chum salmon **Pacific salmon** species (*Oncorhynchus keta*) found in coastal waters and rivers along the Pacific coasts of North America and Japan. Flesh has highly regarded **flavour** and **texture**; occurs in pink or white forms. Mainly canned but also sold fresh, dried-salted, smoked, and frozen. **Roes** are utilized in **caviar substitutes**.

Chungkook-jang Traditional Japanese and Korean fermented product made from **soybeans**. Also known as chunggugjang.

Churning Process used in **buttermaking**. Agitation or churning of cream breaks down the **milk fat globule membranes**, allowing individual **milk fat globules** to coalesce into grains which eventually separate from the **buttermilk**.

Chutneys Fruit or vegetable **pickles**, containing ingredients including **spices** and **sugar**. Originally an Indian delicacy.

Chymosin EC 3.4.23.4. Broad specificity similar to that of pepsin A. Also known as rennin. Component of **rennets**, it initiates the **clotting** of **milk** by cleavage of the Phe105-Met106 bond in the κ-chain of **casein**. Found in the fourth stomach of calves although microbially-produced **recombinant enzymes** are now widely available. Used extensively in **cheesemaking**.

Chymotrypsin EC 3.4.21.1. A serine proteinase produced as an inactive precursor. Cleaves peptide bonds immediately after a Tyr, Trp, Phe or Leu residue.

Chymotrypsin inhibitors Molecules, generally **proteins**, which inhibit the activity of **chymotrypsin** (EC 3.4.21.1, a serine proteinase). These inhibitors occur naturally in a range of plant foods, particularly seeds, where they play a role in plant defence against **pests** and **pathogens**. However, they can also act as **antinutritional factors** in plant foods, reducing the **digestibility** and **nutritional values** of these foods for humans. **Cooking** and other processing treatments can reduce levels of chymotrypsin inhibitors in plant foods. Efforts are also being made to breed plants with reduced levels of these compounds.

Cider In the UK, **alcoholic beverages** made by **fermentation** of **apple musts**. In the USA, this alcoholic beverage is termed hard cider, and the term cider refers to unfermented **apple juices**.

Cider apples Cultivars of **apples** grown for use in **cider** production.

Cider vinegar Fruit-flavoured **vinegar** made by refermenting **cider** or **apple wines**. Used widely as a table vinegar, especially in the USA and apple growing regions of Europe.

Cider yeasts **Yeasts** used for **fermentation** of **apple musts** to produce **cider**.

Ciguatera **Food poisoning** caused by consumption of tropical **marine fish** containing a neurotoxin (**ciguatoxin**) produced by certain **dinoflagellates**. Symptoms include abdominal pain, nausea and vomiting and multiple, varied neurological disorders. Ciguatera poisoning is the most common nonbacterial, fishborne poisoning in the USA (mainly Hawaii and Florida) and is a significant health concern in tropical areas worldwide. Species of fish most frequently implicated in ciguatera outbreaks include **grouper**, **amberjack**, red snappers, **eels**, **sea bass**, **barracuda** and **Spanish mackerel**.

Ciguatoxin Neurotoxin produced by **dinoflagellates** associated with coral reefs, which can accumulate in **fish** and cause **ciguatera** poisoning in consumers. *Gambierdiscus toxicus* is the dinoflagellate most notably responsible for production of ciguatoxin, although other species have been identified recently. At least five types of ciguatoxin have been identified and are noted to accumulate in larger and older fish higher up the food chain.

Cimaterol β-Adrenergic agonist used to enhance growth rates and improve feed efficiency and lean meat content of animals. Use as a growth-promoting agent in farm animals is not permitted in many countries.

Cineole Member of the **terpenes** class of **flavour compounds**, which occurs in many **spices** and **essential oils**.

Cinnamaldehyde Member of the phenolic **aldehydes** class of **flavour compounds**, characteristic of **cinnamon** but also occurring in other foods. Has antimicrobial properties.

Cinnamic acid Member of the phenolic acids class of **flavour compounds** which occurs in a wide range of foods. Cinnamic acid esters are also important flavour compounds. Cinnamic acid and its **esters** have antimicrobial activity.

Cinnamon Widely-used aromatic spice obtained from the dried inner bark of trees belonging to several species of *Cinnamomum*. True cinnamon (also known as Ceylon cinnamon) is *C. zeylanicum,* while much of the cinnamon sold in North America is actually cassia (*C. cassia*). Cinnamon is used in stick (quill) or ground form for flavouring both sweet and savoury foods, including **confectionery**, **meat** dishes and **cola beverages**.

Cinnamon oils **Essential oils** obtained from either **cinnamon** bark or cinnamon leaves. Cinnamon leaf oil has a high **eugenol** content and is used as an alternative to clove oils in seasoning blends. Cinnamon bark oil is characterized by a high **cinnamaldehyde** content and is used as the source of cinnamon **essences** for cooking.

CIPC Alternative term for the herbicide **chlorpropham**.

Ciprofloxacin Fluoroquinolone antibiotic used for treatment and control of gastrointestinal and respiratory infections in farm animals.

Circular dichroism Phenomenon (usually abbreviated to CD) that is observed when optically active matter absorbs left and right hand circular polarized light differently. CD is a function of wavelength and is measured using a CD spectropolarimeter. CD spectra vary according to secondary structure and can be analysed to give information about the secondary structure of biological macromolecules such as **peptides**, **proteins** and **nucleic acids**.

Citral Member of the terpene **aldehydes** class of **flavour compounds**. Occurs in a wide range of plant foods, especially **coriander**, **pepper**, **lemon peel** and **ginger**.

Citrates Salts of **citric acid** which occur naturally in many foods, and may be used as **acidulants** in foods and **beverages**.

Citrate synthases EC 2.3.3.1 (citrate (*si*)-synthases) and EC 2.3.3.3 (citrate (*re*)-synthases). These two enzymes exhibit opposite stereospecificities and are involved in the formation of **citric acid** from acetyl-CoA and **oxaloacetic acid**.

Citric acid Commercially important, versatile organic acid, widely used, along with its salts (**citrates**), in the food and beverage industries. Highly soluble in water and used in **acidulants**, **antioxidants**, **flavourings**, **antimicrobial compounds** and **chelating agents**. Usually obtained commercially by extraction from **citrus fruits** (it is the predominant acid in **lemons**, **oranges** and **limes**), or by **fermentation** of

sugar or fruit processing wastes by *Aspergillus niger*. Isomer of **isocitric acid**.

Citric fermentation The process by which certain organisms produce **citric acid**. *Aspergillus niger* is the organism mostly used in industrial processes. Substrates include **molasses** and **starch hydrolysates**.

Citrinin Yellow pigmented mycotoxin produced by *Penicillium citrinum* and a few species of *Aspergillus*. Used as an antibacterial agent against **Gram positive bacteria**.

Citrobacter Genus of Gram negative, rod-shaped coliform **bacteria** of the family **Enterobacteriaceae**. Occur in the intestines of humans and other vertebrates, water, sewage and soil. Species may be found in **dairy products**, raw **shellfish**, raw **poultry meat** and fresh raw **vegetables**. *Citrobacter freundii* and *C. diversus* may cause diarrhoea in humans.

Citronella Tropical Asian grass (*Cymbopogon nardus*). Lemon-scented leaves are used in **flavourings** in **cooking** and as a **tea**. Source of **essential oils** that are used in commercial flavourings as well as in perfumery and insect repellents.

Citronella essential oils Yellow aromatic **oils** obtained from lemon-scented tropical grasses of the *Cymbopogon* genus (particularly *C. nardus*). Used in the food industry and as an aromatic/deodorizer in perfumes, cosmetics, soaps and insect repellents. Contains **geraniol**, **citronellol** and **citronellal**.

Citronellal Member of the terpene **aldehydes** group of **flavour compounds**, which occurs in **essential oils** of **citrus fruits** and a wide range of **spices**.

Citronellol Member of the terpene **alcohols** group of **flavour compounds**. Occurs in a wide range of plant foods, including **fruits**, **essential oils**, **ginger** and **wines**.

Citrons Long fruits produced by *Citrus medica*, with thick peel and acid flesh. Used in production of candied **peel**, preparation of which involves fermenting immature fruit in **brines** and then soaking in a strong sugar solution. The candied peel is used in **confectionery** products.

Citrulline One of the non-essential **amino acids**, which does not occur in **proteins**.

Citrus beverages **Beverages** based on **citrus juices** and/or whole homogenates of **citrus fruits**.

Citrus essential oils **Essential oils** obtained from **citrus fruits**, e.g. **bergamot oils**. Typically produced by pressing the oil from **citrus peel**, although leaves, fruit or juice may also be used as the source. Applications include as **flavourings** for **soft drinks**, **ice cream**, **chewing gums** and **puddings**. **Limonene** and other **terpenes** are major components.

However, these are frequently removed prior to use of the oils, due to their susceptibility to **off flavour** production as a result of oxidation. Alternative term for **citrus oils**.

Citrus fruits Fleshy and juicy fruits produced on trees of the genus *Citrus*. Include **oranges**, **lemons**, **grapefruit**, **limes**, **tangerines**, **satsumas**, **mandarins** and many hybrid varieties. All are rich in **vitamin C**.

Citrus juice concentrates **Citrus juices** which have been concentrated. May be diluted to produce normal strength citrus juices or used in manufacture of other **beverages** or foods.

Citrus juices Juices extracted from **citrus fruits**; important types include **orange juices**, **lemon juices**, **lime juices** and **grapefruit juices**.

Citrus oils **Essential oils** obtained from **citrus fruits**, e.g. **bergamot oils**. Typically produced by pressing the oil from **citrus peel**, although leaves, fruit or juice may also be used as the source. Applications include as **flavourings** for **soft drinks**, **ice cream**, **chewing gums** and **puddings**. **Limonene** and other **terpenes** are major components. However, these are frequently removed prior to use of the oils, due to their susceptibility to **off flavour** production as a result of oxidation. Also called **citrus essential oils**.

Citrus pectins **Pectins** extracted from **citrus fruits**. **Citrus peel** is one of the main commercial sources of pectins.

Citrus peel Outer skin of **citrus fruits**, consisting of the outer coloured flavedo (also called the epicarp or zest) and the white inner pith (also called the albedo or mesocarp). The flavedo is the source of **citrus essential oils**, while the albedo is used as a source of **pectins**. Peel is also rich in **fibre** and phytochemicals. Often candied and used in **baking**, or used in making **flavourings**.

Citrus red Dye used to improve the colour of **orange peel**.

CJD Abbreviation for **Creutzfeldt-Jakob disease**.

Cl Chemical symbol for **chlorine**.

Cladosporium Genus of **fungi** of the class Hyphomycetes. Occur on **fruits** and **vegetables**. *Cladosporium herbarum* may cause **spoilage** of chilled **meat**. Other species may be responsible for spoilage of **butter**, **margarines**, **stone fruits**, **eggs** and **grapes**.

Clams General name given to a wide range of bivalve **molluscs**; typically marine **bivalves** with equally sized valves that burrow in mud or sand. Many clams are valued as **sea foods** and are eaten in a variety of ways, including baked, fried, stewed, stuffed, raw on the half shell, and in chowders and **soups**.

Clarification Process in which sediment and impurities are separated out of a liquid to make it clearer. Rendered **fats** can be clarified by adding hot water and boiling. The mixture is then strained and chilled. The resulting top layer of fat should be almost entirely clear of residue. Other products to which clarification is applied include **fruit juices**, **wines** and **beer**.

Clarifiers Equipment used for the process of **clarification**, in which sediment and impurities are separated out of a liquid to make it clearer.

Clarity **Optical properties** relating to the extent to which an item is clear and transparent.

Clavaria Genus of edible wild club **fungi** of the class Hymenomycetes.

Claviceps Genus of **fungi** of the order Clavicipitales. Typically parasitic to grasses. Causes plant diseases such as **ergot**, a disease of **rye**.

Clays Sticky impermeable earth that can be moulded when mixed with water and baked to make **containers**. Clay is plastic when moist and becomes permanently hard and retains its shape when baked or fired. Of widespread importance in industry, clays consist of a group of hydrous alumino-silicate minerals. Individual mineral grains are microscopic in size and shaped like flakes. This makes their aggregate surface area much greater than their thickness and allows them to take up large amounts of water by **adhesion**, giving them plasticity and causing some varieties to swell. Clays are effective **filter aids** and are used during adsorption **bleaching** of **oils**.

Cleaning To make a surface free from dirt, pollutants or harmful substances.

Cleaning agents Agents, such as **disinfectants**, used in the **cleaning** process.

Cleaning in place A process in which processing equipment is cleaned using an in-place cleaning system that is usually computer controlled. Cleaning in place (CIP) systems are useful for equipment that is not easily accessible to the operator, and when opening the equipment would be harmful to the operators or the environment, and detrimental to product quality.

Clean in place Alternative term for **cleaning in place**.

Clean room technology Technology that incorporates use of a sterile, dust-free environment. Objectives of a clean room are to isolate a controlled area from the outside, and to control movement of materials and personnel. Parameters requiring control in a clean room are temperature, **relative humidity**, **water activity**, pressure, noise and lighting. Sources and parameters of potential contamination include air quality, type and geometry of air intake systems, personnel, machinery

and equipment, waste produced and **packaging materials**.

Clementines Citrus fruit regarded as a cultivar of **tangerines** or a hybrid of tangerines and sweet **oranges**. Rich in **vitamin C**.

Clenbuterol One of the β-agonist **drugs** which is used in some countries as a growth promoter in slaughter animals. There is concern that residues in **meat** may present a health hazard.

Cloning technology Use of various genetic techniques for producing copies of single **genes** or segments of **DNA** by insertion in **cloning vectors** (e.g. **plasmids** or **viruses**). These **vectors** can then be introduced into recipient cells and propagated. The term also involves production of genetically identical cells (clones) from a single ancestor. In plants, the term refers to natural or artificial vegetative propagation.

Cloning vectors Autonomously replicating **DNA** molecules (e.g. **plasmids**, viral genomes and yeast artificial **chromosomes**) into which foreign DNA fragments can be inserted. They can then be inserted into host cells, propagated and, in the case of expression vectors, used for production of homologous or heterologous proteins.

Clostridium Genus of Gram positive, anaerobic rod-shaped **bacteria**. Occur in soil and in the intestinal tracts of humans and other animals. Some species are **pathogens**, e.g. *Clostridium botulinum*, the causal agent of **botulism**, and *C. perfringens*.

Closures Devices or packaging components used for closing or sealing of containers. Include **caps**, **corks**, crown corks, lids, **stoppers** and **tamper evident closures**.

Clotting The process of **coagulation** to produce a thick mass of cohesive material, e.g. formation of **curd** upon coagulation of **milk**.

Cloud **Turbidity** or **haze** within a product, usually applied to **beverages**.

Cloudberries Fruits produced by *Rubus chamaemorus*. Orange-yellow with an appearance similar to **raspberries** and a flavour like **apples**. Usually eaten stewed or as **jams**.

Cloudiness Extent to which an item is turbid, i.e. hazy in appearance. Usually applied to liquids such as **beverages**.

Clouding agents Substances used to impart the appearance of **turbidity** to foods and **beverages**. **Soy proteins** and citrus fruit processing wastes are frequently used as clouding agents in **citrus beverages**.

Cloves Pungent, aromatic **spices** obtained from the dried, unopened flower buds of the tropical evergreen tree *Syzygium aromaticum* (syn. *Eugenia caryophyllata, E. caryophyllus*). Used whole or ground in a range of foods and **beverages**, including **cakes** and **biscuits**, bread **sauces**, **curries** and mulled **wines**.

Clovis   Alternative term for **carpet shells**.

Cloxacillin   Semisynthetic penicillin antibiotic used principally to treat staphylococcal **mastitis** in cattle. Residues in milk are normally undetectable at 5 days following final dose.

Cluster beans   Alternative term for **guar beans**, seeds of *Cyamopsis tetragonolobus*. Immature pods are eaten as vegetables. **Galactomannans** are extracted from the seeds to make **guar gums**, which are used as **stabilizers** and **thickeners** in foods.

CMC   Abbreviation for **carboxymethylcellulose**.

CO₂   Chemical formula for **carbon dioxide**.

Coagulants   Substances or agents that causes separation or precipitation of solids from a solution, a process known as **coagulation** or **clotting**.

Coagulation   Precipitation of solids from a solution, usually upon addition of specific agents, producing material of a solid or semi solid state. Coagulation is a process particularly applicable to **cheesemaking**. Also known as **clotting**.

Coagulum   Formed by precipitation of **casein** by the action of **acids** or **rennets**, as in **cheese curd**.

Coalfish   **Marine fish** species (*Pollachius virens*) from the cod family (Gadidae) found in the northern and western Atlantic and Barents Sea. This species is often used in production of **fish cakes**, but is also marketed fresh, dried/salted, smoked, canned and frozen. Also known as **pollock** and **saithe**.

Coal tar dyes   **Artificial colorants** originally obtained from coal tar hydrocarbons. The term is now used to refer to any artificial organic dyes or **pigments**, regardless of source. Also known as aniline dyes.

Coating   Covering food with a layer of coating material. For example, chicken pieces may be dipped or rolled in seasoned **breadcrumbs** or **flour** prior to **cooking**. The food can be dipped into beaten **eggs**, **milk** or **beer** before being coated with the dry mixture, to aid adhesion of the **coatings** to the food. Coating food in this manner usually precedes **frying** or **baking**. Products such as **mayonnaise** or **sauces** can also be used to coat food.

Coatings   Materials which form thin continuous layers or coverings over the surface of foods. Used to enclose and/or protect the food, and may be eaten along with the food or removed before consumption. Include **batters**, **breadcrumbs**, **breadings**, **carnauba wax**, **chocolate coatings**, **shellac** and **wax coatings**.

Cobalamins   Term that covers several chemically related compounds, members of the **vitamin B group**, that are essential for cell division in tissues where this process is rapid, e.g. in formation of red blood cells. Deficiency leads to pernicious anaemia when immature red blood cells are released into the bloodstream, and there is degeneration of the spinal cord. This typeof anaemia is the same as seen in **folates** deficiency.

Cobalt   Mineral, chemical symbol Co, which is widespread in foods. An essential nutrient, but toxic in excess.

Cobnuts   Alternative term for **hazelnuts**.

Coca-cola   A proprietary brand of **cola beverages**.

Cocci   Spherical, or near spherical, bacterial cells.

Coccidiosis   Infestation of the **gastrointestinal tract** with parasitic coccidia protozoa. Affects many animals (including cattle, swine, sheep and poultry), but rarely humans. Typically contracted via the faecal-oral route and may vary from mild to fatal. Characterized in animals by diarrhoea, tenesmus, anorexia and nausea. Usually asymptomatic in humans.

Coccidiostats   **Drugs** used for control of pathogenic protozoa (from class Coccidia) responsible for coccidiosis and other parasitic diseases. Normally used prophylactically in feeds for poultry, swine, cattle and sheep. Examples include **dimetridazole**, **nicarbazin** and **salinomycin**.

Cochineal   Water-soluble natural red colorant obtained from the dried bodies of South American insects (*Coccus cacti*). The red colour is due to **carminic acid**, whose aluminium lake is known as **carmine**.

Cockles   General name used for several species of marine bivalve **molluscs**; characterized by a shell having convex radial ribs. Commonly eaten species include *Cerastodermum edulis* (common cockle), *Cardium corbis* and *C. aculeatum* (spiny cockle). Marketed in a variety of ways, including fresh, salted, bottled in **vinegar** and canned in **brines**.

Cockroaches   Common name for orthopteran **insects** of the family Blattidae, which possess flat wide bodies, and long slender segmented antennae. Widespread **pests** in human dwellings and food factories. May be pests of stored foods and act as vectors for **pathogens**.

Cocktails   **Alcoholic beverages** generally based on a mixture of **spirits** with **flavourings** or other ingredients.

Cocoa   Small tropical American tree (*Theobroma cacao*) of the family Sterculiaceae (or Byttneriaceae), seeds of which (**cocoa beans**) are rich in **theobromine** and, after **fermentation** and **roasting**, are used to make **cocoa**, **chocolate** and their products. Sometimes refers to highly concentrated **cocoa powders** made by grinding and removing most of the fats (**cocoa butter**), or a **milk**-based beverage made with such powders. Also known as cacao.

Cocoa beans Seeds or fruits of the cocoa tree (*Theobroma cacao*) that are rich in **theobromine**. After **fermentation** and **roasting**, cocoa beans are used to make **cocoa**, **cocoa powders**, **chocolate** and their products. Also known as cacao beans.

Cocoa beverages Beverages based on **cocoa** (*Theobroma cacao*) solids.

Cocoa butter Edible vegetable fat obtained by pressing or solvent extraction of ground, roasted dehulled **cocoa beans** (*Theobroma cacao*). Composed of symmetrical disaturated oleic glycerol esters resulting in brittleness at room temperature and a sharp **melting point** at 31-35°C. Used primarily in the food industry for manufacturing **chocolate**.

Cocoa butter equivalents Vegetable fats with similar **triacylglycerols** composition and physicochemical properties to **cocoa butter**. Used for partial or complete replacement of cocoa butter.

Cocoa butter extenders Vegetable fats that may be mixed with **cocoa butter** to a limited degree without significantly affecting its **physicochemical properties**.

Cocoa butter replacers Alternative term for **cocoa butter substitutes**.

Cocoa butter substitutes Fractionated **fats** based on various oils (palm, palm kernel, coconut or hydrogenated soybean) designed to replace **cocoa butter** in **confectionery** applications. Also known as cocoa butter replacers.

Cocoa mass Produced by grinding of **cocoa beans** nibs (beans from which the shell or husk has been removed) to release the **cocoa butter** from the cells. Used in the manufacture of **chocolate** and chocolate products. Also called cocoa liquor, **chocolate liquor** and chocolate mass.

Cocoa powders Products obtained by extracting a predetermined amount of **cocoa butter** from **chocolate liquor** using hydraulic presses, and grinding the resulting press cake. Cocoa powders produced are classified according to fat contents.

Cocona Fruits produced by *Solanum topiro* or *S. sessiliflorum*. Orange to maroon in **colour**, with white to pale yellow flesh. Rich in iron; good source of **vitamin A**, **vitamin C** and **niacin**. Used in **salads**, cooked with **fish** or in meat stews, sweetened in **sauces** and **pies**, pickled, candied or in **jams** and jellies. Often processed as a nectars or juices. Leaves are also cooked and eaten as a vegetable.

Coconut butter Alternative term for **coconut oils** when in its semi-solid state.

Coconut cream Product similar to **coconut milk**, but richer. Relatively high fat content, with level varying among commercial brands. For the canned product, coconut milk is filtered, mixed with **emulsifiers** and **stabilizers**, and emulsified to give a creamy consistency, before **pasteurization** and **canning**. Used in the same way as **cream** in many recipes and also in **beverages**.

Coconut milk Liquid prepared by squeezing freshly grated coconut endosperm through **sieves**. Relatively high in fat, level varying among commercial brands. Used in products such as **curries** and **confectionery**.

Coconut oils Semi-solid white **fats** or pale yellow to colourless **oils** extracted from **copra**, the dried pulp of **coconuts**, *Cocos nucifera*. Rich in **lauric acid** and **myristic acid** and used extensively in the food industry. Also known as copra oils.

Coconuts Common name for *Cocos nucifera*. Fruits of the coconut palm consisting of an outer skin, a fibrous region and a hard shell enclosing the commercially used **nuts**. The white endosperm (meat) found inside the shell has a cavity in the centre which contains a watery liquid (**coconut water**, a popular tropical beverage). Endosperm may be eaten fresh, or dried to make **copra**, from which **coconut oils** are extracted, or desiccated coconut. Freshly grated endosperm is squeezed to make **coconut milk**.

Coconut toddy Alcoholic beverages made by **fermentation** of the sap of coconut palms (*Cocos nucifera*).

Coconut water The liquid enclosed within the kernels of **coconuts** (*Cocos nucifera*), which may be used in beverages.

Cocos Genus of **palms**. In some, including the coconut palm (*Cocos nucifera*) and *C. yatay*, fruits, buds and inflorescences are eaten or used in making foods and **beverages**.

Cocoyams Starchy corms of *Xanthosoma sagittifolium* (new cocoyams, also called tannia or yautia) or *Colocasia esculenta* (old cocoyams, alternative term for **taro**) that form part of the staple diet in African countries. Eaten roasted, boiled or baked; the flour prepared from the corm is used as a food ingredient and in making **fufu**.

Cod Name given to several **marine fish** species from the family Gadidae. The principal cod species is *Gadus morhua* (Atlantic cod) which is widely distributed in the north Atlantic and Barents Sea and in commercial terms is the most important food fish in northern Europe. Flesh is lean, firm and white. Other cod species include *G. macrocephalus* (Pacific cod), *G. ogac* (Greenland cod) and *Boreogadus saida* (Arctic cod).

Marketed fresh, frozen, smoked. Often processed as a battered product for frying, grilling or baking.

Codex Alimentarius A food code which provides an opportunity for all countries to join the international community in formulating and harmonizing food standards and ensuring their global implementation. The Codex Alimentarius Commission, established in the early 1960s by the **Food and Agriculture Organization** (FAO) and the **World Health Organization** (WHO), is the body responsible for compiling the standards, codes of practice, guidelines and recommendations that constitute the Codex Alimentarius. Membership of the Commission is open to all Member Nations and Associate Members of FAO and WHO; the Commission meets every two years. National delegations are led by senior officials appointed by their governments. Countries that are not yet members of the Commission attend in an observer capacity.

Cod liver oils Pale yellow **oils** derived from the **livers** of Atlantic **cod** (*Gadus morhua*) and other species of the family Gadidae. Have a typical fish-like flavour which is intensified on exposure to light. Rich in **vitamin A** and **vitamin D**. Contain saturated, monoenoic and **polyunsaturated fatty acids**, such as **eicosapentaenoic acid** and **docosahexaenoic acid**.

Cod livers **Livers** from members of the cod family (Gadidae) which are an important source of **fish oils**. Cod livers are also used in the production of cod liver pastes, which contain **spices** and other **flavourings** and are marketed canned and in the form of **sausages**.

Coeliac disease Life-long **intolerance** to **wheat gluten**, characterized by inflammation of the proximal small intestine. The disease is often manifested as persistent diarrhoea, malabsorption and **malnutrition**. Aetiological mechanisms include genetic predisposition, dietary exposure to **wheat** and immunological factors; prevalence of the disease is high in geographical areas where wheat is a dietary staple. Management of the condition involves consumption of a gluten free diet, which has been facilitated by the development of **gluten free foods**, especially **gluten free bread**.

Coenzymes Low molecular weight non-protein organic molecules, whether freely dissociable or firmly bound, necessary for the activity of certain **enzymes**.

Coertuek Common name for *Echinophora tenuifolia* subsp. *sibthorpiana* which is used as a spice and source of **essential oils** in Turkey.

Coextrusion The process of producing continuous multilayer products in sheet, film, tubing, filament, or other forms, and for production of filled foods. Separate polymer or ingredient streams are fed from different extruders to a die feed block, where they are combined in the die, emerging in combined form as a continuous multilayer extrudate.

Cofermentation **Fermentation** of two or more substrates by a single microorganism or fermentation of a single substrate by two or more **microorganisms**.

Coffee **Beverages** prepared from ground roasted **coffee beans** (*Coffea arabica* and *C. canephora*).

Coffee bags Ground **roasted coffee** packaged in portion-size bags for easy infusion to produce **coffee beverages**.

Coffee bars **Restaurants** serving coffee and light refreshments.

Coffee beans Seeds of the coffee bush (*Coffea arabica* or *canephora*) which are used to prepare **coffee beverages**. As grown, coffee beans are enclosed in soft fruits; these are fermented, and the seeds (coffee beans) are separated from the soft tissue. Raw coffee beans are roasted and ground before use in preparation of **beverages**.

Coffee beverages **Beverages** prepared by infusion of ground roasted **coffee beans** in hot water by a variety of processes. Optionally consumed with addition of other substances, commonly **milk**, **cream** or **sugar**. Types of coffee beverages include **espresso coffee**, **cappuccino coffee** and cafe latte.

Coffee cream In Germany, cream with a minimum fat content of 10% marketed also with fat contents of 12 and 15%. Also called drinking cream. Whitening power in **coffee** is increased by **homogenization**; further processing is performed to increase stability of higher fat products in hot coffee.

Coffee essences Concentrated **coffee extracts**.

Coffee extracts Liquid **extracts** from **coffee beans**, containing active ingredients and **flavour compounds**; coffee extracts may be used for preparation of **coffee beverages**.

Coffee granules Dried **coffee extracts** presented in the form of **granules**.

Coffee grounds Roasted **coffee beans** which have been ground ready for use in preparation of **coffee beverages**.

Coffee oils Volatile, water soluble substances formed during roasting of **coffee beans** so that the sugars and carbohydrates within the bean become caramelized. Contribute to the **flavour** and **aroma** of **coffee**.

Coffee powders Dry **coffee extracts** in the form of powders.

Coffee substitutes Materials for preparation of **beverages** with **sensory properties** resembling those of **coffee**. Commonly based on roasted plant materials, e.g. grains or **chicory** roots.

Coffee whiteners **Whiteners** used in **coffee** and **tea** beverages as an inexpensive alternative to **milk**. Typically made from **vegetable fats, casein, carbohydrates, emulsifiers** and **stabilizers**. Available in liquid or powdered (shelf stable) forms.

Cognac A high quality **brandy** manufactured in a defined district in the Charente and Charente Maritime regions of France.

Cognitive performance **Behavioural effects** relating to acquisition and use of knowledge (perception, attention, memory, speech and language, and reasoning). Some foods can affect cognitive performance.

Coho salmon **Pacific salmon** species (*Oncorhynchus kisutch*) found in rivers and coastal waters along western and eastern Pacific coasts. High fat, firm-textured flesh is somewhat lighter in **colour** than that of other Pacific salmon. Marketed fresh, dried/salted, smoked, canned, cured and frozen.

Cohumulone **α-Acids** fraction present in **hops** and hop products and contributing to the bittering action of hops in **beer**.

Cola beverages **Soft drinks** flavoured with extracts of **cola nuts**.

Cola nuts **Nuts** produced by *Cola nitida* and *C. acuminata*. Used in manufacture of **soft drinks** such as **cola beverages**, or chewed as a stimulant. High **caffeine** content. Also known as kola nuts.

Colby cheese Semi-soft washed-curd **cheese** from the USA, made from **cow milk**. Ripens in 4 months. It has a sweet and mild **flavour** and must be eaten soon after purchase to prevent drying out and loss of flavour.

Cold boning Cutting of **meat** (muscle) from animal **carcasses** that have been refrigerated at 1-2°C for 48 h *post mortem*.

Cold shock proteins Protein fractions which are synthesized in various **bacteria** in response to cold shock, and which contribute to cold tolerance and psychrophilic properties of these bacteria.

Cold shortening Contraction of muscle fibres in raw **meat** at low temperatures. Related to **toughness** in the meat once cooked.

Cold storage **Storage** of foods at **refrigeration** temperature in order to extend **shelf life**.

Cold stores Refrigerated rooms or cabinets used for **storage** of foods at low temperatures, to extend **shelf life**.

Coleslaw Salad of shredded **vegetables**, principally **cabbages**, dressed with **mayonnaise** or an alternative creamy dressing.

Colicins **Bacteriocins** produced by members of the family **Enterobacteriaceae** (e.g. strains of *Escherichia coli* and *Shigella sonnei*) which are often lethal to other susceptible bacterial strains within this family.

Coliforms Gram negative, anaerobic, lactose-fermenting, rod-shaped **bacteria**, typically found in the gastrointestinal tracts of humans and animals (e.g. species of the genera Citrobacter, Enterobacter, Escherichia and Klebsiella). May loosely refer to any Gram negative, rod-shaped enteric bacteria.

Coliphages **Bacteriophages** that infect *Escherichia coli*.

Colitis Inflammatory disease of unknown cause which affects some or all of the colon. Takes various forms, e.g. ulcerative colitis, mucus colitis or ischaemic colitis, which differ in symptoms and effects, and tends to vary in intensity over time. Sensitivity to diet depends on the individual, but in general foods and **beverages** which act as bowel irritants should be avoided. These include **tea, coffee, alcoholic beverages, vinegar**, spicy foods, **fried foods, sugars** and salty foods.

Collagen Insoluble **animal proteins**, with high contents of the amino acids **glycine, hydroxyproline** and **proline**. Collagen is the main fibrous component of **skin**, tendons, **connective tissues** and **bones**. Networks of collagen are also present in tissues and organs including the muscles. Thermal denaturation of collagen occurs between 60 and 90°C. When collagen is boiled it is converted into soluble **gelatin**. Collagen is important in relation to meat **texture**. Collagen crosslinks link together molecules and fibrils of collagen, increasing its tensile strength; thus, the greater the number of crosslinks the tougher the meat. In cooked meat, the presence of collagen crosslinks contributes to shrinkage and tension development, with a subsequent increase in meat **toughness**. Collagen is used to form edible, biodegradable films and **coatings** for the packaging of foods.

Collagenases Digest **collagen** in the triple helical region, hydrolysing peptide bonds prior to a Gly residue. These **enzymes** are useful for **meat tenderization**, while proteolytic digestion products of collagen are useful as **seasonings** and ingredients.

Collagen sausage casings Edible and inedible **sausage casings**, which are regenerated from **collagen** extracted from animal hides and skins. Edible collagen casings have very uniform physical characteristics and are much stronger than natural casings; they are mainly used to prepare fresh pork sausages and frankfurters. Inedible collagen casings must be removed before sausages are eaten; their advantages include strength, uniformity and shrinkage characteristics.

Collards Leaves of a smooth-leaved variety of **kale**. Used as a vegetable. Good source of **vitamin A**, **vitamin C**, calcium and iron.

Colletotrichum Genus of **fungi** of the class Coelomycetes. Some species are important plant **pathogens**. *Colletotrichum gloesporioides* causes **anthracnose** of **tropical fruits** (e.g. **mangoes** and **papayas**), and *C. musae* causes **anthracnose** and crown rot of **bananas**.

Colloidal stability A measure of the longevity of **colloids**; the ability to maintain the suspension of one material in another.

Colloids Mixtures containing small particles of one material suspended in another, often of a different phase. Colloidal particles are generally 1 to 100 nm in size and so are larger than the individual solution molecules, but smaller than particles found in precipitates, which can be removed by **filtration**. Examples of colloids include **aerosols**, **foams**, **emulsions** and **gels**. A colloid containing solid particles suspended in a liquid is more accurately called a sol.

Collybia **Edible fungi**, some species of which are members of the genus **Flammulina** or **Lentinus**.

Colony counting Enumeration of cell colonies in a given sample cultured on a solid medium.

Colony counts Numbers of cell colonies in a given sample cultured on a solid medium.

Colorants Substances that impart **colour**, such as **dyes** or **pigments**. Added to foods to improve visual appearance, replace colour lost during processing and ensure colour consistency. Broadly classified into **natural colorants** and **artificial colorants**, depending on whether they are substances extracted from natural sources or manufactured for use as a food additive.

Colorectal cancer Malignant diseases of the large intestine. Dietary factors suggested to be associated with reduced risk for this cancer include increased intakes of **dietary fibre**, **fruits** and **vegetables** and reduced intakes of **meat** and iron.

Colorimeters Instruments that measure the contents of components in a sample solution by comparison of **colour** with that of standard solutions.

Colorimetry Analytical technique based on comparison of the **colour** of a solution with that of a standard solution.

Colostrum Mammary secretion produced during the first 4-5 days *post partum*. Differs from mature milk mainly in the high content of **immunoglobulins**, which provide passive immunization of the suckling infant or animal. Other differences in composition include increased contents of **milk fats**, short-chain **fatty acids**, **lactoferrin**, **minerals**, most **vitamins**, some **hormones** and some **organic acids** in colostrum, and reduced contents of medium-chain fatty acids, **lactose**, **orotic acid**, and some vitamins and hormones. Sale of colostrum is prohibited.

Colour **Optical properties** relating to the subjective appearance of the wavelength or wavelengths present in a beam of light perceived by the eye. Although it is actually continuous, the visible spectrum is usually split into seven major colours - red, orange, yellow, green, blue, indigo and violet - in order of decreasing wavelength. Colour of foods not only helps to determine quality, but is also an index of **ripeness** or **spoilage**. Various types of spectrophotometers or **colorimeters** can be used for colour measurement.

Column chromatography **Chromatography** technique in which a column or tube is used to hold the stationary phase.

Colupulone **β-Acids** fraction present in **hops** and hop products.

Colza oils Alternative term for **rapeseed oils**.

Comamonas Genus of Gram negative, rod-shaped motile **bacteria**, type species *Comamonas terrigena*. Some species, including *C. acidovorans*, accumulate poly-β-hydroxybutyrate and some, e.g. *C. acidovorans* and *C. testosteroni*, produce biotechnologically useful enzymes such as esterases and quinohaemoprotein alcohol dehydrogenases. *C. terrigena* and *C. testosteroni* were previously members of the genus **Pseudomonas**, while *C. acidovorans* is synonymous with *Delftia acidovorans*.

Combustion In general terms, process in which a substance reacts with **oxygen** or other oxidant giving off heat and light. As an analytical technique, a method for determination of **nitrogen** and crude protein in a sample by burning in oxygen and measurement of the nitrogen gas produced.

Comfrey Common name for the herb *Symphytum officinale*, a member of the **borage** family. Leaves and stems can be eaten, usually boiled or fried; the plant is also used in **flavourings** and **health foods**. Contains pyrrolizidine **alkaloids**, which can be toxic.

Comminution Reduction of a food to minute particles or fragments.

Common Agricultural Policy Common Agricultural Policy (CAP) was established by the 1957 Rome treaty that created the **European Economic Community** (EEC). CAP was intended to stabilize agricultural markets, improve productivity, and ensure a fair deal for both farmers and consumers. It has three major elements: a single market for agricultural products with a system of common prices to producers across the EU; preference for EU producers through a common levy on all agricultural imports from abroad;

and shared financial responsibility for guaranteeing prices.

Common beans Seeds produced by *Phaseolus vulgaris*. Vary in size, shape and **colour**. Eaten fresh, canned or frozen, and available dried. Also used as ingredients in many dishes. Due to the presence of **antinutritional factors**, dried seeds must be soaked and cooked well before consumption. Beans produced by this species have been given a variety of names, including **French beans**, **kidney beans**, **haricot beans**, **snap beans**, **string beans**, cannellino beans and **pinto beans**.

Common millet Cereals belonging to the genera *Panicum* and *Setaria*.

Complexometry Technique in which a substance is measured by the extent to which a complex is formed with an agent. Used to indicate the end point of a **titration** by formation of a coloured complex.

Composite flours Products made by blending **wheat flour** with flour of other origins. Often used to make **bakery products** that are conventionally made with wheat flour alone.

Compotes Fruit products made by stewing fruits with **sugar** or in **syrups**; eaten hot or cold.

Compressibility One of the **rheological properties**, and a measure of the degree to which matter can be squashed or crushed by an externally-applied force. Indicates the **hardness**, **firmness** or sponginess of a material.

Compressimeters Apparatus used for determining **compressibility**.

Compression Flattening of an item by pressure.

Computerized data processing Analysis and organization of data by the repeated use of one or more computer programs.

Comte cheese French **hard cheese** made from **cow milk**. Very creamy, with piquant, yet sweet, **flavour**. Has eyes that vary in size from that of a pea to that of a cherry. Requires a long **ripening** period.

Conalbumin Iron-binding protein found in **egg whites**. Also known as **ovotransferrin**.

Conarachin One of the main **proteins** in **peanuts**. Present in two forms (I and II) that differ in size. Along with **arachin**, these make up 75% of the total protein in peanuts.

Concanavalin A **Lectins** fraction extracted from **jack beans** (*Canavalia ensiformis*) which binds to **glycoproteins** with α-glycoside or α-mannoside groups. Agglutinates many cell types, and is mitogenic.

Concentrated milk Product resulting from removal of a considerable proportion of the water from **milk**. Includes **evaporated milk** and **condensed milk**. A vacuum is applied to reduce the **boiling point** of milk and thus maintain its quality during **evaporation**. Evaporated and unsweetened condensed milk are sterilized by heat. Usually sold in cans in a range of fat contents. Can be reconstituted by addition of water in amounts stated on the packaging.

Concentrated rectified musts **Grape musts** (generally made from **winemaking grapes**) which have been purified (e.g. by **ion exchange**) and concentrated. May be used as a source of added **sugar** in **winemaking**, to increase alcohol concentration in the **wines** or to increase **sweetness** of wines.

Concentration The process by which the strength of a solution or substance is increased. Achieved by a variety of means, including **evaporation**, **filtration** and **dialysis**.

Conching The final step in **chocolate** manufacture, in which machines with rotating blades slowly blend heated **chocolate liquor**, ridding it of residual moisture and volatile acids. Conching continues for 12 to 72 hours (depending on the type and quality of chocolate), while small amounts of **cocoa butter** and sometimes **lecithins** are added to give chocolate its smooth **texture**.

Condensation The conversion of a vapour or gas to a liquid. In physics, condensation is the process of reduction of matter into a denser form, as in the liquefaction of vapour or steam. Condensation is the result of the reduction of temperature by removal of latent heat of evaporation, the liquid product being known as condensate. Condensation is an important part of the process of **distillation**. In chemistry, condensation is a reaction involving the union of atoms in the same or different molecules. The process often leads to elimination of a simple molecule such as water or alcohol to form a new and more complex compound, often of greater molecular weight.

Condensed milk **Milk** thickened by **evaporation** of a considerable amount of its water content. Usually sold in cans and may be sweetened or unsweetened. Unsweetened milk is similar to **evaporated milk**, i.e. sterilized by heat, sold in a range of fat contents and may be reconstituted by addition of water in amounts stated on the packaging. Sweetened condensed milk contains **sugar** in amounts high enough to act as a preservative, and is used in **baking** and to make sweet products such as **confectionery**, **puddings** and **pies**.

Condiments Distinctly flavoured products used to season foods. Refers in particular to items that are added to foods at the table immediately prior to consumption (e.g. **sauces**, **relishes** and **mustard**), rather than items added during **cooking**.

Conductimetry Technique in which the concentration of a substance in solution is measured by conductance of that solution when an alternating current is applied. Changes in conductance can be used to indicate the end point of a **titration**. Alternative spelling is conductometry.

Conductivity Alternative term for **electrical conductivity** or **thermal conductivity**. The former is a measure of the ability of a material to carry an electric current. The latter is a thermophysical property relating to the rate of conduction of heat through a material.

Confectionery Generic term for sweetened food products. **Sugar confectionery** refers to products such as **sweets**, **candy** and **chocolates**, while **bakers confectionery** refers to **bakery products** such as **cakes** and **pastries**.

Confectionery bars Sugar confectionery products formed into bars. Examples include **chocolate bars**.

Confectionery cream Water-in-oil or oil-in-water emulsions used mainly as **fillings** for **bakery products**.

Confectionery fillings Products such as **fondants** or **cremes** which may contain **nuts**, **flavourings** or other ingredients and are used to fill **sugar confectionery** or **bakery products**.

Confectionery pastes Products containing ingredients such as **glucose syrups**, **sugar**, **fats**, **colorants** and **flavourings** that are used in the production of extruded **sugar confectionery**.

Confections Sweet food products, particularly **sugar confectionery**.

Conger eels Marine **eels** within the family Congridae, which includes several species targeted for consumption. Important species include *Conger conger* (from the Eastern Atlantic Ocean around Europe), *C. oceanicus* (from the Western Atlantic) and *C. verrauxi* (caught around the coasts of Australia and New Zealand). Conger eels are often consumed smoked or semi-preserved in jelly.

Conglycinin One of the main **soy proteins**, present as β-conglycinin, which is composed of α, α′ and β subunits. The 7S β-conglycinin and 11S glycinin together account for approximately 70% of the **storage proteins** in **soybeans**.

β-Conglycinin One of the main **soy proteins**, composed of three subunits (α, α′ and β). A 7S globulin which, along with **glycinin**, makes up approximately 70% of the **storage proteins** in **soybeans**.

Conjugases Alternative term for **γ-glutamyl hydrolases**.

Conjugated linoleic acid Group of **linoleic acid isomers** having conjugated double bonds, especially the isomer *cis*-9,*trans*-11-octadecadienoic acid. Occurs naturally in ruminant fats, especially **milk fats**. Thought to have beneficial effects on health, especially **anticarcinogenicity**, anti-atherosclerosis activity, enhancement of immune function, and improvement of **lipids** metabolism and body composition. Commonly abbreviated to CLA.

Connectin Elastic protein found in muscle foods (**meat** and **fish**), in which it is important for **texture** and **tenderness**. Similar to **titin**.

Connective tissues Tissues that connect, bind, support or separate other organs or tissues. Connective tissues include cartilage, ligaments, tendons, **adipose tissues**, the non-muscular structures of blood vessels and the matrix of bones. In **fish**, **collagen** connective tissues separate the myotomes (muscle segments). In animals, connective tissues consisting of both **collagen** and **elastin** bind muscle fibres into bundles and support the blood vessels. **Toughness** of **meat** is correlated with connective tissue content. On cooking, the collagen component of connective tissues is converted into **gelatin**, thereby making meat more tender; however, the elastin component is unchanged on heating. **Frying** or **roasting** have little effect on meat **tenderness**, but tough meat is made more tender by **stewing**.

Conophor nuts Common name for seeds produced by *Tetracarpidium conophorum*. They have a bitter taste when raw, but are palatable when boiled and are a popular snack in Nigeria. Also used as a source of **oils**.

Consistency **Texture** term relating to the degree to which a product, usually a thick liquid, is viscous or dense. The simplest method to determine consistency is to measure the time it takes for the food to run through a small hole of a known diameter. Alternatively, measurements can be made of the time it takes for more viscous foods to flow down an inclined plane using Bostwick **consistometers**. These devices might be used with **tomato ketchups**, **honeys** or **sugar syrups**.

Consistometers Instruments used to measure the uniformity and **consistency** of a manufactured material, such as a food product. The Bostwick consistometer is widely used to evaluate consistency of food suspensions. The Bostwick measurement is the length of flow recorded in a specified time.

Consommes Clear **soups** made by clarifying **broths**, usually fish or meat based. May be served hot or cold. Sometimes used as bases for **sauces**. The term double consomme refers to one which has been reduced to half its original volume, thereby increasing the **flavour**.

Consumer acceptability Extent to which a commercial product is considered satisfactory by consumers. In the case of foods and **beverages**, overall acceptability is judged on the basis of a number of factors, including **sensory properties**, **physical properties** such as **colour**, **appearance** and **texture**. Evaluation of consumer acceptability is important in development and marketing of new products.

Consumer complaints Expression of dissatisfaction made by consumers regarding a commercial product or service. With respect to foods and **beverages**, the term covers complaints made at a local level, e.g. regarding the **acceptability** of a meal served in a restaurant, through to those reported to official agencies, e.g. regarding contamination of products with foreign objects or **food poisoning** incidents.

Consumer education Provision of a variety of forms of training to consumers so as to increase their knowledge of a product or service.

Consumer information Information, such as guidelines and details of use, given to consumers so as to increase their awareness of products and services.

Consumer panels Groups of consumers employed during **sensory analysis** tests who are not specifically trained but can provide a good insight into consumer preference.

Consumer preference Extent to which consumers like one commercial product more than others. In the case of foods and **beverages**, preference is governed by a range of factors, including **sensory properties**, **appearance**, **physical properties**, **texture**, health concerns, price and type of **packaging**. Evaluation of consumer preference variables is important in development and marketing of new products.

Consumer research Any form of marketing research undertaken using the final consumers of a product or service from which to gather data. For example, consumer preference data are obtained and information is gathered on the way in which consumers in a free market choose to divide their total expenditure in purchasing goods and services.

Consumer response Behaviour that consumers exhibit when provided with information in areas such as product purchase, new product development and product labelling. Covers concepts such as consumer attitudes, consumer awareness, consumer choice, consumer complaints, consumer expectations and consumer preference.

Consumer surveys Marketing research tools used to gather data on consumer response to a particular product or service.

Consumption As well as being the action or process of eating foods, this term also means the using up of goods created by production in an economic sense.

Contact Materials Term applied to any material or article coming into contact with foods or **beverages**. Includes **packaging materials** and equipment, cooking utensils, cutlery, preparation surfaces and processing equipment.

Containers Receptacles for holding, storing or transporting substances such as foods. Of many different types, and made from a variety of materials. The term is also used to describe large, portable, standard-sized metal boxes, which are used in the transportation of cargo on lorries or ships.

Contaminants Agents that contaminate. May be undesirable substances (e.g. residues of **pesticides**, **fungicides**, **herbicides** or **fertilizers**) or undesirable or harmful **microorganisms**.

Contamination Process of introducing **contaminants**, or the presence of **contaminants**.

Continuous processing Automated processing systems which operate in a continuous fashion. Such systems allow improved product consistency and reduced manufacturing costs, and are designed to meet the demand for high output.

Contraction Decrease in volume or length of an object. Includes muscle contractions generally, and the temperature-related **cold shortening** of muscle fibres in **meat** during **chilling**. Also used to describe changes in **dough** volume during **proofing**, and temperature-related changes in solution volumes.

Controlled atmosphere packaging Packaging technique in which specified concentrations of gases, including water vapour, are maintained throughout storage to achieve the desired atmosphere. Used to extend the **shelf life** of foods, particularly fresh **fruits** and **vegetables**.

Controlled atmosphere storage Storage of **fruits** and **vegetables** in sealed warehouses where temperature and humidity are closely controlled, and the composition of gases in the atmosphere is altered to minimize spoilage. Usually, the concentration of oxygen is reduced, the concentration of **carbon dioxide** (CO_2) is increased, and **ethylene**, a gas naturally produced by plants that accelerates **ripening**, is removed from the atmosphere. This controlled environment helps slow the enzymic reactions that eventually lead to decomposition and decay, and may increase the time that produce can be stored by several months. Ripening rooms, in which ethylene gas is added to the atmosphere, also help produce higher quality **fruits** and **vegetables**. This technology enables produce to be picked before it is ripe, for easier handling, and then ripened quickly and uniformly under controlled conditions.

Control systems Systems in which inputs and outputs are progressively altered in a well-planned way to cause a process or mechanism to conform to some specified behaviour under a set of given constraints. Computer-based control systems for large industrial plants can involve control of hundreds or thousands of individual variables. Recent developments in control engineering include self-tuning and adaptive control systems, in which controller settings are modified automatically in response to changing process and/or disturbance conditions, and the application of neural networks and artificial intelligence techniques, which mimic the actions of skilled human operators.

Convenience foods Processed foods that can be quickly and easily prepared by the consumer. Examples include **ready to eat meals**, cooked sliced meat, **sauces** for **pasta** and **pizzas** and **microwaveable foods**.

Conveying Process by which items are transported or carried to a particular place.

Conveyors A continuous moving band used for transporting objects from one place to another. Conveyors include simple chutes, unpowered roller conveyors, and a range of powered systems in which materials are carried along by belt, bucket, screw, trolley, or other arrangement. Pneumatic conveyors are tubes in which goods - usually in a finely divided form are moved along by blowers.

Convicine Member of the pyrimidine glucoside class of antinutrients occurring in **faba beans**, **broad beans** and other **legumes**, which causes the haemolytic disease favism in susceptible people.

Cook chill foods Foods, particularly **ready meals**, produced by **cook chill processing** and kept at low temperatures ($<5°C$) from manufacture to point of sale. The minimal processing involved results in high-quality **convenience foods** with a short **shelf life**. Generally packaged in **plastics** trays, as either **ready to eat foods** or easily reheatable products.

Cook chill processing A method of **catering** that involves cooking of foods in batches to a just done status followed by immediate, fast **chilling** (using blast chilling or water bath chilling techniques) to just above **freezing point**. Products are then stored for reheating at a later time. The cook chill process offers a cost effective means of providing quality food while reducing overhead costs.

Cookers Appliances for **cooking** foods, domestic cookers typically consisting of an oven, hob and grill.

Cookies US term for **biscuits**. Compared with other types of biscuit, cookies tend to be larger, with a softer, chewier **texture**. In some parts of the UK, the term refers to sweet **buns** that are filled with cream or topped with **icings**.

Cooking Process of preparation of foods by **mixing**, combining and **heating** the ingredients. Heat-activated cooking methods take five basic forms. Food may be immersed in liquids such as water, **stocks**, or **wines** (**boiling**, poaching, **stewing**); immersed in **fats** or **oils** (**frying**); exposed to vapour (**steaming** and, to some extent, **braising**); exposed to dry heat (**roasting**, **baking**, **broiling**); and subjected to contact with hot fats (**sauteing**).

Cooking fats Fatty substances such as **butter, margarines** and vegetable **shortenings** which are solid at room temperature and are used to moisten, enrich, tenderize and flavour foods during cooking.

Cooking oils Fatty substances which are liquid at room temperature and have usually been refined, bleached and deodorized. May be used for **deep frying**, or in **baking**, **frying** and **grilling** of foods.

Cooking properties Ability of a food product to have acceptable properties upon **cooking**, particularly relating to **texture**, **flavour** and **colour**.

Coolers Devices or containers for making or keeping items cool.

Cooling Process by which the temperature of items is lowered, usually after some form of **cooking**.

Cooperatives Business organizations that are owned and run jointly by their members, with profits or benefits shared among them. For example, farmers have formed cooperatives for many purposes, including marketing of produce, purchasing of production and home supplies, and provision of credit. Farm marketing associations are the most important type of agricultural cooperative. Farm purchasing cooperatives rank second in importance.

Coppas Italian raw, fermented **pork sausages**. Traditionally prepared from entire swine neck muscles, which are deboned and sliced before curing and ripening in casings. However, some coppa may include ingredients such as air-dried neck of pork, swine skin and cartilage. At least three groups of microorganisms (lactic acid bacteria, Micrococcaceae and yeasts) are active in the ripening of coppa. Coppas are usually served in slices.

Copper Mineral, chemical symbol Cu, which is an essential nutrient, but toxic in excess. Copper and its alloys (brass, bronze) may be used in construction of food processing equipment. Copper ions are prooxidative, and may cause **taints** in **wines**.

Copra Dried white flesh of **coconuts** (*Cocos nucifera*) from which **coconut oils** are extracted.

Copra oils Alternative term for **coconut oils**.

Coprinus **Edible fungi**, commonly consumed species including *Coprinus comatus* and *C. cinereus*. Also known as ink cap mushrooms, some species being used in manufacture of ink.

Cordials Term which refers to two types of product: concentrated and sweetened fruit-based **beverages**; and sweet **liqueurs**.

Corers Utensils used to remove the tough central part of various fruits and vegetables, particularly **apples**. Corers are usually made of **stainless steel** and come in different shapes for different uses. An all-purpose corer has a medium-length shaft with a circular cutting ring at the end. An apple corer shaped like a spoked wheel with handles not only cores the apple, but cuts it into wedges as well. A corer for **pineapples** is a tall, arch-handled utensil with two serrated, concentric cutting rings at the base. After the top and bottom of the pineapples are sliced off, the corer is inserted from the top and twisted downward. The tool not only removes the core, but also the outer shell, so producing pineapple rings.

Coriander Common name for *Coriandrum sativum*, an umbelliferous plant cultivated for its aromatic seeds and foliage. Seeds are a major component of **curry powders** and are also used as a **pickling** spice and in **flavourings** for **meat** products and other food products. Seed **essential oils** are also widely used for flavouring purposes, as are the fresh leaves, which are added to foods to impart a delicate, citrus- and parsley-like **flavour**. Also known as cilantro.

Coring Process by which the tough central part of various **fruits** and **vegetables** is removed, often using **corers**.

Coriolus Genus of **fungi** of the class Hymenomycetes. Species may be used in production of **laccases** (e.g. *Coriolus versicolor* and *C. hirsutus*), or in the **decoloration** of brown food **pigments**. *C. versicolor* may also be used in the **bioremediation** of **food factories effluents** and **food factories wastes**.

Corks **Closures** for **bottles**, particularly wine bottles, or jars. Made from cork, as it is a material which can be compressed to a smaller size and resists absorption of liquids. A sensory defect, known as corking, may occur in **wines** due to growth of **microorganisms** on corks. There is also the risk of release of substances such as **trichloroanisole, tannins** and **peroxides** from the corks into the food or beverage contained in the bottle or jar. Synthetic closures made from **plastics** have been developed as an alternative to natural cork closures for wine bottles. **Crown corks** are closures made from metal.

Cork spots Plant disorder affecting **apples** and **pears**. Characterized by large brown spots in the fruit flesh and fruit deformation, with a pitted appearance. Caused by low levels of calcium in affected fruits due to any of a number of factors.

Cork taint **Aroma** and **flavour** defect of **wines** caused by contamination with 2,4,6-trichloroanisole, which may be present in the **corks** used to close the **bottles**.

Corn Grains, also known as maize, from any of numerous varieties of a tall, annual cereal plant (*Zea mays*), which are borne on large ears. Low in **tryptophan**. **Niacin** is present in bound form, making development of the deficiency disease pellagra a possibility in those eating corn as a staple. Corn is processed into a great many products, including **corn oils, corn starch, corn syrups, flour** and **corn masa**. It is also used in making some kinds of **beer, whisky** and **gin**. Some types of corn have a hard endosperm and kernels that burst on heating; these are used to make **popcorn**.

Corn bran Outer protective coating of **corn** kernels; removed during **milling**.

Corn bread **Bread** in which the main cereal component is **corn flour**. May also contain **wheat flour** or, for **gluten free bread, rice flour**. **Arepas** are flat corn bread products popular in South America. Generally spelt cornbread in the USA, where various sweet or savoury **flavourings** may be added, and shape, **texture** and **thickness** of the products are varied according to taste.

Corn chips Crisp **snack foods** made from **corn masa batters**, shaped into flat triangles and fried.

Corn cobs Flowering organs of *Zea mays*, composed of inner and outer leaf-like organs (glumes) which fold around the **corn** kernels, a woody ring of lignified conducting tissue, and an inner pith. Often used as a source of **furfural**. Waste corn cobs are used as substrates for production of various fermentation products including **enzymes, ethanol** and **sugar**.

Corn dogs **Sausages**, especially **frankfurters**, coated in a heavy **corn flour** batter and cooked by **frying** or **baking**. Usually served on a stick.

Corned beef In the UK, corned beef describes **beef** that is brined, chopped, pressed and sold in tins. Conversely, in North America, corned beef describes beef brisket that is cured in seasoned brine and boiled; it is usually served cold.

Cornelian cherries Fruits of the wild dogwood (*Cornus mas*). Eaten fresh or made into **preserves** and **marmalades**. Also used to make alcoholic and non-alcoholic beverages.

Corn fibre oils By-products of the **corn** processing industry which are rich in cholesterol-lowering **phytosterols** and of potential use in the manufacture of **nutraceutical foods**. Produced during **wet milling** of corn.

Cornflakes **Breakfast cereals** made from **corn**, often enriched with **vitamins**.

Corn flour **Flour** often ground from a variety of **corn** with large, soft grains and friable endosperm, from which the **germ** and outer hull are first removed. Also known as maize meal. Distinct from cornflour, which is used as an alternative term for **corn starch**.

Cornflour Alternative term for **corn starch**.

Cornish pasties **Meat pies** traditionally made in Cornwall, United Kingdom. Each pasty consists of a folded pastry case filled with seasoned **meat** and **vegetables**, especially **potatoes**.

Corn masa Dried **corn flour** or **dough** made from this. Produced by cooking and steeping corn grains in unslaked lime followed by rinsing, grinding and drying. Used in manufacture of **tortillas** and related products.

Corn oils Pale yellow **oils** derived from wet grinding the kernels of **corn** (*Zea mays*). Typically bland in **flavour** and widely used as a cooking oil and salad oil and in the manufacture of **margarines**. **Palmitic acid**, **oleic acid** and **linoleic acid** are the major **fatty acids**. **Oxidative stability** is high, despite the highly unsaturated nature. Also known as maize oils.

Corn starch **Starch** isolated from **corn**. Common substrate for manufacture of **syrups**.

Corn steep liquor One of the by-products of processing **corn** to manufacture **corn starch**. The brown, syrupy liquid is rich in **lactic acid** and **phytic acid** as well as containing a range of **amino acids**, **proteins**, **peptides**, **carbohydrates**, **vitamins** and **minerals**. It is used as a **fermentation** substrate in production of substances such as **enzymes**, **polysaccharides** and **antibiotics** by **microorganisms**.

Corn syrups Nutritive **sweeteners** manufactured by partial hydrolysis of **corn starch**. Corn syrups are a mixture of **glucose**, **maltose** and **maltodextrins** produced by acid hydrolysis of corn starch at >100°C, followed by enzymic hydrolysis if required. Degree of hydrolysis required depends on the final application of the syrup; commonly, 40-60% of glycosidic bonds are hydrolysed, i.e. the corn syrups have a dextrose equivalent of between 40 and 60, although corn syrups of 24-80 dextrose equivalents are also produced. Corn syrups are approximately half as sweet as **sucrose**, although **sweetness** increases with increased hydrolysis. In addition to their use as sweeteners, corn syrups are used as **thickeners**, **humectants**, carbon sources for microbial **fermentation** and to provide body to **soft drinks** and **beer**.

Coronary heart diseases Diseases of the heart resulting from narrowing of the coronary arteries resulting in myocardial ischaemia and/or infarction. Narrowing or occlusion of the arteries can be a consequence of **atherosclerosis** or thrombosis. Risk factors for coronary heart diseases have been identified, and include dietary, genetic and lifestyle factors.

Corrinoids Group of compounds containing four reduced pyrrole rings joined into a macrocyclic ring (the corrin nucleus). Includes **vitamin B_{12}**. Many corrinoids have a cobalt atom in the centre of the macrocyclic ring. The B_{12} vitamins are found in **animal foods** but not **plant foods**. Corrinoids are also produced by **bacteria**.

Cortisol Steroid which occurs in animal tissues. May be used as an indicator of stress in slaughter animals, and of **meat** quality.

Corynebacteriaceae Obsolete family of rod-shaped **bacteria** of the order Eubacteriales, the genera of which have been reclassified.

Corynebacterium Genus of Gram positive, aerobic or facultatively anaerobic, rod-shaped **bacteria** of the family Corynebacteriaceae. Occur in soil and vegetable matter. Some species are involved in the **spoilage** of vegetable and meat products. *Corynebacterium glutamicum* is used in the commercial production of **glutamates**, **glutamic acid** and **lysine** for the food industry. *C. manihot* may be used in **starters** for production of **gari**.

Costmary Common name for *Balsamita major* (syn. *Chrysanthemum balsamita*), a perennial herb related to **tansy**. The fragrant mint-flavoured leaves are used for seasoning various foods, including **salads**, **cakes**, meat dishes and **herb tea**. Formerly used to flavour **ale**.

Costus Alternative term for **kuth** (*Saussurea lappa*).

Cottage cheese Soft **white cheese** made from **cow milk**. An acid curd cheese, made without **rennets**. Ripens in 1 or 2 days and has a fat content of only 5-15%.

Cotton candy Fluffy mass of spun **sugar** that is formed from thin threads. Often served around a stick. Also called **candy floss** in the UK.

Cottonseed meal Residue that remains when **oils** have been extracted from **cottonseeds**. Sometimes used as a speciality ingredient in manufacture of **cookies** and frequently in livestock feeds.

Cottonseed oils Pale yellow **oils** derived from **cottonseeds** (*Gossypium* spp.). Used as a salad oil and cooking oil and, on **hydrogenation**, in the manufacture of **margarines**. Rich in **palmitic acid, oleic acid** and **linoleic acid; gossypol** is a minor constituent.

Cottonseed proteins **Proteins** derived from **cottonseeds** (*Gossypium* spp.). Classified as storage and non-storage proteins, both of which show desirable **functional properties** that make them suitable for use as **food supplements**.

Cottonseeds Seeds of the cotton plant (*Gossypium* spp.) which are commercially important for their **oils**. Also a source of **cottonseed meal** and **cottonseed proteins**.

Cough drops Medicated and sweetened **lozenges** that are taken orally to relieve a cough or sore throat.

Coulometry Technique in which the amount of an analyte in solution is measured by converting it from one oxidation state to another, the end point of the reaction being measured with an indicator also present in the solution. A constant current source is used to deliver a measured amount of charge. An intermediate reagent, generated electrochemically from a precursor is often used to cause chemical oxidation. Analyte concentration is calculated from the amount of charge required to cause complete conversion.

Coumaphos One of the **acaricides** used for control of parasitic **mites** of the species *Varroa jacobsoni* in beekeeping. May occur as residues in **honeys**.

Coumaric acid Member of the **hydroxycinnamic acid** class of **phenols**, which occurs in a wide range of plant foods. Has **antioxidative activity** and antimicrobial activity.

Coumarin Constituent of the **flavour compounds** of a wide range of foods, including **cinnamon**. Shows **cytotoxicity** and anticancer activity, and is toxic in excess. Also known as 2-hydroxycinnamic acid lactone.

Coumestrol Member of the **isoflavonoids** class of **phytoalexins**, present in **soybeans** and other crops, which has **phytoestrogens** activity; may be genotoxic.

Countercurrent chromatography Form of liquid partition **chromatography** in which no solid support is required and two immiscible solvent phases are used. Partition takes place in an open column in which one phase (the stationary phase) is retained and the other (the mobile phase) passes through continuously. The stationary phase is retained in the column as a result of column configuration and gravitational or centrifugal force fields. The technique is used in the food industry for preparative separation of food constituents, such as **polyphenols** from **tea**, and **anthocyanins** from **fruits** and **vegetables**, and for analysis of food components and contaminants.

Couplers Devices for connecting or combining items.

Courgettes Small, dark green cucumber-shaped vegetable **marrows** cut from the plant when young. Also known as zucchini in the USA and Canada.

Couscous Granular product originating from North Africa, made either from **durum wheat semolina** or **millet** flour. Also refers to a North African dish made with this product, which is steamed and traditionally served with a stew of **lamb, chicken meat, chick peas** and **vegetables**.

Cowberries Red, acid **berries** produced by *Vaccinium vitis-idaea*; similar to, but smaller than, American **cranberries**. Contain high levels of **benzoic acid**. Used in **jams** and jellies. Also known as mountain cranberries, lingberries or **lingonberries**.

Cow cheese **Cheese** made from **cow milk**.

Cow milk **Milk** produced by dairy **cattle**.

Cow milk cheese **Cheese** made from **cow milk**.

Cowpea meal Flour produced from **cowpeas**. Used as an ingredient in foods.

Cowpeas Seeds of *Vigna unguiculata*, a plant from which young pods and leaves are also consumed. Vary in **colour**, white ones having pigment at the eye only, leading to their alternative names, blackeyed beans or **blackeyed peas**. Rich in protein and carbohydrate.

Cows Adult female bovine animals belonging to the genus *Bos*. In farming, the term cows is used to describe female **cattle** which have borne more than one calf.

Coxiella Genus of Gram negative, anaerobic, rod-shaped **bacteria** of the Coxiella group family. *Coxiella burnetti*, the causal agent of Q fever in humans, may be transmitted from infected animals to humans via their **milk**. Sheep, cattle and goats may act as reservoirs for the disease.

Coypu Large semi-aquatic South American rodents (*Myocastor coypus*), also known as nutria. Coypu are hunted and farmed for their fur and **meat**. Coypu meat is similar in protein content to **poultry meat** and **game meat**, it has favourable composition of **fatty acids** and **amino acids**, and low contents of **fats** and **cholesterol**.

Crab legs Legs from species of large marine **crabs**, which are consumed as sea food delicacies. Commercially important species for crab legs include red king crabs (*Paralithodes camchatica),* blue king crabs, (*P. platypus*) golden king crabs, (*Lithodes aequispinus*), dungeness crab (*Cancer magister*) and edible crab (*C. pagurus*).

Crab meat Edible flesh from the body and legs of **crabs**; in the sea food industry, the term crab meat designates canned white meat from crabs. Usually white in appearance, although leg meat often has red coloration. The most important crab species for meat production are edible crabs (*Cancer pagurus*), dungeness crabs (*C. magister*), **blue crabs** (*Callinectes sapidus*) and king crab species, such as the red king crab (*Paralithodes camchatica*).

Crabs Crustaceans from the order Decapoda, having 10 pairs of legs with the first pair usually modified as pincers. Approximately 4500 crab species occur worldwide, most inhabiting marine or estuarine waters. Many crab species are commercially valuable **sea foods**; marketed in a variety of forms, including fresh cooked whole crab, cooked leg meat, canned meat and pastes.

Crackers Thin, crisp **wafers** or **biscuits** (e.g. water biscuits, cream crackers, wholemeal crackers) made from unsweetened **dough** made with **wheat flour**, fat and **sodium bicarbonate**.

Cracking Breaking of an item with little or no separation of the component parts. Can be used to refer to damage to a commodity, e.g. freeze cracking of foods, or a processing step, e.g. cracking of **eggs** to remove them from their **shells**.

Crambe seeds Seeds from plants belonging to certain species of the genus *Crambe* which belong to the mustard family. **Oils** extracted from the seeds are rich in **erucic acid** and are similar to **rapeseed oils**.

Cranberries Red, acid **berries** produced by *Vaccinium oxycoccus* (large, or American cranberries produced by *V. macrocarpon*). Acidity is due to high levels of a number of acids, including **citric acid**, **quinic acid**, **benzoic acid** and **malic acid**. Because of the acidity, consumption of cranberries can be beneficial in cases of urinary tract infections and some types of kidney stones. Cranberries are used in a range of products, including **sauces**, jellies, **relishes** and **beverages**.

Cranberry juices **Fruit juices** prepared from **cranberries** (*Vaccinium oxycoccus* or *V. macrocarpon*). Thought to have a protective action against urinary tract infections.

Crates Re-usable, slatted, wooden or plastics containers used for transportation of goods, including various foods. Crates subdivided into units are used for holding individual items, such as **bottles**. The term is also used to describe containers for the transportation of live animals, particularly poultry.

Cravings Behavioural term relating to a strong desire or longing for a specific item. The most commonly craved food is thought to be **chocolate**.

Crawfish General name used for marine **lobsters** species within the genera *Palinurus* and *Panulirus* (also called **spiny lobsters**) and *Jasus* (also called rock lobsters). Marketed in a variety of ways, including fresh whole, shelled meat, canned and in pastes. Ground crawfish is often used in **soups**.

Crayfish General name used for various freshwater lobster-like **crustacea** found in lakes, rivers and swamps around the world. Several species are valued as foods, particularly *Cambarus* spp. and *Astacus* spp. from North America and Europe. Usually marketed live and boiled prior to consumption.

Cream Fatty product prepared from **whole milk** by centrifugation. Marketed in a range of types differing in fat content. In the UK, **half cream** contains approximately 12% fat, **single cream** and extra thick single cream 18%, **whipping cream** 34%, **double cream** and extra thick double cream 48%, and clotted cream 55%. In the USA, light cream contains 20-25% fat and heavy cream 40% fat.

Cream cheese **Soft cheese** made from **cow milk**. An acid curd cheese, but unlike **Cottage cheese**, made using **cheese starters**. Generally mild and velvety, but **whey** powder can be added to produce a more grainy texture. Eaten as a cheese and also used in making **cheesecakes** and in **baking**.

Creameries Premises in which **dairy products** are manufactured. Also called **dairies** or **dairy factories**.

Cream horns Baked hollow **puff pastry** products, which are horn-shaped and filled with **whipped cream** and sometimes **jams**.

Creaminess **Consistency** term relating to the extent to which a product is creamy, i.e. smooth, glossy and uniform. As an attribute, creaminess has viscous, flavour and taste aspects.

Creaming Natural formation of a layer of milk fat on the top of milk left to stand for some time. This happens because of the lower **specific gravity** of milk fat and is dependent on the size of the **milk fat globules**. Clustering of the milk fat globules is also affected by **globulins** in the **milk fat globule membranes**. Creaming can be controlled by **homogenization** of milk and **heating** to cause **denaturation** of the **globulins**.

Cream liqueurs **Liqueurs** in which **cream** is combined with alcohol to produce a thick and shelf stable blend. Cream is homogenized to break down the **milk fat globules** to a size suitable to encase the alcohol molecules, preventing the cream from being curdled and the product from separating. The milk fat globules are made as small as possible to give a smooth taste and long shelf life. Cream liqueurs are generally packed in dark glass bottles to protect from **UV radiation** and are best kept refrigerated after the bottle has been opened.

Cream milk **Milk** from which no fat has been removed. Also called full cream milk or whole milk.

Cream puffs Baked **puff pastry** products, which are hollow, and filled with **whipped cream**.

Creatine One of the **nitrogen compounds** which play an important role in muscle metabolism. Occurs in muscle foods such as **meat**, **fish** and **shellfish**. Concentrations of creatine and its anhydride metabolite **creatinine** may be used as indicators of the condition and quality of meat and meat products.

Creatinine Anhydride metabolite of **creatine**. Found in a range of foods.

Creep One of the **rheological properties**, describing a deformation with time of materials under continual stress. An important parameter in a wide range of foods, including **fruits** and **vegetables**, **bakery products**, **dairy products** and extruded **starch**-based foods, as well as in **gels** and films, and **packaging materials**.

Cremes Creams or **custards** used in **fillings** and **desserts**. For example, creme caramel, creme brulee and confectionery cremes.

Cremoso Argentino cheese Argentinian **soft cheese** made from **pasteurized milk**. Has a particularly high moisture content. Its elastic **texture** and delicate **flavour** make it especially suitable for topping **pizzas** and vegetable dishes.

Crepes Thin **pancakes** originating from France made from **batters** containing flour, eggs, salt, milk and water. May be served with sweet or savoury **fillings**. Crepes suzette, traditional **desserts**, are made by warming crepes in orange-butter **sauces**, pouring over orange **liqueurs** and flaming before serving.

Crescenza cheese Italian **cheese** made from **cow milk**. **Texture** and **flavour** can vary from smooth with fresh, clean **acidity** to rubbery and mushy with a sour taste. Best ripened for no longer than 10 days and eaten soon after.

Cresols Methylphenol **flavour compounds** found in a range of foods, especially **smoked foods**. Cresols residues from lacquers applied to **cans** may occur as **contaminants** in **canned foods**. Cresols and cresol derivatives may cause **taints** in foods and **beverages**.

Cress Pungent leaves of seedlings from numerous plants of the family Cruciferae. Used as a salad vegetable, spice and in **soups**. Garden cress, or peppercress, is *Lepidium sativum*.

Creutzfeldt-Jakob disease Abbreviated to CJD. One of the **prion diseases**, this one affecting humans. After a long incubation period (not yet defined), it is characterized by progressive degeneration of the central nervous system. Symptoms include mood swings, aggression, slurred speech, hallucinations, problems in swallowing and ataxia. The initial form of Creutzfeldt-Jakob disease was only observed in subjects >40 years of age. In 1996, however, variant Creutzfeldt-Jakob disease (vCJD), affecting subjects as young as 12 years of age, was formally identified in the UK. The first known vCJD death, identified retrospectively, occurred in the UK in 1995. vCJD is believed to be transmitted to humans in foods (**beef**, meat products and **offal**) derived from **cattle** infected with BSE. In 1997, experiments in mice provided convincing evidence for the link between **BSE** and vCJD. The full extent of vCJD will not be discernible for many years to come because of the long incubation period of the disease; however, many thousands of BSE infected cattle are thought to have entered the UK food chain during the late 1980s, before the first BSE control measures were introduced.

Crispbreads Thin cracker-like products having a considerably lower water content than **bread**. Made from **rye flour** or **wheat flour** and water. Commonly eaten as an alternative to bread by those on a weight reducing diet.

Crispness Term relating to perception of product **texture** in the mouth; the extent to which a product is brittle when bitten.

Crisps Popular savoury bagged **snack foods** comprising very thinly sliced vegetables or extruded cereals that have been fried and flavoured, e.g. **potato crisps**. Also known as **chips** in some countries.

Critical control points Points, steps or procedures at which **quality control** can be applied and a **food safety** hazard prevented, eliminated, or reduced to acceptable levels. The selection of critical control points (CCP) is aided by the use of a CCP Decision Tree, which is designed to help determine what should be used in a Hazard Analysis Critical Control Point (**HACCP**) plan to control hazards.

Croakers General name used for **marine fish** species within the family Sciaenidae, which occur worldwide; also widely referred to as drum. Principal food fish within the family include Atlantic croaker (*Micropogon undulatus*), black croaker (*Cheilotrema saturnum*) and yellow croaker (*Umbrina roncador*).

Crocetin Dicarboxylic carotenoid pigment derived from **saffron** that is used as a natural food colorant. Forms red rhomboid crystals. Slightly soluble in water and organic solvents, and soluble in pyridine and dilute sodium hydroxide.

Crocin Yellow water-soluble carotenoid pigment which is found in the fruits of **gardenia** (*Gardenia jasminoides* Ellis) and in the stigmas of **saffron**. Purified crocin (purity >99.6%) has **antioxidative activ-**

ity comparable to that of **BHA** at some concentrations. Used in **colorants** for foods, e.g. smoked **haddock** and **cod**, and has potential for use in **antioxidants**.

Crocodile meat Meat from **crocodiles**, that is often considered to be a by-product of crocodile skin production. The tail (approximately 63.3% of which is lean meat) is the major carcass component marketed; it represents approximately 20% of carcass weight. **Colour** varies between meat cuts, with meat from the tail and neck appearing white to pink, and leg meat being darker.

Crocodiles The common name given to about 12 species of the genus *Crocodylus* in the family Crocodylidae. Crocodiles are hunted or farmed to produce **crocodile meat** and skins. Crocodile (*C. niloticus*) farming is popular in southern African countries, e.g. Zimbabwe. In the Northern Territory of Australia, crocodile (*C. porosus* and *C. johnstoni*) farming is a rapidly growing industry.

Crohns disease Inflammatory disease or set of diseases of the gastrointestinal tract the cause of which is unclear. Proposed causes include genetic, microbial and environmental factors. Due to the similarity of Johne's disease in cows and other animals to Crohns disease in humans, an association with the causative agent of Johne's disease, *Mycobacterium* avium subsp. *paratuberculosis*, has been suggested. Possible routes of transmission of *M. avium* subsp. *paratuberculosis* from animals to humans include through infected **dairy products** and **meat**.

Croissants Rich, crescent-shaped **rolls** made by laminating butter into a fermented **dough**; often served at **breakfast** with butter and **jams**. Also eaten stuffed with sweet or savoury **fillings**.

Crops Plants, including **cereals**, **vegetables** and **fruits**, which are cultivated commercially for the produce that they yield.

Crops rotation Practice of growing a sequence of different crops on a given piece of land to maintain the fertility of the soil.

Croquant Chopped, roasted **almonds** which are cooked in caramelized **sugar**. Also known as krokant.

Croutons Cubed or shaped pieces of seasoned toasted or fried **bread** used for garnishing **soups** or **salads**. May be flavoured, e.g. with **herbs** or **garlic**.

Crowberries Small black **fruits** produced by *Empetrum nigrum*. Best eaten cooked, as this enhances the **flavour**. Often eaten mixed with other **berries**. Used in **pies**, **soups** and jellies, and to make **wines**.

Crown corks Metal closures for **bottles**. Comprise preformed **caps** and a sealing pad. These are placed over the mouth of the container to be sealed, and the edges are crimped to secure them to the containers.

Commonly used for sealing bottles containing **soft drinks** or **beer**.

Crucian carp A **freshwater fish** species (*Carassius carassius*) from the **carp** family (Cyprinidae) widely distributed in lakes and rivers in Europe and northern and central Asia. Rarely regarded as a high value food fish, but is an important source of protein in some regions of the world. Cultured in parts of Europe and Asia. Normally marketed fresh and frozen.

Cruciferae Family of plants in which the flowers have 4 petals arranged in the shape of a cross. Includes brassicas, mustards and cress.

Cruciferins Globulin seed **storage proteins** which are found in **rapeseeds**.

Crude fibre The indigestible matter left in foods after successive digestion with ether, acids and alkalies, and subtraction of ash. Commonly determined as part of the proximate composition of foods. Values are related to, but not equivalent to, the **dietary fibre** component of foods.

Crumbing Process by which foods are coated in crumbs, usually fresh or dried **breadcrumbs**. Alternatively, breaking a food down into crumbs.

Crumbliness **Texture** term relating to the extent to which a product is brittle, fragile and liable to break up into fragments (crumbs).

Crumbling The process by which goods fall apart into small fragments, or the process by which foods are broken up (usually with the fingers) into small pieces.

Crumpets Small round **bakery products** made with **flour**, water and **milk** with added **sodium bicarbonate**. The resulting **batters** are leavened with **yeasts** and baked on a griddle. Usually served toasted and spread with butter.

Crunchiness **Texture** term related to the extent to which a product is crunchy, i.e. hard and crisp.

Crushing To deform, squash or pulverize an item by compressing forcefully. When applied to foods, this can result in products such as crumbs, pastes or powders. Crushing is often accomplished with a pestle and mortar, or with a rolling pin.

Crust Crisp, outside portion of **bakery products**, e.g. **bread**, that has been caramelized or dehydrated during **baking**.

Crustacea A subphylum of invertebrates containing approximately 30,000 species. Most are aquatic; of these, the majority are marine, but some are found in freshwater. Members of Crustacea include **lobsters**, **crabs**, **crayfish**, **shrimps**, copepods, barnacles and several other groups of organisms.

Cryogenics Branch of physics concerned with the production and effects of very low temperatures. Cryogenic temperatures are achieved either by the rapid evaporation of volatile liquids or by the expansion of gases.

Cryopreservation **Preservation** of foods using very low temperatures.

Cryoprotectants Compounds used to protect **frozen foods** from quality deterioration (**cryopreservation**). Include **sucrose**, which prevents muscle protein from **denaturation** during frozen storage of **surimi**, and **sorbitol**, **starch** or **starch hydrolysates**, which can be used to restrict undesirable changes in the **functional properties** of meat proteins. **Glycerol** is also commonly used as a cryoprotectant.

Cryoscopes Instruments used for measuring the **boiling point** and **freezing point** of liquids.

Cryoscopy Technique for determining the molecular weight of a substance by measuring the amount by which the **freezing point** of a solvent drops upon addition of a known quantity of that substance.

Cryovac Trade name for a process for packaging of **meat**, in which it is sealed in **plastics films** and the air is removed by vacuum pump.

Cryptococcus Genus of yeast **fungi** of the class Hymenomycetes. Occur on plants and in soil. *Cryptococcus neoformans* is often associated with **meat** and meat products where it may cause **spoilage**. *C. albidus* may be responsible for the spoilage of certain **fruits**.

Cryptosporidiosis Enteric disease caused by infection with *Cryptosporidium parvum*. Commonly transmitted through ingestion of food or water contaminated with animal faeces. Characterized by severe diarrhoea, abdominal cramps, fever and headache. May be asymptomatic.

Cryptosporidium Genus of protozoan parasites of the family Cryptosporidiidae. Occur in the intestinal tracts of vertebrates. Some species are pathogenic to humans and other animals. *Cryptosporidium parvum* is the causative agent of **cryptosporidiosis** in humans.

Cryptoxanthin Garnet-red carotenoid pigment with **vitamin A** activity which occurs naturally in **egg yolks**, **butter**, blood serum and in many plants. Slightly soluble in ethanol and methanol, and soluble in chloroform and benzene. It has many nutritional and medical uses. Also known as **provitamin A**.

Crystallization Formation of crystals, particularly used to purify a material or extract it from solution. Extensively used in **sugar processes**, and also in the processing of **butter** and **margarines**, **chocolate** and **ice cream**. Also used to prepare proteins, including **enzymes**, for structural analysis by X-ray diffraction **spectroscopy**.

Crystallography Measurement of the shape and structure of **crystals**.

Crystals Solid materials formed by **crystallization**, in which the atoms are arranged in a single regular arrangement called a lattice. **Sugars** and **fats** readily form crystals under favourable conditions such as temperature and concentration. **Starch** can also crystallize, and in **bread** this **retrogradation** process is associated with **staling**. The presence of crystals in foods strongly influences their **texture**.

Cs Chemical symbol for **caesium**.

Cu Chemical symbol for **copper**.

Cuartirolo argentino cheese Alternative term for Argentinean Quartirolo **cheese**.

Cucumber pickles **Cucumbers** pickled in **brines**. Alternative term for pickled cucumbers.

Cucumbers Fruits produced by *Cucumis sativus*. Contain approximately 95% water and 2% **sugar**, with some **carotenes** in the skin. Usually eaten raw in **salads**, but also used to make **cucumber pickles** and added to **yoghurt** to make raita, commonly eaten with **curries**. Seed kernels may be eaten as a snack food. Small ridge cucumbers are sometimes referred to as **gherkins**.

Cucumber seeds Kernels derived from *Cucumis sativus* which are rich in **proteins** and **oils** and may used as a source of these compounds.

Cucumisins EC 3.4.21.25. Serine **proteinases** occurring in the sarcocarp of **muskmelons** (*Cucumis melo*). Highly homologous with microbial **subtilisins**. Catalyse hydrolysis of **proteins** with broad specificity. Have potential for use in the food industry, including as **milk clotting enzymes** in **cheesemaking**.

Cucurbitaceae Family of food plants including **cucumbers**, **gherkins**, **gourds**, **marrows**, **melons**, **pumpkins** and **squashes**. Both the fruits and other parts of the plants may be consumed. Fruits contain mostly water, with good levels of **vitamin C** and sometimes **carotenes**.

Cucurbitacins Oxygenated tetracyclic triterpenoids produced by plants of the family Cucurbitaceae, such as **gourds** and **cucumbers**. Among the most bitter compounds known to man. Include cucurbitacin A, B and C and momordicoside A. Found in all plant parts except the **seeds**. Accumulation is generally not very high in **fruits**, but varies from season to season and according to location. Although perceived as bitter by humans, cucurbitacins are attractive to some insects and are used in baits.

Cultivar Commercial or cultivated varieties of given species of **plants** or **fungi**. Abbreviated to cv.

Cultivation From agriculture, a general term encompassing the processes associated with growing of crops prior to **harvesting**.

Cultured buttermilk Commercial product made as a substitute for **buttermilk** produced during **churning** as a by-product of **buttermaking**. Made by adding **lactic acid bacteria** to **skim milk**. The **lactic acid** produced during **fermentation** gives the product a tangy **flavour** similar to that of churned buttermilk, but composition of the two products differs. Used as a beverage and as an ingredient in **baking**.

Cultured cream Alternative term for **fermented cream**.

Cultured dairy products Alternative term for **fermented dairy products**.

Cultured foods Alternative term for **fermented foods**.

Cultured milk beverages Alternative term for **fermented milk**.

Cultured milk products Alternative term for **fermented dairy products**.

Cultured milks Alternative term for **fermented milk**.

Culture media Liquid or solid preparation of nutrients specifically designed to support the growth or maintenance of **microorganisms** or other cell types.

Cumin Common name for *Cuminum cyminum*, an umbelliferous herb grown for its aromatic, spicy seeds. These are used whole or ground as a flavouring ingredient in **curry powders**, and a range of other products including **chilli**, **pickles**, **sausages**, **bakery products** and **liqueurs**. Unrelated to **black cumin**.

Cuminaldehyde Aldehyde which is predominant amongst the **carbonyl compounds** in **cumin** (*Cuminum cyminum* L.) seed **essential oils**, representing the major flavour compound in cumin. Colourless liquid, insoluble in water but soluble in ethanol. Has **antimicrobial activity**, particularly against **fungi** and **yeasts**. Also a potent inhibitor of mushroom **tyrosinases**.

Cunninghamella Genus of **fungi** of the class Zygomycetes. Occur as saprotrophs on decaying vegetable matter, soil and dung, or as **parasites** or **pathogens** of plants or animals. *Cunninghamella echinulata* is used in the industrial production of γ-**linolenic acid**.

Cuphea Genus of plants belonging to the family Lythraceae which is being developed as an oilseed crop. **Seeds** of many species, e.g. *Cuphea lanceolata* and *C. viscosissima*, contain **oils** rich in medium chain **saturated fatty acids**. Such species are a potential source of these **fatty acids**, which have beneficial health and nutrition effects in humans and can affect fat quality when fed to animals.

Cupuacu **Fruits** produced by the cupuacu tree (*Theobroma grandiflorum*), which grows in the Amazonian rainforest. The exotic tasting pulp is used in making a range of products including **fruit juices**, **ice cream**, **jams** and **candy**. The seeds, which constitute approximately 20% of the fruit, contain a fat resembling **cocoa butter** and develop a chocolate-like aroma if roasted. They have been used to make a chocolate alternative which is free of **caffeine**.

Curculin Sweet-tasting protein which occurs in the pulp of *Curculigo latifolia* fruits. Also has **flavour** modifying activity which causes organic and inorganic **acids** to taste sweet after tasting the protein.

Curcuma Genus of plants, rhizomes of which are used as **spices** and sources of **essential oils** and **colorants**. Commercially important species include *Curcuma longa* (**turmeric**), *C. aromatica* (wild turmeric) and *C. zedoaria* (**zedoary**). The name is also applied to a natural colorant used to colour foods and textiles and as an indicator in **analytical techniques**. This colorant is sometimes called turmeric, curry or Indian saffron, and is commonly used in **curries**.

Curcumin Phenolic pigment which exists as a yellow-orange powder or needles and is derived from rhizomes of plants of the genus *Curcuma*, e.g. **turmeric** (*Curcuma longa* L). Insoluble in water but soluble in ethanol. Curcumin is a powerful antioxidant, and shows **antitumour activity** in animal studies and **anticarcinogenicity** *in vitro*. Used as a food dye, a biological stain and an analytical reagent. Also known as turmeric yellow.

Curd Protein (**casein**) gel formed by **coagulation** of milk, e.g. during **cheesemaking**. Other **milk proteins** are retained in the liquid portion (**whey**).

Curd cheese A semi-soft **cheese** with a creamy **texture** and mild **flavour**. A white cheese used especially in **cooking**.

Curdlan Extracellular microbial polysaccharide composed entirely of 1→3-β-D-glucosidic linkages which is produced by *Agrobacterium* spp. (formerly *Alcaligenes faecalis* subsp. *myxogenes*). Used as a food additive, particularly in formulation aids, processing aids, **fat substitutes**, **stabilizers thickeners** and **texturizers**. Can undergo both thermo-reversible and thermo-irreversible **gelation**.

Cured meat Meat preserved with the aid of **salt** and **colour** fixing ingredients, e.g. sodium nitrate and/or some sodium nitrite. Other **curing agents** may be added to accelerate **curing** (reducing agents), to modify **flavour** (e.g. **sweeteners**), to modify **texture**, to retard development of oxidative **rancidity**, and to increase **water binding capacity** and decrease shrinkage during subsequent processing. The curing process may involve: rubbing dry curing ingredients into the meat; immersion of meat in curing brines; or injection of the meat with solutions of curing ingredients. In the past, curing was used primarily to preserve meat, but with increases in the use of refrigeration and freezing, the major purpose of curing has changed. Meat curing ingredients are now mainly used to impart unique colour, flavour, **palatability** and texture properties to cured meat products. During the curing process, **nitrates** are converted into **nitrites**. Nitrosomyoglobin, formed from myoglobin and nitric oxide during curing, is responsible for the red colour of cured meat. Health concerns relating to use of nitrites and NaCl in cured meat have led to reductions in use of both ingredients.

Curing Preservation of foods such as **meat**, **cheese** and **fish** by **salting**, **drying**, **pickling** or **smoking**. Smoking can be carried out by the cold smoking method (in which the food is smoked at 70 to 90°F) or by the hot smoking method (which partially or totally cooks the food at 100 to 190°F). Pickled foods are soaked in flavoured, acid-based **brines**. Cheese curing can be undertaken by methods such as injecting or spraying the cheese with specific **bacteria** or by wrapping the cheese in flavoured materials.

Curing agents Ingredients used in the **curing** of foods. Examples include **salt** (sodium chloride; NaCl), **nitrates** and **nitrites**, sugar and **spices**.

Curing brines Brines used for curing of foods, such as meat products. Curing brines are often injected into **meat** to produce the final **cured meat** products.

Currants Term used in two different ways. Firstly, applied to dried **seedless grapes** of Mediterranean origin, similar to **raisins** and used in cooking, mainly in **bakery products**. Alternatively, small acid fruits produced by plants of the genus *Ribes*, including **blackcurrants**, **redcurrants** and **whitecurrants**, which are made into **jams**, jellies, **sauces** or **beverages** as well as being eaten as **desserts**.

Curries Spicy dishes of Indian origin, usually served with rice and/or Indian bread. Based on **meat**, **sea foods** or **vegetables** in piquant **sauces**. Curries can vary in **pungency** from mild to very hot, depending on the added **spices**. **Curry powders** are blends of powdered spices specially prepared for making curries.

Curry leaves Common name for leaves of *Murraya koenigii*, which resemble narrow **bay leaves**, but are more aromatic. Emit a strong warm curry aroma when rubbed or bruised. Used particularly in Indian and Sri Lankan dishes and **sauces** to enhance the **flavour**. For best results, fresh leaves are removed from the branches and added to dishes immediately before serving. May be stored in a refrigerator for short lengths of time, but frozen storage is recommended.

Curry powders Blends of powdered **spices** used for preparing **curries**. Various spices can be used in order to impart a particular **flavour** and/or **pungency**. Popular spice ingredients include **cumin**, **coriander**, **ginger**, **cloves**, **cardamom**, **fenugreek**, **chilli** and **turmeric**.

Curvularia Genus of **fungi** of the class Dothideomycetes. Occur in soil and on plants. Some species (e.g. *Curvularia lunata*) may cause **spoilage** of stored grains (e.g. **sorghum**, **corn**, **rice** and **wheat**).

Custard apples Fruits of any of several plants of the genus *Annona*. Round to heart-shaped with a white to yellow edible pulp. Flesh is eaten as a dessert or used as an ingredient in products such as **fruit salads**, **sherbet**, **ice cream**, **yoghurt** and **milkshakes**. Rich in **vitamin C**. The name has been applied to a number of species, including **cherimoya** (*A. cherimola*), **sugar apples** or sweet sop (*A. squamosa*), **soursop** (*A. muricata*) and the hybrid atemoya.

Custards Cooked or baked **sauces** made from **milk**, **eggs** and **sugar** and thickened with **corn starch**.

Cutability The ease with which an item can be divided into pieces using sharp objects such as **knives** or cleavers.

Cutin Waxy water-repellent biopolymer found in epidermal cell walls, e.g. **fruit peel**, and in the cuticle of plant leaves and stems. Composed of a mixture of oxidized and condensed **fatty acids**, soaps and **esters**. A component of **dietary fibre** and potential source of hydroxy fatty acids.

Cutinases **Lipolytic enzymes** able to hydrolyse **cutin**, the insoluble lipid-polyester matrix covering the surface of plants, and a variety of esters and **triacylglycerols**. Since these enzymes also exhibit **interesterification** and **transesterification** activities, they have a number of potential uses in the food industry, e.g. in production of **flavour compounds**.

Cutlassfish General name used for **marine fish** species within the family Trichiuridae, but particularly refers to *Trichiurus* spp. The most important species commercially are *T. lepturus* (Atlantic cutlassfish) and *T. nitens* (Pacific cutlassfish). Marketed salted/dried and also frozen. Flesh is regarded as having excellent **flavour** when fried or grilled; also used for production of **sashimi** when fresh.

Cutting Process by which an opening or incision is made in an item, or by which a slice is taken from an item, using sharp objects such as a **knives** or cleavers.

Cuttlefish Marine squid-like cephalopod **molluscs**, having calcareous internal shells; occur in deeper oceanic waters. Commercially important species include *Sepia officinalis* (cuttlefish), *Sepiola rondeleti* (lesser cuttlefish) and *Rossia macrosoma* (Ross cuttle). Mantle flesh from cuttlefish is usually marketed in frozen or canned forms.

cv Abbreviation for **cultivar**.

Cyanazine Selective systemic herbicide used for general weed control (pre- and post-emergence) in barley, wheat, corn, rapeseed, potatoes, soybeans and sugar cane. Classified by WHO as moderately toxic (WHO II).

Cyanides Group of compounds containing the -CN group which are salts or **esters** of **hydrogen cyanide**. Can be extremely toxic.

Cyanidin One of the **anthocyanidins**, a pigment often present as a glycoside, which is found in many **fruits** and **vegetables**.

Cyanobacteria Photosynthetic **bacteria** containing **chlorophylls** and other **pigments**. Genera include *Anabaena*, *Nostoc* and *Synechococcus*. Previously better know as **blue green algae**.

Cyanocobalamin Synonym for **vitamin B$_{12}$**. Member of the **vitamin B group**, found in foods of animal origin such as **livers**, **fish** and **eggs**. Vitamin B$_{12}$ is the coenzyme for methionine synthase (EC 2.1.1.13), an enzyme important for the metabolism of **folic acid**, and methylmalonyl coenzyme A mutase (EC 5.4.99.2). **Absorption** of this vitamin requires the presence of an intrinsic factor. Failure of absorption, rather than dietary deficiency, is the major cause of pernicious anaemia.

Cyanogenic glycosides **Cyanogens** which are capable of liberating large amounts of toxic **cyanides**, which can be metabolized to **goitrogens** (thiocyanates). Include **linamarin**, linustatin and neolinustatin. Occur naturally in many plants, including **cereals**, **pulses**, **fruits**, root crops, **nuts** and **oilseeds**, usually in parts that are not eaten or at such low concentrations that they do not present a health risk to consumers. However, in **cassava**, they occur in high levels both in the edible roots and leaves. Readily detoxified by appropriate processing of plant materials.

Cyanogens Colourless flammable highly toxic gases with pungent odours. Produced synthetically by oxidizing **hydrogen cyanide**, but some (e.g. **cyanogenic glycosides**) occur naturally in plants. Starting materials in the manufacture of complex thiocyanates, which are used as **insecticides**.

Cycad seeds Seeds produced by gymnosperms of the genus *Cycas*, especially *C. circinalis*, the false sago palm. Contain a toxic principle, cycasin, which causes a neurological disorder when untreated seeds are consumed.

Cyclamates Salts of **cyclamic acid**, also known as sulfamates. Used as non-nutritive **sweeteners** in foods, usually in the form of calcium cyclamate or **sodium cyclamate**. Cyclamates are approximately 30 times as sweet as **sucrose** and display good solubility and heat stability characteristics for a variety of food applications, such as **low calorie foods**. Use of cyclamates was banned in the USA, UK and Canada due to concerns about possible **carcinogenicity**. However, later studies have failed to confirm this and use is still permitted in many countries.

Cyclamic acid Organic acid used as a sweetener for foods, usually in the form of metal salts (**cyclamates**). Also known as cyclohexanesulfamic acid.

Cyclodextrin glucanotransferases Alternative term for **cyclomaltodextrin glucanotransferases**.

Cyclodextrins **Dextrins** containing at least six **glucose** units in the form of a ring. Can associate with a range of substances and are therefore used as complexing agents, particularly in the β-cyclodextrin form. Used in the food industry as **emulsifiers**, **stabilizers** and masking agents for **off odour** and **off flavour**.

Cyclohexanesulfamic acid Alternative term for **cyclamic acid**.

Cyclohexylamine Amine which exists as a liquid with a strong fishy **aroma**. Miscible with water and common organic solvents. Used in organic syntheses and in the manufacture of **plasticizers**, **dyes**, **emulsifying agents**, dry-cleaning soaps, corrosion inhibitors and rubber chemicals.

Cyclomaltodextrinases EC 3.2.1.54. Hydrolyse cyclomaltodextrins to linear **maltodextrins**. Can also hydrolyse linear maltodextrins, and may hydrolyse **starch**, **pullulan**, **amyloses** and **amylopectins**. They may also exhibit transglycosylation activity.

Cyclomaltodextrin glucanotransferases EC 2.4.1.19. Cyclize part of 1,4-α-D-glucan chains by formation of 1,4-α-D-glucosidic bonds. Cyclomaltodextrins of 6, 7 or 8 **glucose** molecules, known as α-, β- and γ-cyclodextrin, respectively, are formed reversibly by the action of the enzyme on **starch** and **dextrins**. These **cyclodextrins** are useful for **encapsulation** of **volatile compounds**, such as **aroma compounds**. The **enzymes** will also disproportionate linear **maltodextrins** without cyclizing.

Cyclones Processing equipment used for separation of solids from air. Consists of a conical chamber into which the air and solid, such as a food powder, is added tangentially at high speed producing a whirl or cyclone. Particulate matter is forced to the sides of the chamber, decelerates and drops down to the conical end of the chamber from which it is removed. The air stream remains in the central region of the cyclone.

Used for separation of powders, e.g. **milk powders**, from air after **spray drying**.

Cyclopiazonic acid Mycotoxin produced by *Penicillium* species (e.g. *Penicillium verrucosum* and *P. griseofulvum*) and *Aspergillus* species (e.g. *A. flavus* and *A. oryzae*). Formed during fungal growth on food such as **corn**, **peanuts** and **cheese**. Toxic to certain animals (e.g. chickens), but no definite health risk for humans.

Cycloserine Broad spectrum antibiotic that is particularly active against *Mycobacterium* spp. Used to treat mycobacterial infections (such as tuberculosis) in animals.

Cyclospora Genus of parasitic coccidian protozoa of the family Eimeriidae. *Cyclospora cayetanensis* may be transmitted to humans through ingestion of water or food contaminated with **oocysts**.

Cyclosporiasis Disease caused by infection with *Cyclospora* species (especially *C. cayetanensis* in humans). It is characterized by watery diarrhoea, loss of appetite, substantial weight loss, bloating, flatulence, abdominal cramps, nausea, vomiting, muscle aches, low-grade fever and fatigue. Some infected persons are asymptomatic.

Cycocel Alternative term for the plant growth regulator chlormequat. Classified by WHO as slightly toxic (WHO III).

Cymene Volatile, combustible, aromatic hydrocarbon consisting of benzene rings carrying one methyl and one isopropyl group. Exists as a colourless, transparent liquid with an aromatic aroma. Three isomers are known, i.e. *ortho-*, *meta-* and *para-*cymene. *para-*Cymene occurs naturally in several **essential oils**, e.g. **oregano** *(Origanum vulgare* L.). Uses include synthetic resin manufacture, metal polishes, solvents and organic syntheses. *para-*Cymene can be used to produce pure **carvacrol** and *para-*cresol.

Cypermethrin Non-systemic insecticide used to control a wide range of insects in fruits, vegetables, cereals, rapeseed and coffee; also used in animal rearing facilities. Classified by WHO as moderately toxic (WHO II).

Cystatins **Proteins** which inhibit cysteine **proteinases**. These **proteinases inhibitors** are present in many plant seeds, including **legumes** and **cereals**, and are also found in animals tissues, including **meat**, **eggs** and **fish**. Have a potential role in the regulation of **proteolysis** during meat processing as they can inhibit **calpains** and **cathepsins**, and could also be used to maintain the quality of fresh fish and **surimi**.

Cysteine Crystalline sulfur-containing amino acid. In the human diet, cysteine is a conditionally essential amino acid; thus, it may be required in the diet unless abundant amounts of its precursors, **methionine** and **serine**, are available for cysteine synthesis at a nutritionally significant rate.

Cysteine sulfoxides A group of organic sulfur compounds found predominantly in *Allium* and *Brassica* species, where they are important precursors for **flavour compounds** produced by lyases such as **alliin lyases**.

Cysticercosis Infestation with the larvae (cysticerci) of the tapeworm *Taenia solium*. May be caused by ingestion of tapeworm eggs in food and water. Normally, the cysticerci develop in the animal host (swine), and humans are infected with the adult form through eating undercooked infected **meat**.

Cysticercus Larval forms of *Taenia* species of tapeworm. *Cystericercus cellulosae* is the larval form of *T. solium* found in swine, while *C. bovis* is the larval form of *T. saginata* found in cattle.

Cystine White crystalline amino acid; oxidized dimeric form of **cysteine**. In healthy individuals, it is produced from **methionine** or **homocysteine** and is not an essential amino acid.

Cystoseira Genus of **seaweeds** found in low intertidal and subtidal shores of warm and temperate waters around the world. Some species are utilized as food or a source of **phytochemicals**.

Cytidine Nucleoside composed of one molecule of **cytosine** and one molecule of D-ribose. Also known as cytosine riboside.

Cytochalasins **Mycotoxins** produced by certain fungal species (e.g. *Aspergillus*, *Helminthosporium* and *Phomopsis* species). Formed during fungal growth on grains and grain products.

Cytokines Humoral mediators produced by components of the immune system including the interferon and interleukin families and tumour necrosis factor-α (TNF-α). Cytokines are involved in regulation of **immune response** and inflammation and aberrant production is associated with certain **allergies** and inflammatory diseases. Modulation of cytokine status may been one mechanism by which **functional foods**, such as **probiotic foods**, may enhance immunity and health.

Cytokinins Class of **plant growth regulators** which occur naturally in plants and are also applied exogenously to influence the quality of **fruits** and **vegetables**. Particularly active in stimulating growth and cell division.

Cytophaga Genus of Gram negative, anaerobic, rod-shaped **bacteria** of the family Cytophagaceae. Occur in soil, decomposing matter and aquatic habitats. Some species may cause the **spoilage** of **refrigerated foods**, especially **fish** and **shellfish**.

Cytosine Pyrimidine base, which is a constituent of **DNA** and **RNA**.

Cytotoxicity Quality or degree of being cytotoxic (exerting a toxic effect on cells).

Cytotoxins **Toxins** which exert a toxic effect on cells.

D

2,4-D Alternative term for **2,4-dichlorophenoxyacetic acid**.

Dab Marine **flatfish** species (*Limanda limanda*) which occurs abundantly around the northeast Atlantic. Flesh has firm **texture** and a sweet **flavour**. Marketed fresh, dried/salted, smoked and frozen.

Daconil Alternative term for the fungicide **chlorothalonil**.

Dahi **Fermented milk** product popular in India. Dahi made from **buffalo milk** is generally preferred to that made from **cow milk**. A sweet variety of dahi, misti dahi, is prepared by adding **cane sugar** to **milk** during **heating**, giving a caramelized **flavour** and brown **colour**.

Daidzein One of the two **isoflavones** of particular importance in **soybeans**, the other being **genistein**. Both compounds are structurally similar to oestrogenic steroids and possess both oestrogenic and anti-oestrogenic activities, the principal functions responsible for the health benefits associated with consumption of soybeans and soy products.

Dairies Premises in which **dairy products** are manufactured. Also called **creameries** or **dairy factories**.

Dairies effluents **Waste water** released from **dairies**.

Dairies wastes **Wastes** remaining after processing of **dairy products**.

Dairy beverages Drinks based on **milk** or other **dairy products**, e.g. **whey**.

Dairy desserts Ready to eat **desserts** based on **dairy products**, such as **cream**, **milk** or **yoghurt**. Available as chilled, frozen and shelf-stable products. Include **mousses**, **custards**, **fromage frais**, **milk puddings** and **ice cream** products.

Dairy factories Premises in which **dairy products** are manufactured. Also called **creameries** or **dairies**.

Dairy-lo One of the **fat substitutes**, based on whey protein concentrate which has been subjected to controlled thermal **denaturation**, resulting in a functional protein with fat-like properties. Used mainly in **dairy products**, **bakery products** and **salad dressings**.

Dairy products Products manufactured from **milk**. Include as major product groups, **cheese**, **yoghurt**, **butter**, **cream**, **fermented milk**, **ice cream** and **whey** products. Also called **milk products**.

Dairy science Division of **food science** dealing with the characteristics, manufacture and quality of **dairy products** as well as the production, management and distribution of dairy animals such as cows, goats and sheep.

Dairy spreads Spreads based on **milk fats** and containing other, sometimes non-dairy, ingredients to give a lower fat content than **butter**.

Dairy starters Microbial cultures used in manufacture of **fermented dairy products**, including **fermented cream**, **fermented milk** and **cheese**.

Dalia Type of **porridge** made from **wheat grits**.

Daminozide Plant growth regulator (the active component in Alar) which has been widely used in the cultivation of **apples**. Concern arose in the 1980s over the safety of Alar when it was identified as a possible carcinogen. Daminozide is also known by a number of other names, including *N*-dimethylaminosuccinamic acid, kylar and SADH.

Damsons Purple plum-like **fruits** produced by *Prunus damascena*. Eaten cooked or used to make **jams** or damson cheese, a solid preserve of damsons and **sugar**.

Danbo cheese Danish semi-soft **cheese** made from **cow milk**. Has a smooth, dry, yellow rind and is sometimes coated with red wax. Ripened for 6 weeks to 5 months.

Dandelions Common name for *Taraxacum officinale*. All parts of the plant are consumed. The root is used to make **beverages** that smell like **coffee** but have the flavour of **chicory**, the leaves are used in **salads** or as vegetables, and the flower heads are used in **winemaking**.

Danish pastries Sweet **bakery products** made from laminating **yeasts**-fermented **dough** with **butter** or **margarines** and filled with **nuts**, **fruits** or **custards**. Often glazed with thin sugar/water icing.

Dark cutting defect A defect of **beef**, often associated with bull beef. Dark cutting meat, also known as black beef or dark cutter beef, has a darker **colour**, and poorer **flavour** and **texture** than normal beef; moreover, the high **pH** value of dark cutting meat encourages the growth of spoilage bacteria and reduces shelf life. Physiological stress and exhaustion pre-slaughter deplete muscle glycogen stores, ultimately increasing the pH of meat and leading to the development of dark cutting defect. In young bulls, incidence of dark cutting defect can be decreased by low stress handling and prevention of bull behaviour (mounting, mock fighting and butting) in abattoir pens prior to slaughter.

Darkening **Discoloration** of a substance by becoming dark or darker. Red colour is often used by consumers as an indicator of the **freshness** of meat. Darkening of the product, which occurs during storage due to pigment shifts, is perceived as being a negative event, even though this is not a true indicator of wholesomeness or nutritional value. Because of consumer concerns, **packaging films** are designed to protect meat **colour**, largely by controlling diffusion of oxygen. Darkening is also a problem during repeated use of **frying oils**.

Dark firm dry defect Commonly abbreviated to DFD defect, a condition associated with **pork** in which meat has a high **pH** value and darker than normal lean **colour**. The defect results from a decreased **glycogen** content in swine muscles prior to slaughter; it is often associated with pre-slaughter **stress**. In beef, the term dark cutting defect or dark cutter is used to refer to the same condition.

Databanks Large stores of data held on computers.

DATEM Anionic oil in water **emulsifiers** used as **improvers** in **breadmaking**. Acronym for diacetyl tartaric acid esters of mono- and diglycerides.

Date marking Marking of food or beverage containers with a date that may be the date of manufacture, the sell-by date and/or the use-by date (expiry date). The sell-by date is the date by which the manufacturer recommends that a perishable product should be sold. Use-by dates are chiefly used in the UK instead of sell-by dates, and indicate the recommended date by which a perishable product should be eaten or used, after which it is no longer deemed to be safe, desirable or effective. Date marking is often required by law, particularly on packs of foods which should be maintained at low temperature, e.g. **cheese**, **pates** and **ready meals**, and on foods in which **spoilage** organisms are likely to multiply or cross contaminate other foods, e.g. fresh **meat** and **fish**. Other foods, such as **bread** and **cakes**, which tend to deteriorate in quality rather than safety do not require date marking by law, but are often labelled voluntarily by the manufacturer or retailer.

Dates Fruits of the date palm (*Phoenix dactylifera*). Vary in colour, shape and size, and may be soft, dry or semi-dry. Contain high levels of sugar, amounts and individual types of sugars varying among cultivars, but small amounts of **vitamins**. **Vitamin C** content is relatively high in fresh fruits, but is reduced to trace amounts by **drying**. Served as dessert fruits and incorporated into many food products, especially **cakes** and **biscuits**. In addition, in Arab countries, dates are also used in preparation of **syrups**, **vinegar** and **sugar substitutes**.

Date shells Marine **bivalves** (*Lithophaga lithophaga*) occurring along shores of the Mediterranean Sea and eastern Atlantic, which bore into rocks using a secreted acid. Consumed as a table delicacy in some Mediterranean regions.

Dating Process of marking a product or its outer packaging with date information, such as date of manufacture or date by which the product should be consumed to ensure quality.

Davana Common name for *Artemisia pallens*, a plant used as the source of aromatic **herbs** and **essential oils** with a characteristic fruity odour. Used in **flavourings** for **cakes**, **pastries** and value-added **beverages**.

Dawadawa Fat- and protein-rich **fermented foods** from West and Central Africa, traditionally made from **African locust beans**. Seeds are cooked, fermented and formed into balls, which can be used to flavour **soups** and stews. The fermented products can be stored for long periods and are a good source of **linoleic acid** and **vitamin B$_2$**. Also known as **iru** in Nigeria.

DDD Persistent non-systemic organochlorine insecticide used to control a wide range of **insects**; usage on **crops** has generally been displaced by less persistent **insecticides**. Often occurs as a degradation product of **DDT**. Classified by WHO as moderately toxic (WHO II).

DDE Persistent non-systemic organochlorine insecticide used to control a wide range of **insects**; usage on **crops** has generally been displaced by less persistent **insecticides**. Often occurs as a degradation product of **DDT**. Classified by WHO as moderately toxic (WHO II).

DDT Persistent non-systemic organochlorine insecticide used to control a wide range of **insects**. Usage on **crops** has generally been displaced by less persistent **insecticides**. Classified by WHO as moderately toxic (WHO II).

Deacidification Neutralization process whereby the acidity of a substance is reduced. Deacidification is often used in conjunction with the processing of **apple juices**, **cider**, **vegetable oils**, **wines** and **grape musts**. Deacidification of grape musts is crucial for the production of well-balanced wines, especially in colder regions of the world. **Malolactic fermentation** is widely used to reduce the acidity of **grape juices**. Young wines can also be deacidified with calcium carbonate and potassium hydrogen carbonate. Deacidification of vegetable oils (such as **rice bran oils** and **corn oils**) can be carried out using solvent extraction and membrane processing. **Nanofiltration** has been used for deacidifying and demineralizing cottage **cheese whey**, ready for use in **ice cream** and other frozen **dairy desserts**.

Deaeration Removal of air or oxygen from a solution, for example by bubbling with an inert gas. Also known as degassing.

Deaminases **Hydrolases** that act on carbon-nitrogen bonds other than peptide bonds, removing amino groups from compounds. **Ammonia** is produced in the process.

Debaryomyces Genus of yeast **fungi** of the class Saccharomycetes. *Debaryomyces hansenii* may be used as a starter in the manufacture of **fermented sausages**, and has been responsible for the **spoilage** of **fruit juice concentrates** and **yoghurt**.

Debittering Removal of **bitter compounds** from foods such as **citrus fruits**, **chocolate**, **soybeans** and cruciferous **vegetables**, and **beverages** such as **wines**, **fruit juices**, **cider** and **beer**, to make them more palatable. Debittering can be achieved biologically, using **enzymes** or immobilized bacteria. Lactone hydrolases are used commercially for debittering triterpenes present in citrus juices. Correction of excessive **naringin bitterness** in citrus fruits can be achieved through use of **adsorbents** or **cyclodextrins** to form less bitter inclusion complexes. Deliberate aeration of the pulp during apple juice extraction for cidermaking promotes the removal of bitter and astringent **flavonoids** through their binding to the pomace. Fining with **gelatin** decreases contents further still by coprecipitation. Proline-specific **aminopeptidases** can be used for debittering of food protein hydrolysates. Enzymic hydrolysis of **oleuropein** by β-glucosidase from *Lactobacillus* plantarum offers a potential alternative to chemical debittering treatments for table **olives**.

Deboning A process for cutting of **meat** from the **bones**, which can be done either manually or mechanically.

Debranching enzymes Alternative term for **α-dextrin endo-1,6-α-glucosidases** and **isoamylases**.

Debranning Process of **bran** removal from **cereals**. May be achieved by **milling** or by soaking in a solution of an alkali such as sodium hydroxide. Used to enhance milling performance of cereals as well as to provide by-products with potential as food ingredients. However, debranning may also affect the nutritional quality and **functional properties** of the cereal and subsequent products.

Decaffeinated coffee Coffee from which **caffeine** has been removed by a solvent extraction process using aqueous, organic or supercritical solvents.

Decaffeinated tea Tea from which **caffeine** has been removed by a solvent extraction process using aqueous, organic or supercritical solvents.

Decaffeination Removal of **caffeine** from a substance such as **coffee** or **tea**. Caffeine is removed from coffee by soaking **coffee beans** in chemical solvents or water. The resulting decaffeinated product contains approximately 3 mg caffeine per 150 ml cup, compared with 75,150 mg for normal coffee.

γ-Decalactone Flavour compound found in various foods, including **fruits** and **alcoholic beverages**. Microbially synthesized γ-decalactone is used in food **flavourings**.

Decanal One of the aldehyde **flavour compounds**, which occurs naturally in a wide range of foods and **beverages** and is used in **flavourings** for processed products.

Decanoic acid Synonym for **capric acid**. Member of the medium chain-length **saturated fatty acids** with 10 carbon atoms. Found in a range of animal and **vegetable fats** and **vegetable oils**, and, in its free form, contributes to the **flavour** of foods and **beverages**.

Decanol Alcohol with 10 carbon atoms. Along with some of the other higher **alcohols**, contributes to the **flavour** of foods and **beverages**, especially **alcoholic beverages**, and is also widely used as a solvent.

Decanters Stoppered glass **containers** into which **wines** or **spirits** are decanted.

Decarbonation Removal of **carbon dioxide** from a sample. Required for sample preparation prior to **beer** analyses, such as determination of **original gravity** and alcohol content.

Decarboxylases **Lyases** belonging to subclass EC 4.1.1 that remove carboxyl groups from a molecule, especially amino acids and proteins. When acting on single substrates, a molecule of CO_2 is eliminated leaving an unsaturated residue.

Decarboxylation Chemical modification involving the removal of carboxyl groups from **organic compounds**, generating CO_2. Can be due to the influence of **enzymes** (**decarboxylases**) or other **catalysts**, or can occur spontaneously. Several **aroma compounds**, including **diacetyl**, are formed by decarboxylation reactions.

Dechlorination Process of removing residual **chlorine** from a substance. In the food and **beverages** industries, **chlorination** usually cannot be considered without the added expense of dechlorination, as residual chlorine must be removed to prevent chemical changes affecting **flavour**, **aroma** and **colour** of the final product. **Activated carbon** is usually used in the beverages industry to dechlorinate and remove trace levels of outside **flavour compounds** from water to be used in producing **beer** and **soft drinks**. A non-chemical means of dechlorination involves use of a high energy ultraviolet system. This cost effective process reduces free chlorine levels by up to 99%.

Decoction A liquor containing the concentrated essence of a substance, produced as a result of **heating** or **boiling**.

Decoloration Removal of the **colour** from an item. Also known as decolorization.

Decolorization Alternative term for **decoloration**.

Deep freezing A method for preservation of foods by rapid **freezing** and storage at -18°C. Freezing preserves foods by preventing **microorganisms** from multiplying. **Enzymes** in the frozen state remain active, although at a reduced rate. Commercial freezing is usually undertaken by one of the following methods: blast freezing, where air is circulated at -40°C; contact freezing, in which **refrigerants** are circulated through hollow shelves; immersion freezing, where, for example, fruit is frozen in a solution of sugar and glycerol; and cryogenic freezing, using, for example, liquid nitrogen spray. Rapid freezing avoids structural change that would affect **flavour** or appearance of foods, as in the shrinkage and distortion of cells by formation of enlarged ice crystals in the extracellular spaces. Some quick **frozen foods** require thawing before use, and cooking must then be prompt. This method of preservation is widely used for a great variety of foods, including **bakery products** (both ready to eat, and to be cooked when desired), **soups**, and precooked complete **meals**.

Deep frying **Cooking** of foods in an amount of hot **fats** or **oils** sufficient to cover them completely during **frying**.

Deer Common name given to approximately 41 species of even-toed, hoofed, ruminant mammals belonging to the family Cervidae. The term is used specifically to describe any of the small- or medium-sized species of the Cervidae family, as being distinct from other large-sized species such as elks or moose. Deer are farmed or hunted for their meat (**venison**).

Deer meat Alternative term for **venison**.

Defeathering Removal of feathers from the **carcasses** of meat-producing birds, such as **poultry**, during processing. If defeathering is not performed properly, carcasses can be mechanically damaged or microbially contaminated, both of which are of economic importance to the poultry industry.

Defecation Removal of impurities, usually applied to the stage of **purification** of **sugar juices** during **sugar** manufacture. Defecation involves **clarification** of sugar juices by heat and lime. The lime is added to neutralize the **organic acids** present, after which the temperature is raised to approximately 95°C. This lime and heat treatment forms a heavy precipitate of complex composition, which contains insoluble lime salts, coagulated albumin, and varying proportions of **fats**, **waxes** and **gums**. The flocculant precipitate carries with it most of the finely suspended material of the juice that has escaped mechanical screening. Separation of this precipitate from the juice is undertaken using a juice clarifier. Degree of clarification has a great bearing on the boiling house operations, and on yield and refining quality of raw sugar.

Deficiency diseases Conditions arising due to the absence of a dietary nutrient, such as one of the essential **vitamins** or **minerals**. Include various types of anaemia, rickets, scurvy, pellagra, beriberi and goitre. Strategies to counteract these disorders and improve nutrition often combine direct dietary intervention (provision of **food supplements**, food **fortification**, dietary diversification) with agricultural measures (development of foods of improved **nutritional values** and **bioavailability**, development of improved agricultural practices) and economic measures for improving food security.

Defoaming agents Substances, often silicon-based, used to minimize formation of **foams** during food processing. These foams would otherwise cause problems for both the processing operation and final product quality. Typical applications where foaming problems occur include **freeze drying**, **sugar processes** and manufacture of fruit and dietetic **soft drinks**. Similar to **antifoaming agents**.

Defoliation Removal of leaves from plants. Can affect fruit growth and quality.

Deformation Persistent change in shape or size of a substance in response to an externally applied force. Routinely determined for foods during analysis of **rheological properties**, and can include puncture

deformation, torsional deformation, breaking deformation and maximal (peak) deformation.

Defrosting Thawing of **frozen foods**, or alternatively the freeing of an item, e.g. **freezers**, of accumulated ice.

Degassing Alternative term for **deaeration**.

Degreening Process of **ripening** or improvement of skin or **peel** colour, usually by application of **ethylene** to **citrus fruits** (such as **satsuma mandarins** and **lemons**), **bananas**, **rapeseeds** and **mustard seeds**. Decay tends to be more severe in degreened fruit because the degreening process itself promotes decay, and because packaging line fungicide treatments have to be delayed until after degreening. Uneven degreening of bananas is a ripening disorder characterized by either partial or delayed yellowing or by permanent greenness after treatment with exogenous ethylene. Green seed is a significant economic problem in rapeseeds because the **rapeseed oils** extracted from such seed contains chlorophyll-type pigments. Seed crushers can remove the green colour from rapeseed oil with bleaching **clays**, but this involves an added expense and poses an environmental problem.

Degumming The first stage in the purification of crude **oils**, which involves removal of **phospholipids** and colouring materials. Degumming is necessary to prevent separation and settling of **gums** (sticky, viscous oil-water **emulsions** stabilized by phospholipids) during transportation and storage of crude oils, to reduce oil losses in the subsequent phases of refining, and to avoid excessive darkening of the oils in the course of high-temperature **deodorization**. Degumming agents, such as phosphoric acid, may be used together with a **flocculation** agent such as alumina. During water degumming, **phosphatides** in seed oils are removed by centrifugal separation, after precipitation with water. Acid degumming involves removal of gums and impurities via centrifugal separation after precipitation with acid and water. By-products of the degumming process are known as **lecithins**.

Degumming agents Processing aids used to remove **phospholipids**, trace metals and mucilaginous gums during the initial (**degumming**) stage of **oils** and **fats refining**. Examples include water, phosphoric acid and **citric acid**.

Dehairing Removal of the hair from hides and fleece of animal **carcasses**, usually by scalding, singeing or chemical methods. Carcasses are dehaired as an intervention to reduce microbial load and improve visual cleanliness prior to **dressing**.

Dehulling Removal of the **hulls** from **fruits** or **seeds** prior to consumption. Also called **hulling** or

husking. This term also relates to removal of the cluster of leaves from the tops of **strawberries** prior to consumption.

Dehydrated foods Alternative term for **dried foods**.

Dehydration Alternative term for **drying**.

Dehydroacetic acid Organic acid used in **preservatives** to inhibit microbial growth in foods and **beverages**.

Dehydroascorbic acid Oxidized form of **vitamin C**, which together with **ascorbic acid** (the reduced form), makes up the total vitamin C activity in a substance. Present in many food materials, where it has been implicated in **browning** or **discoloration** reactions in certain matrices, such as **citrus juices**. In **breadmaking**, dehydroascorbic acid is formed from ascorbic acid (used in **bakery additives**) and acts as an oxidizing agent, promoting formation of disulfide bonds (important for **dough** strength).

Dehydrogenases Enzymes that oxidize substrates by transferring hydrogen atoms to an acceptor that is either NAD/NADP or a flavin enzyme.

Dekkera Genus of yeast **fungi** of the class Saccharomycetes. *Dekkera bruxellensis* and *D. anomala* are responsible for the **spoilage** of **beer** and **wines**.

Delicatessen foods Speciality **ready to eat foods** purchased from delicatessen shops or departments. Examples include **delicatessen salads**, imported cooked **meat** products and speciality **cheese**. Also known as deli foods in the USA.

Delicatessen salads Ready to eat chilled **salads** (frequently **mayonnaise**-coated) obtained from delicatessen shops or departments. Examples include **coleslaw**, **potato salads** and **herring** salads.

Delphinidin One of the **anthocyanidins**, often present as a glycoside, and found in many **fruits** and **vegetables**.

Deltamethrin Non-systemic insecticide used to control insect **pests** on a wide range of fruits, vegetables and cereals; also used in stored cereals and as a dip or spray for cattle, sheep and swine. Classified by WHO as moderately toxic (WHO II).

Demineralization Removal of minerals from substances. Includes processing steps in food manufacture, such as for **sugar syrups**, **drinking water**, **musts**, **whey** and food factories effluents. Processes used to achieve demineralization include **electrodialysis**, **reverse osmosis** and **nanofiltration**. Also covers the undesirable removal of selected minerals from previously healthy tissues such as bone and tooth enamel, which may be caused by a variety of factors including nutritional imbalance and excess acidity, respectively.

Denaturation Structural change, especially in **proteins** or **nucleic acids**, in response to extreme conditions of temperature, pH, pressure or salt concentration, which renders the molecule incapable of performing its original biological function. Used in food processing to inactivate detrimental **enzymes**, or to alter the **gelation** properties of **proteins** such as **gelatin** or **whey proteins**. However, can also be deleterious, leading to impairment of **functional properties** such as **water holding capacity** in proteinaceous foods, and to reduced product yields in enzyme catalysis.

Denitrification Process of removing **nitrogen** or **nitrogen compounds** from a substance, or alternatively the liberation of elementary nitrogen from nitrogenous compounds in the soil by **bacteria**.

Densitometry Technique for measuring the optical density of a material by recording transmission of light.

Density One of the **physical properties** of a substance, defined as the mass contained in a given volume. Routinely determined for a wide range of foods, including **fruits** and **vegetables** (sometimes related to **ripeness** and composition), **fats** and **oils**, foods produced by **extrusion**, and **cereals**. Density determinations can also be used as **process control** steps in food processing.

Dental caries Disease in which cavities are formed in the teeth resulting ultimately in dental pain and tooth loss. Caries formation is associated with the action of oral *Streptococcus mutans* strains. Cavity formation is increased by the consumption of **sugar**-containing foods, as the sugar is metabolized by the **bacteria** to form **acids**, which destroy the tooth enamel and subsequently the dentine. Increasing oral saliva production, achieved by various means such as chewing **chewing gums**, can buffer bacterial acid production and reduce cavity formation. Sometimes known as caries.

Dental health Measure of the physical condition of an individual's teeth and gums, or factors influencing their condition. Cariogenic foods, including many with a high **sugar** content, promote development of **dental caries** (decay), whilst cariostatic or anticariogenic foods or ingredients reduce these processes. **Fluoridation** of **drinking water** is undertaken with the aim of improving dental health, and **oligosaccharides** with cariostatic properties are being developed for use as **sweeteners**.

Dentex Genus of **marine fish** containing several species of **sea bream**.

Deodorization Removal or concealment of an unpleasant smell in an item. Deodorization is usually the last step in edible oil refining, involving vacuum-steam distillation at elevated temperature, during which free **fatty acids** and odoriferous **volatile compounds** are removed in order to obtain a bland and colourless product. Deodorization can be conducted under continuous, semi-continuous or batch conditions.

Deoxycholate Salt of deoxycholic acid (a secondary bile acid). Used in **surfactants** and selective media for **cell culture**, such as deoxycholate-citrate agar. Also known as **desoxycholate**.

Deoxynivalenol Trichothecene produced by *Fusarium* species. Also known as **vomitoxin**.

Deoxyribonucleases **Enzymes** that cleave the phosphodiester bonds between nucleotide subunits in **DNA**.

Deoxyribonucleic acid Commonly abbreviated to **DNA**.

Depositors Devices for laying down a body of accumulated matter. In the food industry, they may be used to place such substances as **fillings**, **toppings**, **batters** and **mixes** in position.

Depuration To make or become free from impurities using controlled **purification** systems employing sterilized water. Systems can be flow-through or recirculating types, and water sterilization treatments involve the use of chlorine, UV light, **ozone**, membrane filters or **iodophors**. Depuration is usually applied to purification of **shellfish**, such as **oysters** and **mussels**. Post-harvest depuration in controlled waters can increase the safety of shellfish by reducing the number of **pathogens** present following harvesting from moderately polluted water.

Dermatitis Inflammation of the skin. Atopic dermatitis may be associated with other atopic diseases such as **asthma** and type I **allergies**, including those in response to foods.

Desalination Removal of **salt**, e.g. desalination of **sea water**.

Desalting Removal of **salt**.

Desaturases Includes EC 1.3.1.35 and members of subclass EC 1.14.99. These **enzymes** have a number of uses in the food industry, e.g. fatty acid desaturases introduce double bonds into fatty acyl chains and are useful for production of **polyunsaturated fatty acids**, genetic modification of desaturases in **plants** and **microorganisms** can be used to modify contents of **fatty acids**, and cholesterol desaturase can be used to reduce the **cholesterol** content of foods.

Desaturation Process by which a substance is made less saturated. In the case of **organic compounds**, e.g. **fatty acids**, this involves removal of hydrogen atoms from adjacent carbon atoms, thereby forming double bonds and increasing the degree of **unsatura-**

tion. Such reactions are catalysed by **desaturases**. In the food industry, introduction of double bonds into fatty acyl chains in this way is useful for production of **polyunsaturated fatty acids**, intake of which can have beneficial effects for risk of **cardiovascular diseases** development.

Descaling Removal of deposits of scale from an item, particularly removal of limescale from heating elements in kettles and boilers. For removal of fish scales, the alternative term **scaling** is used.

Desiccated coconut Product prepared from coconut endosperm by shredding and drying. Used in manufacture of **sugar confectionery** and **bakery products**.

Desiccation Alternative tem for **drying**.

Designer foods **Functional foods** targeted towards a certain purpose such as the prevention of certain diseases, or provision of tailored health benefits.

Desmosterol Member of the **sterols** group, found in a variety of animal and plant foods including **goat milk**, **sea urchins** and wild **palm oils**. It has also been detected in **human milk**.

Desmutagenicity Specific type of **antimutagenicity** relating to the ability of a chemical to counteract the **mutagenicity** of another chemical. This attribute has been demonstrated for several foods or isolated food components, and contributes to their associated health benefits. Foods and components displaying this property include tea **polyphenols**, extracts of **seaweeds**, **cheese** and **fermented milk**. Some **microorganisms** used in food fermentations have also been shown to have desmutagenic activity, including **Bifidobacterium** species and some **lactic acid bacteria**.

Desorption Physical or chemical **sorption** process by which a substance (gas, liquid or solid) that has been adsorbed or absorbed by a liquid or solid material is removed from the material. Desorption isotherms of foods during **drying** are commonly studied to quantify reductions in moisture content. An O_2 adsorption-desorption process has been observed in **dough** during **breadmaking**. A thermal desorption step is used in analyte separation during GC analyses.

Desoxycholate Synonym for **deoxycholate**.

Dessert mixes Dried **instant foods** used to prepare **desserts**, typically by adding water or milk. Also called **pudding mixes**.

Desserts Sweet foods usually served as the last course of a meal. The term encompasses many different types of food, including dairy- and fruit-based products, cooked or raw. Available frozen, chilled or shelf-stable, as well as in the form of **dessert mixes**.

Popular desserts include **cheesecakes, mousses, gateaux**, fruit products and **ice cream** products.

Dessert wines Sweet **wines** of varying alcohol content usually drunk in small amounts as an accompaniment to the dessert course of a meal. May also refer to **fortified wines**.

Desulfitation Removal of salts of **sulfurous acid**, usually **sulfites**, and SO_2. Microbes can be used for desulfitation of **waste water** (effluent) from food factories. **Wines** for **distillation** can be desulfited using $CaCO_3$. **Musts** that are preserved by heavy sulfitation, and used for adjustment of sweetness of wines, require desulfitation before use. In the Brimstone **winemaking** system, clarified **grape juices** are preserved with high levels of SO_2 (1200-2000 mg/l) and then desulfited just before **fermentation**.

Desulfovibrio Genus of Gram negative, obligately anaerobic rod-shaped **bacteria**. Occur in fresh and salt water sediments, gastrointestinal tracts of animals and faeces. Capable of **sulfates** reduction.

Detergents **Surfactants**, such as soaps, used for **cleaning** purposes.

Detoxicants Substances which inactivate, neutralize, or render harmless **toxins** or poisons.

Detoxification Process of removing poisons or **toxins** (e.g. from foods), or process of inactivating, neutralizing or rendering harmless toxins or poisons. Can be effected by the use of solvents, chemical reactions, enzyme systems or microbial action.

Dewatering Process of removing excess water from a substance, e.g. after washing of a food. Used in processing of foods and in treatment of **wastes**. In the case of foods, water can be removed by various procedures including passing over vibrating screens, using specially designed rotary screens or **centrifugation**.

Dewaxing Process in which **solvents** are used to dissolve **waxes** from oil solutions. During the procedure, the wax solution is chilled and removed by **filtration**.

Dewberries Blackberry-like fruits produced by a number of *Rubus* species, including *R. caesius* in Europe, and *R. hispidus* or *R. canadensis* in America. Similar in appearance to **blackberries**, but smaller, with a slight whitish bloom.

Dextran Polysaccharide obtained from bacterial cell walls. Formed by **fermentation** of **sugars**. Used as a substitute for blood plasma and intravenously as a plasma expander under emergency conditions.

Dextranases EC 3.2.1.11. Catalyse the endohydrolysis of 1,6-α-D-glucosidic linkages in **dextran**, producing **isomaltose** and isomaltotriose. Useful in the sugar industry for degrading any contaminating dextran that may be present, which can interfere with **fil-**

tration and **clarification** of **sugar juices**. May also hydrolyse **starch, amyloses** and **amylopectins**.

Dextransucrases EC 2.4.1.5. Catalyse the synthesis of **dextran** from **sucrose**. Can also synthesize **oligosaccharides**, e.g. leucrose (a potential sweetener), in the presence of appropriate sugar acceptors, e.g. **maltose** (a strong acceptor) and **fructose** (a weak acceptor).

Dextrinases Alternative term for **α-dextrin endo-1,6-α-glucosidases**.

α-Dextrin endo-1,6-α-glucosidases EC 3.2.1.41. Hydrolyse 1,6-α-D-glucosidic linkages in **pullulan, amylopectins** and **glycogen**, and in the α- and β-amylase limit dextrans of amylopectins and glycogen. Also known as limit dextrinases, debranching enzymes, pullulanases and 6-glucanohydrolases.

Dextrins General term used for a range of water-soluble **polysaccharides** formed by partial hydrolysis of **starch**, including **maltodextrins** and **cyclodextrins**. Used for various applications in the food industry, such as prevention of **crystallization** or as thickeners. Their sticky consistency also makes them suitable for use as edible adhesives. Cold-water soluble dextrins are used as carriers for **flavourings** in products such as dry **mixes, soups** and gravy.

Dextrose Name given to the dextrorotary stereoisomer of **glucose** (D-glucose).

DFD defect Abbreviation for **dark firm dry defect** of **pork**.

Dhal Term used in two ways. In India, it is used to denote split **pulses** of a number of varieties, including **grass peas** and **lentils**. It also refers to a spicy dish based on lentils or other pulses that may be pureed and served with curries. Alternative spellings include dal, dahl and dhall.

Dhokla Popular **fermented foods** of India. Typically prepared by soaking **meal** from **chick peas** or other **legumes** in water with **buttermilk** or curds for several hours, seasoning with **ginger** and **chillies**, and steaming the batter. The steamed cake is cut into squares, garnished with grated coconut and **coriander** and served hot.

Diabetes Group of two diseases (diabetes mellitus and diabetes insipidus) of disparate pathology, both characterized by excessive urine production. Diabetes mellitus, the key feature of which is raised blood sugar levels or impaired glucose tolerance, is classified into two types: type 1, juvenile-onset or insulin-dependent diabetes; and type 2, maturity-onset or non-insulin dependent diabetes. Type 1 disease is a result of **insulin** deficiency and type 2 disease is due to insulin resistance. Control of blood sugar levels can be achieved by dietary manipulation in some cases, particularly in mild forms of type 2 disease, by reducing consumption of foods with high **glycaemic index values**. Diabetes insipidus is due, in general, to reduced ability of the kidney to concentrate urine, possibly caused by an impairment in the hypothalamus/antidiuretic hormone system.

Diabetic foods **Dietetic foods** manufactured specifically for individuals suffering from **diabetes**. Generally formulated to be low in absorbable **carbohydrates**, e.g. by replacing **sucrose** with **fructose, sorbitol** or other **sweeteners** that do not induce a large increase in blood glucose level.

Diacetoxyscirpenol Trichothecene produced by **Fusarium** species. Also known as anguidine.

Diacetyl Yellow, flammable liquid with a strong **aroma** and buttery **flavour** derived from **fermentation** of **glucose**. Soluble in water and alcohol. Used as an aroma carrier in foods and beverages.

Diacetyl tartaric acid esters of mono- and diglycerides **Emulsifiers** known by the acronym **DATEM**.

Diacylglycerols **Glycerides** composed of a molecule of glycerol bonded to two **fatty acids**. Possess **emulsifying capacity** and are used as **additives** in foods, including **shortenings**. Also known as **diglycerides**.

Diafiltration Extension of the **ultrafiltration** process in which water is added back to the extract during the concentration process. During diafiltration, both diffusive and convective mass transfer take place simultaneously as a result of two driving forces: a concentration gradient and a transmembrane pressure gradient. This is useful in selectively removing lower molecular weight materials from a mixture, and offers a useful alternative process to **ion exchange** or **electrodialysis** for removal of **anions, cations, sugars, alcohol** or **antinutritional factors**. Diafiltration is an accepted method for production of alcohol free, low calorie and **low alcohol beer**.

Diallyl disulfide Organic sulfur compound which is a major component of **garlic** and **garlic oils** and a major contributor to their **aroma**. In addition to its sensory properties, the compound also possesses health benefits including **antitumour activity** and protection against the risk of **cardiovascular diseases**.

Dialysis Separation of particles in a liquid on the basis of differences in their size and thus ability to pass through a membrane. **Membranes** are chosen that will allow small particles to pass through, but retain larger particles. The process can be used to remove unwanted particles and enrich or concentrate a solution.

Diamine oxidases Alternative term for **amine oxidases**.

Diarrhoea Disorder characterized by loose watery stools which are often evacuated at increased frequency. Diarrhoea may be an indicator of many diseases of the gastrointestinal tract, including **foodborne diseases**, **food poisoning**, **gastroenteritis**, **food intolerance**, colitis and **colorectal cancer**.

Diarrhoetic shellfish poisoning Food poisoning resulting from consumption of marine **bivalves** containing certain **toxins** produced by dinoflagellate **algae** (such as **okadaic acid**). Symptoms include nausea, intestinal pain, diarrhoea and memory loss.

Diarrhoetic shellfish toxins Toxins produced by certain marine dinoflagellate **algae** which are responsible for causing **diarrhoetic shellfish poisoning**. The most important of these toxins are dinophysistoxin-1, **okadaic acid** and derivatives of these compounds.

Diastases Alternative term for **α-amylases**.

Diastatic activity Total activity of **starch** degrading **enzymes** in grain malts. An important quality characteristic for **malting** and **brewing**.

Diatomaceous earths Powdery natural materials formed from the microscopic skeletons of diatoms, deposited in most cases during the Cenozoic era. Diatomaceous earth is fine in texture and grey or white in colour; when pure, diatomaceous earth is composed almost entirely of silicon dioxide or silica, but it is often found mixed with clay or organic matter. The material is used in **fining agents** and **filtration** materials in the food industry, among many other varied and wider fields of application.

Diatoxanthin One of the **carotenoids** detected in several types of **fish** and **shellfish** and also in brown **seaweeds**.

Diazepam Sedative drug that exhibits antihypertensive and myorelaxant properties. Normally used as a feed intake and growth promoting agent. Use to reduce **stress** in animals during transport to **slaughterhouses** is not permitted. Undergoes extensive and complex metabolism in animals.

Diazinon Non-systemic insecticide and acaricide used for control of sucking and chewing **insects** and **mites** on a wide range of fruits, vegetables, cereals, sugar cane, cocoa, coffee and tea; also used as veterinary ectoparasiticide. Classified by WHO as moderately toxic (WHO II).

Diazocyclopentadiene One of the **plant growth regulators**. A competitive inhibitor of **ethylene** that can be used to control ethylene-induced developmental responses in **fruits** and **vegetables**.

Dicamba Selective systemic herbicide used to control annual and perennial broad-leaved weeds and brush species in cereals, asparagus and sugar cane. Often used in combination with other herbicides. Classified by Environmental Protection Agency as slightly toxic (Environmental Protection Agency III). Also known as banvel.

Dichlofluanid Fungicide used for control of scab, brown rot and other fungal diseases in pome fruits, stone fruits and various vegetables; also has a suppressive effect on spider and rust **mites** on fruits. Classified by Environmental Protection Agency as slightly toxic (Environmental Protection Agency III). Also known as euparen.

Dichlorobenzene Insecticide and acaricide used primarily in beekeeping for control of the mite *Varroa jakobsoni*. Also a component of some **packaging materials**, but may migrate into foods and cause **taints**.

2,4-Dichlorophenoxyacetic acid Selective systemic herbicide used for post-emergence control of annual and perennial broad-leaved weeds in cereal crops, orchards, some vegetable crops and sugar cane. Classified by WHO as moderately toxic (WHO II). Also known as 2,4-D.

Dichlorprop Selective systemic herbicide used for post-emergence control of annual and perennial broad-leaved weeds in cereals. Also acts as a plant growth regulator (inhibits formation of abscission zone); used to prevent fruit fall in pome fruits. Classified by WHO as slightly toxic (WHO III).

Dichlorvos Organophosphorus insecticide and acaricide used for control of insect **pests** and **mites** in stored fruits, vegetables and cereals; also used as an anthelmintic in animals. Classified by WHO as highly hazardous (WHO Ib). Also known as vapona.

Dicing **Cutting** of materials, such as foods, into small cubes.

Dicloxacillin Semisynthetic penicillin antibiotic used to treat a range of bacterial infections in animals, particularly those caused by staphylococci.

Dicofol Non-systemic acaricide used for control of **mites** on a wide range of fruits and vegetables. Classified by WHO as slightly toxic (WHO III). Also known as kelthane.

Dieldrin Persistent organochlorine insecticide used for control of a wide range of insect **pests** in crops. Usage on **crops** has generally been replaced by less persistent **pesticides**.

Dielectric constant One of the **electrical properties**, describing the ability of a material to store electrostatic energy when a unit voltage is applied. Also known as relative permittivity. Dielectric constants have been used to determine changes in foods, such as moisture content changes in **sugar confectionery**, or degradation of **frying oils**, and also to monitor processing steps such as the use of **microwaves** in **thawing** and **cooking**.

Dielectric heating   Heating of electrically non-conducting materials, such as foods, by subjecting them to high frequency **electromagnetic fields**. The material to be heated is placed between two electrodes, to which a source of high-frequency energy is connected. In homogeneous materials, the resultant heating occurs throughout.

Dielectric properties   Electrical properties of dielectric materials, i.e. non-conducting materials which can sustain electric fields and act as insulators. These properties include the **dielectric constant**, dielectric relaxation and dielectric loss. Examples of their use in food analysis include assessment of the stability of **dough** during frozen storage, and comparison of the quality of **musts** from different cultivars of **winemaking grapes**.

Diet   Selection by individuals or population groups of **foods** and **beverages** for consumption. Dietary composition is the major factor affecting nutrition status and can have profound effects on health and risks for a range of diseases.

Dietary fibre   Complex mixture of **carbohydrates**, also known as non-starch polysaccharides, derived from plant cell walls and resistant to digestion in the intestinal tract. Little nutritional value in its own right, but considered to have beneficial effects on health. High-fibre diets can help control obesity and constipation, reduce the risk of cancer development and lower blood cholesterol. Fibre-rich foods include wholegrain products, **wholemeal** cereal products, **fruits** and **vegetables**.

Dietary reference values   Usually abbreviated to **DRV**. Set of UK standards detailing the amounts of each nutrient needed to maintain good health. In the case of most **nutrients**, the measured average need plus 20% is satisfactory for the requirements of the majority of the UK population; this is termed the Reference Nutrient Intake.

Dietary supplements   Alternative term for specific types of **food supplements** usually taken in tablet or capsule form as a supplement to the normal diet, with the aim of increasing an individual's intake of a specific **nutrients**, e.g. **vitamins** or **minerals**.

Dietetic foods   Products intended for consumption by individuals with **metabolic disorders** or **allergies**, such as **diabetic foods** or **gluten low foods**. Also used to refer to foods providing specific nutritional benefits to healthy individuals with particular dietary requirements, such as infants or athletes.

Diethylamine   Amine, which exists as a colourless, highly flammable, toxic liquid with an **aroma** of **ammonia**. Miscible with water, alcohol and most organic solvents. Uses include in **pesticides**, resins, polymerization inhibitors, rubber chemicals, pharmaceuticals, electroplating and corrosion inhibitors.

Diethylene glycol   Glycol, which exists as a colourless, viscous, combustible, extremely hygroscopic, non-corrosive liquid. Almost odourless, but has a sweetish **flavour**. Miscible with water, acetone, ether and ethylene glycol, but does not mix with benzene or toluene. When added to water, it lowers the **freezing point**. Used in the manufacture of **corks**, polyurethane and unsaturated polyester resins, **plasticizers**, **surfactants**, dyes, textiles and paper products. Also used in antifreeze solutions and in **humectants** for **casein**.

Diethylnitrosamine   One of the volatile **nitrosamines** with mutagenic activity, synonym *N*-**nitrosodiethylamine**. Occurs predominantly in **meat**, but also detected in other foods, including **cheese** and **fermented foods**. Synthesis has been associated with addition of **nitrates** and **nitrites** to foods during processing.

Diethylpyrocarbonate   Preservative used to prevent the growth of **microorganisms** in wines, other **alcoholic beverages** and **soft drinks**. Also a **histidine** modifying reagent.

Diethylstilboestrol   Synthetic, non-steroidal anabolic agent based on **oestrogens**; currently banned worldwide for use in animals produced for food (has **genotoxicity** and potential **carcinogenicity**). Previously used widely as a growth-promoting agent, principally in cattle but also in sheep and swine.

Differential scanning calorimetry   Technique in which a sample and thermally inert reference material at the same temperature are heated using a temperature programme and the rate of heat flow is measured independently for each. The differential heat flow is monitored as a function of temperature. Can be used to measure heat capacity. Usually abbreviated to DSC.

Differential thermal analysis   Technique in which the difference in temperature between the sample and a reference is measured as heat is applied to the system.

Diffraction   Generally used to describe changes in the direction of waves caused by obstacles. Used specifically in terms of **optical properties** to describe the bending of light when it passes through an obstruction. X-ray diffraction patterns are used to analyse the structure of crystals, including **proteins**, **carbohydrates** and **nucleic acids**. Laser diffraction can be used to analyse the size distribution of particles. White light, electron and neutron diffraction patterns have also been determined during the analysis of foods.

Diffusers Devices assisting in the travel or spread of gas or liquid by **diffusion**.

Diffusion Spontaneous and random movement of molecules or particles in a fluid (gas or liquid) from a region of high concentration to a region of low concentration. Once a uniform concentration (or dynamic equilibrium) is achieved, net diffusion ceases and motion is random throughout the fluid. Diffusion rates are widely used in food analyses, and two common examples include moisture diffusion, which is routinely determined in foods during **drying**, and salt diffusion, which will affect the **curing** rate of foods.

Diffusivity Measure of the ability of a substance to diffuse. Includes **thermal diffusivity**, which describes the diffusion of heat through a material.

Diflubenzuron Selective, non-systemic insecticide used for control of a wide range of leaf eating **insects** and their larvae in fruits and vegetables; also used as an ectoparasiticide on sheep. Classified by Environmental Protection Agency as slightly toxic (Environmental Protection Agency III)

Digestibility Nutrition term relating to the proportion of a food absorbed from the digestive tract into the bloodstream. True digestibility is measured as the difference between intake and faecal output, with allowance being made for that part of the faeces that is not derived from undigested food residues. Apparent digestibility is an approximate measure, which is simply the difference between intake and output.

Digestion Human physiology term relating to the breakdown of large polymeric molecules into their monomeric constituents, achieved chemically or enzymically. In particular, the term is applied to the breakdown by digestive enzymes of complex food molecules, e.g. proteins to amino acids, starch to glucose, fats to glycerol and fatty acids, so that they may be absorbed through the gut lining. Digestion can include the mechanical processes, such as chewing, churning and grinding of food, as well as the chemical action of digestive **enzymes** and other substances such as bile. Chemical digestion begins in the mouth with the action of saliva on food, but most takes place in the stomach and small intestine, where the food is subjected to gastric juices, pancreatic juices and succus entericus.

Digitonin Saponin derived from foxglove (*Digitalis purpurea*) seeds. Unlike digitoxin, the major glycoside obtained from the foxglove, it has no apparent effect on the heart. Used as a reagent in analytical techniques to determine levels of free **cholesterol**.

Diglucosides Compounds which include two molecules of **glucose**.

Diglycerides Glycerides composed of a molecule of glycerol bonded to two **fatty acids**. Possess **emulsifying capacity** and are used as **additives** in foods, including **shortenings**. Also known as **diacylglycerols**.

Dihydrochalcones Class of minor **flavonoids** mainly found in **apples** and apple products such as **cider**. Biochemically related to **flavanones** and **chalcones**. **Neohesperidin dihydrochalcone** is used as a sweetener.

Dihydrostreptomycin Aminoglycoside antibiotic active mainly against **Gram negative bacteria**. Used for treatment of enteric infections in animals and **mastitis** in cows; also used as a topical treatment.

Dihydroxyacetone Ketone which exists as a colourless, hygroscopic, crystalline solid. Soluble in water and alcohol. Used in **emulsifiers, humectants, plasticizers** and **fungicides**.

3,4-Dihydroxyphenylalanine One of the antinutrients of the **amino acids** group. It occurs in **faba beans** (*Vicia faba*) and in some underutilized **legumes** (including *Mucuna pruriens* var. *utilis* and *Cassia hirsuta*). Abbreviated to L-DOPA.

Dika nuts Seeds of *Irvingia gaboensis* (also called wild mango or African mango) or *I. barteri* (also called dika bread and gaboon chocolate). Source of fat (dika butter), and **hydrocolloids** that are used as **thickeners** in foods. Seeds from *I. gaboensis* (**African mango seeds**) are ground to a paste and mixed with spices to make dika bread, a staple food in some African regions.

Diketones Ketones with two carbonyl groups.

Dilatometry Measurement of thermal expansion or dilation of solids or liquids.

Dill Common name for the umbelliferous aromatic herb *Anethum graveolens* cultivated for its aromatic seeds and leaves (dill weed). Used in **flavourings** for products such as **pickles, bread, dressings**. **Essential oils** obtained from leaves and seeds can be used to add **flavour** to pickles, **confectionery** and **chewing gums**.

Dill ether Monoterpene ether ((3R, 4S, 8S)-3,9-epoxy-l-p-menthene) found in **essential oils** extracted from **dill** leaves. Organoleptically considered the most important constituent of dill oils.

Dilution Making a solution less concentrated by adding water or another solvent.

Dimethoate Systemic organophosphorus insecticide and acaricide used for control of a wide range of **insects** and **mites** in fruits, vegetables, cereals, tea and coffee; also used for control of **flies** in animal rearing facilities. May cause **russeting** in some apple varieties. Classified by WHO as moderately toxic (WHO II).

Dimethylamine Amine, which exists as a flammable, anhydrous gas with the **aroma** of **ammonia**. Soluble in alcohol and ether. Uses include the manufacture of solvents, antioxidants, dyes, pharmaceuticals, and acid gas absorbents.

N-Dimethylaminosuccinamic acid Alternative term for **daminozide**.

Dimethylarsinic acid One of the arsenical **herbicides**. Commonly included as one of the organic arsenic species determined as contaminants in foods, particularly **sea foods**. Also known as cacodylic acid.

Dimethyl disulfide One of the volatile **organic sulfur compounds**, and a characteristic flavour and aroma component of many foods and **beverages**, including **mussels**, fermented soy products, **cheese**, **whiskey** and **Brassica** species such as **broccoli**. Also occurs as an **off flavour** compound in **skim milk**.

2,5-Dimethyl-4-hydroxy-3(2*H*)-furanone Chemical name for the flavour compound **furaneol**.

Dimethylnitrosamine One of the volatile **nitrosamines**, which possesses carcinogenic activity. Has been detected in a range of foods, including **cured meat**, **fried foods**, **malt** and **beer**.

Dimethylpolysiloxane An antifoaming agent added to **oils** and **fats** to prevent spattering and foaming during **heating**. Also used to prevent formation of **foams** during other food and beverage processing applications, including **winemaking**, **sugar processes**, and manufacture of fruit and dietetic **soft drinks**.

Dimethyl sulfide Synonym for **methyl sulfide**. Organic sulfur compound, in the form of a colourless liquid, which is commonly used as a solvent. Also occurs naturally in foods and **beverages**, generally as an **off odour** from bacterial metabolism of sulfur-containing **amino acids**.

Dimethyl sulfoxide Organic sulfur compound which exists as a colourless, combustible, hygroscopic liquid. Almost odourless, but has a slightly bitter **flavour**. Soluble in water and alcohol. Used in industrial cleaners, pesticides, paint strippers, human and veterinary medicines, pharmaceutical products, solvents for polymerization and cyanide reactions, and in spinning of synthetic fibres, e.g. polyacrylonitrile.

Dimetridazole Coccidiostat traditionally used for treatment and prevention of histomoniasis in turkeys and chickens, trichomoniasis in cattle, and dysentry in swine. A suspected carcinogen; use in food animals has been banned in the European Community since 1995.

Dim sum Traditional Chinese dish consisting of small portions of different foods, including steamed or fried **dumplings** with various **fillings**.

Dinners Term usually applied to the main meal of the day, served in the evening or at midday. May also refer to frozen and chilled **convenience foods** that comprise a whole meal, such as TV dinners.

Dinoflagellates A group of microscopic (between 20 and 150 m long), generally single-celled organisms (commonly regarded as **algae**); characterized by two flagella that impart a distinctive spiral swimming motion. Abundant in both fresh- and marine waters. Some dinoflagellates produce water-soluble or lipid-soluble small molecular weight compounds (**dinoflagellate toxins**) toxic to humans and other vertebrates.

Dinoflagellate toxins Toxins produced by marine dinoflagellate **algae** which can accumulate in filter feeding **bivalves** and **fish**; consumption of **sea foods** containing these toxins can cause various types of **food poisoning**.

Diols Alcohols which include two hydroxyl groups.

Dioscorin Major storage protein of **yams** (*Dioscorea batatas* Decne and *D. cayenensis*). Possesses **radical scavenging activity**, indicating health benefits for people consuming yam tubers.

Dioxins Polychlorinated hydrocarbons which are very persistent environmental contaminants. Released into the environment as unwanted by-products of manufacturing processes (e.g. manufacture of industrial chemicals and during combustion and incineration processes). Many are carcinogenic, teratogenic and mutagenic. May contaminate food, especially **dairy products**, **meat**, **fish** and **shellfish**.

Dioxygenases Members of EC 1.13 and EC 1.14. **Oxidoreductases** that incorporate two oxygen atoms from O_2 into the compound(s) oxidized.

Dipeptidases EC 3.4.13-EC 3.4.15. **Peptidases** that cleave the peptide bond in **dipeptides**, either specifically or non-specifically (EC 3.4.13). Dipeptidyl-peptidases and tripeptidyl-peptidases (EC 3.4.14) release di- and tri-peptides, respectively, from the N-terminal ends of polypeptide chains, while peptidyl-dipeptidases (EC 3.4.15) release dipeptides from the C-terminus of polypeptide chains. Certain dipeptidases are important for **flavour** development in fermented **meat** and **dairy products**.

Dipeptides Peptides consisting of two amino acid residues.

Dipeptide sweeteners Sweeteners based on dipeptides or their derivatives. Usually more sweet, more stable and lower in calories than conventional sweeteners.

Diphenol oxidases Alternative term for **catechol oxidases**.

Diphenyl Alternative term for the fungicide **biphenyl**.

Diphenylamine Fungicide used as a post-harvest protectant and scald inhibitor on **pome fruits**.

Diphyllobothrium Genus of parasitic tapeworms of the class Cestoda. Occurs in the gastrointestinal tracts of fish, birds, humans and animals. Infection in humans usually occurs through eating raw or undercooked **fish** which is contaminated with the larvae of *Diphyllobothrium latum*.

Dipicolinic acid Substance that occurs as a calcium salt in **bacterial spores**, and that may play a role in increasing the **heat resistance** of spores.

Diplazium esculentum Green fern, the young leaves of which are eaten as a vegetable mainly in India and Indonesia.

Diplococcus Obsolete bacterial genus which included species currently assigned to various other genera.

Diplodia Genus of fungi of the class Dothideomycetes. *Diplodia natalensis* is responsible for stem end rot of **citrus fruits**.

Dipping Process of submerging a food into **sauces** (e.g. **dips**) or **coatings** (e.g. **batters**). Chemical or hot water dipping treatments are also used to decontaminate foods.

Dips Sweet or savoury **sauces** into which accompanying foods (e.g. breadsticks, **crisps**, vegetable crudites) are dipped. Many savoury dips are based on **sour cream**, **cream cheese** or **mayonnaise**. The term may also be applied to chemical and **hot water dips** used to decontaminate foods.

Dipyridyl Organic nitrogen compound formed from **pyridine**. Exists as two **isomers**. Chelating agent able to bind iron.

Diquat Non-selective contact herbicide used for control of broad-leaved weeds in **fruits** and **vegetables**. Also used for pre-harvest desiccation of **cereals**, **legumes** and **sugar beets** and for inhibition of tassle formation in **sugar cane**. Classified by WHO as moderately toxic (WHO II).

Disaccharides **Sugars**, e.g. **maltose**, **sucrose** or **lactose**, which consist of two linked monosaccharide molecules. Dietary source of carbohydrate. Some individuals show disaccharide intolerance, e.g. **lactose intolerance**, and are unable to absorb disaccharides due to an enzyme deficiency.

Discoloration Alteration or **spoilage** of the **colour** of an item.

Diseases Abnormalities of the structure or physiological function of an organism which are regarded as being detrimental to its health.

Dishwashers Kitchen appliances that automatically wash, rinse and dry crockery, cutlery, pans and other utensils.

Disinfectants Chemical agents used for **disinfection**, including **quaternary ammonium compounds**, alcohols, phenols, halogens (chlorine and iodine), halogen compounds, and mercury compounds.

Disinfection Destruction, inactivation or removal of **pathogens** or **spoilage microorganisms**. Commonly refers to the use of **disinfectants** for the treatment of inanimate objects and surfaces (e.g. surfaces in food processing plants and kitchens).

Disinfection by-products By-products of **drinking water** disinfection. **Trihalomethanes** are associated with **chlorination**, while **chlorites** and **chlorates** are associated with **chlorine dioxide** disinfection. **Ozonation** may cause formation of bromates. May be responsible for an increased risk of kidney and bladder cancer in humans and other long term health effects.

Disinfestation Destruction of insect **pests** and other **parasites** of animals or plants. Generally involves the use of **insecticides**, applied either topically or as a spray.

Dispensers Devices that supply or release a product, such as foods and **beverages**, by **dispensing**.

Dispensing The process of supply or release of a product, such as foods and **beverages**, sometimes from special devices (**dispensers**).

Dispersibility Measure of the ability of materials to form **dispersions**, in which one substance is suspended in a second material. Often determined for **dried foods** or ingredients such as powders to illustrate how well they can be rehydrated.

Dispersions Two-phase systems consisting of particles (the disperse phase) suspended in a second substance (the continuous or bulk phase) which generally present in relative excess. Includes **colloids**, **emulsions** and **aerosols**.

Display cabinets Units in which items, including **foods**, are displayed in an appealing manner. Food should be displayed such that its quality is maintained (e.g. lighting and temperature are optimum), and so that it is protected from **contamination** and is attractive to potential customers.

Distillates **Spirits** or their intermediate products manufactured from ethanol-containing mashes or other materials by **distillation**.

Distillation Process for manufacture of **spirits** by condensation of vapour (containing **ethanol** and fla-

vour compounds) liberated from ethanol-containing **mashes** during heating in a still.

Distilleries Factories used for manufacture of **spirits** by **distillation**.

Distilleries effluents **Waste water** produced by **distilleries** during processing.

Distillers grains Alternative term for **distillers spent grains**.

Distillers spent grains Waste product from **distilleries** where **cereals** are used as the raw materials, comprising grain solids remaining after extraction of soluble material in the **mashing** process.

Distillers yeasts Yeasts (*Saccharomyces* spp.) used for **fermentation** of **mashes** to be distilled in manufacture of **spirits**.

Distribution The physical movement of commodities, including **foods**, into the channels of trade and industry. Can involve distributors, wholesalers, retailers, dealers and agents.

Disulfides **Sulfides** which contain two atoms of sulfur.

Diterpenes **Terpenoids** which include four isoprene units and thus contain 20 carbon atoms and 4 branched-methyl groups. Occur in foods, e.g. **coffee beans**, **marjoram** and **rosemary**. Those in rosemary, carnosol and carnosic acid, have **antioxidative activity** in foods. **Coffee** diterpenes show **anticarcinogenicity** in animal studies, but may have hypercholesterolaemic effects in humans.

Dittany Common name for *Origanum dictamnus*, a herb native to Crete. Also known as dittany of Crete. Used as a substitute for **oregano** or **marjoram** and in some Mediterranean dishes. Flowers are used to make **herb tea**. Extracts display high **antioxidative activity**, while **essential oils** have **antimicrobial activity**.

Diuron Systemic herbicide which inhibits photosynthesis. Used for selective control of germinating grasses and broad-leaved weeds in **fruits**, **cereals** and **legumes**. Classified by Environmental Protection Agency as slightly toxic (Environmental Protection Agency III).

Diverticulosis Disease of the large intestine, particularly the distal portion, which is prevalent in older individuals. The wall of the colon forms blind out pockets or diverticulae which can become inflamed (diverticulitis) resulting in acute abdominal symptoms, such as pain, and potentially in severe complications such as peritonitis. Reduced risk for diverticulosis has been associated with increased consumption of **fruits** and **vegetables** and **dietary fibre**.

Djenkol beans Seeds produced by *Pithecellobium lobata*. Contain djenkolic acid, a toxic sulfur-containing amino acid that causes kidney disorders.

DM Abbreviation for **dry matter**.

DNA Abbreviation for deoxyribonucleic acid, a nucleic acid consisting of linked deoxyribonucleotides, each of which contains one of four nitrogenous bases (**adenine**, **thymine**, **cytosine** and guanine), a phosphate group and the pentose sugar deoxyribose. DNA is the genetic material of most organisms and usually exists as a double-stranded molecule in which two antiparallel strands are held together by hydrogen bonds between adenine and thymine and between guanine and cytosine.

DNA-directed DNA polymerases EC 2.7.7.7. Catalyse synthesis of **DNA** from deoxyribonucleoside triphosphates in the presence of a nucleic acid primer. Grouped into two families on the basis of structural similarities, family A (largely bacterial polymerases) and family B (generally polymerases from higher **eukaryotes**). Required for DNA replication and repair; possess 3'-exonuclease activity. Used in **PCR** analysis of **genes**, which has a number of applications in food science. Also known as DNA polymerases and DNA nucleotidyltransferases (DNA-directed).

DNA-directed RNA polymerases EC 2.7.7.6. En-zymes which utilize ATP, GTP, CTP and UTP to synthesize **RNA** from a **DNA** or RNA template. Include enzymes responsible for **transcription** of DNA and primases which synthesize the RNA primers required for replication of DNA. Also known as RNA polymerases.

DNA fingerprinting Technique that allows discrimination at the genomic level, utilizing the fact that the genetic material of an individual is unique. DNA sequences from regions of the genome of related organisms are isolated and analysed either by **Southern blotting** or **PCR** in order to determine the organism's unique DNA fingerprint. Can be used to differentiate between different species, strains or cultivars of **bacteria**, **yeasts**, **fungi** and **plants**.

DNA hybridization Formation of double-stranded **DNA** or DNA/**RNA** sequences by base-pairing between complementary single-stranded sequences. Can be carried out in solution or with one component immobilized on a matrix (e.g. nitrocellulose). The latter is known as **Southern blotting**. Hybridization can also be performed *in situ* using fluorescently-labelled DNA molecules to localize **genes** to specific **chromosomes**.

DNA polymerases Alternative term for **DNA-directed DNA polymerases**.

DNA probes DNA or RNA sequences that have been labelled with radioactive isotopes, dyes or enzymes and are used to detect complementary sequences in DNA or RNA molecules by hybridization.

DNases Alternative term for **deoxyribonucleases**.

Docosahexaenoic acid One of the ω-3 or *n*-3 **polyunsaturated fatty acids** (PUFA), with 22 carbon atoms and 6 double bonds. Only the (all-*Z*)-4,7,10,13,16,19-isomer occurs naturally, and is found principally in **fish oils**. Suggested health benefits associated with docosahexaenoic acid and its related *n*-3 PUFA **eicosapentaenoic acid** include reduced risks of **coronary heart diseases** and cancer, and improved **immune response** and neural development in infants.

Docosapentaenoic acid One of the ω-3 **polyunsaturated fatty acids**, containing 22 carbon atoms and 5 double bonds. Rich dietary sources include **fish oils**, especially **herring** oils, and **cattle livers**. Important for the development of the central nervous system. Consumption also gives protection against **coronary heart diseases**.

Docosenoic acid Monounsaturated fatty acid, which exists as a combustible solid with low toxicity. Insoluble in water, but soluble in alcohol and ether. Occurs naturally as a minor component of many plant seeds and is obtained from plant seed oils, particularly hydrogenated **mustard seed oils** and **rapeseed oils**. Used for the manufacture of **waxes**, **plasticizers**, water-resistant **nylon** and **stabilizers** and as an additive in **polyethylene films**. Also known as **erucic acid**.

Dodecanoic acid Fatty acid which exists as a colourless, combustible solid. Occurs naturally as a glyceride in many **vegetable fats** and as a flavour compound in various foods, including **honeys**, **guavas** and **krill**. Insoluble in water, but soluble in ether and benzene. Uses include in alkyl resins, detergents, **food additives**, **insecticides** and wetting agents. Also known as **lauric acid**.

Doenjang Fermented soybean **pastes**, used as a base for many Korean dishes. Reported to have **antitumour activity** and **antimutagenicity**. Also known as doenzang or tenjan.

Doenzang Alternative term for **doenjang**.

Dogfish General name used for a number of small **sharks** belonging to three different families: Squalidae (spiny dogfish); Scyliorhinidae (catsharks); and Triakidae (smooth hounds). Several dogfish species are utilized as food fish, including *Squalus acanthius* (pickled dogfish), *S. blainvillei* (northern dogfish) and *Mustelus manazo* (smooth dogfish).

Dolphinfish Commercially important **marine fish** species (*Coryphaena hippunus*) belonging to the family Coryphaenidae. Widely distributed in tropical and sub-tropical water throughout the world, and also produced commercially by **aquaculture**. Marketed fresh and frozen. Also known as **mahimahi** or variations of this name, including mahi-mahi and mahi mahi.

Dolphins Marine mammals belonging to the order Cetecea; widely distributed around the world. Dolphins are not commercially exploited on a large scale; however, some species are utilized as a source of meat and oils.

Domiati cheese Egyptian brine ripened **cheese** made from **buffalo milk** or **cow milk**. It is consumed fresh or after **ripening** for three to six months. Sometimes called Damiati or Breda cheese.

Domoic acid Naturally-occurring amino acid found in some marine **algae**. Responsible for **amnesic shellfish poisoning** in humans when filter-feeding molluscan **shellfish** (e.g. **clams**, **mussels**, **scallops** and **oysters**) which feed on the algae are consumed.

Doner kebabs Turkish meat products traditionally prepared from spiced lamb cooked on a spit. Doner kebabs may also be prepared from beef, veal, chicken meat or turkey meat, or meat mixtures. Slices of the spiced meat are usually served with slices of onion in **pita bread**.

Dongchimi Fermented radish root product popular in Korea.

DOPA Amino acid produced in the sympathetic nervous system and the adrenal gland by hydroxylation of **tyrosine**. Occurs in several types of **beans**, e.g. **velvet beans** (*Mucuna pruriens* L.), and can be made synthetically. DOPA is a precursor of dopamine and an intermediate product in the biosynthesis of adrenaline and noradrenaline. Abbreviated form of 3,4-dihydroxyphenylalanine.

Dopamine Important neurotransmitter in both the central and peripheral nervous systems, and metabolic precursor in adrenaline and noradrenaline biosynthesis. Also occurs in **bananas** where it is largely responsible for **enzymic browning**.

Dosa Alternative term for **dosai**.

Dosai Traditional Indian fermented **pancakes** made with **rice** and **black gram**. Also known as dosa.

Dosimeters Instruments used to measure doses of ionizing radiation such as X-rays.

Double cream **Cream** with a high fat content (approximately 48%).

Dough A thick, plastic mixture of **flour** and liquid (e.g. water or milk) that may contain **yeasts** or **baking powders** as leavening agents. May be shaped, kneaded, rolled and baked to make **bakery products**.

Dough conditioners Ingredients added to yeast **dough** to improve its processing characteristics and/or the quality of the finished **bakery products**.

Doughnuts Rounded **bakery products** made from rich, sweetened **dough** leavened with either **yeasts** or **baking powders** and deep fried. May be either ring-shaped with a hole in the centre or filled with **jams**, **whipped cream** or sweet pastes. Often coated with **sugar** or topped with **icings**.

Downstream processing The processing steps involved in **separation** and **purification** of the products of **fermentation** processes and **bioconversions**. Can be performed either simultaneously with the process or after its completion.

Doxycycline Semisynthetic tetracycline antibiotic used to treat a range of bacterial infections in cattle, swine, sheep, goats, poultry and farmed fish. Readily disperses throughout tissues, but excreted relatively slowly. Residues in **kidneys** and **livers** may remain for up to 14 days following withdrawal.

Dracunculiasis Infection transmitted through **drinking water** containing microcrustaceans of the genus *Cyclops*, which harbour infective larvae of the nematode parasite *Dracunculus medinensis* (guinea worm). Infection is initiated with liberation of the larvae in the stomach where they mature and reproduce. Fertilized female worms then migrate to the subcutaneous tissues, usually the extremities, where they form an ulcer. This is accompanied by intense pain, fever, nausea and vomiting.

Dragees Small hard **candy** pieces with hard **sugar** or sugared **chocolate coatings**.

Draught beer **Beer** which is dispensed from **barrels**, **kegs** or other bulk **containers**, rather than packaged in **bottles** or **cans**.

Dressing Post slaughter process, including the steps of skinning, **evisceration**, trimming and washing, that follows the stage in which animals are bled. The head, feet, hides (in the case of sheep carcasses and cattle carcasses), excess fat, viscera and **offal** (edible and inedible) are separated from the bones and **meat**. With automated dressing lines, a series of mechanical devices stun the animals, remove the pelt (first from the brisket, then completely), eviscerate the carcass and process the head. Other devices debone the loin and thoracic regions. With respect to dressed meat, **video image analysis** can be successfully applied to grading for speedy online determination of the fat/lean ratio, and fibre optic probes permit objective prediction of such textural defects in the meat as excessive paleness or darkness. The term can also be applied to the act of applying **coatings** to foods, such as **fish** and **salads**.

Dressings **Condiments** used to coat and add **flavour** to foods prior to consumption. Most common types are **salad dressings**.

Dried dairy products **Dairy products** dried to a low **moisture content**, giving powders with a long **shelf life**. Packaged in materials that are impermeable to **water vapour**, oxygen and light to protect them during storage.

Dried egg products Powders made by **drying eggs** or egg components. Include dried **egg whites**, dried **egg yolks** and dried whole eggs. Utilized in the manufacture of foods where fresh eggs would be used, such as **bakery products**, **bakery product mixes**, **mayonnaise**, **salad dressings** and egg **noodles**. Their long **shelf life** and *Salmonella*-free status make them ideal for use by food manufacturers and caterers.

Dried eggs **Eggs** which have been dehydrated, usually by **spray drying**, to form powders. Also called egg powders. May be used in a range of foods, including **bakery products**, **bakery product mixes**, **mayonnaise**, **salad dressings**, **confectionery**, **ice cream**, **pasta** and **convenience foods**. Their long **shelf life** and *Salmonella*-free status make them ideal for use by food manufacturers and caterers.

Dried figs **Figs** (**fruits** of *Ficus carica*) from which the majority of the water content has been removed, usually by sun **drying**. A rich source of **dietary fibre** and **iron**. Also contain high levels of other minerals, such as **calcium**, **potassium** and **magnesium**, and **polyphenols**, but are low in **sodium** and free of **fats** and **cholesterol**. Eaten as **snack foods**, mixed with **vegetables** or other fruits, or used as ingredients of **bakery products**, **meat** dishes or **fish** dishes. **Purees** prepared from dried figs are used as **fat substitutes** or **sweeteners**.

Dried fish **Fish** subjected to **drying** processes which remove sufficient moisture to inhibit the growth of **microorganisms**, resulting in increased storage life. **Air drying**, sun drying and **freeze drying** are common processes for obtaining dried fish products. Many fish are marketed in dried form.

Dried foods Foods in which the majority of water present has been removed by **drying**, resulting in lighter weight products of extended **shelf life**, e.g. **dried eggs**, **dried fruits**, **dried milk**, mixes and powders. Sometimes rehydrated before consumption, although some, such as dried fruits, are consumed in their dried state. Rehydration properties are affected by the type of drying process used. Also known as dehydrated foods.

Dried fruits Fruits preserved by **drying** (final moisture usually less than 25%). **Sweetness**, and **flavour** in general, are concentrated by the drying process, but nutrients, especially **vitamins**, can be lost. SO_2 spraying can be used before drying to preserve **colour** and nutrients. Eaten out of hand or used in **cooking**. Can be used dried or reconstituted in liquids such as water and **alcoholic beverages**.

Dried meat Meat preserved by **drying**, a process that reduces water activity and so limits bacterial activity and enzymic changes. Traditional drying methods for meat include sun drying, air drying, oven drying and dry curing. A high surface to volume ratio is needed to allow effective drying of meat. Drying produces changes in nutritional value, particularly the vitamin content, and sensory properties of the meat. Today, only a small proportion of meat is preserved by drying. It is mainly prepared when a light weight, high protein product with a good shelf life is required. Some speciality products are, however, prepared. For example, dried beef prepared by dry curing or sweet pickling followed by air drying is an expensive product, usually prepared from very lean beef. It may be sold by the piece or pre-sliced; it is frequently used for hors d'oeuvres.

Dried meat products **Dried foods** produced from **meat**. They include: **pemmican**, produced by sun drying strips of lean meat; and **biltong** and **charqui**, both produced by a combination of brining and air drying. Dried meat products differ considerably from fresh meat and are generally of lower **eating quality**.

Dried milk **Whole milk** dried to a low **moisture content**, giving a powder with a long **shelf life**. Also called **milk powders**.

Dried peas Peas preserved by **drying**. Reconstituted in water before cooking as a vegetable, stir-fried or added to dishes including **soups**, stews and **sauces**.

Dried skim milk **Skim milk** dried to a low **moisture content**, giving a powder with a long **shelf life**. Also called **skim milk powders** and non-fat dried milk.

Dried vegetables Vegetables preserved by **drying**. Commonly used types include **peas**, **carrots**, **peppers** and **onions**. Often reconstituted in water before use or added to dishes such as **soups** and stews.

Dried whey Whey dried to a low **moisture content**, giving a powder with a long **shelf life**. Also called **whey powders**.

Dried yeasts Active **yeasts** preserved by **drying** for ease of handling, transport and storage. Used in **baking**, **brewing** and **winemaking**, and as ingredients of **soups**, **health foods**, **sauces** and **gravy**.

Driers Machines or devices for drying items such as foods. Alternative spelling is dryers.

Drinking chocolate **Chocolate** preparations which are mixed with hot water or **milk** to form **chocolate beverages**. In addition to chocolate, may contain other ingredients such as milk solids, **sugar** and

Drinking straws Hollow tubes, generally made of **plastics** or **paper**, through which **beverages** or liquid foods are sucked into the mouth.

Drinking water **Water** that is suitable for drinking, particularly in terms of its purity, and sensory and hygienic qualities.

Drinking yoghurt Yoghurt with a viscous **consistency** rather than a set curd, prepared by stirring during cooling to 7-8°C before packaging.

Drinks Alternative term for **beverages**.

Drip The liquid that is lost when foods, e.g. **fish** and **meat**, that have been frozen are thawed.

Dripping Unprocessed fat originating from lipid-rich tissues or bones of sheep or cattle. Also the rendered fat produced from roasted meat. Used in **cooking** or as a spread.

Dropsy Alternative term for **oedema**.

Drugs Chemical substances which affect the functioning of living things and the organisms (such as **bacteria**, **fungi** and protozoa) that infect them. Predominant application relevant to the food industry is in animal husbandry, where they are used to cure or prevent diseases in animals, to increase feed efficiency and/or growth rate, and to sedate animals in order to minimize the effects of **stress**. Major classes include **antibiotics**, **anthelmintics**, **anabolic agents** and **barbiturates**. Potential presence of drug residues in animal foods represents a health hazard to consumers.

Drum Alternative term for **croakers**.

Drum drying A process in which **drying** is undertaken continuously on the external surface of an internally steam heated rotating cylinder. A thin film of the product to be dried is applied at one location and removed at another, usually after less than one complete revolution of the cylinder. These **driers** may be atmospheric or vacuum types, and are classified as single drum, double drum, or twin drum (in which the two drums function almost as single drums).

Drums Cylindrical **containers** used for storage and transportation of liquids.

DRV Abbreviation for **dietary reference values**.

Dry beans Type of **common beans** (*Phaseolus vulgaris*).

Dry cured ham Ham which is cured by rubbing **curing agents** in dry form over the surface. Some are cooked after curing, e.g. York ham, whilst others are dried and eaten raw, e.g. Parma ham. For large hams, the curing agents must be applied several times during the curing period. Costs of producing dry cured ham tend to be high because dry curing is slow and requires large amounts of hand labour. However, dry curing of ham can be accelerated through production techniques such as tumbling, blade tenderizing, microbial inoculation, use of nitric oxide, and processing as skinned and/or boneless legs.

Dryers Alternative spelling of **driers**.

Dry ham Raw ham that is dry cured and then dried, either by air **drying** or mechanical means. It is a highly valued speciality product. Some dry ham is smoked. The ham is soft in **texture** and when freshly sliced is pink or red in **colour**. It has a high content of **salt**. Factors affecting quality of dry ham include: genetic type, age, weight, sex, feeding and slaughter of the swine; pH value and water holding capacity of the raw ham before drying; and composition, particularly lipid and protein contents, of the raw ham before drying. Known as prosciutto crudo in Italy and rohschinken in Germany. Varieties include Corsican, Bayonne, Parma, Italian country, Serrano and Iberian hams.

Drying Removal of moisture or liquid from an item to a level of <5%, a process also known as dehydration. Three basic methods of drying are applied to foods: sun drying, a traditional method in which foods dry naturally in the sun; hot **air drying**, in which foods are exposed to a blast of hot air; and **freeze drying**, in which frozen food is placed in a vacuum chamber to draw out the water.

Dry matter Measure of the proportion of a material remaining after the removal of water, also referred to as dry weight. Commonly abbreviated to DM.

Dryness **Sensory properties** relating to the extent to which a product is perceived as being dry.

Dry sausages **Sausages** which are dried during preparation. Often the sausages are hung in a drip room for 2-10 days at 21-27°C before they are transferred to a dry room, which has lower temperature and relative humidity levels, for a further 10-120 days. Natural casings are often used as they shrink and thereby remain in close contact with the surface of the sausage as it loses moisture.

DSC Abbreviation for **differential scanning calorimetry**.

Duck eggs Eggs produced by **ducks**. Consist of approximately 13% protein and 14.8% lipids, and have a mean weight of 70 g. **Egg shells** may be a variety of colours (e.g. white, bluish, greenish, cream, light brown) with speckled or mottled patterns.

Duck livers **Livers** from ducks; part of edible **offal**. Duck livers are cooked by **sauteing**, **frying** or **grilling**, or are used to make **pates** or mousses. In France, the livers of specifically fattened ducks are used to prepare **foie gras**.

Duck meat Meat from **ducks**. Duck **carcasses** have higher fat contents, thicker skin and contain a lower proportion of meat than other poultry carcasses; however, duck meat has a very rich **flavour**. It has a higher **collagen** content and is darker in **colour** than **chicken meat**. Compared with farmed duck meat, wild duck meat has a lower content of fat and a different fatty acid profile.

Ducks The common name given to various domesticated and wild, small water fowl of the family Anatidae; there are many species. Many kinds of ducks are domesticated and are reared for production of **duck meat** and/or **duck eggs**. Wild ducks are hunted for their meat. Different gender and age groups of ducks are known as drakes (adult entire males), ducks (adult females) and ducklings (in general, sexually immature young birds with down rather than feathers).

Dudh churpi Traditional Indian shelf stable dairy product made from partially defatted **yak milk**, **cow milk** or **milk** from crosses between the two animals. Milk is coagulated using acid and heat, and the **curd** is cooked to remove moisture, cut into pieces and dried. Partially dried product (prechurpi) is cooked or dipped in a milk-**sugar** solution, and dried. The final product is chewed. It is a rich source of energy, proteins and minerals.

Dulce de leche Milk-based **confectionery** products popular in Latin America, particularly Argentina. Prepared by condensing a mixture of **milk** and **sugar** to a syrup that is then slightly caramelized by heating and flavoured with **vanilla**.

Dulche de leche Alternative term for **dulce de leche**.

Dulcin Non-nutritive artificial sweetener (4-ethoxyphenylurea), approximately 200-250 times sweeter than **sucrose**. Most countries have banned its use in foods, due to concerns about **toxicity**.

Dulcitol Synonym for **galactitol**. Polyol comprising 6 carbon atoms, produced by **isomerization** of **sorbitol**. Has approximately 0.1 times the **sweetness** of **sucrose**. Present in dulcite (Madagascan manna, *Melampyrum nemorosum*).

Dulse Marine red **algae** (*Rhodymenia palmata*) found along shores of the north Atlantic and northwest Pacific. Eaten as a delicacy in dried form. Also used in **flavourings** for stews and **soups** and in food **thickeners**.

Dumplings Small balls of leavened **dough**, formed from **flour** or meal bound with egg, which are boiled, steamed or baked. Frequently cooked in and served with **soups** and stews. Dessert dumplings are made with sweet dough stuffed with **fruits** and served with **sauces**.

Dunaliella Genus of unicellular halotolerant green **algae** of the family Dunaliellaceae. *Dunaliella salina* and *D. bardawil* are important natural sources of **β-carotene**.

Durian Fruits produced by *Durio zibethinus*. Emit a characteristic sulfurous odour. Often sold in a ready-to-eat form, packaged as the whole edible pulp or in segments, or preserved by **drying**, **fermentation**, salting or **deep freezing**. Used as a source of flavour in **ice cream** and **cookies**. Rich in **sugars** and **vitamin C**.

Dursban Alternative term for the insecticide **chlorpyrifos**.

Durum wheat Species of hard **wheat** (*Triticum durum*), the **flour** of which is glutinous and yellow and used to produce **semolina** from which **pasta** is made.

Dust explosions Explosions caused by clouds of flammable particles at an appropriate concn. coming into contact with an ignition source. Dust and powders present a potential explosion hazard in processing plants, including those in the food industry. Common processes generating explosible dust inc!ude flour and provender milling, sugar grinding, spray drying of milk and instant coffee, and conveyance/storage of whole grains and finely divided materials. Parameters influencing explosions include nature of the combustible material, reactivity, particle dimensions, powder concentration, humidity, ignition energy and presence of inflammable gas. Powders can be classified on the basis of their explosion hazard, with explosible dusts including flour, custard powder, instant coffee, sugar, dried milk, potato powder and soup powder. Methods to control dust explosions include containment, suppression, inerting and venting.

Dyes Natural or synthetic colorants used in foods. In contrast to **pigments**, dyes can usually be solubilized using an appropriate solvent or binder.

E

Earthworms Segmented, burrowing invertebrates of the class Oligochaeta, especially those of the genus *Lumbricus*. Earthworms, such as the red worm *Eisenia foetida*, are used as foods in some areas of the world, including China and the Philippines. They serve as a readily available source of **proteins** and **minerals**.

Eating habits **Consumer response** term relating to the pattern of consumption of foods by particular population groups.

Eating quality The extent to which a food is assessed as being edible, i.e. possessing acceptable **sensory properties**.

Eau de vie French generic term for **brandies** and other **spirits**.

EC Abbreviation for **European Community**.

Echinococcus Genus of tapeworm of the class Cestoda. Infection in humans with *Echinococcus granulosus* may occur after ingestion of water or vegetation contaminated with larval cysts.

Echinoderms A group of exclusively marine invertebrates in the phylum Echinodermata, which contains five classes: Asteroidea (starfish); Ophiuroidea (brittle stars); Echinoidea (**sea urchins**); Crinoidea (feather stars); and Holothuroidae (**sea cucumbers**). Some echinoderms are edible, including the sea urchin species *Loxechinus albus*, *Paracentrotus lividus* and several sea cucumber species.

Echinoids Alternative term for **sea urchins**.

Eclairs Finger-shaped **bakery products** made with choux pastry which is baked and filled with **whipped cream** or **custards** and topped with fondant icing, usually flavoured with **chocolate** or **coffee**. Also a name given to **confectionery** products comprising **toffees** filled with chocolate.

Ecology Biological science, involving the study of interactions of organisms with their environment, including interrelationships between organisms.

e-commerce Buying and selling of products and services transacted electronically via the Internet. Includes dealings among businesses and between companies and consumers. Also called electronic commerce.

Edam cheese Dutch semi-hard **cheese** made from cow **skim milk** or **semi skimmed milk**. Usually coated with red wax, but cheese matured for 17 weeks or longer is coated with black wax. Mainly eaten young for an elastic and supple **texture** and a smooth **flavour**.

Edestin One of the **vegetable proteins** present in certain plant seeds, including **barley** and **hemp seeds**.

Edible containers Holders for foods which are intended to be consumed along with the food they contain. Mainly made from **dough**. Examples include **ice cream cones** and **taco shells**.

Edible fungi Alternative term for **mushrooms**.

Edible oils Lipid-rich substances which are liquid at room temperature and are used in preparing foods. Usually have a high content of **triacylglycerols** and those of plant origin can be a source of bioactive **phytochemicals**. Should be of high quality, pale in colour, free from **off odour** and **off flavour**, and of high **nutritional values**. Includes **vegetable oils** and **marine oils**.

Edible packs Packages for foods made from films and **coatings** that are suitable for consumption along with the products they enclose. The films and coatings are made from natural ingredients such as proteins, carbohydrates or lipids, or their combinations.

EDTA Abbreviation for ethylenediaminetetraacetic acid. Commercially available in the form of sodium and calcium salts, EDTA is one of the best known **sequestrants** and **chelating agents**, controlling the reaction of trace metals present in foods, and thus providing a variety of functions in foods. Applications include prevention of discoloration in canned **corn**, avoidance of crystals formation in canned **sea foods** and prevention of **rancidity** and microbial **spoilage** in **mayonnaise** and fatty **spreads**.

Edwardsiella Genus of Gram negative, facultatively anaerobic, rod-shaped **bacteria** of the family **Enterobacteriaceae** which occur in the intestines of mammals, fish and reptiles. *Edwardsiella tarda* may be an opportunistic pathogen in humans. Infection usually occurs through the ingestion of faecally-contaminated food or water, resulting in diarrhoea.

EEC Abbreviation for **European Economic Community**.

Eels General name used for a number of unrelated **fish** species belonging to the order Apodes and the family Anguillidae; characterized by elongate serpentine bodies lacking scales or pelvic fins. Most species are marine (including moray, snipe and **conger eels**) or have a marine phase. Species within the genus *Anguilla* are particularly valued as food fish, including *A. anguilla* (European eel), *A. rostrata* (American eel) and *A. japonica* (Japanese eel). Flesh tends to be firm, with a rich, sweet **flavour**. Marketed in a variety of forms; smoked, jellied and pickled products are especially popular.

Efficient consumer response Efficient consumer response (ECR) is about change and continued improvement in the grocery supply chain. Four major strategies have been defined within ECR, each of which creates value by satisfying consumer needs for product, convenience and price: Efficient Store Assortment - addresses how many items to carry in a category, what type of items and in what sizes/flavours/packages, and how much space to give to each item; Efficient Replenishment - focuses on reducing and eliminating costs in the order cycle, starting with accurate point-of-sale data; Efficient Promotion - addresses inefficient promotional practices that tend to inflate inventories and practices; and Efficient New Product Introduction - addresses improving the entire process of introducing new products, which is subject to high failure rates, thereby bringing extra costs into the system.

Effluents Liquid **wastes** (**waste water**) discharged into a river or the sea, usually from a factory or plant.

EFTA Abbreviation for European Free Trade Association. EFTA is a trading bloc that was established in 1960 by Austria, Denmark, the UK, Norway, Portugal, Sweden and Switzerland. The aim of EFTA was to work for the removal of trade barriers among its members and to promote closer economic cooperation between EFTA and the rest of Western Europe. EFTA membership expanded when Finland became an associate member in 1961 and a full member in 1986. Iceland and Liechtenstein joined the organization in 1970 and 1991, respectively. However, with the growing success of the EU in the 1970s and 1980s, many members left EFTA to join the EU, and, in late 1993, the only remaining EFTA countries were Norway, Liechtenstein, Iceland and Switzerland. By 1994, EFTA states were concerned that the success of the EU could affect their own economies negatively. The EFTA states negotiated with the EU to establish a broader common market called the European Economic Area (EEA). The EEA comprises all the members of the EU and EFTA, with the exception of Switzerland, which declined to join. The headquarters of EFTA are in Geneva, Switzerland.

Egg nog Alcoholic beverage made using sweetened **milk**, **eggs** and **sherry** and/or **spirits**, e.g. **brandy** or **rum**.

Egg pasta **Pasta** which contains **eggs** as an ingredient.

Egg plants Alternative term for **aubergines**.

Egg powders Alternative term for **dried eggs**.

Egg proteins Proteins found in **eggs**, such as **ovalbumins**, **ovomucoid** and **conalbumin**.

Eggs External reproductive structures produced by the females of certain animals, such as birds, reptiles and fish. The term is used without qualification usually to refer to eggs laid by hens, although eggs produced by other birds, some reptiles (e.g. turtles) and fish (**roes**) are also eaten. Generally composed of **egg yolks** and **egg whites** surrounded by hard **egg shells**. Eaten raw or cooked in a variety of ways, e.g. scrambled, fried, poached or boiled. Also incorporated into a range of foods and beverages, and can be used as **thickeners**, **emulsifiers**, **binding agents** and **foaming agents**.

Egg shells Exterior hard coverings of **eggs**, which are composed mainly of calcium carbonate. Vary in colour according to breed and species of bird. Responsible for permitting gaseous exchange, conserving water, inhibiting microbial penetration and providing mechanical protection.

Eggs lysozymes Alternative term for **egg whites lysozymes**.

Egg whites Portions of **eggs** which surround the **egg yolks**. Composed mainly of water and **albumins**. Form foams upon incorporation of air during **whipping**. Used in this form to make light products such as **meringues** and **sponge cakes**. Also known as albumen.

Egg whites lysozymes Lysozymes found in **egg whites** with good **foaming properties** and **emulsification properties**, particularly after modification or thermal treatment. Their antibacterial properties make them useful for preventing spoilage in foods and **beverages** (e.g. in **meat**, **dairy products** and **beer**). Also potentially useful as **sweeteners**.

Egg yolks Portions of **eggs** which are surrounded by the **egg whites**. Usually yellow in colour. Composed mainly of water, protein and fat. **Colour** may be enhanced by incorporation of pigmented feeds (e.g. yellow corn, alfalfa meal, corn gluten meal, dried algae meal and marigold petal meal) which contain carotenoid **xanthophylls** (e.g. **lutein**, **zeaxanthin**, **carotenes** and **cryptoxanthin**) into the poultry diet.

Separated egg yolks may be used as **emulsifiers** in **mayonnaise** and **salad dressings**.

Egusi Type of watermelon (*Citrullus lanatus, C. vulgaris* or *Colocynthis citrullus*) cultivated mainly in West Africa for its seeds. Dried seeds are rich in **oils** and represent a good source of group B vitamins. They are commonly added to **rice** and legume based dishes, or ground to make a **meal**. The meal is used as a thickener in **soups** and stews, also adding **flavour** and increasing protein contents, or used in preparation of meat-like **patties**.

Eicosapentaenoic acid One of the ω-3 or *n*-3 **polyunsaturated fatty acids** (PUFA), with 20 carbon atoms and 5 double bonds. The most important isomer is the (all-*Z*)-5,8,11,14,17-isomer, and rich sources of this important dietary fatty acid include **fish oils** and marine **algae**. Suggested health benefits associated with eicosapentaenoic acid and its related *n*-3 PUFA **docosahexaenoic acid** include reduced risks of **coronary heart diseases** and cancer, and improved **immune response** and neural development in infants.

Eicosatetraenoic acid One of the ω-3 or *n*-3 **polyunsaturated fatty acids** (PUFA), with 20 carbon atoms and 4 double bonds. An important component of the human diet and a precursor of a range of physiologically active compounds such as **prostaglandins**. Occurs in esterified form as a major component of membrane **phospholipids**. Intermediate in formation of **eicosapentaenoic acid**.

Eicosenoic acid One of the **monounsaturated fatty acids** with 20 carbon atoms; the major isomers are the Δ9 (*n*-11) and Δ11 (*n*-9) forms. Found in a range of foods, including **fish oils**, **peanuts**, **olives** and *Brassica* seeds.

Einkorn Species of **wheat** (*Triticum boeoticum* or *T. monococcum*) grown in arid regions as a livestock feed and one of the first **cereals** grown for food. Ancestor of modern wheat varieties.

Eiswein Sweet **dessert wines** made from **grapes** which have been allowed to freeze, traditionally on the vine. The grapes are hand picked and pressed while still frozen, producing highly concentrated **grape juices** which are rich in **acids**, **sugars** and **aroma compounds**. Produced mainly in Germany, Austria, Switzerland and Canada.

Ekalux Alternative term for the insecticide **quinalphos**.

Elaeis Genus of oil palm, the most common species of which is *Elaeis guineensis*. Seeds are the source of oils similar to **coconut oils** that are used in manufacture of **margarines**, **shortenings** and **cocoa butter substitutes**.

Elaeis oils Alternative term for **palm oils**.

Elaidic acid The *trans* form of an unsaturated fatty acid, which in its *cis* form is **oleic acid**. Exists as a combustible, white solid, which is insoluble in water, but soluble in alcohol and ether. **Hydrogenation** of **fats** for use in **margarines** and **cooking fats** creates *trans* **fatty acids**, including elaidic acid. Elaidic acid occurs in foods, including **butter**, **margarines**, **cereal products** and **snack foods**. As with other *trans* fatty acids, high levels of dietary elaidic acid may have negative **lipaemic activity**.

Elands Large **antelopes** (*Tragelaphus oryx* or *Taurotragus oryx*) found widely distributed in scrub, grasslands and savannah woodland of southern Africa. Hunted as **game**. Attempts have also been made to farm small herds in South Africa and Ukraine for their **meat** and rich **milk**. Antelope meat is red, has a low fat content, and is tender and juicy when cooked. Pot **roasting** is the favoured method of **cooking** eland meat, but it can also be used in place of **beef** in many dishes.

Elasticity **Rheological properties** relating to the ability of a substance to return to its original size and shape after being deformed. The deforming force is known as a stress, and the resulting deformation is the strain. A body is elastic only below a certain stress; above this point, known as the elastic limit, the body is permanently deformed. The point at which the material begins to give is called the yield point.

Elastin One of the **animal proteins** present in mammalian **connective tissues**, and thus a component in **meat** and **meat** products. Particularly rich in **glycine** residues and also contains high levels of **proline**, **alanine** and **valine**.

Elderberries Small purple-black **berries** produced by the elder, *Sambucus nigra*, or American elder, *S. canadensis*. Used in **wines**, juices and other **beverages**, and also in **pies** and **jams**. Rich in **vitamin C**. Contain high levels of **anthocyanins**, making them suitable for use in natural food **colorants**.

Elderberry juices Juices extracted from **elderberries** (*Sambucus nigra*).

Elderflowers Flowers of the elder, *Sambucus nigra*, or American elder, *S. canadensis*. Used to make **wines** and **cordials**; also used in **preserves**, **syrups**, **sorbets**, **ice cream** and **fritters**.

Electrical conductivity Ability of a substance to transmit an electric current. One of the **electrical properties** commonly determined in food analyses. It can be used, for example, as an indicator of *post mortem* changes in **meat** quality and to monitor the composition of **food factories effluents**.

Electrical properties Generalized term for the **physical properties** of a food relating to its ability to conduct electricity. Includes **capacitance, dielectric properties**, conductivity/resistance and **electrostatic interactions**.

Electrical resistance One of the **electrical properties** commonly determined in food analyses, electrical resistance is a measure of the extent to which a material withstands passage of an electric current. Inversely related to **electrical conductivity**. Heat is generated as a consequence of resistance and this characteristic is exploited in some cooking or heating methods, an example being **ohmic heating**.

Electrical stimulation Controlled application of an electrical current to **animal carcasses** immediately after **slaughter**. It is used to increase meat **tenderness**, and also to give meat a lighter, brighter **colour**. In particular, it is used to achieve accelerated conditioning (ageing) of animal carcasses, and to decrease **cold shortening** and subsequent **toughness**, which accompany very rapid **chilling** of **meat**. Electrical stimulation of carcasses breaks cross-linkages between actin and myosin filaments in the muscles, increases enzyme activity and causes some tissue damage; all of these effects increase meat **tenderness**. It may considerably improve the quality of **beef, veal, lamb** and **goat meat**, but has negative or negligible effects on the quality of **pork**. Electrical stimulation is well established in lamb slaughtering practice and has also been widely used in deer slaughtering.

Electrical stunning A form of **stunning**, which is used during **slaughter** to immobilize animals and **birds** before bleeding. It is widely used during the slaughter of swine, sheep and poultry, but can also be used effectively during cattle slaughter. Before consciousness returns, bleeding can be carried out humanely and effectively. As well as improving animal welfare during slaughter, the method has beneficial effects on **meat** quality; for example, it reduces the incidence of the **PSE defect** in **pork**. There are two basic types, namely high voltage and low voltage. Electrical stunners include: pillar types; electrically charged knives; stunning tongs; and electrified water baths.

Electric fields A region of space characterized by the existence of a force generated by electric charge. The magnitude of the electric field around an electric charge depends on how the charge is distributed in space. Each point in space has an electric property associated with it, the magnitude and direction of which are expressed by the value of the electric field strength. The value of the electric field has dimensions of force/unit charge. In the SI system, units are Newtons/Coulomb, equivalent to Volts/Metre.

Electrocution To kill by electric shock. Electrocution may be used to **slaughter chickens** or **fish**. Some evidence indicates that, in comparison with **electrical stunning**, electrocution may reduce faecal loads on poultry **carcasses** under commercial slaughtering conditions. Electrocution is also used as a method to efficiently control **insects** and **mites** in food industry premises.

Electrodes Conductors through which current is applied to or extracted from an electric circuit or system. Usually made of metal. Used as integral parts of instruments employed in detection of sample components.

Electrodialysis Technique in which dialysis is accelerated by application of a potential across the compartments of the apparatus.

Electrolysed water Salted **water** which has been passed through an oxidizing unit, causing it to undergo ionic changes. Depending on which electrode the water is passed over, either acidic or alkaline electrolysed water is formed. Acidic water is lethal to foodborne **microorganisms** and is considered more efficient for washing food, especially **fruits** and **vegetables**, during preparation than using chlorine-containing solutions or, in some cases, heat treatment. Its use has little effect on food **sensory properties**. Alkaline water is useful as a sanitizer, as it functions like a soap to remove substances from food preparation surfaces.

Electrolytes Liquid or solid compounds which, when dissolved in or in contact with water, will dissociate into ions and conduct electricity. In physiological use, the term refers to certain inorganic compounds, e.g. those containing sodium, potassium or calcium, which dissociate into ions that conduct electrical currents and play an important role in controlling body fluid balance.

Electromagnetic fields Fields of force associated with electric charge in motion, having both electric and magnetic components and containing a definite amount of electromagnetic energy. The mutual interaction of electric and magnetic fields produces an electromagnetic field, which is considered as having its own existence in space apart from the charges or currents with which it may be related. Under certain circumstances, this electromagnetic field can be described as a wave transporting electromagnetic energy. In the food industry, electromagnetic fields are utilized in **dielectric heating**.

Electronic noses Apparatus, consisting of arrays of semiconductor metal sensors coated with polymers, used for characterization of **aroma compounds**. The polymers in the sensors adsorb **volatile compounds** from aromas, vapours and gases. Each polymer adsorbs a different combination of ingredients, so that conductivity changes and variations may be processed electronically to produce visual fingerprints.

Electronic tongues Apparatus, consisting of arrays of lipid/polymer membrane based sensors, which can quantify the taste of substances such as amino acid mixtures, **foods** and **beverages**. The lipid/polymer membranes are fitted onto a multichannel electrode, and electric signals from the sensors are fed into a computer; voltage differences between the multichannel electrode and a reference electrode are measured. Output from the sensors varies for chemical substances with different taste qualities but is similar for substances with similar tastes. The sensor array detects the five types of taste quality, i.e. **sourness**, **saltiness**, **bitterness**, **sweetness** and **umami**.

Electron microscopy **Microscopy** technique which utilizes extremely short wave radiation from electrons in a vacuum tube to give high resolution. Commonly abbreviated to EM.

Electron paramagnetic resonance **Spectroscopy** technique for studying the structure and bonding of a paramagnetic substance based on microwave-induced transitions between the energy levels of unpaired electrons. Synonym for electron spin resonance.

Electron spin resonance Alternative term for **electron paramagnetic resonance**.

Electrophoresis Technique in which charged electrical species are separated by migration in an electrolyte through which a current is passed, with cations moving towards the cathode and anions to the anode. Separated species are identified by staining or radioactive labelling. Usually conducted on paper or in a gel (**gel electrophoresis**), although faster methods using capillary columns (**capillary electrophoresis**) have been developed that have other advantages, such as the possibility of on-line detection of separated species.

Electroporation Method for transformation of **DNA** into host cells in which high voltage pulses of electricity are used transiently to permeabilize cell membranes.

Elements Fundamental chemical units of which all matter is composed. Cannot be broken down into simpler substances by ordinary chemical means. For a given element, all atoms have the same number of protons and electrons; however, atomic weight may differ because the number of neutrons in the nucleus differs between isotopes.

Elephant yams Plants of the genus *Amorphophallus* grown for their edible roots. Roots of *A. rivieri* or *A. konjac*, also known as konjac, konjaku or konnyaku, are the source of **konjac glucomannans** which are used as a gum. *A. campanulatus* is the Asian elephant yam.

ELISA Abbreviation for enzyme linked immunosorbent assay, a very sensitive immunological technique which can be used to detect and measure the presence of **antigens** or **antibodies** in a wide variety of biological samples. In the assay, protein antigens or antibodies are labelled with **enzymes**, after which one of the reactants is immobilized onto a support material. As soon as the immunochemical reaction has taken place, unbound substances are washed out and the bound material is quantified by measuring the activity of the enzyme by **spectroscopy**. The immobilization is preferentially performed in the wells of polyvinylchloride or polystyrene microtitre plates, and the colour forming enzymes used are normally **peroxidases**, **alkaline phosphatases** or **glucose oxidases**.

Elk meat Meat from **elks**. Forequarter to hindquarter ratio in elk **carcasses** is similar to that for beef cattle carcasses. Elk carcasses include a high percentage of lean and a low percentage of fat. Amino acid composition is similar to that of **beef**; however, the **physicochemical properties** of elk meat are generally inferior to those of beef. Compared with elk bull meat, elk cow meat requires less **ageing** (conditioning) to attain acceptable **tenderness**.

Elks Large northern deer (*Alces alces*) belonging to the Cervidae family. Wild elks are hunted for their meat. In some countries, e.g. the Union of Soviet Socialist Republics, elks have been domesticated and are used to produce **elk meat** and elk milk. Elk meat is sometimes referred to as **venison**. In popular use, the term is also used to describe North American moose.

Ellagic acid Phenolic organic acid, which in pure form exists as yellow crystals. Only very slightly soluble in water and alcohol. Can be isolated from **tannins** in plant materials, e.g. oak galls, **tea** and some **fruits** and **nuts**. Occurs also in wood aged **alcoholic beverages**. *In vitro*, it shows **antioxidative activity**, whilst in animal studies, it has **antitumour activity** and **anticarcinogenicity**.

EM Abbreviation for **electron microscopy**.

Emamectin Insecticide belonging to the **avermectins** group used to control a range of **insects** including **mites**, leaf miners, **aphids**, **moths** and **bees**. Also used as a parasiticide, effective against sea lice in **fish**.

Emmental cheese Swiss **hard cheese** made from unpasteurized **cow milk**. A difficult cheese to produce due to intricacies of the **fermentation** process required to form the characteristic walnut-sized holes.

Emodin Naturally occurring anthraquinone present in the roots and bark of numerous **plants** of the genus *Rhamnus*. Extracts from the roots, bark, and/or dried leaves of some of these plants, e.g. buckthorn, senna, cascara, **aloe**, frangula and **rhubarb**, are widely used in the preparation of herbal laxative preparations.

Emu eggs **Eggs** produced by **emus**. Consist of approximately 11.9% protein and 16.0% lipids, and have a mean weight of 610 g. **Egg shells** are dark green in **colour**.

Emulsification Process for forming fine dispersions (**emulsions**) of minute droplets of one liquid in another in which it does not dissolve or form a homogeneous mixture.

Emulsification properties Functional properties relating to the ability of food components to form **emulsions**, suspensions of small globules of one liquid in a second liquid with which it will not mix.

Emulsifiers Substances which aid the uniform dispersal (**emulsification**) of one immiscible liquid in another and thereby help in formation of **emulsions**. Widely used in the food industry, where applications include manufacture of **bakery products, confectionery, ice cream, mayonnaise** and **margarines**. Types of emulsifiers used in foods include **carrageenans, lecithins** and **glycerides**.

Emulsifying agents Alternative term for **emulsifiers**.

Emulsifying capacity Functional properties relating to the extent to which food components can form **emulsions**.

Emulsions Types of **colloids** or **dispersions** composed of a mixture of immiscible liquids in which one forms droplets suspended in the other. **Processed foods** based upon emulsions include **sauces, salad dressings, soups, spreads, coatings, mayonnaise, sausages** and some **dairy products**. Emulsions display variable stability, and most require the addition of **emulsifiers** to maintain emulsion structure.

Emu meat Meat from **emus**. Emus have a lower percentage of hot carcass weight and total fat to body weight, but a higher proportion of lean meat to carcass weight than ostriches or rheas. The meat is generally taken from the underbelly and thighs as there is not much meat on the breast. Meat cuts commonly prepared from emu **carcasses** include the side, forequarter, strip loin, neck, hindquarter, thigh, drum, fore saddle and hind saddle. Fat content of emu meat is low and colour is an intense red (pigment content increases with increasing age). **Collagen** content, **colour** and **tenderness** vary between muscles; some muscles are sufficiently tender for **roasting** or **grilling**.

Emus Large, flightless, swift-running Australian birds (*Dromaius novaehollandiae*), which are farmed for the production of **emu meat**, emu eggs, feathers, hides and emu oils.

Enamels Semi-transparent or opaque **ceramics** substances applied as protective or decorative **coatings** to the surface of metals, pottery or glass. Often applied to the surfaces of food **containers**, e.g. **cans** and cooking pots. Enamelled objects that come into contact with food or **beverages** may release lead or cadmium, posing a health risk. Also used to describe paints or varnishes which become smooth and hard when dried.

Enantiomers Stereoisomers of a compound which are mirror images of each other. The left- and right-handed forms of these chiral isomers are optically active and generate a racemate when mixed in equal proportions. Chirality may affect the **biological activity** and **functional properties** of the compound; for example, D-**amino acids** but not L-amino acids are useful as **sweeteners**.

Enantioselectivity Preferential formation of one enantiomer over another in a chemical reaction, expressed quantitatively as enantiomer excess. **Enantiomers** formed may affect the **biological activity** and **functional properties** of the product, e.g. D-**amino acids** but not L-amino acids are useful as **sweeteners**.

Encapsulation A technology that allows sensitive ingredients to be physically enveloped in a protective matrix or wall material in order to protect these ingredients or core materials from adverse reactions, loss of **volatile compounds**, or nutritional deterioration. Spray drying is a **microencapsulation** technique readily used in the food industry. **Carbohydrates**, such as **maltodextrins, starch** and corn syrup solids, and acacia **gums** are widely used examples of encapsulating agents.

Endives Common name for *Cichorium endivia*. Leaves are used fresh in **salads** or blanched to reduce **bitterness**. Common form used is the curled endive; other type is the **Escarole** group, which has broad flat leaves. May have red pigmentation. Similar nutritionally to **lettuces**.

Endo-1,3(4)-β-glucanases EC 3.2.1.6. Enzymes that hydrolyse the 1,3- and 1,4-β-D-glucosidic bonds in β-glucans, which are typically found in **oats, barley**, some **fruits** and certain **microorganisms**. Also known as laminarinases, these enzymes are useful in the brewing industry where β-glucans can cause difficulties during **clarification** of **worts** and **filtration** of **beer**. Also useful in the **winemaking** industry where *Botrytis* contamination is a problem.

Endomyces Genus of **fungi** of the class Saccharomycetes. Occur in soil and plant debris. Some species are plant pathogens (e.g. the mushroom parasite *Endomyces scopularum*). *E. fibuliger* may be responsible for the spoilage of **bread** and other **bakery prod-**

ucts, and is also used in the commercial production of β-glucosidases.

Endomycopsis Obsolete name for a fungal genus whose species have been reclassified into other genera, such as *Saccharomycopsis*, **Endomyces**, **Yarrowia** and **Pichia**.

Endonucleases EC 3.1.21-EC 3.1.31. **Enzymes** that cleave **nucleic acids** at positions within their chains, producing poly- or oligo-nucleotides. Most act specifically on either **DNA** or **RNA**, while some (e.g. S1 nuclease from **Aspergillus**) can act on both DNA and RNA.

Endopeptidases EC 3.4.21-EC 3.4.24 and EC 3.4.99. **Enzymes** that hydrolyse proteins by cleaving specific peptide bonds within protein molecules. These enzymes are classified on the basis of their catalytic mechanism and can be serine, cysteine, aspartic or metalloendopeptidases, although a number cannot yet be assigned to any of these subclasses. Examples include **chymotrypsin**, elastase, **pepsins**, **thermolysins** and **trypsin**. These enzymes have numerous applications in food processing.

Endopolygalacturonases Alternative term for **polygalacturonases**.

Endosulfan Non-systemic insecticide and acaricide used to control a variety of sucking, chewing and boring **insects** and **mites** on a wide range of **fruits**, **vegetables**, **cereals**, **coffee** and **tea**. Classified by WHO as moderately toxic (WHO II). Also known as thiodan.

Endothia Genus of fungi of the class Sordariomycetes. **Microbial rennets** derived from *Endothia parasitica* are used in **cheesemaking**.

Endotoxins Lipopolysaccharide **toxins** of Gram negative **bacteria**, or any microbial toxins which are released only upon cell lysis.

Endo-1,3-β-xylanases Alternative term for **xylan endo-1,3-β-xylosidases**.

Endo-1,4-β-xylanases EC 3.2.1.8. Enzymes that catalyse the endohydrolysis of 1,4-β-D-xylosidic linkages in xylans, yielding **xylose** and **xylooligosaccharides**. Produced by a number of **bacteria** and **fungi**, these **enzymes** can be used for improving the handling and stability of **dough**, degradation of lignocellulosic materials and production of novel **oligosaccharides**.

Endpoint temp. Temperature to which a food product, particularly **meat**, needs to be heated to ensure destruction of **pathogens**.

Endpoint temp. indicators Indicators showing the adequacy of **heating** of foods, particularly **meat** and meat products, in relation to destruction of **pathogens**. The bovine catalase test and tests based on

protein solubility, **enzymes** activity, **colour**, **electrophoresis** patterns of **proteins**, **differential scanning calorimetry** (DSC) of muscle proteins, near infrared spectroscopy (**NIR spectroscopy**) and enzyme linked immunosorbent assays (**ELISA**) can be used for this purpose.

Endrin Persistent organochlorine insecticide used to control a wide range of **insects**; use on crops has generally been displaced by less persistent **insecticides**. Classified by WHO as highly toxic (WHO I).

Energy conservation Planned management of energy supplies by various means. One type of energy conservation is curtailment (doing without). A second type is overhaul (for example, using less energy-intensive materials in production processes, and decreasing the amount of energy consumed by certain products). Another type involves the more efficient use of energy and adjusting to higher energy costs (for example, capturing waste heat in factories and reusing it).

Energy drinks **Soft drinks** containing ingredients intended to enhance or maintain physical energy of the consumer.

Energy foods **Health foods** designed for people, such as sportsmen and sportswomen, requiring a source of high energy. Energy foods are frequently available in the form of carbohydrate-rich energy **food bars**. **Energy drinks** and **isotonic drinks** are popular for the same purpose.

Energy values Alternative term for **calorific values**.

English muffins Thick, round **bread** products which are rapidly fermented using **yeasts** and are well aerated. Baked on a hot plate or griddle and often split and toasted before being eaten, sometimes with sweet or savoury **fillings**, such as **jams**, **bacon** or **cheese**.

Enokitake Alternative term for the **edible fungi** *Flammulina velutipes*.

Enrichment Improvement of the quality or nutritional value of a food, usually by addition of **nutrients**.

Enrichment techniques Procedures which specifically promote the growth of a particular microorganism, thereby increasing its proportion in a mixed population.

Enrobing **Coating** of a centre material, for example **nougat**, biscuit, **fondants** or **caramel**, in **chocolate**. It is necessary to use tempered chocolate for enrobing processes. The centres for coating are placed on a continuous moving wire chain belt, which transports them underneath a flow of chocolate. Below the belt is a bottoming trough that retains the chocolate that falls through the chain belt and recirculates it, forming a layer of chocolate on the undersides of the centres. Sometimes two chocolate streams are used in enrobers;

this is particularly useful when the product to be enrobed has an uneven surface. The first coating flows into all the crevices and provides a good moisture barrier to the product. The second coating gives the chocolate a more rugged appearance. Products finally pass through a cooling tunnel to set the chocolate.

Enrofloxacin Broad-spectrum semisynthetic fluoroquinolone antibiotic used to treat local and systemic infections in animals and poultry. Active against a wide range of **Gram negative bacteria** and also some **Gram positive bacteria**. Metabolized in the liver, the main product being ciprofloxacin, which is detected along with the parent compound in tissues, **milk** and **eggs** of treated animals and poultry. Residues persist longest in poultry skin, and **livers** and **kidneys** of animals and birds.

Entamoeba Genus of protozoan parasites of the family Entamoebidae. Infects humans and other vertebrates. *Entamoeba histolytica* may be responsible for **amoebiasis**.

Enteric viruses **Viruses** that live in the **gastrointestinal tract**. Human enteric viruses may exist as commensals or may be **pathogens** which can cause **gastroenteritis** (particularly members of the families Adenoviridae, Astroviridae, Caliciviridae and Reoviridae). Usually transmitted via the faecal-oral route.

Enterobacter Genus of Gram negative, facultatively anaerobic, rod-shaped **bacteria** of the family **Enterobacteriaceae**. Occur in soil, water, gastrointestinal tracts of humans and animals, and foods (e.g. **dairy products**, raw **shellfish** and raw **vegetables**). Some species may cause opportunistic infections in humans (e.g. *Enterobacter cloacae*).

Enterobacteria **Bacteria** of the family **Enterobacteriaceae**.

Enterobacteriaceae Family of Gram negative, facultatively anaerobic, rod-shaped **bacteria**. Members occur in soil, **water**, **plants** and the gastrointestinal tracts of humans and animals. May occur as **pathogens** in vertebrates (e.g. species of the genera *Escherichia*, *Klebsiella*, *Citrobacter*, *Salmonella*, *Yersinia* and *Enterobacter*) or food **spoilage** bacteria (e.g. species of *Serratia*, *Proteus* and *Erwinia*).

Enterocins **Bacteriocins** produced by *Enterococcus* spp.

Enterococci Term which can be used in two ways. It is used to refer to members of the bacterial genus *Enterococcus*. Alternatively, it can be used loosely with reference to any streptococcal **bacteria** found in the human gastrointestinal tract, including species of *Enterococcus* and *Streptococcus*.

Enterococcus Genus of Gram positive, facultatively anaerobic, coccoid **lactic acid bacteria** of the family Enterococcaceae. Occur in the gastrointestinal tracts of humans and animals. *Enterococcus faecalis* may be an opportunisitc pathogen in humans.

Enterotoxicity Quality or degree of being capable of exerting a toxic effect on the **gastrointestinal tract**.

Enterotoxins Bacterial **toxins** (e.g. **cholera toxin**) which, upon ingestion or production by **microorganisms** within the **gastrointestinal tract**, cause disturbances of the gastrointestinal tract. Diarrhoea is a common symptom.

Enteroviruses **Viruses** of the genus Enterovirus (e.g. coxsackieviruses, polioviruses and echoviruses) which may be pathogenic in humans. Commonly transmitted via contaminated food and water.

Enthalpy Measure of energy (heat) commonly used to study the thermodynamics of chemical reactions. Changes in the structure of food macromolecules, such as **denaturation**, **gelatinization** and **crystallization**, are often associated with changes in enthalpy.

Entoleters Machines used in **disinfestation** of **cereals** and other **foods**. Food is fed to the centre of a high-speed rotating disc which bears studs. The impact of the food being thrown against the studs kills **insects** and destroys their eggs.

Entrees In Europe, a term applied to dishes served before the meat (main) course. In the USA, the term is usually applied to main **meals**.

Environmental protection Ecology term describing measures taken to limit the impact to the environment of human activities. Examples within the food industry include **bioremediation** processes which decrease the chemical and biological value of **effluents** and other **wastes** released into the environment, and the use of readily degradable **packaging materials**.

Environment friendly packaging materials Materials developed for packaging of products including **foods** and **beverages**, with special consideration given to **biodegradability** and **recycling**.

Environment friendly processes Processing procedures that are not harmful to the environment.

Enzyme electrodes Type of **ion selective electrodes** in which the electrodes are coated with a layer containing an enzyme that reacts with the analyte to form a product to which the electrodes respond. Commonly used examples include **glucose** sensitive electrodes, which are coated with **glucose oxidases**.

Enzyme immunoassay Immunoassay (often abbreviated to EIA) in which **antibodies** used to bind to the antigens to be measured are attached to an enzyme as a marker. Antibody-antigen complexes formed are measured on the basis of catalytic activity of the enzyme. **ELISA** is a type of enzyme immunoassay.

Enzyme inhibitors Substances which reduce the activity of **enzymes** and, when present in foods, may act as **antinutritional factors**. Certain **proteinases inhibitors** such as **calpastatins** and **cystatins** play a role in development of **meat tenderness** and also may be useful for maintaining the quality of **fish** and **surimi** by inhibiting **proteolysis**. However, **trypsin inhibitors** and **chymotrypsin inhibitors** present in plant foods, particularly **legumes**, can reduce the **digestibility** and **nutritional values** of these foods.

Enzymes Proteins that act as highly efficient and specific biological **catalysts**. Increase the rate of reactions by decreasing the activation energy but do not alter the equilibrium constant. Divided into six main groups: **oxidoreductases, transferases, hydrolases, lyases, isomerases** and **ligases**.

Enzymic browning Formation of brown coloration of cut **fruits** and **vegetables** due to the action of **catechol oxidases** (polyphenol oxidases). In the presence of oxygen, the **enzymes** break **phenols** down into **quinones**, which polymerize to form brown coloured **melanins**.

Enzymic techniques **Analytical techniques** in which enzyme reactions form a major part.

Epicatechin One of the **catechols** found in **green tea** and **black tea**. Present in lower amounts than **epigallocatechin**. Also found in other plant sources. Displays **antioxidative activity** and, along with other catechols, is associated with the health benefits attributed to green tea consumption, e.g. **anticarcinogenicity** and **antimutagenicity**.

Epicatechin gallate One of the **catechols** found in **green tea** and **black tea**. Present in lower amounts than **epigallocatechin gallate**. Also found in other plant sources. Displays **antioxidative activity** and, along with other catechols, is associated with the health benefits attributed to green tea consumption, e.g. **anticarcinogenicity** and **antimutagenicity**.

Epidemiology Study of the incidence, distribution and causative factors of diseases that are associated with a particular environment or way of life, and of their control and prevention. Epidemiology is fundamental to preventive medicine and public health.

Epidermal growth factors Polypeptide **hormones** which stimulate and sustain epidermal cell proliferation. Synthesized by several glands and organs in the human body. Have numerous beneficial physiological effects on the intestinal mucosa and marked effects on epithelial turnover and microvillous ultrastructure. Epidermal growth factors present in **human milk** af-

fect **gastrointestinal tract** development in infants. *In vitro* and animal studies indicate a role in protection of the gastrointestinal tract against colonization with pathogenic **bacteria**, but epidermal growth factor and its receptors are also involved in many aspects of the development of carcinomas.

Epidermin One of the **lantibiotics** group of polypeptide **antibiotics**. Epidermin is synthesized by *Staphylococcus epidermidis* and displays inhibitory activity towards many **Gram positive bacteria**.

Epigallocatechin One of the major **catechols** found in **green tea** and **black tea**. Also found in other plant sources. Displays **antioxidative activity** and, along with other catechols, is associated with the health benefits attributed to green tea consumption, e.g. **anticarcinogenicity** and **antimutagenicity**.

Epigallocatechin gallate Member of the **catechols**, and a characteristic component of **green tea** and **black tea**. Also found in **seaweeds** and other plant foods. Has **antioxidative activity** and, along with other catechols, is associated with several health benefits attributed to green tea consumption.

Epimerases **Enzymes** that include members of EC 5.1. Catalyse the reversible conversion of an epimer into its counterpart form. Can act on **amino acids**, hydroxy acids, **carbohydrates** and derivatives of these compounds. Useful for preparation of rare **sugars**, and for altering the physical and immunological properties of polymers such as **alginates**.

Epinephrine Alternative term for adrenaline.

Epoxides **Organic compounds** containing a cyclic ether (epoxy) substituent comprising an oxygen atom directly attached via single covalent bonds to two carbon atoms, which may be adjacent or non-adjacent and cyclic or linear. A number of **plastics** used in food **packaging materials** contain an epoxide group.

Eremothecium Genus of **fungi** of the class Saccharomycetes. *Eremothecium ashbyii* is used in the commercial production of **riboflavin**.

Ergocalciferol Synonym for **calciferol** and **vitamin D₂**; one of the group of **sterols** which constitute **vitamin D**. Synthesized by **irradiation** of the plant provitamin **ergosterol**. Alternative recommended name is ercalciol.

Ergosterol Sterol which occurs naturally in **algae**, **bacteria**, **fungi**, **yeasts**, higher plants and animals. When exposed to **UV radiation** it is converted into **vitamin D₂** (**ergocalciferol**), a potent antirachitic substance. Used in synthesis of **oestradiol**.

Ergot A fungus (*Claviceps purpurea*) that attacks mainly **rye**, but also other **cereals**, replacing one or more of the kernels in the mature grain head with a mass called a sclerotium. As well as reducing crop yields, ergot contamination is a health hazard for man and animals eating the grain or its products.

Ergotamine One of the **alkaloids** produced by the **ergot** fungus *Claviceps* *purpurea*, which attacks **cereals**, predominantly **rye**. Also a secondary metabolite of some strains of *Penicillium*, *Aspergillus* and *Rhizopus*. Can cause poisoning (ergotism) if contaminated grain is used for food, but modern grain cleaning and **milling** procedures remove most of the ergot, leaving low levels of ergotamine in **flour**. **Baking** and **cooking** usually cause destruction of remaining alkaloid. Ergotamine is commonly used, in combination with **caffeine**, for treatment of **migraine**.

Erucic acid Monounsaturated fatty acid, which exists as a combustible solid with low toxicity. Insoluble in water, but soluble in alcohol and ether. Occurs naturally as a minor component of many plant seeds and is obtained from plant seed oils, particularly hydrogenated **mustard seed oils** and **rapeseed oils**. Uses include manufacture of **waxes**, **plasticizers**, water-resistant **nylon** and **stabilizers** and as an additive in **polyethylene films**. Alternative term for **docosenoic acid**.

Erwinia Genus of Gram negative, facultatively anaerobic, rod-shaped **bacteria** of the family **Enterobacteriaceae**. Occur on **plants**. Species (e.g. *Erwinia amylovora* and *E. carotovora*) may be responsible for plant diseases (dry necroses, vascular wilts and soft **rots**) and storage rots of **vegetables** (e.g. **potatoes** and **carrots**).

Erysipelothrix Genus of Gram positive, facultatively anaerobic, rod-shaped **bacteria**. *Erysipelothrix rhusiopathiae* occurs as a parasite in humans, mammals, birds and **fish**. Infection in humans is rare, and usually occurs through the handling of contaminated fish and **meat**.

Erythorbic acid Alternative term for the antioxidant **isoascorbic acid**.

Erythritol Tetrahydric polyol, which in pure form exists as combustible, white crystals. Can be made synthetically, but occurs naturally in animal and plant tissues, and in *Protocaccus vulgaris* and other **lichens** of *Rocella* species. Also found in **fermented foods**. Insoluble in water, but slightly soluble in alcohol. Has approximately 75% of the sweetness level of **sucrose**, and also low hygroscopicity, high endothermic reaction and easy crystallization. Used in **sweeteners** for sugar-free or low sugar products. Also known as erythrol.

Erythromycin Macrolide antibiotic used to treat bacterial infections (particularly those caused by staphylococci) in **cattle**, **swine**, **sheep** and **poultry**. Readily

disperses throughout tissues. Residues remain for relatively long periods of time after administration.

Erythrosine Artificial red colorant used for colouring **cherries**, **meat** products, **candy** and **confectionery**. Also known as **FDC red** 3.

Escarole Group of cultivars of **endives** with broad, flat leaves that may have red pigmentation due to the presence of **anthocyanins**.

Escherichia Genus of Gram negative, facultatively anaerobic, rod-shaped **bacteria** of the family **Enterobacteriaceae**. Occur in the **gastrointestinal tract** of humans and animals, and in soil and water (as a result of faecal contamination). *Escherichia coli* infection may occur through the ingestion of contaminated food or water.

Esculetin Synonym for 6,7-dihydroxycoumarin. Metabolite of coumarin found in a range of **plants**. Displays a variety of properties including **anticarcinogenicity**, **antioxidative activity** and inhibition of **lipoxygenases**.

Espresso coffee **Coffee** beverages made by a process based on steam extraction of **ground coffee** in a special apparatus. Usually very dark in **colour** and strong.

Essences **Extracts** which contain at least 1 constituent that defines the quality of the source material, particularly in terms of **flavour**. Extracts may be of natural origin (e.g. **essential oils**) or may be synthetic.

Essential oils Volatile aromatic oils of complex composition extracted from plant material, usually by distillation, although supercritical CO_2 extraction and cold pressing may also be used. Widely used as **flavourings**, either by adding their characteristic **flavour** to an end product or in the creation of natural flavouring blends. Some of the most widely used essential oils are **citrus essential oils**, **peppermint essential oils** and **cinnamon oils**.

Esterases EC 3.1. **Enzymes** that hydrolyse ester linkages resulting in formation of acids and **alcohols** or **thiols**. Esterases are sub-divided into carboxylic esterases, thiolesterases, phosphoric monoesterases, **phosphatases**, phosphodiesterases, triphosphoric monoesterases, sulfatases, diphosphoric monoesterases and phosphoric triesterases. Also includes exonucleases and **endonucleases**.

Esterification The process by which **acids** and **alcohols** react to form alkyl or aryl derivatives. Can be catalysed enzymatically by **esterases** or chemically by mineral acids.

Esters Organic compounds which are formed by combination of an acid with an alcohol. Some esters have a pleasant, generally fruity, **aroma** and occur in plant **essential oils**. Uses vary widely according to type of ester, but include synthesis of **flavourings** and perfumes.

Estragole   Phenol (1-allyl-4-methoxybenzene) which occurs widely in **essential oils** of **herbs** and **spices**. Used in **flavourings** for a wide range of **foods**, and displays **antioxidative activity**. Concerns exist over possible **hepatotoxicity** and **carcinogenicity** associated with chronic consumption. Also known as methylchavicol.

Ethanal   Aldehyde (systematic name for **acetaldehyde**) which in pure form exists as a volatile, colourless liquid with a pungent, fruity **aroma**. Produced by oxidation of **ethanol** and soluble in water and alcohol. **Fruits** and **vegetables** produce ethanal during **ripening**. It is also produced during **fermentation**, and is present in **foods** such as **fermented dairy products** and **alcoholic beverages**. Used in food **flavourings** and in the manufacture of **acetic acid**. Also known as acetic aldehyde.

Ethanol   **Alcohol** which constitutes a major component of **alcoholic beverages**. Formed by **fermentation** of **sugars** by **yeasts**. Synonym for alcohol.

Ethanolamine   Amine which in pure form exists as a colourless, combustible, hygroscopic liquid with an **aroma** of **ammonia**. A member of the **biogenic amines** group, which occurs in various **foods**, including **wines** and **cheese**. Synonym for aminoethanol.

Ethanolic fermentation   The process by which certain **yeasts**, **fungi** and **bacteria** metabolize **sugars** anaerobically to produce **ethanol**. In this process, **glucose** is converted to **pyruvic acid**, which is decarboxylated to **acetaldehyde**. The acetaldehyde is subsequently reduced to ethanol. Synonymous with **alcoholic fermentation**.

Ethephon   White, solid plant growth regulator which is highly soluble in water. By promoting the release of **ethylene**, it promotes the flowering of **plants** and increases the rate of **ripening**. Uses include as a flowering agent in **pineapples** and as a ripening agent in **sugar cane**. Also known as chloroethylphosphonic acid, (2-chloroethyl)phosphonic acid or ethrel.

Ethion   Non-systemic acaricide and insecticide used to control a range of **insects** (especially **mites** and **aphids**) on **fruits**, **vegetables** and **cereals**. Classified by WHO as moderately toxic (WHO II).

Ethiopian mustard   Common name for *Brassica carinata*. Eaten as a green leafy vegetable in Africa. Its potential as an oilseed crop is decreased by the high levels of **glucosinolates** in the **seeds** and of **erucic acid** in the oil.

Ethnic foods   Foods belonging to the traditional cuisine of other ethnic groups. For example, Chinese, Indian and Mexican foods are all popular ethnic foods in the UK and USA. There is an increasing tendency for consumers to try foods from other countries as cultural diversity increases. This is reflected in the continuing increase in international sales of ethnic foods, including ethnic **ready meals**, **flavourings** and **take away foods**.

Ethoxyquin   Used as an antioxidant to prevent pigment discoloration in **paprika** and **chilli** powder. Also used as a herbicide and to prevent superficial **scald** in **fruits**. Alternative term for santoquin.

Ethrel   Alternative term for **ethephon**.

Ethyl acetate   Ester which in pure form exists as a flammable, colourless, volatile liquid with a fruity **aroma**. Slightly soluble in water and soluble in alcohol. Used as a solvent, and in **flavourings** and perfumes.

Ethyl alcohol   Alternative term for **ethanol**.

Ethylamine   Amine which in pure form exists as flammable, colourless, volatile liquid with a strong **aroma** of **ammonia**. Soluble in water and alcohol.

Ethyl butyrate   Ester which in pure form exists as a flammable, colourless liquid with a pineapple-like **aroma**. Virtually insoluble in water, but soluble in alcohol. Occurs as one of the **flavour compounds** in many **fruits**, e.g. **apples**. Used in **flavourings** and perfumes.

Ethyl caproate   Fatty acid ester which in pure form is a combustible, colourless or yellowish liquid with a pleasant **aroma**. Insoluble in water, but soluble in alcohol. Used in organic synthesis and in the manufacture of artificial fruit **essences**. Alternative term for ethyl hexoate.

Ethyl carbamate   Organic nitrogen compound derived from **urea**, which in pure form is a white or colourless crystalline solid. Soluble in water, alcohol and ether, and slightly soluble in oils. A possible carcinogen that is used in **pesticides** and **fungicides**. Formed in **wines**, other **alcoholic beverages** and **fermented foods** during processing or storage. Synonym for **urethane**.

Ethylcarbamate   Alternative spelling of **ethyl carbamate**.

Ethylene   Highly flammable, colourless hydrocarbon gas with a sweetish **aroma** and **flavour**. Slightly soluble in water and alcohol. Occurs in natural gas and coal gas, and is produced by **fruits** and **vegetables** during **ripening**. Removal of ethylene from food packages is used to delay ripening of fruits. As a plant growth regulator, ethylene has many horticultural uses, e.g. as a fruit ripening accelerator.

Ethylenediamine   Amine which exists as a toxic, colourless, alkaline gas or liquid with an **aroma** of

ammonia. Soluble in water and alcohol, and readily absorbs CO_2 from air. Uses include in the manufacture of **chelating agents**, such as **EDTA**, and in **emulsifying agents**.

Ethylene dibromide Colourless, non-flammable liquid with a sweetish **aroma**. Toxic and carcinogenic. Slightly soluble in water and miscible with most organic solvents and thinners. Used in **fumigants** for grain and tree crops, as a general solvent and as a water-proofing preparation.

Ethylene glycol One of the **glycols** or **polyols**. A colourless, viscous, hygroscopic liquid commonly used as a solvent, osmotic solute, antifreeze or plasticizer. Has been used as an additive in edible films.

Ethylene oxide Highly flammable, colourless gas which liquefies at temperatures below 12°C. Soluble in organic solvents and miscible with water and alcohol. It has sporicidal and viricidal activities, and is probably carcinogenic. Sometimes used in **fumigation** of **spices**. Also known as epoxyethane or oxirane.

Ethylenethiourea Primary degradation product of ethylene-bisdithiocarbamate **fungicides** (such as **maneb** and **zineb**), which are used on a wide range of crops. A suspected carcinogen.

Ethyl vanillin Artificial flavouring, approximately 2 to 4 times stronger than **vanillin**. Synthesized from **eugenol**, isoeugenol or **safrole**. Used to enhance fruit and chocolate **flavour** notes in **ice cream**, **beverages** and **bakery products**.

EU Abbreviation for **European Union**.

Eubacteria Former name for a superkingdom of **prokaryotes**, now known as **Bacteria**.

Eucalyptol Monocyclic terpene distributed widely in plants. Occurs as a colourless liquid with a characteristic **aroma** and pungent **flavour**. Major food sources include **eucalyptus** oils, **spices** including **sage**, **rosemary** and **basil**, and **essential oils** extracted from **herbs** and spices. It is used in **flavourings** for **foods** and **beverages**. Cough candy contains particularly high levels of eucalyptol due to a high content of eucalyptus oil.

Eucalyptus Genus of trees found mainly in Australia. Leaves of some species are the source of **essential oils** that are used mainly for medicinal purposes, but can in some cases be used as food **flavourings**. Major floral source for **honeys** in Australia.

Eucaryotes Alternative spelling for **eukaryotes**.

Eucheuma Genus of red **seaweeds** occurring abundantly along shores in the southwest Pacific and Indian Ocean. Several species, such as *Eucheuma cottonii* and *E. spinosum*, are a commercially important source of **carrageenans** used by the food industry. The Philip-

pines, Indonesia and Malaysia are the largest producers of these seaweeds.

Eugenol Combustible, colourless or pale yellow phenol with a spicy **aroma** and **flavour** which is derived from oil of **cloves** and **cinnamon oils**. Only very slightly soluble in water, but soluble in alcohol, ether and volatile oils. Used in **flavourings**, perfumes, essential oil preparations, as a dental analgesic and local anaesthetic, and in the manufacture of isoeugenol for production of **vanillin**.

Eukaryotes Organisms in which the cells have a distinct nucleus containing the genetic material (**DNA**). Includes all organisms except **bacteria**. Alternative spelling is eucaryotes.

Euparen Alternative term for the fungicide **dichlofluanid**.

Euphorbia Plant genus characterized by its members producing a milky juice. Its seeds are of potential use as **oilseeds**, being a rich source of oil which contains high levels of vernolic acid.

European Community In July 1967, three organizations (the **European Economic Community** (EEC), the European Coal and Steel Community (ECSC), and Euratom) fully merged as the European Community (EC). The basic economic features of the EEC treaty were gradually implemented, and, in 1968, all tariffs between member states were eliminated. A meeting of leaders of the member states in December 1969 paved the way for creation of a permanent financing arrangement for the EC based on contributions from the member states, development of a framework for foreign policy cooperation among the member nations, and the opening of membership negotiations with Britain, Ireland, Denmark and Norway. In 1972, it was agreed that the four applicant countries would be admitted on 1 January 1973. Britain, Ireland and Denmark joined as scheduled; however, in a national referendum, the people of Norway voted against membership.

European Economic Community In 1957, the participants in the European Coal and Steel Community (ECSC) signed two more treaties in Rome, one of which created the European Economic Community (EEC, often referred to as the Common Market). The EEC treaty allowed for gradual elimination of import duties and quotas on all trade between member nations and for the institution of a common external tariff. Member nations agreed to implement common policies regarding transportation, agriculture, and social insurance, and to permit the free movement of people and funds within the boundaries of the community.

European Union The European Union (EU) is an organization representing European countries dedicated to increasing economic integration and strengthening cooperation among its members. The EU was formally established on 1 November 1993, and its headquarters are in Brussels, Belgium. The EU is the most recent in a series of European cooperative organizations that originated with the European Coal and Steel Community (ECSC) of 1951, which became the **European Community** (EC) in 1967. The members of the EC were Belgium, Britain, Denmark, France, Germany, Greece, Ireland, Italy, Luxembourg, the Netherlands, Portugal and Spain. In 1991, governments of the 12 member states signed the Treaty on European Union (commonly called the Maastricht Treaty), which was then ratified by the national legislatures of all the member countries. The Maastricht Treaty transformed the EC into the EU. In 1994, Austria, Finland and Sweden joined the EU, bringing the total membership to 15 nations. The EU primarily works to promote and expand cooperation among its members in areas such as economics and trade, social issues, foreign policy, security, and judicial matters. Another goal was to implement Economic and Monetary Union (EMU), which established a single currency for EU members.

Eurotium Genus of xerophilic **fungi** (order Eurotiales) commonly found in soil and concentrated or **dried foods**. Have anamorphic states in the form genus **Aspergillus**. Cause **spoilage** in some **foods** and **beverages**, including stored **grain**, **fruit juices** and **bakery products**.

Evaporated milk Milk concentrated by partial removal of water with the aid of a vacuum to reduce the **boiling point** and thus maintain the quality of the milk during the process. May have a range of fat contents depending on the **concentration** ratio used. After **evaporation**, the product is homogenized, mixed with **stabilizers** and sterilized in cans, or is UHT (ultra high temperature) treated combined with **aseptic packaging** in cartons. May be reconstituted by addition of water in amounts stated on the packaging.

Evaporation Gradual change of state from liquid to gas that occurs at a liquid's surface. The average speed of particles within a liquid depends on the liquid's temperature. Fast-moving particles striking other particles near the liquid's surface may impart enough speed, and therefore enough kinetic energy (energy of motion), to cause the surface particle to leave the liquid and become gas atoms or molecules. As particles with the most kinetic energy evaporate, the average kinetic energy of the remaining liquid decreases. Because a liquid's temperature is directly related to the average kinetic energy of its molecules, the liquid cools as it evaporates.

Evaporators Equipment used in turning a liquid into a vapour by **evaporation**.

Evening primrose oils Plant **oils**, extracted from seeds of members of the genus *Oenothera*, which are rich in γ-linolenic acid and **linoleic acid**. Used mainly in **dietary supplements**.

Evening primrose seeds **Oilseeds** produced by plants of the genus *Oenothera*. Used in the food industry as a source of **evening primrose oils**.

Evisceration The process of disembowelment, the cutting open and removal of the inner organs or entrails of animal **carcasses**. Similar to **gutting** of **fish**.

Ewe cheese **Cheese** made from **ewe milk**. Well-known examples include **Manchego cheese**, **Pecorino cheese** and **Roquefort cheese**. Also known as ewe milk cheese, sheep milk cheese or sheep cheese.

Ewedu Common name for *Corchorus olitorius*. Leaves are used as a pot-herb in West Africa, eaten as a **spinach** substitute in other parts of the world. Also known as moroheiya, Jew's mallow, Egyptian mallow and bush okra.

Ewe milk **Milk** produced by dairy **ewes**. Differs from **cow milk** in having significantly higher protein and fat contents. Most **minerals** and **vitamins** are also present in higher amounts in ewe milk than in cow milk, the notable exception being **carotenes**, contents of which are much lower in ewe milk. Often used in **cheesemaking**. Also known as sheep milk.

Ewe milk cheese Alternative term for **ewe cheese**.

Ewes Mature female **sheep**. The term may also be used to describe adult females of various related animals including **goats** and the smaller **antelopes**.

Exopolysaccharides Extracellular **polysaccharides** synthesized and secreted by **microorganisms**. Includes polysaccharides produced during **fermentation** of foods, and which influence the **viscosity** of the finished product, such as those produced by **lactic acid bacteria** in **yoghurt** or **fermented milk**, and also isolated microbial polysaccharides such as **gellan** which have food applications.

Exotic fruits Fruits from another part of the world or introduced from another country.

Exotic vegetables Vegetables from another part of the world or introduced from another country.

Exotoxins Potent extracellular **toxins** secreted by certain species of **bacteria** (e.g. *Clostridium botulinum* and *Staphylococcus aureus*).

Exo-1,4-β-xylosidases Alternative term for **xylan 1,4-β-xylosidases**.

Expert systems Computer application programs that make decisions or solve problems in a particular field by using knowledge and analytical rules defined by experts in the field. In an expert system, a knowledge base provides specific facts and rules about the subject, and an inference engine provides the reasoning ability that enables the expert system to form conclusions.

Exports Goods or services that are domestically produced but are sold abroad.

Expresso coffee Alternative term for **espresso coffee**.

Extensibility Extent to which a material can be distorted or stretched without breaking. It is often expressed as a proportion of the material's original size. A decrease in extensibility resulting from shortening of muscles has traditionally been used to define *rigor mortis*. Also commonly measured during assessment of the **rheological properties** of **dough**.

Extensographs Instruments used to investigate the **physical properties** of **dough**. Similar to **alveographs**.

Extractive fermentation Simultaneous extraction of **fermentation products** during **fermentation** processes. Organic solvents are frequently used for extraction, and the process often results in higher yields since it eliminates the problem of product inhibition.

Extracts Term usually applied to concentrated **flavourings** obtained by solvent extraction or supercritical extraction of substances such as **herbs**, **meat**, **yeasts** or **fruits**. May also more generally apply to any product obtained by extraction.

Extrudates Items which have been shaped by forcing them through a die (**extrusion**).

Extruders Die equipment that is used to shape items during **extrusion**.

Extrusion Process of shaping items by forcing them through a die.

Extrusion cooking The process of **cooking** of foods by forcing them through a die.

F

Faba beans Seeds produced by *Vicia faba*. Vary in shape, colour and size. Immature seeds are eaten cooked, canned or frozen, while mature seeds are dried. Immature pods are also eaten. Types of faba beans include **broad beans, horse beans, field beans**, tick beans and Windsor beans. Also known as fava beans. In some individuals, an inborn error of metabolism renders them susceptible to a component of faba beans that causes favism, a type of haemolytic anaemia.

σ Factors **Proteins** present in **bacteria** which bind to DNA-directed RNA polymerases, promoting initiation of transcription at promoters of a specific class. Involved in response of the cell to heat shock or other types of stress.

Fagara seeds Seeds produced by plants of the genus *Fagara* or *Zanthoxylum*, some of which are used as the source of **oils** used in **cooking**.

Falafel Fried croquettes of ground **chick peas** and **faba beans** seasoned with **sesame seeds**.

Falling number Indicator used to measure the activity of **α-amylases** in cereal flours. In **wheat**, a low falling number may signal reduced grain quality and poor **breadmaking properties**.

FAO Abbreviation for **Food and Agriculture Organization**.

Farina A fine **flour** or **meal** which is prepared from **cereals**, particularly **wheat**, or other plant foods with a high **starch** content. Can be used in the manufacture of foods such as **pasta**.

Farinographs Instruments used to investigate the **physical properties** of **dough**.

Farmed fish Fish produced by **fish farming** for food purposes. A wide range of fish species are farmed worldwide. Major farmed fish of commercial importance include **Atlantic salmon, rainbow trout, carp, channel catfish, tilapia** and **yellowtail**.

Farmed shellfish Shellfish produced for food purposes by **aquaculture**. A wide range of shellfish species are produced by this process worldwide. These include **mussels, clams, oysters, scallops, shrimps** and **lobsters**.

Farm milk **Milk** collected directly from the producer.

Farnesene One of the sesquiterpenoid volatile **aroma compounds**. Isomers include α-farnesene, which is synthesized in **apples** and is related to the development of **scald**, and β-farnesene which, along with α-farnesene, is a constituent of **essential oils** in several plants including **hops** and citrus species.

Farnesol Terpenoid alcohol which exists as a combustible, colourless liquid with a delicate floral **aroma**; it has low toxicity. Occurs naturally in many **essential oils** and flowers. Used in **flavourings** and perfumes.

Fasciola Genus of parasitic flatworms of the class Trematoda. *Fasciola hepatica* is the causative agent of fascioliasis, which is of great economic importance in cattle and sheep. Human fascioliasis may result from eating raw or improperly cooked **watercress**.

Fast foods Prepared foods obtained from **restaurants** and other catering establishments, where the aim is to provide a fast service and rapid customer turnover at reasonable prices. Examples of fast foods include **burgers, pizzas, sandwiches** and **French fries**.

Fat mimetics Alternative term for **fat substitutes**.

Fat replacers Alternative term for **fat substitutes**.

Fats Non-volatile, water insoluble substances that are usually solid at room temperature and are greasy to the touch. Composed of **esters** synthesized by reaction of **fatty acids** with **glycerol** in a ratio of 3 to 1 to form **triacylglycerols** or **triglycerides**. Arrangement and type of the fatty acids in the glycerol molecule affect the **physical properties** of the fat.

Fat substitutes Substances of various types and origins that show similar properties to **triacylglycerols** in that they have a creamy and fat-like texture but have low **calorific values**. Used in the complete or partial replacement of **fats** in foods, e.g. **low fat foods**. Also known as fat mimetics or fat replacers.

Fattening Feeding of domesticated animals to produce a desirable body weight and body composition for slaughter.

Fatty acid esters **Esters** formed between **fatty acids** and a range of other compounds including **sugars, alcohols, polyols, carotenoids** and **sterols**. Fatty acid methyl esters are commonly prepared from **triglycerides** for GC analysis of fatty acid composition.

Fatty acids **Organic acids** consisting of a chain of alkyl groups containing between 4 and 22 or more carbon atoms with a terminal carboxyl group. In **saturated fatty acids**, e.g. **butyric acid**, **palmitic acid** and **stearic acid**, the carbon atoms of the alkyl chains are connected by single bonds. **Unsaturated fatty acids**, e.g. **oleic acid**, **linoleic acid** and **linolenic acid**, contain at least one double bond. Fatty acids occur naturally and are derived from **animal fats**, **fish oils** and **vegetable fats**. Some, e.g. linoleic, linolenic and **arachidonic acid**, are essential nutrients (essential fatty acids) that are not synthesized in the human body and must be obtained from the diet.

ω-3 Fatty acids **Polyunsaturated fatty acids** having double bonds in the ω-3 position; found in oily fish. May have beneficial effects on health, especially resistance to **cardiovascular diseases**.

ω-6 Fatty acids **Polyunsaturated fatty acids** having double bonds in the ω-6 position. Found in **vegetable oils**. May have beneficial effects for health, especially reducing the risks for **cancer**, **stroke** and **coronary heart diseases**. Include **arachidonic acid**, **linoleic acid** and γ-linolenic acid.

Fatty acid synthases EC 2.3.1.85. Catalyse the synthesis of long chain **fatty acids**. Studies have shown that dietary **polyunsaturated fatty acids** can reduce the activity of this enzyme in animal models. Also thought to be involved in the biosynthesis of **aflatoxins** in *Aspergillus*.

FDA Abbreviation commonly used for the US **Food and Drug Administration**.

FDC blue **Artificial colorants** certified under the US Food Drug and Cosmetic Act. FDC blue 1 (also known as **Brilliant Blue** FCF) and FDC blue 2 (**indigotine**) are currently permitted for food use in the USA.

FDC colours **Artificial colorants** certified by the US Food & Drug Administration as suitable for use in foods under the Food, Drug and Cosmetic (FD&C) Act. FDC colours currently permitted for food use in the USA include **FDC red** 3, red 40, blue 1, blue 2, yellow 5, yellow 6 and green 3.

FDC red **Artificial colorants** certified under the US Food Drug and Cosmetic (FD&C) Act. Red colorants currently certified for food use in the USA are FDC red 3 (**erythrosine**) and FDC red 40 (**Allura red AC**).

FDC yellow **Artificial colorants** approved for food use under the Food, Drug and Cosmetic (FD&C) Act. FDC yellow 5 (**tartrazine**) and FDC yellow 6 (**sun-set yellow** FCF) are currently permitted for food use in the USA.

Fe Chemical symbol for **iron**.

Feathers Flat appendages growing from the skin of birds, consisting of a partly hollow horny shaft fringed with vanes of barbs. Poultry feather **wastes** accumulate during **poultry** processing; they represent an underutilized protein resource, and their disposal carries pollution concerns. *Bacillus* licheniformis secretes keratinase, a proteolytic enzyme which is active on whole feathers, with the ability to hydrolyse collagen, elastin and feather keratin; this enzyme has potential in the **bioremediation** and management of poultry wastes. Feather lysate is a digestible protein source that can be used in animal feeds.

Feeds Materials available for feeding domestic animals, which may be classified loosely into four major groups, namely: green forages; succulent feeds, roots and tubers (e.g. turnips); coarse fodder (e.g. hay) derived from grasses; and concentrates (e.g. cereal grains, oilseeds and various animal by-products).

Feijoa Dark green tropical fruits with white flesh, which are produced by *Feijoa sellowiana*. Used mainly in jellies and **preserves**. Also known as pineapple guavas and guavasteens.

Fenbendazole Anthelmintic used for treatment and control of gastrointestinal roundworms, lung worms and tapeworms in cattle, sheep, pigs and goats. Normally undetectable 7 days after final treatment in all animal tissues except **livers**, where residues may remain for longer periods.

Fenitrothion Non-systemic insecticide with cholinesterase inhibitory activity, used for control of chewing, sucking and boring **insects** in **fruits**, **vegetables** and **cereals**. Also used for control of insects in animal rearing facilities and in stored cereals. Classified by WHO as moderately toxic (WHO II).

Fennel Common name for the plant *Foeniculum vulgare*. Florence or Florentine fennel is eaten as a vegetable. The edible part, eaten raw or cooked, is a false bulb formed by the leaf bases. Has an aniseed **flavour**, and is a good source of potassium and selected **vitamins** and **minerals**. **Fennel seeds** are also harvested for use as a spice and for their **essential oils**.

Fennel seeds Liquorice-flavoured **seeds** from *Foeniculum vulgare*. Used for seasoning **bakery products**, **cheese**, and a number of **meat** and vegetable dishes. Seed oils are used in **liqueurs** and fragrances.

Fenthion Insecticide used for control of chewing, sucking and boring **insects** on fruits, vegetables, tea and sugar cane. Also used for insect control in animal rearing facilities and control of animal ectoparasites. Normally eliminated rapidly from plants and animals.

Classified by WHO as moderately toxic (WHO II). Also known as baytex.

Fenugreek Common name for the leguminous plant *Trigonella foenumgraecum*. Fenugreek **seeds** are used as **spices** in **curry powders**, chutneys and imitation **maple syrups**. The plant itself is rich in **carotenes** and is consumed as a vegetable.

Fenvalerate Non-systemic insecticide and acaricide used on crops for control of a wide range of **insects**, including those resistant to **organochlorine pesticides**, **organophosphorus pesticides** and **carbamate pesticides**. Also used for insect control in animal rearing facilities and as an ectoparasiticide. Classified by WHO as moderately toxic (WHO II).

Fermentation Energy-yielding process in which **organic compounds** are metabolized, usually under anaerobic or microaerobic conditions, to simpler compounds without the involvement of an exogenous electron acceptor. Commonly refers to processes carried out by **microorganisms**, regardless of whether fermentative or respiratory metabolism is involved. Used frequently in the food industry, e.g. for production of **alcohols**, **bread**, **vinegar**, **flavour compounds**, and a wide variety of **fermented foods** and **fermented beverages**.

Fermentation products Products of microbial **fermentation** processes, e.g. **alcohols**, **flavour compounds**, **food additives**, **surfactants** and **organic acids**.

Fermentation technology Technologies and methods used for production of specific products by means of microbial **fermentation**.

Fermented beverages Beverages whose manufacture involves a **fermentation** process, generally **alcoholic fermentation** and/or **lactic fermentation**.

Fermented cream Cream acidified naturally or artificially by the action of **lactic acid bacteria**. **Lactose** in the cream is converted to **lactic acid** by **fermentation**. Types of fermented cream include **sour cream**, **ripened cream** and **smetana**. Used in **cooking** and **baking**, and in **dips**.

Fermented dairy products Produced by **fermentation** of liquid **dairy products** by **lactic acid bacteria** (**starters**). During fermentation, **lactose** is converted into **lactic acid** and sometimes **flavour compounds** such as **diacetyl**, depending on the organisms used and fermentation conditions. Fermentation is allowed to proceed until the required **acidity** is achieved. In some cases, where **yeasts** are also present, **alcohol** is formed in the final product, e.g. **kefir**, **koumiss**. Fermented dairy products include **fermented cream**, some types of **butter**, **cheese**, cul-

tured **buttermilk**, and **fermented milk**, a popular type of which is **yoghurt**. Many traditional fermented dairy products exist throughout the world. Consumption of fermented dairy products, especially those containing specific organisms or **probiotic bacteria**, can enhance intestinal health.

Fermented foods Foods subjected to **fermentation** by beneficial **microorganisms** in order to bring about desirable changes. These changes are mainly concerned with **preservation** (e.g. manufacture of **cheese** and **yoghurt** from **milk**), enhancement of nutritional value (e.g. removal of **antinutritional factors** from **legumes**), or alteration of **flavour** and **texture** (e.g. manufacture of **soy sauces** from **soybeans**). Many different types of fermented foods and **fermented beverages** are available, and play a major part in the human diet. Fermentation is favoured as an inexpensive method of preservation in developing countries and many items, such as **fermented dairy products**, are attracting increasing attention as **functional foods** due to the beneficial actions of the microorganisms and/or enzymes involved in fermentation. Also known as fermented products or cultured foods.

Fermented milk Produced by **fermentation** of **milk** (of various species) by **lactic acid bacteria** (**starters**). During fermentation, **lactose** is converted into **lactic acid**, **aroma compounds** are formed and **milk proteins** are partly decomposed to **peptides** and free **amino acids**, improving **digestibility** of the milk. If **yeasts** are included in the starter mixture, **alcohol** is also present in the final product. Consumption of fermented milk may have many health benefits including alleviation of the symptoms of gastrointestinal disorders. Many types of fermented milk are produced throughout the world, including **yoghurt**, **kefir**, **dahi**, **shubat** and **shrikhand**.

Fermented products Alternative term for **fermented foods**.

Fermented sausages Traditionally produced by chance contamination of **sausages** with local **microorganisms**. In modern practice, however, **starters** are usually added to sausage emulsions in order to produce a more uniform product. Lactic starters are often included, but other microorganisms, particularly those with good nitrate reducing abilities, are also used. As well as affecting sausage **colour** and **consistency**, **fermentation** has major effects on **flavour**. Raw fermented sausages are prepared from unheated raw meat, which is fermented and then held at a controlled temperature and relative humidity until the desired degree of dryness is obtained. In contrast, heat-treated fermented sausages are pasteurized after fermentation and then dried, usually for a brief period.

Fermenters Vessels in which aerobic or anaerobic **fermentation** processes can be carried out, in either batch culture or continuous culture. Typically vertical, closed, cylindrical steel vessels that can range in volume from less than one litre to several thousand litres. Usually have means for ensuring adequate **heat transfer**, **mixing** and **aeration**.

Ferns Non-flowering plants often used as a food. Young shoots (**fiddleheads**) and rootstocks of wild and cultivated species are consumed in a number of countries.

Fertilizers Natural or synthetic substances supplied to plants via the soil or in water to enhance their growth or yield of produce. Commonly contain nitrogen, phosphorus and/or potassium. Concerns about environmental and health hazards associated with their use have led to increasing popularity of organic foods that are cultivated without the use of fertilizers.

Ferulic acid A phenol which in pure form exists as colourless needles and is soluble in water and alcohol. Occurs naturally in plant cell walls and is an *in vivo* substrate for plant **peroxidases**. Displays **antioxidative activity** and is used in food **preservatives**. Microbial and enzymic transformations of ferulic acid can be used to produce useful **aromatic compounds**, including **flavour compounds**, such as **vanillin**.

Feta cheese Greek **soft cheese** made originally from **ewe milk** or a mixture of ewe and goat milks. Pure white, with a crumbly **texture** and high salt content. Saltiness can be reduced by soaking in cold water or milk for a few minutes before consumption.

FIA Abbreviation for **flow injection analysis**.

Fibre Term with a variety of meanings. Sometimes used to mean **dietary fibre**, a complex mixture of **carbohydrates**, also known as non-starch polysaccharides, derived from plant cell walls and resistant to digestion in the intestinal tract. Crude fibre is the indigestible part of foods, determined from the residue remaining after extraction under specified conditions. Soluble fibre is the part of dietary fibre that is soluble and forms a gel in water. Also refers to long thin thread-like structures, such as muscle, glass, nylon and nerve fibres.

Fibreboard Strong material made from wood or other plant fibres which are compressed, with or without binders, into boards. Used to make containers, such as **crates**, and panelling. May be corrugated to improve cushioning characteristics. Compared with plastics, fibreboard has both cost advantages and environmental benefits, particularly as it may be made from recycled fibres.

Fibre optics Application of superfine glass fibres as light conduits in sensors where measurement is based on light transmission. Used in analytical procedures and also in monitoring food processing operations.

Fibrin Insoluble animal protein which is produced on hydrolysis of **fibrinogen** by thrombin. Forms a network of fibres during blood clotting.

Fibrinogen Soluble animal protein which is secreted into blood plasma by the liver. Converted into **fibrin** by the action of thrombin during blood clotting.

Fibrinolysin Alternative term for **plasmin**.

Fibrobacter Genus of Gram negative, obligately anaerobic rod-shaped or coccoid **bacteria**. Occurs in the rumen of ruminants. *Fibrobacter succinogenes* produces **endonucleases** which are capable of degrading **celluloses** and **xylan**.

Ficains EC 3.4.22.3. Also known as ficins, these enzymes are the major proteolytic components of fig latex. They are cysteine **proteinases** that cleave preferentially at Tyr and Phe residues, and that can act on a wide variety of protein substrates. Used as **milk clotting enzymes** during production of teleme, a traditional Turkish dairy product with **yoghurt**-like **texture** and sweet taste.

Ficins Alternative term for **ficains**.

Fiddleheads Edible tightly-curled tips of young fern fronds taking their name from their resemblance to the end of a violin, or fiddle. Deep green in **colour** with a chewy **texture**. Eaten cooked as a starter or side dish, or raw in **salads**. Rich in **vitamin A** and **vitamin C**.

Field beans Type of **faba beans**.

Field peas Variety of peas grown specifically for **drying**, usually not requiring **soaking** before **cooking**. Called split peas when split along the natural seam. Can be yellow or green.

Fig juices **Fruit juices** prepared from **figs** (*Ficus carica*).

Figs **Fruits** produced by *Ficus carica*. Consumed fresh or preserved by **canning** or **drying**. Used as ingredients in **bakery products**. During the process of drying, sugar content of figs increases to about 50%, potassium content increases approximately 5-fold, but the already low **vitamin C** level is halved.

Filberts Alternative term for **hazelnuts**.

Filefish **Marine fish** from the family Balistidae, chiefly found in tropical ocean waters. Generally characterized by a flat body and rough spiny scales. Some species are caught as food fish, including Alutera monoceros (unicorn filefish) and *A. schoepfi* (orange filefish). Usually sold fresh with skin removed.

Fillers Devices used to transfer products into **containers** or **casings** for storage or retail. These are alternatively referred to as **filling equipment**. Also refers to substances, e.g. **buttermilk** powders, **whey protein concentrates** or **corn starch**, used to extend meat batters or **emulsions** to add bulk or **functional properties**.

Filleting Removal of the **bones** from a piece of **meat** or **fish**, so creating a fillet.

Filling Process of transferring products, e.g. foods and **beverages**, into **containers** or **casings** for storage or retail. Also refers to the developmental stage in cereals in which kernel **dry matter** increases.

Filling equipment Devices used to transfer products into **containers** or **casings** for storage or retail. Also sometimes called **fillers**.

Fillings Sweet or savoury ingredients used to fill a cavity within or between layers of a food, e.g. **confectionery fillings**, **pie fillings** and **stuffings**.

Filter aids Materials such as **bentonite, clays, diatomaceous earths** and **kieselguhr** employed to facilitate the course of **filtration**.

Filters Porous devices for removing solid particles from a liquid or gas passed through them.

Filth Food **contaminants** such as hairs, rodent and bird faeces, and insect and feather fragments.

Filth tests Microscopic examinations of food for the presence of **contaminants** (e.g. hairs, rodent and bird faeces, insect and feather fragments). Used as an index of hygienic food processing and handling.

Filtration Process of removing suspended solids from a liquid by straining it through a porous medium that can be penetrated easily by liquids.

Fimbriae Short, thin, hair-like appendages, composed mainly of the protein pilin, which extend from the surface of certain bacterial cells, mainly Gram negative **bacteria**. May be involved in a variety of functions, including bacterial **adherence** to other cells and substrates. Can act as important **virulence factors**. Also known as pili.

Finfish A term used to separate true fish from other **sea foods**, such as **shellfish, crayfish** and **jellyfish**.

Finger millet Edible cereal from *Eleusine coracana* of importance in India and Africa. Used in **porridge** and **gruel**, and to make **beer**. Also known as kurakkan and **ragi**.

Fining **Clarification** of **beer** or **wines** by removal of the minute floating particles that prevent these products from being clear.

Fining agents Substances used for **clarification** of **wines** or **beer**, including **gelatin, isinglass** and **diatomaceous earths**.

Finometers Instruments used to measure the **tenderness** of foods, especially **peas** and **beans**, by determining mechanical resistance.

Fior di Latte cheese Italian **fresh cheese** similar to **Mozzarella cheese**.

Firmness **Texture** term relating to the extent to which a product is dense and firm.

Fish Any of a variety of cold-blooded vertebrate animals found in the fresh and salt waters of the world, ranging from the primitive, jawless **lamprey**, through the cartilaginous **sharks, skate** and **ray**, to the abundant and diverse bony fishes, which include the majority of food fish.

Fish balls Fish products consisting of flesh from white **fish** (such as **cod** or **haddock**) mixed with **milk**, fish stock, **flour** or other binding ingredients, and **seasonings**, which are then shaped into balls and cooked. Marketed as semi-preserved, canned or frozen products. Alternatively known as fish dumplings.

Fishburgers Fish products consisting of minced **fish** flesh, **seasonings** and **preservatives**. Often coated with **batters**, and sold in pre-cooked, frozen form.

Fish cakes Cooked fish products made from fresh or salted **fish**, mixed with **potatoes** and **seasonings**; sometimes **eggs, butter** and **onions** are added. Fish content may range from 35 to 50% by weight. A variety of fish are used, including **cod, haddock, coalfish** and **salmon**.

Fish crackers Fish products popular as **snack foods** in some Asian countries (known as keropok in Malaysia). Commonly made by mixing minced fish flesh with **sago** flour, **tapioca** flour, **salt** and **monosodium glutamate**. The mixture is then moulded into cylinders, steamed, cooled, sliced and sun-dried.

Fish farming Production of **fish** (usually referring to **finfish**) under controlled or semi-controlled conditions for food or industrial purposes. Major **farmed fish** in commercial terms include **Atlantic salmon, rainbow trout, carp, channel catfish, tilapia** and **yellowtail**.

Fish fillets Strips of **fish** flesh cut parallel to the backbone, starting just behind the head of the fish; fins, bones and discoloured flesh are normally removed, but skin may remain.

Fish fingers Fish products consisting of rectilinear portions cut from a block of frozen **fish** flesh; typically the length is about three times the breadth and product weight is approximately 18 g. Often coated with **batters** or **breadcrumbs** and fried in oil.

Fish hydrolysates Products formed from minced or comminuted **fish** (often processing wastes) after treatment with hydrolytic **enzymes**, filtration and drying. Average product consists of 85% hydrolysed protein (mainly small peptides and free **amino acids**), 10% inorganic material and 5% water. Mainly used as **flavourings** in **soups** and in animal feeds.

Fish in juices **Fish** stored in their own juices.

Fish in marinades **Fish**, especially **herring**, soaked in **marinades** (seasoned liquids containing **acetic acid**, **vinegar**, **olive oils**, or **brines**, with or without **spices**) to retard the action of **bacteria** and **enzymes**. Have a characteristic **flavour** and an extended, but limited, **shelf life**.

Fish in oils **Fish** (e.g. **tuna**, **mackerel**) stored in edible **oils**, especially **vegetable oils** such as **soybean oils** or **sunflower oils**. Usually sold as canned products.

Fish in sauces **Fish** (e.g. **sardine**) stored in **sauces**, such as **tomato sauces**. Usually sold as canned products.

Fish liver oils Lipid extracts from **fish livers**, which are rich in *n*-3 **fatty acids**, **vitamin A** and **vitamin D**. Include shark, dogfish, halibut and **cod liver oils**.

Fish livers Livers from some **fish** species which are utilized as foods. Fish valued for their livers include **cod**, **halibut**, **tuna**, certain **sharks** and **mackerel**. Marketed in fresh, frozen, salted and canned forms; also mixed with **fish oils** and **spices** to make pastes.

Fish meal Dried, powdered or granular product obtained from cooked whole **fish** or fish processing **wastes**; fish species frequently used for its production include **anchoveta**, **capelin**, sand eel, **herring** and **mackerel**. Constitutes a valuable ingredient of animal feeds. Sold on the basis of its protein content and rated according to the percentage of protein contained in the product.

Fish mince **Fish** flesh finely cut or crushed into small particles. Often used to make a variety of fish products such as **kamaboko**, **fish fingers** and **surimi**.

Fish noodles Fish products consisting of minced fish flesh mixed with **wheat flour** (or other cereal flour), **cassava starch** and various **additives**; the mixture is extruded through tubular holes to form long strands which are dried in hot air. Products are boiled prior to consumption. A popular meal accompaniment is some parts of South East Asia.

Fish nuggets Fish products comprising pieces of **fish** flesh (not minced) formed into small irregular shapes. May be formed from fillets, fillet pieces or fish blocks; normally occurs in breaded form. Marketed frozen.

Fish oils Lipids obtained from muscle, livers or other organs of fish, particularly **herring**, **menhaden**, **anchovy**, **sardine** and **cod**. Contain *n*-3 **polyunsaturated fatty acids**, the principal ones being **eicosapentaenoic acid** and **docosahexaenoic acid**, which are reported to protect against heart disease. Used in manufacture of **margarines** and **cooking oils**.

Fish pastes Fish products consisting of minced **fish** flesh mixed with **salt**, with or without **spices** and **flavourings**, and ground to a fine consistency with reduced moisture content and often added fat. Frequently marketed as a sandwich spread. Legally, fish pastes should contain not less than 70% fish.

Fish pates Fish products consisting of finely ground **fish** flesh (having reduced moisture content), with added **seasonings** and **flavourings**. Available in spreadable or sliceable forms.

Fish preserves Fish products consisting of **fish** (whole, headed or filleted) preserved in **oils**, **brines**, **sauces** or pickling solutions and stored in cans or sealed glass containers. Fish with high fat content in the flesh (such as **mackerel**, **sardine** and **tuna**) are often used to make preserves.

Fish protein concentrates Dried fish products prepared from ground whole **fish**, which contain enhanced protein content (around 80%) and are marketed in powdered and granular forms; used as food ingredients. Commonly abbreviated to FPC. Occur in two forms, A and B; type A is odourless and colourless, while type B has an odour and flavour associated with its higher fat content.

Fish proteins **Proteins** extracted from **fish** bodies, often from processing wastes such as fish heads and **offal**. Some fish proteins form useful food ingredients due to their functional properties. The major components of **fish protein concentrates** and **fish hydrolysates**.

Fish sauces **Sauces** prepared by fermenting salted **fish** with endogenous **enzymes** for long periods at elevated temperatures until solubilization is achieved. A rich source of certain **amino acids**, especially **lysine** and **methionine**. Popular in South East Asia.

Fish sausages Fish products consisting of ground **fish** flesh, often **tuna**, mixed with small amounts of **fats**, **seasonings** and sometimes a cereal product. The mixture may be cooked or smoked before being packed in **sausage casings**. Marketed in skinless or skinned forms, semi-preserved or canned.

Fish soups **Soups** made from **fish** or other marine animals; usually contain **seasonings** and may contain pieces of fish flesh. Marketed in canned, dried or bottled forms.

Five fingers Fruits produced by *Averrhoa carambola*. The waxy fruits are yellow and juicy, and star-shaped in cross-section. Rich in **vitamin C**, with moderate amounts of sugar. Eaten raw or preserved, and used in

beverages. Also known as **carambolas** and **star fruit**.

Flagellins Protein subunits that make up the filaments of bacterial flagella.

Flaking Process of breaking an item up into small, flat, very thin pieces.

Flame photometry **Spectroscopy** in which a solution of the substance to be analysed is vaporized by introduction into a flame. Spectral lines resulting from a light source going through the vapours are analysed for characteristic bands, the intensity of which are related to the quantity of analyte present in the sample.

Flaming A method of food presentation. Warmed **spirits**, such as **brandy** or **rum**, are sprinkled over foods and ignited just before the product is served.

Flammulina Genus of **edible fungi**, the most commonly consumed species being *Flammulina velutipes* (winter mushroom), an alternative term for *Collybia velutipes*.

Flan cases Bases made of shortcrust **pastry** or **sponge cakes** which are frequently baked blind (i.e. without any **fillings**) and then filled with sweet **fillings**, such as **fruits**, **custards** or **cream**, or savoury mixes of **meat**, **vegetables** or savoury custard, after **baking**.

Flans Open **tarts** containing sweet or savoury fillings such as **custards**, **fruits** or **cheese**.

Flatfish Any **fish** belonging to the order Heterosoma, e.g. **halibut**, **turbot**, **sole**, **plaice** or **flounders**. Most are marine species and many are commercially important food fish.

Flavanols **Flavonoids** which contain a hydroxyl group. In pure form they exist as yellow, needle-like crystals. They include the hydroxyflavones, chrysin, fisetin and **quercitrin**. Occur in many plant foods and **wines**, and are associated with **bitterness** and **astringency**. They may be used to reduce the perceived **sweetness** of foods and **beverages**.

Flavanones One of the major groups of **flavonoids** derived from flavone. In pure form they are colourless, crystalline solids. Found in the tissues of higher plants, including **fruits**, particularly **citrus fruits** and **apples**, and **vegetables**, either in free form or as glucosides. Flavonones occurring in foods include **hesperidin**, **naringenin** and eriodictyol.

Flavins Naturally occurring yellow **pigments** which have a tricyclic aromatic molecular structure. Soluble in water. Include **riboflavin** and its products, e.g. flavin mononucleotide (FMN) and flavin adenine dinucleotide (FAD).

Flavobacterium Genus of Gram negative, aerobic, rod-shaped **bacteria** of the family Flavobacteriaceae. Occur in soil, water, raw meat and milk. Species may be responsible for **spoilage** of **meat**, **fish**, **milk** and **dairy products**. Some species are opportunistic **pathogens** of humans.

Flavomycin Aminoglycoside antibiotic used primarily as a growth-promoting agent in cattle, swine and poultry. Rarely absorbed in the gut of animals and normally excreted rapidly. No withdrawal period is required.

Flavones Flavonoid **pigments** which in pure form exist as colourless, crystalline solids. Insoluble in water. Occur in higher plants, including **fruits** and **vegetables**, and are responsible for ivory and yellow colours in plants and flowers. Some plants contain high levels of flavones, e.g. **parsley** (*Petroselinum crispum*) contains high levels of **apigenin**. Dietary flavones are believed to have various health benefits, e.g. flavones in **tea** and **red wines** may protect against **cancer** and **cardiovascular diseases**. Flavones are used to derive various yellow **dyes**.

Flavonoids Large group of aromatic, oxygen-containing, heterocyclic **pigments**. Include various subgroups of compounds, such as catechins, **flavanols**, **flavanones**, **flavones**, **flavonols**, **anthocyanins** and **leucoanthocyanidins**. Occur widely in higher plants and are responsible for the majority of yellow, red and blue colours in **fruits**, **vegetables** and flowers (with the exception of colours produced by **carotenoids**). Major dietary sources are **apples**, **onions**, **red wines** and **tea**. Believed to protect against **cancer** and **cardiovascular diseases**. Mechanisms of these inhibitory effects are not fully understood, but they are thought to involve inhibition of low density lipoprotein oxidation.

Flavonols **Flavonoids**, distinct from **flavanols**, which contain a hydroxyl group. Include **kaempferol** and **quercitrin**. Occur naturally in plants and are responsible for the ivory and yellow colours of many flowers. Dietary sources include **fruits**, **vegetables**, **red wines**, **green tea** and **black tea**. Believed to protect against **cardiovascular diseases** and **cancer**.

Flavour **Sensory properties** of foods. The tongue can distinguish five separate tastes (sweet, salt, sour, bitter and savoury/**umami**) due to the stimulation of the taste buds. The overall flavour of foods is a combination of these components, together with **astringency** in the mouth, **texture**, and **aroma**.

Flavour compounds Compounds present in substances that give foods their characteristic **flavour**; components capable of stimulating the sense of taste.

Flavoured beverages **Beverages** with added natural or synthetic **flavourings**.

Flavoured milk **Milk** containing **flavourings**.

Flavoured yoghurt Yoghurt containing **flavourings**, usually fruit-based.

Flavour enhancers **Flavourings** used to enhance the original **flavour** and/or **aroma** of a food, without imparting a characteristic taste or aroma of their own. Include **monosodium glutamate** and **ribonucleotides**. Similar to **flavour modifiers**.

Flavourings Substances whose primary purpose as food additives is to impart **flavour** and/or **aroma** when added to foods. Include **natural flavourings**, such as **essential oils** and **spices**, **artificial flavourings**, **seasonings**, **condiments** and extracts. Also known as aromatizing agents or flavours.

Flavour modifiers **Flavourings** used to modify the original **flavour** and/or **aroma** of a food. Similar to **flavour enhancers**.

Flavours Alternative term for **flavourings**.

Flavour thresholds Term used in **sensory analysis** relating to the levels at which perception of increasing concentrations of **flavour compounds** begins.

Flax seed oils Amber to yellow coloured **oils** derived from the cotyledons of **flax seeds** (*Linum usitatissimum*). Use of hydraulic pressure extraction results in pale coloured oils which are bland in **flavour**, while heat and pressure extraction produces darker oils with a bitter flavour and **off odour**. Rich in α-linolenic acid and often used as a food oil. Also known as **linseed oils**.

Flax seeds Seeds from plants belonging to the species *Linum usitatissimum* which are used principally in the production of **flax seed oils** or **linseed oils**. The flax seed meal which remains when the oil has been extracted from flax seeds is used as a livestock feed. Dried flax seeds are also sometimes used in medicinal preparations.

Fleischwurst Ring-shaped German **sausages** which are a type of **bruehwurst** and have a high percentage fat content. Preparation involves **smoking** and scalding.

Flexible packs Packs which are capable of bending easily and repeatedly without breaking.

Flies Common name for species of **insects** of the order Diptera (e.g. blowflies, fruit flies and midges) characterized by one pair of wings and another pair of modified wings used for equilibrium (balancers). May specifically refer to the common housefly, *Musca domestica*, which can carry **bacteria**, inside its body or on its body hair, that can contaminate foods and thereby spread diseases.

Floc A small clump or mass of colloidal particles formed in a fluid by the process of **flocculation** - **aggregation** of fine suspended particles.

Flocculants Substances that induce the formation of a mass of colloidal particles (**floc**) in a dispersion of solids in a liquid.

Flocculation Formation of a mass of colloidal particles (**floc**) in a dispersion of solids in a liquid, or alternatively, removal of suspended solids by coalescence.

Florentines Chewy, thin **biscuits**, often **chocolate**-coated on one side, containing **nuts** and **dried fruits** or **candied fruits**.

Florfenicol Broad spectrum bacteriostatic antibiotic which is a fluorinated analogue of **thiamphenicol**. Similar range of activity to that of **chloramphenicol**, including many **Gram negative bacteria** and **Gram positive bacteria**. Used mainly in treatment of **cattle** and **salmon**. Not labelled for use in lactating cattle or veal calves in some countries.

Flotation Technique in which different types of solid particles in a liquid are separated out, the principle being that some particles will absorb water while others will not.

Flounders General name used for a number of marine **flatfish** species, especially those in the order Pleuronectidae; many species are highly valued food fish, having fine-textured flesh with a delicate **flavour**. Commercially important flounders include *Platichthys flesus* (European flounder) and *Limanda ferruginea* (yellowtail flounder).

Flour Powders made from finely ground, sifted cereal grains. Used as a basic ingredient for **bakery products** and many other products. Unless specified otherwise, the term usually refers to the product of **wheat** grains.

Flour improvers Substances added to milled **flour** to improve its **colour** and/or **baking properties**. These include **oxidizing agents** to accelerate flour ageing (e.g. **potassium bromate**, **ascorbic acid**) and **bleaching agents** such as **chlorine dioxide**.

Flour mills Machines or devices for grinding grain into **flour** and other **cereal products**.

Flow Rheological property concerned with the characteristics of movement of a substance. Flow behaviour affects processing properties of a substance and **texture** of the final product. It is affected by properties such as cohesion and internal friction.

Flow cytometry Technique for sorting, selecting or counting individual cells in a suspension as they pass individually through a small hole or tube in a flow cytometer. May refer specifically to such a technique which involves the detection of a cell-bound fluorescent or fluorochrome label.

Flowers Part of a plant where the fruit or seed develops. Usually brightly coloured to attract insects. Not all flowers are edible, but those that are may be used as a garnish or integral part of a dish. Edible types include nasturtiums, pansies, violas, roses, chrysanthemums and **marigolds**. Flowers may also be candied or crystallized and used to decorate **desserts** or **cakes**.

Flow injection analysis Automated technique in which liquid samples are injected into a stream in which they are mixed with reagents and carried to a sensor or detector for measurement. Compatible with many detection systems, including **spectroscopy**, electrochemical apparatus and **immunoassay**. Commonly abbreviated to FIA.

Flow meters Devices for measuring the flow of a gas or liquid through pipes or other types of equipment.

Fluidization Process in which finely divided solids (e.g. **catalysts**) are made to behave in the same way as fluids by suspending them in moving gases or liquids. The principle is used in **fluidized beds**.

Fluidized beds Novel class of heat transfer media. A fluidized bed is produced by passing a stream of gas or liquid upwards through a bed of particles at sufficient velocity to suspend the particles (**fluidization**). In this state, the mixture of particles and fluid behave like a liquid having density equal to the bulk density of the particles. Circulation of particles in the bed, particularly by the vigorous mixing action of bubbles rising through the bed, results in large **heat transfer** rates between the bed and immersed surfaces. In some instances, the heat transfer rate may be orders of magnitude greater than achieved using the same fluid flow conditions in the absence of particles. Fluidized beds are used in many chemical engineering processes where small solid particles must be brought into intimate contact with a gas stream. Examples are **drying** of finely divided solids, **adsorption** of solvent vapours from air, and heterogeneous catalytic reactions.

Flukes Common name for parasitic flatworms belonging to the class Trematoda.

Flumequine A quinolizine carboxylic acid derivative belonging to the **quinolones** group of synthetic **antibiotics** used to combat intestinal infections in livestock. Residues may occur in the tissues or **milk** of treated animals.

Fluoranthene Polycyclic (tetracyclic) aromatic hydrocarbon which in pure form exists as coloured, needle-like crystals. It is moderately toxic and carcinogenic, insoluble in water, but soluble in organic solvents. Found in coal tar and petroleum. It may occur in foods as a result of **contamination**, **migration** from **packaging** or certain food processing practices, e.g. traditional wood **smoking** of **fish** or **cheese**.

Fluorene Polycyclic (tricyclic) aromatic hydrocarbon which in pure form exists as small white crystalline plates. A mutagen which is insoluble in water, but

soluble in alcohol or ether. May occur in foods as a result of certain food processing practices, e.g. can be formed in wood smoked **cheese** or barbecued **meat**. **Contamination** of foods may also result from pollution.

Fluorescence Absorption of radiation to produce radiation of a longer wavelength, a phenomenon exploited in **fluorescence microscopy**.

Fluorescence microscopy **Microscopy** in which samples are illuminated with UV or blue light causing them to emit light of longer wavelengths.

Fluorescent light Visible light produced by **fluorescence**, especially that from a discharge tube in which a phosphor on the inside of the tube is made to fluoresce by ultraviolet light from mercury vapour. Fluorescent light can accelerate oxidative deterioration of foods such as **oils**, **nuts** and **milk** during storage.

Fluoridation Addition of traces of **fluorides**, particularly to **water supplies** and toothpastes as a means of preventing tooth decay. Fluoridation of water supplies is a controversial issue due to possible health hazards associated with long-term ingestion of high levels of **fluorides**.

Fluorides **Salts** which contain **fluorine**. Often added to toothpastes or **drinking water** in order to reduce the incidence of **dental caries**. Fluoridation of **water supplies** is a controversial issue due to possible health hazards associated with long-term ingestion of high levels of fluorides.

Fluorimetry Alternative term for **fluorometry**.

Fluorine Non-metallic member of the **halogens** family, of which it is the most electronegative and the strongest oxidizing agent. Has the chemical symbol F and atomic number 9. Exists as a pungent, pale yellow gas or liquid. Normally present in bone; both deficiency and excess can lead to skeletal disease. Used to manufacture **fluorides**.

Fluorodensitometry Technique, often combined with **TLC** or **HPLC**, in which concentration of an analyte is determined on the basis of its **fluorescence**.

Fluorometry Technique used to identify a substance from the wavelength of the light that it emits during **fluorescence**. Also called fluorimetry.

Fluorosis Diseases typified by damage to teeth (dental fluorosis; characterized by brown mottling of the enamel) and bones, caused by an excessive intake of **fluorides**. Incidence of dental fluorosis increases when the level of fluoride in the water supply is above a certain limit. The mottled enamel is resistant to dental caries. When the level of fluorides rises still further, systemic fluorosis may occur, with calcification of ligaments.

Flushing Cleansing by passing large quantities of water through an object.

Fluted pumpkins Common name for *Telfaira occidentalis*, a plant cultivated for its leaves, which are used as a vegetable, and its seeds, which are eaten or used as a source of **oils**.

Fluvalinate Insecticide and acaricide used for control of a wide range of **insects** on **fruits**, **vegetables** and **cereals**. Also used for control of **mites** in beehives. Classified by WHO as slightly toxic (WHO III).

Flying fish Any of several **marine fish** species in the family Exocoetidae; very fast swimming fish that can propel themselves out of the water with the aid of specially developed caudal and pectoral fins. Widely distributed in warmer oceanic regions. Species utilized as food fish include *Exocoetus volitans*, *Cypselurus* spp. and *Prognichthys* spp. Marketed fresh or dried; particularly popular in Japan.

Foamed plastics Lightweight **plastics** made by solidifying plastic foams. Plastic foams are produced from liquid plastics, and contain many small bubbles.

Foaming Formation of a mass of small bubbles on or in a liquid (**foams**).

Foaming agents Substances that promote **foaming**.

Foaming capacity **Functional properties** relating to the extent to which an item is able to form **foams**.

Foaming properties **Functional properties** relating to the ability of food components to be formed into **foams**.

Foams Light textured colloidal dispersions of a gas, such as air, in a liquid or solid, typically achieved by **whipping** or frothing. Foams are unstable, requiring the presence of **stabilizers** to form the gas bubble membranes. **Egg whites** have good **foaming capacity** and are used to produce foamed foods such as **meringues** and **souffles**. **Gelatin** and modified **milk proteins** are also widely used to produce foams, e.g. in manufacture of foamed confectionery.

Foie gras A smooth rich paste prepared from fatted goose **livers** or **duck livers**. It is traditionally made in France, where it is a speciality of the Alsace and Perigord regions. It is valued highly for its silky, melting texture. Ideally, it has a delicate rose **colour** with beige mottlings. When aged, it develops a rich **flavour**. Foie gras prepared from goose livers has a richer flavour and is more expensive than foie gras prepared from duck livers. **Ducks** and **geese** reared for foie gras production are force-fed and prevented from exercising, so that they develop hugely enlarged, fat-infiltrated livers. This force feeding raises **animal welfare** concerns and is banned in many countries. After the fattened birds are killed, their livers are removed and often soaked overnight in **milk**, **port** or water, before draining and **marination** in Armagnac, Madeira or port with a mixture of **seasonings**. The livers are then cooked in their own fat and pressed to prepare foie gras. In contrast, pate de foie gras is prepared from a high proportion of pureed goose or duck livers, but usually contains other ingredients such as **swine livers**.

Foils Thin, flexible metallic sheets or strips. Commonly, thickness is specified as being less than a given amount. Used widely as **packaging materials**, e.g. **aluminium foils**.

Folacin Obsolete term for **folic acid** or any derivative exhibiting the vitamin activity of folic acid.

Folate conjugases Alternative term for **γ-glutamyl hydrolases**.

Folates Synonym for pteroylglutamates. Compounds exhibiting the vitamin activity of **folic acid**.

Foliar sprays Used to apply fertilizers or **plant growth regulators** to the leaves of plants.

Folic acid Water soluble member of the **vitamin B group**. In its active form, **tetrahydrofolate**, it is a coenzyme in various reactions involved in the metabolism of **amino acids**, **purines** and **pyrimidines**. Synthesized by intestinal bacteria and widespread in food, especially green **leafy vegetables**. Deficiency causes poor growth and nutritional anaemia. Daily intake should be increased prior to conception and during the first three months of pregnancy to prevent **neural tube defects** (e.g. spina bifida) and other congenital malformations (e.g. cleft lip and cleft palate) in the foetus.

Folpet Fungicide used for control of fungal diseases (including powdery mildew, leaf spot disease and scab) in a range of **fruits** and **vegetables**. Classified by Environmental Protection Agency as not acutely toxic (Environmental Protection Agency IV). Also known as phaltan.

Fomes Genus of **fungi** of the class Hymenomycetes. Occur on trees and on wood. Some species may be responsible for plant diseases. *Fomes fomentarius* produces **enzymes** which are capable of degrading **celluloses**, **xylan** and **lignin**.

Fondants Low moisture **sugar syrups** which are made by boiling concentrated sugar solution, adding **glucose syrups** or inverting agents, and cooling rapidly while mixing, to produce fine sugar crystals in a saturated sugar solution. Used to make fondant **sweets**, as **fillings** in **chocolates** and **biscuits**, and as **toppings** for **cakes**.

Fondues Traditional Swiss dishes made using blends of **cheese** types, such as Gruyere, Emmental and Raclette, which are melted together with **white wines**, **flour** and **seasonings**. The mixture is kept hot by placing it in a pot over a burner and is eaten on cubes of bread which are dipped into the pot. The term is also applied to a **meat** dish in which cubes of raw meat are dipped into a pot of oil which is kept hot over a burner, and consumed once **cooking** is complete.

Fonio   Type of **millet** (*Digitaria exilis* or *D. iburua*) grown in Africa. Utilization is hampered by difficulties in removing the husks from the grain. Also known as hungry rice.

Fontina cheese   Italian **soft cheese** made from **cow milk**. Dense and smooth in **texture** with small round holes. Ripens in about 3 months. Genuine Fontina comes from the Val d'Aosta region of Italy in the Alps near the French and Swiss borders. The primary ingredient of Italian fonduta.

Food additives   **Additives** used specifically in foods.

Food and Agriculture Organization   The Food and Agriculture Organization (FAO) was founded in October 1945 with a mandate to raise levels of nutrition and standards of living, to improve agricultural productivity, and to better the conditions of rural populations. Today, FAO is the largest autonomous agency within the United Nations (UN) system, with 180 Member Nations plus the EC (Member Organization) and more than 4300 staff members around the world. FAO works to alleviate poverty and hunger by promoting agricultural development, improved nutrition and the pursuit of food security. The Organization is active in land and water development, plant and animal production, forestry, fisheries, economic and social policy, investment, nutrition, food standards, and commodities and trade; it also plays a major role in dealing with food and agricultural emergencies. FAO aims to meet the needs of both present and future generations through programmes that do not degrade the environment and are technically appropriate, economically viable and socially acceptable. FAO offers direct development assistance, collects, analyses and disseminates information, provides policy and planning advice to governments, and acts as an international forum for debate on food and agriculture issues.

Food and Drug Administration   US agency within the Department of Health & Human Services which was formed in 1927 by division of the Bureau of Chemistry (established 1862) into the Food, Drug and Insecticide Administration (name shortened in 1930) and the Bureau of Chemistry and Soils. The name is commonly abbreviated to the FDA. It is a scientific, regulatory and public health agency including under its jurisdiction most foods, animal and human **drugs**, therapeutic agents of biological origin, medical devices, radiation emitting devices, cosmetics and animal **feeds**. The FDA evaluates applications for new drugs, foods, **food additives**, **infant formulas** and medi-

cal devices, as well as monitoring manufacture, import, transport, storage and sale of these products. Its mission is to promote public health by helping safe and effective products to reach the market in a timely manner, and monitoring products for continued safety once in use. With respect to foods, the agency aims to ensure that they are safe, wholesome, sanitary and properly labelled.

Food antioxidants   **Antioxidants** used specifically in foods.

Food bars   Hand-held **snack foods**, usually in the shape of a rectangular block, e.g. **cereal bars**, **chocolate bars**, **ice cream bars** and meal-replacement bars.

Foodborne diseases   Diseases whose causative agents are transmitted through food.

Food colorants   **Colorants** used specifically in foods.

Food emulsifiers   **Emulsifiers** used specifically in foods.

Food emulsions   Colloidal suspensions in which a substance is dispersed in another, e.g. oil in water **emulsions**. Emulsions can be formed from immiscible water and oil phases with the aid of **emulsifiers**; **stabilizers** are used to maintain structure. Examples of food emulsions include **milk**, **cream**, **margarines** and **mayonnaise**.

Food enrichment   Addition of nutrients, e.g. **vitamins** and **minerals**, to **processed foods**, such as **cereal products**, to correct for losses occurring during processes such as **milling**. The term **fortification** is used to indicate addition of nutrients to foods to overcome dietary deficiencies in specific population groups.

Food factories effluents   Liquid wastes (**waste water**) often discharged into a river or the sea from food factories.

Food factories wastes   Solid **wastes** generated in food factories during food processing operations.

Food flavourings   **Flavourings** used specifically in foods.

Food handlers   Personnel involved in the preparation, processing or handling of foods.

Food intolerance   Group of diseases in which there is inability to digest a particular food or food constituent properly, often resulting in malabsorption syndromes. Examples include **lactose intolerance**, resulting from lack of a gastrointestinal tract brush border enzyme, and **coeliac disease**, in which an immunological response to **wheat gluten** results in histopathological changes to the intestinal mucosa. Exclusion of the appropriate food from the diet can result in elimination of the symptoms of the disease, and also, in cases such as coeliac disease, reversal of intestinal pathology.

Food poisoning Human disease that results from ingesting food contaminated with **toxins** or **pathogens**. May range in severity from mild to life-threatening.

Food policy A broad term used to encompass those programmes, usually governmental, that most directly affect the food chain. Issues encompassed by this term include the role of food in international trade, agricultural pricing policies, food security, food aid, nutrition planning and food control.

Food powders Alternative term for **dried foods** or **powders** made for use as foods.

Food preservatives **Preservatives** used specifically in foods.

Food reference materials Reference samples comprising food materials of certified composition (e.g. bovine liver and skim milk powder) that are used as standards in analytical procedures.

Foods Substances which can be ingested by living organisms. Usually solid and contain a range of **nutrients** which can be metabolized to produce energy, and sustain life and growth.

Food safety Encompasses activities and policies which are essential for ensuring that food will not cause injury or illness upon consumption.

Food science Study of the characteristics of foods, including chemical properties, biochemical properties, **physical properties**, **physicochemical properties** and biological properties, and effects of these on the quality of products. Also covers application of this information to development of new products and efficient food processing techniques.

Food security Access (both physical and economic) by all people at all times to sufficient food for an active, healthy life.

Foods service **Catering** systems, which supply prepared foods to large groups of consumers. Food is typically prepared and packaged in a central location and then transported and served to the consumer. Examples include **school meals**, **airline meals**, **fast foods** operations and **vending machines** servicers. Also known as food service and foodservice.

Food stabilizers **Stabilizers** used specifically in foods.

Food Standards Agency UK governmental body established in 2000 with the aim of protecting the interests of consumers in relation to foods, and in particular guarding against public health risks arising from the consumption of food.

Food supplements Substances consumed as extra sources of specific **nutrients** in the **diet** with a view to improving nutritional status or health. May be added to foods during processing, e.g. calcium or iron fortification of cereal products, as a form of **food enrichment**. Alternatively, taken separately from foods in the form of tablets, capsules, liquids or oils, depending upon the nature of the dietary constituent. Such preparations may be single or multi-component. Dietary components commonly consumed in this form include **vitamins**, **minerals**, **proteins**, **oils** (e.g. **fish oils** and **evening primrose oils**), **phytochemicals** and plant extracts, such as those from **garlic**.

Food technology Application of a diversity of scientific and practical disciplines, including chemistry, biology, physics and engineering, to the development of food products and to their worldwide distribution.

Food thickeners **Thickeners** used specifically in foods.

Foot and mouth disease Highly infectious viral infection affecting a wide range of animals, including cattle, swine, sheep, goats and buffaloes. Endemic in some parts of the world. Outbreaks in countries previously free of the disease can have serious economic implications due to restrictions on movement of livestock and the widespread slaughter of infected and at-risk susceptible animals. The virus may be present in **milk** of an infected animal before clinical signs of disease appear and is capable of surviving milk **pasteurization**. In meat products, the virus is inactivated by **cooking**.

Forcemeat **Stuffings** made from a seasoned mixture of finely chopped or minced ingredients, such as **meat mince**, **onions**, **breadcrumbs** and **herbs**.

Formaldehyde Aldehyde, which exists as a colourless gas with a pungent aroma. Soluble in water and alcohol. Used as a disinfectant and germicide, and, when in solution, as a preservative for biological specimens and as an embalming fluid. Commercially available as a 37-50% aqueous solution, **formalin**. Also known as **methanal**.

Formalin Solution of **formaldehyde** in water. Also known as **formol**.

Formic acid Organic acid which exists as a combustible, colourless, fuming liquid with a penetrating **aroma**. Soluble in water, alcohol, ether, acetone and benzene. Occurs naturally in **pine needles**, stinging **nettles** and certain **insects**. Used in **preservatives**, **fumigants**, **insecticides**, **refrigerants**, as an antiseptic in **brewing** and in vinyl resin **plasticizers**. Also known as methanoic acid.

Formol Solution of **formaldehyde** in water. Alternative term for **formalin**.

Fortification   Increasing the nutritional quality of a food by addition of **nutrients** such as **vitamins** and **minerals**.

Fortified wines   **Wines** to which **ethanol** has been added, as **spirits** or neutral **alcohol**. Important types include **sherry** and **port**.

Fossa cheese   Italian **hard cheese** made from raw or pasteurized **ewe milk**, **cow milk** or a mixture of both. The name derives from the practice of **ageing** ripened **cheese** for up to 3 months in underground pits dug into tuffaceous rock. During this period, anaerobic **fermentation** takes place, and the cheese develops a unique **flavour** and **aroma**. The final product is white to straw in **colour**, with an irregular shape. After removal from the pit, the cheese can be stored under vacuum for up to a year without loss of its characteristic properties.

Fouling   Accumulation of unwanted materials on the surfaces of processing equipment. The fouling layer has a low **thermal conductivity**, so increasing resistance to **heat transfer** and reducing the effectiveness of **heat exchangers**. Chemical reaction fouling involves deposits that are formed as the result of chemical reactions at the heat transfer surface. This kind of fouling is a common problem in chemical process industries, oil refineries and dairy plants. Biological fouling is the development and deposition of organic films consisting of **microorganisms** and their products. Effective fouling control methods should involve: prevention of foulant formation; prevention of foulants from adhering to themselves and to heat transfer surfaces; and removal of deposits from the surfaces.

Fourier transform IR spectroscopy   Type of **IR spectroscopy** that utilizes the Fourier transform mathematical technique, in which samples are irradiated with polychromatic radiation and the entire range of frequencies is recorded at the same time, giving an interferogram. Fourier transformation then sorts the interferogram into its components, which can be represented as a traditional spectrum. Advantages over conventional IR spectroscopy include increased speed and sensitivity. Usually abbreviated to FTIR spectroscopy.

Fowl   Any **birds** kept for production of **meat** and/or **eggs**, particularly domesticated birds such as **chickens**, **turkeys** and **ducks**. In most commercial production of fowl, hybrids have largely replaced pure- and crossbred fowl. The term may also be used in the names of birds that resemble domestic fowl, e.g. spurfowl; additionally, it may be used for birds collectively, particularly those which are hunted.

Foxtail millet   Cereal plant belonging to the species *Setaria italica*, that is an important food crop in China and other Asian countries.

f.p.   Abbreviation for **freezing point**.

FPC   Abbreviation for **fish protein concentrates**.

Fractionation   Separation of the components of a mixture into fractions, using techniques such as **gel filtration** and **electrophoresis**. This term also relates to precipitation and phase-separation methods used to determine the molecular weight distribution of polymers; these techniques are based on the tendency of polymers of high molecular weight to be less soluble than those of low molecular weight.

Fracture properties   Mechanical properties governing the way in which and conditions under which a structure will break down when an external force is applied.

Franchising   Authorization granted by a government or company to an individual or group enabling them to carry out specified commercial activities, for example to market a company's goods or services, in a designated territory. In return for a specified fee and usually a share of the profits, the franchiser provides the product, the name, and sometimes the plant and advertising.

Frangipans   **Pastry** products or **flans** made with a pastry similar to choux pastry often filled with forcemeat. Also, an almond flavoured cream or paste that is used as **toppings** or **fillings** for **cakes** and **pastries**.

Frankfurters   Mild flavoured, smoked, cooked **sausages** originally produced in Frankfurt, Germany. Varieties include **hot dogs** and **wieners**. They can be made from **beef**, **chicken meat**, **pork**, **turkey meat** or **veal**; typically, they are prepared from a blend of 40% pork and 60% beef. Frankfurter **seasonings** include **coriander**, **garlic**, **mustard**, **nutmeg**, **salt**, **sugar** and **white pepper**. They tend to have high contents of fat and salt. Some are retailed in natural **sausage casings**, but most are prepared in cellulose casings, which are later removed. Most commonly, frankfurters are about 15 cm long, but they are produced in a wide range of sizes. When traditionally made, frankfurters are smoked over hardwood, in order to improve **colour** and **flavour**; however, now **smoke flavourings** are mostly applied as a paint. Despite being precooked, frankfurters taste better after reheating; usually, they are boiled, fried, grilled or steamed immediately before serving.

Frappe   **Confectionery** products made by dissolving **egg whites** in water, adding **sugar syrups**, and whipping to form an aerated foam. Used in **nougat** and **fondants**.

Free radicals   Highly reactive molecular entities, containing one or more unpaired electrons, that are usually short-lived and capable of initiating or mediating a wide variety of chemical reactions. Often formed by the splitting of a molecular bond.

Freeze concentration Concentration of a liquid by **freezing** out pure ice, leaving a more concentrated solution. This process requires less input of energy and causes less loss of **flavour** than concentration by **evaporation**; it is used primarily in the concentration of **fruit juices**, **vinegar** and **beer**. Limitations of freeze concentration are its high cost, the difficulty in separation of ice from solid, and the degree of concentration that can be achieved.

Freeze dried foods Foods dehydrated by **freeze drying**. Used to make various types of products, including **instant soups**, dried **herbs**, **instant coffee** granules and meat products. **Dried foods** obtained in this manner are light, porous, easy to rehydrate and tend to have better shape and colour retention than foods obtained by other **drying** processes.

Freeze driers Apparatus for **preservation** of foods by applying rapid **freezing** followed by a high vacuum which removes ice by sublimation (**freeze drying**).

Freeze drying Preservation of foods by rapid **freezing** followed by subjection to a high vacuum, which removes ice by sublimation. Adequate control of the processing conditions contributes to satisfactory subsequent rehydration, with substantial retention of **nutrients**, and **colour**, **flavour** and **texture** characteristics.

Freezers Refrigerated cabinets or rooms for preserving **frozen foods** at very low temperatures. Foods are usually frozen to an internal temperature of -18°C in freezers; the food must be maintained at this temperature or slightly lower during transport and storage. Commercial freezers include the following types: blast freezers, where air is circulated at -40°C; contact freezers, in which **refrigerants** are circulated through hollow shelves; immersion freezers, where, for example, fruit is frozen in a solution of **sugar** and **glycerol**; and cryogenic freezers, which use, for example, liquid nitrogen spray.

Freezing Method of **preservation** in which **microorganisms** are prevented from multiplying by application of freezing temperatures. Foods are usually frozen to an internal temperature of -18°C in **freezers**; the food must be maintained at this temperature or slightly lower during transport and storage. During freezing, a proportion of the water in the food changes from liquid to solid to form ice crystals, so lowering its **water activity**. Because the process does not kill all types of **bacteria**, those that survive reanimate in thawing food and often grow more rapidly than before freezing. **Enzymes** in the frozen state remain active, although at a reduced rate. Freezing does, however, cause the water in foods to expand, and tends to disrupt the cell structure by forming ice crystals. With quick-freezing, however, the ice crystals are smaller, producing less cell damage than with slowly frozen products. Freezing has been a key technology in bringing **convenience foods** to homes and **restaurants**; it causes minimal changes in quality of food in terms of size, shape, **texture**, **colour**, **flavour** and microbial load.

Freezing point Temperature at which the liquid and solid forms of a substance exist together in equilibrium. Value varies according to pressure and is affected by purity of the substance. Also known as the **melting point**. Freezing point measurement can be used to detect the **adulteration** of **milk** with water, since the value increases when water has been added.

French beans Type of **common beans** (*Phaseolus vulgaris*). Both the pods and seeds are eaten.

French dressing Popular salad dressing made from **vinegar**, **oils** and **seasonings**. Also known as vinaigrette.

French fries Potato products made by cutting **potatoes** into thick or thin strips, soaking in cold water, **drying** and **deep frying** in oil. Also called **chips** (UK), **pommes frites** (France), fries or French-fried potatoes.

Freons Series of nonflammable, nonexplosive fluorocarbons (FC) or **chlorofluorocarbons** (CFC) widely used as **refrigerants**.

Fresh cheese Low-fat **cheese** high in moisture and mild in **flavour**.

Freshness Extent to which a product is fresh and of good **eating quality**.

Freshwater clams Any of a number of **bivalves** inhabiting rivers and lakes.

Freshwater fish Fish that inhabit inland waters (lakes and rivers). Some fish species, e.g. **salmon**, occur in both freshwater and marine phases at different stages during their life history.

Friabilimeters Devices used in assessment of **malt** quality on the basis of friability, a measure of the breakdown of malt endosperm cell wall components. Malt samples are subjected to an abrasive action for separation of hard and ripe constituents which are then weighed. The presence of hard and glassy components, which will cause problems with **brewing**, is detected in this way. May be used in quality control of the **malting** process or to determine malt quality in **breweries**.

Friabilins Water-soluble proteins which control **wheat** kernel hardness and are located on the surface of **wheat starch** granules.

Fricadelles Flattened, round meat products prepared from **meat mince**. They include a high percentage of meat, often **beef** and **pork** in equal quantities, and also binders, such as **bread**, **spices** and **onions**. Varieties include boulettes, bratklopse and bratlinge. They are often served in **sauces** or **gravy**.

Fried foods Foods fried in **fats** or **oils**, e.g. **French fries**, **fritters** and **doughnuts**. Often coated in **batters** or breadings prior to **frying**.

Frigate mackerel Name applied to two species of **marine fish** (*Auxis rochei rochei*, also called bullet tuna, and *A. thazard thazard*, also called frigate tuna). Both species are popular game fish of high commercial importance and belong to the family Scombridae (mackerels, tunas, bonitos). Widely distributed in the Atlantic, Pacific and Indian oceans, where they are highly migratory. Marketed fresh, frozen, dried-salted, smoked and canned.

Fritters Pieces of food (e.g. **fruits**, **meat**, **fish**) that have been dipped in **batters** and deep fried.

Frogs Insectivorous amphibians of the family Ranidae with a short tailless body, smooth moist skin and long hind legs designed for hopping. The species normally consumed is *Rana esculenta*, a large type of frog, but sometimes other species are used as food. Usually, only the hind legs are eaten. **Frogs legs** are particularly popular in France, but are also eaten in many other parts of the world.

Frogs legs Tender, white **meat** from the hind legs of **frogs**. The meat has a very low fat content and a delicate sweet **flavour**; it is particularly popular in France. Ideally, frogs legs are cooked briefly with very little seasoning.

Fromage frais French **soft cheese** of variable fat content, traditionally made from a mixture of ewe and goat milks and eaten soon after production. Moist, creamy and white, the cheese is unripened and made from milk coagulated by **lactic fermentation**; **lactic acid bacteria** are active in the cheese when sold.

Frostings Alternative term for **icings**, used particularly in the USA and Canada. Used more specifically in the UK to refer to soft icings made with **sugar** and **egg whites**.

Frothing Process of forming a mass of small, light bubbles in a liquid by **agitation** or **fermentation**.

Frozen beverages **Beverages**, generally **soft drinks** or **fruit beverages**, which have been frozen. May be served and consumed in a soft-frozen (slush) state.

Frozen desserts **Desserts** preserved by **freezing** and requiring frozen storage. These are often premium quality products, such as ice cream products, **gateaux** and **cheesecakes**. Some require cooking before consumption, but others can be eaten immediately after **thawing** or while still frozen. Unlike many other **frozen foods**, **texture** is not usually compromised by freezing.

Frozen dough **Dough** prepared at a lower temperature than conventional dough (in order to minimize **fermentation** activity of **yeasts**), followed by immediate **freezing** to extend its **shelf life**. Used for production of **bakery products** in in-store bakeries.

Frozen foods Foods preserved by **freezing**, and requiring frozen storage. Usually of higher quality than **canned foods** or **dried foods**, with any losses in quality being due to **texture** deterioration. A wide variety of foods can be frozen, either cooked (**ready meals** and some **desserts**) or uncooked (vegetables, fish fillets and poultry). Some products are thawed before use, while others can be cooked/reheated directly from the freezer.

Frozen meals **Frozen foods** in the form of complete dishes. Usually reheated directly from frozen form prior to consumption. Common types include **pizzas**, **ready meals**, **entrees**, **vegetarian foods** and savoury **pies**.

Frozen storage Storage of foods at freezing temperatures (below $0^{o}C$).

Frozen yoghurt Fermented low-fat **dairy desserts** served in a similar manner to **ice cream**.

Fructans Group of **oligosaccharides** and **polysaccharides** which consist of **fructose** residues attached to a single **glucose** molecule. Depending on the source, chain lengths can range from 3 to 50 residues. In **cereals**, shorter fructans predominate, while **Jerusalem artichokes** contain high levels of **inulin**, a fructan of about 35 residues. **Onions**, **garlic** and **asparagus** are other dietary sources of fructans. In the stomach and small intestine, hydrolysis of fructans is negligible; any trisaccharides which are absorbed directly are usually excreted in urine. The majority of dietary fructans reach the large intestine where **fermentation** occurs.

β-Fructofuranosidases EC 3.2.1.26. Hydrolyse terminal non-reducing β-D-fructofuranoside residues in β-D-fructofuranosides. Substrates include **sucrose**, which is hydrolysed to a mixture of **glucose** and **fructose**. The resultant monosaccharide mixture is called **invert sugar**. Also known as invertases and saccharases, these enzymes can also be used in the production of **marzipan**, soft chocolate **fillings** and artificial **honeys**.

Fructokinases EC 2.7.1.4. Phosphorylate D-fructose in the presence of **ATP** to form D-fructose 6-phosphate and **ADP**. Play an important role during the development of tomatoes, where they are key enzymes in the **fructose** to **starch** biosynthetic pathway.

Fructooligosaccharides **Oligosaccharides** composed of **fructose** monomers used as functional ingredients in **prebiotic foods**. Fructooligosaccharides are found particularly in **chicory** roots as **inulin** polymers and also in *Allium* species. Individual fructooligosaccharides include **kestose** and **nystose**.

Fructose Monosaccharide ketose sugar comprising six carbon atoms. Constituent of **sucrose** which occurs naturally in **fruits** and **honeys**. Commercially produced from **glucose** by **isomerization**, a reaction catalysed by glucose isomerases. May be crystallized from **fructose syrups** by addition of an organic solvent, such as **ethanol**. Fructose is the sweetest natural saccharide and is approximately 1-1.5× as sweet as sucrose. It is also known as laevulose and fruit sugar.

Fructose-bisphosphate aldolases EC 4.1.2.13. **Enzymes** that catalyse the conversion of D-fructose 1,6-bisphosphate to glycerone phosphate and D-glyceraldehyde 3-phosphate, and are involved in the **ripening** of **strawberries**. Also known as aldolases.

Fructose high corn syrups Syrups containing 40-90% **fructose** produced from **glucose syrups** which have been manufactured by hydrolysis of **corn starch**. **Glucose** can be converted to fructose by the action of **glucose isomerases** to produce syrups composed predominantly of glucose and fructose. Higher purity **fructose syrups** are produced using **gel filtration chromatography** to separate fructose from glucose and other **sugars** present. Applications for these syrups include **soft drinks**, **marmalades**, **jams**, canned **fruits**, **fruit juices**, **dairy products** and **bakery products**.

Fructose syrups Aqueous solutions, containing predominantly **fructose**, which are used as **sweeteners**.

Fructosyltransferases Group of glycosyltransferases (EC 2.4) which catalyse the transfer of fructosyl groups to various substrates, commonly **sucrose**. Include inulosucrases, **levansucrases** and sucrose 1^F-fructosyltransferases. May be used in the synthesis of **fructooligosaccharides**.

Fruit beverages **Beverages** derived from **fruit juices**, **fruit extracts** or fruit homogenates.

Fruit brandies Spirits manufactured by **distillation** of fermented fruit mashes.

Fruit bread Bread made by adding up to 50% (flour basis) **raisins** to the **dough** mixture. May also contain other **dried fruits** such as **currants**, **dates** or **bananas**.

Fruit compotes Desserts made from **fruits** stewed in sugar or cooked in **syrups**. Eaten hot or cold.

Fruit cordials Term referring to **fruit juice beverages**, often presented as concentrates for dilution, or to sweet fruit-based **liqueurs**.

Fruit desserts Desserts based on **fruits**. Include **fruit salads**, fruit cocktails, fruit **compotes**, **mousses**, **flans** and **sorbets**.

Fruit extracts Preparations obtained from fruits by a variety of means that can be used as **flavourings** in foods and **beverages**.

Fruit flies Common name for species of **insects** of the family Tephritidae, especially those of the genus *Drosophila*. Serious plant pests whose larvae feed on fruit or decaying vegetable matter.

Fruit gums **Sugar confectionery** products made with **sucrose**, **glucose**, fruit **flavourings** and **gum arabic** either alone (to produce hard gums) or mixed with **gelatin** (to produce soft gums).

Fruitiness Extent to which a product has the **aroma** or **flavour** of fruit.

Fruit jellies Semi-solid foods with an elastic consistency, made either by setting of **fruit juices** containing **pectins** or **gelatin**, or by addition of gelatin to fruit juices.

Fruit juice beverages **Beverages** containing **fruit juices**, together with other ingredients such as water, **sugar** or **flavourings**.

Fruit juice concentrates **Fruit juices** which have been concentrated by **evaporation**, membrane processes or **freezing**. May be diluted to make reconstituted juices, or used as ingredients in a wide range of foods and **beverages**.

Fruit juices Juices extracted from **fruits** consumed as drinks or used as ingredients in a wide range of foods and **beverages**.

Fruit leathers Products made from fruit purees, sometimes sweetened with **sugar** or **honeys**, that are spread in a thin layer and dried. The dried sheets may be cut into strips or rolled into cylinders.

Fruit liqueurs **Liqueurs** made from or flavoured with **fruits**.

Fruit nectars **Beverages** manufactured from **fruit juices** by addition of water and/or **sugar**, optionally with addition of other ingredients.

Fruit peel Rind or skin of fruits. May be removed before consumption of the fruits or eaten at the same time. Rich in fibre. Some types are removed and used in garnishes or as ingredients of various dishes. Peel most commonly used in cooking is that from **citrus fruits**.

Fruit pies Dishes, usually served as **desserts**, having one or more crusts and fruit-based **fillings**. Crusts, generally made from **pastry**, can be on the bottom or top of the dish only, or on both the bottom and top.

The fruit fillings can be prepared from a single fruit or a combination of several **fruits**.

Fruit preserves Prepared by cooking pieces of **fruits** with **sugar** and sometimes **pectins**. Similar to **jams**, except that the fruit pieces tend to be larger in preserves.

Fruit pulps The soft, succulent part of fruits or a preparation made from them by mashing and concentration. Used in the manufacture of a range of foods and **beverages**, including **syrups**, **milkshakes**, **fruit juice beverages** and **ice cream**.

Fruit purees Fruit flesh that is mashed to a smooth, thick consistency by various means, such as forcing through sieves or blending in food processors. Used as garnishes and side dishes or as the base of many types of product, including **beverages**, parfaits, **ice cream**, **mousses** and **souffles**.

Fruits Seed-bearing parts of plants, formed from the ovary after flowering. May be dry or fleshy. The term is commonly restricted to fleshy fruits, which are of economic importance to humans. When other parts of the flower contribute to the structure, they are called false fruits.

Fruit salads **Desserts** comprising a mixture of **fruits** cut into pieces and covered with **syrups** or **fruit juices**. Eaten fresh or available canned.

Fruit syrups **Syrups** produced by **concentration** of **fruit juices**. Used as **flavourings** and **sweeteners**.

Fruit tea **Tea**-type infusion **beverages** made by hot water extraction of soluble constituents from materials derived from **dried fruits**.

Fruit wines Wine-like **alcoholic beverages** made by **fermentation** of fruit **musts** or **mashes**.

Fruit yoghurt **Yoghurt** containing pieces of fruit, **fruit pulps** or **fruit purees**, either as a separate layer or stirred in to give a homogeneous product.

Frying Cooking of foods in hot **fats** or **oils** over a moderate to high heat. In **deep frying**, the foods to be cooked are immersed in the fats or oils.

Frying fats Lipids which are usually solid at room temperature and used as a medium in which to cook foods by **frying**. **Heating** of the fat results in it acting as a thermal transfer agent, with some of it remaining in the **fried foods**. Repeated use of the fat for frying may result in its degradation by means of autoxidation, cyclization or polymerization.

Frying oils Lipids which are usually liquid at room temperature and used as a medium in which to cook foods by **frying**. **Heating** of the oil results in it acting as a thermal transfer agent, with some of it remaining in the **fried foods**. During frying, the heated oil may undergo several degradative changes.

Frying properties Ability of foods to maintain or develop acceptable properties upon application of **frying** procedures.

FTIR spectroscopy Abbreviation for **Fourier transform IR spectroscopy**.

F2 toxin Mycotoxin produced by *Fusarium graminearum*, *F. culmorum* and other ***Fusarium*** species. May be formed when the fungus grows on damp cereal grain (e.g. wheat, barley and corn) used as animal feeds. Has oestrogenic activity and can cause hyperoestrogenism in swine, cattle and poultry. Also known as **zearalenone** and **zearalenol**.

Fucoidans Sulfated **fucose**-containing **polysaccharides** produced by brown **seaweeds**. Potential uses include applications as natural **food antioxidants** and as components of **functional foods**.

Fucose One of the **reducing sugars** found in plant foods and animal foods. Both D- and L- forms occur naturally. Synonym 6-deoxygalactose.

Fucus Genus of **seaweeds** found in lower intertidal zones along rocky shores. Some species, e.g. *Fucus vesiculosus* and *F. spiralis*, are utilized for foods and animal feeds, usually in dried form. **Alginates** are often extracted chemically from dried *Fucus* spp. for use as **bulking agents**, **gelling agents** or **stabilizers** in foods such as **cheese** and **ice cream**.

Fudges Toffee-like **sugar confectionery** products made with **sugar**, **butter** and **milk**, and formed either by rapid agitation or addition of a small quantity of **fondants**, causing sugar crystallization.

Fufu Unfermented or fermented product usually made from **cassava**, but also from other tubers and corms, such as **taro**, **yams** or **cocoyams**. The unfermented form is prepared by boiling or steaming and pounding the vegetables, either individually or in combination. Fermented fufu is prepared from roots which have been soaked for 3-4 days before being formed into pastes. In some areas, fufu is sold as a convenience food. Usually served as an accompaniment to dishes with sauce, such as stews.

Fumarases Alternative term for **fumarate hydratases**.

Fumarate hydratases EC 4.2.1.2. Catalyse the reversible conversion of **fumaric acid** to L-malic acid. Since **organic acids** are essential **flavour compounds** in **alcoholic beverages** and **fermented foods**, modification of fumarate hydratase levels in **yeasts** and **bacteria** can potentially be used for development of fermented foods and beverages with distinctive **flavour**.

Fumaric acid *Trans* isomer of **maleic acid**, used in **acidulants** and **flavourings** in the food industry. Its lack of solubility and nonhygroscopic nature make it particularly suitable for powdered food and beverage mixes. Also used to improve **dough** machineability and **functional properties** of **tortillas**. Can act as a synergistic antioxidant with **BHT** and **BHA** in oil- and lard-based foods. Although less expensive than some acidulants, use in some foods is limited by its strong acid taste.

Fumigants Gaseous **pesticides** used for **fumigation**.

Fumigation Use of gaseous **pesticides** to rid an area of insect **pests**.

Fumonisin B$_1$ Hepatotoxic and nephrotoxic in animals. Although acute **toxicity** in humans is believed to be low, there is evidence to suggest that it may be responsible for oesophageal cancer.

Fumonisin B$_2$ Analogue of **fumonisin B$_1$** with a similar toxicological profile, though it is not as toxic.

Fumonisin B$_3$ Analogue of **fumonisin B$_1$** with a similar toxicological profile, though it is not as toxic as fumonisin B$_1$ or **fumonisin B$_2$**.

Fumonisins **Mycotoxins** produced by *Fusarium* species. (e.g. *F. moniliforme* and *F. proliferatum*) growing on **corn**.

Functional foods Term originally introduced in Japan to mean foods with a physiological function or activity. Used for products containing biologically active components (such as nutrients, **bioactive peptides** or **phytochemicals**) at levels that may confer specific health benefits. Examples include bifidus **yoghurt**, **eggs** incorporating **ω-3 fatty acids** and fibre-enriched **breakfast cereals**. Also known as **nutraceutical foods**. Similar terms include **designer foods**, **medical foods** and **probiotic foods**.

Functional properties Characteristics of a substance that affect its behaviour and that of products to which it is added. Influence potential applications of a substance in the food industry, as a particular functional property may be especially useful for the manufacture and stability of specific types of foods. Include a wide range of characteristics, such as **buffering capacity**, **emulsification properties**, **foaming properties**, **gelling capacity**, **water binding capacity** and **whipping properties**.

Fungal decay Decay caused by the action of **fungi**.

Fungal proteins Fungal mycelia which are used as foods or food ingredients, e.g. **Quorn** (produced by the continuous **fermentation** of *Fusarium graminearum*). Also know as **mycoprotein**.

Fungal spores **Spores** produced by **fungi**, e.g. ascospores, basidiospores, chlamydospores, sporangiospores and zygospores.

Fungi Eukaryotic microorganisms of the kingdom Fungi, that possess cell walls and lack chlorophyll. Some species are **pathogens** of humans, animals and plants. Certain fungi are used commercially (e.g. in the production of **enzymes** and **fermented foods**). Species such as *Penicillium* and *Aspergillus* are important agents of food **spoilage**, while other species (e.g. *Penicillium camemberti* and *P. roqueforti*) are desirable and essential in the ripening of certain types of **cheese**.

Fungicides Chemical substances used to kill or inhibit the growth of **fungi** that cause diseases of plants and animals. Most are applied as sprays or dusts and either have a systemic or protectant effect. Examples of fungicides commonly applied to crops include **benomyl**, **captan**, **dichlofluanid**, **maneb** and **zineb**. Residues in foods and the environment can represent a health hazard.

Fural Alternative term for **furfural**.

Furaldehyde Alternative term for **furfural**.

Furaneol Synonym for **2,5-dimethyl-4-hydroxy-3(2H)-furanone**. One of the main **flavour compounds** in **strawberries**, and also present in **pineapples** and **roasted foods** such as **coffee**. May be used in food **flavourings**.

Furanones Important **flavour compounds** of **strawberries**, **pineapples** and various other fruits; also present in **roasted foods** such as **roasted coffee**, roasted almonds and **popcorn**. Include **2,5-dimethyl-4-hydroxy-3(2H)-furanone** (**furaneol**).

Furans Any of a group of unsaturated **heterocyclic compounds** that occur as colourless, volatile liquids and are composed of a ring of four carbon atoms and one oxygen atom. May also refer to the simplest of these compounds, C_4H_4O, which is used as an organic intermediate.

Furazolidone Synthetic antibiotic used for treatment of *Salmonella* infections in cattle, swine, sheep, poultry and farmed fish. A suspected carcinogen. Also used as an additive to feeds for growth-promoting purposes. It is rapidly metabolized in animals, with tissue residue levels depleting rapidly. Systemic use in animals reared for food purposes is prohibited in the EU and USA.

Furcellaran **Gums** produced from the red alga *Furcelleria fastigiata*, also known as Danish agar. Used as **thickeners** and **gelling agents** in **flans**, **jellies** and **dairy desserts**. Some similarities to **carrageenans**.

Furcellaria Genus of **seaweeds** widely distributed along rocky shores at lower intertidal zones. Some species, e.g. *Furcellaria lumbricalis*, are utilized as a source of **carrageenans** in the food industry.

Furfural Viscous, colourless volatile liquid aldehyde, which has a distinct **aroma** and is unstable in air, exposure to which results in polymerization to a reddish brown **colour**. Composed of a furan ring and aldehyde side chain, it is derived from the thermal breakdown of **pentoses** from cornstalks, **corn cobs** and **bran distillation**. Often used as a solvent. Alternative terms include **fural** and **furaldehyde**.

Furosine Amino sugar generated during acid hydrolysis of fructosyl-lysine. May be a useful indicator of the extent of damage that occurs during the early stages of the **Maillard reaction**.

Furylfuramide Mutagenic nitrofuran compound (2-(2-furyl)-3-(5-nitro-2-furyl)acrylamide). Previously permitted as a food additive in Japan. Also known as AF-2.

Fusaproliferin Mycotoxin produced by *Fusarium* species, especially *F. subglutinans* and *F. proliferatum*. May be produced, often in association with **fumonisin B$_1$** and **beauvericin**, in *Fusarium*-infected **cereals**.

Fusarenon X Trichothecene produced by *Fusarium* species during growth on foods.

Fusarin C Mycotoxin produced by *Fusarium* species during growth on foods. Strongly mutagenic and possibly carcinogenic in humans.

Fusarium Genus of **fungi** which occur in soil and decaying organic matter. Some species may cause plant diseases, and the spoilage of stored **fruits** and **vegetables**. May also cause diseases in humans and animals through the production of **mycotoxins** on foods and feeds.

Fusel oils Colourless viscous liquids with an unpleasant **aroma** and **flavour**. Composed of a mixture of **amyl alcohol** with higher alcohols and traces of other components. Present in distilled **spirits** as by-products of **alcoholic fermentation**. More toxic than **ethanol**.

Fusidium Obsolete name for a fungal genus whose species have been reclassified into other genera.

Fusion proteins Proteins containing **amino acids** sequences from two distinct proteins, formed by expression of a recombinant gene in which two coding sequences have been joined together in-frame. Fusion of proteins with affinity tags can be used to facilitate purification, while fusion with signal peptides can be used to facilitate secretion of proteins from cells.

Fuzzy control Control processes based on the theory of fuzzy logic, an artificial intelligence concept used in **expert systems** for estimating the degree of certainty of conclusions.

Fuzzy logic process control A form of logic used in **process control** in which statements can be given fractional values rather than simply true or false.

Fynbo cheese Danish semi-hard cheese made from pasteurized **cow milk**.

G

Galactanases EC 3.2.1.89 (arabinogalactan endo-1,4-β-galactosidases) and EC 3.2.1.90 (arabinogalactan endo-1,3-β-galactosidases). **Enzymes** that catalyse the endohydrolysis of 1,4- and 1,3-D-galactosidic linkages, respectively, in **arabinogalactans**. Can be used for production of **galactooligosaccharides** by virtue of their transglycosylation activity.

Galactans **Galactose** polymers found in **agar**, **carrageenans**, **pectins** and **hemicelluloses**. Complete hydrolysis of galactans results in the production of galactose only, whilst incomplete hydrolysis generates **galactooligosaccharides**.

Galactitol Synonym for **dulcitol**. Polyol comprising six carbon atoms, produced by **isomerization** of **sorbitol**. Has approximately 10% the **sweetness** of **sucrose**. Also known as dulcitol due to its presence in dulcite (Madagascan manna, *Melampyrum nemorosum*).

Galactolipids **Glycolipids** which contain **galactose** residues and/or *N*-acetylgalactosamine and are found in nervous tissue. Also known as **cerebrosides**.

Galactomannans Polymers of D-**galactose** and D-**mannose** found in **bacteria**, **yeasts** and **legumes**, possibly as storage polysaccharides.

Galactooligosaccharides **Oligosaccharides** that consist mainly of **galactose** residues. Produced by action of **β-galactosidases** on **lactose**. Present naturally in **human milk** and thought to be the main carbon source for *Bifidobacterium* in the neonatal **gastrointestinal tract**. Added to probiotic preparations, e.g. **fermented milk** and **yoghurt**, as a source of carbon for the **probiotic bacteria**. Galactooligosaccharides have approximately 0.2× the **sweetness** of **sucrose**. They are useful in food processing as they have greater **thermal stability** and acid resistance than sucrose. For this reason, they are often included in **jams** and **bread**.

Galactosamine Derivative of **galactose** in which the hydroxyl group of the carbon-2 atom is replaced by an amino group. Found in **glycolipids**, **mucopolysaccharides** and chondroitin sulfate.

Galactose Monosaccharide with six carbon atoms which occurs naturally as a component of many complex plant-derived **polysaccharides**, such as **pectins** and **gums**. Constituent of **lactose**, from which it may be produced by hydrolysis. Has approximately 40% the **sweetness** of **sucrose** and is used in **sweeteners**.

Galactosidases **Glycosidases** that exist as **α-galactosidases** and **β-galactosidases**.

α-Galactosidases EC 3.2.1.22. Hydrolyse terminal, non-reducing α-D-galactose residues in α-D-galactosides, including **galactose oligosaccharides**, **galactomannans** and **galactolipids**. Can also hydrolyse α-D-fucosides. Useful for hydrolysis of oligosaccharides in **soymilk**, making it more suitable for human consumption. Also known as melibiases.

β-Galactosidases EC 3.2.1.23. Hydrolyse terminal non-reducing β-D-galactose residues in β-D-galactosides. Used for production of low lactose **dairy products** (suitable for individuals who suffer from **lactose intolerance**) where they hydrolyse lactose to **glucose** and **galactose**. Lactose hydrolysates are significantly more soluble than **lactose** and can be used as **syrups** for the manufacture of baked goods, in the production of **ice cream** and **desserts**, and in **dressings** and **lemonade**. These enzymes are also useful for utilization of **whey**-containing wastes.

Galactosides **Glycosides** formed from mixing **galactose** with an **alcohol**; on hydrolysis, galactose is produced.

Galacturonic acid Member of the **uronic acids** derived from D-**galactose** by oxidation of the alcohol group of the carbon-6 atom to form a carboxyl group. Found in **pectins**, plant **gums** and bacterial cell walls.

Galangal Rhizomes from the zingiberaceous plant, *Alpinia galanga* or *A. officinarum* (lesser galangal). Similar to **ginger**, and used as a spice in South East Asia and some other regions in **flavourings** for products such as **curries**, **vinegar** and **wines**. Also reported to have medicinal properties.

Galgals Type of **lemons** produced by *Citrus pseudolimon*, which are indigenous to and cultivated on a commercial scale in India. Used in manufacture of **pickles** and as a source of **fruit juices**, **peel**, **pectins** and **essential oils**.

Gallic acid Organic acid with **antioxidative activity** which occurs naturally as a component of **tannins**, e.g. in **tea**. Gallic acid esters, such as octyl gallate and **propyl gallate**, are used as **antioxidants** in the food industry.

Game The collective name for **birds** and animals which normally live in the wild and are hunted for sport or **game meat**. In many countries, game may only be killed by people possessing a Game Licence and a licence is also needed to sell game. Legislation may also specify close seasons when game must not be shot or open seasons when particular types of game may be shot. Game is regarded as a valuable asset on many farms. If wild game is managed carefully, it is possible to produce a regular crop of game birds and animals which can be culled to provide game meat. A high level of consumer demand for game meat has led to farming, including ranch-raising, of wild game; for example, red deer have been farmed successfully in Scotland, eland in Zimbabwe and reindeer in the north of Scandinavia. The majority of commercially available game meat is from farmed game.

Game birds Heavy bodied, ground-nesting birds which are farmed or hunted for their **meat**. They belong to the order Galliformes and include grouse, **guinea fowl**, **partridges**, **pheasants** and **quails**.

Game meat The meat of wild or farmed **game birds** or animals. Game meat has a characteristic flavour and dark red colour. The flavour and aroma of game meat may be very strong; to decrease these characteristics, game meat is often marinated before cooking. Game meat tends to have a low fat content, which is attractive to consumers, but may make it difficult to cook. Meat from game animals, such as wild boars, dogs, rats and bears, may be infested with *Trichinella spiralis* larvae; such meat must be cooked thoroughly to avoid the risk of trichinosis.

Gamma irradiation Exposure of foods to **gamma rays**, generated by radioactive decay of cobalt-60 (^{60}Co) or caesium-137 (^{137}Cs). Used for **sterilization** or **preservation** purposes. **Irradiation** delays **ripening** of **fruits** and **vegetables**, inhibits **sprouting** in bulbs and **tubers**, causes disinfestation of grain, **cereal products**, fresh and **dried fruits** and vegetables, and destroys **bacteria** in fresh **meat**. Continued consumer concern over the safety of irradiation and **irradiated foods** has limited its full-scale use.

Gamma rays Penetrating electromagnetic radiation of shorter wavelength than **X-rays**. For food **irradiation**, sources used for generation of gamma rays include cobalt-60 (^{60}Co) and caesium-137 (^{137}Cs).

Gammon The thigh and adjacent parts, including the hind leg, of a side of **bacon**, usually cured while still part of the swine carcass. Preparation involves **brin-**ing of the meat as if it were to become bacon and then draining for about one week. Some gammon is cold-smoked before being sold, whilst other gammon is unsmoked and is also known as pickled pork. Gammon is usually sold uncooked, but cut into small portions or sliced as gammon steaks. It is commonly cooked by baking or pan frying, or is cooked in a casserole with **vegetables** or **pulses**.

Gangliosides **Glycolipids** composed of a fatty acid (most often stearic acid) and an oligosaccharide, containing hexose and **sialic acid** residues, attached to a sphingosine. Highest concentrations are found in central nervous system tissues.

Ganoderma **Edible fungi** used in health foods and medicines, especially in China and Japan. Most common example is *Ganoderma lucidum*.

Garbanzo beans Seeds produced by *Cicer arientinum*. Used in stews and **casseroles**, and as the major component of **humous** and **falafel**. Rich in copper and **folic acid**, with relatively high contents of protein, **vitamin A** and iron. Also known as **chick peas**.

Gardenia Genus of flowering plants. Fruits of *Gardenia jasminoides* are used as a source of food **colorants**. The colorants are primarily composed of yellow **carotenoids**, **crocin** and its congeners, and iridoid **glycosides** such as geniposide.

Gari Meal produced by roasting and drying fermented **cassava** mash. Major food source in West Africa. Protein content is low. May contain potentially toxic levels of residual **cyanogens**, depending on the processing techniques used.

Garlic Pungent, edible bulbs of *Allium sativum*. One of the world's most widely used **spices**, used to flavour many different dishes. Each bulb comprises a number of cloves, which release a characteristic **aroma** when peeled and crushed. This aroma is due to the presence of **allicin**, which is believed to play a key role in the beneficial health effects reported for garlic. As well as being used fresh, much of the crop is further processed to yield garlic powder, garlic salt or **garlic oils**.

Garlic oils Highly pungent **essential oils** obtained from **garlic**. Used in spice mixes and other **flavourings**. Major constituent is allyl sulfide.

Garnishes Decorative and edible accompaniments to sweet or savoury dishes, usually added just before serving. May be placed on the plate beside the dish or applied to the surface of the food. Vary greatly in size and content, including sprigs of **parsley** or other **herbs**, **salad vegetables**, **croutons**, slices of fruit, whole **fruits** and **chocolate** shapes. Garnishes often indicate the main ingredient or **flavour** of a dish.

Gas chromatography **Chromatography** technique, usually abbreviated to GC, in which the sample is vaporized and injected into a carrier gas (mobile phase) that moves through a column, the inner surface of which is coated with a stationary phase. Sample components are separated on the basis of their affinity for the stationary phase, and identified by the time they are retained by the stationary phase. A range of detection techniques can be used in combination with gas chromatography, including mass spectroscopy (GC MS).

Gases Substances which have no fixed shape, low density and viscosity and no fixed volume, but which will adopt the volume of the space available, irrespective of the amount present. Composed of widely separated molecules which may be easily compressed and has the ability to diffuse readily. Distinct from the solid and liquid states.

Gas liquid chromatography Chromatography technique in which the mobile phase is a gas and the stationary phase is a liquid adsorbed on a porous solid in a tube or on the inner surface of a capillary column. Usually abbreviated to GLC. Components of the sample are partitioned between the gas and liquid phases, the rate at which they are eluted from the column depending on their partition coefficients. They are identified by the time taken to reach the detector for the system.

Gassericins Bacteriocins produced by *Lactobacillus gasseri*.

Gastritis Inflammation of the stomach. Causes can include consumption of corrosives and irritants (such as **alcoholic beverages**) and infection with *Helicobacter pylori*.

Gastroenteritis Inflammation of the mucous membranes of the stomach and intestines. Major causes include a range of **pathogens** that may be ingested via contaminated foods and **water supplies**. These include species of *Salmonella*, *Shigella*, *Campylobacter*, and *Vibrio*, and *Escherichia coli*, rotaviruses and **small round structured viruses**.

Gastrointestinal tract The organ commencing at the mouth and finishing at the anus, including the stomach and intestines, into which foods are taken and digested, and from which nutrients and non-nutrients are absorbed into the body, and waste is excreted.

Gastropods Common name for **molluscs** within the class Gastropoda; characterized by a single muscular foot. Includes **snails** (aquatic and marine), **limpets** and **sea slugs**.

Gateaux French word for **cakes**. Can refer to plain or fancy cakes, e.g. made from layers of **sponge cakes** filled and topped with **fruits**, **jelly** or **cream**.

GATT Abbreviation for **General Agreement on Trade and Tariffs**.

Gauges Instruments that measure and give a visual display of amounts, levels or contents.

Gayal meat Meat from **gayals**, which are a type of semi-domesticated oxen.

Gayals Large, semi-domesticated oxen found in India. Semi-domesticated form of the gaur (*Bos gaurus*) that is sometimes classified as *Bos frontalis*. Raised for their **meat** and **milk**. Bulls are crossed with English cattle breeds to produce good quality beef cattle.

Gazelle meat Meat from **gazelles**, which include several species of small, slender antelope, many of which belong to the genera *Gazella* of the family Bovidae. The meat is usually obtained from wild game animals and is appreciated for its **tenderness**.

Gazelles Any of several species of small, slender **antelopes**, many of which belong to the genus *Gazella* of the family Bovidae. Found wild in Africa, the Middle East and Asia. Source of **gazelle meat**.

Gbure Common name for *Talinum triangulare*, a leafy vegetable consumed in West and Central Africa. High in **fibre** and rich in essential **amino acids**.

GC Abbreviation for **gas chromatography**.

GC MS Abbreviation for **gas chromatography** combined with **mass spectroscopy**.

Geese The common name for any of numerous domesticated or wild waterfowl belonging to the family Anatidae, in which they comprise several genera (e.g. *Anser* and *Branta*). Most domesticated geese are kept in small flocks under free-range conditions for production of goose eggs and **goose meat**. A male goose is called a gander, whilst the female is a goose and the sexually immature young (with down rather than feathers) is a gosling.

Gelatin Soluble protein extracted from animal **collagen**, bones or **connective tissues** using hot water and acid or alkaline treatment. Widely used in the food industry in **gelling agents**, e.g. in **aspic**, **jellies**, **ice cream**, **yoghurt** and canned **meat**, and can also act as an emulsifier or stabilizer, e.g. in **marshmallows** and **confectionery fillings**. Lacks the essential amino acid **tryptophan**, but is a source of several **amino acids** lacking in plant foods such as **wheat**, **oats** and **barley**. Alternatively spelt gelatine.

Gelatinization Process involving disruption of molecular order within **starch granules** as a result of heating in water. Occurs over a temperature range and is also affected by granule size. Alterations caused include irreversible swelling, loss of birefringence, leaching of amylose and reduced crystallinity. Prolonged heating of the starch granules will eventually lead to total disruption.

Gelation Process of **gels** formation by **coagulation** of sols or **aggregation** of particles. Formed in a variety of ways according to the type of material concerned. In the case of polymer molecules, gelation is caused by formation of intermolecular crosslinks during heating or cooling. Aggregation of particles may be induced by a variety of stimuli including changes in **pH** or **ionic strength**. Also called gelling.

Gel electrophoresis **Electrophoresis** technique in which separation is performed in a gel, usually comprising agarose or polyacrylamide.

Gel filtration **Chromatography** technique in which separation is based on the size and shape of molecules. Samples are applied to a column of gel, e.g. polyacrylamides, cross-linked dextrans, large polysaccharides, and components are separated on the basis of their ability to penetrate the pores of the gel beads while being washed through with a liquid mobile phase. The technique may be used to estimate relative molecular mass (M_r). Also known as **gel permeation chromatography**.

Gelidium Genus of red **seaweeds** known as **onigusa** in Japan. Eaten in some Asian countries, also a source of **agar**.

Gellan Exopolysaccharide produced by ***Sphingomonas*** *paucimobilis* ATCC 31461 (formerly *Pseudomonas elodea*) which is composed of a tetrasaccharide repeating unit comprising **glucose**, **rhamnose** and **glucuronic acid** in the ratio 2:1:1, and with acyl substitution of one of the glucose residues. Forms **gels** in the presence of cations and used as **gelling agents** and **thickeners** in foods.

Gellan gums **Gums** containing **gellan**, a microbial polysaccharide produced by ***Sphingomonas*** *paucimobilis* ATCC 31461 (formerly *Pseudomonas elodea*), which form transparent and heat- or acid-resistant **gels**. Widely used as **thickeners** and **gelling agents** in foods.

Gelling Alternative term for **gelation**.

Gelling agents **Additives** used to promote **gelation**. Used in manufacture of **jellies** and other food **gels**. Commonly used gelling agents include **pectins**, **agar**, **guar gums** and **gellan gums**.

Gelling capacity One of the **functional properties** of a substance concerned with its ability to form a gel.

Gel permeation chromatography **Chromatography** technique in which separation is based on the size and shape of molecules. Samples are applied to a column of gel, e.g. polyacrylamides, cross-linked dextrans, large polysaccharides, and components are separated on the basis of their ability to penetrate the pores of the gel beads while being washed through with a liquid mobile phase. The technique may be used to estimate relative molecular mass (M_r). Also known as **gel filtration** chromatography.

Gels Solid or semi-solid jelly-like **colloids**, such as those formed when **gelatin** is mixed with hot water and allowed to cool. Products such as **pectins** and **agar** are well known for their gel-forming ability. Gels, including **agar gels**, are widely used as food **stabilizers** and **thickeners**.

Gene cloning Insertion of **DNA** sequences containing **genes** into **vectors** (e.g. **plasmids** or **viruses**) that can then be propagated in a host organism, thus producing multiple copies of the gene of interest.

Gene disruption Use of both *in vitro* and *in vivo* recombination to replace wild type **genes** or **DNA** sequences with a mutant version.

Gene expression The process by which proteins are produced from their coding **genes** by means of **transcription** followed by **translation**.

Gene libraries Collections of cloned **DNA** fragments in which the inserted sequences together represent entire **genomes** of organisms (genomic libraries). Alternatively, the cloned DNA may be composed of cDNA molecules formed from an mRNA template (cDNA libraries), thus representing only the expressed portions of genomes.

Gene probes Molecules that have been labelled with radioactive isotopes, fluorescent dyes or enzymes that bind selectively to specific **genes**, thus allowing identification or isolation.

General Agreement on Trade and Tariffs The General Agreement on Trade and Tariffs (GATT) was a treaty and international trade organization in existence from 1948 to 1995. GATT members worked to minimize tariffs, quotas, preferential trade agreements between countries, and other barriers to international trade. In 1995, GATT's functions were taken over by the World Trade Organization (WTO), an international body that administers trade laws and provides a forum for settling trade disputes among nations. GATT members sponsored eight specially organized rounds of trade negotiations. The last round of negotiations, called the Uruguay Round, began in 1986 and ended in 1994. At the end of the negotiations, the members of GATT, as well as representatives from seven other nations, signed a trade pact that will eventually cut tariffs overall by about one-third and reduce or eliminate other obstacles to trade. The pact also took steps toward opening trade in investments and services among member nations and strengthening protection for intellectual property. Throughout 1995, GATT and the WTO coexisted while GATT members sought their governments' approval for WTO membership. After the transition period, GATT ceased to exist. All of the 128 nations that were contracting parties to the 1994 GATT agreement eventually transferred membership to the WTO.

Genes Units of inheritance that occupy specific loci within nucleic acid molecules (e.g. **chromosomes**, **plasmids**). Consist of specific **DNA** sequences that code for functional polypeptides or **RNA** molecules (e.g. rRNA, tRNA). Eukaryotic genes often consist of coding units (exons) separated by one or more non-coding unit (introns).

Genetically engineered foods Foods that have been modified or that have been prepared with agents, e.g. enzymes, or contain ingredients that have been modified using genetic techniques. Used to confer new properties such as enhanced **nutritional values** and prolonged **shelf life**. More commonly referred to as **genetically modified foods** or GM foods.

Genetically modified foods Commonly abbreviated to GM foods. Foods that have been modified or that have been prepared with agents, e.g. **enzymes**, or contain ingredients that have been modified using genetic techniques. Used to confer new properties such as enhanced **nutritional values** and extended **shelf life**. Less commonly referred to as **genetically engineered foods**.

Genetically modified microorganisms Microorganisms that have been modified by **genetic techniques** to enhance their properties or confer upon them new properties. Abbreviated to GM microorganisms.

Genetically modified organisms Organisms that have been modified by **genetic techniques** to enhance their properties or confer upon them new properties. Abbreviated to GM organisms or GMO.

Genetic disorders Deleterious effects caused by alterations in the genetic material of organisms that may or may not be inherited in a Mendelian fashion.

Genetic engineering General term covering various **genetic techniques** for *in vitro* manipulation of genetic material. Can be used for construction of new **genes** or novel combinations of genes, usually for insertion into host cells, placing genes under the control of different regulatory systems or introducing specific **mutations** into DNA molecules.

Genetic fingerprinting Process by which different **DNA** samples are compared to determine if they are from the same individual. Uses **PCR** or **Southern blotting** to compare the characteristic polymorphic patterns of certain regions of genomic DNA. Also **DNA fingerprinting**.

Genetic mapping Process by which the relative positions of **genes** on **DNA** molecules (usually **chromosomes**) and the distances between them are determined.

Genetics The study of heredity and variation, i.e. the patterns of inheritance of specific traits.

Genetic techniques Methods used in the study of **genetics** and for the manipulation of genetic material.

Genetic variants Organisms and cells that differ in **phenotype** due to differences in **genotype**, rather than to environmental factors.

Gene transfer General term for insertion of foreign **genes** into cells or organisms.

Geniposide Iridoid glucoside which is found in the fruits of **Gardenia** *jasminoides* and is a constituent of gardenia yellow, a natural colorant used in a range of foods. Can be transformed into a blue pigment, which is also of potential use as a food colorant, by enzymic hydrolysis to genipin and reaction with **amino acids**.

Genistein Yellow isoflavone which occurs in free or glucosidic form and has a weak oestrogenic effect. Found in **soybeans**, **chick peas**, **lucerne** and clover.

Genomes The genetic material of an organism or cell, comprising the complete set of **genes**.

Genomics The study of **genomes**.

Genotoxicity Quality or degree of being capable of exerting a damaging effect on the **DNA** that forms **genes**.

Genotype The genetic constitution of an organism or cell that determines the expression of specific traits.

Genotyping **Genetic techniques** (e.g. repetitive DNA sequence analysis, **RAPD** analysis, gene sequence analysis, **PCR**, **pulsed field gel electrophoresis** and **RFLP**) used to determine and compare the genetic constitution of organisms and cells.

Gentamicin Aminoglycoside antibiotic used to treat a range of bacterial infections in farm animals. Used especially in swine for treatment of colibacillosis and swine dysentry; also used for treatment of **mastitis** in cattle. Depletes relatively slowly from tissues, particularly **kidneys**. Relatively long withdrawal periods are required for animals intended for consumption.

Gentiobiose Disaccharide reducing sugar produced by reaction of two molecules of **glucose** in the presence of **β-glucosidases**. Systematic name is 6-*O*-β-D-glucopyranosyl-D-glucose.

Geobacillus Genus of Gram positive aerobic thermophilic **bacteria**. Several species of *Bacillus* have been transferred to this new genus, including *B. stearothermophilus* and *B. thermoleovorans*.

Geosmin Heterocyclic volatile compound which naturally occurs in fresh water and imparts musty and earthy **flavour** and **aroma taints** to **beverages** such as **drinking water** and **wines**, as well as to **freshwater fish**.

Geotrichum Genus of **fungi** of the class Hyphomycetes. Occur in soil, water, dairy products and grains. Species may be involved in the production of **fer-**

mented foods, or may cause food **spoilage**. *Geotrichum candidum* imparts **flavour** and **aroma** to many types of **cheese** and assists in the **fermentation** of **cocoa**. However, it can also cause spoilage of dairy **cream** and **butter**, sour rot of **citrus fruits** and **peaches**, and watery soft rot of **vegetables**.

Geranial Structural *trans* isomer of **citral** derived from oxidation of **geraniol** and present in **lemongrass oils**.

Geraniol Colourless or pale yellow unsaturated monoterpene alcohol which has a rose-like aroma and occurs in geranium and rose **essential oils**. Used in **flavourings**.

Geranyl acetate Volatile compound occurring as one of the natural **flavour compounds** in the **essential oils** of many **herbs** and **spices**. Extracted as a colourless liquid by fractional distillation of selected essential oils or prepared by **acetylation** of **geraniol**. Used in **flavourings** for foods and **beverages**.

Germ Germinating portion or embryo of a cereal grain which is extracted and discarded when the grain is milled to make white **flour**. High in **fats** and several **vitamins**.

Germicides Antimicrobial chemical agents used for **disinfection**, antisepsis or **sterilization**.

Germination Sprouting of a seed, spore or other reproductive body. Influenced by a number of factors, including temperature, light and oxygen supply. Used commercially in preparation of **cereals** for manufacture of **alcoholic beverages**, and in production of **mushrooms**.

Germination capacity Ability of a seed to germinate.

Gesatop Alternative term for the herbicide **simazine**.

Gestagens Steroid **hormones** which induce progestational effects in the uterus.

Ghee Product made from **butter**; originally produced in India but now more widespread. Butter is melted at a high temperature, during which moisture is evaporated. **Proteins** are then removed from the melted butter by **centrifugation**.

Gherkins West Indian gherkins are fruits produced by *Cucumis anguria*. Usually 4-5 cm long, and used mainly in **pickles**. In Europe, the term gherkins usually refers to small ridge **cucumbers**.

Giardia Genus of protozoan parasites of the family Hexamitidae. Occur in water and the **gastrointestinal tract** of humans and other mammals. *Giardia lamblia* is the causative agent of **giardiasis** in humans.

Giardiasis Disease caused by infection with *Giardia lamblia*. Commonly transmitted through ingestion of food or water contaminated with cysts. Characterized by watery diarrhoea, abdominal cramps, nausea and flatulence. Infection may be asymptomatic.

Gibberellic acid Plant growth regulator belonging to the **gibberellins** group which may be obtained commercially by culture filtration of the fungus *Gibberella fujikuroi*.

Gibberellins Any of a group of **plant growth regulators** originally produced by *Gibberella fujikuroi*. Promotes processes such as stem elongation, **germination** and flowering. Often used to stimulate germination of dormant grain such as **barley** during **malting**.

Giblets Edible **offal** from the carcasses of poultry and game birds. Giblets include the **livers**, **hearts**, **gizzards** and necks of the **birds**; they are usually removed before the birds are cooked. Giblets, with the exception of livers, are often used to make **gravy**, **stocks** or **soups**.

Gigartina Genus of **seaweeds** found on rocky shores around the world. Some species are used as a source of **carrageenans** and nutraceuticals in the food industry.

Gin Spirits made by flavouring rectified **ethanol** with **juniper** and other plant ingredients, usually by redistillation of the spirits with the **flavourings**.

Gingelly oils Alternative term for **sesame oils**.

Gingelly seeds Alternative term for **sesame seeds**.

Ginger Rhizomes from *Zingiber officinale*. Used fresh or dried as **spices** in a number of foods and **beverages**, including **gingerbread** and **ginger beer**. **Pungency** is due to the presence of **gingerols**. May also refer to related *Curcuma* species such as *C. xanthorrhiza*.

Ginger ale **Ginger**-flavoured sweetened **carbonated beverages**, often added to **spirits** such as whisky or brandy prior to consumption.

Ginger beer Effervescent **ginger**-flavoured slightly-alcoholic **beverages** made by **fermentation** of a **sugar** medium containing ginger and other **flavourings**.

Gingerbread Dark **molasses**-based **cakes** or **biscuits** flavoured with ground **ginger** and other **spices**. Often cut into shapes, decorated and glazed.

Gingerols Phenolic **ketones** which are the major **pungent principles** of fresh **ginger**.

Ginjoshu Type of **sake**.

Ginkgo nuts Seeds produced by *Ginkgo biloba* (maidenhair tree), a plant grown in Asia. Fresh nuts are soaked in hot water to loosen the skin. Also available dried and canned in **brines**. Used widely in Japanese cooking (turn green when cooked) and in Chinese medicines.

Ginseng   Root of the plant *Panax ginseng*, used for preparation of **beverages**. Widely considered to have health-promoting properties, possibly related to the presence of **saponins (ginsenosides)**.

Ginseng saponins   Alternative term for **ginsenosides**.

Ginsenosides   Complex mixture of **saponins** which are believed to be the active components of **ginseng**, *Panax ginseng*, and are thought to be responsible for the reported health benefits associated with this plant. Also known as ginseng saponins.

Girdling   Removal of a strip of bark from the circumference of a tree, with the intention of improving growth or quality of fruits.

Gizzards   Muscular, thick-walled stomachs of **birds**, which lie between the proventriculus and the upper limit of the small intestine; poultry gizzards form a part of edible **offal**. In birds, the function of the gizzard is to grind food, typically with swallowed grit and small stones.

Glass   Brittle, usually transparent or translucent material used widely to make bottles and other containers. Manufactured by fusing sand (silica and silicates) with soda and lime. Also refers to individual drinking vessels made from glass.

Glass bottles   Bottles made from **glass** which are commonly used as containers for **beverages** and other liquids. Available in a range of shapes, capacities and **colour**.

Glass containers   Containers made from **glass** which may be used to store or package a range of foods. Include **glass bottles**, **beakers**, jars and pots.

Glassine   Smooth, thin, glossy transparent or semi-transparent **paper** made primarily from chemical wood pulps. Has a high resistance to transmission of air and is grease resistant. To make it impervious to water vapour, some glassine is lacquered, laminated or waxed. Used for wrapping food.

Glassiness   **Optical properties** relating to the extent to which a product appears to have the surface properties of glass, i.e. smoothness, uniformity, shininess and glossiness.

Glass transition   Reversible sudden transition of an amorphous polymer from a glassy condition to a flexible condition when it is heated to a specific temperature range (**glass transition temp.**). Due to a change in the arrangement of the polymer molecules from a coiled and motionless state to one where they are free to move.

Glass transition temp.   Temperature range at which the **glass transition** (change from a glassy to a flexible condition) of polymers takes place. Value varies according to the polymer and the range is relatively small.

Glazes   Substances, such as **milk**, beaten **eggs** or thin **jams**, which are used to create a shiny appearance or provide protective **coatings** on foods. Also, smooth, glossy, glass-like materials fused onto the surface of pottery, where they form hard, impervious decorative coatings.

Glazing   Application of a liquid, such as **milk** or beaten **eggs**, to hot or cold foods to produce a smooth, shiny coating after setting. For example, milk or beaten eggs can be brushed onto **pastry** before **baking** to add **colour** and shine.

GLC   Abbreviation for **gas liquid chromatography**.

Gliadins   **Cereal proteins** from the endosperm of **wheat** or **rye**. The elastic constituent of **gluten**.

Gliocladium   Genus of **fungi** of the class Sordariomycetes. Species may be capable of being antagonists of other fungal species (e.g. Rhizoctonia species). *Gliocladium roseum* is an effective **biocontrol** agent for managing fungal diseases of certain crops (e.g. **strawberries**).

Gliotoxin   Mycotoxin produced by *Trichoderma viride* and species of *Aspergillus*, *Gliocladium* and *Penicillium*. Inhibits replication of certain **viruses** (e.g. polioviruses) and also exhibits **antibacterial activity**, **antifungal activity** and **antitumour activity**.

Globe artichokes   Common name for *Cynara scolymus*. Plant has a large, thistle-like flower head with edible fleshy leaves and heart. Generally eaten cooked, either hot or cold, and can be canned. Small, immature flower heads may also be consumed, cooked and preserved in **olive oils**. Globe artichokes are high in **fibre**, low in fat and calories, and rich in **vitamin A** and **vitamin C**.

Globins   **Animal proteins** that contain some arginine and tryptophan, are rich in histidine and are deficient in isoleucine. They often form the protein portion of conjugated proteins, e.g. the globins in **haemoglobin** or **myoglobin**.

Globulins   Any of a class of spherical or globular shaped high molecular weight proteins which are relatively insoluble in water and soluble in dilute salt solutions. Found widely throughout nature; they include **lactoglobulins**, serum globulins and **immunoglobulins**. Subdivided into α-, β- and γ-globulins.

γ-Globulins   **Globulins** occurring in animal tissues and products derived from them; includes **immunoglobulins**.

Gloss   **Optical properties** relating to the surface lustre or sheen on a product. Gloss is important to the attractiveness of specific products such as **gelatin desserts** and buttered vegetables.

Glucagon Polypeptide hormone secreted by pancreatic cells in response to a decrease in serum **glucose**. Acts by promoting the conversion of liver **glycogen** into glucose, thereby increasing the level of blood sugar, and has an opposite effect to that of **insulin**.

Glucanases General term for **enzymes** that hydrolyse **glucans**. Include **glucan endo-1,3-β-D-glucosidases**, **endo-1,3(4)-β-glucanases** and **licheninases**.

β-Glucanases Alternative term for **endo-1,3(4)-β-glucanases**.

1,4-α-Glucan branching enzymes EC 2.4.1.18. **Enzymes** which transfer a segment of a 1,4-α-D-glucan chain to a primary hydroxyl group in a similar glucan chain. These enzymes convert **amyloses** into **amylopectins**. The recommended name requires a qualification depending on the product, glycogen or amylopectin, e.g. glycogen branching enzyme, amylopectin branching enzyme (often referred to as Q-enzyme).

Glucan endo-1,3-β-D-glucosidases EC 3.2.1.39. **Enzymes** which hydrolyse 1,3-β-D-glucosidic linkages in 1,3-β-D-glucans. Also known as **laminarinases**. Important in the **malting** and **brewing** industries, and potentially useful for production of functional **oligosaccharides**. May be involved in the **ripening/softening** of **fruits** and in plant defence.

Glucan 1,4-α-glucosidases EC 3.2.1.3. Also known as glucoamylases, these **enzymes** hydrolyse terminal 1,4-linked α-D-glucose residues successively from non-reducing ends of chains, releasing β-D-glucose. Can also hydrolyse α-D-1,6-glucosidic linkages, although at a slower rate. Useful for degradation of **starch** (**saccharification**) for production of **sugar syrups** and for conversion of residual **dextrins** to fermentable sugars during production of low calorie **beer**.

4-α-Glucanotransferases EC 2.4.1.25. **Enzymes** which transfer a segment of a 1,4-α-D-glucan to a new position in an acceptor, which may be **glucose** or another 1,4-α-D-glucan. Involved in **starch** metabolism in plants and can catalyse the synthesis of cycloamylose. Together with maltogenic **amylases**, these enzymes may also be useful for the synthesis of **isomaltooligosaccharides**, useful as low calorie **sweeteners** and in **probiotic foods** as bifidogenic agents. Also known as disproportionating enzymes, dextrin glycosyltransferases and D-enzymes.

Glucans Soluble, indigestible **polysaccharides** composed predominantly of D-**glucose** residues and found in **cereals** such as **oats**, **barley** and **rye**.

β-Glucans Soluble but undigested polysaccharide complexes composed of D-glucose in either straight or branched chains with glycosidic linkages. Present in most cereal grains but occur in larger amounts in **barley** and **oats**.

Glucides Subclass of **saccharides**.

Glucitol Alternative term for **sorbitol**.

Glucoamylases Alternative term for **glucan 1,4-α-glucosidases**.

Glucobrassicin Alternative term for 3-indolylmethyl glucosinolate, one of the major **glucosinolates** found in **vegetables** of the genera **Brassica** and *Raphanus*.

Glucocorticoids Any of a group of corticosteroids secreted by the adrenal cortex that controls carbohydrate and protein metabolism by promoting **glycogen** deposition in the liver. Glucocorticoids have **anti-inflammatory activity**.

Glucokinases EC 2.7.1.2. Catalyse the transfer of a phosphate group from **ATP** to D-glucose to form D-glucose 6-phosphate, the first reaction of the glycolytic pathway. Also used as a trivial name for **hexokinases** (EC 2.7.1.1).

Glucomannans Viscous **polysaccharides** comprising **glucose** and **mannose** which occur naturally in the food reserves of some plants, such as konjac (**elephant yams**). Like **galactomannans** they form thermally reversible **gels** with **xanthan**.

Gluconic acid Organic acid which is soluble in water and alcohol and is formed by oxidation of glucose in which the CHO group has been converted to COOH. Predominant acid found in **honeys**.

Gluconobacter Genus of Gram negative, aerobic, rod-shaped acetic acid **bacteria** of the family Acetobacteraceae, formerly known as *Acetomonas*. Occur in soil, plants, **fruits** and flowers. Species are used in the production of **vinegar**, and may cause **spoilage** of **beer**.

Glucono-δ-lactone Lactone that forms **gluconic acid** when dissolved in water. Used as an acidulant and hence to induce **gelation** in a range of foods, and as a **leavening** agent in **bakery products**. Also frequently used as an additive in **cheese**, **meat** and **sausages**.

Glucosamine Crystalline amino derivative of **glucose** and the principal component of **chitin**, mucoproteins and **mucopolysaccharides**.

Glucose Monosaccharide with six carbon atoms. Free glucose is present naturally in **fruits** and **honeys** and it is the monomer unit from which **starch** and **cellu-**

loses are synthesized; commercial manufacture of glucose is by hydrolysis of starch. It is the main energy source for living cells. Glucose is a constituent of **sucrose** and is used in **sweeteners**. Free glucose has 0.7-0.8× the **sweetness** of sucrose. The D-stereoisomer of glucose is known as **dextrose**.

Glucose isomerases EC entry for these enzymes has been deleted. Activity now attributed to **xylose isomerases** (EC 5.3.1.5) or, in the presence of arsenate, to glucose-6-phosphate isomerases (EC 5.3.1.9). The term glucose isomerases is still used widely, however. Catalyse the **isomerization** of **glucose** to **fructose** and used for large scale production of **fructose high corn syrups**.

Glucose oxidases EC 1.1.3.4. Flavoenzymes that oxidize β-D-glucose to β-D-gluconic acid and H_2O_2. Used for stabilizing foods and **beverages** by removing **glucose** and O_2, and preventing deterioration of **colour** and **flavour** (e.g. **liquid egg whites, fruit juices** and **beverages, beer, wines, mayonnaise, bread dough** and **cheese**). Also used in **biosensors** for determination of **glucose** levels and for removal of O_2 in food packaging, thus protecting against oxidative deterioration.

Glucose-6-phosphate Primary metabolite of **glucose** in living cells. Formation of glucose-6-phosphate from glucose is the first reaction in **glycolysis** and is catalysed by **hexokinases**.

Glucose-1-phosphate adenylyltransferases EC 2.7.7.27. Also known as ADP-glucose pyrophosphorylases, these **enzymes** transfer a phosphate group from **ATP** to α-D-glucose 1-phosphate to form pyrophosphate and ADP-glucose. They are key enzymes in the **starch** biosynthetic pathway in plants.

Glucose syrups **Syrups** consisting predominantly of **glucose**. Produced commercially by hydrolysis of **starch**; **corn starch** is the most commonly used substrate.

Glucosidases Alternative term for **glycosidases**.

α-Glucosidases EC 3.2.1.20. **Enzymes** which hydrolyse terminal, non-reducing 1,4-linked α-D-glucose residues of **oligosaccharides** and α-D-glucosides. **Polysaccharides** are hydrolysed only slowly. These enzymes are specific with respect to the **glucose** residues of **glucosides** but less specific with respect to the **aglycones**. Can be used to hydrolyse **maltose** and maltose derivatives, **sucrose**, and aryl- and alkyl-α-D-glucosides. Also known as maltases and glucoinvertases.

β-Glucosidases EC 3.2.1.21. **Enzymes** which hydrolyse terminal non-reducing β-D-glucose residues with the release of β-D-glucose. Substrates include **cellobiose**, cellooligosaccharides and aryl-β-

glucosides. These enzymes have a number of uses in the food industry, including **debittering** of table **olives**, increasing the **aroma** of **wines**, enrichment of **genistein** in **soy protein concentrates** and release of **phenols** from phenolic glycosides in plant tissues.

Glucosides A range of **glycosides** found mainly in plants, the sugar component of which is **glucose**. These compounds may be useful as aroma precursors, **pigments** and **surfactants**, and may exhibit **antioxidative activity**. However, cyanogenic glucosides found in several plants are a potential source of **cyanides** and are therefore potentially toxic.

Glucosinolases Alternative term for **thioglucosidases**.

Glucosinolates Class of toxic **glucosides** which are found in **Brassica** (e.g. **broccoli, cabbages, radishes**). Degraded by **thioglucosidases** to produce mustard oils, accounting for the pungent **flavour** of these compounds. May have anticancer properties, since they increase the rate at which potential **carcinogens** are excreted.

Glucosyltransferases Members of subclass EC 2.4; synonymous with **transglucosylases**. **Enzymes** that transfer glucosyl groups from a donor molecule to an acceptor.

Glucuronic acid Uronic acid derived by oxidation of the carbon-6 atom to a carboxyl group. Potential **toxins** are conjugated with glucuronic acid in the liver to form glucuronides before being excreted. Glucuronic acid is also found in **mucopolysaccharides**.

α-Glucuronidases EC 3.2.1.139. **Enzymes** which hydrolyse α-1,2-glycosidic bonds of the 4-O-methyl-D-glucuronic acid side chain of **xylan**. Act as part of an array of **xylan degrading enzymes** produced by a range of **microorganisms**. Such enzymes are useful in food processing procedures involving degradation of plant cell walls, e.g. extraction of juices and **essential oils**, or **clarification** of **wines**, and in production of modified xylans for use as **bulking agents**.

β-Glucuronidases EC 3.2.1.31. **Enzymes** which can be used to hydrolyse glycyrrhizin β-glucuronide, modifying its sweetness. The enzymes present in intestinal bacteria are thought to be responsible for production of toxic **aglycones** and **carcinogens**, a process which may be induced by dietary carbohydrates. Determination of β-glucuronidase activity may, therefore, be useful for evaluating dietary-mediated colon cancer risk.

Glutamate decarboxylases EC 4.1.1.15. Involved in synthesis of γ-amino-n-butyric acid (GABA), which is reported to have **antihypertensive activity** and is found in high amounts in foods such as red mould **rice**

and **tea**. These **enzymes** are involved in acids resistance in **bacteria** and are produced by **cheese starters** during ripening.

Glutamate dehydrogenases **Dehydrogenases** which catalyse the conversion of glutamate to 2-oxoglutarate and ammonia, using NAD^+ (EC 1.4.1.2), $NAD(P)^+$ (EC 1.4.1.3) or $NADP^+$ (EC 1.4.1.4) as acceptor. Useful for determination of glutamate content of foods. Production of these enzymes by **lactic acid bacteria** may be important for **flavour** development in certain **dairy products**, including **cheese**.

Glutamate oxidases Alternative term for **L-glutamate oxidases**.

L-Glutamate oxidases EC 1.4.3.11. Convert L-glutamic acid to 2-oxoglutaric acid. Used for determination of L-glutamic acid levels in foods and, in combination with other **enzymes**, for detection of **aspartame** and **aspartic acid**. Have also been used for detection of the neurotoxin β-N-oxalyl-α,β-diaminopropionic acid in **grass peas**, and are potentially useful for preparing α-keto acids from a racemic mixture of acidic **amino acids** and for resolution of racemates.

Glutamates Salts of **glutamic acid** used as **flavourings**, e.g. the flavour enhancer **monosodium glutamate**.

Glutamic acid Amino acid which is believed to play a part in thehigh-quality **flavour** of young fresh vegetables and in the enhancement of other flavours in general. Salts of glutamic acid (**glutamates**) are widely used as **flavourings**.

Glutaminases EC 3.5.1.2. Convert L-glutamine to L-glutamic acid. Can be used to increase the **glutamic acid** content and, hence the **sensory properties**, of foods, e.g. **soy sauces**, **protein hydrolysates** used as **flavourings** and certain meat products.

Glutamine Non-essential amino acid which is a monoamide of **glutamic acid**. Abundant in plants such as **beets**, **carrots** and **radishes** and important in cell metabolism.

γ-Glutamyl hydrolases EC 3.4.19.9. **Carboxypeptidases** that hydrolyse γ-glutamyl bonds. Used in analysis of **folates** in foods and **beverages**, and also known as conjugases and folate conjugases.

γ-Glutamyltransferases EC 2.3.2.2. Also known as glutamyl transpeptidases, these **enzymes** catalyse the hydrolysis and transpeptidation of γ-glutamyl compounds, such as **glutathione**. Activity of these enzymes in **onions** has been linked to the formation of a variety of **organic sulfur compounds** active as defence molecules and as **flavour compounds**. The enzymes have also been used as markers for the **pasteurization** of **milk**.

Glutathione Tripeptide widely distributed in cells and composed of **glutamic acid**, **cysteine** and **glycine** residues. Functions as a redox agent and a coenzyme for some **enzymes**. Also shows **antioxidative activity** in the protection of sulfhydryl groups in enzymes and other proteins.

Glutelins Group of **globulins** present in the seeds of **wheat**, **rice** and **barley**. Soluble in dilute acids or alkalies and insoluble in water, they are a constituent of **gluten**.

Gluten Water insoluble protein complex found in the endosperm of **wheat** and **rye** and composed predominantly of **gliadins** and **glutenin**. When mixed with water, forms cohesive, elastic, cross-linked molecules. These confer elasticity to **bread dough**, allowing the dough to trap **carbon dioxide** during **breadmaking** and causing the **bread** to rise.

Gluten free bread **Bread** formulated to contain no **gluten** by excluding **wheat** and **rye** proteins to make it suitable for consumption by people suffering from **coeliac disease**.

Gluten free foods Foods formulated to contain no **gluten** by excluding **wheat** and **rye** proteins to make them suitable for consumption by people suffering from **coeliac disease**.

Glutenin Glutelin found in the endosperm of **wheat** and one of the major components of **gluten**.

Gluten low bread Alternative term for **gluten free bread**, a **bread** made with the exclusion of **wheat** and **rye** proteins to make it suitable for consumption by people suffering from **coeliac disease**.

Gluten low foods Alternative term for **gluten free foods**, foods made with the exclusion of **wheat** and **rye** proteins to make them suitable for consumption by people suffering from **coeliac disease**.

Glycaemic index values **Nutritional values** relating to the ability of carbohydrate to increase the level of **glucose** in the blood compared with the same amount of glucose.

Glycans Alternative term for **glucans**.

Glycation Modification involving nonenzymic reaction of **sugars** with **proteins** (or sometimes **lipids**), as in the **Maillard reaction**. Results in alterations in physicochemical, biological and functional properties, such as **foaming properties**, **emulsification properties** or **antioxidative activity**, of proteins.

Glycerides Synonym for **acylglycerols**. **Fatty acid esters** of **glycerol**, such as **monoglycerides**, **diglycerides** and **triglycerides**. Major components of natural **fats** and **oils** (particularly as triglycerides); also used as **emulsifiers**.

Glycerin Synonym for **glycerol**; alternative spelling glycerine.

Glycerine Synonym for **glycerol**; alternative spelling glycerin.

Glycerol Clear, odourless, sweet-tasting, viscous, hygroscopic liquid produced by fat **saponification**. Functions as a humectant to prevent **candy** and other foods from drying out, as a solvent for **flavourings** and **colorants**, as an emulsifier, and as a **texture** improver for **cakes**. Also used to control crystallization and in the formulation of **fat substitutes**. Synonym for glycerin and glycerine.

Glycerolipids Alternative term for **glycolipids**.

Glycerol monolaurate Alternative term for **monolaurin**.

Glycerol monostearate Synonym for **glyceryl monostearate**, an ester formed by reaction of **stearic acid** with **glycerol**. Used in food **emulsifiers**, in the manufacture of products such as **coffee whiteners** and **ice cream**. Included as **bakery additives** in manufacture of **bread** and other **bakery products** due to the anti-staling properties of the glyceride component.

Glyceryl monostearate Ester formed by reaction of **stearic acid** with **glycerol**. Used in food **emulsifiers**, in the manufacture of products such as **coffee whiteners** and **ice cream**. Included as **bakery additives** in manufacture of **bread** and other **bakery products** due to the anti-staling properties of the glyceride component. Also called **glycerol monostearate**.

Glycine Non-essential achiral amino acid, structurally the simplest of the **amino acids**. Sweet-tasting and used to retard the onset of **rancidity** in **fats**, as well as being a nutrient. **Gelatin** is a particularly rich source of glycine.

Glycine betaine One of the soluble **nitrogen compounds** and a derivative of **betaine** occurring in a range of foods, especially **sugar beets**, **spinach** and **molasses**; also found in some **shellfish**, where it is important for **flavour**. An effective osmoprotectant, glycine betaine is also synthesized by **microorganisms** living at very high osmotic pressures. Accumulation of glycine betaine in some pathogens, e.g. *Listeria monocytogenes*, allows them to survive under conditions of extreme temperature, leading to food safety problems. The compound may also be added to increase thermal tolerance and osmotolerance in **bacteria** used in food manufacture.

Glycinin One of the main **soy proteins**. An 11S storage protein that, along with **β-conglycinin** (7S globulin), makes up approximately 70% of **storage proteins** in **soybeans**.

Glycoalkaloids Natural **toxins** which are synthesized by plants of the family Solanaceae, including **potatoes**. Consist of **alkaloids** with one or more **sugar** residues attached. Include α-**solanine** and α-**chaconine**.

Glycogen High molecular weight branched polysaccharide comprising D-glucopyranose residues (**glucose** in the ring conformation). Formed predominantly in muscle and liver tissues and is the main store of energy in animals and humans.

Glycolic acid Colourless, hygroscopic chemical intermediate of the conversion of **glycine** to **ethanolamine**. Constituent of **cane sugar juices** and unripe **grapes**.

Glycolipids Compounds consisting of lipid moieties which are glycosidically linked to one or more monosaccharide residues. Also known as glycosphingolipids.

Glycols General term for **diols**, **organic compounds** with two alcohol groups. Include **ethylene glycol** and **1,2-propanediol (propylene glycol)**.

Glycolysis Series of reactions which take place in most living cells by which **glucose** is converted into **pyruvic acid** and then to **lactic acid**.

Glycomacropeptides Low molecular weight **whey proteins** produced during **cheesemaking** when κ-**casein** is treated with **chymosin**. Show **biological activity** and are potential ingredients for **functional foods**.

Glycopeptides Compounds in which a carbohydrate is covalently linked to an oligopeptide composed of D- and/or L-amino acid residues.

Glycoproteins Conjugated **proteins** composed of polypeptide backbones to which **carbohydrates** are covalently attached. Present in **ovalbumins**, **mucins** and fish **antifreeze proteins**.

Glycosidases EC 3.2. **Enzymes** that hydrolyse **carbohydrates** such as **starch**, **celluloses** and **pectins**. Able to cleave short-chain **oligosaccharides** as well as **polysaccharides** with various structures. They are used in all areas of the food industry, but their major application is in starch processing. Subdivided into enzymes that hydrolyse *O*-, *N*- or *S*-glycosyl compounds (EC 3.2.1, 3.2.2 and 3.2.3, respectively). Some can also transfer glycosyl residues to oligosaccharides, polysaccharides and other alcoholic receptors.

Glycosides Compounds occurring abundantly in plants in which a sugar is combined with a non-sugar entity (aglycone); this may be an alcohol, phenol or sterol, and replaces the hydroxyl group on the carbon-1 atom. Often found in fruit **pigments**, e.g. **anthocyanins**.

Glycosyltransferases EC 2.4. **Enzymes** that catalyse the transfer of glycosyl groups to an acceptor.

Usually, other **carbohydrates** act as acceptors, although inorganic phosphate can also be an acceptor, such as in the case of **phosphorylases**. Some of the enzymes in this group also catalyse hydrolysis, which can be regarded as the transfer of a glycosyl group to water. The subclass is subdivide further, according to the nature of the sugar residue being transferred, into hexosyltransferases (EC 2.4.1), pentosyltransferases (EC 2.4.2) and those transferring other glycosyl groups (2.4.99).

Glycyrrhizic acid　Glycoside extracted from **liquorice** (*Glycyrrhiza*) which has an intensely sweet **flavour**.

Glycyrrhizin　Sweet-tasting glycoside derived from **liquorice** root. Used in high-intensity **sweeteners** for foods.

Glyoxal　Dicarbonyl compound found as an aroma precursor/compound in **wines**. Also one of the **Maillard reaction products** in **nonenzymic browning**.

Glyphosate　Non-selective systemic herbicide used for control of deep-rooted perennial plants and annual and biennial grasses and broad-leaved weeds in a wide range of crops. Also used for pre-harvest drying of **cereals** and **legumes** and for control of sucking **insects** on fruit trees. Classified by Environmental Protection Agency as slightly toxic (Environmental Protection Agency III).

GM foods　Abbreviation for **genetically modified foods**.

GM microorganisms　Abbreviation for **genetically modified microorganisms**.

GMO　Abbreviation for **genetically modified organisms**.

GM organisms　Abbreviation for **genetically modified organisms**.

GMP　Abbreviation for **guanosine monophosphate**.

Gnathostoma　Genus of parasitic **nematodes** of the family Gnathostomatidae. Occur in the **gastrointestinal tract** of dogs, cats and wild animals, and also in **fish**, **shellfish**, **meat** and water. *Gnathostoma spinigerum* is the causal agent of **gnathostomiasis** in humans.

Gnathostomiasis　Disease in humans caused by infection with ***Gnathostoma*** *spinigerum*. Commonly transmitted by consuming raw or undercooked contaminated **fish**, **shellfish** or **meat**, or drinking contaminated water. Characterized by a skin condition (creeping eruption) in which migrating larvae create tunnels under the skin that develop into abscesses. The larvae may also migrate through the eye or brain causing severe damage.

Gnocchi　Small **dumplings** made from a **dough** of **potatoes** and **flour**. Cooked in boiling water and served as a side dish or main course with a savoury sauce. **Eggs**, **cheese** or chopped **spinach** may also be added to the dough.

Goa beans　Seeds produced by *Psophocarpus tetragonolobus*. Rich in protein. As well as the seeds, immature green pods, leaves and root tubers are eaten. Also known as **winged beans** and **asparagus peas**.

Goat cheese　**Cheese** made from **goat milk**. Usually has a slightly harsh and piquant **flavour** and **aroma**.

Goat meat　**Meat** from **goats**; also known as **chevon**, particularly in India. It resembles mutton, but includes very little intermuscular fat. During the dressing process, goat **carcasses** nearly always become tainted with the typical aroma of goat, which transfers from the goat skin. The most tender meat comes from young goats, also known as kids, capretto or cabrito; meat from older goats is tougher. Goat meat is widely consumed in North Africa and the Middle East. It is often produced from goats managed traditionally, as free-foraging herds; consequently, goat meat tends to be fairly lean.

Goat milk　**Milk** produced by dairy **goats**. Similar in composition to **cow milk**, but with slightly higher contents of **calcium**, **niacin** and **vitamin A**, and significantly lower concentrations of **folic acid** and **vitamin B$_{12}$**. Goat milk contains almost no **carotenes**. Often used in **cheesemaking**.

Goat milk cheese　**Cheese** made from **goat milk**. Usually has a slightly harsh and piquant **flavour** and **aroma**.

Goats　The common name given to a number of hardy, mainly domesticated ruminant mammals in the genus *Capra*. Goats are related closely to sheep. They are reared world wide as a source of **goat milk**, **goat meat**, hair and hides. Different gender and age groups of goats are known as bucks or billys (adult entire males), does or ewes (adult females), goatlings (generally, sexually mature females to the end of their first pregnancy) and kids (generally, sexually immature animals which are less than one year old).

Godulbaegi　Common name for *Ixeris sonchifolia*, used as a vegetable and component of **kimchies** in Korea. Also known as Korean lettuce.

Gofio　Powdery cereal product made by **milling** of toasted grain, including **wheat**, **corn**, **barley**, **rye** and mixtures of these **cereals**. Consumed widely in the Canary Islands, being served with **milk** at **breakfast**, mixed with **soups**, as an ingredient of main course dishes or in **desserts**, combined with **honeys** and **almonds**.

Goitrogens Compounds found in foods (especially *Brassica* species, **peanuts**, **cassava** and **soybeans**) that can cause goitre, especially when dietary intake of iodine is low, by inhibiting synthesis of thyroid hormones (**glucosinolates**) or uptake of iodide into the thyroid gland (thiocyanates).

Goldenberries Alternative term for **cape gooseberries**.

Gonyautoxins **Paralytic shellfish toxins** which are sulfonated derivatives of **saxitoxin** and neosaxitoxin. Produced by *Gonyaulax* species (e.g. *Gonyaulax catenella* and *G. tamarensis*) and other red tide **dinoflagellates**. Responsible for **paralytic shellfish poisoning** in humans due to consumption of molluscan **shellfish** (**clams**, **mussels**, **oysters** and **scallops**) which filter feed on these dinoflagellates.

Good Manufacturing Practice Part of **quality assurance** which ensures that products, including foods, are consistently produced to the quality standards appropriate to their intended use and as required by the marketing authorization or product specification. Often abbreviated to GMP. Concerned with both production and **quality control**. It contains the following ten principles: writing procedures; following written procedures; documenting for traceability; designing facilities and equipment; maintaining facilities and equipment; validating work; job competence; cleanliness; component control; and auditing for compliance.

Gooseberries Green, yellow or red fruits produced by *Ribes grossularia* or *R. uva-crispa*. Consumed fresh or cooked. Also used in **jams**, jellies, juices, **syrups** and as the base of **flavourings**.

Goose meat **Meat** from **geese**. Goose meat is dark in **colour** and has a high fat content. Meat from young geese (goslings) is more tender and more expensive than meat from older birds. Wild goose meat has a much stronger flavour and is tougher than domesticated goose meat. In many European countries, goose meat is particularly popular at Christmas.

Gorgon nuts Seeds produced by *Euryale ferox*, an aquatic plant grown in Asia. Rich in carbohydrates, proteins and minerals. Popped or fried kernels may be eaten as **snack foods**. Also used in milk-based preparations in India, where they are known as makhana.

Gorgonzola cheese Italian soft **blue cheese** made from **cow milk**. Mould is added to the **cheese milk** and after about 4 weeks of **ripening** spread of the mould is encouraged by piercing with thick needles. Ripening lasts 3 to 6 months and the finished cheese is wrapped in foil to keep it moist. **Flavour** ranges from mild to sharp, depending on age.

Gossypol Yellow, potentially toxic phenolic substance composed of four benzene rings attaching to isopropyl, hydroxyl, aldehyde or ketone side chains. Occurs in some varieties of **cottonseeds** from which it is removed during the refining process for **cottonseed oils**.

Gouda cheese Dutch semi-hard **cheese** made from **cow milk**. Usually coated with yellow wax, mature cheese ripened for 18 months or more is coated with black wax.

Goulash A rich stew from Hungary. Usually made with meat and vegetables and highly seasoned with **paprika**.

Gourds Fruits of the family Cucurbitaceae containing mainly water, but a relatively good source of **vitamin C** and, in some cases, **carotenes**. Types include balsam pears or **bitter gourds**, snake gourds, bottle gourds (grown mainly for the outer shell which is used as a container), wax gourds or ash gourds, and fig leaf or Malabar gourds. Eaten as a vegetable or used in **pickles** and **curries**.

Gourd seeds Seeds produced by plants belonging to certain members of the family Cucurbitaceae, which bear fruits that have a hard rind. Contain high quantities of **oils** and **proteins**. Potential sources of **edible oils**.

Gouter Light meal eaten in the afternoon in France, traditionally at the end of the school day. Usually includes some kind of **bread** with a sweet or savoury spread, and sometimes cake. Also called the fourth meal.

Gracilaria Genus of red **seaweeds** containing several species of high commercial importance, particularly as a source of **agar**, which has many uses in the food industry. Commercially cultivated in parts of Asia, South America and southern Africa.

Grading Establishing the degree or rank of an item within a scale. In the food industry, grading is the classification of a food by variables such as quality, size and **colour**.

Grain Collective name for seeds of cereals such as **wheat**, **oats** and **corn**.

Grain alcohols **Spirits**, commonly with a neutral **flavour** and **aroma**, made by **distillation** of fermented **mashes** derived from grain (commonly unmalted).

Grain amaranth Seeds produced by *Amaranthus* varieties cultivated as pseudocereal plants (also **amaranth grain**).

Graininess **Consistency** term relating to the extent to which a product is grainy, i.e. granular, sandy and gritty.

Gram negative bacteria **Bacteria** that, following staining with crystal violet, are decolorized by organic solvents (e.g. ethanol or acetone) but stain red with the counterstain (safranin) in the Gram stain procedure. Their cell walls are composed of a thin layer of peptidoglycans covered by an outer membrane of **lipoproteins** and **lipopolysaccharides**.

Gram positive bacteria **Bacteria** that resist decolorization by organic solvents (e.g. ethanol or acetone) to retain their original purple crystal violet stain in the Gram stain procedure. Their cell walls are composed of a thick layer of peptidoglycans with attached teichoic acids.

Grams Alternative term for **legumes**.

Grana cheese Italian hard grating **cheese** made from **cow milk**. Various types are made, including **Grana Padano cheese**. Most are aged for up to 4 years.

Granadillas Alternative term for **passion fruits**.

Grana Padano cheese Italian **hard cheese** made from unpasteurized **cow milk**. Similar to **Parmesan cheese**, with a hard, thick rind and a grainy, crumbly interior. **Ripening** lasts 12 to 48 months.

Granola **Breakfast cereals** composed of rolled **oats** mixed with **dried fruits**, **brown sugar** and **nuts**.

Granulated sugar Crystalline solid comprising at least 99.8% **sucrose**. Granulated sugar is produced by **crystallization** or graining of concentrated **sugar syrups** and is the most pure form of **sugar** manufactured from **sugar beets** and **sugar cane**.

Granulation Processing of a food into small compact particles (granules). Granulators are often used in cane and beet **sugar** manufacture to remove unbound moisture from the sugar by **driers** and coolers. Moisture is removed in driers by blowing hot air through a stream of cascading sugar or through a bed of wet sugar.

Granules Small particles or grains. **Starch** exists as granules which are insoluble in cold water but form a viscous solution when heated. Some food ingredients and **instant foods** are provided in the form of dried granules, which are reconstituted with water before use or consumption.

Granulometry Technique for measuring particle or granule size distribution.

Grapefruit **Citrus fruits** (*Citrus paradisi*) formed as a hybrid of **pomelos** and sweet **oranges**. Cultivars may be white, pink or red fleshed and seedless or seeded. Contain less total sugar and **vitamin C** than some other citrus fruits and approximately 1% citric acid. Also contain the bitter compound **naringin**.

Grapefruit juices **Fruit juices** prepared from grapefruit (*Citrus paradisii*). Good source of **vitamin C**. Widely consumed as **beverages**, sometimes sweetened, due to their **bitterness**.

Grapefruit peel Outer skin of **grapefruit**. Used to make candied peel, as a garnish, or as an ingredient in **bakery products** and a range of dishes. Also used as a source of **essential oils** and **pectins**.

Grape jams **Jams** made from **grapes**, usually of specific varieties, e.g. Concord or Catawba.

Grape juice concentrates **Grape juices** which have been concentrated. May be diluted to produce reconstituted grape juices, or used in **winemaking**.

Grape juices **Fruit juices** extracted from **grapes** (*Vitis* spp., especially *V. vinifera*). May be drunk as **beverages**, or fermented to produce **wines**.

Grape marc By-products from **wineries**, comprising the **pomaces** remaining after **grapes** have been pressed and the **musts** separated.

Grape musts **Grape juices**, especially those to be fermented in **winemaking**.

Grape pomaces Residue, including **grape skins** and **grape seeds**, remaining after separation of **musts** from pressed **grapes**.

Grapes Seeded or seedless fruits produced by the genus *Vitis*, the most important species of which is *V. vinifera*. The majority are cultivated for production of **wines**, significant amounts also being grown as **table grapes** or for preparation of **dried fruits** (**raisins**, **currants**, **sultanas**); some are used to prepare **grape juices**. Contain 15-25% sugar as well as **tartaric acid** and **malic acid**, but little **vitamin C**. The skins of red and black varieties contain **anthocyanins**. Table grapes tend to have firmer flesh and lower acidity than **winemaking grapes**.

Grape seed oils Unsaturated **oils** extracted from **grape seeds**, large quantities of which are produced as by-products in **winemaking**, and the manufacture of **grape juices** and seedless **raisins**. Rich in **palmitic acid**, **stearic acid**, **oleic acid** and **linoleic acid**. Used in **cooking**.

Grape seeds Seeds or pips found in the centre of some types of **grapes**. Produced in large amounts as by-products of **winemaking**. Contain up to approximately 17% oil and 15% protein. Used as a source of **grape seed oils** and full-fat or defatted flour.

Grape skins Outer **peel** of **grapes**. Contain a number of **flavonoids** and **phenols**. Sometimes fermented along with the **grape juices** and **grape seeds** during **winemaking** to influence the **flavour** and **colour** of the finished product. Source of **antho-**

cyanins, which may be extracted and used as food **colorants**.

Grape spirits **Spirits**, such as **brandy**, derived from fermented **mashes** based on **grapes**.

Grappa Italian **spirits** distilled from a fermented **grape marc mashes**.

Grass carp **Freshwater fish** species (*Ctenopharyngodon idellus*) distributed across eastern Europe and Asia. Flesh is not highly prized for **flavour**, but grass carp is extensively cultured as a food fish in some parts of Asia. Also known as **white amur**.

Grass peas Seeds produced by *Lathyrus sativus*. Eaten after boiling in water or used to make **dhal**. Flour is used to make **chapattis**, paste balls and **curries**. Rich in proteins and carbohydrates, but prolonged consumption can cause lathyrism, a neurological disease resulting in weakness or paralysis of the legs. Also known as chickling vetch.

GRAS status Designation awarded to **food additives** by the US Food and Drug Administration (FDA) to indicate that they are not considered to be a health risk for consumers. GRAS is an acronym for generally recognized/regarded as safe.

GRAS substances **Food additives** which have been granted **GRAS status**.

Grating Reduction of a piece of firm food, e.g. **cheese** or **vegetables**, to small shreds by rubbing on a coarse, serrated surface, usually a kitchen utensil called a grater.

Gravimetry Technique based on weighing of the sample. Examples of its use include weighing of a sample before and after heating to indicate the content of volatile compounds and quantitative analysis of a substance following precipitation.

Gravity In a broad sense, the gravitational force which acts on any object within the earth's gravitational field. Also, the attraction between two massive bodies. The force of gravity acting on a body determines its mass. The centre of gravity of an object is the point at which the total weight acts. **Specific gravity** is a number equal to the ratio of a substance's weight to that of an equal volume of water.

Gravy **Sauces** produced using **fats** and juices that exude from **meat** during **cooking**. A **roux** is produced from the meat fats and **flour**, then liquid (e.g. **wines**, **cider**, **stocks**) is added and the mixture is thickened by heating. **Browning agents** may also be added to colour the sauce if required.

Gravy granules **Instant foods** in the form of free flowing granules that produce a ready to serve **gravy** when reconstituted with boiling water.

Gravy powders **Instant foods** in the form of a powder from which a ready to serve **gravy** is produced on addition of boiling water.

Greasiness **Sensory properties** relating to the extent to which a product is greasy, i.e. smeared or covered with grease, slippery or fatty, or is perceived to have a greasy quality in the mouth.

Great sturgeon Alternative term for **beluga**.

Green beans Type of **common beans** (*Phaseolus vulgaris*). Both pods and seeds are eaten. Also known as **string beans**, **French beans** and **snap beans**.

Green coffee **Coffee beans** which have been fermented but not roasted.

Greengages Green plum-like **fruits** produced by *Prunus italica* or *P. domestica*. Eaten raw or cooked and used in **jams** and jellies. Relatively low content of **vitamin C**.

Green gram Alternative term for **mung beans**.

Greenland halibut **Marine fish** species (*Reinhardtius hippoglossoides*) of high commercial importance belonging to the family Pleuronectidae. Found in the Arctic, North Pacific and North Atlantic Oceans. Marketed dried-salted or frozen and usually cooked by **steaming** or **frying**.

Green-lipped mussels Alternative term for **green mussels**.

Green mussels Common name for mussels of the species *Perna canaliculus* or *P. viridis*. Also known as green-lipped mussels.

Green peas Immature seeds of *Pisum sativum*. Contain more moisture, but less proteins, fats and carbohydrates, than mature seeds. Source of **vitamin A**, **vitamin C**, niacin and iron. Eaten fresh, canned or frozen. In some cultivars, e.g. mangetout, sugar peas and snow peas, the seeds and pods can both be eaten. Also called English peas and garden peas. Tiny, young green peas are known as petit pois.

Green peppers **Sweet peppers** or **bell peppers** (*Capsicum annuum*) picked while young and unripe; if left on the vine, they become **red peppers**. Vary in size and shape. Mildly flavoured, and rich in **vitamin C**. Eaten raw, in **salads** or as crudites, or cooked, sometimes stuffed with **meat** or other vegetables.

Green tea **Tea** made from **tea leaves** which have not undergone **fermentation** before **drying**.

Green vegetables Plants with edible green leaves. Good sources of a range of **vitamins** and **minerals**. Include **lettuces**, **cabbages**, **spinach** and **kale**.

Grenadier General name used for deepwater **marine fish** species in the family Macrouridae, having a distinct body shape (long slender body with large head). The most important commercial species are *Macrourus* *berglax* (rough head grenadier), *Coryphaenoides rupestris* (roundnose grenadier) and *C. acrolepis* (Pacific grenadier). Marketed fresh and frozen.

Grifola frondosa **Edible fungi**, also known as maitake, with medicinal properties. Used in **health foods** and **beverages**.

Grilling **Cooking** of food on a grill, using radiated heat. Considered by some consumers a healthier way of cooking than **frying**, as no **fats** or **oils** are needed.

Grinding Reduction of a food to small particles or powders by crushing in grinders. Grinding can be undertaken to varying degrees, producing food that is fine, medium or coarse in **texture**, as desired.

Grits Hulled, de-germinated and coarsely ground grain, especially **corn**. Often boiled and served at **breakfast** or as a side dish. Also called hominy grits.

Grittiness **Mouthfeel** term relating to the extent to which a product is perceived to be grainy or sandy.

Groats Husked but unflattened grain, especially **oats**, used to make **gruel** and **porridge**.

Grocery stores Shops or businesses, including **supermarkets**, that sell groceries, i.e. merchandise including foods and household goods.

Groundbeans Seeds of *Macrotyloma geocarpa*, a plant native to Africa. Used to prepare protein enriched **weaning foods** and **tempeh**. Also known as Kersting's groundnuts.

Ground beef Alternative term for **beef mince**.

Ground chicken Alternative term for **chicken mince**.

Ground coffee **Coffee beans** which have been ground ready for use to prepare **coffee beverages**.

Groundfish General name used for **marine fish** species which normally occur on or close to the sea bed, such as **cod**, **flatfish**, **haddock**, **hake** and **pollack**.

Ground meat Alternative term for **meat mince**.

Groundnut oils Pale yellow oils extracted from peanuts (*Arachis hypogaea*). Rich in **palmitic acid**, **oleic acid** and **linoleic acid** with good **oxidative stability**. Due to the desirable **flavour**, often used in **cooking** and as a substitute for **olive oils** and other **edible oils**. Also known as arachis oils or peanut oils.

Groundnuts Alternative term for **peanuts**.

Ground pork Alternative term for **pork mince**.

Ground turkey Alternative term for **turkey mince**.

Groundwater Water held in the soil or crevices in rocks, especially below the water table. May be treated and used as **drinking water**.

Grouper Group of **marine fish** species of the genus *Epinephilus* belonging to the family Serranidae, some of which are of commercial importance. Widely distributed in the Atlantic and Pacific oceans. Marketed and consumed in a variety of ways.

Growth factors **Proteins** involved in control of growth and differentiation of cells or organisms. Examples include epidermal growth factor, nerve growth factor, insulin-like growth factors and fibroblast growth factor. Levels of growth factors in serum can be affected by intake of specific foods such as soy products, with consequent effects on neoplastic transformation and bone metabolism.

Growth hormone Alternative term for **somatotropin**.

Growth promoters Organic compounds of plant and animal origin which stimulate growth. Plant growth promoters include a variety of **plant growth regulators**, such as **auxins**, **gibberellins** and **cytokinins**. Animal growth promoters include **hormones**, **antibiotics** and **β-adrenergic agonists**. Use of some animal growth promoters is banned in certain countries.

Growth stimulators Alternative term for **growth promoters**.

Gruel Thin, watery **oatmeal porridge** made by soaking oatmeal in water or **milk** before **cooking**. The solids can be removed before consumption.

Gruyere cheese Swiss **hard cheese** made from unpasteurized **cow milk**. Rind is hard and dry with tiny holes. **Texture** is more dense and compact than that of **Emmental cheese**. Salted in **brines** for 8 days and usually ripened for 2 months at room temperature. **Curing** lasts for 3 to 10 months.

Guacamole Dish made from mashed **avocados** mixed with **lemon juices** or **lime juices** (to prevent discoloration) and **seasonings**. Finely chopped vegetables are sometimes added. Eaten as side dishes, **dips** or **sauces**.

Guaiacol Member of the **phenols** group. Guaiacol has an antiseptic or medicinal-type **aroma** and is present as one of the **aroma compounds** in **beer**, **wines** and **whisky**. Also occurs as a taint in **fruit juices** caused by bacterial **spoilage**, and is used as a substrate for analysis of **peroxidases**.

Guanidine Strongly alkaline member of the **organic nitrogen compounds**, commonly used in the hydrochloride form for the denaturation of proteins. Synonyms include imino-urea and aminomethanamidine.

Guanosine Member of the **nucleosides** group, formed from guanine linked to a ribose molecule. Occurs as a component of **nucleotides** and **nucleic acids**. Often found as a mono- di- or tri-phosphate.

Guanosine monophosphate Member of the **nucleotides** group, commonly abbreviated to GMP.

Several chemical forms of GMP exist, including a cyclic form (cGMP) with a role in animal cell metabolism. Guanosine monophosphates, especially guanosine 5'-monophosphate, are used in foods, particularly savoury foods, as **flavour enhancers**. GMP can be purified from **yeast extracts** or produced by **fermentation** of **microorganisms**. Also known as guanylic acid.

Guanylic acid   Alternative term for **guanosine monophosphate**.

Guarana   Paste made from seeds of the Brazilian plant *Paullinia cupana*. Contains methylxanthine **alkaloids**, and is used in **flavourings** and as a stimulant in **soft drinks**.

Guar beans   Seeds of *Cyamopsis tetragonoloba*. Immature pods are eaten as vegetables. **Galactomannans** are extracted from the seeds to make **guar gums**, which are used as **stabilizers** and **thickeners** in foods. Also known as **cluster beans**.

Guar gums   High viscosity **gums** isolated from ground endosperms of the legume *Cyamopsis tetragonoloba*, also known as **guar beans** or **cluster beans**. Composed of repeating (1→4)-β-D-mannopyranosyl units with branches of α-D-galactopyranosyl units linked via (1→6) linkages. Mannose:galactose ratio is 2:1. These gums are used as **thickeners** (thickening capacity is approximately 8-fold that of **starch**), **stabilizers** and **emulsifiers** in foods, e.g. in **low fat foods** and **ice cream**.

Guava juices   **Fruit juices** prepared from **guavas**.

Guava pulps   Soft mass prepared from the flesh of **guavas**. Used in a range of products including **beverages**, **ice cream**, **yoghurt**, **bakery products**, **jams** and jellies.

Guava purees   Smooth creamy preparation made from the flesh of **guavas** by sieving or reducing in a blender or liquidizer. Used as **sauces** or in preparation of products such as **fruit juices**, **fruit nectars**, **bakery products**, **ice cream**, **yoghurt** and **jams**.

Guavas   **Fruits** produced by *Psidium guajava*. Variable in shape, with a yellow-green skin, and white to red flesh containing a great number of seeds. Good source of **vitamin A** and **vitamin C** (content of latter depends on cultivar and environment), and contains relatively high amounts of **niacin** and **carotenes**. Eaten fresh (sprinkled with **lime juices**) or canned; also used in various products, including **preserves**, **jams** and jellies, or made into **guava juices** or nectars.

Guinea fowl   The common name given to medium-sized, ground-dwelling African **game birds** of the sub-family Numidinae. Wild guinea fowl are hunted for their **meat**. In captivity, they are reared world wide to produce guinea fowl meat and guinea fowl eggs.

Guizotia   Genus of plants cultivated as an oilseed crop. The clear, edible **oils** produced from the seeds may be used in foods and are often used as substitutes for **olive oils**.

Gulabjamans   Popular Indian sweet prepared by addition of **sugar** or **jaggery** to **khoa**. Also called gulabjamun.

Gum acacia   Dried exudates from African species of the genus *Acacia*, particularly varieties of *A. senegal* and *A. verek*. Forms low viscosity aqueous solutions that are used as **thickeners** and **coatings** for products such as **confectionery**, **jellies** and **chewing gums**, **stabilizers** of beer **foams** and **flavourings**. Synonym for **gum arabic**.

Gum arabic   Dried exudates from African species of the genus *Acacia*, particularly varieties of *A. senegal* and *A. verek*. Forms low viscosity aqueous solutions that are used as **thickeners** and **coatings** for products such as **confectionery**, **jellies** and **chewing gums**, **stabilizers** of beer **foams** and **flavourings**. Also called **gum acacia**.

Gum balls   Alternative term for **chewing gums**.

Gum confectionery   Collective terms for **chewing gums** and **bubble gums** and their products.

Gum ghatti   Low viscosity **gums** obtained as stem exudates from *Anogeissus latifolia*. Major components are **arabinose**, **galactose**, **mannose**, **xylose** and **glucuronic acid** in a ratio of 10:6:2:1:2. Uses include as food **coatings** and **stabilizers**.

Gum guaiac   Alcohol soluble **gums** obtained as a resin from *Guajacum officinale* or *G. sanctum* wood. Predominantly composed of α- and β-guaiaconic acids with guaiacic acid and **vanillin**. Primarily used as **antioxidants** in **chewing gums**.

Gum kondagogu   Exudates of *Cochlospermum gossypium*, a tree native to India. Classified as a variety of **karaya gums**, although the **gums** do differ in composition, **functional properties** and **physical properties**. Used as a food additive and as a substitute for **gum tragacanth**.

Gums   High molecular weight **polysaccharides** that form viscous solutions or **gels** when dissolved or dispersed in a solvent, usually water. Obtained from plant exudates and **seaweeds** or produced as **exopolysaccharides** by **bacteria**. Gums have many applications in the food industry: low viscosity gums, i.e. **gum arabic** and **gum ghatti**, are used as water binding agents for prevention of **syneresis** and as encapsulating agents for **flavourings**; medium viscosity gums, i.e. **gum tragacanth** and **alginates**, provide body and are useful **emulsifying agents**, e.g. in

Hepatitis Inflammation of the liver which can be a result of infections or non-infectious pathology. Certain causes of infection, such as hepatitis A virus, can be borne in foods and **water supplies**.

Hepatitis viruses Viruses labelled A to E that cause inflammation of the liver (**hepatitis**). Hepatitis A and E viruses can be transmitted through faecal contamination of food or water.

Hepatotoxicity Quality or property of having a poisonous or destructive effect on liver cells.

Hepatotoxins **Toxins** that act specifically or primarily on the liver.

Heptachlor Non-systemic insecticide used for control of termites, **ants** and soil-dwelling **insects** in a wide range of crops. Classified by WHO as moderately toxic (WHO II).

Heptachlor epoxide Primary degradation product of the insecticide **heptachlor**. Occurs more commonly in animal tissues than does the parent compound.

Heptadecanoic acid Carboxylic acid with 17 carbon atoms, member of the **saturated fatty acids**, with a melting point of 59-61°C. Synonyms include **margaric acid**, margarinic acid and *n*-heptadecylic acid. Occurs as a free fatty acid and lipid component of **animal fats** and **vegetable fats**.

Heptanoic acid Member of the **saturated fatty acids** with seven carbon atoms. Important in the **flavour** and **aroma** of many foods and **beverages**, including **beer**, **wines**, **tea**, **fruits** and **cereal products**.

2-Heptanone Methylketone which is one of the important **flavour compounds** in foods, especially **cheese** and other **dairy products**.

Heptenal Aldehyde identified in a variety of foods. Several isomers exist and have been associated with fishy or boiled potato-like **aroma**.

Herbal beverages **Beverages** in which herbal material is a significant source of **flavour** and/or active ingredients.

Herbal tea Alternative term for **herb tea**.

Herbicides Chemical substances used to kill or inhibit growth of unwanted plants around crops. Most are applied as sprays and have either a systemic or contact effect. Examples of herbicides commonly applied to crops include **atrazine**, **diuron**, **glyphosate** and **propham**. Residues remaining in foods and the environment can represent a health hazard.

Herbs General term for flowering plants, parts of which are used predominantly as **flavourings** rather than as foods.

Herb tea Tea-type infusion **beverages** prepared from dry plant material other than **tea leaves** (*Camellia sinensis*).

Heritability The capacity to be transmitted from one generation to another. The hereditary or genotypic variance expressed as a percentage of the total variance in the feature examined.

Herpesviruses Enveloped DNA **viruses** of the family Herpesviridae. Occur in humans and cold-blooded vertebrates and invertebrates. Usually transmitted through contact.

Herrgard cheese Swedish semi-**hard cheese** made from pasteurized **cow milk**. Similar to **Gruyere cheese**, but more supple and softer. The natural pale rind is often covered with yellow wax. The cheese melts easily and has a mild nutty **flavour**. As well as the full-fat version (45% fat), a low-fat type (30% fat) is made using **skim milk**.

Herring Generally refers to the **marine fish** species *Clupea harengus*, an abundant fish caught in huge numbers in the north Atlantic and north Pacific Ocean. May also be used as a general name for several small pelagic marine fish species within the family Clupeidae. A wide range of herring products are marketed, including **kippers** (smoked herring) and salted, cured, dried and canned herring.

Hesperidin Flavonoid glucoside found in **citrus fruits**.

Heterocyclic amines **Amines** with a cyclic molecular structure containing atoms of at least two different elements in the ring or rings. Formed particularly in **meat** and **fish** during **grilling** or **frying**. Some are of concern because of their **mutagenicity** or **carcinogenicity**.

Heterocyclic aromatic amines **Heterocyclic amines** containing ring structures with conjugated double bonds and delocalized electrons. Formed particularly in **meat** and **fish** during **grilling** or **frying**. Some are of concern because of their **mutagenicity** or **carcinogenicity**.

Heterocyclic compounds **Organic compounds** having a closed chain or ring which contains more than one type of atom. Commonly include nitrogen, sulfur or oxygen atoms in place of carbon atoms. Examples include **aniline**, **heterocyclic amines**, **lactones** and **pyrazines**.

Hexachloran Alternative term for the insecticide **HCH**.

Hexachlorobenzene Selective fungicide used in **fumigants** for control of common bunt and dwarf bunt in wheat. A suspected carcinogen. Commonly abbreviated to HCB. Classified by WHO as extremely hazardous (WHO Ia).

Hexachlorobiphenyl One of the **polychlorinated biphenyls** (PCB) used for a variety of industrial purposes, including manufacture of capacitors, transformers, **plasticizers**, adhesives, pesticide extenders, paints and water-proofing compounds. Although use has been discontinued since 1977, it is very persistent in the environment. Associated with the yusho **food poisoning** incident (caused by ingestion of rice oils contaminated with PCB on the Japanese island of Kyushu in 1968).

Hexachlorophene Organochlorine compound, 2,2′-methylenebis(3,4,6-trichlorophenol), with disinfectant activity.

Hexadecanoic acid Straight chain, C16 member of the **saturated fatty acids**, synonym, **palmitic acid**. A major component of **animal fats** and **vegetable fats**. Synthetic precursor of several **unsaturated fatty acids**.

Hexadecenoic acid A C16, straight chain member of the monounsaturated **fatty acids** containing one double bond. Most common forms include the 9Z-isomer (synonym, **palmitoleic acid**) and the 11Z-isomer, but 3E-, 6E-, 6Z- and 9E-isomers are also found. Occurs as a component of **animal fats** and **vegetable fats**.

Hexanal Member of the **aldehydes** group of **aroma compounds**, synonym caproaldehyde. Imparts a green, fruity **aroma** in many foods, but also occurs as a fatty acid oxidation product in lipid-containing foods where it is associated with **rancidity**.

Hexanoic acid Synonym for **caproic acid** or capronic acid. A C6 member of the **carboxylic acids** (**fatty acids**) family of aliphatic compounds. Contributes to the **flavour** and **aroma** of many foods, including **cheese**.

Hexanol A C6 alcohol which occurs as a **flavour** and **aroma** component in many foods and **beverages**.

Hexenal A C6 aldehyde with one double bond. Several isomers are found in foods, including *trans*-2-hexenal and *cis*-3-hexanal. Associated with green, fresh **aroma** characters and occurs in many **fruits** and other foods and **beverages**.

Hexokinases EC 2.7.1.1. **Enzymes** that catalyse the transfer of a phosphate group from **ATP** to D-hexoses to form D-hexose 6-phosphates. **Glucose, mannose, fructose, sorbitol** and **glucosamine** can act as acceptors. Hexokinases are of use in a variety of analytical applications including measurement of glucose, fructose, mannose, **ATP** and creatine kinases. Since they allow for measurement of glucose in the presence of fructose, they are used widely for detection of **adulteration** in **wines** and **fruit juices**.

Hexosamines Amino sugars comprising six carbon atoms. Examples include **glucosamine** and **galactosamine**.

Hexoses General term for **sugars** comprising six carbon atoms, e.g. **glucose, mannose, galactose, fructose, sorbose** and **tagatose**.

Hexylamine One of the **biogenic amines**, identified in **milk, cheese** and **sake**.

4-Hexylresorcinol Phenolic inhibitor of polyphenol oxidases (**catechol oxidases**), used to control **browning** of **fruits**, and **melanosis** in **shrimps**.

Hg Chemical symbol for **mercury**.

Hickory nuts Nuts produced by trees of the genus *Carya*, the most popular of which are **pecan nuts**, produced by *C. pecan*. Common hickory nuts are small with a very hard shell and are produced by the shagbark hickory tree (*C. ovata*); these are used in **bakery products**, often as a substitute for pecan nuts.

Hickory smoke Natural flavourings produced by extraction of condensed smoke produced by burning of wood from hickory trees (*Carya* species).

High amylose corn starch Starch manufactured from hybrid corn plants that have been selected for the high **amyloses:amylopectins** ratio of their starch. Amylose content in high amylose corn starch is usually ≥55%. Due to the high amylose content, the starch produces firm **gels** on heating.

High calorie foods Any foods that have a high calorie content in relation to bulk, such as **peanut butter** or chocolate syrup. Also includes **dietetic foods** and **energy foods** which have been specifically manufactured to give an increased calorie content. These are designed for weight gain and may be targeted at individuals with specific nutritional requirements, e.g. athletes, invalids, low birth-weight infants. Lightweight, calorie-dense foods are also used as **space flight foods** and military rations.

High density polyethylene Polyethylene of high-density grade. Used as a packaging material in many food and beverage applications. Commonly abbreviated to HDPE.

High gravity brewing Brewing process in which **worts** of higher than normal concentration are fermented, and the resulting high-concentration **beer** is diluted to normal beer strength.

High performance liquid chromatography Column chromatography technique with a liquid mobile phase in which high column inlet pressure, narrow bore columns and small particle size stationary phases are used to achieve rapid separation. Usually abbreviated to HPLC. Can be applied to separation of a wider range of compounds than is possible with **gas chromatography**. Also called high pressure liquid chromatography.

High pressure liquid chromatography Column chromatography technique with a liquid mobile phase in which high column inlet pressure, narrow bore columns and small particle size stationary phases are used to achieve rapid separation. Usually abbreviated to HPLC. Can be applied to separation of a wider range of compounds than is possible with **gas chromatography**. Also called high performance liquid chromatography.

High pressure processing Nonthermal **preservation** technique used to inactivate vegetative **microorganisms** in foods by isostatic pressure **pasteurization** (1000-9000 atmospheres). High pressure processing affects only noncovalent bonds, enabling phase transitions, permeabilization of **biological membranes**, **denaturation** of proteins, **gelatinization** of proteins and **starch**, increasing reaction rates, and compacting of materials. **Bacterial spores** are considerably more resistant to high pressure processing than vegetative or germinating cells.

Hijikia Genus of **seaweeds** including the edible species *Hijikia fusiforme*; a dried form of this seaweed is used as a food ingredient in Japan and other parts of Asia.

Hilsa Fish species (*Tenualosa ilisha*) from the herring family that is distributed around the northern part of the Indian Ocean. Migrates into river systems during part of its life cycle. Popular food fish in India. Marketed fresh or as a dried/salted product. Sometimes spelt hilsah.

Hilsah Alternative spelling for **hilsa**.

Himegai Japanese name given to **mussels**.

Hiochi bacteria *Lactobacillus* species which cause **spoilage** of **sake**.

Hippuric acid Member of the **organic acids**, synonyms include N-benzoyl glycine, benzoylamino acetic acid and benzamido acetic acid. Contributes to the **flavour** of several **dairy products**, including **cheese**, **yoghurt** and **kefir**. Often converted to **benzoic acid** during microbial **fermentation**.

Hispanico cheese Spanish semi-**hard cheese** made from raw or pasteurized **cow milk**, or a mixture of cow and **ewe milk**.

Histamine One of the **biogenic amines**, synonym 2-(4-imidazolyl)ethyl amine. Formed by decarboxylation of **histidine**. Present naturally in a wide range of foods, including **yeast extracts**, **cheese**, **red wines** and **fish**. Histamine poisoning (**scombroid poisoning**) has occurred after consumption of **fish** (commonly **mackerel**, **tuna** and **bonito**) due to the presence of high levels of histamine as a result of microbial **spoilage**. Histamine is potentially toxic at high levels, and is not destroyed during **cooking**. Symptoms of histamine toxicity include violent headaches, flushing, rashes, sweating, cramps and diarrhoea.

Histidine One of the non-essential **amino acids**, occurring in animal and plant proteins. Precursor of **histamine**.

Histidine decarboxylases EC 4.1.1.22. Convert L-histidine to **histamine**. Production of histamine in foods and **beverages** by **bacteria** can result in **spoilage** and may represent a serious health problem.

Histochemistry Study of chemical components of cells and their distribution by means of chemical reactions. Methods used include **microscopy**, radiography and **chromatography**.

Histology Study of the microstructure of cells.

Histones Group of low molecular weight, basic nuclear proteins found in eukaryotes, which are involved in packaging of nuclear **DNA** into chromatin. Histones are commonly rich in **lysine** or **arginine** residues.

H₂O₂ Chemical formula for **hydrogen peroxide**.

Hogs Castrated male **swine** (also known as bars or barrows) reared for **pork** production.

Hoki Marine fish species (*Macruronus novaezelandiae*) from the hake family. Distributed in the southwest Pacific Ocean around South Australia and New Zealand. A commercially important food fish marketed fresh or frozen as fillets; often processed into fish blocks for reprocessing into **fish fingers** or other ready-to-cook fish products.

Hominy Hulled, de-germinated and dried coarsely ground **corn** kernels used to prepare various foods, including **puddings** and **bread**. More finely ground kernels are referred to as hominy **grits**.

Homocysteine One of the sulfur-containing **amino acids**. Precursor of **methionine**. Plasma homocysteine levels are frequently determined as an independent risk factor for **cardiovascular diseases**, and have been studied in relation to **coffee** and **alcohol** consumption.

Homogenization Creation of **emulsions** by reducing all the particles to the same size. For example, in **homogenized milk**, the **milk fat globules** are emulsified, preventing the **cream** from separating out. Commercial **salad dressings** are also often homogenized.

Homogenized milk Milk treated in a homogenizer to break up the **milk fat globules** and reduce **creaming**, thus increasing **shelf life**. Modifications to **casein** structure improve **digestibility** of the milk; smaller milk fat globules and increased surface area increase contact with the taste buds, giving a fuller **flavour**. Homogenized milk has a greater whitening power in coffee. It is more sensitive to light-induced

off flavour but less sensitive to development of **flavour** defects caused by **oxidation**.

Homogenizers Apparatus used in **homogenization** of foods, such as **milk**.

Honey beverages **Beverages** in which **honeys** are major constituents, as **sweeteners**, **flavourings** or sources of fermentable material.

Honeybush tea **Herb tea** prepared from fermented leaves of South African plants of the genus *Cyclopia*.

Honeycombs Storage units of beehives constructed from **beeswax** by worker honeybees. Honeycombs are formed from a framework of hexagonal shaped cells. The cells are used to store **honeys**, and insect eggs and larvae.

Honeydew honey **Honeys** produced from honeydew, a sweet substance secreted by plant lice usually onto trees, e.g. beech honeydew, which is gathered by honeybees. Honeydew honeys are considered to be of inferior quality to honeys produced from nectar.

Honeydew melons Melons (*Cucumis melo*) which when ripe have a creamy skin **colour** and pale green, juicy, sweet flesh. Tend to be large and have a long **shelf life**. A source of **vitamin C**, **potassium** and some trace **minerals**.

Honeys Natural **syrups** produced by honeybees predominantly from nectar but also from honeydew and **fruit juices**. Honey consists of approximately 20% (w/w) water and 80% **sugars**, mostly **fructose** and **glucose**. Honeys also contain the **flavour compounds** and **aroma compounds** present in the nectar or fruit juices collected, composition of which is dependent on its botanical origin, and it is these minor components that give honeys their individual **flavour**. Honeys are collected from honeycombs, where they are stored, and may be used directly as both foods and **sweeteners**.

Hop essential oils **Essential oils** prepared from hops (*Humulus lupulus*). Major components present are the **bitter acids humulones** and **lupulones**, and a terpenoid, **humulene**. The highest concentrations of **flavour compounds** are contained in the lupulin glands of hop leaves, thus lupulin essential oil is used as a concentrated source of hop **flavour** for **beer brewing**.

Hop extracts Extracts of the active ingredients α-**acids**, β-**acids**, **resins**, **essential oils**) of **hops**. Used in **brewing**.

Hop pellets **Hops** which have been comminuted and compressed into pellets. Used in **brewing**.

Hoppers Large containers for **grain**, typically those that taper downwards and discharge their contents through valve-like openings at the base. In general, used as temporary receptacles for grain.

Hopping Process used in brewing. It is the addition of **hops** to fermenting **worts** to impart **flavour** and **bitterness**. Hops may also be added to the finished **beer** (dry hopping) to enhance hop flavour.

Hops Dry cones of the hop plant (*Humulus lupulus*). Used as **flavourings** and bittering agents in **beer**.

Hop substitutes Substances used in place of **hops** to impart **flavour** and **bitterness** in **beer**. Required particularly in situations where climatic and economic considerations prohibit the use of conventional brewing materials, e.g. in Nigeria where malted or unmalted **sorghum** has been used instead of malted **barley** to produce **lager**. Materials which have been used successfully as hop substitutes include **seeds** from *Garcinia kola* and extracts from bitter leaf (*Vernonia amygdalina*).

Horchata Spanish **beverages** made from aqueous extracts of **chufa nuts** (*Cyperus esculentus* tubers).

Hordein Prolamin found in **barley**.

Hordenine One of the **biogenic amines**. Found in germinated **barley**, **sorghum** and **millet**, and in **malt** and **beer**.

Hordeumin High molecular weight anthocyanin/polyphenol complex formed during **ethanolic fermentation** of uncooked **barley bran**. Exists as a purple pigment at low pH values and has potential for use in **colorants** for foods.

Hordothionins Antifungal proteins which occur in **barley** kernels.

Hormones **Organic compounds** synthesized in minute quantities in specialized tissues such as the endocrine organs and then transported in the bloodstream to a target organ, which it stimulates. Definition now extended to include regulatory compounds in lower animals and plants and synthetic **growth promoters**. Hormones are routinely fed to animals to stimulate productivity, but are also of concern due to accumulation of residues and inappropriate or illegal usage.

Horse beans Type of **faba beans** (*Vicia faba*).

Horse gram Seeds produced by *Dolichos biflorus*, used as a pulse crop in India, where it is also known as kulthi. In Burma, dry seeds are processed in a similar way to **soybeans** to make fermented **sauces**.

Horse mackerel Name given to a number of **marine fish** species from the mackerel family (Carangidae) within the genera *Trachurus* and *Decapterus*. Important species include *Trachurus trachurus* (Atlantic horse mackerel), *T. japonicus* (Pacific horse mackerel) and *Decapterus macarellus*. Marketed fresh and frozen, dried-salted, smoked and canned. Also known as **jack mackerel** and scad.

Horse meat **Meat** from **horses**. Horse **carcasses** have a high dressing out percentage. Other benefits of horse meat include rapid ageing post-slaughter, good tenderness, low contents of fat (with high proportion of unsaturated fatty acids) and **cholesterol**, and high protein and iron contents. When freshly cut, horse meat is dark red or bluish in **colour**, but, after several hours, it develops a rusty colour. **Aroma** of the meat is sweet. It is not marbled with fat. Often, the intermuscular fat resembles beef fat, but some horse meat has yellow, soft and greasy fat. Large amounts of horse meat are eaten in continental Europe, Japan and Russia. Horse meat is processed into a wide range of meat products including brined horse meat, horse **sausages** and cured smoked ham-type products. In some countries, horse meat is eaten raw (e.g. as steak tartare in France) and consumption may be associated with **foodborne diseases**, such as **trichinosis**.

Horse milk **Milk** produced by horses. Also called **mare milk**.

Horseradish Common name for *Armoracia rusticana*, vegetables of the Brassicaceae family. **Spices** of horseradish root have a pungent **flavour** and are used as **flavourings**, e.g. of horseradish sauce, a traditional accompaniment to roast **beef** in the UK. Distillates from horseradish root possess **antimicrobial activity**. Source of horseradish **peroxidases**.

Horses Herbivorous, solid-hoofed, quadruped mammals belonging to the Equidae family; there are several species, including *Equus caballus*. Both domesticated and wild horses are used for the production of **horse meat**.

Horticultural products Products of horticulture, such as **fruits**, **vegetables** and **flowers**.

Horticulture Cultivation of **fruits** and **vegetables** for human consumption, and of **flowers** and other plants for ornamental purposes. Practiced on a small scale as a pastime (gardening) or on a larger, commercial scale (also market gardening).

Hot boning Cutting of **meat** (muscle) from animal carcasses that have first been conditioned at 16°C for varying time periods *post mortem*.

Hot dogs Hot **frankfurters** served in long, soft **bread rolls**, with added **mustard**, **tomato ketchups** or other **condiments**. Hot dogs are particularly popular in the USA.

Hot peppers Fruits produced by various members of the *Capsicum* genus. Vary in size, shape and **colour**, but always with numerous seeds. Very pungent, due to the presence of **capsaicin** in the seeds and veins. Include **chillies**. Rich in **vitamin A** and **vitamin C**; good source of **vitamin E**, potassium and **folic acid**. Used as a dried powder in many dishes, such as stews, and to make hot sauces.

Hotrienol Member of the **terpenoids**, synonym 3,7-dimethyl-1,5,7-octatrien-3-ol. **Aroma** constituent present in several plants, including **elderflowers**.

Hot water dips Treatment used to protect **fruits** and **vegetables** from conditions such as **chilling injury**, **pests** infestation and decay during cold storage.

HPLC Abbreviation for high performance/pressure liquid chromatography.

H$_2$S Chemical formula for **hydrogen sulfide**.

Hsian-tsao Common name for *Mesona procumbens*, a perennial plant the leaves of which yield a gum which is used in making **desserts**. Leaf extracts exhibit **antioxidative activity** and **antimutagenicity**. Leaves are also used to prepare **beverages**.

H$_2$SO$_4$ Chemical formula for **sulfuric acid**.

HTST pasteurization High temperature, short time (HTST) **pasteurization** treatment used widely in the food industry, but particularly applied to liquid foods such as **raw milk** and **fruit juices** to reduce substantially the total bacterial count for improved **shelf life** and to eliminate any **pathogens**. For milk, heat treatment is accomplished using plate **heat exchangers**. Cold raw milk held in a cool storage tank is pumped into **pasteurizers**, where it is heated to a temperature of at least 72°C. The milk, at pasteurization temperature and under pressure, flows through the holding tube where it is held for at least 16 seconds. At the end of the tube is an accurate temperature-sensing device that checks if any of the heated milk has not reached the pasteurization temp. If any milk has not, a diversion device is activated, and the product is made to flow back through the heat exchanger. Properly heated milk continues to flow through the system and is cooled to 4°C or less. Cold, pasteurized milk passes through a vacuum breaker then on to a storage tank filler for packaging.

HTST processing Alternative term for **HTST pasteurization**.

Huckleberries **Berries** produced by plants of the genus *Gaylusacia*, commonly *G. baccata*. Resemble **blueberries** in appearance, but have harder seeds inside, a thicker skin and slightly more astringent **flavour**. Eaten raw or in **bakery products** such as pies.

Huitlacoche Parasitic fungus (*Ustilago maydis*) that infects ears of corn, causing kernels to swell and darken. It is an edible fungus and was originally consumed only in Mexico, though huitlacoche is now considered as a delicacy internationally. **Flavour** is a cross between those of **corn** and **mushrooms**. Sold canned and frozen, it may be used in any dish which calls for cooked mushrooms. Also known as corn smut, maize mushroom and cuitlacoche.

Hulling Removal of the **hulls** from **fruits** or **seeds** prior to consumption. Also called **dehulling** or **husking**. Also, removal of leaves from the tops of **strawberries** prior to consumption.

Hulls The outer (usually fibrous) coverings of some fruits or seeds, that are removed by **dehulling** or **hulling** prior to consumption. Also known as husks or shells.

Hulupones Oxidation products of **β-acids** found in **hops** and hop products.

Human immunodeficiency viruses Retroviruses also known as HIV which are responsible for the disease acquired immunodeficiency syndrome (**AIDS**) in humans. There is concern over the risk of virus transmission to infants from infected mothers during breast feeding.

Humanized milk **Milk** in which the **nutrients** composition is adjusted to that of **human milk** as far as possible, making it suitable for feeding to infants.

Human milk **Milk** produced by women during human lactation. Composition differs considerably from that of **cow milk**. Although fat contents of human and cow milks are similar, **fatty acids** composition varies. Human milk contains less protein than cow milk; proportions of individual **proteins** and **amino acids** also differ. Contents of **lactose, oligosaccharides** and some **vitamins**, and activities of some **enzymes** are higher in human than in cow milk, while human milk contains a lower amount of **minerals** in total. Also called breast milk or mothers' milk.

Human milk substitutes Preparations for feeding to infants and young children as a replacement for **human milk**, designed to meet their specific nutritional requirements. Also called **infant milk formulas**. May be based on **cow milk** or **soymilk**.

Humectants Ingredients added to increase or maintain the **water activity** of foods. Examples of humectants include, **gums**, which possess water binding activity, and **NaCl, glycerol** and **sucrose**, which increase water activity by altering the osmotic pressure of foods.

Humic acids Complex **organic acids** of polyphenolic structure formed in soils and peat which can form adsorption complexes with minerals. Present in many natural water sources, requiring removal during purification for **drinking water** production.

Humicola Genus of **fungi** of the class Hyphomycetes. Species may produce various enzymes, e.g. **cellulases** (*Humicola insolens*), **lipases** (*H. lanuginosa*) and acid **proteinases** (*H. lutea*).

Humidification Process whereby the level of moisture in the air is increased. By circulating air of higher humidity, the moisture content of hygroscopic products can be increased. This process, known as conditioning, is applied to some grain prior to **milling** or other processing.

Humidity Moisture content of the atmosphere. **Relative humidity** (abbreviated to RH) is the moisture content of the air at a given temperature as a percentage of the level required to cause saturation at that temperature.

Humous Dish made from **chick peas** pureed with **garlic, lemon juices** and **olive oils** or **sesame oils**. It may also contain **tahini**. Served as **dips**, often accompanied by **pita bread**, or **sauces**. Alternative spellings include hummus, hoummos and houmous.

Humulene Sesquiterpene **aroma compounds** present in **essential oils** of **hops**.

Humulinic acid Intermediate product in **isomerization** of **humulones** to **isohumulones** during boiling of hopped **worts** or manufacture of isomerized **hop extracts**.

Humulones Fractions of the **α-acids** group of **bitter compounds** in **hops** and hop products. Important **bitter compounds** in hops and **beer**.

Hurdle technology Food processing technique employing a combination of preservation procedures or hurdles to inhibit growth of **microorganisms** in the product. These include manipulation of factors such as temperature, **water activity** and **acidity**, as well as processes such as gas packaging and **high pressure processing**. The aim is to interfere with several different mechanisms within microorganisms simultaneously. This multi-targeted approach allows effective use of mild techniques.

Hurum Expanded waxy **rice** product consumed especially in India. Preparation involves **soaking, parboiling** and **flaking** of waxy rice, addition of fat and expansion in sand.

Husbandry The breeding, care and cultivation of **crops** and animals. It may also include the management and conservation of plant or animal resources.

Husking Removal of the husks from **fruits** or **seeds** prior to consumption. Also called **dehulling** or **hulling**. Also relates to the removal of husks from the tops of **strawberries** prior to consumption.

Husks The outer (usually fibrous) coverings of some **fruits** or **seeds**, that are removed by **husking** prior to consumption. Alternatively, the circle of leaves on the tops of **strawberries** where they were attached to the plants. Also known as hulls or shells.

Hyacinth beans Alternative term for **lablab beans**.

Hybridization Formation of double-stranded nucleic acid molecules by base-pairing between complementary single-stranded molecules. Used to detect specific sequences and for determining the degree of sequence identity, and can be carried out in solution or with one component immobilized on a suitable matrix (e.g. nitrocellulose). Hybrids can be detected by **EM** or by labelling one of the components, e.g. fluorescently or radioactively. Hybridization can also be performed *in situ* using fluorescently-labelled **DNA** molecules to localize **genes** to specific **chromosomes**.

Hybrids The offspring of two parents differing in at least one genetic characteristic (trait). Also, heteroduplex DNA or DNA-RNA molecules.

Hydnocarpus Genus of tree, the seeds of which are used to obtain an oil which contains **palmitic acid** and small quantities of **phytosterols**.

Hydnum **Edible fungi**, the most commonly consumed species being *Hydnum repandum* (*Dentinum repandum*). Best eaten cooked as it is bitter when raw.

Hydration Process by which water is added to a dried substance to hydrate it.

Hydrocarbons Any **organic compounds** that contain only carbon and hydrogen.

Hydrochloric acid Solution of hydrogen chloride gas in water, chemical formula HCl. Strong mineral acid widely used in the food industry as a processing aid.

Hydrochlorofluorocarbons Organic compounds (abbreviated to HCFC) consisting of carbon, hydrogen, chlorine and fluorine. Used as **refrigerants**. HCFC are less destructive to the ozone layer than **chlorofluorocarbons** (CFC). HCFC are currently used as replacements for CFC, but their use is to be phased out, as specified by the amended Montreal Protocol, when they are expected to be replaced by **hydrofluorocarbons** (HFC).

Hydrocolloids High molecular weight polymers of animal, plant or microbial origin that form viscous solutions or **gels** on addition of water, e.g. **gums** and **gelatin**.

Hydrocooling Precooling method for heat sensitive products, such as certain **fruits** and **vegetables**. During hydrocooling, fruits and vegetables are cooled by direct contact with flowing cold water, which absorbs heat directly from the produce. Hydrocooling allows the grower to harvest produce at optimum maturity with greater assurance that it will reach the consumer at maximum quality. Hydrocooling benefits the produce by slowing the natural deterioration that starts shortly after harvest, slowing the growth of decay organisms and reducing wilt by retarding water loss.

Hydrocyanic acid Toxic, colourless gas with a boiling point of 26°C. Synonym **hydrogen cyanide** and chemical formula HCN. Occurs as a hydrolysis product of **cyanogenic glycosides** in a range of foods, especially **cassava**, but also including **edible fungi**,

flax seeds and **wines**. Used as a fumigant to control pests in stored foods.

Hydrocyclones **Cyclones** used for **clarification** of liquids, such as for removal of dust and soil particles from thin **sugar juices** and extraction of **casein** particles from **whey**. Liquid is added tangentially at high speed to a conical chamber to produce a spinning motion (the cyclone). Particulate matter is forced to the sides, decelerates and falls to the bottom of the chamber from which it is collected. A liquid column is formed in the centre of the cyclone and rises to an outlet at the top of the chamber.

Hydrofluorocarbons Hydrofluorocarbons (HFC) are organic compounds that contain hydrogen, carbon and fluorine. HFC, which do not contain chlorine, are not harmful to the ozone layer, and so are suitable replacements for **chlorofluorocarbons** (CFC) in **refrigeration**.

Hydrogen Odourless, colourless diatomic gas, chemical symbol H_2, which combines readily with other organic and inorganic elements. Extremely abundant, being present in water and all organic compounds. The hydrogen isotopes deuterium and tritium are **radioelements**.

Hydrogenated fats Oils from an animal or vegetable source that have been subjected to **hydrogenation**, which hardens and stabilizes the oil by reducing unsaturated double bonds in the **fatty acids**.

Hydrogenation Chemical reaction in which molecular hydrogen reacts with **hydrocarbons** or unsaturated **fatty acids**, usually in the presence of **catalysts**. Often used to harden oils, which also improves their **oxidative stability**. In this **hardening** process, hydrogen reduces carbon atoms linked by a double bond, decreasing the level of saturation of the **fatty acids**. Often used in the manufacture of **margarines**.

Hydrogen azide Colourless liquid, chemical formula HN_3, with strong reducing activity. One of a range of **disinfectants** used in the food industry.

Hydrogen cyanide Toxic, colourless gas with a boiling point of 26°C. Synonym **hydrocyanic acid** and chemical formula HCN. Occurs as a hydrolysis product of **cyanogenic glycosides** in a range of foods, especially **cassava**, but also including **edible fungi**, **flax seeds** and **wines**. Used as a fumigant to control pests in stored foods.

Hydrogenomonas Obsolete genus of rod-shaped **bacteria**, the species of which have been reclassified into other genera (including *Aquaspirillium* and *Pseudomonas*).

Hydrogen peroxide Strong oxidizing agent and antimicrobial compound with chemical formula H_2O_2. Used in foods at low concentrations (e.g. maximum limit is 0.05% in **milk**) as **preservatives, dough conditioners, bleaching agents**, and for artificial ageing of **wines** and **spirits**, and **refining** of **fats and oils**. At concentrations greater than those used in foods and **beverages**, it is used in **disinfectants**.

Hydrogen sulfide   Toxic, colourless gas, chemical formula H_2S, with a distinctive odour of rotten eggs. Formed by reduction of **organic sulfur compounds**, including proteins, during microbial **fermentation** and can occur in **musts** and **worts** as an undesirable by-product of **alcoholic fermentation** by **yeasts** giving rise to sulfide **taints** in the resulting **wines** and **beer**. Also produced by **spoilage bacteria** during decomposition of high-protein foods such as **meat** or **fish**. Dietary protein from meat is an important substrate for hydrogen sulfide generation by bacteria in the human large intestine and this process has been implicated in the development of ulcerative **colitis**.

Hydrolases   EC 3. **Enzymes** which catalyse the hydrolysis of various bonds, including **esters, glycosides**, ethers, **peptides** and **amides**.

Hydrolysed lactose syrups   **Syrups** manufactured by acid or enzymic hydrolysis (treatment with **β-galactosidases**) of **lactose** syrups or **whey**. Consist of an aqueous solution of **glucose** and **galactose**; whey-derived hydrolysed lactose syrups also contain salts and **oligosaccharides**.

Hydrolysed starches   Alternative term for **starch hydrolysates**.

Hydrolysed starch syrups   **Syrups** manufactured by acid and/or enzymic hydrolysis of **starch** slurries. The starch may be derived from any source, although commonly **corn starch** is used due to advantages of cost and availability. Examples of hydrolysed starch syrups include **corn syrups, glucose syrups** and **maltose syrups**.

Hydrolysis   Reaction in which a substance is split into two or more component parts by the action of water in the presence of **catalysts** such as **enzymes, acids** or **alkalies**, acting at specific points within the molecules. Types of hydrolysis include **proteolysis**, in which **proteins** are broken down to component **peptides** or **amino acids**, **lipolysis**, in which **lipids** are broken down into constituent **fatty acids**, and **saponification**, in which lipids are hydrolysed in the presence of alkalies to form soaps.

Hydrometry   Measurement of specific gravity of a liquid or strength of alcoholic beverages. Usually performed using a sealed graduated tube weighted at one end, which sinks in the liquid to a depth that indicates the specific gravity.

Hydroperoxide lyases   Involved in production of **flavour compounds** in higher plants. Cleave 9- and 13-hydroperoxides of **linoleic acid** and **linolenic acid** into volatile C6- or C9-aldehydes and C12- or C9-oxoacids, respectively. The C6- and C9-**volatile compounds** are useful for production of natural food **flavourings**.

Hydroperoxides   **Organic compounds** in which one hydrogen atom of a hydrocarbon is replaced by an -O-OH group. Lipid hydroperoxides are formed by **lipoxygenases** during oxidation of lipids and these are further degraded enzymically or thermally to produce acids and aldehydes which can be associated either with **flavour** and **aroma** development or with decreases in lipid quality in **fats** and **oils**.

Hydrophobicity   State in which a substance has low affinity for water. Extent to which molecules are insoluble in water.

Hydroponics   Cultivation of plants in a nutrient solution rather than soil.

Hydroquinone   Member of the **phenols** group of **aromatic compounds** with **antioxidative activity**. Synonyms include 1,4-benzenediol, *p*-dihydroxybenzene and quinol. Occurs naturally in several foods and **beverages**, including **fruits, vegetables, grain, coffee, tea** and **beer**. Can also include any member of the aromatic *p*-diols derivable from *p*-quinones or any compound with a quinol nucleus.

Hydrothermal processing   Application of heat and moisture treatments, such as steam infusion processes used for **cooking, puffing** or **flaking** of foods.

Hydroxybenzoic acid   Crystalline derivative of **benzoic acid** containing one hydroxyl group per molecule. **Esters** of *p*-hydroxybenzoic acid (**parabens**) are used as **food preservatives** and **artificial flavourings**.

Hydroxybenzoic acid esters   **Esters** of organic **alcohols** (usually **ethanol, butanol** or **propanol**) and *p*-hydroxybenzoic acid. Uses include as food **preservatives** and **artificial flavourings**. Also called **parabens**.

3-hydroxy-2-butanone   Chemical name for the flavour compound **acetoin**.

Hydroxybutyric acid   One of the short chain **fatty acids**, with four carbon atoms. Synonym, hydroxybutanoic acid. Not widely identified as a lipid component of foods, but does occur in an esterified form as an aroma compound in **sake** and **cheese**. 3-Hydroxybutyric acid has been used as a marker for fertile incubated **eggs** in which the embryo has died, and which are not permitted to be used in foods.

Hydroxycinnamic acid   One of the aromatic **phenols** widely distributed in plant foods including **fruits**

and **cereals**, and plant-derived **beverages** including **fruit juices**, **wines**, **whisky** and **sake**. Three isomers exist, including 4-hydroxycinammic acid (synonym **coumaric acid**). Also more widely used as a general term to describe hydroxy-substituted forms of **cinnamic acid**, including **ferulic acid** (4-hydroxy-3-methoxycinnamic acid) and **caffeic acid** (3,4-dihydroxycinnamic acid).

5-(Hydroxymethyl)-2-furaldehyde Synonym for **hydroxymethylfurfural**. Member of the heterocyclic **organic compounds** composed of a furan ring with aldehyde and hydroxymethyl substituents. Found as a natural component in **honeys** and as a thermal breakdown product of **sugars** in heat-treated products such as **UHT milk** and pasteurized **fruit juices**. Often determined chemically as a marker of **nonenzymic browning**.

Hydroxymethylfurfural Member of the heterocyclic **organic compounds** composed of a furan ring with aldehyde and hydroxymethyl substituents. Synonyms include **5-(hydroxymethyl)-2-furaldehyde** and 5-(hydroxymethyl)furfural. Found as a natural component in **honeys** and as a thermal breakdown product of **sugars** in heat-treated products such as **UHT milk** and pasteurized **fruit juices**. Often determined chemically as a marker of **nonenzymic browning**.

Hydroxyproline Member of the **amino acids** with eight possible structural isomers, of which only the L-isomers are known to occur naturally. Found in several animal **proteins** including **collagen** and **gelatin**, and in extensin, a plant protein.

Hydroxypropylcellulose Non-ionic ether of cellulose that forms a viscous liquid when solubilized in water. Uses in foods include as **emulsifiers**, **stabilizers**, encapsulating agents and **thickeners**.

Hydroxystearic acid A C18 member of the **fatty acids** family of aliphatic compounds, synonym hydroxyoctadecanoic acid. Produced in microbial **bioconversions** of **oleic acid** as an intermediate in the formation of **lactones**.

5-Hydroxytryptamine Chemical name for **serotonin**. A biogenic amine which may occur in foods. In the body, functions as a neurotransmitter, and is toxic at excessive concentrations.

Hygiene Science of health and its preservation, or a practice or condition that is conducive to the preservation of health.

Hygienic quality Extent to which something is conducive to health.

Hygrometers Instruments used to measure the **humidity** of the atmosphere.

Hygromycins **Antibiotics** which exhibit relatively poor **antibacterial activity**, but are effective **ant-**helmintics. Used to control parasitic worm infections in swine and poultry.

Hygroscopic properties Extent to which a substance absorbs moisture from the atmosphere without dissolving in the moisture. Highly hygroscopic substances, e.g. **silica gels**, can be used as desiccants.

Hyperactivity Psychiatric condition, also known as ADHD (attention deficit hyperactivity disorder) characterized by inattention, restlessness and impulsiveness. Has been linked anecdotally to consumption of refined **sugar** and **food additives**, particularly **colorants** such as **tartrazine**. However, scientific evidence indicates that most cases are not a result of dietary factors.

Hypercholesterolaemia Condition in which abnormally high levels of **cholesterol** are present in the blood. A very high cholesterol level is a known risk factor for **coronary heart diseases** and **stroke**. Blood cholesterol levels may be controlled by diet or **functional foods** containing cholesterol-lowering constituents, such as **stanol esters**.

Hyperlipaemia Group of diseases characterized by elevated levels of plasma lipids, such as **cholesterol** or **triacylglycerols**. Dietary factors that have been proposed to reduce hyperlipaemia include a favourable dietary profile of **fatty acids**, increased **dietary fibre** content, consumption of **soy proteins** and **isoflavones**, and use of **functional foods**, such as **spreads** enriched with **phytosterols**, and **probiotic foods**.

Hypermarkets Very large self-service shops selling foods and household goods, and sometimes clothing.

Hypertension Prevalent disease in which blood pressure is elevated. In the majority of cases the cause is unknown; a rare cause is excessive consumption of **liquorice** rich in **glycyrrhizic acid**. High blood pressure is a risk factor for other diseases such as **cardiovascular diseases** and has been shown to be improved by reduction in body mass index. The association between hypertension and consumption of **salt** is controversial.

Hyphaene Genus of **palms**. **Fruits** of some species are eaten or made into **beverages**.

Hypochlorites Salts of hypochlorous acid (HClO), such as sodium hypochlorite. Widely used as **disinfectants**.

Hypocholesterolaemic activity Ability of a food, nutrient or diet to produce hypocholesterolaemia, a state wherein blood **cholesterol** level is abnormally low, or to lower high cholesterol levels (as in hypercholesterolaemia) to within the normal range. Reduction of blood cholesterol levels are associated with reduced risk of **cardiovascular diseases**. Dietary components possessing hypocholesterolaemic activity include **dietary fibre**, some **plant proteins**, some **fatty acids**, **phytosterols** and **probiotic bacteria**. Included as a specific type of **hypolipaemic activity**. Alternative spelling hypocholesterolemic activity.

Hypolipaemic activity Ability of a food, nutrient or diet to reduce the fasting and/or postprandial levels of plasma lipids, including **cholesterol** and **triacylglycerols**. Reductions in certain plasma lipid parameters, such as fasting levels of total cholesterol and cholesterol within low density lipoproteins (LDL), and postprandial triacylglycerol concentrations, are associated with reduced risk for **cardiovascular diseases**. Dietary components demonstrating hypolipaemic activity include certain **fatty acids**, **phytosterols** and phytosterol-enriched **margarines**, **probiotic bacteria** and **dietary fibre** fractions.

Hypoxanthine Member of the **purines** group, synonym 6-hydroxypurine. Combines with ribose to form **inosine**, one of the **ribonucleosides**. Produced as a breakdown product from adenine nucleotides, and is often determined as a marker of **freshness** in **fish**. Used in combination with **xanthine oxidases** in a chemical assay for free **radical scavenging activity**.

Hyssop **Spices** from *Hyssopus officinalis*. Hyssop has a warm, **camphor**-like **aroma** and a warm, sweet and slightly burning **flavour**.

I

IAA Abbreviation for **indol-3-ylacetic acid**.

Iberian ham A variety of high-quality, dry cured ham, traditionally produced in the Iberian peninsula (Spain and Portugal). These hams are usually produced from Iberian or Iberian x Duroc **swine**. Traditionally the hams are subjected to long periods of **ageing**, during which intense enzymic action helps to develop their distinctive **flavour**. Increasingly, however, shorter curing periods are being used to reduce costs; this practice results in reduced flavour intensity of the product. Iberian ham of a superior quality is produced from swine fed on acorn-based feeds.

Ice Solid form of water, used for numerous food processing applications, including **chilling** and **glazing** of foods (e.g. **fish**). Small pieces of ice, e.g. ice cubes or crushed ice, may be added to **beverages** to cool them, while ice crystal characteristics play an important role in determining the quality of **frozen foods**. Flavoured ice is consumed in the form of **ice lollies** and **water ices**.

Ice cream Frozen dairy product with creamy, smooth and crystalline consistency. In addition to **milk** and **dairy products** such as **cream**, **milk powders**, **butter** and sweetened **condensed milk**, also contains **sugar**, **flavourings** and **additives** such as **emulsifiers** and **stabilizers**. The ingredient mix is processed in an ice cream freezer where it is frozen by contact with the refrigerated wall, blades scraping the mixture from the walls while whipping air into the ice cream. The soft-serve ice cream produced can be hardened further by placing in a suitable freezing apparatus.

Ice cream bars Confectionery snack products containing **ice cream** covered with **chocolate** or other **coatings**. May be stick novelties, wafer products or cone products.

Ice cream cones Thin, slightly sweetened, **wafers** baked on a waffle iron and curled before cooling to form a cone shape. Used to hold one or more scoops of **ice cream**.

Ice cream mixes Commercial products used in manufacture of **ice cream**. Contain all the main components of the final product, including **milk**, **cream**, **sugar**, **flavourings** and **emulsifiers**.

Ice cream wafers Thin, slightly sweetened, waffle-textured **wafers** that are usually triangular or rectangular and served as an accompaniment to **ice cream** or used to make an ice cream sandwich.

Iced coffee Chilled **coffee beverages**.

Iced tea Chilled **tea beverages**.

Ice lollies Portions of **ice cream**, flavoured **water ices** or coated ice cream products. Generally served on a stick.

Ice milk Low-fat **ice cream**.

Ice nucleation activity Promotion of the formation of ice crystals. Agents displaying ice nucleation activity include small particles, such as food particles, and large molecules, such as ice nucleating proteins.

Ices Term sometimes used for **ice cream**.

Icings **Toppings**, usually for **cakes** and **biscuits**. Basic formulations for icings consist of **icing sugar** mixed with water. Other ingredients that may be used include **butter/margarines**, **egg whites** and **colorants**.

Icing sugar Powdered **granulated sugar** used as an ingredient of **fondants** and **icings** that require **sweetness** and a smooth **texture**. **Anticaking agents**, usually **starch** or tricalcium phosphate, are commonly added to icing sugar.

ICPAES Abbreviation for **inductively coupled plasma atomic emission spectroscopy**.

ICPMS Abbreviation for **inductively coupled plasma mass spectroscopy**.

IDF Abbreviation for **International Dairy Federation**.

Idiazabal cheese Spanish **hard cheese** made from unpasteurized **ewe milk**. Has a compact but not crumbly **texture** and characteristic smoky **flavour**.

L-Iditol 2-dehydrogenases EC 1.1.1.14. Catalyse the interconversion of L-iditol and L-sorbose, although they can also act on D-glucitol (giving D-fructose) and other closely related **sugar alcohols**. Useful in the analysis of **sorbitol**. Also known as sorbitol dehydrogenases.

Idli A steamed, naturally fermented cake-type product widely consumed as a breakfast food or snack in India. Prepared by fermenting a slurry of ground **rice** and

legumes (usually **black gram dhal**) and steaming the resulting **batters** to give products with a soft, sponge-like **texture** and good **digestibility**.

Ika shiokara Traditional Japanese **sea foods**, consisting of **squid** flesh fermented with squid liver contents and salt.

Illipe butter Naturally occurring vegetable fat derived from **nuts** of various species of the genus *Shorea* (**illipe nuts**). Shows similar **fatty acids** composition and **melting** profile to **cocoa butter** and is thus used in **cocoa butter substitutes**.

Illipe nuts Nuts produced by various species of the genus *Shorea*, which yield **fats** (**illipe butter**) with similar properties to **cocoa butter**.

Image analysis Analysis of a sample on the basis of its structure, as determined by non-destructive techniques such as **microscopy**. Parameters of interest in the image can be both classified and quantified using the human eye or computer programs.

Image processing Technique that can be used with **image analysis** in which the image of the sample is processed in some way to make it easier to perform further interpretation. Thus, the image quality is improved but no analysis or quantification is performed.

Imaging **Analytical techniques** used to produce images of objects or substances which will allow their structure to be studied. Includes **magnetic resonance imaging** and thermal imaging.

Imazalil Systemic fungicide which inhibits **ergosterol** biosynthesis. Used to control a wide range of fungal diseases on **fruits** and **vegetables**. Particularly active against fungal strains resistant to **benzimidazole**. Also used as a seed dressing for control of fungal diseases affecting **cereals**. Classified by WHO as moderately toxic (WHO II).

Imbibition Process of **absorption**, by soaking up a liquid.

Imidacloprid Widely used systemic insecticide related to nicotine and used for control of chewing and sucking **insects** (e.g. **aphids**, **thrips**, some **beetles** and soil-dwelling insects) in **cereals**, **fruits** and **vegetables**.

Imidan Alternative term for the insecticide **phosmet**.

Imidazoles **Heterocyclic compounds** containing a 5-membered imidazole ring with two nitrogen atoms. These **organic nitrogen compounds** are present in **histidine**, **histamine** and imidazole **alkaloids**, and imidazole, a weak base, has been used in the extraction of **fats**, **proteins** and **polysaccharides** for chemical analysis.

Imitation cheese Product with the appearance and **sensory properties** of **cheese**, but which is differ-

ent from genuine cheese in composition. May be based on **soybeans** rather than **milk**.

Imitation crabmeat Product resembling flesh from **crab legs** which is actually derived from flesh of marine fish (usually a mild-flavoured **fish** such as **pollock**). The fish is processed by rolling 'sheets' of fish and adding **colorants** to give it the required appearance; crab sticks or flaked imitation crabmeat are commonly produced. The resulting product is lower in **cholesterol** than real **crab meat**.

Imitation cream Product with the appearance and **sensory properties** of **cream**, but which differs from genuine cream in composition. Usually prepared with **vegetable proteins** and **vegetable fats** as substitutes for milk-based components. Also called non-dairy cream.

Imitation dairy products Substitutes for **dairy products**, with vegetable-based components (often soy products) usually replacing all or part of the milk constituents. Products have the appearance and **sensory properties** of dairy products, but differ in composition. Nutritional properties of the imitation products may not match those of the dairy products they are intended to replace. Commonly produced types include **imitation cheese**, imitation milk and **imitation cream**.

Imitation foods Alternative term for **simulated foods**.

Immobilization Process by which microbial, plant and animal cells, and macromolecules (e.g. **enzymes**) are attached to solid surfaces or entrapped within gels. They can then be used in applications such as **bioconversions** and **biotransformations**, **affinity chromatography** and **biosensors**.

Immobilized cells Microbial, plant and animal cells that have been attached to solid surfaces or entrapped within gels. Can be used in **bioconversions** not possible with isolated **enzymes** and in **biosensors**. Entrapment is the most commonly used method for immobilization; gels used include **agar**, **alginates**, **carrageenans**, polyacrylamides and polyurethane.

Immobilized enzymes Enzymes that have been attached to solid surfaces or entrapped within gels. **Immobilization** methods include covalent attachment or ionic binding to solid carriers or supports (e.g. **celluloses**, synthetic polymers and DEAE-cellulose), cross-linking with bifunctional reagents, **encapsulation** (e.g. in liposomes) and entrapment within gels. Immobilized enzymes often offer a number of advantages over free enzymes, such as ease of reuse and increased stability.

Immune response Reaction of the body to foreign substances (**antigens**). **Antibodies** produced by lymphocytes in response to the antigens can destroy the antigens directly or label them in a way that makes them susceptible to attack by white blood cells. White blood cells specific to the antigens (T-cells) may also be produced. Synonymous with **immunological response**.

Immunoaffinity chromatography **Chromatography** technique in which the stationary phase (immunosorbent) is prepared by immobilizing **antibodies** specific to the analytes of interest onto the surface of a rigid or semi-rigid support. Used as a clean up or preconcentration step in an analytical procedure as well as a separation technique.

Immunoassay Analytical technique in which substances are measured using specific antibodies that bind to the corresponding antigens. Binding is measured by use of antibodies labelled with radioactive isotopes, enzymes (**enzyme immunoassay**) or fluorescent dyes.

Immunochemical analysis **Analytical techniques** in which specific immune reactions are employed in the investigation.

Immunodiffusion Immunological technique in which **antigens** are detected by precipitation reaction with specific **antibodies** in agar gel. Antigens diffuse out from wells cut into the gel to react either with antibodies diffusing from a central well or antibodies incorporated in the gel.

Immunoelectrophoresis Technique combining separation of sample components by **electrophoresis** with immunological identification of the separated substances using specific **antibodies**.

Immunofluorescence Immunological technique in which **antibodies** labelled with a fluorescent dye are used to detect **antigens** in the samples.

Immunogenicity Extent to which a substance can cause an **immune response**. Affected by a number of factors, including nature of the substance, dose and previous exposure of the host.

Immunoglobulin A One of the 5 major classes of **immunoglobulins**, commonly abbreviated to IgA. Produced predominantly against ingested **antigens**, and found in external secretions of mammals, such as saliva, sweat and tears. Also present in colostrum, providing a valuable source of immunity for suckling animals and infants. Provides local immunity against infection in the gut or respiratory tract, preventing attachment of **microorganisms** to epithelial cells.

Immunoglobulin E One of the five major classes of **immunoglobulins**; commonly abbreviated to IgE. Helps to protect against parasitic infections. On binding **antigens**, IgE molecules trigger **histamine** release from circulating **leukocytes**. Following sensiti-

zation, however, these **antibodies**, when present in foods, can also be responsible for certain **allergies**, such as type I hypersensitivity and **anaphylaxis**.

Immunoglobulin G The most abundant of the major classes of **immunoglobulins** in the bloodstream; commonly abbreviated to IgG. Produced by B lymphocytes following previous exposure to a given antigen.

Immunoglobulins Proteins (commonly abbreviated to Ig) also known as **antibodies**, which are produced by white blood cells in response to foreign **antigens**. Capable of binding the antigens as part of the body's **immune response**. There are 5 main classes of immunoglobulins (IgG, IgE, IgM, IgA and IgD), each of which has distinct roles in the immune system.

Immunoglobulin Y One of the biologically active substances found in hens **egg yolks**. Active against a wide range of **bacteria**. Uses of this class of **immunoglobulins** include as a therapeutic agent and potentially in **food preservatives**.

Immunological effects Influence of exposure to a substance on a body's immune system.

Immunological response Alternative term for **immune response**.

Immunological techniques **Analytical techniques** in which **antigens** are detected using **antibodies**. Include **agglutination tests**, **ELISA**, **radioimmunoassay**, **immunoelectrophoresis** and **immunodiffusion**.

Immunology Science concerned with the way in which the body reacts to foreign substances. Includes immunity, components of the immune system and diagnosis of disease.

Immunomagnetic separation Technique in which a substance is separated from a sample using magnetic beads coated with specific **antibodies**. After allowing interaction of the beads with the analyte of interest, they are removed from the sample using a magnetic particle separator. Often used as an enrichment stage in isolation and detection of **microorganisms**.

Immunomodulation Process of influencing the functioning of a body's immune system.

IMP Abbreviation for **inosine monophosphate**.

Impala Swift-running, medium-sized, graceful antelopes (*Aepyceros melampus*). Impala are hunted for their **meat**.

Impedance Opposition to the flow of current in an electrical circuit.

Impellers Devices for driving an item forwards, employed in food processing.

Impingement drying **Drying** technique originally used for paper and textiles but more recently applied to foods. Gas jets are arranged in such a way that the gas, e.g. superheated steam or hot air, impinges perpendicularly on the food to be dried. The gas is directed at high velocity, removing moisture from the surface of the food. Processing time is reduced compared with that required for other types of drying.

Imports Goods or services that are produced abroad but purchased for use in the domestic economy.

Improvers Additives that improve the quality of the final product. Used predominantly in the bakery industry. Includes **flour improvers** which enhance the **breadmaking properties** of **flour**.

Incaparina Low cost protein-rich **food supplements** introduced by the Institute of Nutrition of Central America and Panama (INCAP) to combat protein **deficiency diseases** in infants and others at risk from **malnutrition**. The original formulation is based on **cottonseed meal** and **corn** and has a nutritional value similar to that of **milk**. Other formulations have been developed based on **soybeans** and low-cost local **vegetables**.

Indian mackerel Marine fish species (*Rastrelliger kanagurta*) from the mackerel family (Scombridae) which is mainly found in the Indo-west Pacific region. Marketed fresh, frozen, canned, dried-salted and smoked; also made into **fish sauces**.

Indian mustard Annual plant (*Brassica juncea*), related to rapeseed, grown for its seeds which are a source of **vegetable oils**.

Indian shad Marine fish species (*Tenualosa ilisha*) of the family Clupeidae and of minor commercial importance. Found in the Indian Ocean. Marketed fresh or dried-salted. Also known as hilsa shad.

Indigo carmine Disodium salt of 5,5'-ingotin disulfonic acid. Artificial deep blue **food colorants**. It has low solubility in water but is heat stable. Added to products including **beverages**, dessert powders and **confectionery**. Also known as **indigotine**.

Indigotine Alternative term for deep blue **artificial colorants** known as indigo carmine.

Indole acetic acid Alternative term for **indol-3-ylacetic acid**.

Indoles Group of nitrogen-containing **heterocyclic compounds** based on the 2,3-benzopyrrole (indole) skeleton. Indole-containing **organic compounds** include **tryptophan, skatole**, indole **alkaloids, indol-3-ylacetic acid** and indole-3-carbinol, a **glucobrassicin** derivative isolated from cruciferous vegetables with possible **anticarcinogenicity**. Indole has an animal-like **aroma** and has been identified as a volatile constituent in several foods and **beverages**.

Indol-3-ylacetic acid One of the **auxins** group of **plant growth regulators**, with the synonym indole acetic acid and the abbreviation IAA. Controls plant

growth and differentiation thereby affecting the yield and quality of **fruits** and **vegetables**.

Induction heating Heating, e.g. of foods, by production of an electric or magnetic state by the proximity (without contact) of an electrified or magnetized body.

Inductively coupled plasma atomic emission spectroscopy One of the 2 main inductively coupled plasma **spectroscopy** techniques (the other is **inductively coupled plasma mass spectroscopy**) which can be used to detect very small amounts of most elements in solid or aqueous samples. Usually abbreviated to ICPAES. Samples are nebulized and passed through a tube in an inert gas (e.g. argon) atmosphere. The tube is heated by radiofrequency radiation to produce a plasma with an extremely high temperature. When the sample flows into the plasma, atoms are excited and emit energy at characteristic wavelengths, which is detected usually by photographic emulsion detectors or photoelectric transducers.

Inductively coupled plasma mass spectroscopy Mass spectroscopy technique utilizing inductively coupled plasma. Usually abbreviated to ICPMS. Samples are nebulized and passed through a tube in an inert gas atmosphere. The tube is heated by radiofrequency radiation to produce a plasma with an extremely high temperature, which is then analysed by mass spectroscopy.

Infant foods Foods designed to meet the nutritional needs of infants, such as **infant formulas** and **weaning foods**. A wide range of processed infant foods is available in industrialized countries, including **rusks**, pureed **ready meals**, fruit drinks and cereal-based dishes. Foods are typically fortified with minerals and vitamins, and designed to be low in sugar and salt.

Infant formulae Alternative term for **infant formulas**.

Infant formulas Liquid foods for infants used as a substitute for **human milk**. Usually take the form of modified **cow milk** products (**milk infant formulas**), which aim to mimic the composition of human milk. Formulas may also be based on **milk** from other species, **soymilk** or other products in order to meet the nutritional needs of infants suffering from **intolerance** to cow milk.

Infant milk formulas Preparations for feeding to infants and young children, intended to satisfy their specific nutritional requirements. May be based on **cow milk** or **soymilk**. Also called **human milk substitutes**.

Infectivity Ability of **pathogens** to become established within or on the tissues of a host, or the capability of pathogens to be transferred from one organism to another.

Infestation Condition in which a host is occupied or invaded by **parasites**, e.g. ticks, lice or **mites** which may live on the surface of a host, or worms which may live within the organs of a host.

Information processing Evaluation of data using a computer, to generate usable information.

Infrared Section of the electromagnetic spectrum of which the radiation has lower energy than the visible spectrum and wavelengths ranging from 750 nm to 1 mm.

Infrared irradiation Application of **infrared radiation** to foods to extend **shelf life**. Alternatively known as IR irradiation.

Infrared radiation Electromagnetic radiation having a wavelength just greater than that of red light but less than that of **microwaves**, emitted particularly by heated objects.

Infrared spectrophotometry Alternative term for infrared spectroscopy or **IR spectroscopy**.

Infusions Extracts produced by soaking a substance, usually of plant origin, e.g. **spices**, **teas** or fruits, in a solvent, usually water. Solvent-soluble components, including **flavour compounds** and **aroma compounds**, leach out from the material into the solvent.

Inheritance Transmission of a trait from a parent to its offspring.

Injera Flexible, spongy, pancake-like flat unleavened **bread** prepared from spontaneously fermented **millet** flour **dough**.

Ink jet printers Non-impact printers in which the print image is formed by minute jets of ink. The jets of ink pass through an electrical field and this directs droplets of ink precisely onto the surface. Uses include production of high quality print food labels.

Ink jet technology Printing technology that involves spraying droplets of ink through computer-controlled nozzles.

Inn breweries Small-scale **breweries**, integrated with **pubs**, inns or **restaurants** where the **beer** is served.

Inorganic acids **Acids** which do not contain the carboxylic acid moiety common to **organic acids**. Includes the mineral acids **hydrochloric acid**, **nitric acid**, **phosphoric acid** and **sulfuric acid**.

Inorganic compounds Chemical compounds that do not contain C-H bonds.

Inosine Ribonucleoside formed from **hypoxanthine** (6-hydroxypurine) linked to a ribose molecule. Unlike other **ribonucleosides**, does not occur as a component of **nucleic acids**, but is used in synthetic oligo-

nucleotide probes. Often found as the ribonucleotide **inosine monophosphate** (IMP). Inosine levels can be used as indicators of fish **freshness**.

Inosine monophosphate Member of the **ribonucleotides**, commonly abbreviated to IMP. Occurs as a flavour compound in foods and is particularly associated with **umami** flavour. Content in **meat** and **fish** is used as a **freshness** indicator.

inositol Common name for the cyclitol *myo*-inositol. A polyol which occurs widely in foods as the free form, as **inositol phosphates** or as a component of **phosphatidylinositol**. Participates in cell signalling as a part of a membrane secondary messenger system and can also act as an antinutritional factor.

Inositol phosphates **Antinutritional factors** found in foods, especially **cereals** and **legumes**, which can compromise the absorption of minerals from the gastrointestinal tract. May be present in a range of forms, from bis phosphates up to hexaphosphates (also known as **phytates**). To improve the nutritional value of foods, both exogenous and endogenous **phytases** can be utilized to hydrolyse the higher inositol phosphates into lower **phosphates**, which generally have lower capacities to bind minerals.

Insect foods **Insects** that are eaten as foods in many parts of the world, including China, Japan and rural areas of Africa and South America, where they can serve as a valuable and readily available source of **proteins** and **minerals**. Types of insect consumed include grasshoppers, crickets, locusts, **bees** and **ants**. Most species are roasted, fried or boiled prior to consumption, although a few are eaten live. Insect foods are generally regarded as taboo in the Western World, although some insect products are available as novelty foods.

Insecticides Chemical substances used to kill **insects**. Used primarily to control **pests** that infest crops or to eliminate potential disease-carrying insects in specific areas. Classified into several groups, the most important of which are carbamate insecticides, fumigant insecticides, **organochlorine insecticides**, **organophosphorus insecticides** and **pyrethroid insecticides**. Residues persisting in foods and the environment can represent a health hazard.

Insects Members of the class Insecta, such as **flies**, **ants** and **beetles**. May generally refer to any other arthropods which resemble insects, such as spiders. Typically have a segmented body with an external chitinous covering, three pairs of legs, and, in most groups, two pairs of wings. Some species may be consumed as **insect foods**, while others may act as **pests** of crops and stored foods.

Insertion sequences Small, simple **transposons** (mobile units of **DNA**) usually ranging in size from 700 to 1500 base pairs. Possess short repeated nucleotide sequences at either end and carry no genetic information other than that required for their **transposition**. When inserted into bacterial DNA, insertion sequences (often abbreviated to IS) inactivate the gene, but activity is restored upon removal. IS transfer events are important mediators of genetic polymorphisms in both prokaryotes and eukaryotes, and have been widely studied in both **pathogens** and beneficial **microorganisms** within the food industry.

Instant beverages Dried **beverages** formulated and processed in a manner giving rapid solubility in water or other liquids.

Instant cocoa **Beverage mixes** containing **cocoa powders** that are usually reconstituted with hot **milk** or water to make **cocoa beverages**.

Instant coffee Dried (generally freeze dried) **coffee extracts** processed to a form which dissolves rapidly in water.

Instant foods Processed foods that have undergone **instantization**, so that they can be easily and rapidly reconstituted by bringing them into contact with a liquid such as milk or water. Common instant foods include **gravy granules**, **instant noodles**, **milk powders**, **instant coffee** and **tea powders**.

Instantization Processing of **dried foods** in a way that facilitates preparation or reconstitution of the final product. Common techniques used in instantization include **agglomeration** of particles and lecithination.

Instant noodles **Noodles** that have been pre-cooked and reconstitute rapidly when hot water is added to them.

Instant soups Dried **soup mixes** that are designed to rehydrate rapidly upon addition of water. Often prepared by **freeze drying**. Typically sold as convenience **snack foods/beverages** in single serving sachets.

Instant tea Dried (generally freeze dried) **tea** extracts processed to a form which dissolves rapidly in water.

Insulin One of the mammalian endocrine **hormones**. This polypeptide is synthesized in the pancreas in response to elevated blood glucose levels. Deficiencies in secretion of insulin or physiological responses to insulin occur in type I (insulin-dependent) and type II (non-insulin dependent) **diabetes** mellitus, respectively. Diet can be used to control type II diabetes, and information regarding postprandial blood insulin and glucose responses to foods (their insulinaemic and **glycaemic index values**) is useful in dietary control of this disease.

Integrated pest management Approach to crop **pests** control that uses a combination of various physical, chemical and biological pest control tactics in an attempt to reduce reliance on chemical **pesticides**, and hence minimize harmful residues in crops and pollution of the environment. Pest control tactics employed include biological control, use of conventional plant breeding or **genetic engineering** to improve crop resistance to pests, use of agricultural practices that lessen the degree of pest damage (e.g. mixed cropping, time of planting), and selective use of **insecticides** or other chemical agents (e.g. insect growth regulators).

Interesterification The process by which fatty acyl residues are interchanged between **triglycerides** in a mixture of **lipids**. Can be catalysed by **lipases**, and may be used to modify the composition and properties of **fats** and **oils**.

Interfacial tension Attractive force between molecules at an interface.

Interferometry Analytical technique based on differences in refractive index between the sample under investigation and a standard. Measurements are made on an interferometer, an optical instrument in which a beam of light is split and subsequently reunited after traversing different paths, producing interference.

Intermediate moisture foods Semi-moist foods, which do not require **refrigeration** and can be eaten without further preparation. Foods are preserved by limiting **water activity** to a level unable to support microbial growth, e.g. by addition of **humectants**. Examples of intermediate moisture foods include **dried fruits**, beef **jerky** and semi-dried **sausages**.

Intermittent warming Warming of commodities, such as **fruits** and **vegetables**, to room temperature at intervals during storage to prevent **chilling injury** symptoms from developing. Chilling injury is a problem in most crops of tropical or subtropical origin. Symptoms of chilling injury, such as pitting, **discoloration**, internal breakdown and decay, can result in large postharvest losses during marketing. Intermittent warming may, however, cause undesirable **softening**, increase decay, and cause condensation to form on the product.

International Dairy Federation Organization comprising more than 40 member countries throughout the world which aims to form a centre for collection and dissemination of information for the dairy sector, and to serve as a link between the dairy sector and organizations representing other sectors. Each member country has a national International Dairy Federation (IDF) committee representing the dairy sector in that country, covering the full range of dairy activities. As well as organizing events at which experts can report progress in various areas of research, the IDF also publishes technical and scientific findings and works closely with the **Codex Alimentarius** in many areas, including the provision of draft standards for **milk** and **dairy products**.

International Organization for Standardization Commonly abbreviated to ISO. A network of the national standards institutes of some 130 countries, with a central office in Geneva, Switzerland, that coordinates the system and publishes finished standards.

Intestines The portion of the **gastrointestinal tract** which extends from the lower opening of the stomach to the cloaca or anus. Intestines of slaughtered animals form a part of edible **offal**; after cleaning they may be used as **casings** for the production of meat products, e.g. **sausages**.

Intolerance Group of diseases in which there is inability to digest a particular dietary constituent properly, often resulting in malabsorption syndromes. Examples include **lactose intolerance**, resulting from lack of a gastrointestinal tract brush border enzyme, and **coeliac disease**, in which an immunological response to **wheat gluten** results in histopathological changes to the intestinal mucosa. Exclusion of the relevant component from the diet can result in elimination of the symptoms of the disease, and also, in cases such as coeliac disease, reversal of intestinal pathology.

Introns Sequences of **nucleotides** interrupting the coding sequence of a gene. These are transcribed into **RNA** but are removed by splicing before translation of the RNA into the protein product. The remaining sequences, which together code for the product, are called exons.

Inulases Alternative term for **inulinases**.

Inulases II Alternative term for **inulin fructotransferases (depolymerizing)**.

Inulin Polysaccharide composed mainly of fructofuranose residues (**fructose** in the ring conformation) although it also contains a glucopyranose residue. Inulin occurs naturally in some plants, e.g. **Jerusalem artichokes** and **chicory**, where it replaces **starch** as an energy store.

Inulinases EC 3.2.1.7. Catalyse the endohydrolysis of 2,1-β-D-fructosidic linkages in **inulin**, a linear, β-2,1-linked polymer of **fructose** which serves as an energy reserve in many plants. The product, **fructose**, has a high degree of sweetness and is important in the dietetic foods and **beverages** industries. Enzymic hydrolysis of inulin using inulinases offers an alternative to the standard procedure for production of fructose which uses **starch** as the source material.

Inulin fructotransferases (depolymerizing) EC 2.4.1.93. Enzymes which remove successive terminal D-fructosyl-D-fructofuranosyl groups from **inulin** as the cyclic 1,2':2,3'-dianhydride, leaving a residual di- or trisaccharide. The product, di-D-fructofuranose dianhydride (DFA III), is half as sweet as **sucrose** and is of interest for use in low calorie **sweeteners**.

Invertases Alternative term for **β-fructofuranosidases**.

Invert sugar Syrup composed of a mixture of **glucose** and **fructose** manufactured from **sucrose** by acid hydrolysis or action of **β-fructofuranosidases** (invertases). Uses include as a substrate for manufacture of **sorbitol** or **mannitol**, and in **sweeteners** and **humectants**.

Iodates Salts containing an IO_3 anion. Include potassium iodate **oxidizing agents**, which are added to **wheat dough** during **breadmaking**. Iodates are also added to table **salt (NaCl; sodium chloride)** and **infant formulas** for iodine **fortification** of the diet.

Iodides Salts that contain an I anion or other compounds containing **iodine** with an oxidation state of -1. Potassium iodide (KI) is added to table **salt (NaCl; sodium chloride)** and **infant formulas** for iodine **fortification** of the diet.

Iodine One of the **halogens**, chemical symbol I. Occurs naturally in the diatomic form I_2, and is a bluish-black solid which sublimes to form a bluish irritant gas. An essential dietary mineral which is accumulated in the thyroid gland and used to synthesize the thyroid hormones, including **thyroxine**, which are important for normal growth and development. Foods particularly rich in iodine include **seaweeds** and **marine fish**. Low dietary intakes of iodine can cause hypothyroidism and associated iodine **deficiency diseases** such as goitre. **Fortification** of the diet with iodine in the form of **iodates** or **iodides** is common.

Iodine values Measure of the unsaturation of **fats** or **oils**, based on the amount of iodine absorbed in a given time. Also known as iodine number.

Iodized salt Ordinary **salt** (NaCl) fortified with 0.01% potassium iodide. Added to foods in order to prevent **iodine** deficiency.

Iodometry Redox analysis technique based on reaction with iodine/iodide. Strong reducing agents are determined by titration with iodine while strong oxidizing agents react with iodide to form iodine. Iodine is titrated with a standard solution of thiosulfate, using a starch solution as an indicator.

Iodophors Complexes of iodine and certain high molecular weight surfactant compounds (e.g. **polyvinylpyrrolidone** and **quaternary ammonium com-**

pounds). Used in the food industry as **disinfectants** and **detergents**.

Ion chromatography **Chromatography** technique allowing simultaneous determination of anions and cations in a sample by using a sequence of a cation exchange resin column, a detector, an anion exchange column and another detector.

Ion exchange Reversible process in which substitution of ions for others of the same charge occurs. Solution containing ions is passed through a molecular network containing groups that can be ionized. Ions in the solution attach to the network, releasing free or mobile ions from the network. The reaction is classified according to the nature of the substituent groups in the network, i.e. cation exchange or **anion exchange**. Substances acting as ion exchangers or ion exchange resins include aluminosilicates, cross-linked polymers and celluloses. This process is the basis of separation by **ion exchange chromatography**.

Ion exchange chromatography **Chromatography** technique in which separations are carried out on **ion exchange** resins. Ions from the sample solution that pass into the exchangers are displaced by varying the pH, concentration or ionic strength of the eluting liquid, usually using a gradient. Separation is based on **anion exchange** or cation exchange depending on the type of resin used.

Ionic strength Parameter which is a function of the charge and concentration of **ions** in a solution.

Ionization Process by which a neutral substance becomes charged, forming ions. The conversion is due to the addition or removal of electrons induced by various means, including heating, chemical reaction, exposure to ionizing radiation or passage of an electric current.

Ionol Alternative term for the antioxidant **butylated hydroxytoluene**.

Ionones Volatile aroma compounds found particularly in **fruits**, **wines** and **tea**. One of the major ionones, β-ionone, has a violet-like aroma.

Ions Electrically charged atoms or groups of atoms. Positively charged **cations** result from the loss of electrons and negatively charged **anions** from their acquisition.

Ion selective electrodes **Electrodes** used to determine concentrations of specific ions, including metal ions and **salts**, e.g. **nitrates**, in aqueous solutions.

Ioobai **Fruits** produced by *Myrica nagi*. Kernels are eaten traditionally in China and the surrounding area.

Ipomeamarone One of the toxic **phytoalexins** formed in **sweet potatoes** as a result of mechanical injury or fungal infection.

Iprodione Contact fungicide with protective and curative action; used for control of a wide range of fungal diseases on **fruits**, **vegetables**, **cereals** and **oilseeds**. Sometimes used as a postharvest dip or as a seed treatment. Classified by the Environmental Protection Agency (EPA) as not acutely toxic (Environmental Protection Agency IV). Also known as rovral.

IR Abbreviation for **infrared**.

IR analysis Alternative term for **IR spectroscopy**.

Iridaea Genus of red **seaweeds** occurring on rocky shores around many parts of the world. Commercially important source of **carrageenans** used to make **thickeners**, **gels** and **stabilizers** for the food industry. Some species are cultivated commercially.

IR irradiation Alternative term for **infrared irradiation**.

Iron Group 8 metal, chemical symbol Fe. Forms salts in either the ferric (iron(II)) or ferrous (iron(III)) oxidation states. One of the essential minerals, iron is required to synthesize ferritin, **lactoferrin**, **haemoglobin**, cytochromes and other **haemoproteins**. Iron deficiency in the diet can lead to anaemia, and non-soluble iron salts such as ferrous sulfate are used for **fortification** purposes. Good sources of iron include **meat** and meat products, **cereals** and **green vegetables**. **Bioavailability** of iron in the diet is influenced by the presence of other chemicals such as calcium and **phytates**.

Irpex Genus of **fungi** of the class Homobasidiomycetes. Occur on felled timber and living trees. **Proteinases** produced by *Irpex lacteus* are used as **milk clotting enzymes** in **cheesemaking**.

Irradiated foods Foods subjected to **irradiation** to improve **shelf life** and eliminate harmful **bacteria**, **insects** and other **pests**. Types of food that can be successfully irradiated include **poultry meat** and red **meat**, **fruits**, **vegetables** and **cereals**. Regulations vary between countries as to which (if any) foods may be irradiated. Irradiated **spices** are currently the only irradiated foods licensed for sale in the UK.

IR radiation Alternative term for **infrared radiation**.

Irradiation Application of various forms of **radiation**. In food processing, this can be exposure of items to low doses of high-frequency energy from **gamma rays**, **X-rays** or accelerated electrons with the aim of extending **shelf life**. These rays contain sufficient energy to break chemical bonds and ionize molecules that lie in their path. The two most common sources of high-energy radiation used in the food industry are cobalt-60 (^{60}Co) and caesium-137 (^{137}Cs). For the same level of energy, gamma rays have a greater penetrating power into foods than high-speed electrons. The unit of absorbed dose of radiation by a material is denoted as the gray (Gy), one gray being equal to **absorption** of one joule of energy by one kilogram of food.

Irrigation Artificial supply of water to land by such means as ditches and pipes for the purpose of nourishing plants.

IR spectra Absorption patterns resulting from **IR spectroscopy** analysis of samples. Serve to analyse the composition of samples, and identify impurities.

IR spectroscopy Abbreviation for infrared spectroscopy. Analytical technique in which samples are identified on the basis of absorption of light of infrared wavelength.

Iru Traditional Nigerian fat- and protein-rich **fermented foods** made from **African locust beans**. Seeds are cooked, fermented and formed into balls, which can be used to flavour **soups** and stews. The fermented products can be stored for long periods and are a good source of **linoleic acid** and **vitamin B₂**. Similar to **dawadawa**, a product made in West and Central Africa.

Ishiru Traditional Japanese **fish sauces** usually made from **squid livers** (ika-ishiru) or sardine (iwashi-ishiru). Production involves a long natural **fermentation** period. Used as **seasonings** in a range of dishes.

Isinglass Inner lining of swim bladders of fish, commonly from **hake** or **sturgeon**. Used as a substitute for **gelatin** and in **clarification** of **wines** and **beer**. Also termed fish glue.

ISO Abbreviation for **International Organization for Standardization**.

Iso-α-acids **Bitter compounds** formed from **hops**-derived **α-acids** during boiling of **worts** or preparation of isomerized **hop extracts**. Important bitter compounds in **beer**.

Isoamyl acetate Ester with banana-like odour. One of the natural **aroma compounds** found as a result of yeast **fermentation** in **beer**, **sake** and **wines**, and also occurs naturally in fruits such as **apples** and **bananas**. Widely used as an added flavour compound in processed foods. Can be produced in microbial fermentations and also enzyme **bioconversions**.

Isoamyl alcohol One of the aliphatic **alcohols**, with a characteristic odour and pungent taste. Synonyms include isopentanol, **methyl butanol** and isopentyl alcohol. Used as an **esterification** substrate for production of isoamyl **esters**. Also identified as one of the **aroma compounds** present in **wines**, **cider** and **beer** as a result of yeast **fermentation**.

Isoamylases EC 3.2.1.68. Hydrolyse 1,6-α-D-glucosidic branch linkages in **glycogen**, **amylopectins** and their β-limit **dextrins**. 1,6-Linkages are hydrolysed only if at branch points. Although both are also known as debranching enzymes, isoamylases are distinguished from **α-dextrin endo-1,6-α-glucosidases** (EC 3.2.1.41) by their inability to attack **pullulan**, their limited action on α-limit dextrins and their complete action on glycogen.

Isoascorbic acid Isomer of L-**ascorbic acid**. Exhibits **antioxidative activity** and **antimicrobial activity** and thus has uses in food **preservatives**. It is also added to **meat** and **meat** products to stabilize **colour** and **flavour**. Also called **erythorbic acid** and γ-lactone.

Isobutanol One of the aliphatic **alcohols**, with a mild alcoholic, sweet odour. Synonyms include **isobutyl alcohol** and **methyl propanol**. One of the **aroma compounds** produced during **fermentation** in **alcoholic beverages** including **wines**, **beer** and **cider**.

Isobutyl alcohol One of the aliphatic alcohols, with a mild alcoholic, sweet odour. Synonyms include **isobutanol** and **methyl propanol**. One of the **aroma compounds** produced during **fermentation** in **alcoholic beverages** including **wines**, **beer** and **cider**.

Isobutyric acid One of the short-chain **fatty acids**, with four carbon atoms. Has a pungent **aroma** and has been identified in carob, **wines** and **beer**. Synonymous with 2-methylpropanoic acid.

Isocaproic acid Member of the short-chain fatty acids. Identified as one of the **aroma compounds** present in **meat** and **fish**. Synonymous with 4-methylvaleric acid and 4-methylpentanoic acid.

Isochlorogenic acid One of the **phenols** present in **fruits** and **vegetables**. Synthesized in response to damage or wounding.

Isocitrate dehydrogenases **Dehydrogenases** which catalyse the conversion of isocitrate to 2-oxoglutarate and carbon dioxide, using either NAD^+ (EC 1.1.1.41) or $NADP^+$ (EC 1.1.1.42) as the acceptor molecule. Isozyme profiles of isocitrate dehydrogenases can be used in **species identification** in **meat** and cultivar differentiation in **fruits** and **vegetables**. Other applications of the enzyme include determination of isocitrate levels in **fruit juices** and **vegetable juices**.

Isocitric acid One of the **organic acids**, produced as an intermediate in the tricarboxylic acid and glyoxylate cycles. Found in many **fruits** and fruit products, including **fruit juices**. Also formed as a by-product during microbial fermentation to produce its isomer, **citric acid**.

Isoelectric focusing **Electrophoresis** in which a pH gradient is incorporated into the gel diffusion medium. Sample components migrate through the gel until they reach the point where the pH is equal to their **isoelectric points**. Commonly abbreviated to IEF.

Isoelectric points   pH at which the net charge on a molecule is zero. At their isoelectric points, proteins will not migrate in an electric field.

Isoenzymes   Multiple forms of **enzymes** that catalyse the same reaction but which differ in characteristics such as primary structure, kinetics, electrophoretic mobility and immunological properties.

Isoflavones   Subclass of the **flavonoids**, sharing a basic structure of two benzyl rings joined by a three carbon bridge which may or may not be closed into a pyran ring. Isoflavones differ from **flavones** in that the benzyl B ring is joined at position 3 instead of position 2. These phytochemicals are more restricted in occurrence than other **flavonoids**, but can be found in several **legumes**, including **soybeans**, **lentils**, **peas** and **mung beans**. Soybeans and soy products provide a major dietary source of isoflavones, including **daidzein** and **genistein**, which display activity as **phytoestrogens**.

Isoflavonoids   A subclass of the **flavonoids** which includes **isoflavones**.

Isoglucose   **Fructose** sweetener prepared from **starch**. Starch is dispersed in water and hydrolysed to produce **glucose syrups**, and the glucose is then isomerized to fructose via a reaction catalysed by **glucose isomerases**. When produced from **corn starch**, isoglucose preparations are known as **fructose high corn syrups**.

Isohumulones   Components of the **hops**-derived **iso-α-acids** fraction in **worts** and **beer**. Formed by **isomerization** of **humulones** during boiling of **worts** or preparation of isomerized **hop extracts**. Important **bitter compounds** in beer.

Isoleucine   One of the essential **amino acids**. A common protein constituent and free amino acid in many foods.

α-Isolupanine   *Cis,cis*-lupanine. Alkaloid occurring in **lupin seeds** (*Lupinus* spp.).

Isomalt   Polyol or dissacharide alcohol produced by reduction of **sucrose** and composed of a mixture of glucomannitol and glucosorbitol. Isomalt has approximately half the sweetness of sucrose and half the calorific value. It has low solubility in water and is used in the manufacture of hard **sugar confectionery**, **chewing gums**, **preserves** and **bakers confectionery**.

Isomaltooligosaccharides   **Oligosaccharides** produced from **starch** or **corn** powder. Used in the form of **syrups** as low calorie **sweeteners**. Effective in stimulating the growth of ***Bifidobacterium*** species and also beneficial in preventing dental caries,

improving intestinal function and enhancing immune response in humans. Potentially useful ingredients for **functional foods**.

Isomaltose   Isomer of **maltose** with 2 molecules of **glucose** linked by an α-1,6-glycosidic bond rather than an α-1,4- bond as in maltose.

Isomaltulose   Disaccharide with the systematic name 6-*O*-α-D-glucopyranosyl-D-fructofuranose (hydrolysis produces **glucose** and **fructose**). It occurs naturally and is present in **honeys** and **cane sugar juices**. Manufactured by bacterial transglucosylation of **sucrose** and marketed under the name Palatinose. Has approximately half the **sweetness** of **sucrose** but is more resistant to hydrolysis, digestion and microbial degradation, thus it is thought to have potential as a sweetener for **low calorie foods**.

Isomerases   EC 5. **Enzymes** that catalyse geometric or structural changes within a molecule to form a single product. Reactions do not involve a net change in the concentrations of compounds other than the substrate and product. Subdivided into **racemases** and **epimerases** (EC 5.1), *cis-trans*-isomerases (EC 5.2), intramolecular oxidoreductases (EC 5.3), intramolecular transferases (mutases; EC 5.4), intramolecular lyases (EC 5.5) and other isomerases (EC 5.99).

Isomerization   Reaction in which the structure of a molecule is altered so that it is converted into one of its **isomers**.

Isomers   Series of compounds that have the same molecular formula but which differ in structure (structural isomers) or orientation (**stereoisomers**).

Isoniazid   Common name for isonicotinic acid hydrazide, an antibiotic used for treatment of **tuberculosis** in humans and animals. Banned in some countries for use in the treatment of dairy **cattle** as it can interfere with the tuberculin test and the presence of residues in **milk** is a potential hazard to human health.

Isoprene   Branched five-carbon chain hydrocarbon that forms a recognizable structural component of **isoprenoids, terpenoids** and other compounds derived from isopentenylpyrophosphate, the biosynthetic isoprene unit. Synonym 2-methyl-1,3-butadiene.

Isoprenoids   **Organic compounds** based on the **isoprene** hydrocarbon structural unit. Include a large range of chemicals, such as **carotenoids, steroids, terpenoids** and **tocopherols**. Many substances contain both isoprenoid and non-isoprenoid components.

Isospora   Genus of protozoan **parasites** of the class Coccidia. Occur in the intestines of birds, amphibians, reptiles and mammals, including man. Infection with

Isospora belli in humans may occur due to the ingestion of contaminated food or water.

Isosyrups **Fructose syrups** prepared by hydrolysis of **corn starch** followed by treatment with **glucose isomerases** to convert the **glucose** in the hydrolysate to **fructose**.

Isotachophoresis **Electrophoresis** technique in which separation of sample components is based on their ionic mobility. The ions separated are sandwiched between an electrolyte of higher mobility and one of lower mobility.

Isothiocyanates **Organic compounds** containing a nitrogen-carbon-sulfur unit. Structural isomers of thiocyanates. Many isothiocyanates are pungent **volatile compounds** released upon damage to tissues, for example in *Brassica* species. **Allyl isothiocyanate**, which has **antimicrobial activity**, contributes to the pungency of **mustard**, **watercress**, **horseradish** and **wasabi**.

Isotonic drinks **Beverages** which are isotonic with normal human body fluids, and contain components such as **electrolytes** and **sugars**. Claimed to enhance performance and recovery from physical exertion.

Isovaleraldehyde Aldehyde; synonym **3-methylbutanal**. Volatile flavour compound identified in **vinegar**, **coffee** and **tomatoes**, and also as an **off flavour** in **sake** and chlorinated **drinking water**.

Isovaleric acid One of the short, branched-chain **fatty acids**. Volatile aroma compound in **Cheddar cheese** and **Swiss cheese**; also identified in **whisky**, **wines** and **beer**, and as an off odour compound in **natto**. Synonymous with 3-methylbutanoic acid.

Isozymes Alternative term for **isoenzymes**.

Itaconic acid Alternative term for 2-methylenebutanedioic acid or methyl succinic acid. An organic acid produced commonly during **fermentation** of **sugars** such as **glucose** or **molasses** by *Aspergillus* *terreus*. Also a pyrolysis product of **citric acid**. Used in **resins** and **plasticizers**.

Ivermectin One of the **anthelmintics** used widely for treatment of **nematodes** and arthropod **parasites** infections in cattle, sheep, goats and swine. Excreted relatively slowly from animals so that residues may persist for long periods in tissues, particularly in **livers** and **adipose tissues**.

Izvara Bulgarian product made from **cow milk** coagulated with **rennets** and fermented with 1-5% **butter starters**.

J

Jaboticaba Purple, grape-like **fruits** produced by trees of the genus *Myrciaria* (*M. cauliflora, M. jaboticaba* or *M. trunciflora*). Eaten fresh or used to make jellies and **alcoholic beverages**.

Jack beans Seeds of *Canavalia ensiformis*. Mature seeds must be boiled in water before consumption because of the presence of toxic constituents. Immature seeds and pods (**snap beans**) are also eaten. When roasted, seeds are used as **coffee substitutes**.

Jack fruits Alternative term for **jak fruits**.

Jack mackerel Alternative term for **horse mackerel**.

Jaggery Unrefined brown coloured **sugar** produced mainly in India by **evaporation** of **sugar cane juices**. Also known as **gur**.

Jak fruits Fruits produced by *Artocarpus heterophyllus* (*A. integrifolia*) and related to **breadfruit** and **figs**. One of the largest cultivated fruits, weighing up to 100 pounds. When ripe, jak fruits are eaten raw, while flesh and seeds of green fruits are eaten cooked, commonly in **curries**. Also known as jack fruits.

Jalapeno peppers Small smooth-skinned **chillies** originating in Mexico. Usually about 2 inches long and 0.75 inches in diameter. **Colour** varies from dark green to bright red when ripe. Range in spiciness from hot to very hot, but the extremely hot veins and **seeds** are easy to remove. Available fresh, canned or dried. Used to add spiciness to Mexican dishes, sauces and other dishes, or served stuffed and deep fried. Smoked jalapenos are known as chipotles. Also used in jalapeno cornbread.

Jams Conserves made by boiling whole fruits with **sugar** to form **fruit pulps**. Called jelly in the USA.

Japanese apricots Small yellow **fruits** produced by the ornamental tree *Armeniaca mume* (*Prunus mume*). Eaten raw or used to make **fruit juices** and **pickles**. Also known as ume or mei.

Japanese chestnuts Large fruits produced by *Castanea crenata*. The flesh is creamy and sweet, but the outer peel is difficult to remove.

Japanese flounders Marine **flatfish** species (*Paralicthys olivaceus*) from the flounder family (Paralicthyidae), which occurs in the western Pacific Ocean. Highly prized as a food fish in Japan. Usually marketed fresh. Also known as hirame and **bastard halibut**.

Japanese pears Oriental pears produced by *Pyrus serotina* or *P. pyrifolia*. Also referred to by many other names, including **Asian pears**, Chinese pears and sand pears.

Japanese pepper Common name for *Xanthoxylum piperitum* or sansho. The leaves are used in **seasonings** or as spicy **vegetables** in Japanese cooking.

Japanese plums Large, yellow to red **fruits** produced by *Prunus salicina*. Alternatively, another name for **loquats**, small yellow fruits produced by *Eriobotyra japonica*.

Japanese radishes Oriental type of *Raphanus sativus* with long, mild flavoured roots of up to 20 kg in weight. Traditionally used in **soups** and **sauces** or cooked with **meat**. Sold in the UK as mouli or rettich. Also known as daikon.

Jarlsberg cheese Norwegian **hard cheese** made from **cow milk**. It has a similar **consistency, texture** and hole formation to **Emmental cheese**, but a more nut-like and sweeter **flavour**. The cheese is golden yellow in **colour**, and contains holes of various sizes. It is used as a table, dessert or sandwich cheese.

Jasmine Natural **flavourings** with warm, spicy characteristics derived from flowers and leaves of jasmine (*Jasminus* spp.). Predominant **flavour compounds** and **aroma compounds** include jasmonates, jasmones, benzyl acetate, indol and **eugenol**.

Jasmonic acid Jasmonic acid and methyl jasmonate, collectively referred to as jasmonates, are naturally occurring **plant growth regulators** involved in various aspects of plant development and responses to biotic and abiotic stresses. Used to regulate the yield and quality of **fruits** and **vegetables**.

Jellied milk **Milk** to which is added **sugar, flavourings**, thickening agents and **gelling agents**. Also known as jellified milk.

Jellies Small, soft sweets, usually fruit flavoured, of gelatinous texture, made in various shapes and often coated with sugar. The singular term, **jelly**, is used to refer to jam-like products, usually clear, that are made from strained **fruits** containing **pectins** which are boiled with **sugar**. Also refers to soft, semi-transparent foods prepared from **gelatin** which are sweetened, flavoured, cooled in a mould and eaten as **desserts**.

Jelly In the UK, a term applied to fruit-flavoured sweetened **desserts** set with **gelatin** (**table jellies**) and also to clear **jams** made from boiled, sweetened **fruit juices**. In the USA and Canada, the term is synonymous with any type of jam. Also used for savoury products with a jelly like consistency and set with gelatin, e.g. calf's foot jelly.

Jelly babies **Jelly confectionery** products formed into stylized shapes resembling babies.

Jelly confectionery Collective term for **confectionery** products made with **jelly**.

Jelly figs **Fruits** produced by *Ficus awkeotsang*. Seeds are used in Taiwan to make jelly **cakes** and jelly **desserts**. Also used in manufacture of **soft drinks**.

Jellyfish Common name used for any free-swimming marine and freshwater invertebrates from the phylum Cnidaria. Some species are consumed in dried form.

Jelly rolls US term for **swiss rolls**, thin **sponge cakes** which are covered on one side with **jams** and rolled into cylinders.

Jeotgal Traditional Korean salted and fermented sea food **sauces** prepared from waste tissues of **fish** or **shellfish**, such as the internal organs of **whelks**, a by-product of their processing.

Jerky Meat products prepared by drying long, narrow strips of **meat**, commonly beef. Also known as jerked meat. World wide, various types of jerky are produced. For example, in South Africa, a spicy version of jerky, known as **biltong**, is produced, often using game meat, and in the Caribbean, strips of meat are soaked in a spicy marinade and then dried to produce a version of jerky known as tasajo. The chewy strips of dried meat do not require refrigeration and, thus, are popular **snacks**. The major disadvantage of jerky prepared from red meat is that it has high contents of salt and fat; in comparison, turkey jerky is a healthier alternative as it has lower contents of fat and salt.

Jerusalem artichokes Stem tubers of *Helianthus tuberosus*. White to yellow or red to blue in **colour**; irregular and knobbly in shape. Consumed boiled or baked. Rich source of **inulin**.

Jessenia Genus of **palms**, the most common species being *Jessenia bataua*. Seeds are a source of **palm oils**; the sweet pericarp is also eaten.

Jicama Common name for the tropical legume *Pachyrrhizus erosus* or *P. tuberosus*. Young pods are eaten as a vegetable, but the mature seeds are poisonous. Large, turnip-like tubers are thinly sliced and eaten raw, cooked in stews and **soups** or pickled. Tubers are used as substitutes for water chestnuts or **yams**. Source of a **starch** similar to **arrowroot**. Also known as **yam beans** and Mexican potatoes.

Jobs tears Edible seed kernel from the wild grass *Coix lacryma-jobi* used as a cereal food in parts of East Asia and the Philippines. Also known as adlay.

Jointing Cutting of animal **carcasses** into joints.

Jojoba oils Liquid wax **esters** of long chain **fatty acids** (e.g. **eicosenoic acid** and **erucic acid**) with long chain **alcohols** (e.g. eiconsanol and docosanol) derived from seeds of *Simmondsia chinensis*. Show high **oxidative stability** and may be used as **coatings** for **dried fruits**.

Jojoba seeds Seeds produced by the plant *Simmondsia chinensis*, native to south western USA and northern Mexico. Similar in **colour** and shape to **coffee beans**, and rich in **tocopherols**. Source of **jojoba oils**, uses of which include **coatings** for **dried fruits**. Contain **simmondsin**, a cyanide-containing glycoside produced as a by-product in the manufacture of jojoba oils, and with potential as an **appetite** suppressant.

Jowar Indian name for **sorghum** (*Sorghum vulgare*). Also known as great millet, kaffir corn and guinea corn.

Juiciness **Sensory properties** relating to the extent to which products, such as **fruits**, **vegetables** and **meat**, are juicy or succulent. Dependent on the amount of cell sap released during fracture. Uncooked cell walls fracture across to reveal the internal structure of the cell, thus releasing sap. In contrast, cooked cells tend to separate along the middle lamella, so cell sap is not released in the same way.

Jujubes **Fruits** produced by *Zizyphus jujuba* or *Z. mauritiana*. Similar to dates in appearance and **flavour**. Relatively high sugar content. Eaten in a number of ways, including fresh, dried, boiled with rice, smoked, pickled, stewed and baked. Also called ber fruits and Chinese dates.

Juniper **Spices** of ripe berries from the common juniper tree, *Juniperus communis*. Berries are fermented and distilled to produce **gin**. Juniper berry extracts are used as **flavourings**, e.g. in **ice cream**, **sugar confectionery** and **bakery products**.

Junket **Desserts** prepared from sweetened and flavoured **curd**.

Jute Rough fibre made from the inner bark of tropical plants belonging to the genus *Corchorus*, especially *C. olitorius* (in India) and *C. capsularis* (in China). Jute fibre is used to make jute board, a strong flexible **cardboard** often used to make shipping **cartons**.

Also woven into sacking, and used for making wrapping paper and twine.

Jute seeds Seeds from either of two Asian plants, *Corchorus capsularis* or *C. olitorius* which may be used as **oilseeds**.

K

Kachkaval cheese **Hard cheese** popular in the Balkan countries sometimes made from raw **ewe milk**. Kachkaval has a smooth dry rind and an amber-coloured interior with a moderately firm **texture** and no holes. **Flavour** is piquant and slightly salty.

Kaempferol Member of the **flavonoids** group. Present in a range of foods including many **fruits** and **vegetables**. Displays **antioxidative activity**. Synonymous with 3,5,7,4′-tetrahydroxyflavone.

Kafirins **Prolamins** found in **sorghum**.

Kahweol Member of the **terpenoids**. One of the diterpenes found in **coffee** which, along with **cafestol**, may be associated with increases in plasma cholesterol levels.

Kajmak cheese Yugoslav fresh **cream cheese** made from cow, ewe or goat milk.

Kakdugi Radish **kimchies**.

Kaki figs Alternative term for **persimmons**.

Kaki fruits Alternative term for **persimmons**.

Kalakand Sweetened dairy product that is popular in India. Made by evaporating acidified **buffalo milk**.

Kalamansi juices **Fruit juices** extracted from **fruits** of *Citrus microcarpa*.

Kale Non-heading **cabbages** with large leaves that have a mild, cabbage-like **flavour**. Leaves vary in **colour** and form according to variety, and are a good source of **vitamin A**, **vitamin C**, calcium, **folic acid** and iron. Used in **salads** and **soups**, or as a side vegetable. Varieties with curled and crimped leaves (curly kale) are most popular for human consumption. Also known as borecoles and collards.

Kamaboko Japanese name for fish products consisting of processed, stabilized **fish mince** having a firm, elastic or rubbery **texture**; often used as a general name for all **surimi** products made in Japan. Various types of kamaboko are produced and classified into several categories according to heating method, shape or ingredients used.

Kanamycin Aminoglycoside antibiotic active against many **pathogens**. Used in treatment of respiratory diseases, **mastitis** and other infectious conditions. Residues may persist in **milk**, but are normally eliminated by 3 days after the end of treatment.

Kangaroo meat **Meat** from **kangaroos**, herbivorous marsupials belonging to the genus *Macropus*. Kangaroo carcasses have high lean and low fat contents; the proportion of high-value meat in kangaroo carcasses is greater than in sheep carcasses. The usual meat cuts available are fillet, loin and rump (the prime roasting and grilling cuts), topside, tail and chopped meat. As kangaroo meat oxidizes rapidly on contact with air, the majority is sold sealed, either vacuum packed or under plastic film. Kangaroo meat is dark coloured and lean; it has a distinctive texture and flavour, a low fat content and contains predominantly **polyunsaturated fats**. In addition to its nutritional benefits, kangaroo meat is associated with a low incidence of **pathogens** and a low potential for transmission of **zoonoses**.

Kangaroos Large, herbivorous marsupials belonging to the genus *Macropus* of the Macropodidae family; there are several species. In Australia, three species of kangaroo are harvested commercially for **kangaroo meat** production, namely the red kangaroo (*M. rufus*), the eastern grey kangaroo (*M. giganteus*) and the western grey kangaroo (*M. fuliginosus*).

Kanjan Alternative term for **kanjang**.

Kanjang A Korean style soy sauce produced by **fermentation** of meju (soybean paste). Also known as kanjan.

Kanji Traditional Indian beverage made from black **carrots**. Peculiar to the northern plains of India, black carrots are black on the outside but a rich red colour under the skin. The carrots are parboiled in water with **salt** and other **flavourings** such as ground **mustard seeds** and **chilli** powder. The mixture is then left to ferment in the sun, resulting in a sour and spiced red drink which is consumed as an accompaniment to **meals**.

Kapok oils Yellow-green **oils** obtained from seeds of the kapok tree (family Bombacaceae) which contain cyclopropene acids and are used in both food and soap manufacture.

Kapok seeds By-products of kapok fibre production; used for extraction of **kapok oils**.

Karaya gums Exudates of *Sterculia urens*, a tree that is native to India; hence, the **gums** are also known as

Indian tragacanth. Used as food **thickeners, stabilizers, emulsifiers** and **texturizing agents**.

Kareish cheese Egyptian brine-ripened **cheese** made from cow or buffalo **raw milk**. Slightly acidic and salty **flavour**.

Kashar cheese Turkish semi hard or **hard cheese** generally made from raw **ewe milk**, alone or mixed with raw **goat milk**. Similar to Kashkaval or **Kachkaval cheese** popular throughout Balkan countries.

Kasseler Cured pork products prepared from the loin of **swine**. Kasseler is cured and drained, and is then smoked and cooked again in a process similar to that used in **ham** production. It is a more delicate product than ham, and cannot tolerate being tumbled or massaged in order to increase take up of additional water.

Kasseri cheese Greek pasta filata type cheese made from **ewe milk** or a mixture of ewe milk and **goat milk**. Traditionally made from **raw milk**, as processing procedures are considered to inhibit harmful **microorganisms**. Rindless, but with a white crust. Interior is pale yellow in colour with a springy texture. Flavour is salty and buttery, with an underlying sweetness. Kasseri cheese is used as an alternative to **mozzarella cheese** in local dishes.

Katemfe Common name for *Thaumatococcus daniellii*. **Fruits** of this plant are the source of the protein sweetener **thaumatin**.

Katsuobushi Fish product consisting of dried flesh of **skipjack tuna** (**bonito**) which has been boiled, defatted and shaped into a stick-like form. Normally used as a condiment.

Katyk Fermented product prepared from cream of **ewe milk**.

Kava Beverage made by aqueous extraction of powders prepared from rhizomes of kava kava (*Piper methysticum*). Used in the south Pacific region as a narcotic/stimulant. Also used in treatment of anxiety and a range of disorders. The pharmacologically active components are lactones. Non-addictive, but can have adverse effects, such as muscle weakness, drying of the skin, and liver damage, if consumed over a long period of time or in high amounts.

Kawal Strong-smelling pastes prepared by **fermentation** of leaves of the legume *Cassia obtusifolia*. Rich in protein. Used as **meat substitutes** in **soups** and stews.

KCl Chemical formula for potassium chloride. One of the chlorides widely used in food processing at varying levels to replace salt (NaCl), for example in **brines**, in order to reduce Na levels in foods, and specifically to produce **low sodium foods** and **salt substitutes**.

Kebabs Pieces of **meat**, **fish** and/or **vegetables** grilled or roasted on skewers or spits.

Kecap An Indonesian soy sauce prepared by fermentation of **black soybeans** in a 2-stage process involving a **solid state fermentation** and a brine **fermentation**.

Keeping quality Alternative term for **shelf life**.

Kefalograviera cheese Greek **hard cheese** made usually from **ewe milk**.

Kefalotyri cheese Greek **hard cheese** made from whole raw **ewe milk** or a mixture of ewe and **goat milk**. The **colour** varies from white to yellow and it has a tangy **flavour** and sharp **aroma**. It ripens in 2-3 months and is generally served grated over cooked dishes. Also produced in Romania.

Kefir Alcoholic **fermented milk** product made traditionally by addition of **kefir grains** to **milk**. The traditional product contains **alcohol** and CO_2 in addition to **lactic acid**, making it foaming and viscous. Since this can cause blowing of packs, **starters** with few or no **yeasts** and lactobacilli are used in industrial production of kefir. Commercial kefir tends to contain much lower amounts of alcohol than traditionally prepared products. Kefir is generally more digestible than milk and more easily tolerated by lactose-intolerant individuals. It is marketed with various fat contents.

Kefiran One of the **exopolysaccharides** produced by *Lactobacillus kefiranofaciens* and found in **kefir grains**. Potentially useful as a food additive due to its **gelation** properties. Also displays **antitumour activity**.

Kefir grains Traditionally used in the culture of **milk** during manufacture of **kefir**. An irregularly shaped mass of microbial polysaccharide, mesophilic streptococci, leuconostocs, mesophilic and thermophilic lactobacilli, and **yeasts**. Yeasts comprise the grain core, with yeasts and rod-shaped **bacteria** immediately outside, and rod-shaped bacteria predominantly at the periphery. Due to problems with blowing of packages, kefir grains are not used in industrial manufacture of kefir, **starters** consisting almost entirely of streptococci, with few or no yeasts and lactobacilli, being used instead. The product then lacks the foaming characteristics of traditional kefir.

Kegs Small **barrels**, often used for transportation or storage of **alcoholic beverages**, especially **beer**. May be made from wood, but are commonly made from plastics or metal.

Kelp Alternative term for **seaweeds** of the genus *Laminaria*.

Kelthane Alternative term for the acaricide **dicofol**.

Kenaf seeds Seeds produced by *Hibiscus cannabinus* which are used as a source of **edible oils**.

Kenkey Corn-based product from Ghana that is made by boiling fermented **dough**.

Keratin One of the structural fibrous **animal proteins**, found in vertebrate skin and specialized epidermal structures, including **feathers**, nails, hair, hooves, horns and quills. Keratin-degrading **microorganisms** and serine **proteinases** (keratinases) are of interest for **bioremediation** of wastes from slaughterhouses and food factories wastes from meat and carcass processing.

Kesari dhal Alternative term for **grass peas**.

Kestose Fructooligosaccharide comprising 2 **fructose** residues and a **glucose** residue. Produced by hydrolysis of **inulin** or via the action of fructosyltransferases using **sucrose** as substrate.

Keta salmon **Pacific salmon** species (*Oncorhynchus keta*) found in seawater and rivers along north Pacific coasts. Mostly marketed as a canned product but also sold fresh, dried-salted, smoked, and frozen.

Ketchups Synonym for **catsups**. Originally, a spicy pickled fish condiment, nowadays the term refers to various thick piquant **sauces** containing **sugar**, **spices**, **vinegar**, and other ingredients such as **tomatoes**, **mushrooms**, **nuts** or **fruits**. **Tomato ketchups** are one of the most well known types of ketchup and are a popular accompaniment for **French fries**, **burgers** and many other foods.

Ketjap Alternative term for **kecap**.

Ketones Types of **carbonyl compounds** in which the carbonyl substituent is bound to two carbon atoms. Many ketones are important volatile **aroma compounds** in foods and **beverages**.

Ketoses **Nonreducing sugars** in which, in the acyclic form, the actual or potential carbonyl group is nonterminal (ketonic). Many of these sugars, which are **monosaccharides**, have the suffix '-ulose'. Examples include **xylulose** and arabino-2-hexulose (**fructose**).

Kettles Metal or plastic **containers** with a lid, spout and handle for boiling **water**. Also metal containers for heating any liquids. Fish kettles are long **pans** specially designed for cooking **fish**.

Khoa Heat-concentrated milk product usually prepared from **buffalo milk** and popular in India. Used as the base material for a number of Indian sweets, such as **burfi**, **peda** and **gulabjamans**.

Khurchan **Concentrated milk** product popular in India. Prepared by simmering whole **milk** and adding **sugar**.

Kicap Alternative term for **kecap**.

Kidney beans Type of **common beans** (*Phaseolus vulgaris*) with kidney-shaped seeds. Red kidney beans form an integral part of the Mexican dish chilli con carne. Due to the presence of **antinutritional fac-**

tors, such as **lectins**, beans must be well soaked in water and cooked prior to consumption.

Kidneys Paired abdominal excretory organs, which form a part of edible **offal**. Kidneys from young animals, particularly lambs and veal calves, are more tender than those of older animals; those from milk-fed animals are palest in colour. Lamb and calf kidneys have the most delicate flavour, whilst cattle and swine kidneys have a stronger **flavour**.

Kieselguhr Soft, crumbly sedimentary material used as a filtering agent and in other industrial applications.

Kilka Brackish and **freshwater fish** species (*Clupeonella cultriventris*) from the herring family (Clupeidae), found in the Black Sea (northwestern parts), Sea of Azov and Caspian Sea; also occurs in lakes in Turkey and Bulgaria. Often marketed as a dried, salted product. Also known as **black sea sprat**.

Killer yeasts **Yeasts** (including **brewers yeasts**, **wine yeasts** and **sake yeasts**) which secrete protein or glycoprotein **toxins** able to kill sensitive yeast strains. This may be disadvantageous, if desirable yeast strains are killed, or beneficial if wild yeasts or contaminating yeasts are eliminated.

Kilning Final stage of **malting**, in which steeped germinated **malting barley** is heated and dried to a specified moisture content. This halts metabolism and enzyme activity in the **malt**. Kilning temperature and duration may be selected to give malts with a range of **colour** and **flavour**.

Kilns Furnaces or **ovens** for burning, **baking** or **drying**. An oast is a kiln used to dry products such as **hops** and **malt**.

Kimchies Fermented vegetable products, made mainly from **cabbages** or **radishes**, eaten widely in Korea. Prepared vegetables are soaked in **brines** for several hours before mixing with **flavourings** and **fermentation** by **microorganisms** present in the raw materials. Rich in **vitamin C**.

Ki-mikan Common name for **citrus fruits** produced by *Citrus flaviculpus*, which are eaten in Japan. Also known locally as ogon-kan.

Kinases **Enzymes** that transfer a phosphate group from one compound, such as **ATP**, to another. The acceptor may be an alcohol group (EC 2.7.1), a carboxyl group (EC 2.7.2), a nitrogenous group (EC 2.7.3) or a phosphate group (EC 2.7.4). The pyrophosphokinases are in subclass EC 2.7.6. The dikinases (EC 2.7.9) transfer 2 phosphate groups from a donor such as ATP to 2 different acceptors.

Kinema Traditional Indian product made by **fermentation** of cooked **soybeans**, usually with *Bacillus subtilis*. Rich source of protein, with a stringy **texture**

and characteristic **flavour**. Consumed as **meat substitutes**, usually in a side dish with cooked **rice**.

Kinetin Member of the **cytokinins** group of **plant growth regulators**. Kinetin (6-furfurylaminopurine) occurs naturally in plants and is an important determinant of growth and development. Also added exogenously to **fruits** and **vegetables** during cultivation.

Kingklip Eel-like **marine fish** species (*Genypterus capensis*) primarily caught off the coast of southern Africa. Normally marketed in frozen form, but also sold fresh.

King salmon Alternative term for **Chinook salmon**.

Kinins Alternative term for the **cytokinins**, a group of **plant growth regulators**. Also, in animal physiology, refers to a class of linear polypeptide hormones including bradykinin and angiotensin.

Kinnow mandarins Variety of **mandarins** with very sweet flesh and numerous seeds.

Kippers Fish products consisting of boned and gutted **herring** which are split open along the back, lightly brined and cold smoked; sometimes artificially coloured. Marketed chilled, frozen or canned; ground flesh is made into kipper paste.

Kirsch **Fruit brandies** distilled from **cherries**, commonly in the presence of the cherry stones.

Kishk Dried mixture of **fermented milk** and **cereals** originating from the Middle East. Typically, made from **bulgur** wheat fermented with **yoghurt** and then dried and ground to a powder. Easy to store and can be reconstituted with water to make soups. Known as **tarhana** in Turkey or **trahanas** in Greece.

Kisra Fermented thin pancake-like leavened **bread** made from whole **sorghum** flour.

Kissel Russian jelly-type **desserts** made from sweetened **fruit purees**. Typically made with **cranberries** and thickened with **arrowroot**, **corn starch** or **potato meal**. Served hot or cold.

Kiwano **Fruits** produced by *Cucumis metuliferus*. A spiky cross between **cucumbers** and **melons** containing white seeds in a bright green, jelly-textured pulp. Also known as horned melons.

Kiwifruit **Fruits** produced by *Actinidia deliciosa* or *A. chinensis*. Rich in **vitamin C**; also contains a range of **minerals** and B vitamins. The flesh is bright green near the surface, with a ring of black seeds near the centre and a core of lighter green flesh. Eaten fresh, often in **fruit salads**, or used to top **desserts** and in garnishes. Also known as Chinese gooseberries.

Kiwifruit juices **Fruit juices** extracted from **kiwifruit** (*Actinidia chinensis*).

Kjeldahl nitrogen Total nitrogen in a substance, determined by digesting the sample with sulfuric acid and a catalyst. Kjeldahl nitrogen is used extensively for determination of protein levels in foods. In these cases, the nitrogen measured is converted to the equivalent protein content by use of an appropriate numerical factor.

Klebsiella Genus of Gram negative, facultatively anaerobic, rod-shaped coliform **bacteria** of the family **Enterobacteriaceae**. Occur in the gastrointestinal and respiratory tracts of humans and animals, soil, **dairy products**, raw **shellfish** and fresh raw **vegetables**. *Klebsiella pneumoniae* may be responsible for **gastroenteritis** in humans due to consumption of contaminated food. *K. aerogenes* is responsible for early **blowing** in **cheese**.

Kloeckera Genus of yeast **fungi** of the class Saccharomycetes which are anamorphs of **Hanseniaspora** species. Occur on **fruits** and in soil. *Kloeckera apiculata* is used in **winemaking**.

Kluyveromyces Genus of yeast **fungi** of the class Saccharomycetes. Occur in foods, **beverages**, plants, soil, insects and sea water. *Kluyveromyces marxianus* var. *marxianus* is used in the production of **fermented milk** (e.g. **koumiss** and **kefir**), *K. marxianus* var. *bulgaricus* in used in the production of **yoghurt** and *K. marxianus* var. *lactis* is used in the production of **buttermilk**, Italian **cheese** and fermented milks. *K. fragilis* may be responsible for the **spoilage** of **yoghurt**, while *K. marxianus* may cause **cheese** spoilage.

Knackwurst Cooked, smoked **sausages**, traditionally made in Germany. Knackwurst are prepared from similar ingredients to **bologna** and **frankfurters**, including coarsely textured **pork**, **beef** and **veal**; however, knackwurst additionally include **garlic**, which gives them a stronger **flavour**. They may also be known as garlic sausages or knoblouch. Some are prepared in wide diameter edible natural casings, whilst in other types, casings are removed before retail. Although they are cooked sausages, they are recooked before eating; commonly, they are simmered with **sauerkraut**, served like **frankfurters**, or added to stews and **soups**.

Kneading Working of **dough**, usually with the hands or by machine, in order to form a cohesive, smooth and elastic mass. The network of **gluten** strands stretches and expands during kneading, so enabling dough to retain gas bubbles formed by the actions of the leavening agent. When done by hand, kneading is performed by pressing down into the dough with the heels of both hands, then pushing away from the body. The dough is then folded in half, given a quarter turn, and the pressing and pushing action is repeated.

Knives Sturdy and well balanced **cutting** instruments consisting of a blade fixed into a handle, or blades on a machine for cutting, **peeling**, **slicing** or spreading.

Most knife blades are made of steel or ceramic zirconia, a hard material that doesn't rust, corrode or interact with food. Knife handles are usually made of wood, plastic, horn or metal. Preferably, the end of the blade should extend to the far end of the handle, where it should be anchored by several rivets. Knives are tailored for specific applications. For example, a chef's knife has a broad, tapered shape and fine edge, which is ideal for **chopping vegetables**, while a slicing knife with its long, thin blade cuts cleanly through cooked **meat**. Knives with serrated edges are good for slicing softer foods such as **bread**, **tomatoes** and **cakes**. The easy-to-handle, pointed, short-bladed paring knife is ideal for **peeling** and **coring fruits**.

Kocho Traditional Ethiopian product made by **lactic acid bacteria fermentation** of ensete (*Ensete ventricosum*; Abyssinian banana), a crop related to **bananas** and **plantains**. The pseudostem, corm and inner leaf sheaths are the plant parts which are fermented. Fermentation lasts for a month to a year, depending on the ambient temperature. The fermented product is then baked.

Kochujang Korean name for chile bean pastes, pastes or **sauces** made from fermented **soybeans**, or sometimes fermented **black beans**, **chillies**, **garlic** and **seasonings**. Popular also in Chinese dishes. Also known as kochu chang.

Kochwurst German **sausages** made from precooked ingredients. The major types include: **liver sausages**, brawn sausages, **blood sausages**, spreadable sausages and aspic sausages.

Kohlrabi Variety of *Brassica oleracea*. Available in white, green and purple types. Rich in **vitamin C** and **potassium**. Leaves are used in **salads** or cooked as a vegetable; the swollen, turnip-like stem is eaten raw or cooked. Also known as cabbage turnips.

Koji **Cereals** or **beans** inoculated with *Aspergillus* or other fungi and used as a starter for a wide range of Oriental **fermented foods** and **fermented beverages**, including **miso**, **sake** and **soy sauces**. Acts as a supplier of various **enzymes**, such as **lipases**, which contribute to the quality and **functional properties** of the products.

Kojic acid Fungal metabolite produced by *Aspergillus oryzae* which shows **antibacterial activity**. Also inhibits the activity of **catechol oxidases** and **tyrosinases**. Used primarily as **browning inhibitors** in foods.

Kokja Starters containing **fungi** and **bacteria** used in manufacture of Korean **takju rice wines**.

Koko Thin, fermented **porridge** made from **corn**, **sorghum** or **cassava** flour, either singly or in mixtures. Often consumed as **infant foods** in Ghana and Kenya. Also known as **uji**.

Kokum Common name for the tropical tree, *Garcinia indica*, **fruits** of which are used in preparation of a spice. The dark purple fruits are picked when ripe, dried and the **peel** removed for use in foods, where it adds **colour** and a sour, slightly astringent **flavour**. Used especially in **curries**, vegetable dishes, **chutneys** and **pickles**. **Fats** prepared from kokum **seeds** have been used in **cocoa butter extenders** suitable for use in **chocolate** and **sugar confectionery**. Kokum is also known by a variety of other names, including cocum, kokam and Goa butter.

Komatsuna Leaf vegetables (*Brassica campestris* or *B. rapa*) that are types of **turnips** developed for their leaves. Rich in **vitamin C** with a relatively high content of **carotenes**. Young shoots are used in **salads**, while leaves are cooked as a vegetable or used in **soups**. Also known as mustard spinach or spinach mustard.

Kombu Japanese name for **seaweeds** of the genus *Laminaria*.

Kombucha Beverages made by **fermentation** of tea infusion with a mixed **bacteria/fungi** culture.

Koningklip Alternative term for **kingklip**.

Konjac Alternative term for some **elephant yams**.

Konjac glucomannans Gums composed of glucose and mannose obtained from **elephant yams**. Used primarily in Japan as **gelling agents**.

Konnyaku Alternative term for some **elephant yams**.

Korn Spirits, produced mainly in Germany and the Netherlands, made by **distillation** of fermented grain **mashes**.

Kosher foods Foods permitted under Jewish biblical law and prepared in accordance with Jewish dietary code. Laws relate not only to the types of foods permitted (e.g. **pork** and **rabbit meat** products are non kosher) but also to the methods of slaughter/preparation, and to food combinations (e.g. **meat** products and **dairy products** may not be mixed). Kosher foods are perceived by many as having been prepared to high standards of wholesomeness and **hygiene**, and are currently attracting a new market of non-Jewish consumers who use kosher certification as an indication of quality.

Kostroma cheese Russian cheese made from **cow milk**.

Koumiss Fermented milk usually made from **mare milk**. Produced using a 2-stage **fermentation** in which **lactic acid bacteria** are added, followed by **yeasts** on completion of **lactic fermentation**. In addition to **lactic acid**, it contains **ethanol** and CO_2, giving a light effervescence.

Krill Any of a number of small, shrimp-like marine crustacean species occurring abundantly in cooler waters. An important commercial species is *Euphausia superba*, which is caught in huge numbers and primarily used to make a protein-rich meal for animal feeds.

Krokant Alternative term for **croquant**.

Kudzu **Legumes** of the genus *Pueraria*. Leaves are used in **salads** and the **tubers** are eaten cooked. The large tubers are also used as a source of **starch** (Japanese arrowroot) that is used in **thickeners**.

Kulfi Concentrated frozen milk product similar to **ice cream** popular in India and Pakistan.

Kumquats Orange or golden-yellow **fruits** of trees of the *Fortunella* species, belonging to the same family as *Citrus* species. Rich in **vitamin C**. Eaten fresh, cooked, candied or preserved in **syrups**. Used in **marmalades**, **chutneys** and **jellies**.

Kunun zaki Traditional Nigerian non-alcoholic fermented beverage which is one of a group of **beverages** called kunu. Commonly made from **millet**, **sorghum**, **rice**, **acha** or **corn**, singly or in mixtures, a combination of sorghum and millet being preferred. Cereal grains are steeped in water and dry or wet milled with **spices** such as **ginger**, red pepper, **black pepper**, **cloves** and **garlic** to impart **flavour**. Saccharifying agents, including malted rice, **sweet potatoes**, **soybeans** and malted sorghum, may also

be added. The finished product is sweet with a potato-like flavour.

Kurakkan Alternative term for **finger millet**.

Kurthia Genus of Gram positive, obligately aerobic coccoid or rod-shaped **bacteria**. Occur in manure and stagnant water. *Kurthia zopfii* is responsible for the **spoilage** of **meat** and meat products.

Kuruma prawns Species of **prawns** (*Penaeus japonicus*) highly valued for its **flavour** and **texture**, particularly in Japan. Occurs in the Indian Ocean and the Southwestern Pacific Ocean from Japan to Australia; cultured in Japan and Australia.

Kusaya Traditional Japanese fish product consisting of dried, brined **mackerel**.

Kusum Seeds from the kernels of the tree *Schleichera trijuga*, **oils** extracted from which are rich in **arachidic acid** and may be used for culinary purposes.

Kuth Common name for *Saussurea lappa* or *Costus*. Used as a source of **inulin** in production of **inulinases** during **fermentation** of ***Aspergillus*** *niger*.

Kvass **Alcoholic beverages** originating in Russia, made by **fermentation** of **mashes** based on mixed **cereals** and **bread**.

Kwoka Non-fermented **corn** product popular in Nigeria.

Kylar Alternative term for the plant growth regulator **daminozide**.

L

Laban Alternative term for **leben**.

Labban Alternative term for **leben**.

Labelling Process of attaching labels to items to make them identifiable, or the information included on the labels. For foods, information may include **bar codes**, brand names, **trademarks**, illustrative matter, and compositional and nutritional details.

Labels Pieces of paper, plastics or fabric which are attached to and provide information about an item. For foods, this information may include **branding** or the **trademarks** of a food company, the geographical origin, **date marking**, compositional details, **health claims**, nutritional values and warnings relating to specific ingredients, e.g. nuts. The content of information on food labels is often governed by legislation.

Lablab beans Seeds of *Lablab niger* or *Dolichos lablab*. Rich in proteins and carbohydrates. Young and mature seeds as well as young pods are consumed. Also known as hyacinth beans and bonavist beans.

Labneh Strained concentrated **yoghurt** product popular in the Middle East. Also called yoghurt cheese and labaneh.

Laccases EC 1.10.3.2. Ligninolytic multicopper-containing **enzymes** that catalyse the oxidation of **phenols** and non-phenols with concomitant reduction of molecular oxygen. Can be used for removing phenols from **fruit juices**, as well as olive oil waste water and other **effluents**, detoxification of lignocellulosic hydrolysates and gelation of sugar beet **pectins** for use in foods. Can cause oxidative **spoilage** of **wines**.

Lacon Traditional Spanish dry cured **pork** foreleg product made by a process similar to that used in production of **dry cured ham**. The specific designation Lacon Gallego is used to indicate its geographical origin, the Galicia region.

Lacquers Liquids consisting of resins, cellulose esters, shellac or similar synthetic substances dissolved in a solvent, such as ethyl alcohol. Dry to form shiny, hard, protective or decorative coatings for plastics, wood, metals and other products.

Lactacins **Bacteriocins** synthesized by *Lactobacillus* spp. that are inhibitory only to other lactobacilli. Lactacin A is produced by *L. delbrueckii subsp. lactis* JCM 1106 and 1107. It has a narrow host range and is heat labile. Lactacin B is produced by *L. acidophilus* N2, and its synthesis is chromosomally linked. This protein forms aggregates of molecular weight 100,000 Da; however, the actual molecular weight of lactacin B is 6000-6500 Da. Lactacin F is produced by *L. acidophilus* 88, and its synthesis is plasmid linked. It has a broader activity range than lactacin B, and forms aggregates of molecular weight 180,000 Da; however, the actual molecular weight of lactacin F is 25,000 Da.

α-Lactalbumin One of the major **whey proteins**, accounting for approximately 20% of total whey proteins in **cow milk**. Rich in **tryptophan** and **cystine**. Found in genetic variants A, B and C that differ in **amino acids** composition and have a bearing on the properties and yield of **milk**.

Lactalbumins **Albumins** present in **milk**. The main protein is **α-lactalbumin**.

β-Lactam antibiotics Large group of **antibiotics** comprising naturally occurring and semisynthetic **penicillins**; the most widely used antimicrobial drugs in veterinary practice. Commonly classified into 4 groups according to antimicrobial properties: aminocillins; **cephalosporins**; carbapenems; and monobactams.

β-Lactamases EC 3.5.2.6. Hydrolases that hydrolyse β-lactam **antibiotics**. Those that act on **penicillins** are sometime known as penicillinases.

Lactarius **Edible fungi** also known as milk cap **mushrooms**. Commonly consumed species include *Lactarius deliciosus, L. helvus, L. trivialis* and *L. sanguifluus*. While most edible types are eaten cooked, some species are dried and used in **condiments**, and others are pickled or salted.

Lactases Alternative term for **β-galactosidases**.

Lactate dehydrogenases EC 1.1.1.27 (L-lactate dehydrogenases) and EC 1.1.1.28 (D-lactate dehydrogenases). Catalyse the conversion of **pyruvic acid** into (*S*)- and (*R*)-lactic acid, respectively. Involved in **lactic acid** biosynthesis and useful for determination of D- and L-lactic acid levels in **beverages**, and for detection of lactic acid spoilage **bacteria** in **beer**.

Lactate 2-monooxygenases EC 1.13.12.4. Flavoproteins that convert (*S*)-lactate to acetate. Have been used to construct **lactic acid biosensors** and for production of D-lactate from a racemic mixture. Also known as lactate oxidases.

Lactate oxidases Alternative term for **lactate 2-monooxygenases**.

Lactates Salts or esters of **lactic acid**. Lactates such as **sodium lactate** are widely used in foods as **preservatives**, whilst calcium or iron lactates can be used in food **fortification**. Lactate concentrations are frequently determined in foods as a measure of lactic acid levels.

Lactation Physiological process involving secretion of **milk** from the mammary gland, usually beginning at the end of pregnancy and controlled by the hormones prolactin and oxytocin. At the beginning of lactation, **colostrum** is produced, mature milk being secreted later. In cows, milk yield as well as composition varies during lactation. Yield increases up to the 2nd month of lactation and decreases thereafter. Milk protein and fat contents are lowest during the 2nd month, then increase. Free **fatty acids** contents and proportions of **stearic acid**, **oleic acid** and **linolenic acid** in milk fat increase as lactation progresses, while proportions of short- and medium-chain fatty acids and **linoleic acid** decrease. **Lactose** content of milk decreases as lactation proceeds. Contents of **immunoglobulins**, **minerals** and **trace elements**, and activities of some **enzymes** increase towards the end of lactation.

Lactation number Value defining the number of lactations undergone by an animal. Can affect **physicochemical properties** and **functional properties** of **milk**.

Lactation stage Measure of the number of weeks of **lactation** that have passed since parturition. Lactation is generally divided into three stages during which three distinct secretions are produced: **colostrum**; transient milk; and mature milk. Colostrum is produced for approximately the 1st week, transient milk for the following 2-3 weeks and mature milk is produced thereafter.

Lactic acid α-Hydroxypropionic acid. One of the **organic acids** present in **sour milk**, **molasses**, **fruits**, **beer** and **wines**. Produced via **lactic fermentation** of **sugars** by **lactic acid bacteria**, a process that is an important step in manufacture of **cheese**, **yoghurt** and other acidic **fermented dairy products**. Also used for acidulating **worts** in **brewing** and in **preservation** of **meat** products, such as **salami** and **pepperoni**.

Lactic acid bacteria **Gram positive bacteria** (e.g. **Lactobacillus**, **Lactococcus**, **Leuconostoc**, **Pediococcus** and **Streptococcus** species) that are capable of **lactic fermentation** of sugar substrates. Used extensively in the food industry as **starters** to initiate lactic acid fermentation in the production of **fermented dairy products** (e.g. **yoghurt** and **cheese**), fermented meat products (e.g. **salami**), and fermented plant products (e.g. **sauerkraut** and **sourdough**).

Lactic beverages Beverages, manufacture of which includes **lactic fermentation**.

Lactic fermentation Process by which certain **bacteria**, such as **lactic acid bacteria**, convert **sugars** entirely, or almost entirely, to **lactic acid** (homolactic fermentation) or to a mixture of lactic acid and other products (heterolactic fermentation). Lactic acid bacteria produce either L(+)- or D(-)-lactic acid or both, depending on the specificity of the NAD-dependent lactate dehydrogenases present.

Lacticins **Bacteriocins** synthesized by **Lactococcus** spp. Lacticin 481 is synthesized by *L. lactis* subsp. *lactis* CNRZ 481, and is inhibitory towards strains of *Lactococcus*, **Lactobacillus**, **Leuconostoc** and **Clostridium** *tyrobutyricum*. Lacticin 481 is produced optimally at pH 5.5. The molecule has a relative molecular mass of 1700, is rich in nonpolar amino acids and contains the unusual amino acid **lanthionine**.

Lactic starters Starters containing **lactic acid bacteria**.

Lactic streptococci Bacteria of the genus **Streptococcus** capable of **lactic fermentation**, and therefore often used as **starters** in the production of **fermented foods**. *S. salivarius* subspecies *thermophilus* is used in **starters** for production of **yoghurt**.

Lactitol Polyol, with the systematic name 4-O-β-galactopyranosyl-β-D-sorbitol, present in **milk**. May be isolated from **whey** or manufactured by **hydrogenation** of lactose. Has approximately 20% of the **sweetness** of **sucrose** and is used in **sweeteners** and **bulking agents** for **sugar confectionery**, **ice cream** and **jams**. Lactitol is not readily absorbed by the **gastrointestinal tract** and thus may be used in **low calorie foods** and foods suitable for consumption by diabetics.

Lactobacillaceae Family of Gram positive, anaerobic or facultatively anaerobic rod-shaped or coccoid **lactic acid bacteria**. Occur in the mouth and **gastrointestinal tract** of humans and animals, in food (e.g. **dairy products**) and in fermenting **vegetable juices**. Includes the genera **Lactobacillus**, **Carnobacterium**, **Leuconostoc**, **Oenococcus** and **Pediococcus**.

Lactobacillus Genus of Gram positive, anaerobic or facultatively anaerobic, rod-shaped **lactic acid bacteria** of the family **Lactobacillaceae**. Occur in foods and **beverages** (e.g. **wines**, **beer**, **fruits**, **meat** products, **dairy products**) and in the mouth and **gastrointestinal tract** of humans and animals. Used as **starters** in the manufacture of **fermented foods** and **fermented beverages** (e.g. *Lactobacillus acidophilus* in the manufacture of **kefir**, *L. plantarum* in the manufacture of **sauerkraut** and *L. delbrueckii* subspecies *bulgaricus* in the manufacture of **cheese**). Some species may be responsible for **spoilage** of beer, **meat**, **milk** and wines.

Lactocins **Bacteriocins** produced by *Lactobacillus* spp. Lactocin 27 is produced *by L. helveticus* LP27, and is a 12,400 Da glycoprotein inhibitor with a narrow spectrum of activity (restricted to *L. helveticus* and *L. acidophilus*). Lactocin 27 exerts a bacteriostatic effect rather than being bactericidal in activity. Lactocin 27 is very heat stable and is not thought to be plasmid encoded. Lactocin S is produced by *L. sake* L45. It has antimicrobial activity against other lactobacilli, *L. mesenteroides* and **Pediococcus** sp., and is moderately heat stable. Lactocin S has an estimated molecular weight of <13,700 Da, and is plasmid encoded.

Lactococcins Plasmid encoded **bacteriocins** produced by *Lactococcus* spp. that are small and heat stable (lactococcins A, B and M). Lactococcins have a narrow host range, against lactococcal strains only. Their mechanism of action is on the bacterial membrane of susceptible organisms. *L. lactis* subsp. *cremoris* 9B4 produces two bacteriocins, one with low potency and one with high potency (lactococcin A).

Lactococcus Genus of Gram positive, facultatively anaerobic coccoid **lactic acid bacteria** of the family Streptococcaceae. Occur in **milk** and **dairy products**. Used extensively as **starters** (e.g. *Lactococcus lactis* subspecies *lactis* and *L. lactis* subspecies *cremoris* strains) in the manufacture of **fermented dairy products** (e.g. **cheese** and **fermented milk**).

Lactoferrin Transferrin found in the **milk** of most mammals. Sometimes called lactotransferrin. In common with other **transferrins**, it binds **iron**, giving it a red tinge. Characteristics of lactoferrin are similar in human and cow milks, but amounts are much higher in **human milk**, where lactoferrin accounts for up to 20% of total protein. **Antibacterial activity** of lactoferrin is attributed to its ability to bind iron, making the nutrient unavailable for microbial growth. Gram negative **bacteria** are inhibited particularly due to high iron requirements. This antibacterial action is particularly valuable in preventing gastrointestinal infections in the newborn. Additional physiological functions suggested for lactoferrin include regulation of iron transport and absorption, and participation along with other proteins, e.g. **immunoglobulins** and **lysozymes**, in local immunity.

β-Lactoglobulin One of the major **whey proteins**, accounting for approximately 50% of total whey proteins in **cow milk**. Small globular protein rich in **methionine**. Exists as a dimer at neutral **pH**, with one free thiol group and two disulfide bridges. Several genetic variants that affect milk properties and yield have been identified in cow milk, but variants A and B are most common. Often used as a surfactant in food **dispersions** such as **emulsions** to stabilize polyphasic systems.

Lactoglobulins Globulins found in **milk**. The main protein is **β-lactoglobulin** which accounts for approximately 50% of the total content of **whey proteins**.

Lactones **Heterocyclic compounds** containing intramolecular cyclic esters formed by a condensation reaction between two hydroxy carboxylic acid substituents. Examples include the acidulant **glucono-δ-lactone** and the **aroma compounds γ-decalactone** and **coumarin**.

Lactoperoxidase systems Antimicrobial systems that occur naturally in **raw milk**, consisting of lactoperoxidases, thiocyanate (the major antimicrobial agent) and H_2O_2. The systems can be activated by addition of exogenous thiocyanate and H_2O_2 in order to increase the storage time of raw milk. They may also be useful for extending the **shelf life** of other foods.

Lactose Disaccharide **sugar** comprising **glucose** and **galactose** residues linked by glycosidic bonds. Predominant sugar in **milk** which may be recovered from **whey** by removal of **whey proteins** and minerals, followed by **crystallization** of lactose monohydrate and, to a lesser extent, anhydrous lactose. Has low **sweetness**, approximately 16-20% that of **sucrose**, and is used in **sweeteners** for **sugar confectionery** and **infant formulas**. Some individuals suffer from **lactose intolerance** due to an inability to digest this sugar. Also known as milk sugar.

Lactose intolerance Impaired ability to digest the disaccharide **lactose** due to lack of lactases (**β-galactosidases**) in the small intestinal mucosa. Undigested lactose remains in the intestinal contents, and is fermented by **bacteria** in the colon, resulting in explosive and watery diarrhoea. Treatment is to omit lactose from the diet.

Lactose synthases EC 2.4.1.22. Catalyse the transfer of **galactose** from UDPgalactose to D-glucose, forming **lactose**. These **enzymes** are complexes of *N*-acetyllactosamine synthases (EC 2.4.1.90) and **α-lactalbumin**. In the absence of α-lactalbumin, the enzymes catalyse the transfer of galactose from UDPgalactose to *N*-acetylglucosamine.

Lactose syrups Syrups consisting predominantly of **lactose**. Manufactured from **whey** by removal of **whey proteins** and minerals using **ultrafiltration** and **ion exchange chromatography**, respectively. Used as **sweeteners** in **dairy products**, **infant formulas** and **sugar confectionery**.

Lactosucrose Oligosaccharide with the systematic name 4-G-β-D-galactosylsucrose, which promotes the growth of **Bifidobacterium** in the human **gastrointestinal tract**. It also inhibits growth of harmful **bacteria**. Made commercially from **sucrose** and **lactose** in a reaction catalysed by **β-fructofuranosidases**. Used as a low-calorie sweetener in foods and **beverages**, including **soft drinks**, **bakery products** and **sugar confectionery**, and as a component of **functional foods**.

Lactulose Nutritive sweetener produced by **isomerization** of **lactose** which has 1.5× the **sweetness** of lactose.

Lactylates **Salts** or **esters** of lactyl lactate. Include **stearoyl lactylates**, which are used in the food industry as **emulsifiers**.

Laevulose Alternative term for **fructose**.

Lager Type of **beer**, originating in Central Europe but now popular worldwide. Made by **fermentation** with **bottom fermenting yeasts**.

Lake water Water derived from lakes. After treatment, it may be used as **drinking water**.

Lamb **Meat** derived from young **sheep**; also animals from which lamb is produced. Lamb is pink in **colour** and has creamy-white fat, which has a firm, dry **texture**. The older the sheep, the coarser the texture and stronger the flavour of lamb. In many cultures, consumption of lamb is associated with festivals and religious ceremonies.

Lambanog **Spirits** made in the Philippines from fermented sap of coconut **palms**.

Lamb chops Thick slices of lamb, usually including an 'eye' of meat, a rib and a layer of subcutaneous fat.

Lamb cutlets Portions of **lamb**, particularly lamb chops from just behind the neck. Lamb cutlets are usually cooked by **grilling** or **frying**.

Lambic Belgian **beer**, made by a slow spontaneous **fermentation** process. Frequently flavoured with **fruits** such as **cherries**, **raspberries** or **peaches**.

Lamb kidneys Paired abdominal excretory organs, which form part of the edible **offal** from lamb **carcasses**. Kidneys from young animals, particularly lambs, are more tender than those of older animals. Lamb kidneys, along with those from calves, have a more delicate **flavour**, whilst **cattle kidneys** and **swine kidneys** have a stronger flavour.

Lamb mince **Meat mince** prepared from lamb. Lamb mince is more delicately flavoured and has a lower fat content than pork mince or beef mince. Also known as ground lamb or minced lamb.

Lamb sausages **Sausages** made from **lamb** or **mutton**. Fresh lamb sausages may be seasoned with **rosemary** or **mint**. Dry fermented lamb sausages are often prepared using lean meat from older sheep.

Laminaria Genus containing several species of large brown **seaweeds**, most of which are utilized for food purposes. Rich source of **minerals**, particularly calcium, potassium, magnesium, iron and trace minerals such as manganese, copper and zinc. Popular in Japan, where some species are used to flavour dashi, a soup stock. Used as a source of **alginates** for the food industry. Usually sold dried, in strips or sheets. Also known as kelp, **kombu** and tangle.

Laminarin β-1,3-Glucan which acts as a reserve polysaccharide in *Laminaria* spp. Located in membrane-bound vesicles. Possesses **hypolipaemic activity** and anticoagulant properties. Used as a substrate for detecting β-1,3-glucanase activity.

Laminarinases Alternative term for **endo-1,3(4)-β-glucanases** and **glucan endo-1,3-β-D-glucosidases**.

Laminates Materials made up of several layers of reinforcing fibres produced by placing layer on layer and bonding the sheets together, usually with heat or pressure. Laminates include fibreglass, plywood and reinforced plastics.

Lamprey **Fish** species (*Lampetra fluviatilis*) of minor commercial importance. Found in coastal waters and rivers and as a freshwater inhabitant of lakes in Europe. Since mucus and serum are poisonous, flesh must be washed thoroughly before consumption. Fresh and smoked fish are eaten fried.

Land snails A large group of creeping terrestrial gastropod molluscs. Several species are harvested from the wild or farmed as a source of **snail meat**.

Langoustines French name for **Norway lobsters** (*Nephrops norvegicus*). Also known as **scampi**.

Langsat **Fruits** produced by *Lansium domesticum*. White flesh is juicy and aromatic. Usually eaten out of hand, but can also be used in **cooking**. Also known as lanzones, lanzons and ayer-ayer.

Lannate Alternative term for the insecticide **methomyl**.

Lanternfish Any of a number of relatively small deepwater **marine fish** species from the family Mictophidae. Widely distributed around the oceans of the world. Some species are utilized as food fish.

Lanthanides Group of **elements** with atomic numbers 57-71, of which cerium is the most abundant. All have similar physical and chemical properties. Used widely in industry, e.g. in alloys and magnets and as **catalysts**; used in **fertilizers** for food crops in some countries. Due to their widespread use, lanthanides can enter the food chain as pollutants, **wild mushrooms** being particularly susceptible to accumulation, although trace amounts are found in many foods. Also known as rare earth elements.

Lanthionine Di(α-amino acid) formed from the **amino acids alanine** and **cysteine**, synonyms include *S*-(alanin-3-yl)-L-cysteine and 2,2′-diamino-3,3′-thiobis(propionic acid). The lanthionine skeleton occurs in **lantibiotics**, a group of polypeptide **bacteriocins** synthesized by **Gram positive bacteria**.

Lantibiotics Plasmid encoded **bacteriocins** produced by *Lactococcus* lactis, consisting of small membrane active peptides (<5 kDa) containing the **amino acids lanthionine**, β-methyl lanthionine and other dehydro residues. Heat sensitive at pH 9.4, and act on a wide host range of **Gram positive bacteria**. Include **nisin** and lacticin 481.

Lanzones Alternative term for **langsat**.

Lao-chao Traditional fermented **rice** product.

Lard Soft, white solid fat traditionally obtained by **rendering** or **melting** the internal **fats** from swine. Rich in a number of **fatty acids**, including *sn*-2 palmitic, stearic, oleic and linoleic acids; contains **cholesterol**. Has a bland **flavour** and odour. Used in **cooking** and **baking**.

Lasagne Rectangular sheets of **pasta**. Usually eaten layered with meat or vegetables and cheese **sauces** and baked.

Lasalocids Polyether **antibiotics** which are **coccidiostats** used widely in the control of **coccidiosis** in poultry; also used as **growth promoters** in cattle. Rapidly metabolized in animals and residues are normally absent from all tissues except **livers** within 7 days post-treatment. Residues may accumulate to relatively high levels in **eggs**.

La Serena cheese Spanish **cheese** made from raw **ewe milk** using **vegetable rennets** prepared from thistles. Has a semi-hard rind, a soft to semi-hard curd and a minimum fat content of 50%.

Lasoda fruit **Fruits** produced by *Cordia myxa*. Harvested green and used in making **pickles**.

Lassi Sweetened **fermented milk** beverage popular in India. Prepared by stirring **sugar**, water and **flavourings** into **dahi**, giving a viscous, white, mild to highly acidic drink.

Laurel Common name for the plant *Laurus nobilis*, leaves and berries of which are used as **flavourings** in **sauces**, **pickles** and **seasonings**. Imparts a sweet, spicy **flavour**. Leaves should not be eaten due to the presence of the carcinogen **safrole**. Also termed **bay**, bay laurel or sweet bay.

Laurencia Genus of red **seaweeds** found on rocky shores around the world. Some species (e.g. *Laurencia obtusa*) show commercial potential as a source of **polysaccharides** for food **thickeners** and **stabilizers**.

Lauric acid One of the medium-chain **saturated fatty acids**. Contains 12 carbon atoms and has a m.p. of 44°C. Synonym, **dodecanoic acid**. Slight odour of **bay** oil. Occurs as a triacylglycerol component of **milk fats** and **vegetable oils** including **rapeseed oils** and **palm oils**, and is a component of several **cocoa butter substitutes**. Identified as an aroma component in **cheese**.

Lautering Separation of **worts** from insoluble material in **brewing mashes** by running off the worts through the perforated bottom of **lauter tuns**, in which the insoluble solids are retained.

Lauter tuns Circular vessels equipped with a perforated or wire mesh base and rotating stirrer arms, used for the **lautering** process.

Lavender Common name for plants of the genus *Lavandula*. Used mainly as a source of **essential oils** which are used medicinally and in aromatherapy, but also a flavouring ingredient in foods. Applications include **sauces**, **dressings**, **cookies** and **herb tea**. Lavender flowers are the botanical source of popular monofloral **honeys**.

Laver Name given to dried, edible **seaweeds** of the genera *Porphyra* and *Ulva*.

Laverbread Product made from red **seaweeds** of the genus *Porphyra*. Prepared by boiling in **brines**, cooling and chopping; often fried prior to consumption.

Lben Alternative term for **leben**.

LC Abbreviation for **liquid chromatography**.

LDPE Abbreviation for **low density polyethylene**.

Lead One of the **heavy metals**, chemical symbol Pb. The main source of lead for humans is dietary; lead can be present as a contaminant in both foods and **beverages**, including water. Following consumption, lead is accumulated predominantly in bones and teeth. Some research suggests an essential role for lead in mammalian nutrition, although further work is needed to define its role precisely. In excess, lead causes a range of **toxicity** problems including anaemia, encephalopathy, neuropathy and renal dysfunction.

Leaf beet Common name for *Beta vulgaris*. Leaves, including the stalk, are eaten as a green vegetable in a similar way to **spinach**. Used raw in **salads**, boiled as a vegetable and in savoury dishes. Also known as **Swiss chard**, chard, white beet, spinach beet and silver beet. Good source of **vitamin A**, **vitamin C** and **iron**.

Leaf proteins Proteins contained in plant leaves, a very good source of protein in the diet.

Leaf vegetables Plants in which the edible parts are the leaves.

Leafy vegetables Leafy plants, the stems and leaves of which are used as vegetables.

Lean The part of **meat** which contains very little fat.

Leavening The process by which **dough** is made to rise due to **fermentation** by **yeasts**.

Leban Alternative term for **leben**.

Lebaycid Alternative term for the insecticide **fenthion**.

Leben A **fermented milk** similar to **yoghurt** produced in North Africa and Asia Minor. Since the **starters** used include **yeasts**, the product contains some **ethanol**. Also known as laban, labban, lben, leban and lebben.

Lecithinases Lecithinases A, C and D, alternative names for **phospholipases** A$_2$, C and D, respectively; lecithinase B, alternative term for **lysophospholipases**.

Lecithins Products comprising **phospholipids**. Composed of phosphate esters of **diglycerides** (mostly **oleic acid**, **palmitic acid** and/or **stearic acid**) esterified to choline via the phosphate group. Due to the presence of both polar and non-polar moieties, the molecule forms micelles and has uses as food **emulsifiers**. Prevalent in **soybeans** and **egg yolks**; by-products in manufacture of **soybean oils**. Lecithin is also called phosphatidylcholine.

Lectins Carbohydrate-binding **proteins** or **glycoproteins**, synonyms include **phytohaemagglutinins** and agglutinins. Lectins are of non-immune origin and agglutinate cells and/or precipitate glycoconjugates. Found in many plant foods and can have detrimental properties as **antinutritional factors** and **toxins**, or possible beneficial properties including **antitumour activity**. Lectins are widely used analytically as specific binding and separating agents.

Leeks Common name for *Allium ampeloprasum* var. *porrum* or *A. porrum*. Lower part is eaten as a vegetable or used as an ingredient in **soups** and stews.

Lees Sediments of **yeasts** and other insoluble material formed at the bottom of **containers** of **wines**.

Legionella Genus of Gram negative, aerobic rod-shaped **bacteria** of the family Legionellaceae. Occur in aquatic habitats, including domestic water systems (e.g. air-conditioning cooling towers, showers and nebulizers), surface waters, moist soils and thermally polluted streams. *Legionella pneumophila* is the causative agent of **legionellosis** in humans.

Legionellosis Medical name for **Legionnaires disease**.

Legionnaires disease Severe contagious disease caused by *Legionella pneumonophila*, characterized by influenza-like symptoms, high fever, chills, headache, pleurisy, pneumonia and sometimes death. Infection occurs through inhalation of contaminated aerosols (e.g. from air-conditioning cooling towers, showers and nebulizers).

Legume meal **Flour** made from seeds of legume plants.

Legume proteins Proteins formed in legume seeds, a very good dietary source of protein.

Legumes **Vegetables** of the family Leguminosae (Fabaceae). The seeds or beans are contained in pods. Edible products include dry seeds (**beans** or **pulses**), immature green seeds, **oilseeds** (such as **soybeans**), green pods, **spices**, shoots, leaves and sprouts. Rich sources of good quality proteins, but generally low in fat (exceptions include **peanuts**, soybeans and **chick peas**). Also good sources of **dietary fibre** and some B vitamins. **Carotenes**, **vitamin C** and **vitamin E** can be obtained from immature seeds, pods, leaves and sprouts. Some seeds also contain **antinutritional factors** or **toxins** that can cause diseases. These can usually be destroyed by careful processing of the seeds.

Legume sprouts Produced by **germination** of legume seeds, commonly **mung beans**, **alfalfa**, **lentils**, **soybeans** and **black gram**. Rich in protein, **vitamins** and **minerals**. Fresh sprouts are crisp and tender, and are often eaten raw. In dishes, they are cooked for a short period only to avoid wilting. Also available canned.

Legumin One of the **storage proteins** formed in **seeds** of **legumes**.

Lemonade Effervescent or still **beverages** made from **lemon juices**, or, more generally, **carbonated beverages** with a lemon **flavour**. May be added to **spirits** before consumption.

Lemon balm **Spices** also called bee balm, melissa, bee herb, balm mint and balm gentle. Lemon balm leaves possess a citrus-like **aroma** and can be added directly to foods without further processing.

Lemon essential oils Distillates of **lemon peel** used as **flavourings**. The active component of lemon oils is **citral**, a mixture of the terpene **aldehydes neral** and **geranial**.

Lemon grass Alternative term for **lemongrass**.

Lemongrass **Spices** made from the grasses *Cymbopogon flexosus* or *Andropogon nardus* (East Indian lemongrass) or *Cymbopogon citratus* (West Indian lemongrass). Used as **flavourings** in Asian cuisine and particularly in Thai dishes. The characteristic flavour compound of lemongrass is **citral**.

Lemon grass oils Alternative term for **lemongrass oils**.

Lemongrass oils **Essential oils** produced by steam **distillation** of fresh **lemongrass**, comprising approximately 65-75% **citral**.

Lemon juices **Fruit juices** prepared from **lemons** (*Citrus limon*). Used in **beverages** and as a flavouring ingredient in **cooking**.

Lemon peel Outer skin of **lemons**. Used to make candied peel, as a garnish and to add **flavour** to a range of sweet and savoury dishes.

Lemons Yellow **citrus fruits** (*Citrus limon*) that are extremely rich in **vitamin C**. Total sugar content is relatively low for a citrus fruit. Its citric acid content of approximately 5% makes it too acidic for eating as a dessert. However, **lemon juices** are widely used as food and beverage **flavourings**, and **lemon peel** is also used in foods.

Lemon tea **Tea beverages** with the **flavour** of **lemons**.

Lentils **Seeds** of the **legumes** *Lens culinaris* or *L. esculenta*, rich in proteins and carbohydrates. Used to make **dhal**, in **soups** or in **snack foods**. **Flour** made from the seeds can be used as an ingredient in **cakes** and **infant foods**. Young pods of the plant are eaten as a vegetable.

Lentinus **Edible fungi**, the most commonly consumed example being shiitake or Japanese black forest **mushrooms** (*Lentinus edodes*, renamed *Lentinula edodes*).

Lepiota Genus of **fungi** including a number of edible species.

Leptospira Genus of Gram negative, obligately aerobic spiral-shaped **bacteria** of the family Leptospiraceae. Occur in fresh water, salt water and soil, and as **parasites** in domestic and wild animals and humans. Serotypes of *Leptospira interrogans* are the causative agents of **leptospirosis** in humans. Transmission to humans is usually via direct contact with animals or animal **carcasses**, or via exposure to water contaminated with the urine of infected animals. People at risk of infection include those whose work brings them into contact with animals (e.g. farmers, meat handlers, veterinarians), and those exposed to urine-contaminated streams, rivers and standing water.

Leptospirosis Disease affecting animals and humans caused by infection with various serotypes of ***Leptospira*** *interrogans*. Carried by a wide range of animals, infection in humans being caused by contact with the animals, **carcasses**, fluids such as **milk** or water contaminated with animal urine.

Lettuces Common name for *Lactuca sativa*. Generally used as a salad plant, but sometimes eaten as a vegetable. Good source of **fibre**, potassium, **β-carotene**, **vitamin E** and **vitamin C**. Some cultivars have red pigmentation.

Leucaena Genus of **legumes**. Seeds of some species, mainly *Leucaena leucocephala* and *L. glauca*, are used as food and as a source of **gums**; leaves and pods are also eaten. However, proteins in leaves, pods and seeds contain the toxic amino acid mimosine, which can be destroyed by heating.

Leucine One of the essential **amino acids**. A common protein constituent and free amino acid in many foods. Leucine is also a precursor of several **aroma compounds** and participates in the **Maillard reaction**. Produced industrially by **fermentation** of ***Corynebacterium*** *glutamicum*.

Leucoanthocyanidins **Anthocyanidins** found in a range of plant foods.

Leucoanthocyanins **Anthocyanins** found in a range of plant foods, and also in **wines**. In a polymerized form, constituents of **polyphenols** and condensed **tannins**.

Leucocyanidin Anthocyanidin found in plant foods and derived products including **beer**. Present in a polymerized form in **polyphenols** and condensed **tannins**.

Leucocytes Alternative spelling of **leukocytes**.

Leuconostoc Genus of Gram positive, facultatively anaerobic coccoid **lactic acid bacteria** of the family Streptococcaceae. Occur in **dairy products** and in fermenting **vegetables** and **fermented beverages**. Species may be used as **starters** in the production of **fermented foods**. *Leuconostoc mesenteroides* subspecies *cremoris* strains are used as starter cultures in the production of **fermented dairy products** (e.g. **fermented cream**, **cheese**, **kefir**, **buttermilk**).

Leukocytes White, nucleated blood cells that lack **haemoglobin**, which are found in blood and lymph. Formed in lymph nodes and bone marrow. Can produce **antibodies** and move through the walls of vessels to migrate to the sites of injuries, where they surround and isolate dead tissue, foreign bodies and bacteria. There are two major types: those with granular cytoplasm (granulocytes), which include basophils and neutrophils; and those without granular cytoplasm, such as lymphocytes and monocytes.

Levanases EC 3.2.1.65. Catalyse the random hydrolysis of 2,6-β-D-fructofuranosidic linkages in 2,6-β-D-fructans (**levans**) containing more than 3 **fructose** units. Useful for production of **fructooligosaccharides**.

Levans Fructose-based **polysaccharides**, these β2→6 fructans are synthesized by **bacteria**, commonly *Zymomonas mobilis*. As high molecular weight polymers, levans have similar potential applications in foods as **dextran**; they are also used in the production of **fructooligosaccharides**, including **kestose**.

Levansucrases EC 2.4.1.10. Transfer a fructosyl group from **sucrose** to 2,6-β-D-fructans (**levans**), increasing the chain length by one fructosyl unit. Useful for production of **fructooligosaccharides**.

Lichenases Alternative term for **licheninases**.

Licheninases EC 3.2.1.73. Hydrolyse 1,4-β-D-glucosidic linkages in β-D-glucans containing 1,3- and 1,4-bonds. Act on lichenin and cereal **β-glucans**, but not on β-D-glucans containing only 1,3- or 1,4-bonds. Used in the **brewing** industry to hydrolyse mixed β-glucans during **malting** and brewing, and potentially useful for production of probiotic **oligosaccharides**.

Lichens Composite, plant-like organisms of the division *Lichenes* formed by the symbiotic association of **fungi** and **algae**. Form crusty patches or bushy growths on areas such as tree trunks and rocks. Used mainly as a source of **dyes**, but some species, such as *Parmelia nepalensis*, *Ramalina farinacea* and *Gyrophora esculenta* are eaten.

Life cycle assessment Assessment of the impacts associated with a system, function, product, or service over its entire life cycle. Sometimes considered to include four stages: initiation; inventory; impact analysis; and improvement.

Ligases EC 6. **Enzymes** that catalyse the joining of 2 molecules with concomitant hydrolysis of the pyrophosphate bond in **ATP** or a similar triphosphate. Important in the synthesis or repair of many biological molecules, such as **DNA**. Subdivided into enzymes that form carbon-oxygen bonds (EC 6.1), carbon-sulfur bonds (EC 6.2), carbon-nitrogen bonds (EC 6.3), carbon-carbon bonds (EC 6.4) and phosphoric ester bonds (EC 6.5).

Light Source of illumination that makes objects visible; electromagnetic radiation in the wavelength range 390-740 nm.

Lignans **Cinnamic acid** dimers in which the phenylpropane units are linked tail-to-tail. These **phenols** are present in many plant foods: **flax seeds** are a particularly good source of lignans, but they are also present in **cereals**, **vegetables**, **fruits**, **legumes**, **processed foods** and **beverages**. Lignans are of interest as **phytoestrogens**, and may play a role in the prevention of oestrogen-dependent cancer.

Lignin Random phenylpropanoid polymer component of plants, where it confers strength, rigidity and resistance to degradation. Lignin is one of the most abundant biopolymers, and a major component of insoluble **dietary fibre** in plant foods.

Ligninases Fungal **enzymes** involved in the degradation of **lignin**. Similar to **laccases**, **lignin peroxidases** and **manganese peroxidases**.

Lignin peroxidases **Lignin**-degrading **enzymes** potentially useful for lignin depolymerization, degradation of toxic pollutants and catalysis of difficult chemical transformations (e.g. during the production of **vanillin**). Thought to be required for the decoloration of **olive oil mills effluents** by **fungi**.

Lignocelluloses Complex of **lignin** and **celluloses** found in the cell walls of plants, and a component of **dietary fibre** in plant foods. Plant-derived wastes such as **pomaces** and **bagasse** contain lignocelluloses, and these wastes can be hydrolysed chemically or enzymically to release **sugars** which can be used as microbial **fermentation** substrates, for example for **ethanol** synthesis.

Lima beans Seeds produced by *Phaseolus lunatus*. Variable in size, shape and **colour**. Rich in proteins and a good source of **vitamin A**, **vitamin C**, some of the **vitamin B group**, **fibre** and **potassium**. As well as dried beans and immature beans (often canned or frozen), pods and leaves are also eaten. Mature seeds can contain toxic **hydrocyanic acid**, which is destroyed by **soaking** and **boiling** in water before consumption. Also known as butter beans, sieva beans and Madagascar beans.

Limburg cheese Belgian **soft cheese** made from **cow milk**. Sometimes called Limburger cheese. The washed rind is reddish-brown and the slightly sticky smear interior is yellow. It has a spicy and aromatic **flavour**, and a characteristic **aroma** caused by enzymes breaking down proteins on the cheese surface. Unripened cheese contains some holes, but ripened cheese has only a few, if any, small holes. The cheese ripens in 6-12 weeks and its fat content can be between 20 and 50%.

Lime berries Reddish **fruits** produced by *Triphasia trifolia* or *T. aurantiola*. Also known as limau kiah.

Lime essential oils **Essential oils** from **limes** produced by compression of **peel** or **distillation** of mashed lime pulps or juices. Used as **flavourings**, particularly in **carbonated beverages**, such as **cola beverages**. The predominant flavour compound present is **terpineol** which is produced from **citral** during distillation.

Lime juices **Fruit juices** prepared from **limes** (*Citrus aurantifolia*). Used in **beverages** and as a flavouring ingredient in **cooking**.

Limes Greenish-yellow **citrus fruits** (*Citrus aurantifolia*) which are rich in **vitamin C**. Total sugar content is relatively low for a citrus fruit and they are very acid. Used in **marmalades** and as **flavourings** in products such as **sauces**, **pickles** and **chutneys**. **Lime juices** are used in **beverages** and the peel is a source of **oils**. Cultivated mainly in warmer climates, as the plant is very sensitive to frost.

Liming One of several **sugar processes** used for **purification** of **sugar juices**. Involves addition of some form of lime, e.g. calcium oxide, milk of lime (a slurry of calcium hydroxide) or calcium saccharate, to sugar juices and heating. The lime neutralizes **organic acids** present and forms insoluble lime salts with the impurities. Suspended particles from the **sugar cane** or **sugar beets** that remain after filtration associate with the precipitate formed. Forms of liming include cold liming, hot liming and intermittent liming; these differ with respect to the order in which addition of lime and heating are carried out.

Limit dextrinases Alternative term for **α-dextrin endo-1,6-α-glucosidases** and **oligo-1,6-glucosidases**.

Limoncello Lemon liqueur traditionally made in Italy by soaking zest of **lemon peel** in alcohol, such as **vodka**, and adding sugar syrup. Commonly consumed on its own cold or iced, as an ingredient of longer drinks or poured over **ice cream** or **fruits**.

Limonene One of the monoterpenoid **aroma compounds**, with lemon-like odour. Found in **citrus fruits** and their products, including **citrus juices** and **citrus essential oils**. Also found in **dill** and **caraway** seeds.

Limonin One of the main **bitter compounds** found in **citrus fruits**. Limonin and other **limonoids** are highly oxygenated triterpenoids of interest as anticarcinogenic phytochemicals.

Limonoid glucosides **Limonoids** with carbohydrate (**glucose**) substituents; in contrast to limonoids, the **glucosides** are generally non-bitter. Over 17 different limonoid glucosides have been isolated from **citrus fruits**, and limonoids are mainly accumulated as glucoside derivatives in mature citrus fruit tissues. Along with limonoid aglycones, the glucosides show possible **anticarcinogenicity**.

Limonoids Highly oxygenated triterpenoids found predominantly in **citrus fruits**. Over 35 limonoids have been identified in citrus species, and many are **bitter compounds**. Limonoids possess anticarcinogenic activity and also antifeedant activity against insects and termites.

Limpets Any of a number of marine gastropod **molluscs** having compressed conical shells. Found attached to substrates on rocky shores worldwide. Meat is valued for its **flavour**, but generally has a tough **texture**. Consumed raw or lightly sauteed; meat is often tenderized prior to consumption.

Lin Alternative term for **tench**.

Linalool One of the monoterpenoid **aroma compounds**, with floral/sweet/citrus **aroma** characteristics. Linalool is found naturally in many foods and **beverages**, and is also added as a flavour compound to **processed foods**.

Linalyl acetate Ester with sweet/floral **aroma** characteristics. This flavour compound is found in several plant **essential oils**, including **bergamot oils**, sage oils and **citrus oils**.

Linamarases Alternative term for **β-glucosidases**.

Linamarin One of the cyanogenic glycosides, linamarin is found in **cassava** roots. This toxin has to be removed by processing, generally **fermentation**, before cassava can be eaten safely.

Lincomycin Antibiotic produced by *Streptomyces lincolnensis*. Used for initial treatment of mild to moderate staphylococcal infections in a variety of farm animals. Also used for growth promoting purposes. Distributes widely around the body, but normally reabsorbs completely from muscle tissue.

Lindane Alternative term for the insecticide **HCH**.

Ling **Marine fish** species (*Molva molva*) of high commercial importance belonging to the Lotidae (hakes and burbots) family. Widely distributed in the Atlantic Ocean. Marketed fresh, frozen and dried-salted. Cooked in a variety of ways, including **steaming**, **frying**, **broiling** and **baking**.

Ling cod **Marine fish** species (*Ophisdon elongatus*) of commercial importance belonging to the Hexagrammidae (greenlings) family. Also prized as a game fish. Found mainly in the northeast Pacific Ocean. Marketed fresh and frozen. Cooked in a variety of ways, including **steaming**, **frying**, **broiling**, **boiling**, **baking** and by **microwaves**. Ling cod **livers** are particularly rich in **vitamin A**.

Lingonberries Red, acid **berries** produced by *Vaccinium vitis-idaea*. Contain high levels of **benzoic acid**. Used in **jams** and jellies. Also known as **cowberries** or lingberries.

Linoleates Salts or **fatty acid esters** of **linoleic acid**. Also anionic form of linoleic acid. Often used in model systems to determine lipid oxidation or **antioxidative activity** of selected chemicals.

Linoleic acid One of the **polyunsaturated fatty acids**, synonym octadecadienoic acid. Contains 18 carbon atoms and 2 double bonds at positions 9 and 12. Linoleic acid is an essential nutrient in mammals, and is present in many plant and animal foods. However, there is some concern that excessive consumption of linoleic acid is harmful as it has been associated with increases in **cancer** and **coronary heart diseases**.

Linolenic acid One of the **polyunsaturated fatty acids**, synonym octadecatrienoic acid. Contains 18 carbon atoms and 3 double bonds at positions 9, 12 and 15 (α-linolenic acid) or at positions 6, 9 and 12 (γ-linolenic acid). α-Linoleic acid is an essential nutrient in mammals, and is found in many plant oils, especially **linseed oils**. γ-Linolenic acid is found in several plant oils, particularly in **evening primrose oils**, and is also found at low levels in animal lipids, including those of **human milk**. γ-Linolenic acid is a precursor for **arachidonic acid** and the **prostaglandins**.

Linseed oils Yellow to amber viscous **vegetable oils** obtained from **flax seeds**, *Linum usitatissimum*. Rich in iodine and α-linolenic acid. Polymerize on exposure to air, resulting in thickening. Used as a food oil. Also known as **flax seed oils**.

Linseeds Seeds derived from flax, *Linum usitatissimum*, used as the source of **linseed oils**.

Linuron Selective systemic herbicide used for pre- and post-emergence control of annual grasses and broad-leaved weeds around a range of plants, including **vegetables**, **cereals**, **coffee** and **tea**. Classified by Environmental Protection Agency as slightly toxic (Environmental Protection Agency III).

Lipaemic activity Human physiology term relating to the ability of certain compounds to either increase or decrease levels of lipids in the blood.

Lipases Enzymes that hydrolyse tri-, di- or mono-acylglycerols at a lipid-water interface to form free **fatty acids** and either di- or mono-glycerides, or free **glycerol**. The term usually refers to **triacylglycerol lipases**, which act on **triglycerides**. Can cleave various natural **lipids** and **oils**, such as **olive oils**, **soybean oils**, **coconut oils**, **butterfat**, and pork and beef **fats**, and can show positional-, fatty acid- or stereo-specificity. Useful for enhancing of **flavour** during **cheese ripening** and, due to their **esterification, interesterification** and **transesterification** activities, for production of modified **esters** and lipids, speciality fats and **cocoa butter substitutes**. Lipases are also active in organic solvents.

Lipids Naturally occurring organic chemicals that are characteristically poorly soluble in water but are soluble in organic solvents. Lipids constitute one of the four main classes of compounds found in living tissues, and also one of the major nutrient types, and as a class include **oils**, **fats**, **fatty acids**, long-chain (or fatty) **alcohols**, **triglycerides**, **phospholipids**,

waxes, **steroids**, **terpenoids** and some **hormones** and **vitamins**.

Lipolysis Hydrolysis (splitting) of **lipids** by **lipases** to yield **glycerol** and **fatty acids**.

Lipolytic enzymes Encompasses **lipases, lipoprotein lipases** and **phospholipases**.

Lipopolysaccharides Complexes formed between **polysaccharides** and **lipids**. Lipopolysaccharides are an important component of the outer membrane of **Gram negative bacteria** and are key determinants of **antigenicity** and **toxicity**.

Lipoprotein lipases EC 3.1.1.34. Also known as diacylglycerol lipases, these **enzymes** hydrolyse **triacylglycerols** to form **diacylglycerols** and free **fatty acids**. Can also hydrolyse diacylglycerols and exhibit **esterification** activity.

Lipoproteins Conjugated molecules containing proteins and lipids. The lipid may be a phospholipid, triglyceride or **cholesterol**, or a mixture of these. Serum lipoprotein and lipoprotein-cholesterol profiles are frequently measured as an indicator of diet and health. Also, oxidation of serum low density lipoprotein (LDL) is implicated in the aetiology of **coronary heart diseases**. Certain functional food constituents such as **flavonoids** from **green tea** and **red wines** have the ability to inhibit LDL oxidation due to their **antioxidative activity**.

Liposcelis Genus of **insects** of the family Psocidae common as pests in cereal stores in hot, humid areas. Can feed on **grain**, **oilseeds** and **pulses**. In large numbers, they may cause heating of grain with consequent damage to its quality and value. Also found in food manufacturing premises and domestic situations where conditions are favourable.

Lipovitellins Lipoproteins present in **egg yolks**.

Lipoxidases Alternative term for **lipoxygenases**.

Lipoxygenases EC 1.13.11.12. Found in virtually all **legumes**, these **enzymes** catalyse the oxidation of **unsaturated fatty acids** containing a *cis-cis* penta-1,4-diene unit to the corresponding monohydroperoxide. The preferred substrates are **linoleic acid, linolenic acid** and **arachidonic acid**. Although lipoxygenases are important for synthesis of **flavour compounds** in, e.g. **tomatoes** and **olive oils**, and can be used to introduce new flavours into foods, they can also contribute to food **spoilage** by production of a rancid **off flavour**.

Liqueurs Alcoholic beverages made from **spirits** or neutral **alcohol** with addition of other ingredients such as **sugar** and **flavourings**.

Liquid chromatography Alternative term for **high performance liquid chromatography**. Usually abbreviated to LC.

Liquid egg Pasteurized **egg whites**, **egg yolks** or whole **eggs** in liquid form. The long **shelf life** and **Salmonella**-free status of such products make them suitable for use by food manufacturers and caterers.

Liquid egg whites Pasteurized **egg whites** in liquid form. Processing conditions confer a long **shelf life** and ensure that they are free of **Salmonella** contamination. May be used in the manufacture of **meringues** and **cakes**.

Liquid egg yolks Pasteurized **egg yolks** in liquid form. Processing conditions confer a long **shelf life** and ensure that they are free of **Salmonella** contamination. May be used in the manufacture of **mayonnaise** and **salad dressings**.

Liquid membranes Thin layers of liquid, separating two phases: a process stream and a stripping phase. Impurities, e.g. metal ions, can be extracted almost completely by a carrier that is dissolved in the liquid membrane. On the other side of the membrane, stripping takes place. While the carrier is stripped continuously, the driving force for the extraction remains high. Types of liquid membranes in use include: bulk liquid membranes; emulsion liquid membranes; thin sheet supported liquid membranes; hollow fibre supported liquid membranes; two module hollow fibre supported liquid membranes; and spiral wound membranes.

Liquid nitrogen **Nitrogen** gas (N_2) that has been cooled to a temperature less than or equal to 77.4 K, thus existing in a liquefied state.

Liquid smoke Oil or water extracts of smoke produced from burning woods, often maple, oak or mesquite. Imparts a smoky **flavour** to foods.

Liquid whole egg Pasteurized blend of **egg whites** and **egg yolks** in liquid form. Processing conditions confer a long **shelf life** and ensure that the product is free of **Salmonella** contamination. May be used in the manufacture of **doughnuts**, **cookies**, **mayonnaise**, **salad dressings** and egg **noodles**.

Liquorice **Sugar confectionery** product made from the dried root extract of the Mediterranean plant *Glycyrrhiza glabra*. Contain the triterpenoid glycoside **glycyrrhizin**.

Listeria Genus of Gram positive, aerobic rod-shaped or coccoid **bacteria**. Occur in soil, fresh and salt water, sewage sludge and decaying vegetation. *Listeria monocytogenes*, the causative agent of **listeriosis** in humans, has been associated with foods such as **soft cheese**, **milk**, **ice cream**, raw **vegetables**, prepared **salads**, **cakes**, **fermented sausages**, sliced cold **meat**, and raw and smoked **fish**.

Listeriolysins **Toxins** produced by *Listeria monocytogenes* which lyse cells.

Listeriosis Infection in humans caused by *Listeria monocytogenes*. Usually transmitted by contaminated foods. Pregnant women, babies, the elderly and the immunocompromized are particularly susceptible to infection. Symptoms vary from a mild influenza-like illness with high fever and dizziness to meningitis and meningoencephalitis. In pregnant women, intrauterine or cervical infections may result in spontaneous abortion, stillbirth or premature birth. Gastrointestinal symptoms such as nausea, vomiting and diarrhoea may precede more serious forms of listeriosis or may be the only symptoms exhibited.

Litchis **Fruits** produced by *Litchi chinensis*. A rough, pink-red rind covers the white edible aril that encloses a single seed. The aril is a good source of **vitamins** (B, C, D and E). Available fresh, canned and frozen. Eaten alone or as a component of **sauces** and **compotes**. Also known as lychees, lechees, lichees and litchees.

Lite beverages **Beverages** with a low content of **alcohol** and/or **sugar** compared with conventional beverages of the same general type.

Lite foods Foods that are low (light) in **calories**, **fats**, **cholesterol**, **sugar** and/or **salt**.

Litesse Bulking agent and fat substitute which is only partially metabolized in the body. Derived from **polydextrose**, with small amounts of **sorbitol** and **citric acid**. Used in production of **low calorie foods**, including **bakery products**, **dairy products**, **salad dressings** and **confectionery** products.

Lithium Lithium, chemical symbol Li, is a group 1 alkali metal element, and may be one of the essential minerals, although lithium-dependent **enzymes**, **hormones** or other essential functions are not recognized. Animal studies suggest that lithium is essential for normal growth and reproduction. Lithium is found in a wide range of animal and plant foods, and is particularly rich in **eggs** and **milk**.

Liver flukes Parasitic flatworms of the class Trematoda that invade and cause damage to the livers of vertebrates, e.g. *Fasciola hepatica*.

Liver pates Meat products based on finely comminuted or mashed **livers**, often swine livers or poultry livers. Pre-scalding of the liver tends to increase redness of the pates. Commonly, liver pates are prepared from ingredients including scalded fat and have a spreadable texture; in contrast, when prepared using non-scalded fat they tend to have a sliceable texture.

Livers Large, vascular, abdominal organs. Livers of slaughtered animals and poultry form a part of edible **offal**. They can be cooked by grilling or sauteing, but become tough if overcooked. Livers are often processed to produce liver products including **foie gras**

and **fish liver oils**. Fish livers are particularly rich sources of vitamins A and D. Animal livers are good nutritional sources of **iron, vitamin A, vitamin B$_1$, vitamin B$_2$, niacin** and **vitamin B$_{12}$. Retinols** are stored in the liver and very high concentration occur in the livers of animals fed on retinol-supplemented feeds and among wild animals which eat fatty fish. Because of the teratogenicity of high concentration of retinols, pregnant women are often advised not to eat liver products.

Liver sausages Cooked, ready-to-eat **sausages** prepared from finely minced **swine livers** and other **meat**, and seasoned with **onions** and **spices**. Liver sausages may be prepared using smoked meat, such as **bacon**, or may be smoked after cooking. Their texture ranges from firm and sliceable to smooth and spreadable. Plastic bags or tubes are often used as casings for liver sausages, but other liver sausages are shaped into loaves. Usually, they are used to prepare **snack foods** or **sandwiches**. They are also known as **liverwurst** or leberwurst.

Liverwurst The German term for **liver sausages**, including the famous braunschweiger.

Livestock Domesticated animals reared for production of food (**meat, eggs** or **milk**), other animal products (wool, skins or fur) or for other commercial purposes.

Lizardfish Any of a number of **marine fish** species in the family Synodontidae. Widely distributed in warmer oceanic waters. A few species are fished commercially, principally off the coast of Japan. Marketed fresh and also used to make **kamaboko** products.

Loaf vol. Space occupied by **bread** as it rises during **baking**. Often measured in cubic centimetres. Used as a measure of **breadmaking** quality of **cereals, flour** and **dough**.

Lobsters Common name for several large marine **crustacea** belonging to the families Homaridae (including the large north Atlantic lobsters of the genus *Hommarus*) and Palinuridae (including rock lobsters and **spiny lobsters**). Many species are of high commercial value as they are prized for their flesh.

Locust bean gums **Gums** extracted from young **carob beans** (*Ceratonia siliqua*). Uses include as **thickeners** (water binding agents) and **stabilizers** in foods, such as **soft cheese, bread, sausages** and **ice cream**, where they helps protect against freeze/thaw damage. May also be used as a substitute for **cocoa, coffee** or **chocolate**.

Locust beans Alternative term for **carob beans**.

Loganberries Red, acid **berries** produced by *Rubus loganobaccus*, possibly a hybrid between **dewberries** or **blackberries** and **raspberries**. Contain high amounts of **citric acid** and **vitamin C**. Usually too acid to eat fresh, but are consumed canned or in **jams** or **wines**.

Lokum Alternative term for **Turkish delight**.

Lollipops Large **sugar confectionery** products on wooden or plastic sticks.

Longaniza Dry, cured **pork sausages**, traditionally produced in Spain. Ingredients for these highly seasoned, light-coloured **sausages** include lean **pork**, belly pork, **pimiento peppers, additives** and **condiments**. In Spain, they are often served as an entre with **potatoes** and other **vegetables**, but also make good cooking sausages, **fillings** for **omelettes** or tapas.

Longans **Fruits** produced by *Dimocarpus longan, Euphoria longana* or *Nephelium longana*. The thin, brown rind contains the soft, white edible pulp that surrounds a single seed. Eaten raw, preserved or dried; also available canned. Consumed as **snack foods** or used in **soups**, some savoury dishes and **desserts**. Also known as dragon's eyes.

Long life foods Foods that have a prolonged **shelf life**, usually under ambient conditions. Includes ultra high temperature (UHT) treated and sterilized products, such as **UHT milk**, and shelf stable **bakery products**.

Loquats **Fruits** produced by *Eriobotrya japonica*. Pale yellow to deep orange in **colour**, they are rich in **carotenes** but contain little **vitamin C**. Eaten fresh or used to make **jams, jellies, desserts** and **pies**. Also known as Japanese medlars, **Japanese plums**, Chinese medlars and Chinese loquats.

Lorries Large motor vehicles designed to transport heavy loads. Used in a wide range of applications, including transport of **animals** to **slaughterhouses**, carriage of **cereals** and other raw materials to processing facilities, and transfer of **processed foods** from factories to retail premises. Also known as **trucks**, especially in Canada and the USA.

Los Pedroches cheese Spanish semi-**hard cheese** made from raw or pasteurized **ewe milk**, usually from Merino ewes. **Rind** is yellow and shiny. The ivory white interior is compact, with small holes distributed throughout.

Lotus roots Underground stems, or rhizomes, of the lotus plant (*Nelumbo nucifera*), commonly used in Asian cooking. Rich in **sodium**, the **vitamin B group, vitamin C** and **vitamin E**. Eaten as a vegetable and also in sweet dishes. Lotus root flesh is creamy-white, with the **texture** of raw **potatoes**. **Flavour** is similar to that of fresh **coconuts**. Seeds and leaves of the lotus plant are also consumed.

Loukanka Raw **dry sausages**, traditionally produced in Bulgaria. They are made from **pork**, or pork and **beef** mixtures. Loukanka may be eaten smoked or unsmoked.

Lovage Common name for *Levisticum officinale* Koch, fruits of which are used as spices. Imparts a warm, maple-like **flavour** during **cooking** similar to that of **celery**, however, unlike celery, lovage maintains its flavour after cooking. Lovage leaves and **essential oils** are often included in sweet **sauces**, **gravy**, **pickles** and **seasonings**.

Low alcohol beer **Beer**, the **alcohol** content of which is lower than that normal for the specific type; legal definitions covering the limit differ between countries. Low alcohol beers are made by two general classes of process: formation of lower than normal amounts of alcohol by interrupted **fermentation** or restricted fermentation (using immobilized **yeasts** or low fermentation temperatures); or removal of alcohol from normally-fermented beer (by techniques such as **vacuum evaporation** or **dialysis**). **Sensory properties** of low alcohol beer frequently differ from those of normal beer; defects include a **worts**-like **flavour**, and lack of typical beer **aroma** notes formed during fermentation.

Low alcohol beverages **Beverages**, the **alcohol** content of which is lower than that normal for the beverage type; legal definitions of the limit differ between countries. Low alcohol beverages are made by two general classes of process: formation of lower than normal amounts of alcohol (by restricted or interrupted **fermentation** processes); or removal of most of the alcohol from normally-fermented beverages (generally by **evaporation** or membrane processes). Low alcohol beverages commonly have **sensory properties** which differ, to a greater or lesser extent, from those of normal beverages of the same type.

Low alcohol wines **Wines**, the **alcohol** content of which is lower than that normal for the specific type; legal definitions for limits differ between countries. Low alcohol wines are made by two general classes of process: formation of lower than normal amounts of alcohol (by use of glucose oxidase treated **musts**, early arrest of **fermentation**, aerobic fermentation or use of special **yeasts**); or removal of alcohol from normally-fermented wines (by **distillation** processes, membrane processes, **adsorption** or extraction). Low alcohol wines commonly have **sensory properties** which differ from those of conventional wines of the same type.

Low calorie foods Any foods that are low in **calories**, i.e. those that are naturally low in calories such as **lettuces**, and processed foods that have been manufactured to give a reduced calorie content for a given reference amount, such as **low calorie spreads**. Although originally developed for those with specific health or weight problems, low calorie processed foods are now consumed by many consumers who perceive them as a healthy option. **Sensory properties** of these foods have also improved due to developments of new **sugar substitutes** and **fat substitutes**. Many of these foods can also be classed as **low fat foods**.

Low calorie spreads **Spreads** with reduced **calories** contents.

Low density polyethylene **Polyethylene** of low-density grade. Less rigid and with better resistance to impact than **high density polyethylene** (HDPE). Commonly abbreviated to LDPE.

Low fat foods Foods that are low in **fats**, either naturally or because they have been formulated to contain a reduced fat content compared with a given reference amount. Some of the most popular foods in this sector are low fat **dairy products**, **low fat spreads** and **bakery products**, many of which contain **fat substitutes** as a means of reducing fat content while maintaining acceptable **sensory properties**. Much of the growth in this sector is attributed to consumer perception of these foods as a healthy option. Also classed as **low calorie foods**.

Low fat spreads **Spreads** with reduced **fats** contents.

Low sodium foods Foods containing relatively low levels of **sodium**, and therefore deemed suitable for consumption by those suffering from **hypertension** and certain other diseases. Reduced sodium levels may be achieved by replacement of NaCl with **salt substitutes** containing **potassium** or **magnesium** compounds.

Low sugar confectionery **Confectionery** in which **sucrose** is partially replaced with **sweeteners** (e.g. **polyols**). Suitable for consumption by individuals following a special diet.

Low sugar foods Foods manufactured in such a way that they are low in **sugar**, such as **low sugar confectionery**. Commonly contain **sweeteners** and **bulking agents** as **sugar substitutes**. Such foods may also provide a reduction in calories (**low calorie foods**) and are regarded as a healthy option by the consumer. The reduced sugar contents may also be beneficial for **dental health**.

Lozenges Small flat sweets made from **icing sugar**, **glucose syrups**, **gum arabic/gelatin** and **flavourings**. Sometimes medicated, as in the case of **cough drops**.

LTLT pasteurization Low temperature, long time batch **pasteurization** treatment (also known as the holder method) that is applied to liquid foods, particularly **milk**. A quantity of milk is placed in an open vat, heated to 63°C, held at that temperature for 30 minutes, and then pumped over a plate-type cooler prior to **bottling** or **cartoning**. In addition to destroying common **pathogens**, this heat treatment also inactivates **lipases**, which might otherwise quickly cause the milk to become rancid.

Lubricants Substances, e.g. oil or grease, applied to equipment components to minimize friction.

Lucerne Alternative term for **alfalfa**.

Lukum Alternative term for **Turkish delight**.

Lulo Alternative term for **naranjilla** (*Solanum quitoense* or *S. angulatum*). Orange **fruits** with green-yellow juicy flesh. Rich in **vitamin A** and **vitamin C**. Most commonly used in **beverages**, but also eaten out of hand, as ingredients in **desserts**, or in jellies and **marmalades**. Also known as quito oranges.

Luminescence The emission of **light** from a substance or organism, and which occurs at temperatures below those required for incandescence. Includes photoluminescence, **chemiluminescence**, electroluminescence, **fluorescence** and **phosphorescence**.

Lumpfish Marine fish species (*Cyclopterus lumpus*) belonging to the lumpfishes and snailfishes family (Cyclopteridae). Widely distributed in the western and eastern Atlantic Ocean. Eaten fresh or smoked, especially in Nordic countries. Eggs are used as inexpensive **caviar substitutes**; **roes** are also sold fresh. Also known as lumpsucker.

Lumpiness Texture term relating to product **consistency** and the extent to which an item contains lumps. Lumpy products contain inhomogeneities in structure, which can be present as invisible defects. Lumpiness has a negative effect on the **spreadability** of products such as **margarines**, and hampers the formation of a smooth surface of the spread film.

Luncheon meat A cooked meat product prepared from chopped **pork**, **ham** and/or **beef**. Luncheon meat is available canned or sliced, and is sold in vacuum packaging.

Lungs Paired organs within the ribcage into which air is inhaled during breathing. The lungs of slaughtered animals form a part of edible **offal** and lung mince may be included in cooked **sausages** (e.g. **frankfurters** and **pepperoni**). Mechanical **stunning** used in cattle slaughter may result in brain emboli in the lungs; this is of particular concern in relation to **BSE** and the transmission of **prions** in foods.

Lupanine One of the toxic **alkaloids** present in **lupins**.

Lupin proteins Vegetable proteins extracted from **lupin seeds**.

Lupins Species of *Lupinus*, some of which are used as food. Seeds are rich sources of proteins and **oils**. High levels of **alkaloids** make some seeds too bitter for consumption, but contents may be reduced by washing in water. Varieties selected as grain crops are low in alkaloids (sweet lupins). Seeds have been used as **coffee substitutes** and seed **flour** has been suggested as a substitute for **soy meal**.

Lupin seed oils Vegetable oils derived from seeds from plants of the genus *Lupinus* which have low to intermediate levels of **unsaturated fatty acids**.

Lupin Seeds Seeds from species of the genus *Lupinus*, annual or perennial herbs or shrubs of the family Leguminosae. Rich in protein, with low to intermediate levels of **unsaturated fats**; may be used as **oilseeds** or are roasted, boiled and salted and used as **snack foods**.

Lupulin A fine yellow powder or resin containing high concentrations of the **bitter compounds** and **essential oils** present in **hops**. Occurs in lupulin glands found predominantly on hop cones.

Lupulones Alternative term for the **β-acids** found in **hops** and **beer**. Compared to other **bitter compounds**, these poorly soluble resin constituents have little bittering capacity in beer.

Lutein One of the most widespread naturally occurring **carotenoids**. Found in many foods, and particularly **fruits** and **vegetables**.

Luteolin Member of the **flavonoids**, found in a range of plant foods, including **sage**, **olives**, **lettuces**, **endives** and **citrus fruits**. Has also been found in **honeys**.

Lyases EC 4. **Enzymes** that cleave C-C, C-O, C-N and other bonds by means other than hydrolysis or oxidation. Two substrates are involved in one reaction direction, but only one in the other. When acting on the single substrate, a molecule is eliminated leaving an unsaturated residue.

Lycadex Maltodextrin derived from **corn, potatoes, wheat** and **tapioca**. Used as a fat substitute, texture modifier or bulking agent in **processed foods, bakery products, dairy products** and **salad dressings**.

Lycasin Registered trade name for a polyol manufactured by **hydrogenation** of **starch hydrolysates**. Consists of approximately 50% **maltitol**, 16% maltotriitol and 7% **sorbitol**. It is used as a sweetener and may, in combination with other **sugar alcohols**, be used as a substitute for **glucose syrups**.

Lychees Alternative term for **litchis**.

Lycopene　One of the **carotenoids**, particularly characteristic of **tomatoes**.

Lycoperdon　**Edible fungi** commonly known as puff balls.

Lyes　Aqueous solutions of **alkalies**, generally **sodium hydroxide** or potassium hydroxide, of use in food **processing** treatments such as **peeling** or **sugar processes**.

Lymeswold cheese　British mould-ripened cheese made from **cow milk**.

Lyophilization　Alternative term for **freeze drying**.

Lysine　One of the essential dietary **amino acids**. Present as a free amino acid and protein constituent in a wide range of foods. Cereals such as **rice** and some **wheat** varieties contain low lysine levels, and both conventional plant breeding and genetic engineering techniques have been used in attempts to increase lysine contents of these dietary staples.

Lysinoalanine　Dipeptide formed from **lysine** and **alanine**. One of the cross-linked **peptides** formed in food **proteins** during **thermal processing**, especially in alkaline conditions. Can be used as an indicator of **milk** quality after thermal processing.

Lysins　Members of EC 3.4.24. A group of metalloendopeptidases that includes **collagenases**, **thermolysins** and **autolysins**.

Lysolecithin　Monoglyceride phosphate ester conjugated to choline via esterification with the phosphate moiety. Produced by hydrolysis of **lecithins**. Also called lysophosphatidylcholine.

Lysophospholipases　EC 3.1.1.5. Hydrolyse single fatty acid ester bonds in lysoglycerophosphatidates with the formation of glyceryl phosphatidates and free fatty acids. Also known as **lecithinases** B and **phospholipases** B, these **enzymes** are potentially useful for improving the quality of **wheat** hydrolysates.

Lysophospholipids　**Phospholipids** deacylated at position 1 or 2.

Lysozymes　EC 3.2.1.17. Hydrolyse 1,4-β-linkages between N-acetylmuramic acid and N-acetyl-D-glucosamine residues in peptidoglycans, and between N-acetyl-D-glucosamine residues in chitodextrins. Found in **milk**, particularly **human milk**, and **egg whites** (**egg whites lysozymes**; **eggs lysozymes**). Important antimicrobial **preservatives** since they are able to break down the cell wall of many Gram positive **bacteria**. Used in the production of certain types of **cheese** to kill harmful bacteria, and for cold **sterilization** of certain foods and **beverages**.

M

Mabinlin Sweet protein isolated from seeds of the Chinese plant *Capparis masaikai*. Four homologues of mabinlin have been isolated, mabinlin I to IV. A recombinant mabinlin has been produced that is 400 times sweeter than **sucrose** and used as a sweetener for low calorie **beverages** and **low calorie foods**.

Maca Common name for *Lepidium meyenii*, an Andean crop grown for the roots or **tubers** which are eaten as a vegetable.

Macadamia nuts **Nuts** produced by the Australian species *Macadamia integrifolia* or *M. tetraphylla*, with smooth or rough shells, respectively. Considered among the finest gourmet nuts, they are eaten roasted and salted, or as ingredients in **bakery products, ice cream** and **sugar confectionery**. Also known as Queensland nuts.

Macaroni Hollow tubes of **pasta** which are usually short and curved.

Macaroons Small chewy **cakes** or **cookies** made from ground **almonds**/almond paste or coconut, **sugar** and **egg whites**. Often baked on rice paper.

Mace One of the **spices**, along with **nutmeg**, derived from seeds of *Myristica fragrans*. Mace is produced from the arillodes of *M. fragrans*. These are red-coloured structures, situated on top of the **nuts** of this plant, that resemble a cockerel's comb.

Maceration Softening or breaking up of foods by soaking in a liquid, or the soaking of foods (usually **fruits**) in a liquid in order to absorb the flavour. **Spirits** or **liqueurs** are often used as the macerating liquid.

Machine vision Inspection systems in which samples are examined using a camera, the image from which is analysed by computer using image processing algorithms. Operations which can be performed include defect detection, dimensions measurement, orientation detection, grading, sorting and counting.

Mackerel Any of a number of **marine fish** species in the family Scombridae, many of which are commercially important food fish. Found in temperate and tropical seas around the world. Commercially important species include *Scomber scombrus* (**Atlantic mackerel**) and *Scomber japonicus* (**Pacific mackerel**). Flesh is firm and fatty, with a distinctive savoury **flavour**. Marketed fresh, frozen, smoked, salted, dried and canned. **Roes** of some species are also consumed, often marketed as canned products.

Macrocystis Genus of large brown **seaweeds** (**kelp**) found on rocky coastal substrates in many parts of the world. Some species, such as *Macrocystis pyrifera*, are an important source of **alginates** used by the food industry.

Madeira **Fortified wines** produced in the island of Madeira, characterized by being aged for several months at high temperature in special rooms called estufas. Types include Sercial (the driest), Verdelho and Bual (the sweetest).

Madeirization In the context of **Madeira wines**, the process of development of the characteristic **flavour** as a result of controlled heat treatment. For other wines, a **flavour** defect due to excessive heating and oxidation.

Madhuca seeds Seeds from plants of the genus *Madhuca*, often used as **oilseeds**.

Magnesium One of the essential mineral nutrients, chemical symbol Mg. Widely distributed in plant and animal foods, good sources including **fruits, vegetables** and **dairy products**. Standard Western diets generally contain adequate levels of magnesium, so **fortification** is largely unnecessary. Absorption of dietary magnesium may be affected by other dietary nutrients such as calcium, phosphate and vitamin D, and also by some clinical conditions, including alcoholism and **diabetes**. Magnesium is an important bone constituent and intracellular inorganic cation acting as an essential co-factor in many enzymic reactions. Magnesium deficiency can cause calcification of soft tissues, electrolyte imbalances, gastrointestinal symptoms and personality changes. If taken in excess, magnesium **toxicity** symptoms can include nausea, vomiting, hypotension and neurological changes.

Magnetic fields Regions around a magnet within which the force of magnetism acts. Various applications in the food industry include non-thermal **preservation** techniques.

Magnetic resonance imaging Non-destructive analytical technique based on nuclear magnetic resonance which is used widely in the food industry. Applications include assessment of meat quality, determination of components in foods and measurement of **thermophysical properties**.

Mahewu African lactic fermented, non-alcoholic **beverages** made from **corn**, **sorghum** or **millet**.

Mahimahi Alternative term for the common **dolphinfish** (*Coryphaaena hippunis*), a **marine fish** species of high commercial importance. Widely distributed in tropical and sub-tropical waters throughout the world, and also produced commercially by **aquaculture**. Marketed fresh and frozen. Other forms of the name are mahi mahi and mahi-mahi.

Mahon cheese Spanish **hard cheese** made from **cow milk**. It is produced on the Balearic Island of Minorca. During manufacture, **curd** is piled in the centre of a piece of cheesecloth, the corners of which are knotted and twisted together. The cheese is then pressed and twisted for a few days, resulting in the typical 'cushion' shape of this cheese. The hard, orange rind carries the imprint of the cheesecloth. Although sold at various stages of maturity, Mahon is usually sold young, when it has a smooth and supple texture combined with a sweet and fruity aroma. It must be consumed within 10 days of purchase.

Maida Indian refined white **flour** made from **wheat**.

Maillard reaction Chemical reaction that occurs between **reducing sugars** and the amino groups of **proteins** or **amino acids** present in foods, and is responsible for **nonenzymic browning**. **Maillard reaction products** cause a darkening of **colour**, reduced solubility of proteins, development of bitter **flavour**, and reduced nutritional availability of certain amino acids, such as **lysine**. Rate of Maillard reaction is influenced by many factors, including **water activity**, temperature and pH of foods.

Maillard reaction products Soluble and insoluble polymers produced via the **Maillard reaction** when **reducing sugars** and amino groups of **amino acids** and **proteins** are heated together. Maillard reaction products contribute to the **colour** and **flavour** of foods such as **soy sauces**, **caramels** and **toffees**, **milk chocolate** and **bread**. Maillard reaction products are important functional components of **caramel colorants** (produced by heating **sucrose** in the presence or absence of ammonium ions) that are used in a wide range of foods and **beverages**.

Maize Alternative term for **corn**.

Maize meal Alternative term for **corn flour**.

Maize oils Alternative term for **corn oils**.

Makhana Alternative term for **gorgon nuts**.

Malabar nightshade Alternative term for **Ceylon spinach**.

Malachite green Chemical dye which shows **antibacterial activity**, **antifungal activity** and also properties of **anthelmintics**. Used primarily in **aquaculture** for treatment and control of a range of parasitic and fungal infections in **fish** and **shellfish**. Residues may persist in **aquaculture products**. Suspected mutagen.

Malate dehydrogenases Generic term for a group of **enzymes** including: EC 1.1.1.37 which converts (*S*)-malic acid and NAD$^+$ to **oxaloacetic acid** and NADH; EC 1.1.1.38 and EC 1.1.1.39 which convert (*S*)-malic acid and NAD$^+$ to **pyruvic acid**, CO_2 and NADH; and EC 1.1.1.40 which converts (*S*)-malic acid and NADP$^+$ to pyruvic acid, CO_2 and NADPH. The latter three **enzymes** are also known as malic enzymes. Involved in **malic acid** metabolism and the ripening of certain **fruits**, and can be used for determination of the malic acid content of foods and **beverages**.

Malathion Non-systemic insecticide and acaricide used for control of biting, chewing and sucking **insects** in a wide range of crops, including **pome fruits** and **stone fruits**, **vegetables** and **rice**. Also used to control pests during storage of **cereals**. Classified by WHO as slightly toxic (WHO III). Also known as carbofos.

Malay apples Bright red **fruits** produced by *Syzygium malaccense*, a tree native to Malaysia and India. The white flesh is slightly sweet and juicy. Eaten raw or used to make **preserves** and **wines**. Also known as mountain apples or pomerac.

MALDI-TOF-MS Commonly used abbreviation for matrix-assisted laser desorption/ionization time of flight **mass spectroscopy**. Technique used to determine biomolecular structure of substances such as **proteins**, **sugars** and **oligonucleotides**, including those of food origin. Molecules are embedded in a matrix on a metal surface, desorbed into a gas phase by the force of a laser beam, accelerated by an electric field and fly through a drift tube at high vacuum. They are characterized according to molecular weight, which is indicated by the time taken to pass through the drift tube.

Maleic acid Carboxylic acid which occurs as a colourless, crystalline solid and is used in making synthetic resins. On heating, water is eliminated, forming maleic anhydride, which can be used in modification of **proteins**, particularly **enzymes**, and in preparation of copolymers used in **packaging** of foods. Maleic acid has antibotulinal activity, making it suitable for use in food **preservatives**, and is also a substrate for synthesis of D-**malic acid** by many **bacteria**. A more modern name for maleic acid is *cis-*

butenedioic acid. The more stable *trans* isomer is **fumaric acid**.

Maleic hydrazide One of the **plant growth regulators**. Used particularly to control **sprouting** in **potatoes** and **onions** during storage.

Malic acid Aliphatic dicarboxylic acid, an important metabolic intermediate in the glyoxylate and tricarboxylic acid cycles, and also commonly accumulated in some **fruits** and **vegetables** including **apples** and **grapes**. This organic acid is the substrate for **malolactic fermentation** by **bacteria** which produces **lactic acid** and **carbon dioxide** and reduces the overall acidity of the fermented products, generally **wines**, thereby increasing product quality.

Malic enzymes Alternative term for certain **malate dehydrogenases**.

Malignant hyperthermia Progressive hyperthermia, severe muscular rigidity and acidosis, which occurs in some **swine** in response to **stress**. It is associated with porcine stress syndrome, pronounced **halothane sensitivity** and the **PSE defect** in **pork**.

Mallards Wild **ducks** (*Anas platyrhynchos*) belonging to the Anatidae family, which are hunted for production of **duck meat**.

Mallow seeds Seeds produced by plants belonging to the family Malvaceae, often used as **oilseeds**.

Malnutrition Condition resulting from inappropriate **nutrition**. Includes both inadequate and excessive dietary intakes of **nutrients** and/or **calories**. Insufficient protein intake causes kwashiorkor in children, and a diet deficient in all nutrients causes marasmus. Lack of vitamins causes a wide variety of **deficiency diseases**, including scurvy, rickets, beriberi and pellagra. Malnutrition may result from eating disorders, such as anorexia nervosa and bulimia nervosa. Overnutrition can lead to **toxicity** and **obesity**.

Malolactic fermentation A type of **fermentation** carried out by species of **bacteria** such as *Lactobacillus*, *Leuconostoc* and *Pediococcus*, in which L-malic acid is converted to L-lactic acid and CO_2. In certain fermented products (e.g. **wines** and **soy sauces**), it has the effect of reducing the **acidity**, since lactic acid is a weaker acid than malic acid, and can be used, therefore, to impart desirable acidity on these products.

Malonaldehyde One of the **aldehydes** produced as a result of lipid oxidation, synonym **malondialdehyde**. Traditionally used in the determination of **thiobarbituric acid values**, a measure of lipid **rancidity** or oxidation.

Malondialdehyde One of the **aldehydes** produced as a result of lipid oxidation, synonym **malonaldehyde**. Traditionally used in the determination of **thiobarbituric acid values**, a measure of lipid **rancidity** or oxidation.

Malt Cereals grains which have been steeped, partially germinated, then kilned to terminate **germination**. The malting process includes **starch saccharification** and partially breakdown of **proteins** present in the grain to yield fermentable material; enzyme activity is also increased. Malt is used mainly in **brewing**; small quantities are used in making **bakery products**. Malt is most commonly made from **barley**, but other cereals such as **wheat** and **sorghum** may also be malted.

Maltases Alternative term for **α-glucosidases**.

Malt beverages **Beverages** based on **malt**. May resemble **beer**, but do not comply with national regulations for beer.

Malthouses Industrial premises used for **malting** of **barley**.

Malting Process of conversion of **cereals** (especially **barley**) into malt by controlled **steeping**, **germination** and **kilning** to terminate **germination**.

Malting barley **Barley** (*Hordeum vulgare*) cultivars which have composition and **germination** properties making them suitable for **malting** and **brewing**.

Malting properties Properties of **barley** or other **cereals** which determine suitability for **malting** and quality of the **malt** produced. These include **germination** characteristics, composition, **proteins** and **starch** modification properties, and activity of **enzymes**.

Maltitol Polyol, systematic name 4-*O*-α-glucopyranosyl-D-sorbitol, manufactured by **hydrogenation** of **maltose syrups**. Has 0.6-0.9× the **sweetness** of **sucrose** and is used in **sweeteners**.

Maltodextrins **Dextrins** of varying, but generally intermediate, length (degree of polymerization), containing D-glucopyranose residues with α1→4 linkages, as in **maltose**. Synonym for **maltooligosaccharides**.

Maltohexaose Oligosaccharide consisting of six **maltose** residues linked via α-1,4-glycosidic bonds. Produced by hydrolysis (treatment with acids or α-amylases) of **starch**. Has low **sweetness** compared with **sucrose** (0.1× as sweet) but higher **viscosity**, thus making it useful in **bulking agents**.

Maltol Pyrone with the systematic name 3-hydroxy-2-methyl-4H-pyran-4-one. Used as **flavourings** with **caramel**-like **aroma** that impart a freshly baked **flavour** and **aroma** to **bread** and **cakes**.

Maltooligosaccharides **Oligosaccharides** containing D-glucopyranose residues with α1→4 linkages, as in **maltose**. Synonym for **maltodextrins**.

Maltose Disaccharide comprising two molecules of **glucose** linked by a α-1,4-glycosidic bond which is manufactured by hydrolysis of **starch**. Has 0.4-0.5× the **sweetness** of **sucrose** and is used in **sweeteners** and as a **fermentation** substrate in **brewing**. Also known as malt sugar.

Maltose syrups **Syrups** in which the predominant **sugar** present is **maltose**. Manufactured by hydrolysis of **starch** and may contain up to 90% maltose.

Maltotetraose Oligosaccharide consisting of four **maltose** residues linked by α-1,4-glycosidic bonds which is produced by hydrolysis of **starch**. Has approximately 0.2× the **sweetness** of **sucrose**. Maltotetraose syrups have many applications, including as **sweeteners**, **bulking agents**, **humectants** and glazing agents.

Maltotriose Oligosaccharide consisting of three **maltose** residues linked by α-1,4-glycosidic bonds which is produced by hydrolysis of **starch**. Has approximately 0.3× the **sweetness** of **sucrose**.

Malt vinegar **Vinegar** produced by **fermentation** of **barley malt**. **Starch** is hydrolysed during **malting** and the sugars in the resulting hydrolysate are fermented to produce **acetic acid**. Malt also imparts **flavour** to the vinegar. Malt vinegar is often used for **pickling** and as a condiment, most commonly in the UK.

Malvidin One of the **anthocyanidins**, a pigment commonly found in **grapes** and wines, sometimes as a glycoside. Also found in other **berries**.

Mamey **Fruits** produced by *Mammea americana*. Round and green with a rough leathery skin and pale yellow flesh. Eaten fresh or in **jams**, **preserves** or **sauces**. Pulps are used to make **wines**. Mature fruits contain high levels of **pectins**.

Manchego cheese Spanish **hard cheese** made from pasteurized **ewe milk**, the name indicating that it is made in the La Mancha region of Spain. The cheese has a black, grey or buff rind, and a white to yellow interior, depending on age. The interior contains a number of holes and has a mild, nutty and slightly briny flavour which can have a peppery bite in older cheeses. The finished cheese is usually smeared with olive oil and surface mould is removed. Manchego is sold at various stages of maturity; at 13 weeks of **ripening**, it is described as curado (cured) and, after more than 3 months old it is referred to as viejo (aged).

Mancozeb Fungicide with protective action used for control of many fungal diseases (e.g. **blights**, leaf spot, **rusts** and downy mildew) in a range of **fruits**, **vegetables** and **cereals**. Classified by Environmental Protection Agency as not acutely toxic (Environmental Protection Agency IV).

Mandarin juices **Fruit juices** prepared from **mandarins** (*Citrus reticulata*).

Mandarins Small, loose skinned **citrus fruits** (*Citrus reticulata*). Eaten as a dessert, commonly as canned segments. Relatively high **vitamin C** content. Varieties include **tangerines** and **satsumas**, but the names tend to be used indiscriminately. Used in several citrus hybrids.

Mandoo Korean **dumplings** which are stuffed with a spicy mixture of **vegetables** and/or **meat**. **Fast foods** eaten as a snack or main dish. Cooked by **steaming**, **frying** or **boiling**. Also used in making dumpling **soups**.

Maneb Fungicide with protective action used for control of a range of fungal diseases in **fruits**, **vegetables** and **cereals**. Classified by Environmental Protection Agency as not acutely toxic (Environmental Protection Agency IV).

Manganese A mineral, with chemical symbol Mn. Limited evidence for its role as an essential nutrient in humans, although it is required as a cofactor for several enzymes. However, **deficiency diseases** have been reported in other animals. Widely distributed in foods and **beverages**. Toxicity in humans is generally associated with mining, although manganese levels in foods are often determined along with those of other **heavy metals**.

Manganese peroxidases EC 1.11.1.13. Oxidize Mn(II) to Mn(III). Major ligninolytic enzymes produced by a number of white rot fungi that are important in the potential use of these organisms for **lignin** degradation, degradation of toxic pollutants and decoloration of **olive oil mills effluents**.

Mangoes **Tropical fruits** produced by *Mangifera indica*. Vary in shape, size and **colour**, but the flesh surrounding the large stone is always yellow to orange. Rich in **vitamin C** and **carotenes**, with approximately 14% sugar. Eaten fresh as a dessert, also canned or dried. Used in a range of products, including **jams**, **pickles** and **chutneys**, or as a source of **fruit juices**. The seeds (kernels) inside the stone can also be used as a food or as a source of **flour**, **fats** and **oils**.

Mango jams **Jams** made from **mangoes**, sometimes combined with other **fruits**.

Mango juices **Fruit juices** prepared from **mangoes** (*Mangifera indica*).

Mango kernels Edible seeds found within the stone of **mangoes**. Good source of nutrients for humans in times of food shortages. **Fats** and **oils** extracted from the kernels have been used in foods, e.g. as **cocoa butter substitutes**. Meal prepared from the kernels can be used as a substitute for **wheat flour** in **baking**.

Mango nectars **Fruit juice beverages** made by addition of water, **sugar** and optionally other ingredients to **mango juices**.

Mango pickles Products made by **pickling** pieces of **mangoes** with **spices**, salt and **oils**.

Mango pulps Soft mass prepared from the flesh of **mangoes**. Used in a range of products including **beverages**, **ice cream**, **yoghurt**, **bakery products**, **jams** and jellies.

Mango purees Smooth creamy preparation made from the flesh of **mangoes** by sieving or reducing in a blender or liquidizer. Used as **sauces** or in preparation of products such as **fruit juices**, **fruit nectars**, **bakery products**, **ice cream**, **yoghurt** and **jams**.

Mangosteens **Tropical fruits** produced by *Garcinia mangostana*. Dark purple with a hard rind, juicy flesh and unique, pleasant **flavour**. Sugar content is relatively high, but **vitamin C** level is low.

Manioc Alternative term for **cassava**.

Mannanases Alternative term for **β-mannosidases**.

Mannan endo-1,4-β-mannosidases EC 3.2.1.78. Catalyse the random hydrolysis of 1,4-β-D-mannosidic linkages in **mannans**, **galactomannans** and **glucomannans**. Useful for production of **food additives**, extraction of **vegetable oils** from **legumes** and reduction of the viscosity of **coffee extracts** during the manufacture of **instant coffee**.

Mannans **Polysaccharides** containing a high proportion of **mannose**. Mannans that also contain **glucose** or **galactose** residues are known as **glucomannans** and **galactomannans**, respectively. Mannans are produced by plants, e.g. **konjac glucomannans**, **bacteria** and **fungi**, including **yeasts**. Uses include in **thickeners** and **texturizers**.

Mannases Alternative term for **β-mannosidases**.

Mannitol Polyol consisting of six carbon atoms that occurs naturally in plants, plant exudates and **seaweeds**. Manufactured by reduction of **mannose** or reduction and **isomerization** of **glucose**. Has approximately $0.6\times$ the **sweetness of sucrose**. Uses include as nutritive **sweeteners**, **anticaking agents**, **stabilizers** and **thickeners**. The name is derived from manna, the sweet exudate from the ash tree, from which it has been isolated. Also called manna sugar.

Mannoproteins **Glycoproteins**. Yeast mannoproteins are used in **winemaking** to prevent haze formation.

Mannose Monosaccharide consisting of six carbon atoms (**hexoses**). Has approximately $0.6\times$ the **sweetness** of **sucrose**.

Mannosidases EC 3.2.1.24 (**α-mannosidases**) and EC 3.2.1.25 (**β-mannosidases**). Hydrolyse terminal, non-reducing α- and β-D-mannose residues in α- and β-D-mannosides, respectively. The former are involved in the **ripening** of **fruits** and can be used for synthesis of novel **cyclodextrins**. The latter, also known as mannases and mannanases, can be used for synthesis of alkyl β-mannosides, which are useful as **surfactants**. They can also be used for synthesis of mannooligosaccharides.

α-Mannosidases EC 3.2.1.24. Hydrolyse terminal, non-reducing α-D-mannose residues in α-D-mannosides. Involved in the **ripening** of **fruits** and can be used for synthesis of novel **cyclodextrins**.

β-Mannosidases EC 3.2.1.25. Hydrolyse terminal, non-reducing β-D-mannose residues in β-D-mannosides. Can be used for synthesis of alkyl β-mannosides, which are useful as **surfactants**, and also for synthesis of mannooligosaccharides. Also known as mannases and mannanases.

Manometers Instruments used for measuring the pressure of liquids or gases.

Manometry Measurement of the pressure or tension of gases or liquids.

Maple saps Sweet viscous fluid produced by and tapped from maple trees (*Acer*). Maple trees are native to North America and only sap obtained from the sugar maple or the black maple, species producing the sweetest sap, are used in manufacture of **maple syrups**.

Maple syrups Concentrated sugar solution produced by **evaporation** of **maple saps**. **Sucrose** is the predominant sweet substance, comprising approximately 60% of the syrup by weight; **hexoses** are also present. Maple syrups also contain **flavour compounds**, e.g. syringaldehyde, and **natural colorants**, which provide the characteristic maple syrup **flavour** and amber **colour**.

Marbling Streaks of intramuscular **animal fats** in meat from mammals. Marbling is one of the factors used to assess quality of **meat**, particularly **beef**. For example, good quality beef is marbled with fine strands of fat; this fat bastes the meat as it cooks, thus affecting **juiciness** and **tenderness**. Lower quality beef has either no marbling or thicker marbling; it tends to be tougher after **cooking**.

Marc **Spirits** made by **distillation** of fermented **mashes** based on **grape marc**.

Mare milk **Milk** obtained from horses. Differs from **cow milk** by its lower fat and protein contents (1.5

and 2.4%, respectively) and higher **lactose** content (approximately 6.2%). Levels of most **minerals** are also lower than in cow milk, but contents of iron and copper are higher. **Vitamin A** and most B vitamins are present in lower concentrations in mare milk than in cow milk, but contents of **carotenes** and **niacin** are higher than in cow milk. **Ascorbic acid** is present in a similar amount to that in cow milk.

Margaric acid Carboxylic acid with 17 carbon atoms, member of the **saturated fatty acids**, with a melting point of 59-61°C. Synonyms include **heptadecanoic acid**, margarinic acid and *n*-heptadecylic acid. Occurs as a free fatty acid and lipid component of **animal fats** and **vegetable fats**.

Margarines Water-in-oil **emulsions** usually composed of approximately 80% **animal fats** or hydrogenated **vegetable fats** and 20% water, together with **emulsifiers**, **colorants**, **vitamin A**, **vitamin D** and **flavourings**. Usually solid at room temperature. Used as **spreads**, **butter substitutes**, in **baking** or as **cooking fats**. Low fat products may contain as little as 20% fat.

Maribo cheese Danish semi **hard cheese** made from **cow milk**. Similar in appearance to **Gouda cheese**, with a yellow wax coating and a firm interior containing many eyes. Sometimes flavoured with **caraway** seeds.

Marigolds Bright yellow **flowers** of the genus *Tagetes* used to add **flavour** and **colour** to dishes including **soups** and **salads**. Petals are dried and the powder used as **colorants** for foods. Dried preparations are also added to chicken feeds to enhance pigmentation of **egg yolks**.

Marinades Seasoned liquids used for marination mainly of **meat** or **fish**. Usually contain **oils** mixed with **wines**, **vinegar** or **lemon juices**, and **herbs** or **spices**.

Marination Soaking of foods in **marinades**, mixtures of ingredients such as oils, vinegar and herbs, before cooking, in order to add **flavour** or tenderize. Because most marinades contain acidic ingredients (lemon juices, vinegar or wines), marination should be conducted in glass, ceramic or stainless steel, but not in aluminium, containers.

Marine fish Any **fish** which exist in **sea water** environments. The majority of commercially important food fish are found in sea water.

Marine oils **Lipids** derived from marine animals. Include **fish oils**, **squid oils**, **seal oils** and **whale oils**.

Marjoram Common name for *Origanum majorana*, the leaves and seeds of which are used as spices. Also called sweet marjoram. Leaves of the plant have a warm wood-like **aroma** similar to that of **nutmeg**. Leaf **essential oils** are also used as **flavourings**.

Marker genes **Genes** that confer a readily detectable **phenotype** on cells, either in culture, or in transgenic or chimeric organisms. They may encode reporter **enzymes** or markers conferring **antibiotics resistance**.

Market research The activity of gathering information about customers' needs and preferences. Market research uses surveys, tests and statistical studies to analyse consumer trends and to forecast the quantity and locale of a market favourable to the profitable sale of products or services. The social sciences, for example psychology and sociology, are increasingly utilized to provide clues to people's activities, circumstances, wants, desires and general motivation.

Markets As well as conveying the offering of goods for sale or promotion of products, this term can also cover the regular gatherings for the purchase and sale of food, livestock and other commodities, the outdoor spaces or large halls where vendors sell their goods, or particular areas of commercial or competitive activity.

Marlins Any of a number of large, fast swimming **marine fish** species belonging to the family Istiophoridae. Commercially important species include *Makaira indica* (black marlin), *M. nigricans* (blue marlin) and *Tetrapturus audax* (striped marlin). Marketed fresh or frozen and occasionally smoked; also used in manufacture of **fish sausages** in Japan.

Marmalades **Preserves**, often clear, produced from the pulps and rind of **fruits**, mainly **citrus fruits**.

Marrons glaces **Chestnuts** cooked in **syrups** and glazed.

Marrows **Vegetables** produced by plants of the genus *Cucurbita*, which includes also **squashes** and **pumpkins**. Vegetable marrows are varieties of *C. pepo*. Large cylindrical or round vegetables of various colours, with greenish-white or yellow flesh. Contain mainly water (usually at least 90%), with small amounts of **starch**, sugar, **fats**, **proteins**, **carotenes** and vitamin B, and moderate amounts of **vitamin C**. Eaten boiled or stuffed with meat or other vegetables. Marrows harvested when young are **courgettes** or zucchini.

Marshmallows Soft **aerated confectionery** products made from **corn syrups**, **glucose**, **gelatin** and **egg whites**. Originally manufactured from the root sap of the marsh mallow plant (*Althaea officinalis*).

Marula Plum-size **fruits** produced by *Sclerocarya caffra* or *S. birrea* subsp. *caffra*, a tree native to Africa. Rich in **vitamin C** and several minerals. Beneath a strong, leathery skin are a layer of white flesh similar to **mangoes** and a pit containing a small, tasty kernel.

Eaten out of hand or made into **jams**, **jellies** and a range of **beverages**, including **fruit juices**, **wines**, **beer** and **schnapps**-like **spirits**.

Marzipan Malleable confection made with crushed **almonds** or almond pastes, together with powdered **sugar** and **egg whites**. Often used to decorate **cakes** or as **fillings** in **pastries** and candy.

Mascarpone cheese Italian high-fat **soft cheese** made from **cow milk**. Although not strictly a true cheese, it is described as a **curd cheese**. Mascarpone is made by adding a culture to the cream skimmed from milk used in manufacture of Parmesan cheese. **Tartaric acid** is also used in its production. After addition of the culture, the cream is gently heated and allowed to mature and thicken, after which it takes only a few days to ripen. The white to yellow cheese is spreadable and frequently used in dishes and **sauces**.

Mashes Mixtures of ground **malt**, optionally with other **brewing adjuncts**, with hot water. Heated under controlled conditions to solubilize and extract fermentable constituents and other materials of importance for the **brewing** process and **beer** quality.

Mashing Preparation of aqueous extracts of **malt** (optionally together with **brewing adjuncts**) by heating them in water under a time/temperature regime which will optimize enzymic **solubilization** and extraction of carbohydrates, soluble nitrogen compounds and other constituents of importance for **fermentation** and **beer** quality. Brewing enzyme preparations may be used to enhance the enzymic solubilization process, especially when non-malted adjuncts are used.

Mashua Alternative term for **anu**.

Massecuites Mixture of crystallized **sugar** and **sugar syrups** which is produced during manufacture of sugar. Centrifuged to separate the sugar crystals (which are dried and stored) from the syrup, which undergoes further **crystallization** to improve sugar yield.

Mass spectrometry Alternative term for **mass spectroscopy**.

Mass spectroscopy **Spectroscopy** technique in which separation is based on atomic and molecular mass. Samples are bombarded with electron beams which fragment the molecules. The fragments are accelerated through magnetic fields and sorted on the basis of charge to mass ratio. Usually abbreviated to MS.

Mass transfer Movement of matter from one place to another, usually considered with reference to a defined boundary, as in the transfer of water within or from a wet product during **drying**.

Mastication First stage in the digestion of foods, whereby food taken into the mouth is processed into a form suitable for swallowing. During mastication, foods are chewed, ground and torn with the teeth, and mixed with saliva. Small food particles result which have a large surface area on which saliva can act. Mastication also releases food **flavour** and **aroma**. In conjunction with the action of the tongue, a cohesive food bolus is formed of the correct size to pass through the oesophagus.

Mastitis Inflammation of the mammary gland caused by pathogenic **microorganisms**. In cows, can cause reductions in milk yield and alterations in the composition of milk from infected quarters.

Masu salmon Alternative term for **cherry salmon**.

Mate Infusion **beverages** prepared from dry leaves and twigs of the plant **yerba mate** (*Ilex paraguariensis*).

Matjes Traditional Dutch product of lightly cured **herring**. Herring used for matjes production must have no development of the reproductive system, giving them a high fat content. They are prepared in a special way, cutting into the gills and leaving the pancreas in the fish after **gutting** so that the pancreatic **enzymes** promote maturation of the product. As well as having a high fat content, matjes are rich in ω-3 **fatty acids**.

Matsutake Wild Japanese **mushrooms** (*Tricholoma matsutake*) which are usually exported either in canned or dried form.

Maturation Alternative term for **ageing** and **ripening**.

Maturity Alternative term for **ripeness**.

Mauritia Genus of palm trees that grow in South America. Fruits are used in preparation of **beverages** and in some cases as the source of **oils**. Pulps of the fruits from *Mauritia vinifera* are used as a food. **Wines** and **sago** are produced from stems of *M. flexuosa*.

Mawa Type of **condensed milk** made by heating milk until boiling and then stirring continuously over a low heat until it thickens to the consistency of **cream cheese**. Used in preparation of Indian **desserts** and **sweetmeats**. Also known as khoya.

Mawe **Porridge** made from dehulled and partially germinated white **corn**.

Maximum residue limits Maximum concentrations of pesticide residues, resulting from the registered use of agricultural or veterinary **pesticides**, that are recommended to be legally permitted or recognized as acceptable in or on a food, agricultural commodity or animal feed. Commonly abbreviated to **MRL**.

Mayonnaise **Salad dressings** prepared from **vegetable oils**, **egg yolks**, **vinegar** or other acidifying agent (e.g. **lemon juices**) and **flavourings** (e.g. **mustard**). For manufacture of commercial mayonnaise, oil content must be $\geq$65% (by weight). Commonly 70-80% (by weight) oil is used to give a thicker product that has been shown to be more acceptable to consumers.

MCPA Selective systemic herbicide used for post emergence control of annual and perennial broad-leaved weeds in **cereals**, **potatoes**, **peas** and **asparagus**. Classified by WHO as slightly toxic (WHO III).

Mead **Alcoholic beverages** made by **fermentation** of a medium in which **honeys** are the main source of fermentable **sugars**.

Meadowfoam Flowering plant, *Limnanthes alba*, which yields high quality **oils** from its seeds. 95% of the oil is composed of 20 and 22 carbon **fatty acids**. It shows high **oxidative stability** and may be used as a substitute for **whale oils** or **jojoba oils**.

Meadowfoam oils **Oils** extracted from meadowfoam (*Limnanthes alba*), which contain high proportions of **fatty acids** with more than 20 carbon atoms, including some which are unique to this oil. Display high **oxidative stability** and can improve the stability of other **vegetable oils**. Development of low **erucic acid** lines has enabled food application of meadowfoam oils. They have been used as **plasticizers** in **chewing gums**.

Meal **Flour** prepared from non-cereal plants.

Mealiness **Sensory properties** relating to the extent to which products (usually **fruits** such as **apples**, **peaches** and **nectarines**) are perceived as being mealy, i.e. soft, powdery and floury. Mealiness is the result of breakdown of flesh into small pieces that tend to be dry in the mouth; it is related to an increase in the levels of water-soluble **pectins** and decreases in insoluble pectins during **ageing**. Thus, when eaten, the cells separate easily without the release of cell sap, and the mouth perceives the outside surfaces of the cells rather than the cleaved cells leaking sap.

Meal replacers Products designed for consumption in place of conventional **meals** for a specific dietary purpose, e.g. weight management.

Meals Processed foods eaten at mealtimes and/or designed to be one of the main dishes of the day, e.g. lunches, pub meals, **ready meals**, **school meals**.

Meat Animal tissues which are used as food, including those of domestic mammals, poultry, game birds and game animals. Meat is composed of **lean** muscles, **connective tissues**, **fats**, skin, nerves, blood vessels and water. It can be classified as red or white, based on its **colour** intensity, which results from the proportion of red and white muscle fibres that it contains. Red fibres have a higher **myoglobin** content than white fibres. Composition of meat differs between species and between retail cuts; it depends greatly on

the fat to lean ratio, which determines energy value and concentrations of most nutrients. Water content of meat tends to decrease with increasing fat content. Lean meat includes substantial amounts of high biological value proteins; however, meat is also an important dietary source of fat, high **bioavailability** inorganic nutrients (including Fe, Zn, Cu and Se) and the vitamin B group.

Meat alternatives Alternative term for **meat substitutes**.

Meat analogues **Simulated foods**, comparable in structural and mechanical properties to natural **meat**. They can be produced from various high protein content, raw materials including **beans**, **fish** and **grain**, and also from protein recovered from **offal**. Ingredients such as dry spun protein fibres, e.g. **casein**, may be incorporated into meat analogue mixtures as **texture** imparting materials.

Meat balls Meat products prepared from chopped **meat**, which is formed into balls and then cooked. Ingredients may also include **onions**, **breadcrumbs**, **eggs** and **seasonings**.

Meat emulsions Meat products which include sausage emulsions and emulsions used in the preparation of comminuted meat products. They are composed of a continuous phase (protein and water) and a dispersed phase (fat particles). They are prepared from **meat**, such as **mechanically recovered meat** and **offal**, and other ingredients, such as non-meat proteins (e.g. **sodium caseinate** and soy protein isolates). **Enzymes** may be added to improve the **functional properties** of meat and non-meat proteins in the emulsions. Mechanical treatment during **comminution** has major effects on properties of products prepared from meat emulsions.

Meat extenders Non-meat ingredients used to improve **flavour**, **texture**, appearance and **nutritional values** of **meat emulsions**. In general, they cost less per kilogram than **meat**, and include: **dairy products**, such as **dried skim milk**, **sodium caseinate**, milk coprecipitates, **whey** and whey products, and other milk derivatives; soy protein isolates and concentrates; **oilseeds**; cereal products; and pea meal, chick pea meal and textured navy bean protein concentrate.

Meat extracts Water-soluble extracts of **meat** which are used as **flavourings**. **Meat mince** is immersed in boiling water to leach out the water-soluble extracts; meat extract (no. 1 extract) is produced by concentrating these extracts. Exhaustive extraction of meat produces a direct extract, which contains a high concentration of **gelatin**. Meat extracts are rich nutritional sources of the **vitamin B group**, particularly **vitamin B$_2$**, **vitamin B$_{12}$** and **nicotinic acid**.

Meat loaf Meat products commonly prepared from comminuted **meat**, such as **meat mince**, poultry mince or fish mince. Meat loaf may include **offal**, blood and low value meat, such as **mechanically recovered meat**. Other ingredients may include binders, **onions**, tomato puree, **garlic**, white **bread**, **milk**, **herbs** and **seasonings**. The ingredients are mixed before cooking, usually in a loaf tin; however, meat loaf may also be prepared in casings. Some meat loaf is prepared with **colour** contrasts or patterns; preparation of these products tends to involve traditional high-cost labour-intensive methods. Once cold, meat loaf can be cut into firm slices. Generally, it is served cold.

Meat mince **Meat** cut up or shredded (minced) into very small pieces by the process of **mincing**. Quality depends on the part of the animal carcass that the meat originated from; in particular, it varies with fat and connective tissue contents. Also known as ground meat or minced meat.

Meat pastes Comminuted meat products similar to **pates**, and of intermediate **texture**, commonly with a meat content of approximately 70%. The non meat portion consists of rusk and water, or other suitable filler such as **soy protein concentrates** or **sodium caseinate**. The product is usually heat sterilized after filling into jars or cans.

Meat patties Round, flat cakes of comminuted **meat**. Although they may be prepared from **meat mince**, they may also be reconstituted, e.g. from **mechanically recovered meat**. Some may include **meat extenders**. Varieties include **beef patties**, **chicken patties** and **turkey patties**.

Meat pies Meat products in which chopped meat or **meat mince** is encased in **pastry** and baked. Meat pies often contain **offal** and low value meat, such as **mechanically recovered meat**. They may be prepared in pie dishes that are lined and sealed with **pastry**, e.g. steak and kidney pie. Pasties are a type of meat pie prepared in a folded pastry case, e.g. **Cornish pasties**.

Meat sauces Any **sauces** that contain **meat** as the main ingredient. Meat sauces are usually used as an accompaniment to **pasta** and **rice**, for example bolognese sauces or meat curry sauces.

Meat substitutes **Simulated foods** used as direct substitutes for **meat**. They may be included in meat products or may provide vegetarian alternatives to meat. Meat substitutes include **textured vegetable proteins** (TVP), texturized milk proteins, **quorn** and **tofu**. **Aroma compounds**, **stabilizers** and **colorants** may be included. Also known as meat alternatives.

Mechanical boning Removal of bones from **meat** or **fish**, usually before **cooking**, using specially designed **boning** equipment.

Mechanical harvesting Gathering (harvesting) of crops by mechanical means.

Mechanically recovered meat **Meat** recovered from bone using separation machinery. Mechanical recovery increases the efficiency of separation and thereby allows the recovery of extra meat per carcass; it is also less time consuming than hand **boning** of meat. In many systems, meat and bone are forced against perforated plates or cylinders; the meat passes through, leaving the bone to be removed as waste. Composition of the meat recovered varies between the methods used, but in general consists of comminuted meat, bone marrow, collagen, bone and fat. Bone content is very important and must be minimized. Initial raw materials need to have low bacterial counts; they should be handled at low temperature and treated as promptly as possible. Advanced meat recovery (AMR) systems produce a product which is similar in appearance, texture and composition to meat trimmings and similar hand deboned meat products. Other systems produce a paste- or batter-like meat product, or liquid meat extracts. Mechanically recovered meat is widely used in meat products. It is also known as mechanically separated meat or mechanically deboned meat.

Mechanical properties In relation to foods, **physical properties** associated with the reaction of foods to **stress**. Include parameters such as **hardness**, **viscosity**, **elasticity** and adhesiveness.

Medical foods Foods specially formulated to be consumed by individuals who suffer from disease or health conditions that require special dietary management, because of distinctive nutritional requirements associated with the conditions.

Mediterranean diet Diet eaten in certain Mediterranean countries, in which the populations enjoy the lowest recorded rates of chronic diseases and the highest adult life expectancy. Contains an abundance of food from plant sources, including **fruits** and **vegetables**, **potatoes**, **bread** and grains, **beans**, **nuts** and **seeds**. Emphasis is placed on eating a variety of **minimally processed foods** and, wherever possible, seasonally fresh and locally grown foods. **Olive oils** replace other fats and oils (including **butter** and **margarines**) in the diet. The diet also includes daily consumption of low to moderate amounts of **cheese** and **yoghurt**, and weekly consumption of low to moderate amounts of **fish** and **poultry meat**, and from zero to four **eggs** per week (including those used in cooking and baking). Fresh fruit is used as the typical daily dessert; sweets with a significant amount of **sugar** (often in the form of **honeys**) and saturated fats are consumed not more than a few times per week. Red meat is consumed only a few times per month. There is also moderate consumption of **wines**, normally with **meals**.

Medlars **Fruits** produced by *Mespilus germanica*. Rich in sugar and potassium, but not a good source of **vitamin C**. Palatable only when partially rotten or after exposure to frost, when they become soft. Consumed along with **port** or used in making **jams** and **wines**.

Megasphaera Genus of anaerobic, spheroid **Gram negative bacteria** of the family Veillonellaceae. Some species, especially *Megasphaera cerevisiae*, are responsible for **spoilage** of **beer**.

Megrim **Marine fish** species (*Lepidorhombus whiffiagonis* or *L. boscii*) of high commercial value belonging to the family Scophthalmidea. Found in the north east Atlantic Ocean and western Mediterranean Sea. Flesh tends to be dry and is best eaten fried in fat. Skin is used as a source of **collagen** and **gelatin**.

Meitauza Traditional Chinese food made by fermentation of **okara**.

Meju Product made traditionally from **soybeans** that are malted, formed into blocks and dried. Fermented to produce **soy sauces** and **bean pastes** as by-products.

Melamine Toxic, nonflammable, colourless to white crystalline solid. Insoluble in organic solvents, very slightly soluble in ethanol and slightly soluble in water. When heated, it sublimes slowly before melting at 347°C. Used to make melamine resins and to give wet strength to paper. Melamine resins are made by polymerizing melamine with formaldehyde; they are used in various coatings, insulators and as adhesives.

Melanins High molecular weight **pigments** with reddish-brown to black **colour**, formed by the action of **oxidases** on **phenols**, as in **enzymic browning**. Widely distributed in animals and plants, generally bound to proteins. Although a normal constituent of certain foods and **beverages**, including **black tea**, melanins can sometimes produce an undesirable **discoloration** of foods, such as **mushrooms**, several **fruits** and **shrimps**.

Melanoidins **Pigments** with yellow to brown **colour** and **malt**-like **aroma** formed by reactions between **reducing sugars** and **amino acids** in foods during heating. Formation of these **Maillard reaction products** is important during food **processing** procedures such as **baking** and **roasting**.

Melanosis Darkening in **shrimps** between the shell and tail muscle, which develops as the product deteriorates. Produced by an enzymic reaction affecting naturally occurring **amino acids** when exposed to sun-

light. While they may not be as attractive, affected shrimps are safe to eat, unless **spoilage** characteristics are present. Sulfiting agents are used to prevent melanosis.

Melatonin Hormone produced by the pineal gland in animals where it stimulates colour change in lower vertebrates and plays a role in circadian rhythms of humans. Also present in **insects**, **bacteria** and **plants**. Its activities as a broad-spectrum free radical scavenger and indirect antioxidant suggest health benefits of ingestion.

Melezitose Trisaccharide formed from two molecules of **glucose** and one molecule of **fructose**. Occurs naturally in **honeys** and tree exudates.

Melibiases Alternative term for **α-galactosidases**.

Melibiose Disaccharide formed from a molecule of **galactose** and a molecule of **glucose** linked by a 1,6-glucosidic bond. The dihydrate of melibiose has approximately one third the **sweetness** of **sucrose** by weight.

Melomel Type of **mead** made from **honeys**, water and any **fruits** other than **grapes** or **apples**.

Melons Widely grown **fruits** produced by *Cucumis melo*. Available in a number of types, including winter melons, **cantaloupes**, **muskmelons** and ogen melons, which differ in surface and flesh characteristics. Consumed as a dessert, sometimes sprinkled with **ginger** or **lemon juices**. Flesh contains at least 90% water, relatively high amounts of sugar and **vitamin C** and, in cases where there is a pink or orange colour, high levels of **carotenes**.

Melon seeds Seeds found in the centre of **melons**. Rich in protein and fat. Used in the manufacture of **bakery products** and **confectionery**, as well as in the preparation of **beverages**. Also roasted and consumed as **snack foods**.

Melting Conversion of solid foods (such as **butter** or **chocolate**) into a liquid or semi-liquid state by application of heat.

Melting point Temperature at which a solid changes into a liquid, i.e. the solid and liquid forms exist together in equilibrium. A pure substance at a pressure of 1 atmosphere has a single reproducible melting point. The melting point is a characteristic of a pure substance; the presence of impurities lowers the melting point.

Membrane bioreactors **Bioreactors** in which reaction products are removed through **membranes** by, for example, **ultrafiltration**, **reverse osmosis** and **dialysis**, thus allowing continuous operation. Can be used in processes such as **bioremediation** of **waste water**, **purification** of **drinking water**, **bioconversions** and **biotransformations**. The membranes can also be used as supports for **immobilization** of **enzymes** or cells.

Membranes Solid matrices used for separation of molecules in processes such as **dialysis**, **filtration** and **reverse osmosis**, as supports for **immobilization** of cells and **enzymes**, and in techniques such as **blotting** and **hybridization**.

Menadione Synonym for **vitamin K_3**. Synthetic compound with **vitamin K** activity, used in prevention and treatment of hypoprothrombinaemia, secondary to factors that limit absorption or synthesis of vitamin K. Two to three times more potent than naturally occurring vitamin K.

Menaquinones Synonym for vitamin K_2 series. A variety of metabolites with **vitamin K** activity synthesized mainly by intestinal **bacteria**. Found in **meat**, **livers**, **eggs** and **cheese**. Formerly called farnoquinone.

Menhaden Any of several species of **herring**-like **marine fish** from the genus *Brevoortia*. Found off the east coast of the USA, in the Atlantic and in the Gulf of Mexico. Marketed fresh, salted, canned or smoked; mainly used for production of **oils**, **fertilizers** and **fish meal**.

Menhaden oils Important commercial **fish oils** which are rich in ω-3 **polyunsaturated fatty acids**. Extracted from fish belonging to the genus *Brevoortia*.

Menthol One of the monoterpenoid **aroma compounds** and a secondary alcohol. Characteristic component of **mint oils**. Widely used in mint **flavourings**.

Menthone Member of the monoterpenoid **aroma compounds**, with a ketone functional group. Present in **mint** and **mint oils**, and used in mint **flavourings**.

Mercaptans **Organic compounds**, synonym **thiols**, containing the thiol (-SH) group, also called a mercapto group or a sulfhydryl group. Sulfur analogues of **alcohols** in which the oxygen atom has been replaced by a sulfur atom.

Mercaptophos Alternative term for the insecticide **fenthion**.

Mercosur A regional trade organization formed in 1991 to establish a common market and a common trade policy.

Mercury A heavy metal, chemical symbol Hg, formerly known as quicksilver. Liquid at room temperature, and exhibiting two valencies - mercury(I) and mercury(II). Present in the environment naturally as mercury sulfide, but also as an industrial pollutant, for example as **methylmercury**, and occurs as a contaminant in foods. Accumulation of mercury in **fish** and other **sea foods** is of particular concern. Toxic-

ity symptoms include chronic muscular problems and reduced fertility.

Merguez Highly seasoned fresh **sausages** which are popular in France. They are prepared from **beef**, **pork** or **mutton**. Usually, they are grilled or fried before eating.

Meringues **Confectionery** products made by **whipping egg whites** to a foam, incorporating **sugar** and drying to a crisp finish. The term may refer to small cakes or shells made of this material which have been decorated or filled, e.g. with **whipped cream**, **ice cream** or **fruits**. Also used as **toppings** added to **flans** or **pies**, as in lemon meringue pies.

Merissa Type of **sorghum beer** made and consumed in Africa.

Mesentericins Class II **bacteriocins** produced by *Leuconostoc* *mesenteroides*. Active against species of the genera *Lactobacillus*, *Carnobacterium* and *Listeria*, including *L. monocytogenes*, suggesting a role in food **preservation**.

Mesophiles Organisms, especially **microorganisms**, that grow best at intermediate temperatures. Their optimum growth temperature lies within the generally accepted range of 20 to 45°C.

Mesquite pods Pods produced by the mesquite tree (*Prosopis* species, including *P. juliflora* and *P. glandulosa*, honey mesquite), a plant that grows well in semi-arid climates. The sweet pods are a good source of calcium, manganese, iron and zinc, and are sometimes made into **syrups**. Mesquite meal, made by grinding whole pods, is additionally rich in proteins and is effective in controlling blood sugar levels in diabetics due to the high content of **fructose**. It is used in **flavour enhancers** for a large range of cereal products and other foods. Seeds inside the pods can be eaten whole or ground to a flour that is used in food products such as **bakery products** and **burgers**. They are also a source of **gums**.

Mesquite seed gums **Gums** obtained from seeds of trees of the genus *Prosopis*. Physical and chemical properties of mesquite seed gums resemble those of **gum arabic**, for which they can be used as substitutes.

Metabisulfites Disulfurous acids, the disodium salts of which are used as **preservatives** and **antioxidants**.

Metabisulphites Alternative spelling of **metabisulfites**.

Metabolic disorders Generic term for diseases caused by an abnormal metabolic process. They can be congenital, due to inherited enzyme abnormality (inborn errors of metabolism), or acquired due to disease of an endocrine organ or failure of a metabolically important organ such as the liver.

Metacercariae Mature infectious forms of parasitic trematode larvae.

Metalaxyl Systemic, benzenoid fungicide used for control of a wide range of fungal **diseases** in food **crops**, including **fruits** and **vegetables**. Classified by WHO as slightly hazardous (WHO III).

Metal detectors Electronic devices that give an audible signal when close to metal; used to detect metal foreign bodies or contaminants during food processing.

Metalloenzymes **Enzymes** that contain a bound metal ion as part of their structure. This ion may be required for enzymic activity, either participating directly in catalysis or stabilizing the active conformations of the proteins.

Metallothioneins Cysteine-rich **proteins** which bind divalent heavy metal ions. Widely distributed in animals and **microorganisms**. Metallothionein-like proteins have been identified in plants.

Metals Metals are generally solid, have a metallic lustre, are malleable and ductile, and conduct both heat and electricity. Approximately 75% of known **minerals** are metals. Metal ions can replace the hydrogen in acids to form salts, they also form alloys with each other.

Metanil yellow **Azo dyes** not permitted for use in foods, drugs or cosmetics. Also called CI Acid Yellow 36.

Metaphos Alternative term for the insecticide **parathion-methyl**.

Methallyls Short-chain aliphatic compounds with alcohol, chloride or cyanide substituents. Used as **fumigants** to control pests in stored grain.

Methamidophos Systemic insecticide and acaricide used to control chewing and sucking **insects** and spider **mites** on a range of crops, particularly **pome fruits**, **stone fruits**, **citrus fruits**, **potatoes** and **corn**. Classified by WHO as extremely hazardous (WHO Ib). Also known as monitor.

Methanal Simplest of the **aldehydes**, synonym **formaldehyde**.

Methanearsonic acid Alternative term for the herbicide **methylarsonic acid**.

Methanethiol Smallest of the **thiols**, synonym **methyl mercaptan**. One of the volatile **aroma compounds** found in **cheese** and other foods.

Methanol One of the **alcohols**, methanol contains a single carbon atom, and is a light, volatile flammable, poisonous, sweet-smelling liquid at room temperature. Widely used as a solvent, antifreeze or fuel. Can occur as a fermentation by-product in **alcoholic beverages** and **vinegar**. Synonym for **methyl alcohol**.

Methidathion Non-systemic insecticide and acaricide used for control of a wide range of chewing and sucking **insects** (especially scale insects) and spider **mites** in a wide range of **fruits**, **vegetables** and **cereals**. Classified by WHO as extremely hazardous (WHO Ib). Also known as supracide.

Methional Aldehyde, synonym 3-(methylthio)propionaldehyde, with a boiled-potato like **aroma**. Important aroma compound in **wines**; also identified in many other foods, including **sea foods**, **coffee**, **beer** and **yeast extracts**.

Methionine One of the essential dietary **amino acids**, this thiol-containing amino acid is a common protein constituent in foods. Also a precursor of several **organic sulfur compounds** which are important in food **flavour**.

Methionol A thiol alcohol, synonym 3-(methylthio)-1-propanol. One of the important sulfur **flavour compounds** found in **wines** and fermented soy products.

Methomyl Systemic insecticide and acaricide used to control a wide range of **insects** and spider **mites** on **fruits** and **vegetables**. Also used for control of **flies** in animal houses and **dairies**. Classified by WHO as extremely hazardous (WHO Ib). Also known as lannate.

Methoprene Hormonal insecticide (pheromone analogue) with insect growth regulating activity. Used for control of a range of **insects** in food storage areas and processing and handling establishments. Also used in cultivation of **mushrooms**. Classified by Environmental Protection Agency as not acutely toxic (Environmental Protection Agency IV).

Methoxychlor Insecticide used for control of a wide range of **insects** (particularly chewing insects) in **fruits**, **vegetables** and **cereals**. Also used for insect control in animal houses, **dairies** and food factories.

Methyl alcohol Alternative term for **methanol**.

Methylamine Amine present in a wide range of foods and **beverages**. Often included in with the **biogenic amines** in food analyses.

Methylarsonic acid Selective contact herbicide used for control of grass weeds in **sugar cane** and **citrus fruits** plantations. Classified by WHO as slightly toxic (WHO III).

Methyl bromide Colourless, poisonous gas, synonym bromomethane. Used in the **fumigation** of **fruits** and **vegetables** to control pests, but now largely being replaced with other **fumigants** such as **phosphine**.

3-Methylbutanal Chemical name for **isovaleraldehyde**. May be one of the **flavour compounds** or cause **taints** in various foods, **beverages** and water.

Methyl butanol One of the aliphatic **alcohols**, with a characteristic odour and pungent taste. Synonyms include **isoamyl alcohol**, isopentanol and isopentyl alcohol. Used as an **esterification** substrate for production of isoamyl esters. Also identified as one of the **aroma compounds** present in **wines**, **cider** and **beer** as a result of yeast **fermentation**.

Methyl carbamate Carcinogen that may occur, along with **ethyl carbamate**, in some **fermented foods** and **alcoholic beverages**.

***N*-Methylcarbamate insecticides** Class of **insecticides** sharing carbamic acid as a common base structure. Widely used for control of insect **pests** on crops and in food storage and preparation areas. Generally biodegradable and of low soil persistence. Commonly used examples include **aldicarb**, **carbaryl**, **methomyl** and **propoxur**.

***N*-Methylcarbamate pesticides** Major class of **pesticides** which includes ***N*-methylcarbamate insecticides**. Members share carbamic acid as a common base structure.

Methylcellulose Methyl ester of cellulose. Prepared by alkali treatment of **celluloses** followed by methylation of the alkali cellulose with chloromethane. Due to its ability to absorb water and form viscous colloidal aqueous solutions, methylcellulose is often used as a substitute for **gums**. Food industry uses include as **thickeners**, **stabilizers**, **emulsifiers**, **bulking agents** and **binders**. It is often included as an ingredient of **fillings** for **meat pies** or **fruit pies**.

1-Methylcyclopropene Volatile unsaturated cyclic hydrocarbon which acts as an inhibitor of **ethylene** activity by binding to ethylene receptors. Inhibits postharvest **ripening** and **softening** in **fruits** and **vegetables**, thus extending **shelf life**.

Methylglyoxal Aldehyde present in many foods, but most commonly determined along with other dicarbonyl compounds as a natural component in **beer** and **wines**, and as an **ozonation** by-product in water purification. Synonyms include pyruvic aldehyde and **pyruvaldehyde**. Can be formed as one of the **Maillard reaction products** in **nonenzymic browning**, but is toxic at high levels.

Methylhistidine Amino acid which is frequently determined in **meat** and meat products as an indicator of connective tissue content.

Methyl iodide Organic halogen compound, synonym iodomethane. Used in some **disinfectants** and in **fumigation** of fruits. Also used in several analytical techniques, including methylation treatments.

2-Methylisoborneol Compound of the **terpenoids** group, formed by soil **microorganisms**, which can cause mouldy, musty taints in water and **freshwater fish**.

Methyl jasmonate One of the group of **plant growth regulators** which control growth and development. Particularly involved in plant defence responses. Can be applied exogenously to control fruit development and abscission.

Methyl linoleate Methyl ester of **linoleic acid**. Used widely as a substrate in studies of lipid **oxidation** and **antioxidative activity**.

Methyl mercaptan Smallest of the **thiols**, synonym **methanethiol**. One of the volatile **aroma compounds** found in **cheese** and other foods.

Methylmercury Organomercury compound produced as a result of industrial activity and present environmentally as a pollutant of soils and water, and hence plants and animals. Often measured as an indicator of **mercury** contamination of foods, especially **sea foods** and water.

***S*-Methylmethionine** Synonym for **vitamin U**. A compound found in raw **cabbages**, other **green vegetables**, **beer** and **citrus juices**. A precursor of the **off flavour** compound **dimethyl sulfide**. Used in treatment of ulcers.

Methylobacillus Genus of obligately methanol assimilating **Gram negative bacteria**. Of particular interest as sources of **biomass** and **exopolysaccharides**.

Methylococcus Genus of Gram negative, aerobic coccoid **bacteria** of the family Methylococcaceae. Occur in mud, soil and water. Capable of oxidizing methane. *Methylococcus capsulatus* is used in the production of **single cell proteins**.

Methylomonas Genus of Gram negative, aerobic rod-shaped **bacteria** of the family Methylococcaceae. Obligately methylotrophic (able to metabolize single-carbon compounds as the sole source of both carbon and energy). Some species, e.g. *Methylomonas methanica*, *M. methylotropha* and *M. clara*, are used in the production of **single cell proteins**.

Methylparaben Common name for 4-hydroxybenzoic acid methyl esters used as **preservatives** for foods and **beverages**.

Methylparathion Alternative term for the insecticide **parathion-methyl**.

Methylpentoses General term for **sugars** containing six carbon atoms but only five hydroxyl groups. Examples include **rhamnose** and **fucose**.

Methyl propanol One of the aliphatic **alcohols**, with a mild alcoholic, sweet odour. Synonyms include **isobutyl alcohol** and **isobutanol**. Several isomers exist, including 2-methyl-1-propanol and 2-methyl-2-propanol. One of the **aroma compounds** produced during **fermentation** of **alcoholic beverages**, including **wines**, **beer** and **cider**.

Methyl sulfide Colourless liquid, synonym **dimethyl sulfide**, commonly used as a solvent. Also occurs naturally in foods and **beverages**, generally as an **off odour** resulting from bacterial metabolism of sulfur-containing **amino acids**.

Methylthiophanate Alternative term for the fungicide **thiophanate-methyl**.

Methylxanthines Group of **alkaloids** including **caffeine, theobromine** and **theophylline**, which are commonly found in **tea, coffee, cola beverages, cocoa** and **chocolate**.

Metmyoglobin A brown pigment, formed by **oxidation** of **myoglobin**, in which water is bound to the ligand, and the haem group of myoglobin is in the ferric (Fe^{3+}) state. In **meat**, metmyoglobin produces a brown/grey coloration, which is unattractive to consumers; thus, metmyoglobin formation is a major problem in maintaining a stable display of retail meat. Several approaches may be taken to delay the formation of metmyoglobin in meat, including: production of meat from animals fed on antioxidant supplemented feeds; use of **modified atmosphere packaging** for meat; and treatment of the meat surface with **antioxidants**, such as **vitamin C**.

Metolachlor Selective herbicide used for pre-emergent control of annual grasses and some broad-leaved weeds around **cereals** and vegetable crops. Classified by WHO as slightly toxic (WHO III).

Metribuzin Selective systemic herbicide used for pre- and post-emergent control of many grasses and broad-leaved weeds around **cereals** and vegetable crops. Classified by Environmental Protection Agency as slightly toxic (Environmental Protection Agency III). Also known as sencor.

Metroxylon Genus of **palms**, the trunks of which are a source of **sago**. Main species is *Metroxylon sagu*, but *M. rumphii* is also a sago producer. Young apical shoots of the plants (palm hearts) are consumed as a vegetable.

Mettwurst Fermented **sausages** made from minced, cured **pork** and **beef**; they are a type of German-style **salami**. They are seasoned using ingredients such as **allspice, coriander, ginger** and **mustard**. Mettwurst are smoked and air dried. There are two major types, fresh (raw) and cooked. They are made using starter cultures. Recipes vary widely, and accordingly the characteristics of mettwurst are very diverse. For example, consistency ranges from very finely minced to coarsely chopped, and from spreadable to elastic and sliceable.

Mevalonic acid One of the **organic acids**, synonym 3,5-dihydroxy-3-methyl-valeric acid. Important intermediate in the synthesis of **isoprenoids**.

Mezcal Spirits made in Mexico by **distillation** of the fermented sap of the **agave** plant.

Mg Chemical symbol for **magnesium**.

Micellar electrokinetic chromatography Capillary **electrophoresis** technique in which neutral compounds are separated using surfactant micelles. Usually abbreviated to MEKC.

Microalgae Microscopic **algae**. Particularly unicellular algae such as *Chlamydomonas* and *Chlorella*.

Microbacterium Genus of Gram positive, aerobic rod-shaped **bacteria** of the family Microbacteriaceae. Some species, e.g. *Microbacterium lacticum and M. flavum*, occur in **dairy products** (e.g. spray **dried milk, cheese**), presumably due to improper cleaning of dairy equipment, and cause **spoilage**. *M. thermosphactum* may cause spoilage of vacuum-packaged **meat** and meat products.

Microbial biomass Quantitative estimate of the entire assemblage of **microorganisms** in a given habitat in terms of mass, volume or energy.

Microbial counts Numbers of **microorganisms** in a given sample.

Microbial proteins Proteins produced by **microorganisms**.

Microbial rennets Enzymes sourced from **microorganisms**, commonly **fungi**, that are used as substitutes for **animal rennets** in **coagulation** of milk for **cheesemaking**.

Microbial spoilage Spoilage caused by the activity of **microorganisms**.

Microbial spores Spores of **bacteria** or **fungi**.

Microbicidal compounds Compounds used for killing **microorganisms**.

Microbiological quality Extent to which a substance (e.g. a food) is contaminated with **microorganisms**.

Microbiological techniques Techniques used in **microbiology**, including those used to detect or quantitate **microorganisms** in substances such as foods and **beverages**.

Microbiology Scientific study of **microorganisms** and their interactions with other organisms and the environment.

Microbreweries Small **breweries** making speciality **beer** in small quantities (generally under 15,000 barrels annually). Frequently, the products are sold on the premises.

Micrococcaceae Family of Gram positive, aerobic or facultatively anaerobic coccoid **bacteria**. Consists of the genera *Arthrobacter*, *Kocuria*, *Micrococcus*, *Nesterenkonia*, *Renibacterium* and *Rothia*.

Micrococcus Genus of Gram positive, obligately aerobic coccoid **bacteria** of the family **Micrococca-**

ceae. Occur in soil, water and **raw milk**. *Micrococcus varians* is used as a starter in the ripening of dry **fermented sausages**. Other species may cause **spoilage** of **meat** and **eggs**.

Microcrystalline celluloses Highly crystalline particulate material produced by hydrolysis of **celluloses**. Used in **stabilizers**, **texturizing agents** and **fat substitutes** in foods such as **salad dressings**, **dairy products** and **bakery products**.

Microcystins **Hepatotoxins** produced by some strains of the cyanobacterium *Microcystis aeruginosa*. Exert hepatotoxic effects in humans and animals upon ingestion of contaminated water.

Microcystis Genus of Gram negative, photosynthetic **cyanobacteria** that occur in aquatic environments. Species are planktonic in fresh water, and often form blooms in water (e.g. reservoirs). *Mycrocystis aeroginosa* produces **microcystins** which are hepatotoxic in humans and animals upon ingestion of contaminated water.

Microemulsions **Emulsions** having a droplet diameter that is too small to be seen by the naked eye, typically 10-100 nm. Applications include edible films, **coatings** and delivery systems for **nutrients** and **flavourings**.

Microencapsulation Process in which thin films or polymer coatings are applied to small solid particles, droplets of liquids or gases. Can be used to encapsulate **enzymes**, **microorganisms**, **flavour compounds**, **sweeteners** and food ingredients. Useful for controlled **flavour** release and enhancing the stability of sensitive ingredients. Methods for microencapsulation include spray drying, spray chilling and spray cooling, extrusion, air suspension coating, liposome entrapment, co-crystallization, molecular inclusion and interfacial polymerization.

Microfiltration A method of sterile **filtration** that removes particles of approximately 0.1-10.0 micrometers in size, such as large fat globules, large proteins and suspended particles such as microbial cells. Microfiltration is generally used in the **clarification** and separation of **beer**, **wines** and **soft drinks**, and in the dairy industry for processing of low heat sterile **milk**.

Microflora In a microbiological context, refers to all the **microorganisms** present in a particular habitat. May refer to all the microscopic plants, **bacteria**, **fungi** and **algae** present in a particular habitat in a broader biological context. Also, can be used to describe the plants, bacteria, fungi and algae that are present in a particular microhabitat.

Microfluidization High pressure **homogenization** technique for deagglomeration and dispersion of uni-

form submicron particles and creation of stable **emulsions** and **dispersions**. Combined forces of shear and impact act upon products to create finer, more uniform dispersions and emulsions than can be produced by any other means.

Micrometers Instruments used in conjunction with microscopes or telescopes for measuring small distances.

Micromonospora Genus of Gram positive, aerobic filamentous **bacteria** of the family Micromonosporaceae. Occur in soil, decaying vegetation and water. Some species, e.g. *Micromonospora chalcea* and *M. cellulolyticum*, produce **cellulases**.

Micronization Indirect infrared (IR) heating method that relies on heat that is generated externally being applied to the surface of a food mostly by radiation, but also by convection, and, to a lesser extent, conduction. IR heating is mostly used to alter **eating quality** of foods by changing the surface **colour**, **flavour** and **aroma**. The main commercial application of radiant energy is in **drying** of low moisture foods and in **baking** and **roasting ovens**.

Microorganisms Microscopic organisms which include **algae**, **bacteria**, **fungi**, **protozoa** and **viruses**.

Microsatellite markers Highly polymorphic **DNA** markers comprising mono-, di-, tri- or tetra-nucleotides repeated in tandem arrays and distributed throughout **genomes**. Used as genetic markers in **genetic mapping** studies.

Microsatellites Repetitive stretches of short sequences of **DNA** distributed throughout **genomes**.

Microscopes Apparatus used to make a magnified image of a small sample. Include light microscopes and more complex instruments such as electron microscopes that measure transmission, reflection or emission of electrons from the sample.

Microscopy Analysis of samples using microscopes which produce magnified images. Includes basic light microscopy and more complex techniques such as **electron microscopy**.

Microwaveable containers Containers that may be used safely for microwave **cooking** or **reheating** of foods. Must be made of materials that will not cause damage to **microwave ovens** during operation or allow **migration** of undesirable components into the foods being heated.

Microwaveable foods Foods suitable for heating in **microwave ovens**. **Ready meals** that can be rapidly reheated in this manner and **microwave popcorn** are some of the most popular types of microwaveable products. Further applications have been limited by problems such as lack of **browning** in foods cooked in **microwave ovens**, and arcing during microwave cooking of **foils** packaged foods. However considerable advances have been made in development of **microwave susceptors** and other devices for promoting browning and crisping during microwave heating and cooking.

Microwaveable packaging   **Packs** or wrappings that may remain on foods during microwave **cooking** or **reheating** without causing damage to **microwave ovens** or causing **contamination** of the products with undesirable components.

Microwave ovens   Ovens, ranging in power from approximately 500 to 900 Watts, that use **microwaves** to cook, heat or defrost foods. A microwave oven cooks with high-frequency electromagnetic waves that cause food molecules to vibrate, so creating friction that heats and cooks the food. Microwaves penetrate only a few centimeters into the food, so the centre of most products is cooked by heat conduction. Non metal containers (such as glass and ceramics) need to be used in these ovens, as microwaves can pass through them (unlike metal) and therefore cook the food from all angles at once. As the microwaves pass through these non metal containers, they are able to stay relatively cool themselves while the food they contain becomes hot. However, during longer cooking periods, the containers can become very hot due to heat conduction from the hot food. To assist in administration of an even distribution of microwaves, some microwave ovens have turntables; others have revolving antennae for the same purpose. Factors that affect how quickly food cooks in a microwave oven include: temperature of the food when cooking begins; volume of food being cooked at one time; size and shape of the food; amount of fat, sugar and moisture in the food; bone distribution; and food density. Because of the way in which foods are cooked in microwave ovens, **browning** does not occur in the normal manner. **Microwave susceptors** are used to promote browning.

Microwave popcorn   **Popcorn** made with popping **corn** which has been specially formulated for preparation in **microwave ovens**.

Microwaves   Electromagnetic waves with a wavelength in the range 0.001 to 0.3 m, shorter than that of normal radio waves, but longer than those of **infrared radiation**. Microwaves are used in **microwave ovens** for **cooking**, **heating** and **defrosting** of foods.

Microwave susceptors   Devices used in the form of **active packaging** that cause **browning** and crisping of foods that are prepared in **microwave ovens**. A microwave active metal is lightly deposited on a thermally stable substrate (such as **PET**) and this sheet is laminated onto a back stock that provides rigidity. Once placed in the microwave, these packages will reach temperatures in excess of 170°C almost instantaneously. The high temperatures allow the food to cook quickly, and promote the **Maillard reaction**, thus enhancing browning characteristics.

Migraine   Condition characterized by severe, usually unilateral, vascular headache. Sometimes combined with any of a range of other symptoms such as nausea, vomiting and heightened sensitivity to light or sound. In some cases, attacks are triggered by ingestion of specific foods, **food additives** or **beverages**. Commonly suspected dietary triggers include **alcoholic beverages**, beverages containing **caffeine**, cheese, some **beans**, **cured meat** and **chocolate** based products.

Migration   Movement of undesirable compounds (e.g. **plasticizers** from **packaging materials**) into foods.

Milk   Secretion of the mammary gland of mammals. Composition varies among species, and is affected by many factors, including feeds and season. When used without further clarification, the term milk is generally accepted to mean **cow milk**. Cow milk is sold in various forms that differ in fat content (**whole milk**, **semi skimmed milk** and **skim milk**). Whole milk contains approximately 87% water, 4% fat, 3% protein and 5% **lactose**. It is rich in **calcium** (approximately 1.2 g/l), **riboflavin** (2 mg/l), **vitamin B$_{12}$** and **iodine**. A good source of **vitamin A** and **vitamin B$_1$**. Also contains **folates** and other **vitamins** and **minerals**. Due to risk of contamination with **pathogens** and **spoilage** organisms, milk for drinking is generally sold pasteurized, sterilized or UHT (ultra high temperature) treated, although **raw milk** is sometimes used to make dairy products such as **cheese**.

Milk beverages   Drinks in which **milk** is a major constituent. Include **milkshakes**, **flavoured milk**, carbonated milk beverages, milk mixed with fruit or vegetable juices or pulps, and products enriched with specific nutrients, e.g. **fibre** or **calcium**. Alternative term for milk drinks.

Milk chocolate   Type of **chocolate** made by incorporating **milk powders** along with **sugar**, **chocolate liquor** and **cocoa butter**. More widely eaten than dark chocolate and white chocolate. Compared with dark chocolate, milk chocolate has a creamier **texture** and taste, and tends to be softer.

Milk clotting   Process in which **milk** is separated into **curd** and **whey** by the action of **milk clotting enzymes**, e.g. **rennets**, **lactic acid** produced by **bacteria**, or a combination of both. Used in **cheesemaking**. During clotting (or **coagulation**), κ-casein, which resides on the surface of **casein micelles** and confers stability, is removed by the action of the enzymes, causing the destabilized **casein** to precipitate;

acid acts by destroying linkages between components of the micelle. Curd produced by using enzymes generally has a higher **calcium** content than that formed by **acids**.

Milk clotting enzymes **Enzymes** used in the **clotting** or **coagulation** of **milk** during **cheesemaking**. Most commonly, **rennets** extracted from the stomach of young ruminants have been used traditionally in this process, but other sources of enzyme have been developed in the light of shortages of **animal rennets** and the increasing popularity of vegetarian products. Alternatives include **microbial rennets**, produced by a range of **microorganisms**, and enzymes produced by plants, e.g. **cardoons**.

Milk fat globule membranes Membranes surrounding **milk fat globules**, comprising approximately 60% lipid and 40% protein. **Enzymes** and **trace elements** are associated with these membranes. When broken down, e.g. during **churning** of **cream** in **buttermaking**, fat globules are released and may coalesce.

Milk fat globules Emulsified form in which **milk fats** exist in milk. Surrounded by **milk fat globule membranes**. Fat globules have a diameter of 2-6 μm and a large surface area.

Milk fats **Lipids** present in **milk** mainly in the form of emulsified **milk fat globules**. Mainly **triglycerides**, with small amounts of **monoglycerides**, **diglycerides**, **cerebrosides** and free **fatty acids**. **Milk fat globule membranes** also contain **phospholipids** and **sterols**. Fat content of milk varies greatly among species and among animal breeds. Fatty acid composition of milk fats is governed by many factors, including feed, lactation stage and fat content of milk. **Cow milk** fat contains a great number of fatty acids, principal ones including **palmitic acid**, **oleic acid**, **myristic acid**, **stearic acid**, **linolenic acid** and **linoleic acid**.

Milkfish **Marine fish** species (*Chanos chanos*) widely distributed in the Indo-Pacific; a commercially important food fish in south east Asian countries. Flesh is white and tender. Marketed fresh or frozen; often used to make **fish cakes** and **surimi**.

Milk ice Product similar to **ice cream** but generally containing less milk fat.

Milk infant formulas Preparations for feeding to infants and young children intended to satisfy their specific nutritional requirements. Made from **cow milk** with the nutrient composition adjusted to mirror that of **human milk**. Composition is varied according to the age of the infant to be fed.

Milking Drawing of **milk** from the udder of female mammals. Extraction is performed manually or, where large numbers of animals are to be milked, using equipment specifically designed for the purpose (**milking machines**).

Milking frequency Number of times an animal is milked during a given period. Dairy cattle are generally milked twice daily. Alteration of milking frequency can have effects on **milk** composition and quality.

Milking interval Time elapsing between consecutive milkings. Affects **milk** composition and quality.

Milking machines Devices used to extract **milk** from the udders of mammals. Modern machines operate by suction, utilizing a partial vacuum and a pulsating action to simulate hand milking.

Milk powders Products prepared by drying **whole milk** to a low moisture content, giving a powder with a long **shelf life**. Also called **dried milk**.

Milk products Alternative term for **dairy products**.

Milk protein concentrates Preparations made by **concentration** of milk, usually **skim milk**, by **ultrafiltration**, during which **milk proteins** are separated from other milk constituents, followed by drying. Vary in milk protein content according to manufacturing procedures. Used in a similar way to **skim milk powders** in a variety of foods, including **processed cheese**, **infant formulas**, **beverages**, **bakery products**, **dairy spreads** and diet products.

Milk proteins Proteins found in milk, comprising **casein** (approximately 75% of total protein) and the **whey proteins**, including **α-lactalbumin**, **β-lactoglobulin**, serum albumin and **immunoglobulins**. Used as ingredients in various foods, such as **bakery products**, coffee **whiteners**, nutritional beverages and **imitation cheese**, to modify **functional properties** and **sensory properties**. In some food allergies, **cow milk** proteins act as **allergens**.

Milk puddings **Puddings** made by baking **milk** with a grain, such as **rice**, **semolina** or **tapioca**, **sugar** and sometimes **flavourings**.

Milkshakes **Beverages** made by addition of **flavourings**, often **fruits**-based, to **milk** and agitation by beating or shaking, sometimes with the addition of **ice cream**.

Milk substitutes Multipurpose term covering replacements for mothers' milk, e.g. **human milk**, used in infant feeding (**infant formulas**) or young animal feeding, as well as products prepared for use by individuals unable to tolerate **milk** or not wishing to consume it for other reasons, e.g. vegans. Depending on the intended consumers, the latter category may or may not contain dairy components, e.g. **whey**. Non dairy foods used as the basis of milk substitutes include **soybeans** and **oats**.

Millet Small seeds from any of a number of cereal grasses, including **common millet** (*Panicum miliaceum*), **finger millet** (*Eleusine coracana*), **foxtail millet** (*Setaria italica*), **pearl millet** (*Pennisetum typhoideum*) and **teff** (*Eragrostis tef*). Good sources of many minerals and with good storage properties. Forms the staple diet of much of the world population, especially in Asia and Africa. Consumed like **rice** or made into various products such as **porridge**, **gruel** and **bread**.

Millet oils **Vegetable oils** extracted from **millet** grains.

Milling Grinding in **mills**. For example, grinding of grain to produce **flour**.

Milling properties Ability of solid materials, such as grain, to be ground into powders.

Mills Machinery for **grinding** solid substances, or buildings equipped with such machinery. Include equipment used for grinding grain into **flour**.

Milo Drought resistant grain **sorghum**, especially *Sorghum bicolor*, which is similar to **millet** and is grown in Africa, Asia and the USA.

Milt Gonads from male fish, particularly **herring** and **mackerel**. Often called soft roe. Marketed as fresh or canned products.

Miltone Vegetable-toned milk product developed in India to overcome problems of **milk** shortages. Prepared using a protein isolate from **peanuts** which is added to **cow milk** or **buffalo milk** along with **sugar** and **vitamins**. Suitable for drinking on its own, in tea or coffee, or for processing into **yoghurt**.

Minas cheese Brazilian **fresh cheese** made from **cow milk**. White with a mild **flavour**.

Minas Frescal cheese Brazilian **fresh cheese** made from **cow milk**. White with a mild **flavour**.

Minced beef Alternative term for **beef mince**.

Minced meat Alternative term for **meat mince**.

Mincers Devices used to cut up or shred foods, particularly **meat**, into very small pieces.

Mincing Shredding or **cutting** up of food, particularly **meat**, into very small pieces, usually using devices called **mincers**.

Mineral oils Oils derived from hydrocarbon sources, some of which may be of food grade and may be used as **food additives**. Other mineral oils are not of food grade and may act as food contaminants.

Minerals Solid inorganic elements, including metals and non-metals. Also compounds occurring naturally in the earth's crust. Minerals are not normally volatilized when their organic matrix is ashed to remove car-

bonaceous materials. Many minerals are essential nutrients in that they are necessary in the diet of humans or animals to allow completion of the life cycle.

Mineral waters Natural **spring waters** or similar waters, which are produced and bottled under conditions specified under national regulations. In the UK, mineral waters may also be used as a general term for carbonated **soft drinks**.

Minimally processed foods Foods that are processed using technologies that do not significantly alter their fresh-like attributes, but achieve reliable preservation and control over enzyme activity and microbial growth. Most commonly applied to **fruits** and **vegetables**. Examples of such processing for fruits and vegetables include washing, **sorting**, **cutting**, **trimming**, **slicing** and **dicing**. Other methods of minimal processing include various high temperature short time (HTST) thermal processes in combination with minimal processes such as hermetic **packaging** and **refrigeration**. In order for minimally processed foods to have a reasonable **shelf life**, **modified atmosphere packaging** has become an integral part of many minimal processing procedures.

Minimal processing Limited processing of products to a level where they maintain the characteristics of fresh foods. Examples include industrial processes such as washing, **sorting**, **cutting**, **trimming**, **slicing** and **dicing**. Because a number of minimal processing technologies result in wounding of plant tissues and subsequent acceleration of deteriorative processes, controlled environmental conditions are critical requirements in the transportation, **distribution**, **storage** and retail display of these products. **Modified atmosphere packaging** has become an integral part of minimal processing.

Mint Plants of the genus *Mentha*, leaves of which are used as **spices**. Mint leaves are often added directly to foods and **beverages** or during **cooking** of dishes and impart a cool, fresh **flavour**. The predominant flavour compound of mint is **menthol**. Species with food industry applications include **peppermint** (*M. piperita*), **spearmint** (*M. spicata*) and Japanese mint (*M. arvensis*).

Mint oils **Essential oils** distilled from **mint**. The characteristic fresh, cool **flavour** and **aroma** of mint oils is due to the presence of the terpenoid, **menthol**. Mint oils are used as **flavourings** for **sugar confectionery**, such as **mints** and **chewing gums**, and in **beverages**, e.g. **cordials**.

Mints Sweets or **lozenges** which may be hard, soft or covered in **chocolate** and are flavoured with either **peppermint** or **spearmint**.

Miraculin   Flavour modifiers. Miraculin is a flavourless glycoprotein with a molecular weight of 44 kDa that was isolated originally from miracle fruits, berries of the African shrub *Richardella dulcifica*. Foods and **beverages** having a sour **flavour** that are eaten subsequent to consumption of miraculin are perceived as being more sweet than they actually are.

Mirex   Systemic insecticide used for control of **ants** and chewing **insects** in some crops. Use for pest control purposes has been banned in most parts of the world, although high persistence means residues may still occur in the environment and in animals.

Mirin   Condiments prepared by **fermentation** of steamed **rice**, **koji** and **ethanol** by *Aspergillus* spp. Used predominantly in Japanese cuisine.

Mirliton   Alternative term for **chayote**.

Miso   Pastes made by fermenting usually cooked **soybeans**, but sometimes **barley** or **rice**, and salt. Aged in cedar vats for 1-3 years. Ingredients and length of ageing are varied to produce products differing in **sensory properties**. Used as the base for manufacture of **soups** and **sauces**, and for flavouring other foods.

Mites   Common name for most members of the arthropod order Acarina. Includes many species that are **parasites** of animals or plants.

Mixers   Electric machines or devices for combining food ingredients by **beating**, **mixing** or **whipping**. There are two basic types of mixer - stationary and portable. Stationary mixers tend to be more powerful and can therefore handle heavier mixing jobs; they are usually equipped with an assortment of attachments including dough hooks, wire whisks, paddle-style beaters, and even citrus juicers, ice crushers, pasta makers, sausage stuffers and meat grinders. Portable mixers are small in size, due in part to their small motors, but are consequently easy to store. The term mixers is also applied to **beverages**, such as soda water, tonic water, cola, lemonade or fruit juice that can be combined with **spirits** to make beverages such as **cocktails**.

Mixes   Blends of ingredients in dried form that can be reconstituted in a liquid, sometimes with the addition of other ingredients, to make the desired product. The reconstituted mixture may require cooking or be ready to eat. Products available in this form include **cakes**, **soups** and **desserts**.

Mixing   Combining food ingredients together by hand or using electric machines (**mixers**).

Mixographs   Instruments used to investigate the **physical properties** of **dough**.

Modified atmosphere packaging   Packaging technique used primarily to extend **shelf life** of foods. Gas composition within the package is changed by altering levels of oxygen, nitrogen and carbon dioxide. This inhibits microbial growth, controls enzymic and biochemical reactions, reduces moisture loss and protects against infestation with **pests**. Used for packaging a wide range of foods.

Modified starches   Starch that has been modified by chemical reaction or physical means in order to adapt it for a specific application or improve its general applicability, i.e. to increase stability. Chemical modifications include crosslinking, **acetylation**, phosphorylation, reaction with 1-octenylsuccinic anhydride, hydroxypropylation and oxidation. Physical modifications involve pregelatinzation of starch by **drying** or **heating**.

Modori   Formation of very brittle heat-set **gels** on incubation of fish **surimi** pastes at 50-70°C. Gel structure is irreversibly destroyed and mechanical strength is reduced. Inhibited by addition of **sugars**, such as **glucose**, **fructose** and **sucrose**, dried **egg whites**, **alcohols**, such as *n*-butyl-, *n*-amyl- and *n*-hexyl alcohols, and **proteinases inhibitors**. The phenomenon appears to be associated with endogenous serine **proteinases** and cysteine proteinases.

Moinmoin   Steamed pastes usually prepared from **cowpeas** and popular in Nigeria.

Moisture content   Level of moisture held in a substance, stated as a percentage of the wet or dry weight.

Moisture sorption   Process whereby moisture binds to another substance.

Moisture transfer   Movement of moisture in stored products as a result of moisture and temperature changes.

Molasses   Low purity, thick, brown **syrups** produced as a by-product of **sugar** refining, and including the syrups remaining after **sucrose crystallization** has been exhausted. Uses include as a feedstock for microbial **fermentation**. Molasses from **cane sugar refining** are also known as blackstrap molasses.

Molecular weight   Sum of the atomic weights of the atoms in a molecule, usually measured in Daltons (abbreviation Da).

Molluscicides   Chemical substances used for control of terrestrial and aquatic **molluscs** such as **snails** and **mussels**. Examples commonly used on crops include metaldehyde and thiodicarb.

Molluscs   A diverse group of invertebrate organisms belonging to the phylum Mollusca; includes **gastropods** (e.g. snails), **bivalves** (e.g. oysters) and **cephalopods** (e.g. **squid**). The majority of molluscs are of marine origin, but large numbers of species occupy freshwater and terrestrial habitats. Many species of molluscs are collected or cultivated for human consumption.

Mol. wt. Abbreviation for **molecular weight**.

Molybdenum One of the essential metal minerals, chemical symbol Mo. Component of xanthine dehydrogenase and sulfite oxidase in animals and also nitrogen-fixing **enzymes** in some **microorganisms** and plants. Dietary requirements for molybdenum are very low and **deficiency diseases** are rare. Is found in **vegetables**, **cereals**, **oilseeds**, **fish** and water.

Momoni **Condiments** prepared from fermented fish which are used in parts of Africa, particularly Ghana. Used as **flavourings** in **soups** and stews, and also as a source of protein.

Monascus Genus of **fungi** that produce the red, yellow and purple **pigments**, rubropunctatin and monascorubin, ankaflavin and monascin, and rubropuntamine and monascorubramine, respectively. *Monascus pupureus* ATCC 16365 is commonly used for pigment production. Traditionally, in Asia, *Monascus* spp. are grown on steamed **rice** to produce red coloured foods. *Monascus* **colorants** have not been universally approved for use in foods.

Monellin Protein dimer with a sweet **flavour** isolated from **berries** of *Dioscoreophyllum cumminsii*, a plant native to Africa. Purified monellin is ≥2000× sweeter than **sucrose** on a weight for weight basis. **Sweetness** of monellin is dependent upon correct folding of the protein, therefore, conditions that induce protein **denaturation**, e.g. high temperatures or exposure to strong acids or alkalies, will decrease intensity of the sweet flavour.

Monensin Polyether antibiotic and coccidiostat used to control **coccidiosis** in cattle, lambs and poultry. Also used to treat ketosis in dairy cows and in **growth promoters** for cattle. Residues in **poultry meat** and **eggs** may persist for several weeks post-treatment, but it is excreted more rapidly in cattle.

Monilia Former name for the genus *Candida*.

Moniliformin Mycotoxin produced by several *Fusarium* species, mainly *F. proliferatum* and *F. subglutinans*, growing on **corn**. Acutely toxic for various animal species.

Monilinia Genus of **fungi** of the class Leotiomycetes. *Monilinia fructicola*, *M. laxa* and *M. fructigena* cause postharvest brown rot of **stone fruits**.

Monitor Alternative term for the insecticide **methamidophos**.

Monkfish Name applied to **marine fish** species of the genera *Lophius* or *Lophoides* (angler fish) and *Squatina* (angelsharks). Widely distributed throughout the world. Marketed fresh, frozen or dried-salted and cooked by a variety of methods. May also be used as a source of **fish oils** and **fish meal**.

Monoacylglycerols Types of **glycerides**, synonym **monoglycerides**. Composed of a glycerol molecule in which one of the hydroxyl groups has been acylated with a fatty acid substituent.

Monoamine oxidases Alternative term for **amine oxidases**.

Monoclonal antibodies **Antibodies** derived from a single antibody-producing cell, or produced artificially by a single clone and consisting of identical antibody molecules. Produced by fusing antibody-forming lymphocytes from mouse spleen with mouse myeloma cells. The resulting hybrid cells multiply rapidly and produce the same antibody as the parent lymphocytes. Monoclonal antibodies are widely used to detect and measure the amounts of particular **antigens**, or entities that can act as antigens.

Monocrotophos Systemic insecticide and acaricide used to control a wide range of **pests**, including sucking, chewing and boring **insects** and spider **mites** on **citrus fruits**, **vegetables**, **cereals** and **sugar cane**. Classified by WHO as highly hazardous (Ib). Also known as azodrin.

Monoglycerides **Lipids** composed of **glycerol** esterified to a single fatty acid, such as glycerol monooleate and **glycerol monostearate** (glycerol esterified with **oleic acid** and **stearic acid**, respectively). These compounds are present in naturally occurring **fats** and **oils**. Monoglycerides have many applications in the food industry, including uses as **emulsifiers**, inhibitors of **staling** in **bread dough** and **cake mixes**, encapsulators for **flavourings**, moisture barriers, and in manufacture of **margarines**. Specific properties are determined by the nature of the fatty acid present. Also called **monoacylglycerols**.

Monolaurin **Monoglycerides** formed by **esterification** of **glycerol** with **lauric acid** (dodecanoic acid). Monolaurin exhibits inhibitory activity against several foodborne pathogenic bacteria, including *Listeria monocytogenes*, and is used in food **preservatives**. Other uses include as mild **surfactants** and **emulsifiers**. Also called glycerol monolaurate.

Monooxygenases Members of EC 1.13 and EC 1.14. **Oxidoreductases** that incorporate one oxygen atom from O_2 into the compound oxidized.

Monophenol monooxygenases EC 1.14.18.1. Also known as **tyrosinases**, **phenolases**, monophenol oxidases and cresolases, these enzymes can catalyse the reaction of **catechol oxidases** under certain conditions. Involved in **enzymic browning** in **fruits**, **vegetables** and cereal grains, but are useful for the oxidation of **phenols** during **tea**, **coffee** and **cocoa** processing.

Monosaccharides General term for a single sugar unit comprising five or six carbon atoms in a ring conformation (furanose and pyranose, respectively).

Monosodium glutamate Monohydrate sodium salt of L-glutamic acid used in **flavour enhancers**. Most commercial production of this additive is from glutamate produced as a result of **fermentation** by **bacteria**, including *Micrococcus glutamicus*. Often abbreviated to MSG.

Monoterpenes Monoterpenoids synthesized from isoprene and containing two isoprene units. May be acyclic, monocyclic or dicyclic.

Monoterpenoids Class of **terpenoids** which includes **monoterpenes** and their oxygenated and hydrogenated derivatives. Occur naturally in a wide range of plant foods and derived products, including **wines**, **beer**, **fruit juices** and other fruit products. Many are **aroma compounds** and components of **essential oils** and **flavourings**.

Monounsaturated fatty acids Unsaturated fatty acids containing a single double bond. Examples include **oleic acid**, **palmitoleic acid** and **erucic acid**, significant sources of which include **olive oils**, **fish oils** and **rapeseed oils**, respectively.

Montasio cheese Italian **hard cheese** originally made from **ewe milk** but now made from unpasteurized **cow milk**. Rich and creamy, with a fruity **flavour**. The yellow-brown rind is smooth and springy in young cheese but becomes darker and harder with age. Interior is firm with small holes, becoming granular or brittle in mature cheese. Ripens in 3-18 months.

Monterey sardine Marine fish species (*Sardinops sagax caeruleus*) of high commercial importance belonging to the family Clupeidae. Found in the Pacific Ocean, off the coast of California. Other subspecies of *S. sagax* (common name South American **pilchards**) are found in various areas of the Indo-Pacific region. Marketed fresh, frozen or canned and cooked mainly by **frying** or **broiling**. Also used to make fish meal.

Mood Pattern of behaviour exhibited in relation to a current state of mind. Usually a relatively short-lived and low-intensity emotional state. Can be affected by **diet**, and can affect consumer response to foods as well as **eating habits**.

Mook Traditional Korean food made from **mung beans**, **cowpeas**, **buckwheat** or **acorns**, which takes the form of a **starch** gel.

Moose Large ruminant animals (*Alces american*) belonging to the Cervidae family, which are hunted for their **meat**. Moose often feed in marginal wasteland environments, which may be contaminated with pol-

lutants (e.g. **heavy metals**), so high levels of contaminants may occur in moose **carcasses**, **moose meat** and moose **offal**.

Moose meat Meat from **moose**. Popular in the USA, a commonly consumed **game meat** in Sweden and also a traditional food for some ethnic groups, such as the James Bay Cree Indians in Canada. Since moose often feed in marginal wasteland environments, which may be contaminated with pollutants (e.g. **heavy metals**), high levels of contaminants may occur in moose **carcasses**, meat and **offal**. Moose meat is sometimes referred to as **venison**.

Moraxella Genus of Gram negative, aerobic rod-shaped **bacteria** of the family Moraxellaceae. Occur in soil and water and on plants and animal hides. Responsible for **spoilage** of fresh **meat** and **fish**.

Morchella **Edible fungi** commonly known as morels. Main species is *Morchella esculenta*.

Morello cherries Dark coloured type of **sour cherries** (*Prunus cerasus*) used in **cooking** or canned. Also processed into juices, **liqueurs**, e.g. **kirsch**, and **jams**.

Morels Edible fungi of the genus *Morchella*, particularly *M. esculenta*.

Morganella Genus of Gram negative, facultatively anaerobic rod-shaped **bacteria** of the family **Enterobacteriaceae**. Species, especially *Morganella morganii*, can produce **histamine** in scombroid **fish** and fish products, which can cause **scombroid poisoning** in humans when consumed.

Morin Flavonol, with a structure similar to that of **quercetin**, which is a natural component of many plants and **wines**. Possesses **antioxidative activity** in lipids, **antimicrobial activity** and nitrite scavenging activity, but can act as a pro-oxidant for non-lipid food constituents. Potentially useful as an antioxidant in **oils**.

Moringa Genus of plants native to tropical Asia. **Seeds** of some species, especially *Moringa oleifera*, yield high quality **oils** used as **cooking oils** as well as lubricants. Other plant parts are also used as foods, **fruits** and leaves as **vegetables** and roots as a source of **spices**.

Morning goods Bakery products that are usually, but not always, eaten at **breakfast** (e.g. **croissants**, **brioches**).

Moroccan smen Alternative term for **smen**.

Moromi Fermenting mash mixture of **rice**, **koji**, **yeasts** and water that is pressed to separate **sake** from suspended solids during sake brewing. Moromi is also the mash based on koji (derived from **soybeans** or **cereals**), with the addition of brines, which is fermented to make **soy sauces**.

Mortadella Large, fully cooked semi-dry **sausages**, originally made in the Bologna area of Italy. Mortadella is prepared from very finely chopped, cured **pork** and **beef**, with the addition of cubes of white fat; it is lightly spiced with **aniseed** and **garlic**, and smoked at high temperature before air drying. Other versions include German-style mortadella, which is prepared from high quality, finely minced meat with cubes of pork fat and **pistachio nuts**. Mortadella has one of the highest fat contents of all cooked sausages. Usually, it is sliced and served, but it may be added to **pasta stuffings**, **sauces** or sautes.

Mortierella Genus of **fungi** of the family Zygomycetes. *Mortierella alpina* is used in the production of **arachidonic acid** for use in foods.

Morwong Any of a number of **marine fish** species in the family Cheilodactylidae, widely distributed in the Pacific and Indian Oceans and parts of the Atlantic Ocean. Commercially important species include *Nemadactylus macropterus* (morwong), *N. douglasi* (grey morwong) and *Cheilodactylus spectablis* (banded morwong). Marketed fresh, usually whole, gutted or as fillets; sometimes sold as frozen fillets or in ready-to-cook packs.

Moth beans Seeds produced by *Phaseolus aconitifolius*. Grown as a food source particularly in India, where the seeds are eaten whole or split, often fried in oils; the green pods from the plant are also eaten as a vegetable. Also known as mat beans, matki beans, mout beans and dew gram.

Moths Common name for mostly nocturnal **insects** of the order Lepidoptera. Adults or larvae may be **pests** of plants and stored foods. The potato tuber moth (*Phthorimaea operculella*) is an important pest of **potatoes**, the codling moth (*Cydia pomonella*) is a pest of **walnuts** and **apples**, and the Indian meal moth (*Plodia interpunctella*) is a pest of **flour** and **dried fruits**. Larvae of the emperor moth (*Imbrasia belina* Westwood) are an important food source in southern Africa.

Mottling Blotchy **discoloration** of foods.

Moufflon Small wild **sheep** (*Ovis orientalis*) which are believed to be the common ancestors of all domestic sheep. They are hunted for their **meat**.

Moulding Formation of an object out of a malleable substance. Also use of containers (moulds), usually distinctively shaped, to form food into a specific shape. Moulds can range in size from small, individual candy-size moulds to large pudding moulds and cheese moulds. The foods to be moulded (e.g. a gelatin-based dessert) are poured or packed into the mould and then left until they become firm enough to hold their shape.

Moulds Common alternative term for **fungi**.

Mountain apples Alternative term for **Malay apples**.

Mousses Creamy, frothy **desserts** typically made from **fruit purees**, **whipped cream** and/or beaten **eggs**, and set with **gelatin**. Savoury mousses are similar light-textured dishes made from **meat** or **fish**.

Mouthfeel **Sensory properties** relating to sensations produced in the mouth by foods during **mastication**. Mouthfeel is affected by a wide variety of food properties, including **viscosity**, **flavour** and **foaming**.

Moxidectin One of the broad-spectrum **anthelmintics** used to control infections by **parasites**, predominantly heartworm infections, in cattle, sheep and goats, but not approved for use in dairy animals. Rapidly metabolized and excreted from treated animals.

Mozzarella cheese Italian **soft cheese** made from **buffalo milk**. A plastic, spun-curd cheese made by coagulating **pasteurized milk** at 90°F, cutting the **curd**, treating it with hot water (200°F) and kneading into a shiny lump. Pieces are then taken off, cooled, salted and marketed soon after.

m.p. Abbreviation for **melting point**.

MPN Abbreviation for the most probable number method, a technique for estimating the number of viable **microorganisms** suspended in a liquid. Sets of tubes containing growth medium are inoculated with successively smaller volumes of sample solution. Following incubation, the tubes are examined for microbial growth and the number of cells in the original sample is calculated from the pattern of growth, using probability tables. Also called the multiple tube method.

MRL Abbreviation for maximum residue limits. Maximum concentrations of pesticide residues, resulting from the registered use of agricultural or veterinary **pesticides**, that are recommended to be legally permitted or recognized as acceptable in or on a food, agricultural commodity or animal feed.

mRNA **RNA** molecules derived from **DNA** by **transcription** that function as templates for synthesis of proteins (**translation**) in cells or for synthesis of cDNA.

MS Abbreviation for **mass spectroscopy**.

Mucic acid Member of the **organic acids**, synonym galactaric acid. Occurs as an oxidation product of **galacturonic acid** and is found particularly in **grapes** infected with *Botrytis* cinerea.

Mucilage **Gums** produced as plant exudates, in particular those produced by **seaweeds**.

Mucins **Glycoproteins** secreted by animal mucous cells and glands. Found in saliva, and gastric and intestinal secretions.

Mucoids Glycoproteins or **mucins** with mucus-like properties. Also used to describe gummy or slimy bacterial colonies.

Mucopolysaccharides Any of a group of **polysaccharides** containing **amino sugars**, such as **glucosamine**, and a uronic acid residue. Can have gel-like properties, since their structure allows significant water sorption. Examples include hyaluronidase and chondroitin. Synthesized by certain **microorganisms**, but also present in cartilage from **meat** and **fish**. Can affect **viscosity** of **fermented foods**.

Mucor Genus of **fungi** of the class Zygomycetes. Occur on vegetable matter, soil and dung. *Mucor racemosus* and *M. mucedo* may be responsible for spoilage of **bread** and **meat**, while other species may be parasitic to stored grains. *M. hiemalis* is used in the production of **sufu**, and *M. racemosus* and *M. rouxianus* are used in the production of **pozol**.

Mud crabs Common name for marine and estuarine **crabs** of the genus *Scylla*, especially *S. serrata*. Widely distributed in the Pacific and Indian Oceans; also produced by **aquaculture** in Asia and Australia. Prized for the delicate, sweet **flavour** of the moist meat, found mainly in the claws. Cooked by **steaming**, poaching, pan **frying** or on a barbecue. Meat is eaten on its own or as an ingredient of **soups** or **pasta fillings**. Also known as mangrove crabs.

Muenster cheese Alternative spelling of **Munster cheese**.

Muesli Mixture of untoasted cereal flakes (e.g. **oats**, **wheat** and **rye**), **dried fruits** and **nuts**, often used as **breakfast cereals**. Can be sweetened or unsweetened.

Muesli bars Fibre-rich **cereal bars** based on **muesli** ingredients.

Muffin cakes Round **cakes** which may be leavened with **yeasts** or **baking powders** and sweetened with sugar. May be plain, or flavoured with **fruits**, e.g. **blueberries**, **dried fruits**, **nuts**, **chocolate** or savoury ingredients such as **cheese**.

Muffins Term that has two different meanings. American muffins are small round **cakes** which may be leavened with **yeasts** or **baking powders** and sweetened with **sugar**. They may be plain, or flavoured with **fruits**, e.g. **blueberries**, **nuts** or savoury ingredients such as **cheese**. Often eaten with breakfast or as an accompaniment at dinner. **English muffins** are thick, round **bread** products which are rapidly fermented and well aerated. Baked on a hot plate or griddle and often split and toasted before being eaten, sometimes with sweet or savoury **fillings**, such as **jams**, **bacon** or **cheese**.

Mugwort Plants of *Artemisia* spp., the leaves of which are used as spices. In Asian cooking, leaves are added as an ingredient to **stuffings** and **rice cakes**, and tea **flavourings**. In Western countries, it has uses as flavourings for poultry or pork dishes.

Mulberries Berries produced by plants of the genus *Morus*. The common or black mulberry (*M. nigra*) produces purple fruit similar in appearance to **raspberries**. The white mulberry (*M. alba*) is grown mainly as a food source for silkworms, but also for the fruits, which are dried before consumption; leaves are a potential source of natural food **antioxidants**. Mulberries are eaten as a dessert, added to **tarts** and **pies**, or made into **jams** and **wines**. Rich in potassium and **vitamin C**. The main acid is **citric acid**.

Mulberry leaves Leaves of the white mulberry (*Morus alba*), extracts of which have **antioxidative activity** and are of interest as potential natural food **antioxidants**. Also valued for nutritional and health-promoting activities.

Mulching Covering or surrounding of plants with mulch, comprising organic matter such as leaves, peat or straw. Done to inhibit growth of weeds, and to prevent evaporation of moisture or freezing of the plant roots.

Mullet Any of around 80 estuarine and **marine fish** species in the family Mugilidae, widely distributed in Atlantic and Pacific coastal waters; some species migrate inshore. Many species are important food fish, including *Liza ramada* (thin-lipped grey mullet), *L. aurata* (golden grey mullet) and *Mugil cephalus* (striped mullet; black mullet). Marketed fresh (whole, gutted or fillets) and as smoked or salted products. **Roes** of some species are popular as dry-salted products.

Multipacks Packages containing several individual containers of foods or **beverages** that may be separated before consumption. Commonly used for **dairy desserts**, **snack foods** and **carbonated beverages**.

Mung beans Pulses produced by *Phaseolus aureus* or *Vigna radiata*. Eaten boiled or in **dhal**; flour produced from the **beans** may also be used in **baking** or made into **porridge**. Contain little fat, but high levels of proteins and carbohydrates. Most commonly used pulses for production of **bean sprouts**. Also known as green gram.

Munggo Alternative term for **black gram**.

Munster cheese French **soft cheese** made from pasteurized **cow milk**. The edible skin is sticky and orange in **colour**, while the soft interior has a mild, piquant **flavour** which becomes more pungent as the cheese is washed. **Ripening** occurs from the inside out. American versions of this cheese have a lighter coloured interior and a mild flavour. Alternative spelling **Muenster cheese** is used in some countries.

Muramidases Alternative term for **lysozymes**.

Murex Genus of gastropod **molluscs** resembling **whelks**. Found in tropical and sub-tropical coastal areas. Flesh of some species is consumed.

Muscadine grapes **Grapes** produced by *Vitis rotundifolia* that have a characteristic musky **flavour** and are astringent and lacking in sweetness. Grown mainly as **table grapes**, but some are used in **winemaking**.

Muscles Tissues composed of bundles of specialized cells which are capable of contraction and relaxation to create body movement. There are >600 muscles in an animal carcass; these vary widely in shape, size and activity. There are three types of muscle, namely: skeletal, cardiac and smooth. The largest part of the musculature consists of skeletal muscles and it is this part of animal carcasses that is generally referred to as **meat**; organs comprised of cardiac or smooth muscle tend to be classified as offal. Muscle tissue also contains structural elements (**collagen**, reticulin and elastin).

Mushrooms Fruiting bodies of various species of fungi. Eaten raw or used to add flavour to dishes, **soups** and **sauces**. Many species are gathered wild, but care must be taken as some are poisonous. The most commonly cultivated species is *Agaricus bisporus*; other types of commercial importance include shiitake, **straw mushrooms**, oyster mushrooms and winter mushrooms. Rich in phosphorus, magnesium, potassium, selenium and **riboflavin**, and low in fat.

Muskmelons **Fruits** produced by *Cucumis melo*. Yellow or green skin with a raised network of a lighter shade. Flesh is green to orange, comprising mainly water, but with high levels of sugar, **vitamin C** and **carotenes**. Eaten as a dessert. Also known as netted melons or nutmeg melons, and include Galia melons.

Mussel poisoning Toxic reaction following consumption of contaminated **mussels**. Especially refers to a severe and often fatal intoxication after eating mussels that have fed on red tide flagellates (particularly the **dinoflagellates** *Gonyaulax*) and accumulated certain **alkaloids** in their tissues.

Mussels Any of a large group of marine and freshwater bivalve **molluscs** from the family Mytilidae. Distributed worldwide, but more common in cooler waters. Many species are valued for the delicate, sweet **flavour** and **texture** of their flesh. Important commercial species include *Mytilus edulis* (**blue mussels**), *M. galloproviancilis* (Mediterranean mussels) and *Perna canaliculus* (**green-lipped mussels**), all

of which are cultured. Marketed live (whole with shells), and as fresh, smoked, canned, salted and semi-preserved products.

Mustard **Condiments** prepared from dried ripe seeds (**mustard seeds**, also used to produce pungent **spices**) of *Brassica nigra* (black or brown mustard), *B. juncea* (brown mustard only) or *Sinapis alba* (**white mustard** or **yellow mustard**). For serving, mustard powder is added to water, **salt**, **vinegar** and/or other ingredients, e.g. **wines**.

Mustard greens Leaves of the brown or Indian mustard plant (*Brassica juncea*) eaten as a vegetable and used in manufacture of **kimchies**.

Mustard seed oils **Oils** extracted from **mustard seeds** belonging to the genera *Brassica* or *Sinapis* used in the food and soap industries.

Mustard seeds Globular seeds of black or brown mustard (*Brassica nigra* or *B. juncea*) or white or yellow mustard (*Sinapis alba*), which are odourless when whole and have a pungent **flavour**.

Musts **Fruit juices** (especially those extracted from **winemaking grapes**) intended for **alcoholic fermentation** to produce **wines**.

Mutagenesis Generation of **mutations**.

Mutagenicity Capability of inducing **mutations**.

Mutagens Chemical or physical agents which promote **mutagenesis**.

Mutanolysins Enzymes produced by *Streptomyces globisporus* and *S. mutans* which are similar to **lysozymes**. Used in conjunction with or instead of lysozymes to hydrolyse bacterial cell walls prior to extraction of their contents, e.g. for identification purposes. Cell wall digestion is achieved by cleavage of β-1→4-*N*-acetylmuramyl-*N*-acetylglucosamine linkages of peptidoglycan.

Mutants Include populations, organisms, genes, **chromosomes** that differ from the corresponding wild type by one or more **mutations**.

Mutarotases Alternative term for **aldose 1-epimerases**.

Mutations Detectable and heritable structural changes to the genetic material of a cell or organism, or the results of such changes. May occur by chemical changes to the **DNA**, e.g. substitution of one nucleotide for another, or physical damage such as breakage or rearrangement. Depending on where in the DNA sequence alteration occurs, a mutation may not be detected (silent mutation) or may be apparent from effects on the gene product. Mutations may be random, spontaneous or induced by **mutagens**.

Mutton **Meat** from mature **sheep**, which are over one year old, including meat from ewes, rams, wethers and hoggets. Mutton tends to be cheaper than **lamb**, but also tends to be tougher, darker in **colour**, fattier and less delicately flavoured. It is the preferred meat for Muslims. Also known as sheep meat, sheep muscles, ram meat or ram muscles.

Mutton sausages **Sausages** in which the main meat component is **mutton**.

Mycobacterium Genus of Gram positive, aerobic rod-shaped **bacteria** of the family Mycobacteriaceae. Occur in **dairy products**, soil and water. *Mycobacterium paratuberculosis*, which causes Johne's disease in **cattle**, is suspected of causing **Crohns disease** in humans who consume contaminated **milk**.

Mycoplasma Genus of Gram negative, facultatively anaerobic **bacteria** of variable form of the family Mycoplasmataceae. Species (e.g. *Mycoplasma bovis*) are the causative agents of **mastitis** in **cattle**.

Mycoprotein Commercially produced high-protein **biomass** of **fungi**. A major example of a mycoprotein is **Quorn**, which is produced using *Fusarium graminearum*.

Mycotoxicosis Disease of humans and animals resulting from the ingestion of **mycotoxins** in foods or feeds.

Mycotoxins Toxins, e.g. **aflatoxins** and **ochratoxins**, produced by **fungi**.

Mylar Lightweight but strong film made from **polyethyleneterephthalate**.

Myofibrillar proteins Salt-soluble **proteins**, including **actins** and **myosin**, which are the predominant type of proteins in muscle and are responsible for contraction, **texture** and **water holding capacity**. Degradation of these proteins is important for *post mortem* **tenderization** of **meat**.

Myofibrils Elongated contractile elements contained within skeletal and cardiac muscle fibres. The major **myofibrillar proteins** are **actins**, **myosin**, **actomyosin** and **tropomyosin**. Within myofibrils, thick filaments, consisting almost entirely of myosin, and thin filaments, consisting almost entirely of actin, are aligned parallel to each other; in certain regions these myofilaments overlap and this arrangement produces a characteristic striated appearance in myofibrils. Areas of different densities, which are visible within the light and dark bands of myofibrils, include: the isotropic band (I-band); the anistotropic band (A-band); and the Z-line, a thin dark line, which bisects the I-band. The basic repeating contractile structural unit of myofibrils is called a sarcomere; it extends from Z-line to Z-line. During muscle contraction, actin and myosin within the thick and thin myofilaments interact to form actomyosin and this causes shortening of the muscle fibres. *Post mortem,* there is lateral shrinkage of the myofibrils in **meat**. Fluid is expelled from the spaces between filaments and is drained by gravity, forming **drip**.

Myoglobin Purplish-red protein **pigments** found in muscles (**meat**). Myoglobin has one haem unit and one globin chain. In meat, myoglobin content differs between species and between different muscles. **Colour** lightness of meat is inversely correlated with myoglobin content. Meat colour is affected by oxidation state of myoglobin; the three major myoglobin derivatives are reduced myoglobin (purplish-red), **oxymyoglobin** (bright red) and **metmyoglobin** (brown/grey). Colour changes due to oxygenation of myoglobin are reversible. When meat is cured with nitrite, myoglobin is converted into the bright red pigment nitrosomyoglobin. Thermal denaturation of myoglobin to a brown pigment begins at about 65°C; consequently, the red colour of raw meat changes to brown on **cooking**.

Myosin Myofibrillar **globulins** that are the most abundant proteins in **meat** and the predominant salt-soluble muscle proteins. During muscle contraction, myosin combines with **actins** to form **actomyosin**. Myosin molecules are shaped like elongated rods with thickened regions at one end. Proteolysis of myosin by trypsin results in formation of two myosin fractions, **myosin heavy chains** and **myosin light chains**. Myosin is insoluble in water and only slightly soluble in acids; however, it is soluble in salt solutions or alkalies. Myosin gelation is a principal factor in obtaining good **texture** in meat products.

Myosin heavy chains Heavy chain isoforms of **myosin**. They constitute the head of the myosin molecule, and play an important role in heat-induced gelation of myosin. Also known as heavy meromyosin.

Myosin light chains Light chain isoforms of **myosin**. They form the main structural element of myosin. Although they do not seem to contribute to heat-induced gelation of myosin, they may help to stabilize the gel once it is formed. Also known as light meromyosin.

Myrcene One of the acyclic **monoterpenes**, found in the **essential oils** of a variety of useful plants, such as **lemon grass**. Has a spicy, balsamic **aroma**. Myrcene-containing **essential oils** are widely used for flavouring foods.

Myricetin Member of the **flavones** found particularly in **berries** and **wines**. Myricetin occurs in both glycosylated and aglycone forms and has **antioxidative activity**.

Myristic acid One of the **saturated fatty acids**, with 14 carbon atoms; synonym, tetradecanoic acid. Major component of many **animal fats**, **vegetable fats** and **oils**.

Myristicin One of the alkenylbenzene group of **aromatic compounds**, synonym 5-methoxy safrole. Major aroma compound in **mace** and **nutmeg**, and also found in **parsley**, **dill** and **carrots**.

Myrosinases Alternative term for **thioglucosidases**.

Myrothecium Genus of **fungi** of the class Hyphomycetes. Some species (e.g. *Myrothecium roridum*) may produce **mycotoxins** during growth on foods. Other species (e.g. *M. verrucaria*) may be used in the production of **enzymes** (e.g. **cellulases**, **polygalacturonases** and **xylan degrading enzymes**).

Myrrh **Flavourings** isolated from gum exudates collected from plants of the genus *Commiphora* which are native to Africa and Arabia.

Myrtle Common name for *Myrtus communis*, an evergreen tree with aromatic blue-black **berries**. Fruits and leaves are used as **condiments**, as a source of **essential oils** and in production of **liqueurs**.

N

N_2 Chemical symbol for **nitrogen** gas.

Na Chemical symbol for **sodium**.

NAA Abbreviation for the analytical technique **neutron activation analysis** and the auxin **naphthaleneacetic acid**.

NaCl Chemical formula for **sodium chloride**.

NAD(P) Abbreviation for **nicotinamide adenine dinucleotide (phosphate)**.

Naegleria Genus of **amoebae** of the family Vahlkampfiidae. Occur in damp soil, mud, water and sewage. *Naegleria fowleri*, a water contaminant, is the causative agent of meningoencephalitis in humans.

NAFTA Abbreviation for **North American Free Trade Agreement**.

Nalidixic acid Quinolone antibiotic that exhibits **antibacterial activity** against various **Gram negative bacteria**. Used in Japan for treatment of vibriosis and furunculosis in **salmon** and **trout**. Residues in treated **fish** may persist for 10 days or more post-treatment.

Nan Flat **bread** originating from northwest India made from white **flour**, leavened with **sodium bicarbonate** and baked in a tandoor.

Nanofiltration Form of filtration that uses semipermeable membranes of pore size 0.001-0.1 μm to separate different fluids or ions, removing materials having molecular weights in the order of 300-1000 Da. Nanofiltration is most commonly used to separate solutions that have a mixture of desirable and undesirable components. An example of this is the **concentration of corn syrups**. Nanofiltration is capable of removing ions that contribute significantly to osmotic pressure, and this allows separation at pressures that are lower than those needed for **reverse osmosis**.

NaOH Chemical formula for **sodium hydroxide**.

Naphthalene Aromatic hydrocarbon with a distinctive coal tar-like odour. Used as an insecticide and in the synthesis of dyes.

Naphthaleneacetic acid Auxin; 2-(1-naphthyl)acetic acid. Used to regulate the yield and quality of **fruits** and **vegetables**.

Naphthol Phenol that is a major metabolite of the insecticide **carbaryl**.

2-(1-Naphthyl)acetic acid Chemical name for the plant growth regulator **naphthaleneacetic acid**.

Naphthylmethylcarbamate Alternative term for the insecticide **carbaryl**.

Napins **Storage proteins** of **rapeseeds** (*Brassica napus*).

Naranjilla Orange **fruits** with green-yellow flesh produced by (*Solanum quitoense* or *S. angulatum*). The juicy pulp is used in **beverages** and **sherbet**. Also eaten out of hand, and used as an ingredient in **desserts**, jellies and **marmalades**. Rich in **vitamin A** and **vitamin C**. Alternative term for **lulo** and quito oranges.

Narazuke **Vegetables** pickled in **sake lees**. Originally made from uri, a cross between **cucumbers** and **melons**, but now made using **aubergines**, small melons, **radishes** and cucumbers.

Naringenin Non-bitter flavanone found mainly in **citrus fruits**, but also in other **fruits**, e.g. **tomatoes**.

Naringin Bitter glycoside present in **citrus fruits**.

Naringinases Commercial crude fungal enzyme preparations consisting of **α-L-rhamnosidases** and **β-glucosidases**. Used to degrade **naringin**, a bitter flavonoid found in **citrus fruits**, during extraction of juices in order to reduce the **bitterness** to acceptable levels.

Nata Thick white mucilaginous mat formed by **fermentation** of *Acetobacter* xylinum grown on the surface of **coconut water**, **coconut milk** or other sugary **fruit juices**. Used in production of **desserts**, including nata de coco popular in the Philippines.

Natamycin Polyene antibiotic with **antifungal activity** used topically on skin and mucous membranes for treatment of ringworm in cattle and horses. Absorption through dermal tissues of treated animals is normally negligible. Also used as a fungicide on bulbs such as **onions** and **garlic**. Also known as pimaricin.

Natto Traditional Japanese product made by **fermentation** of **soybeans** with *Bacillus* subtilis (*B. natto*).

Natural colorants **Colorants** that exist in nature.

Natural flavourings **Flavour compounds**, also **essential oils**, extracts and hydrolysates containing flavour compounds, that are derived from natural sources, such as plants, animal foods and edible yeasts. Usually they have little or no nutritive value but are used solely to impart **flavour**.

Natural foods Foods produced by natural farming and using natural processing methods, e.g. without the use of added **salt**, **sugar** and **preservatives**. Similar to **organic foods** but not necessarily totally organic.

Natural sweeteners Sweet-tasting substances that occur in nature. Saccharides, such as **sucrose** (sugar), D-**glucose** (dextrose) and **fructose** (laevulose) are the major natural **sweeteners** used by the food industry. Other natural sweeteners include sweet-tasting proteins (e.g. **thaumatin**), terpenoids (e.g. **glycyr-rhizin**), steroidal **saponins** (e.g. polypodoside A), dihydroisocoumarins (e.g. phyllodulcin) and **flavo-noids** (e.g. **neohesperidin**).

Navy beans Type of **common beans** (*Phaseolus vulgaris*).

N compounds Compounds that contain the element **nitrogen**.

NDGA Abbreviation for **nordihydroguaiaretic acid**.

Near infrared **Infrared radiation** which has a wavelength between 0.7 and 2.5 μm. Near infrared (commonly abbreviated to NIR) is subdivided into very near infrared (0.7-1 micron) and short wave infra-red (1.0-2.5 microns).

Nectarines **Fruits** produced by *Prunus persica* var. nectarina. Similar to **peaches** in composition and **flavour**, but with a smoother skin and richer **colour**. Sweet, juicy flesh varies in colour from white to yellow, depending on variety. Varieties also differ in stone tenacity (clingstone or freestone). Rich in **vitamin A**, **vitamin C** and potassium. Eaten out of hand or in **salads**, and used as a garnish, in **toppings** and in various **desserts**.

Nematocides Chemical substances used for control of **nematodes** that parasitize animals or infest crops. Generally fall into two major classes, **fumigants** and non-fumigants (contact), based on chemical and physical characteristics. Commonly used examples include **methyl bromide** and **oxamyl**.

Nematodes Group of worms which are members of the phylum Nematoda. Occur in soil and fresh and marine waters. Some are **parasites** of humans, animals and plants.

Neocallimastix Genus of anaerobic **fungi** which occur in the rumen of animals. Species (e.g. *Neocalli-mastix patriciarum* and *N. frontalis*) are used in the production of **enzymes** (e.g. **xylan degrading enzymes** and **cellulases**).

Neohesperidin Flavonoid glycoside present in Spanish **oranges** (*Citrus aurantium*). The sugar component is a disaccharide, β-neohesperidose (2-*O*-α-L-rhamnopyranosyl-β-D-glucopyranose). A bitter compound of poor water solubility, it is one of the important **flavour compounds** of **orange juices**.

Neohesperidin dihydrochalcone Non-nutritive **sweeteners** with 1000-1500× the sweetness of **sucrose**. Produced from **naringin**, a flavonoid glycoside extracted from **grapefruit** peel.

Neomycin Aminoglycoside antibiotic produced by **Streptomyces** *fradiae*. Used for treatment of a range of bacterial infections in cattle, sheep, swine and poultry. Relatively persistent in muscle tissues, **eggs** and **milk**. Parenteral use in food-producing animals is not permitted in the EU.

Neopullulanases EC 3.2.1.135. Hydrolyse **pullulan** to **panose** (6-α-D-glucosylmaltose). Useful for production of novel branched **maltooligosaccharides** and phosphoryl **oligosaccharides** for use in foods, the former of which can be used as growth factors for bifidobacteria and as non-cariogenic **sweeteners**, while the latter may increase Ca **bioavailability**.

Neosartorya Genus of **fungi** of the class Eurotiomycetes. *Neosartorya fischeri* may be responsible for the **spoilage** of canned and bottled **fruits**.

Neotame High-intensity, non-nutritive sweetener which is an N-substituted derivative of **aspartame**. Approximately 7,000-10,000× sweeter than **sugar**. A free flowing white crystalline powder which is water-soluble and heat-stable, and can be used in **cooking** as well as in tabletop applications. Although similar to aspartame, neotame is degraded differently in the human digestive system, avoiding problems caused by the presence of **phenylalanine** for people suffering from **phenylketonuria**.

Nephelometry Technique used to determine the size and concentration of cells or particles in a solution by measuring the intensity of scattered light. Light scattering depends on the number and properties of the particles in the solution.

Neral Aldehyde; *cis*-citral. Volatile flavour compound found in plant **essential oils**.

Nerol Monoterpene alcohol. Volatile flavour compound found in many plant **essential oils** and involved particularly in the **flavour** and **aroma** of **grapes** and **wines**.

Neroli oils Yellowish **essential oils** derived from bitter orange blossoms by steam distillation. Have an intense **aroma** of orange blossom.

Net protein ratio Weight gain of a group of animals (e.g. rats) fed a test diet plus the weight loss of a similar group fed a protein free diet, and the total divided by the weight of the protein consumed by the animals on the test diet.

Net protein utilization Net protein utilization, commonly abbreviated to NPU, is an index of the **nutritional values** of **proteins**. This quality ratio indicates the amount of dietary protein retained in the body under specific clinical conditions. Changes in body nitrogen levels following consumption of a dietary protein are compared with those following consumption of a protein-free diet for the same duration, and then the dietary nitrogen retained in the body is expressed as a proportion of nitrogen intake.

Nettles Plants of the genus *Urtica*, including stinging or common nettles (*U. dioica*) and small nettles (*U. urens*). Leaves are rich in **vitamin C** and can be used as a vegetable when young. Also used in herbal preparations and **soups**, and to make **beer**, **wines** and **teas**.

Neural networks Systems of computer programs and data structures which are modelled on the human nervous system and brain. Incorporate large numbers of processors operating in parallel, each with an individual sphere of knowledge which has been fed into it along with rules about relationships. Networks can use this information to recognize patterns in large amounts of data. Used in the food industry to model processes and predict the behaviour of foods under specific conditions. Also known as **artificial neural networks**.

Neural tube defects Congenital malformations of the spinal cord caused by the folds of the ectodermal neural plate failing to close properly in early embryonic development. Failures to close at the top result in anencephaly, which is always fatal; failures to close along the spine result in spina bifida, which can have either a reasonably hopeful or a very poor prognosis depending on location and other characteristics of the opening. Supplements of **folic acid** begun before conception reduce the risk of neural tube defects developing in the foetus.

Neurological shellfish poisoning Food poisoning associated with consumption of **shellfish** containing **neurotoxins** produced by the dinoflagellate **algae** *Pytchodiscus brevis*. Gastrointestinal and neurological symptoms normally occur within 3 to 6 hours of ingestion of contaminated shellfish.

Neurospora Genus of **fungi** of the class Sordariomycetes. Occur on decaying or burnt vegetation. *Neurospora sitophila* is responsible for **spoilage** of **bread**. *N. intermedia* is used as a starter for **ontjom** and in the **fermentation** of bongkrek.

Neurotoxicity Property of being toxic to nervous tissues.

Neurotoxins Toxins that act specifically or primarily on nervous tissues (e.g. **botulotoxins** and **saxitoxin**).

Neutralization Process of making something chemically neutral, with a pH of approximately 7.

Neutron activation analysis Analytical technique in which samples are irradiated with a reactor, accelerator or isotopic neutron source. Radioactive nuclides are produced by the addition of neutrons to nuclei of specific atoms and these nuclei release energy in the form of gamma rays or electrons to convert back to a stable state. The radiation detected is a measure of the energy of the nuclides produced in the sample. Commonly abbreviated to NAA.

NH$_3$ Chemical formula for **ammonia**.

Niacin A member of the **vitamin B group**. Generic descriptor for two compounds in foods which have the biological activity of the vitamin: **nicotinic acid** (pyridine 3-carboxylic acid) and **nicotinamide** (the amide of nicotinic acid). The metabolic function of niacin is in the coenzymes nicotinamide adenine dinucleotide (NAD) and nicotinamide adenine dinucleotide phosphate (NADP), which operate, often in partnership with **thiamin** and **riboflavin** coenzymes, to produce energy within the cells. Niacin is found in animal tissue as nicotinamide and in plant tissues as nicotinic acid; both forms are of equal niacin activity. Rich sources of niacin include **livers**, **kidneys**, lean **meat**, **poultry meat**, **fish**, **rabbit meat**, **cornflakes** (enriched), **nuts** and **peanut butter**. Niacin can withstand reasonable periods of cooking, heating and storage. Canning, drying and freezing result in little destruction of the vitamin. In cereals, niacin is largely present as niacytin, which is not biologically available. Deficiency of niacin leads to pellagra (photosensitive dermatitis), depressive psychosis and intestinal disorders. Previously known as vitamin PP.

Niacinamide Synonym for **nicotinamide**.

Nicarbazin Coccidiostat used prophylactically for prevention of intestinal and caecal **coccidiosis** in chickens. Not used in laying hens due to detrimental effects on production of **eggs**. Withdrawal periods are specified for treated chickens.

Nickel Transition element with the chemical symbol Ni.

Nicotinamide Synonym for niacinamide and nicotinic acid amide. The amide form of **nicotinic acid** which has **niacin** activity as a constituent of 2 coenzymes (nicotinamide adenine dinucleotide (NAD) and nicotinamide adenine dinucleotide phosphate (NADP)); these coenzymes act as intermediate hydrogen carriers in a wide variety of oxidation and reduction reactions. Nicotinamide can be formed in the body from the amino acid **tryptophan**; on average 60 mg of dietary tryptophan is equivalent to 1 mg of preformed niacin.

Nicotinamide adenine dinucleotide (phosphate) A coenzyme derived from **niacin**; commonly abbreviated to NAD(P).

Nicotinic acid A member of the **vitamin B group** found in plant tissues. Contributes, along with **nicotinamide** found in animal tissues, to **niacin** activity. Chemical name pyridine 3-carboxylic acid.

Nicotinic acid amide Synonym for **nicotinamide**.

Nigerose Disaccharide composed of two **glucose** residues linked via an α-1,3-glycosidic bond. Isomer of **maltose**.

Niger seeds Seeds from the plant *Guizotia abyssinica*, which is grown in India and Ethiopia as an **oilseeds** crop.

Nile perch Large **freshwater fish** species (*Lates niloticus*) widely distributed in lakes and rivers around Central Africa; a highly valued food fish. Fresh and frozen fillets are exported from Kenya, Tanzania and Uganda to markets in Europe, Israel and the USA.

Nile tilapia **Freshwater fish** species (*Oreochromis niloticus*, formerly *Tilapia nilotica*) of high commercial importance belonging to the cichlid family (Cichlidae). Widely distributed in rivers and lakes of Africa and also produced by **aquaculture**. Marketed fresh and frozen.

NIR Abbreviation for **near infrared**.

NIR spectroscopy **Spectroscopy** performed at wavelengths in the near infrared (NIR) region.

Nisin Polypeptide **antibiotics** produced by *Lactococcus lactis*. Used in **preservatives** for a variety of foods, such as **processed cheese**, **meat** and meat products, **fish**, and canned fruits and vegetables.

Nitrates Salts of **nitric acid** found in many animal and plant foods as a result of use of nitrate **fertilizers**, the nitrification process in the soil, or use of sodium nitrate or potassium nitrate **food additives**. Health risks are associated with conversion of nitrates into **nitrites** in the **gastrointestinal tract**. Contamination of **drinking water** with nitrates from chemicals used in agriculture is a particular concern.

Nitric acid Strong acid that forms **nitrates** with metals, carbonates, hydroxides or oxides. Powerful oxidizing agent. Used in digestion or extraction of samples during analysis. Commercially utilized in production of **fertilizers**, explosives and dyes.

Nitric oxide Gas (chemical formula NO) produced by **reduction** of **nitric acid**, **nitrates** or **nitrites**, or oxidation of **ammonia**.

Nitrification Conversion of ammonia or other N compounds into **nitrites** or **nitrates**.

Nitrites Salts of **nitrous acid** formed by **reduction** of **nitrates**. Can be **oxidizing agents** or **reducing agents**. Authorized as **food additives** for **preservation** of **meat** and **cheese**. Health risks are associated with formation of **nitrosamines** from nitrites in the presence of **amines**.

Nitrofurazone Broad-spectrum synthetic antibiotic used for prophylactic and therapeutic purposes in swine, cattle, sheep, poultry and fish. Also used for treatment of topical bacterial infections and for growth promoting purposes. A suspected carcinogen; systemic use in animals reared for food purposes is prohibited in the EU and USA.

Nitrogen Colourless and odourless gas that constitutes approximately three-quarters of the Earth's atmosphere by volume. The common form is dinitrogen (chemical symbol N_2). Constituent of **proteins**, **amino acids**, and many other groups of chemicals, e.g. **amines**, **alkaloids** and **purines**.

Nitrogen compounds Compounds that contain the element **nitrogen**.

Nitrogen dioxide Brown gas with the chemical formula NO_2.

Nitrogen monoxide Alternative term for **nitric oxide**.

Nitrogen solubility index **Physicochemical properties** defined as extracted nitrogen as a percentage of sample nitrogen content, determined between pH 2 and 12.

Nitrosamines **Nitroso compounds** with strong **carcinogenicity** formed by reaction of **amines** with nitrogen oxides or **nitrites**.

Nitroso compounds **Organic compounds** containing the nitroso group, many of which are **mutagens**.

***N*-Nitrosodiethylamine** Volatile nitrosamine with mutagenic activity. Found in a range of foods, sometimes as a result of indirect contamination, e.g. **migration** from **rubber** or **packaging materials**, or as a result of formation during **processing**.

***N*-Nitrosodimethylamine** Volatile nitrosamine with mutagenic activity; commonly abbreviated to NDMA. Found in a range of foods, sometimes as a result of indirect contamination, e.g. **migration** from **rubber** or **packaging materials** or as a result of formation during **processing**.

Nitrosomonas Genus of **Gram negative bacteria** of the family Nitrosomonadaceae, occurring in soils and water. The species *Nitrosomonas europaea* is utilized for biological nitrogen removal from **waste water**, including **food factories effluents**.

Nitroso pigments **Pigments** formed during **curing** of **meat** by the reaction of nitric oxide (synthesized by conversion of **nitrites** used in **curing agents**) with **metmyoglobin** or **myoglobin**. Responsible for the pink **colour** of **cured meat**.

***N*-Nitrosopyrrolidine** Volatile nitrosamine with mutagenic activity. May be formed in a range of foods, including **bacon**, during processing.

Nitrous acid Fairly strong acid with the chemical formula HNO_2.

Nitrous oxide Colourless gas with the chemical formula N_2O, also known as dinitrogen oxide. Used as a mild anaesthetic (laughing gas).

Nivalenol Trichothecene produced by *Fusarium* species (e.g. *F. nivale*) during growth on foods such as **wheat**, **rye**, **barley**, **corn** and **millet**.

Nixtamalization Traditional process used to improve the nutritional quality of **corn**. Nixtamalization involves **cooking** and **steeping** corn in a lime solution, washing the corn (nixtamal) and stone grinding nixtamal to form a corn dough or masa. Masa is used to produce nixtamalized products (e.g. corn **tortillas**, **tortilla chips**, **corn chips** and **taco shells**).

N-Lite D Trade name for a waxy **corn** maltodextrin which is used as a fat substitute, particularly in **ice cream** and other **dairy products**.

NMR Abbreviation for **nuclear magnetic resonance**.

NO₂ Chemical formula for **nitrogen dioxide**.

Nocardia Genus of Gram positive, aerobic rod-shaped or filamentous **bacteria** of the family Nocardiaceae. Occur in soil. Some species are causative agents of **mastitis** in **cattle**, while others may be used in the production of **biosurfactants**.

Nocardiopsis Genus of alkalophilic **bacteria** of the order Actinomycetales found in soil. Type species is *Nocardiopsis dassonvillei*. Producers of a number of enzymes, including **proteinases** and **glycosidases**.

N-Oil Tapioca-derived dextrin that exhibits a creamy, fat-like **texture** and is used as a fat substitute in foods such as **salad dressings** and **dairy products**.

Nomograms Graphical plots in the form of line charts which may be used to solve particular types of equations. Scales for the variables involved in the formula are presented in a way such that corresponding values for each variable are on a straight line intersecting all scales. Thus, when values for two variables are known, the value of a third can be read from its scale.

Nonachlor Primary degradation product of the organochlorine insecticide **chlordane**. May accumulate in animal tissues and **milk**, where it persists for long periods.

Nonanal Aldehyde important for the **flavour** and **aroma** of many foods.

Nonanone Methyl ketone that is important for the **flavour** and **aroma** of many foods including **dairy products**, **fruits** and **vegetables**.

Nonenal Aldehyde important for the **aroma** of many foods. Also involved in formation of cardboard **off flavour** in **beer**.

Nonenzymic browning Food **browning** process promoted by heat treatment, which includes a wide range of reactions, such as the **Maillard reaction**, caramelization, chemical oxidation of **phenols** and maderization.

Nono Nigerian **fermented milk** product.

Nonreducing sugars **Sugars** that do not have a free carbonyl group (ketone or aldehyde) and therefore are not able to act as **reducing agents**.

Nonthermal processes Processing techniques that do not require heat. Usually refers to food pasteurization and sterilization treatments that do not employ heat during processing. Examples include: **high pressure processing** (inactivates vegetative **microorganisms**); ultrasonication (inactivates vegetative **bacteria** and reduces heat resistance of **bacterial spores**); high voltage electric pulse treatment (**electroporation**; inactivates vegetative microorganisms); ionizing radiation treatment (inactivates **pathogens**); high intensity light pulse treatment (inactivates vegetative bacteria); and high intensity **magnetic fields** processing (inactivates microorganisms).

Noodles Elongated, ribbon-like **pasta** made with **eggs** and **rice**, **wheat** or **buckwheat** flour. Used in European and Oriental cuisine. Often used to add bulk to **soups** and stews.

Nootkatone Sesquiterpene that is one of the essential **aroma compounds** in **grapefruit**.

Norbixin One of the dicarboxylic carotenoid **pigments** present in **seeds** of the shrub *Bixa orellana*. The main water soluble component of the natural orange colorant, **annatto**.

Nordihydroguaiaretic acid Phenolic lipid soluble lipoxygenase inhibitor that is used mainly in **antioxidants** for **fats** and **oils**. Often abbreviated to NDGA.

Norflurazon Selective herbicide used for preemergence control of annual and perennial grasses and broad-leaved weeds around plants such as those producing **fruits**, **asparagus**, **artichokes**, **nuts** and **soybeans**. Classified by Environmental Protection Agency as not acutely toxic (Environmental Protection Agency IV).

Norharman β-Carboline formed from **tryptophan** during heating. Demethylated analogue of Harman, its co-mutagen.

Nori Dried seaweed product obtained from red **algae** in the genus *Porphyra* (particularly *P. tenera* and *P.*

yezoensis). Popular in Japan, where it is often consumed in toasted form. Good source of **vitamin B₁₂**, **dietary fibre** and certain **minerals**; may possess anticarcinogenicity.

North American Free Trade Agreement The North American Free Trade Agreement (NAFTA) is a pact that calls for the gradual removal of tariffs and other trade barriers on most goods produced and sold in North America. NAFTA, which became effective in Canada, Mexico and the USA on 1 January 1994, built upon a 1989 trade agreement between the USA and Canada that eliminated or reduced many tariffs between the two countries. NAFTA called for immediate elimination of duties on half of all US goods shipped to Mexico and the gradual phasing out of other tariffs over a period of about 14 years. The treaty also protected intellectual property rights and outlined the removal of restrictions on investment among the three countries. Mandates for minimum wages, working conditions and environmental protection were added later as a result of supplemental agreements signed in 1993. Talks began in late 1994 to expand NAFTA to include all Latin American nations (with the exception of Cuba). These talks include plans to create a free-trade zone throughout the Americas in the 21ˢᵗ century. Formal negotiations to include Chile in NAFTA began in 1995.

Northern blotting A method for analysing **RNA**. RNA is separated by electrophoresis, transferred to a chemically reactive matrix (e.g. nitrocellulose) on which it binds covalently in a pattern identical to that on the original gel, and detected by complementary labelled probes (RNA or single-stranded **DNA**) that hybridize to specific RNA sequences.

Norvegia cheese Norwegian semi-hard cheese similar to **Gouda cheese**.

Norwalk viruses **Small round structured viruses** of the family Caliciviridae. Responsible for acute **gastroenteritis** in humans. Transmitted by the faecal-oral route via contaminated water and foods (e.g. **shellfish** and **salads**).

Norway lobsters Marine species of **lobsters** (*Nephrops novergicus*) found in the North Sea, the northeast Atlantic and the Mediterranean sea. Highly valued for their flesh. Marketed fresh (whole, tail meat with shell or shelled cooked or uncooked), frozen, semi-preserved or as a component of pastes and **soups**. Also known as Dublin Bay **prawns**, **langoustines** and **scampi**.

Nostoc Genus of filamentous **cyanobacteria** which occur naturally in damp habitats as green to black gelatinous colonies. Some species produce high levels of phycobiliproteins, making them a potential source of natural **pigments** for use in foods. Several species,

e.g. *Nostoc flagelliforme*, are eaten in various countries, including China. *N. commune* is rich in **dietary fibre**.

Nougat **Aerated confectionery** products made with **honeys** or **sugar**, **egg whites** and **starch syrups**. Often contains **nuts**, **dried fruits** and/or **cherries** and may be either chewy or brittle in consistency.

Novagel Commercial name for cellulose gel consisting of **microcrystalline celluloses** and **guar gums**.

Novel foods Foods prepared using unconventional processes (particularly genetic technology), derived from unconventional sources or offering non-nutritional benefits. Examples include **biotechnologically derived foods**, **designer foods** and **medical foods**.

Novobiocin Narrow-spectrum antibiotic with **antibacterial activity** against many **Gram positive bacteria**. Frequently used in combination with **penicillins** for treatment of cattle **mastitis** and to control cholera and staphylococcal infections in poultry. Residues in muscle tissues normally deplete rapidly following treatment.

NPR Abbreviation for **net protein ratio**.

NPU Abbreviation for **net protein utilization**.

Nuclear magnetic resonance **Spectroscopy** technique based on the magnetic moment of atomic nuclei. An external magnetic field will partially align the axis of spin of spinning nuclei, but some precession about the magnetic field will occur. The precession depends on the magnetic field applied and the magnetic moment of the nucleus (dependent in turn on the chemical state of the atom), and is specific to the type of nucleus. The precession rate, measured by emission or absorption of applied radiofrequency, is used to give details about the composition of the sample. Commonly abbreviated to NMR.

Nuclear power Power generated by nuclear reactors in nuclear power plants or stations. Accidents at nuclear power stations have caused fallout of radiocaesium, and radioactive contamination of growing foods.

Nucleases EC 3.1.11-EC 3.1.16 (exonucleases) and EC 3.1.21-EC 3.1.31 (**endonucleases**). Enzymes that cleave the phosphodiester bonds between nucleotide subunits of **nucleic acids**.

Nucleic acids Polymers of **nucleotides** in which the 3′ position of one nucleotide sugar is linked to the 5′ position of the next by a phosphodiester bond. The two major types are **DNA** and **RNA**.

Nucleosides Compounds of purine or pyrimidine bases with a sugar, usually **ribose**.

Nucleotidases EC 3.1.3.31. Catalyse the dephosphorylation of **nucleotides**, forming a nucleoside and orthophosphate. Exhibit a wide specificity for 2′, 3′

and 5'-nucleotides, and also hydrolyse glycerol phosphate and 4-nitrophenyl phosphate.

Nucleotides Compounds of purine or pyrimidine bases with a sugar phosphate.

Nukazuke Japanese fish product consisting of fermented fish (usually **sardine**) in **rice bran**-based **pickles**.

Nuoc-mam Fermented **fish sauces** produced by fermenting **anchovy** (or other small **marine fish**) in **salt**, **flavourings** and **spices** for long periods. The resulting product is clear amber in colour. Used in Vietnamese and Thai cuisine.

Nuruk Starters (comprising **yeasts** and other **fungi**) for Korean **rice wines**.

Nutmeg One of the **spices**, along with **mace**, derived from seeds of *Myristica fragrans*. Kernels may be used whole, grated or ground. Characteristic **flavour compounds** include α- and β-pinene, **myristicin**, **camphene**, dipentene and sabanene.

Nut oils Oils extracted from **nuts**, such as **almonds**, **hazelnuts** and **walnuts**. Best used uncooked, as heat often destroys their delicate **flavour**.

Nut pastes Pastes made from **nuts** that are used as the base for making **confectionery fillings** or nut **spreads**.

Nutraceutical foods Alternative term for **functional foods**.

NutraSweet Registered brand or trade name of the low calorie sweetener **aspartame**.

Nutrients Essential dietary factors, such as **vitamins**, **minerals**, **amino acids** and **fatty acids**, that are required by the body but cannot be synthesized in the body in adequate amounts to meet requirements, so must be provided by the diet. Nutrient deficiency can cause poor growth, deformity, malfunctioning and sterility. A range of characteristic **deficiency diseases** is recognized in humans.

Nutrition Science of the relationship between foods, **nutrients** and health. A major aspect considered is the way by which an organism absorbs and utilizes food components. The study of nutrition involves identification of individual nutrients that are essential for growth and maintenance of the individual, interrelationships among nutrients within individual organisms, and quantitative requirements of organisms for specific nutrients under various environmental conditions in order to optimize health.

Nutritional labelling Information appearing on **labelling** or packaging of foods relating to energy and nutrients in the food. The information which must or may be given, and the format in which it must appear, is governed by law in most countries.

Nutritional values Indications of the level to which a food contributes to the overall diet. These values depend on the quantity of food ingested and absorbed, and the amount of essential nutrients it contains. Nutritional values can be affected by cultivation conditions, handling and storage practices, and processing.

Nuts **Fruits** consisting of an edible kernel within a shell, the thickness and hardness of which varies among types. Kernels have a high fat content and are often used as the source of **nut oils**. They are also rich in **fibre**, **vitamin E**, **folic acid** and a range of **minerals**. Nuts are generally available shelled or unshelled; shelled nuts are sold in many forms including raw, blanched, roasted and flavoured. They are eaten out of hand or used in a variety of sweet and savoury dishes. Commonly consumed nuts include **walnuts**, **pistachio nuts**, **pine nuts**, **cashew nuts** and **almonds**. Some foods known as nuts are not true nuts, e.g. **Brazil nuts** are really **seeds** and **peanuts** are **legumes**.

Nylon Family of strong, elastic polyamide materials, which vary from moderately flexible to strong, tough and rigid products. Can be shaped when heated into forms such as sheets, bristles and fibres. Resistant to greases and oils. Used widely as a packaging material for foods in **packs** such as **pouches** and boil-in-bags.

Nypa Genus of **palms**. The nipa palm (*Nypa fruticans*) is the source of a **sugar**-containing sap.

Nystose Fructooligosaccharide comprising three **fructose** residues and a **glucose** residue. Produced by hydrolysis of **inulin** or from **sucrose** via the action of fructosyltransferases.

Nyufu Type of fermented **tofu**.

O

O_2 Chemical symbol for **oxygen** gas.

O_2 **absorbers** Abbreviation for **oxygen absorbers**.

Oak Hard, durable **wood**, usually with a distinct grain, obtained from oak trees, which belong to the many species within the genus *Quercus*. Used to impart a distinctive **aroma** and **flavour** to foods by various methods, including **smoking** (e.g. for meat products and **fish**), storing and/or **ageing** in oak **barrels** (e.g. for **wines** and **spirits**), and addition of oak wood supplements or extracts. The term is also used to describe the smoky flavour and aroma characteristics of **wines** and **spirits** aged in oak barrels.

Oat bran Outer layer found under the hull of the oat grain which forms the **milling** fraction.

Oat fibre Indigestible material derived from oat grains, consumption of which is reported to reduce serum **cholesterol**.

Oat flour Ground oat grains from which their outer layers have been removed. Used as an ingredient in **bakery products** and **snack foods**.

Oat gums **Gums** produced from **oats** that are composed predominantly of $(1{\rightarrow}3)(1{\rightarrow}4)$-β-D-glucan (**β-glucans**). Used as **thickeners** in foods.

Oatmeal Rolled or ground **oats**. Also refers to **porridge** made from rolled or ground oats.

Oat oils **Oils** extracted from **oats**. Highly unsaturated and containing high levels of **linoleic acid**.

Oatrim Fibre-rich fat substitute made from **oat flour** and **oat bran** which is used in **bakery products**, **salad dressings**, **confectionery** and **dairy products**.

Oats Edible starchy grain derived from plants belonging to the genus *Avena*, particularly *A. sativa*, *A. steritis* and *A. strigosa*, used as a cereal food. A rich source of **vitamin B₁**; also rich in protein and high in fat.

Oat starch **Starch** isolated from **oats**.

Obesity Condition in which the body's weight is beyond the limit of skeletal and physical requirements due to excessive accumulation of fat. Possible causes include overeating, an inappropriate diet, genetic factors and **metabolic disorders**. Obesity is associated with an increased risk of developing a range of diseases, including adult-onset **diabetes**, **cardiovascular diseases** and certain types of **cancer**. In some cases, obesity can be reversed through measures such as adoption of a low-calorie diet, making changes to lifestyle or consuming **functional foods** designed for this purpose.

Obesumbacterium Genus of Gram negative, facultatively anaerobic rod-shaped **bacteria** of the family **Enterobacteriaceae**. *Obesumbacterium proteus* is responsible for producing **off odour** and **nitrosamines** in fermenting beer **worts**.

Oca Common name for *Oxalis tuberosa*, the stem tubers of which are eaten like **potatoes**. The **oxalic acid** present in some cultivars can be removed by sun drying or **freeze drying**. Also known as iribia, cuiba and New Zealand yams.

Ochratoxin A Most toxic of the known **ochratoxins**.

Ochratoxins Mycotoxins produced by certain species of ***Penicillium*** (e.g. *P. viridicatum*) and ***Aspergillus*** (e.g. *A. ochraceus*) during growth on foods and feeds (e.g. **wheat**, **rye**, **barley**, **oats**, **corn** and **peanuts**). Nephrotoxic and carcinogenic in humans and animals (e.g. **cattle** and **swine**) when ingested in contaminated foods and feeds.

Octadecanoic acid Synonym for **stearic acid**, one of the **fatty acids** that occur naturally in the form of **glycerides** in **animal fats** and **vegetable fats**.

Octadecenoic acid Synonym for **oleic acid**, an unsaturated fatty acid that occurs as the glyceryl ester in **fats** and **oils**. One of the major **fatty acids** in **cow milk**.

Octanal Aldehyde contributing to the **flavour** of many foods. Formed by lipid oxidation.

Octanoic acid Synonym for **caprylic acid**, a saturated fatty acid that occurs in **milk** and **coconut oils**.

Octanol Alcohol with a strong odour. Manufactured by the action of concentrated NaOH on **castor oils**. Has good demulsifying and wetting power; used as a foam reducing agent. Also known as octyl alcohol or capryl alcohol.

Octanone Methyl ketone important for the **flavour** of foods, especially **mushrooms**.

Octopine Guanidino amino acid that is formed as a product of glycolysis in **cephalopods**, where it may be used as an indicator of quality and **freshness**.

Octopus Any of a number of eight-armed cephalopod **molluscs** from the order Octopoda. Widely distributed in shallow marine waters. Many species are consumed, particulary in Japan and Mediterranean countries. Marketed fresh and frozen; also as smoked and canned products. Ink sacs found in all species contain a black liquid that is sometimes used in food **colorants** or **flavourings**.

Odour Alternative term for **aroma**.

Odour activity values Ratio of the concentration of **aroma compounds** present in a product to the **odour threshold values**.

Odour threshold values Levels at which perception of increasing concentrations of **aroma compounds** begins. The concept of odour threshold is useful in defining **aroma** purity, estimating the necessary amount of starting material, serving as a reference point in describing intensity and aroma quality, and evaluating which components present are important in contributing to a characteristic aroma.

OECD Abbreviation for **Organization for Economic Cooperation and Development**.

Oedema Excessive accumulation of tissue fluid in body tissues, leading to swelling. Popularly known as dropsy. Causes of oedema include heart failure, kidney failure, liver failure and **malnutrition**. Diuretic drugs can relieve symptoms by causing the patient to pass more urine. Allergic reactions may be accompanied by local oedema.

Oenococcus Genus of Gram positive, anaerobic, coccoid **lactic acid bacteria** of the family **Lactobacillaceae**. *Oenococcus oeni* is used as a starter culture in **winemaking**, where it carries out **malolactic fermentation**.

Oenocyanins **Anthocyanins** that occur in red **grape skins**. Used as **natural colorants** in foods.

Oenological properties Properties of ingredients such as **winemaking grapes**, **musts** and **wine yeasts** which are of relevance in relation to **winemaking**.

Oestradiol Female sex hormone, one of the major **oestrogens**. Found at varying levels in a range of foods. Implants of oestradiol-containing **growth promoters** can be used to improve growth performance and characteristics of **carcasses** in animals. Sale of **meat** containing synthetic oestradiol is prohibited in some countries.

Oestrogens Group of steroid **hormones** derived ultimately from **cholesterol**, in which carbon atoms 1 to 6 are in the form of an aromatic ring. Natural oes-

trogens are produced predominantly in the ovaries and are responsible for development of female secondary sexual characteristics and regulation of the menstrual cycle. Alternative spelling estrogens.

Oestrone One of the **oestrogens**, systematic name 3-hydroxyoestra-1,3,5(10)-trien-17-one. Produced by reduction of the androgen androstenedione or by oxidation of 17β-**oestradiol**. Alternative spelling estrone.

Offal **Animal foods** described collectively as by-products of animal slaughter. Offal includes all parts of the carcass that are cut away when the carcass is dressed, such as the **intestines**, **kidneys** and **livers**. In many cultures, edible portions of offal are considered as delicacies. Kidneys, livers, **brains**, **hearts** and **sweetbreads** (pancreas and thymus) are often eaten, but other organs may be associated with cultural limits or taboos. Religious traditions often regard offal as unclean and, accordingly, place restrictions on consumption of offal. As carcass organs can form the foci of infection, routine veterinary inspection of offal at slaughterhouses is used to identify and exclude diseased animals from the food chain.

Off flavour **Taints** perceived in the mouth upon tasting a product. Off flavours are negative attributes, and affect the **eating quality** of foods; they may also indicate that a food is spoiled.

Off odour **Taints** perceived in the nose when foods are smelled. Off odours are negative attributes, and affect the **eating quality** of foods; they may also indicate that a food is spoiled.

Ogi Nigerian weaning food made by fermenting **corn gruel** with **lactic acid bacteria**. May also be prepared with **millet** or **sorghum**.

Ogiri West African fermented **condiments** used to season **soups** or stews. Typically made from fermented **castor beans**, **melon seeds** or **sesame seeds**.

Ohmic heating Thermal processing of foods using energy produced in the form of heat when a current passes through an electrical resistance. In this form of electric resistance heating, the food itself acts as a conductor between a ground and a charged electrode. The food may be immersed in a conducting liquid. Heating is accomplished according to Ohm's law, where conductivity of the food determines the current that will pass between the ground and electrode. Ohmic heating can be used as a **cooking** technique, and also for **pasteurization** and **sterilization**.

Oil expellers Equipment used in extraction of **vegetable oils** from **oilseeds**. **Oils** are pressed from the source material using pressure from an auger, which turns inside a barrel. The barrel is closed except for a single hole through which the extracted oil drains. Expellers remove larger proportions of oil from seeds than can be achieved with hydraulic batch presses. Also known as continuous screw presses.

Oil palms Palm trees, *Elaeis guineensis*, native to tropical Africa. Yield **palm oils** from the fleshy endosperm of its seeds and **palm kernel oils** from the seed kernels.

Oils Lipid-rich, viscous substances derived from animal, vegetable or mineral sources that are liquid at room temperature and insoluble in water.

Oilseed proteins **Proteins** derived from **oilseeds**, which have desirable **functional properties** and nutritional characteristics and may reduce the risk of certain diseases. May be used as **food supplements**.

Oilseeds Seeds, e.g. **sesame seeds**, **sunflower seeds** and **soybeans**, from which **vegetable oils** may be extracted. The oilseed cake or meal which remains after oils have been extracted is often used as a livestock feed, since it is rich in protein.

Okadaic acid Polyether toxin produced by certain dinoflagellate **algae** that can accumulate in **bivalves**. Causative agent of **diarrhoetic shellfish poisoning**.

Okara Fibre-rich by-product remaining after extraction of **soymilk** from ground **soybeans**. Also rich in high quality proteins. Used in **soups**, vegetable dishes, **sausages** and **bakery products**.

Okra Common name for *Hibiscus esculentus*. Good source of **vitamin A** and **vitamin C**. Immature pods are eaten as a vegetable, pickled or used to thicken **soups** and stews. Also known as okro, lady's fingers, gumbo and bindi.

Oleandomycin Macrolide antibiotic produced by *Streptomyces* *antibioticus*; used in cattle to treat intermammary infections and for growth promoting purposes. Distributes widely in tissues of animals following administration.

Olefins **Hydrocarbons** containing one or more carbon double bond(s) whose names have the suffix -ene, e.g. ethene, but-1-ene, but-2-ene (the number designates the position of the double bond, between C1 and C2 and between C2 and C3, respectively). Simple olefins have only one double bond. Synonymous with alkenes.

Oleic acid Monounsaturated fatty acid of 18 carbon atoms with the systematic name *cis*-Δ^9-octadecenoic acid. Prepared by hydrolysis of **animal fats**, such as **tallow**, or **vegetable oils**, such as **olive oils**, **sunflower oils** and **soybean oils**.

Olein Common name for 9-octadecenoic acid 1,2,3-propanetriyl ester, **triglycerides** composed of **glycerol** esterified with 3 molecules of **oleic acid**. A

major constituent of **fats** and **oils**. Used as solvents for oil soluble **flavour compounds** and **vitamins**, and as food **stabilizers**.

Oleomargarines Highly unsaturated fractions of **tallow** which have a low **melting point** and are separated by **fractionation**. Used in manufacture of **margarines**.

Oleoresins Extracts of oil-soluble components of plant materials, usually **spices**. Produced by direct contact of the spices with highly hydrophobic organic solvents, e.g. hexane. The organic solvents can then be evaporated to concentrate the extract. Used as **flavourings** by the food industry. Oleoresins are cheaper to produce than **essential oils** and easier to use than spices, but do not have the full flavour profile of essential oils.

Oleosins Alkaline **proteins** found in the oil bodies of plant **seeds**.

Olestra Calorie-free sucrose polyester used as a fat substitute, formed from **sucrose** and **fatty acids** of **soybean oils**, **corn oils**, **cottonseed oils** or **sunflower oils**. Neither metabolized nor absorbed by the body, it absorbs fat-soluble **vitamins** and **carotenoids** and may cause anal leakage. Stable under thermal conditions (e.g. **frying**) and may be used in **bakery products** and **snack foods**.

Oleuropein One of the **phenols** present in **olives** and **olive oils**, responsible for their **bitterness**. Before consumption, olives are debittered by hydrolysis of oleuropein, e.g. by **fermentation** or NaOH treatment.

Olfactometry Measurement of the olfactory properties of a substance. Often used in conjunction with **gas chromatography** in analysis of gaseous components.

Oligo-1,6-glucosidases EC 3.2.1.10. Hydrolyse 1,6-α-D-glucosidic linkages in **isomaltose** and **dextrins** produced from **starch** and **glycogen** by α-**amylases**. Also known as limit dextrinases and isomaltases, some preparations can catalyse the reaction of sucrose α-glucosidases.

Oligonucleotide probes Alternative term for **gene probes**.

Oligonucleotides Short fragments of single-stranded **DNA**, typically up to 20 **nucleotides** in length.

Oligosaccharides Compounds comprising between three and ten **monosaccharides** linked by glycosidic bonds. Synthesis is by limited hydrolysis of **polysaccharides** or via addition of monosaccharides, a process catalysed by **glycosyltransferases**.

Oligouronides **Oligosaccharides** containing residues of **uronic acids**, such as **glucuronic acid** and

galacturonic acid. May be produced by hydrolysis of **pectins** or **polyuronides**.

Olive oil mills effluents Waste water produced during processing of **olive oils** which have high levels of organic aliphatic and **aromatic compounds** and often represent an environmental problem in areas where olive oil is produced.

Olive oils Oils that are rich in monounsaturates and are derived from the mesocarp of fruits of the olive tree, *Olea europaea*. Virgin olive oil is produced from the first pressings of ripe **olives**, while other grades are produced from subsequent pressings and then refined. Used in **cooking** or in **salad dressings**.

Olives Fruits produced by *Olea europea* with fleshy pulp and stony kernels. Change from green to black as they mature. Used as table olives or a source of **olive oils**. Table olives (black and green) are pickled in **brines**. Olives are sometimes pitted (stone removed) and stuffed with vegetables, such as **pimento peppers** and **onions**, or other foods.

Omelettes Eggs which have been beaten, sometimes with **seasonings** and other ingredients such as **milk**, and fried. May be plain or filled with savoury or sweet **fillings**, e.g. **cheese** or **jams**.

Omija Raspberry-like **fruits** produced by *Schizandra chinensis*, also called the five taste tree. Used in oriental medicine and also in fruit punch and **fruit tea**.

Oncom Alternative term for **ontjom**.

Onigusa Japanese name for red **seaweeds** of the genus *Gelidium*. Found in intertidal or subtidal areas in many parts of the world; normally found in greater abundance in exposed coastal areas. Primarily used as a source of **agar** for food and **beverages** processing. Some species are consumed, in fresh, dried, pickled or jelly forms, particularly in Asian countries.

Onions Underground bulbs of *Allium cepa*, composed of fleshy leaf bases, and varying in size, shape, **colour** and **flavour** according to variety. Not rich in nutrients, but a good source of flavour in cooking. Fresh and dried onion products are used in sauces and a variety of dishes, including **soups**, stews and **salads**, and are essential ingredients of **pickles** and **chutneys**. Available also canned, pickled and frozen.

Ontjom A **tempeh**-like fermented product made usually from press cake of **peanuts**, although other starting materials, such as **okara**, can be used. *Neurospora sitophila* is used to ferment ontjom, giving an orange-red covering to the product. Deep-fried slices are consumed as a side dish. Also known as oncom.

Oocysts Spherical cysts which form around two conjugating gametes in the sporozoa of certain protozoans. Extremely resistant to adverse environmental conditions.

Oolong tea Type of tea in which the **tea leaves** have been partially fermented, rather than not fermented (as in **green tea**) or fully fermented (as in **black tea**).

Opacity Degree of obstruction an item produces to the transmission of visible light.

Opalescence A pearly or milky mineral lustre resembling that of opal, resulting from the characteristic internal play of colours, in turn resulting from the reflection and refraction of light passing through adjacent thin layers of different water content.

Opaque 2 Variety of corn bred to contain higher levels of **amino acids** such as **lysine** and lower concentrations of **zein**.

Operators Sites within prokaryotic **operons** where repressor proteins bind to the **DNA**, thereby inhibiting **transcription** of adjacent genes. Typically consist of or contain palindromic sequences, and may lie between promoters and the first structural genes of operons, or may overlap or even occur within promoters. In some operons, two operators may be present.

Operons Groups of contiguous structural **genes** and their associated control elements that are found in prokaryotes and which are transcribed as single **transcription** units from common **promoters**, thereby allowing coordinated regulation. The structural genes within operons may or may not be related in function, e.g. they may encode enzymes of particular metabolic pathways.

Optical density A measure of light absorption of a translucent medium, equivalent to the logarithm of the **opacity**. In the food industry, optical density is used in measurements of various parameters, including **turbidity**, **browning** and bacterial growth.

Optical properties **Physical properties** relating to the appearance of a product, including **clarity**, **colour**, **reflectance**, **turbidity** and **fluorescence**.

Optical rotation Ability of some compounds to rotate a plane of polarized light due to asymmetry of the molecule. If the plane of light is rotated to the right, the substance is dextrorotatory and is designated by the prefix (+); if the plane of light is rotated to the left, the substance is laevorotatory, and the prefix is (-). A mixture of the two forms is optically inactive and is termed racemic. **Sucrose** is dextrorotatory but is hydrolysed to **glucose** (dextrorotatory) and **fructose** which is more strongly laevorotatory; therefore, hydrolysis changes the optical activity from (+) to (-). A mixture of glucose and fructose is termed **invert sugar**.

Orange beverages Beverages based on **orange juices**, orange extracts or comminuted **oranges**.

Orange essential oils Essential oils produced by compression of **orange peel** that are composed predominantly of D-**limonene** but may also contain other **aroma compounds**, including **octanal**, **myrcene**, **linalool**, **decanal**, sinensal, ethyl butyrate and valencene. Composition of the essential oils is dependent on the species of orange from which they are produced.

Orange juice beverages   **Beverages** based on **orange juices** in combination with other ingredients.

Orange juice concentrates   **Orange juices** which have been concentrated; commonly used for preparation of reconstituted **orange juices** or **fruit juice beverages**.

Orange juices   **Fruit juices** extracted from **oranges** (*Citrus sinensis*). Rich in **vitamin C**.

Orange peel   Outer skin or rind of oranges composed of the coloured flavedo (or zest) and the inner white albedo (or pith). Used to make candied peel, as a garnish or to add **flavour** to **bakery products** and a range of dishes.

Orange roughy   Deepwater **marine fish** species (*Hoplostethus atlanticus*), widely distributed in the Atlantic, Pacific and Indian Oceans. Increasingly targeted as a food fish, particularly off the coast of New Zealand. Flesh is prized for its firm **texture** and delicate, shellfish-like **flavour**. Marketed fresh and frozen.

Oranges   **Citrus fruits** of 3 main types - sweet, loose skinned and bitter. The juicy pulp may or may not contain seeds according to cultivar. All are rich in **vitamin C**, some B vitamins and **minerals**. The sweet orange (*Citrus sinensis*) has the highest commercial production and is used for eating fresh and extraction of **orange juices**. These include navel, Valencia and blood oranges. Loose skinned oranges include **mandarins** and **tangerines**. Bitter oranges, also Seville or sour oranges, are too sour to eat raw, and are used in making **marmalades**, **food flavourings**, **liqueurs**, such as curacao, and candied peel.

Orange wines   **Fruit wines** made using **oranges** or **orange juices** as starting material. Many different types of oranges are used. Usually consumed as an aperitif or dessert wine.

Orbignya   Genus of **palms** including the cohune palm (*Orbignya cohune*) and the babassu palm (*O. phalerata*). **Fruits** are used as a source of **palm oils** and food.

Oregano   Common name for *Origanum vulgare* and other members of this genus native to Europe (*O. syriacum*, *O. compactum* and *O. onites* but not *O. majorana* which is the source of the spice **marjoram**) the leaves of which are used as **spices**. Mexican oregano is produced from *Lippia* spp. which are cultivated pre-

dominantly in the Americas. **Carvacrol** is the main aroma compound of oregano.

Oregano oils   **Essential oils** extracted from **oregano**. Rich in **phenols**. Possess **antimicrobial activity** and are used to protect packaged foods, e.g. **fish** or **meat**, from **spoilage**. Also possess **antioxidative activity**.

Oreochromis   Genus of **freshwater fish** belonging to the family Cichlidae, many of which are of commercial importance. Found in lakes and rivers across Africa. Include **tilapia**, with the most important species in commercial terms being *Oreochromis nilotica* (**Nile tilapia**) and *O. mossambicus* (Mozambique tilapia).

Organic acids   **Organic compounds** comprising one or more substituents with the chemical formula -CO(OH). Examples include **fatty acids**, **citric acid** and **acetic acid**. Include **carboxylic acids**.

Organic compounds   Compounds based on a skeleton of one or more carbon atoms. In their simplest forms, carbon atoms are bound to each other and to hydrogen (e.g. **hydrocarbons**); these include **paraffins** and **olefins**. More complex organic compounds have one or more hydrogen atoms substituted with other elements or groups, e.g. halogens, nitrogen, sulfur, hydroxyl groups, as in **organic halogen compounds**, **organic nitrogen compounds**, **organic sulfur compounds** and **alcohols**, respectively. Carbon atoms may form linear structures and ring structures; a hydrocarbon ring comprising six carbon atoms and six hydrogen atoms is known as a benzene ring and organic compounds containing this structure or derived from it are known as arenes or **aromatic compounds**.

Organic foods   Foods produced by organic farming methods, i.e. without the use of chemical **fertilizers** or **pesticides**, and without any additives. The aim is to provide high quality, healthy food free from chemical **residues**. In the case of livestock, strict attention is paid to animal welfare, growth promoters are banned and use of veterinary drugs is kept to a minimum. Organic foods are regarded as a healthy, environmentally friendly option by the consumer, but future market growth is uncertain due to problems associated with high prices and provision of consistent quality.

Organic halogen compounds   **Organic compounds** which contain one or more carbon atoms linked via covalent bonding to one or more halogen atoms (F, Cl, Br, I). This group includes **organochlorine compounds**, **polybrominated biphenyls** and **chlorofluorocarbons**.

Organic nitrogen compounds   **Organic compounds** containing one or more carbon atoms linked via covalent bonding to nitrogen. **Amino acids, purines, pyrimidines** and **alkaloids** are all examples of these compounds.

Organic sulfur compounds   **Organic compounds** which contain one or more sulfur atoms, either linked directly to a carbon atom via covalent bonding or indirectly via an oxygen atom. Examples include **thiols, methionine** and **allicin**.

Organization for Economic Cooperation and Development   The Organization for Economic Co-operation and Development (OECD) is an international organization founded in 1961 to coordinate the economic policies of industrialized nations. The OECD succeeded the Organization for European Economic Cooperation, an agency founded in 1948 to direct reconstruction efforts in European nations devastated by World War II. The OECD provides expertise and supervision on matters concerning the domestic economies of member nations and their interactions within the global economy. It publishes a broad range of statistics and forecasts on agriculture, economic development and aid, education, energy, the environment, trade, health, labour markets, science and technology, and taxation. The OECD hosts annual meetings of finance and economic ministers from its member nations; these meetings lay the groundwork for the annual meetings of heads of states from the Group of Seven, the world's most powerful industrial nations. During the 1990s, the OECD expanded its activities in response to the end of the Cold War in 1991, the increasingly global nature of trade, and the emergence of new issues such as environmental protection. The OECD also represents the industrialized nations in negotiations and discussions with developing nations about development assistance, the global distribution of resources and wealth, and the structure of the international economy.

Organobromine compounds   **Organic halogen compounds** containing one or more carbon-bromine bonds. Include **polybrominated biphenyls** such as **polybrominated diphenyl ethers**, which are flame retardants sometimes found as contaminants in environmental matrices, including **fish**.

Organochlorine compounds   **Organic compounds** which contain one or more carbon-chlorine bonds. Examples include **organochlorine insecticides, organochlorine pesticides** and solvents, such as **chloroform** and methylene chloride.

Organochlorine insecticides   Class of **insecticides** which are used widely for control of **insects** on crops and in food storage areas. May persist for long periods in the environment and in animal tissues. Use of many highly persistent examples, such as **al-**drin, **DDE, DDT** and **endrin**, has been discontinued in most parts of the world.

Organochlorine pesticides   Major class of **pesticides** comprising chlorine-containing **organic compounds**. Includes **organochlorine insecticides**.

Organoleptic evaluation   Alternative term for **sensory analysis**.

Organoleptic properties   Alternative term for **sensory properties**.

Organophosphorus insecticides   Class of **insecticides** which are widely used for control of insects in crops and food storage. Act as inhibitors of **cholinesterases**. Commonly used examples include **chlorpyrifos, dichlorvos, methidathion** and **parathion**.

Organophosphorus pesticides   Major class of **pesticides** comprising phosphorus-containing **organic compounds**. Include **organophosphorus insecticides** and acaricides, and some antifungal agents.

Organotin compounds   **Organic compounds** which contain one or more atoms of tin. Uses include as **pesticides** and **fungicides**. Regarded as contaminants of foods, since some organotin compounds have been shown to be toxic when tested in animal models. Examples include **butyltins** and triphenyltin (fentin).

Original gravity   Amount of extract (soluble material) present in **worts**, as calculated from the amount of non-fermented extract left in the finished **beer**, together with the amount of extract equivalent to the quantity of **ethanol** present in the beer.

Ornithine   Non-protein amino acid derived from L-**arginine** by hydrolysis. Intermediate of the urea cycle in terrestrial vertebrates. Has an amino propane side chain and is also termed 2,5-diaminopentanoic acid.

Orotic acid   Synonym for vitamin B_{13}. An intermediate in the biosynthesis of **pyrimidines**, a growth factor for some **microorganisms**.

Ortanique   Cross between **oranges** and **tangerines** with a distinctive acid-sweet **flavour**, very juicy flesh and thin skin. Flesh is deep orange in **colour** with few or no seeds.

Orthocide   Alternative term for the fungicide **captan**.

Oryzaephilus   Grain beetles of the order Coleoptera. Some (e.g. *Oryzaephilus surinamensis* and *O. mercator*) are pests of stored cereal grains (e.g. **wheat, rice** and **barley**).

Oryzanols   Ferulic acid **esters** of terpene **alcohols** commonly prepared from **rice bran oils** but which have also been extracted from **corn oils** and barley oils. Used predominantly as **antioxidants**.

Oryzenin Glutelin which is one of the main **storage proteins** in **rice**.

O₂ scavengers Abbreviation for **oxygen scavengers**.

Osladin Steroidal **saponins** and the main active sweet component of rhizomes of the fern *Polypodium vulgare*. Osladin is glycosylated with two disaccharide units of 2-*O*-α-L-rhamnopyranosyl-β-D-glucopyranose. Also known as polypodoside A.

Osmolality Concentration of osmotically active particles in a solution, measured in osmoles of solute/kg of solvent.

Osmolarity Concentration of osmotically active particles in a solution, measured in osmoles of solute/litre of solution.

Osmoregulation Regulation of osmotic pressure, especially in the body of a living organism.

Osmosis Passage of water through a differentially permeable membrane, from a region of low concentration of solutes to one of higher concentration. Osmosis stops if the pressure of the more concentrated solution exceeds that of the less concentrated solution by an amount known as the osmotic pressure between them.

Osmotic dehydration Alternative term for **osmotic drying**.

Osmotic drying Water removal **preservation** technique based on the water and solubility activity gradient across a cell's semi-permeable membrane. Involves immersing high moisture foods in an osmotic solution, usually of sugar or salt. Water flows out of the material, and sugar may flow in, depending on several variables. Osmotic drying with osmotic syrup recycling requires two to three times less energy than convection **drying**. At relatively low process temperatures (up to 50°C), it improves product **colour** and **flavour** retention. Application of osmotic drying in the food industry is restricted, as simultaneous solute transfer into the foods can affect product quality.

Osmotic stress Stress exerted on an item when under osmotic pressure.

Osteoporosis Weakening and brittleness of the bones, resulting in them becoming liable to fracture. Generalized osteoporosis occurs most commonly in the elderly, and in women following the menopause; it can also result from long-term steroid therapy, infection or injury. Although there is a net loss of calcium from the body, this is the result of osteoporosis, not its cause, and there is no evidence that calcium supplements affect the progression of the disease. A high calcium intake in early life may be beneficial, since this may result in greater bone density at maturity, but the most important factor is regular exercise to stimulate bone metabolism.

Ostiepok cheese Slovak plasticized, smoked **cheese** made from **ewe milk**.

Ostriches Large flightless fast-running African **birds** (*Struthio camelus*) belonging to the Struthionidae family. In recent years, popularity of **ostrich meat** has increased in many countries; consequently, ostrich farming has expanded greatly and is now popular in many European and Scandinavian countries as well as in Africa.

Ostrich fern Common name for *Matteuccia struthiopteris*. The tightly curled tips of the young fronds are the **fiddleheads**, which are eaten as a vegetable in **salads** or in **soups**. Also called the fiddlehead fern.

Ostrich meat **Meat** from **ostriches**. Ostrich carcasses contain a large proportion of lean meat, the majority of which is found in the 10 major muscles of the legs and thighs. Meat from the thigh region is darkest in **colour** whilst meat from the *iliotibialis cranialis* is a bright red colour. Ostrich meat has a low fat content compared with other red meats. Ostrich meat products include **sausages**, **salami**, **steaks**, minced meat, burgers and **biltong**.

Ouzo **Aniseed** flavoured **spirits** produced in Greece. Usually drunk mixed with, or accompanied by, water.

Ovalbumins Predominant **proteins** in **egg whites** of **eggs** produced by **poultry** including **chickens**, **ducks**, **geese** and **guinea fowl**.

Ovens Enclosed chambers or compartments in which foods are cooked or heated, for example during **baking** and **roasting**.

Overwrapping Packaging technique in which several **packs** or **multipacks** are wrapped together often with **cellophane** or other **plastics films** to form a single unit.

Ovine Affecting, resembling or relating to **sheep**.

Ovomucins Sulfated **glycoproteins** found in **egg whites** which are responsible for their gel structure. Possess antiviral activity and act as **trypsin inhibitors**.

Ovomucoid Heat resistant **glycoproteins** found in **egg whites**. Show activity as **trypsin inhibitors**.

Ovotransferrin **Glycoproteins** found in **egg whites**. Possesses **antimicrobial activity**. Also known as **conalbumin**.

Oxacillin Semisynthetic isoxazolyl penicillin antibiotic used primarily for treatment of staphylococcal **mastitis** in cattle. **Milk** from treated cattle is normally regarded as safe for human consumption by 5 days post-treatment.

Oxalates Salts and esters of **oxalic acid**. Present at high concentrations in **fruits** and **vegetables**, e.g. **potatoes**, **spinach**, **rhubarb**, **plums**, **tea** and some

nuts, where they are regarded as **antinutritional factors**. High concentrations of oxalates in urine are associated with formation of renal stones.

Oxalic acid Organic acid comprising two carboxylic acid groups which has many industrial applications including **clarification** of **fats** and **oils**, and acid hydrolysis of **starch** to produce **sugar syrups**. Present as **oxalates** in **fruits** and **vegetables**, e.g. **spinach**, **beets** and **strawberries**, where they are considered to be **antinutritional factors** due to their involvement in formation of renal stones.

Oxaloacetic acid Organic acid which is an intermediate in the citric acid cycle where its reaction with acetyl-CoA produces citrate-CoA which is hydrolysed to citrate. Also known as oxosuccinic acid.

Oxamyl Contact and systemic insecticide, acaricide and nematocide used to control chewing and sucking **insects**, spider **mites** and **nematodes** in a wide range of **fruits** and **vegetables**. Classified by WHO as extremely hazardous (WHO Ib).

Oxen Adult castrated male **cattle**, particularly those used as draft animals. In broader use, the term is used to describe all domesticated bovine animals kept for draft purposes, and for **meat** or **milk** production.

Oxidants Chemicals that are capable of causing the oxidation of other chemicals, i.e. they donate oxygen or remove electrons.

Oxidases EC 1.1.3. **Oxidoreductases** that catalyse reactions in which O_2 acts as an acceptor.

Oxidation Addition of oxygen to a compound, for example using **oxidizing agents**. Also includes reactions in which atoms in the reacting materials lose electrons, frequently together with the removal of hydrogen ions. Oxidation-reduction reactions always occur simultaneously; if one reactant is oxidized, another must be reduced.

Oxidation reduction potential Alternative term for **redox potential**.

Oxidative stability Extent to which a substance can withstand the stress of oxidation.

Oxidizing agents Chemicals capable of oxidation which are themselves reduced during the process, i.e. they gain electrons. Important oxidizing agents for the food industry include **chlorine dioxide**, which is used as an antimicrobial agent in water and **sea foods**, and **iodates**, which are used as **flour improvers**.

Oxidoreductases EC 1. **Enzymes** that catalyse oxidation-reduction reactions. The substrate that is oxidized is regarded as the hydrogen or electron donor. This group includes **dehydrogenases**, **oxidases** and reductases.

Oxolinic acid Quinolone antibiotic used for treatment of bacterial infections in cattle, swine, poultry and farmed fish. Residues normally deplete relatively rapidly in poultry and swine, but in fish, may persist for long periods depending on water temperature and salinity level.

Oxtail The skinned tail of all categories of **cattle**. Oxtail has a high percentage of bone running through the middle and has a high fat content. The tails of older animals contain greater proportions of meat than those of younger animals. It is usually sold jointed into pieces. Small pieces of oxtail are often used to prepare oxtail **soups** or **stocks**, whilst larger pieces may be cooked by braising. Oxtail requires long, slow cooking to extract the best **flavour**.

Oxygen Element with an atomic weight of 16 and an atomic mass number of 8. Most common form of free oxygen is the diatomic species, molecular oxygen (O_2). Oxygen is the most abundant element of the Earth (air is composed of approximately 20% O_2). Essential for respiration in animals and aerobic **microorganisms**, produced by photosynthesis and is a common substituent of **organic compounds**, including biopolymers, such as **proteins** and **polysaccharides**. Reaction of foods with oxygen (oxidation) is a common cause of food **spoilage**, e.g. oxidation of **fats** and **oils** causes **rancidity**, and presence of oxygen may allow growth of aerobic food spoilage microorganisms.

Oxygen absorbers Materials which reduce the oxygen contents of food **containers** and maintain them at a very low level. This inhibits the growth of **microorganisms** and insects, and oxidative chemical reactions, increasing the stability and **shelf life** of the packaged products. Also referred to as **oxygen scavengers**.

Oxygenases EC 1.13-EC 1.14. **Enzymes** that catalyse the incorporation of molecular **oxygen** from O_2 into the compound oxidized. Dioxygenases (which contain Fe) incorporate two atoms of oxygen, while monooxygenases incorporate only one atom.

Oxygen scavengers Materials which reduce the oxygen contents of food **containers** and maintain them at a very low level. This inhibits the growth of **microorganisms** and insects, and oxidative chemical reactions, increasing the stability and **shelf life** of the packaged products. Also referred to as **oxygen absorbers**.

Oxymyoglobin Bright red **pigments** which represent the reduced form of **myoglobin**. In oxymyoglobin, oxygen is bound to the ligand, and the haem group of myoglobin is in the ferrous (Fe^{2+}) state. When fresh **meat** is cut and a new surface is exposed to oxygen, the surface **colour** changes from dark purple to bright red; this colour change, associated with oxymyoglobin formation is known as **bloom**.

Oxytetracycline Tetracycline antibiotic used for treatment and control of a wide range of bacterial infections in cattle, swine, sheep, poultry and fish. Used particularly to treat intestinal and respiratory diseases in swine and bacterial diseases in farmed fish. Residues normally deplete within 5 days post-treatment in cattle, swine, sheep and poultry, but tend to persist for longer periods in fish.

Oxytocin Peptide hormone (nine **amino acids**) synthesized in the posterior pituitary gland. Stimulates uterine smooth muscle to induce uterine contractions and promote labour. Also induces secretion of **milk** in response to a suckling stimulus.

Oyster mushrooms **Mushrooms** of the genus *Pleurotus*.

Oyster nuts **Seeds** produced by *Telfairia pedata*. Used as a source of **oils** or eaten roasted. Similar in **flavour** to **almonds**.

Oyster plant Alternative term for **salsify**.

Oysters Common name for marine or freshwater bivalve **molluscs** in the family Ostreidae; distributed worldwide. Prized for **flavour** and **texture** of flesh, which ranges from creamy beige to pale grey in **colour**. Many species are commercially important, including *Ostrea edulis* (flat oysters), *Crassostrea gigas* (Pacific oysters) and *C. virginica* (blue point oysters). Marketed live (in shell), fresh (shucked), frozen, dried, smoked and semi-preserved.

Oyster sauces **Sauces** used in Oriental dishes, particularly in Chinese dishes. Prepared by **proteolysis** of **oysters** tissues.

Ozonation Application of **ozone** (O_3), a pungent, toxic form of oxygen with three atoms in its molecule, formed in electrical discharges or by UV light. Ozone is used in the **purification** of drinking water and in **oxidizing agents**.

Ozone Form of oxygen comprising three oxygen atoms. Gas which is a strong oxidizing agent with broad spectrum **antimicrobial activity**. Uses within the food industry include in **disinfectants**; permitted for use on food surfaces. Also known as triatomic oxygen.

P

Pachysolen Genus of yeast fungi of the class Saccharomycetes. *Pachysolen tannophilus* is used in the production of **xylitol** from hardwood hemicellulose hydrolysate, and in the production of **ethanol**.

Pacific hake Marine fish species (*Merluccius productus*, *M. gayi gayi* or *M. gayi peruanus*) of high commercial importance. Widely distributed in the eastern Pacific Ocean. Marketed fresh and frozen and cooked in a number of ways, including **steaming**, **boiling** and **frying**. Also used in **fish meal** production.

Pacific mackerel Marine fish species (*Scomber japonicus*) from the mackerel family (Scombridae); distributed in the Indo-Pacific. Commercially important food fish (especially in Japan). Flesh has high fat content with a strongly distinctive savoury **flavour**. Marketed fresh, frozen, smoked, salted and occasionally canned. Also known as **chub mackerel**.

Pacific ocean perch Marine fish species (*Sebastes alutus*) found in offshore waters of the North Pacific region. Important commercial food fish. Marketed fresh or frozen (whole or fillets); **livers** are used as a source of vitamin-rich **oils**.

Pacific salmon General name referring to any of the six species of **salmon** (cherry, chinook, chum, coho, pink and sockeye salmon) occurring in the North Pacific Ocean. All are highly valued food fish.

Pacific whiting Marine fish species (*Merluccius productus*) from the hake family (Merluccidae), found in the northeastern Pacific Ocean. A commercially important food fish; usually marketed frozen, as flesh quality rapidly deteriorates following capture. Also known as Pacific hake.

Packaging Enclosure or **wrapping** of products. Functions include product containment for handling, transportation and use, **preservation**, optimization of product presentation, hygiene and to facilitate product dispensing and use. The term covers retail (primary), grouped (secondary) and transport (tertiary) forms.

Packaging films Packaging materials in the form of thin sheets which can be wrapped round a product. Films can be made from synthetic materials, such as **plastics**, or natural substances, such as **whey proteins**.

Packaging materials Substances used to make **packs**. Packaging for foods is commonly made from a variety of materials, including **glass**, **plastics**, **rubber**, **wood** and **paper**, which are formed into a range of container types. The type of material chosen depends on the product to be packaged and the intended use.

Packinghouses Establishments in which products are packed.

Packs Containers of varying shapes and sizes made from **paper**, **plastics**, **cardboard** or other materials that are used to enclose items such as food. The term is also used to describe items or groups of items which are packed in containers or enclosed in **packaging materials**.

Paclobutrazol Heterocyclic organochlorine compound which is used as a plant growth regulator and fungicide. Inhibits synthesis of **gibberellins** which consequently retards growth and enhances flowering and fruiting. For this reason, fruit trees are often treated with paclobutrazol.

Pacu Name given to fruit-eating **freshwater fish** species, including *Piaractus mesopotamicus*, found mainly in Brazil and related to the piranha. Commonly produced by aquaculture. Marketed fresh and frozen.

Paddlefish Freshwater fish species (*Polyodon spathula*) of commercial importance belonging to the family Polyodontidae. Found in North American river systems. Used to make **surimi**. Paddlefish **roes** are used as **caviar substitutes**.

Paddy Rice that remains in the **husks**. Refers to rice when still in the field or after threshing. Also refers to a field used for growing rice that is subject to irrigation or flooding.

Padi straw mushrooms Alternative term for the **edible fungi** *Volvariella volvacea*.

Paecilomyces Genus of **fungi** of the class Eurotiomycetes. Occur in soils, foods, **fruit juices** and plant debris. Some species (e.g. *Paecilomyces variotii* and *P. lilacinus*) may be responsible for the **spoilage** of foods (e.g. **oilseeds**, **cereals**, **bread**, **meat** and **cheese**).

Paenibacillus Genus of Gram positive, facultatively anaerobic **bacteria**, members of which can fix nitrogen, produce **antimicrobial compounds** and synthesize hydrolytic **enzymes**. The type species is *Paenibacillus polymyxa*. *P. larvae* is a pathogen of honey **bees** and causes an infectious disease called American foulbrood.

PAGE Abbreviation for **polyacrylamide gel electrophoresis**.

PAH Abbreviation for **polycyclic aromatic hydrocarbons** or **polynuclear aromatic hydrocarbons**.

Pak choi Type of Chinese cabbage (*Brassica chinensis*) cultivated originally in the Far East and South East Asia but becoming popular in Western countries. Used widely in stir fried dishes and soups, eaten as a cooked vegetable or used raw in **salads**. Also known by various other names, including bok choy and white mustard cabbage.

Pakoras Indian **snack foods** consisting of pieces of spiced **meat** and/or **vegetables** enclosed in **batters** and deep fried.

Palatability **Sensory properties** relating to the extent to which a food is palatable, i.e. possessing satisfactory properties or composition to be acceptable for consumption. Palatability of many foods can be enhanced by selective processing. For example, **cereals** may be crushed by **milling** to produce **flour**, a process that removes or breaks down the indigestible outer husk of the cereal seeds. The flour can be made more palatable by making it into **bread** or **pasta**.

Palatinit Low calorie sweetener. Disaccharide sugar alcohol produced from **sucrose** via transglycosylation and reduction reactions. Composed of a mixture of two isomers, namely 6-*O*-(D-glucopyranosyl)-D-sorbitol and 1-*O*-(D-glucopyranosyl)-D-mannitol. Has sweetness of a similar intensity to sucrose, but unlike sucrose it is resistant to enzymic hydrolysis and is therefore a poor substrate for growth of microorganisms, including those responsible for food spoilage and formation of **dental caries**. It also potentiates flavours of aromatic foods and masks the metallic **aftertaste** produced by some **artificial sweeteners**. Also called isomaltitol.

Palatinose Commercial name for the disaccharide **isomaltulose**. Isomer of **sucrose** produced by bacterial transglucosylation.

Pale soft exudative defect Commonly abbreviated to PSE defect, a condition affecting **meat**, especially **pork**. It is often **stress** related and is associated with accelerated *post mortem* muscle metabolism and a low pH value in meat. There is a linear relationship between **myosin** denaturation and drip loss or surface lightness within the PSE quality class. Excessive **colour** variation, poor **water binding capacity** and de-

creased water holding capacity occur in PSE meat, making it unsuitable for further processing. It is a major problem for the swine industry. Approximately 60% of the PSE defect in pork is associated with porcine stress syndrome (PSS), a genetic disorder which enhances susceptibility to stress. Certain breeds of swine, such as Pietrain, are highly susceptible to the PSE defect. **Halothane sensitivity** tests have been used to screen breeding swine for PSS with the aim of preventing further propagation of the PSE defect in breeding herds. However, as about 40% of the PSE defect in pork is associated with stress prior to slaughter, and poor meat handling and storage, removal of the mutant PSS gene from breeding stock will not completely eliminate PSE meat. An inherited condition also underlies much of the PSE defect in **turkey meat**.

Palmarosa *Cymbopogon martini* or East Indian geranium, a plant whose leaves are used as **spices**. Palmarosa **essential oils** are also used as **flavourings**, having a sweet rose-like **aroma** with herbaceous undertones due to the presence of the aromatic alcohols **geraniol** and **nerol**.

Palm hearts Young apical shoots (also called cabbages) of **palms**, used as a vegetable. Long and slender with a delicate, artichoke-like **flavour**. Available fresh in some countries; otherwise, sold canned in water. Used in **salads** or in cooked dishes. Also known as hearts of palm.

Palmitic acid Saturated fatty acid containing 16 carbon atoms. Present as glyceride esters in many **fats** and **oils**, including **palm oils**, from which it is commonly obtained.

Palmitoleic acid Monounsaturated fatty acid comprising 18 carbon atoms and a double bond between atoms 9 and 10. Systematic name is cis-Δ^9-hexadecenoic acid. Component of **fats** and **oils**.

Palm kernel oils Oils produced from the kernels of the fruits of **oil palms**, *Elaeis guineensis*, usually by solvent extraction. Classed as lauric oils. Used in the manufacture of **margarines**, **cooking fats** and **confectionery**.

Palm oil mills effluents Organic waste water produced during processing of palm oils. Have high carbon contents and low nitrogen contents and often represent an environmental problem in areas where palm oil is produced.

Palm oils Oils derived from the fleshy portion of the fruits of **oil palms**, *Elaeis guineensis*. Rich in **carotenes**, which are often removed to give the oil a paler **colour**. In addition to their use as **cooking oils**, they are also used in the manufacture of soaps and candles.

Palm olein Olein isolated from **palm oils**.

Palms Tropical evergreen plants of the family Palmae or Arecaceae with a variety of uses. Products made from plant parts include **palm oils**, **sago**, **starch**, **sugar**, **palm wines** and **spirits**. Fruits and **palm hearts** of some species are eaten. Commercially important examples include date palms (*Phoenix dactylifera*), borassus palms, coconut palms (*Cocos nucifera*) and sago palms (*Metroxylon sagu*).

Palm stearin Stearin isolated from palm oils.

Palm wines **Alcoholic beverages** made by **fermentation** of juices tapped from the stems of several species of **palms**.

Palmyra Species of **palms** (*Borassus flabellifer*) which yields edible fruit and whose inflorescence (complete flower head) is a source of **palm wines**, **sugar** and **vinegar**. Also commonly spelt as palmyrah.

Palmyrah Alternative spelling for **palmyra**.

Palytoxin Polent marine toxin produced by zoanthids of the genus *Palythoa*. It has been detected in a range of **sea foods**, including **fish**, **crabs** and **seaweeds**. Can cause **food poisoning** and even death in people eating contaminated products.

Pancakes Thin, flat **cakes** made by **frying batters** in a pan or on a greased griddle and cooked on both sides until brown.

Pancreas An elongated, tapered organ located in the abdomen; it is mainly composed of exocrine tissue but includes islets of endocrine cells. Animal pancreases form a part of edible **offal**, which is known by butchers as **sweetbreads** or, specifically, as gut sweetbreads.

Pancreatins Mixed **hydrolases** prepared from pancreas. Useful for production of vegetable **protein hydrolysates**, casein **phosphopeptides** and powdered **milk infant formulas** in which the **casein** is pre-digested, and also for liquefaction of **fish proteins** and **meat** residues.

Paneer Indian cheese-like product made by acid **coagulation** of heated **buffalo milk**. White in **colour** with a spongy body and sweet, mildly acidic and nutty **flavour**. Used in the preparation of many products, including **curries**, vegetable dishes and sweets.

Panettone Rich Italian yeast **cakes** made with **candied fruits**, **eggs** and **butter**. Traditionally eaten on festive occasions.

Panning Method used to make coated **sugar confectionery**. Used to make two types of product, i.e. hard centres, such as **nuts** or **dried fruits**, covered with **chocolate**, or chocolate or similar centres coated with **sugar**. In both cases, the **coatings** are applied to the centres while they are tumbled in a pan or drum. Temperature control is used to harden chocolate coatings, while sugar coatings are hardened by moisture reduction.

Panose Oligosaccharide comprising three **glucose** residues, with one glucose residue α-1,6- linked to maltose (α-1,4- linked glucose disaccharide). Produced by hydrolysis of **pullulan** or via the action of **glycosyltransferases** on maltose.

Pans Metal containers, usually broad, flat and shallow, in which foods are cooked. Also, open containers in which **panning** of **confectionery** is performed.

Pantothenic acid Member of the **vitamin B group**. Chemically, pantothenic acid is the β-alanine derivative of pantoic acid, and is required for the synthesis of coenzyme A (involved in the metabolism of fats, carbohydrates and amino acids) and of acyl carrier protein (involved in the synthesis of **fatty acids**). Dietary deficiency is unknown; it is widely distributed in all living cells, the best sources being **livers**, **kidneys**, **yeasts**, and fresh **vegetables**. **Royal jelly** is also a rich source. Approximately 50% of pantothenic acid in grains is lost by **milling**, up to 50% in **fruits** and vegetables is lost during **canning**, **freezing**, and **storage**, and from 15 to 30% in **meat** is lost during **cooking** or canning. Pantothenic acid is reasonably stable in natural foods during storage, provided that oxidation and high temperatures are avoided.

Papads Traditional Asian **snack foods** made from a mixture of **black gram** meal, salt, **oils** and **spices**, which is deep fried or toasted.

Papain EC 3.4.22.2. A cysteine endopeptidase from the latex of **papayas** with broad specificity, but with a preference for a residue bearing a large hydrophobic side-chain at the P2 position. Many other plants contain homologues of papain. Used for **tenderization** of **meat**, **stabilization** of **beer**, **coagulation** of **milk** in **cheesemaking** and hydrolysis of **fish proteins**.

Papaya nectars **Fruit juice beverages** made by addition of water and/or **sugar**, and optionally other ingredients, to papaya juices.

Papayas **Fruits** produced by *Carica papaya*, a member of the pawpaw family. Vary in size, shape and **colour**. Rich in **vitamin A**, **vitamin C** and potassium. Flesh is yellow to orange, with a large number of small black seeds in the centre. Both flesh and **seeds** are edible. Unripe fruits are sometimes eaten as a vegetable; ripe fruits are eaten as **desserts**, or used to make **soft drinks**, **jams**, or **ice cream**. Leaves, stems and fruits of the plant contain the enzyme **papain**, used in **tenderization** of **meat** and **clarification** of **beer**. Also called **pawpaws** in the UK and fruta bomba in Cuba.

Paper Material manufactured in thin sheets from wood pulp or other fibrous substances. Used widely as a me-

Pearl barley Whole barley kernels with the husk and part of the bran layer removed by **polishing**. Often added to **soups**.

Pearling As well as referring to the formation of pearl shaped items, this term relates to the removal of indigestible **hulls**, **aleurone** and **germ** layers from **cereals** by abrasion. With respect to **barley**, three successive pearlings removes all of the hull and most of the bran layer, leaving what is termed pot barley. Two to three additional pearlings, followed by sizing with a grading wheel, produces pearl barley. Also known as attrition milling and abrasive debranning.

Pearl millet **Millet** kernels from which the husk and bran layer have been removed by polishing. Also a type of millet (*Pennisetum typhoideum*).

Pears **Pome fruits** produced by plants of the genus *Pyrus*. Common or European pears are *P. communis*; **Asian pears** are members of the species *P. pyrifolia*. Generally, European pears are bell-shaped with soft flesh and Asian pears are round with crunchy flesh. A great many cultivars are grown commercially. Good source of **dietary fibre**, **vitamin C** and **potassium**. Eaten fresh or canned. Used as dessert fruits, cooked in dishes, in **jams** or processed into **fruit juices** and **fruit nectars**. Juice from some cultivars is fermented to produce **perry**.

Peas Common name for *Pisum sativum*, a widely cultivated legume. Good source of protein and **vitamin C**. Green or immature seeds are cooked as a vegetable, canned or frozen. Dry or mature seeds are cooked, used in **soups** or other dishes, or rehydrated and canned as processed peas. In some cultivars, including snow peas, snap peas and sugar snap peas, the pod is also eaten.

Pea starch **Starch** isolated from **peas**.

Pecan nuts Type of **hickory nuts** produced by *Carya pecan* or *C. illinoensis*. Kernels have a high oil content. Eaten out of hand and also in a range of sweet and savoury dishes, the most famous being pecan pie, one of the popular **desserts** in the USA.

Pecan oils **Oils** extracted from **pecan nuts**. Rich in **unsaturated fatty acids**, with only small amounts of **saturated fatty acids**. Possess the characteristic sweet **aroma** of pecan nuts. Blends with other **vegetable oils** have been suggested as bases for **margarines** and **shortenings**.

Pecorino cheese Name for all Italian hard cheeses made from **ewe milk**. Types include Pecorino Romano from the Rome area, **Pecorino Sardo cheese** from Sardinia and Pecorino Siciliano from Sicily. The rind is pale straw to dark brown in **colour** depending on age, and the interior is white to pale yellow with small eyes. Pecorino Romano is larger than other Pe-

corino cheeses and takes 8-12 months to mature, after which it has a salty **flavour** with a fruity tang. Pepato is spiced with **peppercorns**.

Pecorino Sardo cheese **Hard cheese** made in Sardinia from **ewe milk** (Pecorino is a name given to all Italian hard cheeses made from ewe milk). Rind is pale straw to dark brown in **colour**, depending on age. Interior is white to pale yellow with small eyes. **Flavour** is salty with a fruity tang which becomes stronger as **ripening** proceeds.

Pectate lyases EC 4.2.2.2. Catalyse the eliminative cleavage of pectates to **oligosaccharides** with 4-deoxy-α-D-gluc-4-enuronosyl groups at their non-reducing ends. Can act on other polygalacturonides but do not act on **pectins**. Also known as pectate transeliminases. Thought to be involved in postharvest decay of **fruits** by **bacteria** and **fungi**, causing tissue degradation of cell walls, and softening and rotting of plant tissues.

Pectate transeliminases Alternative term for **pectate lyases**.

Pectic enzymes Group of **enzymes** that catalyse degradation of pectic polymers in the cell walls of plants. These enzymes are involved in the **ripening** of **fruits**, and have a number of uses in the processing of fruits and **vegetables**. The group comprises **polygalacturonases**, **pectinesterases**, **pectate lyases** and **pectin lyases**.

Pectic substances **Pectins** and **polysaccharides** derived from them, such as polygalacturonic acid, polyglucuronic acid and **polyuronides**.

Pectinases Alternative term for **polygalacturonases**.

Pectinatus Genus of Gram negative, obligately anaerobic rod-shaped **bacteria** of the family Selenomonadaceae. *Pectinatus cerevisiiphilus* is responsible for the **spoilage** of **beer**.

Pectinesterases EC 3.1.1.11. Hydrolyse the methyl ester groups of **pectins**, resulting in deesterification. The **enzymes** act preferentially on a methyl ester group of a galacturonate unit next to a non-esterified galacturonate unit. Found in various **fruits**, where they are involved in **ripening**. Used for **clarification** and reduction of the **viscosity** of **fruit juices**, and have also been used for production of low-sugar fig **jams** and detection of **adulteration** in **orange juices**. Also known as pectin methylesterases.

Pectin lyases EC 4.2.2.10. Catalyse the eliminative cleavage of **pectins** to **oligosaccharides** with terminal 4-deoxy-6-methyl-α-D-galact-4-enuronosyl groups. They are the only **pectic enzymes** that can depolymerize pectins without altering their **esterification** levels. Used for **clarification** and reduction of the **viscosity** of **fruit juices**, and for softening the tissues of **fruits** and **vegetables**. Potentially useful in the **bioremediation** of **waste water** from the processing of fruit juices.

Pectin methylesterases Alternative term for **pectinesterases**.

Pectins **Polysaccharides** present in all plant cell walls. Composed of chains of α-(1→4) linked D-polygalacturonate interspersed with (1→2)-L-rhamnose residues, usually found in a partially methyl esterified form. Also has side chains composed of neutral sugars. Major sources of pectins include **citrus peel** and **apple pomaces**. Pectins are **hydrocolloids** and form **gels** via cooling or enzymic action. Used as **gelling agents**, **stabilizers** and **thickeners** in **beverages** and semi-solid foods, such as **jams** and **jellies**.

Pectolytic enzymes Alternative term for **pectic enzymes**.

Peda Indian sweet made using **khoa** as the base material. There are regional variations in its manufacture techniques, with consequent effects on sensory and compositional properties. Generally, khoa and **sugar** are heated to the desired **texture** and then divided into portions (usually round balls) that are packed in paperboard boxes lined with greaseproof paper.

Pediocins **Bacteriocins** produced by several strains of *Pediococcus* spp. that are bactericidal against a wide range of **Gram positive bacteria**. Plasmid encoded pediocin A, synthesized by *P. pentosaceus* (FBB-61 and L-7230), has a wide host range against Gram positive bacteria. Pediocin AcH, synthesized by *P. acidilactici* H, is a plasmid encoded, hydrophobic, inhibitory protein with a molecular weight of 2700 Da. Some **Gram negative bacteria** can be made susceptible to pediocin AcH when they are sublethally stressed. **Antibacterial activity** of pediocin AcH is through destabilization of cytoplasmic membranes. Pediocin PA-1, synthesized by *P. acidilactici* PAC 1.0, is a plasmid encoded protein with a molecular weight of 16,500 Da. Both pediocin AcH and PA-1 are ribosomally synthesized. Bactericidial efficiency of pediocins varies greatly under different conditions; some are due to the physical and chemical properties of the molecules, and some are due to environmental factors.

Pediococcus Genus of Gram positive, facultatively anaerobic coccoid **lactic acid bacteria** of the family **Lactobacillaceae**. *Pediococcus acidilactici* and *P. pentosaceus* are used as **starters** in the manufacture of fermented meat and vegetable products. *P. inopinatus*, *P. dextranicus* and *P. damnosus* may be responsible for **spoilage** of **beer** and **wines**.

Peel **Rind** of **fruits** and **vegetables**. A source of **essential oils** that may be used as **flavourings**, **dietary fibre**, **pectins**, **vitamins** and **minerals**. Peel from some sources, e.g. **citrus peel**, is used in foods and **beverages**, eaten candied or chocolate coated, processed into **marmalades** or incorporated into garnishes. The term also refers to a spade-like device used for moving loaves of **bread** or **pizzas** into or out of **ovens**.

Peeling Removal of the outer covering, or **peel**, from **fruits** or **vegetables** using **knives** or special peelers. Also commonly removal of the shell from hard boiled **eggs**.

Pekmez Traditional Turkish concentrated fruit juice based product usually made from **grape juices**, but also from other **fruit juices**.

Pelargonidin One of the **anthocyanidins**, systematic name 3,4′,5,7-tetrahydroxyflavylium chloride. **Glycosides** of this compound are plant **pigments** which have been identified in crops, including **strawberries**, **radishes** and red fleshed **potatoes**. Name is derived from the flowering plant pelargonium, which is a source of the pigment pelargonin, the 3,5-diglucoside of pelargonidin.

Pelargonium Genus of plants which includes geraniums, **essential oils** from which may be used in foods and beverages as **flavourings** or antimicrobial agents.

Pelmeni **Dumplings** filled with **meat** or **fish** traditionally eaten in Russia.

Pelshenke values Scores that provide estimates of the potential **breadmaking** strength of **wheat** in relation to its **gluten** quality.

Pelt 44 Alternative term for the fungicide **thiophanate-methyl**.

Pemmican Meat products consisting of small, pressed cakes of pounded **dried meat**, fat and **fruits**. The meat is mixed to a paste with melted fat and the other ingredients, before shaping into cakes and drying in the sun. Pemmican was originally made by North American Indians, but has subsequently gained popularity as a useful food for travellers, including Arctic explorers.

Penamellera cheese Spanish semi **hard cheese** made from **cow milk**, **goat milk** or **ewe milk**. A natural rind cheese with a nutty **flavour** and meaty **aroma**. The interior is dense with some small holes.

Penetration Process of entry and permeation into an item. Penetration tests are widely used as a simple way to determine **yield stress** of a product.

Penetrometers Instruments used for measuring the **firmness** of foods, especially **fruits**, on the basis of the depth of penetration of a probe under a known load.

Penetrometry Technique for measuring the **firmness** of foods, especially **fruits**, based on the depth of penetration by a probe under a known load.

Penicillic acid Mycotoxin produced by *Aspergillus ochraceus* and *Penicillium viridicatum*. May occur in a wide range of foods susceptible to **spoilage** by these **fungi**, including **barley**, **corn**, **rice**, **cheese** and **fish**.

Penicillinases β-Lactamases that hydrolyse β-lactam **antibiotics**. Used for detection of antibiotics residues in foods.

Penicillin G Natural penicillin antibiotic produced by *Penicillium chrysogenum*. Active against **Gram positive bacteria**. Used for treatment of bacterial infections in all farm animals, particularly for control of **mastitis** in dairy cows and for treating infections of the gastrointestinal system and urinary and respiratory tract. Residues in **milk** and muscle tissues are rarely detectable beyond 5 days from the final treatment. Also known as **benzylpenicillin**.

Penicillins Group of **antibiotics** widely used to treat bacterial diseases in animals, and constituting the most important group of antibiotics. Classified in three distinct groups: natural penicillins (including **penicillin G**); penicillinase-resistant penicillins (including **cloxacillin** and **oxacillin**); and broad spectrum penicillins (including **amoxicillin** and **ampicillin**).

Penicillium Genus of **fungi** of the class Ascomycetes. Some species, e.g. *Pencillium digitatum*, *P. expansum* and *P. implicatum* can cause food **spoilage** and some are capable of causing food spoilage at **refrigeration** temperatures. Some species produce **mycotoxins**, e.g. **citrinin**, luteoskyrin and **patulin**. Certain species are used in production of **organic acids** and **antibiotics**, while others are used in **cheesemaking**, e.g. *P. camemberti* (**Brie cheese, Camembert cheese**) and *P. roqueforti* (**Roquefort cheese, Stilton cheese**).

Pentanal Synonym for **valeraldehyde**. Organic compound present in many foods that has an unpleasant odour and a low odour threshold value. One of the main compounds that can cause **off odour** in **sake**.

Pentane One of the **paraffins**. Saturated aliphatic hydrocarbon composed of five carbon atoms and used as a solvent.

Pentanedione Synonym for **acetylacetone**. Ketone which occurs in the **flavour compounds** of foods and **beverages**, including **beer**, **coffee** and **fermented dairy products**.

Pentanoic acid Synonym for **valeric acid**. Volatile fatty acid comprising five carbon atoms and a single carboxylic acid group. Contributes to the **aroma** of mature **cheese**. Uses include as a reactant in production of **aroma compounds** and **flavourings**. Also one of the main malodorous pollutants from livestock houses.

Pentanol Synonym for **amyl alcohol**. One of the higher **alcohols**, comprising five carbon atoms and a single alcohol group. Of importance in the **flavour compounds** fraction of **alcoholic beverages**. Forms part of the toxic **fusel oils** fraction of **spirits**. Used as a solvent and as a substrate for production of the flavouring amyl acetate.

Pentosanases **Enzymes** that hydrolyse **pentosans**. Include endo- and exo- arabanases (α-*N*-arabinofuranosidases), which are used in production of **fruit juices**, xylan degrading enzymes and **hemicellulases**. Used in **breadmaking** for improving **dough** properties and **loaf vol.**, and for extending **bread shelf life**.

Pentosans Polysaccharides formed from **pentoses**. Found mainly in fibrous plant tissues, e.g. almond shells and **cereals**. Pentosan composition of cereals, such as **wheat** and **rye**, may influence grain **texture**.

Pentoses Monosaccharides comprising five carbon atoms. Examples include the aldoses, **ribose**, **arabinose** and **xylose**, and the ketose, **xylulose**.

Peonidin One of the **anthocyanidins**, systematic name 3,4′,5,7-tetrahydroxy-3′-methoxyflavylium chloride. Glycosides of this compound are plant **pigments** which are present in red **grapes**, purple-flesh **sweet potatoes** and black **rice**. Name is derived from peonies, plants with violet-red flowers from which peonin, the 3,5-diglucoside of peonidin, has been obtained.

Pepino Fruits produced by *Solanum muricatum*. Vary greatly in size, shape and **colour**, and may be seeded or seedless. Rich in **vitamin C** and potassium, with smaller amounts of **vitamin A**. **Flavour** resembles that of cantaloupe or honeydew **melons**. Used peeled as a dessert or as a component of a number of dishes; **seeds** are also edible. Available dried, canned or bottled. Also known as pepino dulce, melon pepino, melon pear and mellofruit.

Pepper Spices obtained by crushing dried berries from *Piper nigrum* (**black pepper** and **white pepper**) or *Schinus molle* (pink or red pepper). Pepper imparts a warm aromatic **flavour** to foods. The main aroma compound present is **piperine**.

Peppercorns Whole dried berries from *Piper nigrum* or *Schinus molle* (black and red peppercorns, respectively). Used as culinary spices to impart a warm, aromatic **flavour** to foods.

Peppermint Common name for *Mentha piperita*, leaves of which are used as **spices**. When added to foods or **beverages**, peppermint imparts a fresh, cool

flavour. The main active aroma compound of peppermint is **menthol**.

Peppermint essential oils **Essential oils** distilled from **peppermint**. The characteristic fresh, minty notes, produced by **menthol**, are not present in the primary distillate but are formed by further processing or natural ageing of the oils. The oils also contain various quantities of menthofuran, peroxidation of which produces an undesirable **aftertaste**, and thus content of this molecule influences quality of peppermint essential oils.

Pepperoni Highly spiced, ready-to-eat, Italian salami-type **sausages** prepared from **pork** and **beef**. They are seasoned with **black pepper**, **cayenne pepper**, **garlic** and **salt**, and dried slowly to a hard **texture**. They are often sliced thinly and served as an appetizer or added to **pizzas**.

Peppers Fruits produced by plants of the genus *Capsicum* (family Solanaceae), the most important species being *Capsicum annuum* and *C. frutescens*. Vary in size, shape, **colour** and **pungency**, but all are hollow, with many seeds in the centre. All types are rich in **carotenes** and **vitamin C**. According to variety, peppers are used as vegetables or as the source of **flavourings** for foods. Types include **bell peppers**, **sweet peppers**, **red peppers**, **green peppers**, **pimento peppers** or **pimiento peppers**, and **chillies**.

Pepsins EC 3.4.23.1-3. Acid **proteinases** secreted from vertebrate gastric mucous membranes as inactive precursors that are converted autocatalytically to the active enzyme under acidic conditions. Preferentially cleave peptide linkages between two aromatic **amino acids**. Used for preparation of **protein hydrolysates**, **stabilization** of **beer** and form part of the active constituents of **rennets** used in the dairy industry.

Peptidases Enzymes that hydrolyse peptide bonds. Include **aminopeptidases**, **carboxypeptidases** and **endopeptidases**.

Peptides Compounds formed by two or more **amino acids** linked via peptide bonds, e.g. **dipeptides** (two amino acids linked), oligopeptides (several amino acids linked) and **polypeptides** (many amino acids linked).

Peptidyl-dipeptidase A EC 3.4.15.1. Releases C-terminal dipeptides from **polypeptides**, provided proline is not present on either side of the cleavage site. Also known as angiotensin I-converting enzyme. Inhibitors of this enzyme are potentially useful as components of **functional foods**, since they exhibit **antihypertensive activity**.

Peptones Protein hydrolysates produced via the action of pepsin. Peptones are formed in the stomach during digestion of **proteins**.

PER Abbreviation for **protein efficiency ratios**.

Pera Khoa-based dairy product popular in India. Also called **peda**.

Peracetic acid Chemical with the systematic name ethaneperoxoic acid. Strong oxidizing agent which is used as a disinfectant in the food industry.

Perch **Freshwater fish** species (*Perca fluviatilis*) widely distributed throughout Europe. Cooked flesh is normally firm with a mild **flavour**. A popular food fish in some regions of Europe, where it is marketed fresh and frozen.

Performance drinks Non-alcoholic beverages formulated with ingredients claimed to enhance physical or mental performance.

Pergamyn Transparent, **celluloses**-based **paper**. Possesses many properties that make it suitable for packaging of foods, including lack of taste or smell, and its greaseproof nature.

Perilla Genus of plants the green or red leaves of which are used in **salads**, as a vegetable or as a garnish. The most commonly consumed species is *Perilla frutescens*. Also known as green shiso, Japanese basil and red shiso. Grown also for **perilla seeds**, a source of **perilla oils**.

Perilla oils Pale yellow **oils** extracted from **perilla seeds** (produced by the Asiatic mint plant, *Perilla* spp.). Used in the Far East as **cooking oils** and also used in the manufacture of paints and varnishes.

Perilla seeds Oil-rich seeds produced by plants of the genus *Perilla*, especially *P. frutescens*.

Perishability Extent to which an item is perishable, i.e. having a short **shelf life** or deteriorating quickly during storage.

Perishable foods Foods with a short **shelf life**, such as **milk**, **eggs**, **meat**, **fish** and many **fruits** and **vegetables**.

Periwinkles Any of a number of small marine gastropod **molluscs**; abundant on rocky shores along Atlantic coasts. Several species are popularly consumed, including *Littorina littorea* (common or edible periwinkles), *L. obtusata* (smooth periwinkle), *L. irrorata* (gulf periwinkles) and *L. angulifera* (southern periwinkles). Usually marketed fresh (in shell, cooked or uncooked).

Permeability Ability of items such as membranes or other barriers to permit fluids to flow through them. Permeability is an important indicator of membrane functionality, and is expressed as a volume flow of liquid through a unit area of membrane at some defined transmembrane pressure. Permeability of food **pack-**

aging materials is important in relation to product **shelf life**. **Modified atmosphere packaging** of foods can involve use of films with various gas permeability coefficients.

Permeation Passage of fluids through items such as membranes, food **packaging materials** or other barriers, or, in chemical terms, the diffusion or penetration of ions, atoms or molecules through a permeable substance. In the food industry, knowledge of the level of permeation of gases through functional barriers such as packaging materials is important in relation to product **shelf life**.

Permethrin Non-systemic insecticide used for control of a wide range of insect **pests** in **fruits** and **vegetables**; also used to control biting **insects** in animal rearing establishments. Classified by WHO as moderately toxic (WHO II).

Permissible levels Recommended limits for the amounts of particular contaminants (e.g. residues of veterinary **drugs**, **heavy metals**) that may be permitted in certain foods.

Pernod French **aniseed**-flavoured **aperitifs**, originally formulated as a substitute for **absinthe**.

Peroxidases Includes EC 1.11.1.7 and other members of subclass EC 1.11.1. Involved in **ripening** of **fruits**, **enzymic browning** and degradation of **lignin** by white-rot **fungi**. Peroxidases have a number of industrial applications, including use in **time temp. indicators**, such as those used for investigating inhibition of **microorganisms** during the thermal processing of low-acid foods, detection of **phenols** and production of **flavour compounds**. In addition, the degree of inactivation of peroxidases can be used as an indicator of the extent of **blanching** in **vegetables**.

Peroxidation Formation of peroxides as a result of the action of oxygen, especially on **polyunsaturated fatty acids**. **Vitamin E** prevents lipids peroxidation in cells.

Peroxides Compounds containing either the peroxide ion, e.g. sodium peroxide, or covalently bonded dioxygen (R-O-O-R), the simplest being hydrogen peroxide. Organic peroxides may be formed via autoxidation reactions or by direct oxidation, processes involved in the development of **rancidity** of **fats** and **oils**.

Peroxide values Measure of the number of millimoles of peroxide absorbed by 1000 g of oil or fat, used as an indicator of **rancidity**. As fats decompose, peroxides are formed. Chemically, peroxides are capable of causing the release of I from KI. Therefore, the amount of I released from KI added to a fat is a rancidity test. The more peroxide present, the more I released; hence, the higher the peroxide values.

Perry **Cider**-like **alcoholic beverages** made by **fermentation** of **pear juices**, commonly prepared from special cultivars of **pears**.

Persimmon juices **Fruit juices** extracted from **persimmons** (*Diospyros kaki*).

Persimmons **Fruits** produced by *Diospyros kaki*. Contain moderate amounts of **vitamin C**, **carotenes** and **sugars**. Most varieties are orange in **colour** when ripe, with the appearance of **tomatoes**. Some varieties have an astringent taste, especially when unripe, due to high levels of **tannins**. Non-astringent fruits are eaten out of hand, cooked, candied or made into **jams** or **jellies**.

Persipan Product which is often used as an alternative to **marzipan** and is similar in composition, but is made using **apricot kernels** instead of **almonds**.

Persulfates Salts of peroxodisulfuric acid which are strong **oxidizing agents**. Ammonium persulfate and potassium persulfate are permitted **food additives**. Ammonium persulfate is a bakery additive, uses including **bleaching agents** for **starch** and food **preservatives**. Potassium persulfate has uses in **defoaming agents**. Alternative names include peroxosulfates and peroxodisulfates.

Peruvian carrots Alternative term for **arracacha**.

Peruvian parsnips Alternative term for **arracacha**.

Pervaporation Membrane **separation** technique in which a liquid feed mixture is separated by partial vaporization through a non-porous, selectively permeable membrane. A vapour permeate and a liquid retentate are formed. Partial vaporization is achieved by reducing the pressure on the permeate side of the membrane (vacuum pervaporation) or, less commonly, by sweeping an inert gas over the permeate side (sweep gas pervaporation). Vacuum pervaporation at ambient temperature using hydrophilic membranes is used to dealcoholize **wines** and **beer**, whereas hydrophobic membranes are used to concentrate **aroma compounds** such as **alcohols**, **aldehydes** and **esters**.

Pesticides Chemical substances used to kill plants, animals or other organisms that interfere with agricultural production or are harmful to humans. Major groups include **herbicides** (for control of unwanted plants), **insecticides** (for control of insect **pests**), **fungicides** (for control of pathogenic or spoilage **fungi**) and **rodenticides** (for control of rats, mice and other **rodents**). Many are non-specific and may be too toxic to organisms that are not considered pests. Some persist for long periods in the environment and can accumulate in the food chain. Residues in foods may represent a health risk to consumers

Pesto Sauces, often served with **pasta**, the major ingredients of which are **basil**, **garlic**, **nuts** and **olive oils**.

Pests Organisms (typically **rodents**, **insects** and **pathogens**) that are regarded as harmful to humans, animals or plants.

PET Abbreviation for **polyethyleneterephthalate**.

Petitgrain oils **Essential oils** used in **flavourings** for many foods, especially **confectionery** products.

Petroselinic acid Monounsaturated fatty acid comprising 18 carbon atoms obtained from **parsley** seed **essential oils**. Systematic name *cis*-6-octadecenoic acid.

Petunidin One of the **anthocyanidins (flavonoids** and red/blue **pigments**, the **colour** of which is pH dependent), systematic name 3,3′,4′,5,7-pentahydroxy-5′-methyoxyflavylium chloride. Glycosides of petunidin are plant pigments that are present in crops, including red **grapes**, **blackberries**, **blueberries**, purple fleshed **sweet potatoes** and black **rice**. Name is derived from blue petunia, a flowering plant from which the pigment petunin, the 3, 5-diglucoside of petunidin, is obtained.

PFGE Abbreviation for **pulsed field gel electrophoresis**.

P-Fibre **Fibre** extracted from the interior of **peas**. Often used as a fat substitute in **dairy products**.

pH Measure of the degree of **acidity** or **alkalinity** of a substance. pH (an abbreviation for potential of hydrogen) is defined as the negative logarithm of the hydrogen ion concentration. The scale ranges from 0 (very strongly acid) to 14 (very strong alkaline). A neutral solution, such as pure water, at 25°C has a pH of 7.

Phaeophytins Brown **pigments** produced by removal of magnesium ions from **chlorophylls** using limited hydrolysis. Present in **green vegetables** as degradation products of chlorophylls; degradation is accelerated by **cooking** or processing and thus may cause **browning** in vegetables or vegetable products.

Phaffia Genus of yeast **fungi** of the class Heterobasidiomycetes. Species (e.g. *Phaffia rhodozyma*) may be used in the production of **astaxanthin**, which is added to animal feeds to confer a reddish **colour** to **fish** flesh, **poultry meat** and **egg yolks**.

Phages Alternative term for **bacteriophages**.

Phalsa Small, round, dark-purple or nearly black fleshy **fruits** with a pleasantly acidic fibrous flesh. Botanical name is *Grewia subinaequalis* or *G. asiatica*. Native to India and Nepal, but also found in Australia. May be eaten fresh as a dessert, or made into **syrups** for use in the manufacture of **soft drinks**.

Phaltan Alternative term for the fungicide **folpet**.

Phane Product made from caterpillars of the emperor moth (*Imbrasia belina* Westwood) which feed on the mophane tree (*Colophospermum mopane*). Caterpillars are cooked after removal of the stomach contents and then either eaten immediately or as a snack after **salting** and **drying**. Consumed as a delicacy in Botswana and other parts of southern Africa.

Phanerochaete Genus of **fungi** of the class Homobasidiomycetes. *Phanerochaete chrysosporium* is used in the production of ligninolytic **enzymes** (e.g. **lignin peroxidases**).

Phase behaviour Activity of the various components of a mixture; of primary importance for food formulation and processing. For example, examination of the phase behaviour of fat mixtures (**palm kernel oils**, **cocoa butter** and **anhydrous milk fats**) can aid in the understanding of softening and **bloom** formation in compound **coatings**. Information regarding the phase behaviour properties of biopolymer systems may be useful in the design of new low fat foods.

Phaseolins One of the major types of **storage proteins** (7S) of **common beans** (*Phaseolus vulgaris*).

Pheasant meat **Meat** from **pheasants**, medium-sized, long-tailed sedentary game **birds** belonging to the Phasianidae family. Birds are hunted as game and also reared commercially for meat production. Meat is lean and dry, and is marketed fresh and frozen. Meat from female pheasants tends to be juicier and more tender than that from males; **flavour** of wild pheasant meat tends to be stronger than that of farmed birds. For optimum meat flavour and **texture**, it is recommended that pheasants are hung before **cooking**. The main method of cooking is **roasting**, or **braising** for older birds, but meat is also used in stews and **soups**.

Pheasants Medium-sized, long-tailed sedentary game **birds** belonging to the Phasianidae family; there are several species. Pheasants are hunted for their **meat**. They are also reared commercially for **pheasant meat** production. Meat from female pheasants tends to be juicier and more tender than that from male pheasants. **Flavour** of wild pheasant meat tends to be stronger than that of farm-raised birds.

Phenanthrene Polycyclic aromatic hydrocarbon consisting of three condensed benzene rings which is present in coal tar. Used in the manufacture of **pigments**. Detected as a contaminant in foods, including **cheese**, **sea foods** and cooked **meat**.

Phenethyl alcohol Synonym for **phenylethanol**. Aroma compound present in **essential oils** of orange blossom, rose and hyacinth. Used in food **flavourings** for imparting a mildly floral **flavour** to foods such as **ice cream** and **chewing gums**.

Phenobarbital Barbiturate that is used mainly as an anticonvulsant drug for the treatment of all forms of epilepsy (except absence seizures) in animals.

Phenolases Alternative term for **catechol oxidases**, **laccases** and **monophenol monooxygenases**.

Phenolic compounds Alternative term for **phenols**.

Phenols Group of **organic compounds** comprising at least one benzene ring that is covalently bonded to one or more hydroxyl groups. Phenols have wide distribution and applications, and are available in synthetic or natural forms, e.g. **lignin** and **catechols**. Uses include as **disinfectants** (**cresols**), in manufacture of **azo dyes** and **plastics**, and as **flavourings** (**vanillin**), **antioxidants** (**sesamol** and **NDGA**) and **pigments** (**curcumin**). Some phenols, e.g. **chlorophenol**, are also considered to be **toxins**.

Phenotype Observable characteristics of an organism, either in total or with respect to particular traits, resulting from the interaction of the **genotype** and the environment.

Phenylacetaldehyde Aromatic aldehyde which has an **aroma** resembling hyacinths or lilacs. Applications include in **aroma compounds** and **flavour compounds** used in foods, such as **bakery products**, **sugar confectionery** and **ice cream**.

Phenylacetic acid Volatile aromatic organic acid. Used in **flavourings** for foods such as **bakery products**, **ice cream** and **sugar confectionery**. Also a substrate for synthesis of other **flavour compounds**. Alternatively called α-toluic acid and benzeneacetic acid.

Phenylalanine Essential amino acid with an aromatic side chain which is obtained in the diet from **proteins**, such as **ovalbumins**, **lactalbumins** and **zein**. In common with the other **amino acids**, only the L-enantiomer of phenylalanine is utilized significantly by humans. Substrate for manufacture of the dipeptide sweetener **aspartame**. Given the internationally recognized three letter and single letter codes Phe and F, respectively.

Phenylalanine ammonia-lyases EC 4.3.1.5. Deaminate L-phenylalanine to form *trans*-cinnamate and ammonia. The reverse reaction can be used for production of L-phenylalanine, a precursor of **aspartame**. Involved in accumulation of **flavonoids** in **apples** and **enzymic browning** in **fruits**, and are markers of environmental stress in plant tissues, e.g. **chilling injury** and wounding. They may also act on L-tyrosine.

Phenylethanol Synonym for **phenethyl alcohol**. Aroma compound present in **essential oils** of orange blossom, rose and hyacinth. Used in food **flavourings** for imparting a mildly floral **flavour** to foods such as **ice cream** and **chewing gums**.

Phenylketonuria Genetic disease (commonly abbreviated to PKU) in which patients are unable to metabolize the amino acid **phenylalanine**, which is a normal constituent of diet. The amino acid and its derivatives accumulate in the body and prevent proper mental development. The gene responsible for phenylketonuria is recessive, so that a child is affected only if both parents are carriers of the defective gene. Infants with the disease need a special diet that contains little phenylalanine, which should be maintained until adolescence.

Pheromones Substances secreted by a species, which are recognized by members of the same species. Used for intraspecies communication, e.g. for attraction.

Phloroglucinol Phenol with the systematic name 1,3,5-benzenetriol. Used in manufacture of **pigments** and **plastics**.

Phloxine Red xanthene **pigments** used as food **colorants**. Also known as Food Red No. 104.

pH meters Instruments for measuring the **pH** of a solution.

Phoenix Genus of **palms** that includes the date palm *Phoenix dactylifera*, the wild date palm *P. sylvestris* (used as a source of **sugar**) and the **sago** producing palm *P. acaulis*.

Phoma Genus of **fungi**. Some species (e.g. *Phoma herbarum* and *P. sorghina*) may cause **spoilage** of **fruits** (e.g. **melons**, **papayas** and **bananas**), **vegetables** (e.g. **sugar beets**), **cheese** and **cereals** (e.g. **sorghum**, **barley**, **corn** and **rice**).

Phosalone Non-systemic insecticide and acaricide used primarily for control of **insects** on **pome fruits** and **stone fruits**; also used in **grapes**, **rapeseeds**, **potatoes** and **vegetables**. Classified by WHO as moderately toxic (WHO II).

Phosmet Non-systemic insecticide and acaricide with predominantly contact action. Used for control of biting, sucking and chewing **insects** on a range of **fruits**, **vegetables** and **cereals**. Also used to control weevils on **sweet potatoes** during storage and as an animal ectoparasiticide. Classified by WHO as moderately toxic (WHO II). Also known as imidan.

Phosphamidon Systemic insecticide and acaricide used to control sucking, chewing and boring **insects** and spider **mites** on a wide range of crops. Classified by WHO as extremely hazardous (WHO Ia).

Phosphatases Enzymes that hydrolyse phosphomonoesters. Acid phosphatases are specific for a single charged phosphate group and **alkaline phosphatases** for a double charged group. Alkaline phosphatase levels can be used to determine the contents of **tannins** in **grapes** and **red wines**, the degree of **milk pasteurization**, and the extent of heat treatment of **meat** and meat products. Alkaline phosphatase has also been used in amperometric **biosensors** for determination of levels of **phosphates** in **drinking water**. Acid phosphatases produced by **lactic acid bacteria** are involved in **flavour** development during **cheese ripening**, but can also cause food deterioration.

Phosphates **Salts**, condensation products or **esters** of **phosphoric acid**. Phosphates used in the food industry include the food **additives trisodium phosphate** and potassium phosphate, and **polyphosphates**.

Phosphatides **Salts** or **esters** of **phosphatidic acid**.

Phosphatidic acid Simplest phospholipids, composed of glycerol esterified to two **fatty acids** and **phosphoric acid**. Also called diacylglycerol-3-phosphoric acid.

Phosphatidylcholine Alternative term for **lecithins**.

Phosphatidylethanolamine Phospholipid produced by **esterification** of **phosphatidic acid** to **ethanolamine**.

Phosphatidylinositol Phospholipid formed by **esterification** of **phosphatidic acid** to **inositol**.

Phosphatidylserine Phospholipid formed by **esterification** of **phosphatidic acid** to **serine**.

Phosphine Fumigant gas produced from phosphorus or metal phosphides.

Phosphodextrins Phosphate esters of **dextrins**.

Phosphoglucomutases EC 5.4.2.2 (phosphoglucomutases) and EC 5.4.2.6 (β-phosphoglucomutases). The former convert α-D-glucose 1-phosphate to α-D-glucose 6-phosphate. Also catalyse, although more slowly, the interconversion of 1- and 6-phosphate isomers of many other α-D-hexoses, and the interconversion of α-D-ribose 1-phosphate and 5-phosphate. The latter convert β-D-glucose 1-phosphate to β-D-glucose 6-phosphate.

Phosphoglycerides Phosphate esters of **glycerides**.

Phospholipases EC 3.1.1.32 (phospholipase A1), EC 3.1.1.4 (phospholipase A2), EC 3.1.4.3 (phospholipase C) and EC 3.1.4.4 (phospholipase D). These **enzymes** hydrolyse **phospholipids**; phospholipases A1 and A2 hydrolyse sn-1 and sn-2 acyl esters, respectively, while phospholipases C and D cleave sn-3 phosphodiester bonds. Useful for production of **emulsifying agents** and novel **lecithins** with nutritional applications, and for improving the softness of **bread**.

Phospholipase C acts as a virulence factor in certain bacterial **pathogens**. Phospholipase B is included under **lysophospholipases**.

Phospholipids **Lipids** comprising a **glycerol** or sphingosine backbone esterified to two **fatty acids** and **phosphoric acid** or a phosphoric acid ester. Examples include **phosphatidic acid, phosphatidylserine, phosphatidylinositol, phosphatidylethanolamine** and **lecithins**.

Phosphopeptides **Peptides** containing one or more **serine** or **threonine** residue esterified to a phosphate group.

Phosphorescence **Luminescence** that persists after the cause of excitation has been removed.

Phosphoric acid Synonym for orthophosphoric acid. Acid produced by reaction of **phosphates** with **sulfuric acid** or by oxidation of phosphorus followed by addition of water. Permitted food additive that is used to acidify **fruit juice beverages** and **cola beverages**, and as a substrate for **phosphates**.

Phosphorus Mineral element with the chemical symbol P. Forms three different types of crystal structure, termed white, red and black phosphorus which also differ with respect to **physical properties** and reactivity.

Phosphorylases Members of EC 2.4. Enzymes that transfer glycosyl groups from a donor compound to inorganic phosphate.

Phosphotransferases EC 2.7. **Enzymes** that transfer a phosphate group from a donor to an acceptor that may be an **alcohol** (EC 2.7.1), a carboxyl group (EC 2.7.2), a nitrogenous group (EC 2.7.3) or a phosphate group (EC 2.7.4). Diphosphotransferases transfer two phosphate groups to either a single acceptor (EC 2.7.6) or two different acceptor molecules (EC 2.7.9).

Phostoxin Compound used, typically in pellet form, as a fumigant in stored foods (especially **cereals** such as **wheat, barley, corn** and **rice**).

Phosvitin Phosphoproteins found in **egg yolks**. Possesses **antioxidative activity**.

Photobacterium Genus of Gram negative, facultatively anaerobic coccoid or rod-shaped **bacteria** of the family Vibrionaceae. Occur in sea water, the **gastrointestinal tract** of **fish** and marine animals, and the luminous organs of certain fish and cephalopods. *Photobacterium phosphoreum* may be responsible for the **spoilage** of fish and fish products.

Photocolorimetry **Colorimetry** technique in which results obtained using a colorimeter are recorded permanently using photography.

Photodensitometry Technique used to determine the density of a substance by examination of photographic negatives. Used in combination with chromatographic techniques, such as **thin layer chromatography**, and **gel electrophoresis** to quantitate separated components. Also used widely in medicine, where it is known alternatively as radiographic absorptiometry, to assess bone mineral changes.

Photolysis Cleavage of one or more covalent bonds in a molecule due to the absorption of energy from light or some other form of electromagnetic radiation (e.g. **UV radiation**, **X-rays**).

Photometry Science of visual radiation and the theory of its measurement. Luminous quantities can be measured by the human eye, while radiant quantities are measured by devices sensitive to electromagnetic energy. Photometric measurements are performed using photometers equipped with photoelectric cells of various types and sensitivities.

Photooxidation Oxidation reactions initiated by the presence of light.

Phthalic acid Aromatic organic acid with the systematic name benzene-1,2-dicarboxylic acid. Used for manufacture of **pigments** and **phthalic acid esters**.

Phthalic acid esters **Esters** of **phthalic acid** which have uses as **plasticizers**, e.g. in food **packaging materials**. **Migration** from the packaging materials into packaged foods or **beverages** can occur.

Phulka Puffed unleavened Indian **bread** made from **wheat flour** and similar to **tortillas**. Eaten warm as an accompaniment to **curries**.

Phycocyanin Plant photosynthetic protein covalently bound to a green/blue bilatriene pigment (a phycobiliprotein). Produced by **cyanobacteria**, e.g. *Spirulina platensis*, and has uses in **natural colorants**.

Phycoerythrin Plant protein conjugated to a red biladiene pigment (a phycobiliprotein). Involved in photosynthesis and produced by red **algae**. Uses include in **natural colorants**.

R-phycoerythrin Red phycobiliprotein pigment found in certain red **algae** that consists of two polypeptide chains each linked covalently to an open-chain tetrapyrrole chromophore. Phycoerythrins are useful as **colorants** in foods.

Phyllophora Genus of red **seaweeds** occurring on rocky coastlines around the world. Some species are utilized by the food industry as a source of **carrageenans**.

Phylloquinone Synonym for **vitamin K₁**. Fat-soluble vitamin found in all green plants. Especially abundant in **alfalfa** and green **leafy vegetables**. Essential for production of prothrombin, and several other proteins involved in the blood clotting system, and the bone protein osteocalcin. Deficiency causes impaired blood coagulation and haemorrhage. Two groups of compounds have **vitamin K** activity: phylloquinones; and a variety of **menaquinones** synthesized by intestinal **bacteria**.

Phylloxera Genus of plant-eating **insects** of the family Phylloxeridae. *Phylloxera vitifoliae* is a serious grapevine pest.

Physical properties Characteristics of substances that do not involve a chemical change, such as **density**, **electrical properties**, **mechanical properties** and **optical properties**.

Physicochemical properties Characteristics of chemical systems determined by application of physical principles, i.e. the **physical properties** of chemical compounds.

Physics The study of systems and their interactions with one another, in terms of the interrelationship between matter and energy, without reference to chemical change. Traditionally divided into the study of mechanics, electricity and magnetism, heat and thermodynamics, optics and acoustics. More modern aspects include quantum mechanics, relativity, nuclear physics, particle physics, solid-state physics and astrophysics.

Physiological effects Effects that products or their components have on human physiological processes.

Physiology Study of the function of biological processes within living organisms. Broken down into the study of the function of particular organs. The concept of homeostasis, the regulation of the internal environment within certain parameters, is central to this science.

Phytases EC 3.1.3.8 (3-phytases) and EC 3.1.3.26 (4-phytases). **Enzymes** that dephosphorylate **phytates**, **antinutritional factors** found in **beans** and **bran** products. Can be used to increase the nutritional properties of beans and cereal products by increasing the **bioavailability** of phosphorus and other minerals.

Phytates Salts or **esters** of **phytic acid** containing **inositol** and **phosphates** as the base. Especially abundant in the outer layer of **cereals**, in dried **legumes** and some **nuts** as both water-soluble salts (sodium and potassium) and insoluble salts of calcium and magnesium. Phytates may decrease absorption of calcium, zinc and iron from the intestine.

Phytic acid Hexaphosphoric acid ester of **inositol** present mainly in cereal **grain**, **nuts** and **legumes**.

Phytoalexins Organic nitrogen compounds produced by plants in response to infection or injury which exhibit **antimicrobial activity**.

Phytochemicals Physiologically active chemicals produced by plants. Used in **functional foods** and **nutraceutical foods**.

Phytochrome Protein bound pigment of plants which regulates flowering in response to light.

Phytoestrogens Non-steroidal, non-nutrient compounds occurring naturally in plants which possess oestrogenic or anti-oestrogenic activity via binding to oestrogen receptors. Examples include **isoflavonoids** and **lignans**, which are present in **tea**, **coffee**, **cereals**, **fruits**, **vegetables** (especially **soybeans**) and **alcoholic beverages**.

Phytofluene One of the **carotenoids** found in orange and red **fruits** and **vegetables**, especially **tomatoes**.

Phytohaemagglutinins **Lectins** produced by **common beans**. Agglutinate mammalian red blood cells (erythrocytes) and have mitogenic effects.

Phytohormones Chemicals produced by plants which regulate plant physiology. Produced in various parts of the plant and generally have no specific target organs, but act on various plant tissues.

Phytophthora Genus of **fungi** of the order Pythiales. Species responsible for several fruit and vegetable diseases. *Phytophthora infestans* causes late **blights** of **potatoes**, *P. syringae* and *P. cactorum* cause storage rot of **apples** and **pears**, and *P. citrophthora* causes storage rot of **citrus fruits**.

Phytostanols Hydrogenated **phytosterols** found naturally in only small amounts in plants, but produced commercially by **hydrogenation** of naturally occurring phytosterols. Due to their hypocholesterolaemic activity, phytostanols are used in **functional foods** and **beverages**, generally in the form of **stanol esters**.

Phytosterols Steroid **alcohols** present in plants, particularly in **oils** and **waxes**. Have hypocholesterolaemic activity and are thus used in **functional foods**, such as specially formulated **margarines** and **spreads**. Examples include **sitosterol** and **stigmasterol**.

Picante cheese Hard or semi-hard spicy **cheese** made in Portugal from mixtures of **goat milk** and **ewe milk**. Also called Picante da Beira Baixa cheese.

Pichia Genus of yeast **fungi** of the class Saccharomycetes. Occur in tree exudates, tunnels of woodboring **beetles**, **grain**, **flour**, **wines**, faeces and skin. Species may cause **spoilage** of **wines**, **sauerkraut** and **delicatessen foods**. *Pichia farinosa* and *P. fermentans* are involved in **cocoa fermentation**.

Picking Alternative term for **harvesting**. Generally refers to manual, rather than mechanical, gathering of **crops**.

Pickled cheese **Cheese** that is ripened in **brines**. **Curd** is cut into pieces that are put into containers filled with brine or salty **whey** and left to ripen for several months. Examples of this type of cheese include **Feta cheese**, **Domiati cheese**, Brinza cheese and **Kareish cheese**. Also known as brine ripened cheese.

Pickled cucumbers Alternative term for **cucumber pickles**.

Pickled eggs Products prepared by **pickling** hard boiled **eggs** in solutions usually of **vinegar** mixed with **flavourings**. As well as eggs from chickens, **duck eggs** and **quail eggs** are commonly used.

Pickled onions Small **onions** (commonly pearl onions) pickled in **vinegar** mixture or **brines**. Used as a condiment or garnish.

Pickles Foods preserved by **pickling** in a liquid such as **vinegar** or **brines**, usually containing **spices** to enhance **flavour**. Can be made from vegetables, fruits, meat, eggs or nuts. Popular pickled foods include **sauerkraut**, **cucumber pickles** and **chutneys**.

Pickling **Preservation** of foods in a pickling liquid such as **vinegar** or **brines**, often containing **spices**. Foods commonly preserved in this way include **vegetables**, **fruits**, **meat**, **eggs** or **nuts**. **Pickles** can be of various flavours, and can be sweet, savoury or spicy.

Picloram Selective systemic herbicide used for control of many annual and perennial broad-leaved weeds around monocotyledon crops, including some **cereals**. May be used as a plant growth regulator. Classified by Environmental Protection Agency as not acutely toxic (Environmental Protection Agency IV). Also known as Tordon.

Pidan Alkali-treated preserved **duck eggs**. Prepared by storing fresh duck eggs under a mixture of caustic soda, burnt straw ash and slaked lime for several months, until the **egg whites** and **egg yolks** coagulate and become discoloured. Also known as Chinese eggs or thousand year eggs.

Pie fillings Sweet or savoury preparations used as **fillings** for **pies**. Prepared fillings based on ingredients such as **fruits**, **meat** or **vegetables** and containing **seasonings** and other **additives** are available commercially.

Pies **Bakery products** consisting of a **pastry** case filled with either a sweet or savoury mixture and baked until the crust is crisp. May be topped with pastry or an alternative such as mashed **potatoes**, or may have no top.

Pigeon meat **Meat** from wild and domesticated **pigeons**. Among the various cuts from pigeon **carcasses**, breast meat constitutes the largest portion followed by the wings, back, neck, thighs and drumsticks. Pigeon meat is relatively fatty and has a higher energy value than **chicken meat**; it is also darker in **colour** than chicken meat

Pigeon peas Seeds produced by *Cajanus cajan*. Grow in long, twisted pods and are usually greyish in **colour**. Young seeds are eaten as a vegetable, but mature seeds are often dried and split, eaten as **dhal** in India. Green pods may also be used as a vegetable and seeds can be germinated to produce sprouts. May be used instead of **soybeans** to make **tempeh**. Also known as red gram.

Pigeons Various stout bodied, fruit- or seed-eating **birds** belonging to the Columbidae family; there are many species. Pigeons are hunted and farmed for **pigeon meat**. Young unfledged pigeons are called squabs.

Pigging Cleaning of **pipes** or ducts in processing equipment, including that in food factories, by forcing a tightly fitting, flexible object, such as a brush, blade or swab (pig), through the pipeline in order to scrape or push out the residual contents.

Pigmentation **Colour** that a substance exhibits, due to the presence of **pigments**.

Pigments Compounds, usually fine, solid particles, that give **colour** or other properties to a tissue, object or substance. For example, **chlorophylls** impart a green colour to **lettuces** and **peas**, **carotenes** are responsible for the orange colour of **carrots**, **lycopene** gives the red colour to **tomatoes**, **anthocyanins** contribute the purple colour of **grapes** and **blueberries**, and **oxymyoglobin** gives the red colour to **meat**. Pigments are sensitive to chemical and physical effects during food processing, and to chemical change during **ripening**. Pigments may also be added intentionally to foods in the form of **food colorants**.

Pigs Specific types of **swine**. Domesticated omnivorous ungulates, which are related to the wild boar (*Sus scrofa*) with some crossing with the Chinese type (*Sus indicus*). Pigs are kept for **bacon**, **ham** and **pork** production. Commercial farming systems commonly produce four classes of pigs, namely: pork pigs (also known as porkers) usually slaughtered at about 19 weeks of age; bacon pigs (also known as baconers) usually slaughtered at about 24 weeks of age; cutters usually slaughtered at about 23 weeks of age; and heavy hogs usually slaughtered at about 27-28 weeks of age. Pig performance and carcass confirmation are optimized by selective breeding and feeding, especially in bacon pigs.

Pike Any of several **freshwater fish** species in the genus *Esox*; distributed across Europe and North America. Valued as a food fish in some regions, where it is utilized fresh or frozen. Also known as pickerel in North America.

Pilchards Any of a number of small **herring**-like **marine fish** species in the family Clupidae; worldwide distribution. Many species are also referred to as **sardine**; the term pilchards generally refers to larger individuals within the species. Commercially important species include *Sardina pilchardus* (European pilchard), *Sardinops caerulea* (Californian pilchard) and *S. melanosticta* (Japanese pilchard). Marketed fresh, smoked, salted and dried; particularly popular as a canned product in various **sauces** or oils.

Pimaricin Alternative term for the fungicide **natamycin**.

Pimento Alternative term for **allspice**.

Pimento peppers Large **red peppers** (*Capsicum annuum*). Flesh is more aromatic and sweeter than that of **bell peppers**. Available fresh, canned and bottled. Used as a stuffing for green **olives**. Dried fruits are used as a source of **paprika**. Also known as **pimiento peppers**.

Pimiento peppers Large **red peppers** (*Capsicum annuum*). Flesh is more aromatic and sweeter than that of **bell peppers**. Available fresh, canned and bottled. Used as a stuffing for green **olives**. Dried fruits are used as a source of **paprika**. Also known as **pimento peppers**.

Pineapple juices **Fruit juices** extracted from **pineapples** (*Ananas comosus*).

Pineapple nectars **Fruit juice beverages** made by addition of water and/or **sugar**, and optionally other ingredients, to **pineapple juices**.

Pineapples **Fruits** produced by *Ananas comosus*. Good source of potassium and **fibre**, and contain moderate amounts of **sugars** and **vitamin C**. Consumed fresh, dried or canned, and used to make **pineapple juices** and **jams**. Pineapple juices can be further processed into **vinegar** or **spirits**. Fruits and stems of the plant contain **bromelains**, **proteinases** used in the pharmaceutical industry and in **beer** manufacture.

Pine needles Needles produced by plants of the genus *Pinus*, that have possible health promoting properties. Extracts are used in food **flavourings**, especially in Korean foods. Needles are also used to make **beverages**, including **teas**.

Pinenes **Terpenoids** and **flavour compounds** with a **camphor**-like **aroma** frequently present in **essential oils**. Used for manufacture of synthetic pine oil.

Pine nuts **Nuts** produced by *Pinus* species. Found on the woody scales of female pine cones. Removal is labour-intensive, making the nuts expensive. Eaten raw or roasted, or used in savoury and sweet dishes.

Pink salmon Smallest of the **Pacific salmon** species (*Oncorhynchus gorbuscha*); found in rivers and coastal waters along western and eastern Pacific coasts. Mostly sold canned but also utilized fresh, smoked and frozen; **roes** are used as **caviar substitutes**, especially in Japan.

Pinto beans Type of **common beans** (*Phaseolus vulgaris*).

Pipecolic acid Cyclic amino acid, the L-form of which is present in plants and can be produced from L-lysine.

Piperidine Organic nitrogen compound derived from **piperine** by heating. Present in **pepper** in small amounts.

Piperine Alkaloid and flavour compound isolated from **black pepper**, where it is primarily responsible for **pungency**. May also be used in **flavourings** for **brandy**.

Piperonyl butoxide Chemical substance used primarily as a synergist for **pyrethroid insecticides** to enhance their **toxicity**. Also used as a food additive in Japan (as a preservative for **cereals** and **legumes**). Classified by Environmental Protection Agency as not acutely toxic (Environmental Protection Agency IV).

Pipes Tubes of various diameters through which substances, including gases and liquids, can flow. Usually made of metal or plastic. Used to convey ingredients and products during processing of foods.

Pips Alternative term for small seeds, usually applied to those within **fruits**.

Pirimicarb Selective systemic carbamate insecticide. Used to control **aphids** and other **insects** in a wide range of plants, including **cereals**, **fruits** and **vegetables**. Classified by WHO as moderately hazardous (WHO II).

Pirimiphos-methyl Broad-spectrum insecticide and acaricide used to control a wide range of **insects** and **mites** in **fruits**, **vegetables**, **cereals** and **sugar cane**. Also used for pest control in stored grain and in animal houses. Classified by WHO as slightly toxic (WHO III).

Piscicides Chemical substances used for control of undesirable **fish** species, normally non-indigenous species that have been introduced to lakes or river systems. Rarely used in practice, mainly due to their tendency to harm many other organisms in addition to target species. Examples include antimycin and rotenone.

Piscicolins Plasmid encoded nonlantibiotic **bacteriocins** synthesized by ***Carnobacterium*** *piscicola*. Small hydrophobic peptides that are moderately heat stable. Activity is not affected by exposure to pH values in the range 2 to 8. The antibacterial spectrum of the bacteriocins produced by *C. piscicola* includes various genera of **lactic acid bacteria** and generally also includes ***Listeria*** *monocytogenes*. **Gram negative bacteria** are not inhibited.

Pisco South American spirit made by distilling Muscat grape **wines**.

Pistachio nuts **Nuts** produced by *Pistacia vera*. Shells split as the nut matures, making the kernels easy to remove. Kernels are green and have a unique **flavour** that makes them a popular constituent for a range of sweet and savoury dishes. Also eaten raw or roasted and salted in their shells. Rich in calcium, phosphorus, iron, **thiamin** and **vitamin A**.

Pita bread Round or oval, flat **bread** originating from the Middle East. Made from **yeasts**-leavened **dough**, which expands when baked to form a pocket which can be opened and filled to form a sandwich. Also spelt pitta bread.

Pitayos Pink to red **fruits** produced by several species of **cacti**, including *Stenocereus queretaroensis*, native to Mexico. White, juicy flesh is full of tiny seeds. Eaten out of hand, and used in **preserves**, **sherbet** and **beverages**. Also called pitayas.

Pito Traditional African **alcoholic beverages** made by **fermentation** of **mashes** based on **cereals**.

Pitta bread Alternative spelling for **pita bread**.

Pizza dough Yeasts-leavened **dough** used to make the base for **pizzas**.

Pizza fillings Foods used to top **pizzas**. Include **tomatoes**, **Mozzarella cheese**, **salami** and **sea foods**.

Pizzas Baked tarts of Italian origin composed of a flat base of yeast **dough** topped with seasoned **tomato sauces**, **cheese** (usually **Mozzarella cheese**) and other foods such as **salami**, **olives**, **vegetables** and **sea foods**. Traditionally baked rapidly in wood burning **ovens** and served hot.

PKU Abbreviation for **phenylketonuria**.

Plaice Generally refers to the marine **flatfish** species (*Pleuronectes platessus*; European plaice), found in the northeast Atlantic, where it is a highly valued food fish. Other plaice species include *P. quadrituberculatus* (Alaska plaice) and *Hippoglossoides platessoides* (American plaice). Marketed live (on ice), fresh (gutted or fillets), frozen or smoked.

Plankton General name given to animal (zooplankton) or plant (phytoplankton) organisms which float more or less passively in large bodies of freshwater and in oceans. The majority of planktonic organisms are microscopic. Forms the primary food base for larger aquatic organisms.

Plantains **Fruits** resembling **bananas** produced by *Musa paradisiaca*. Larger, firmer and starchier than bananas and always eaten cooked. Rich in **vitamin A**. Eaten fried, baked or boiled. When green, the cooked fruit tastes like **potatoes**, but as it ripens, it becomes sweeter, black skinned fruits being used for dessert recipes.

Plantaricins Bacteriocins synthesized by *Lactobacillus* *plantarum*. Plantaricin A is produced by *L. plantarum* C-11; it is bactericidal towards some **lactic acid bacteria**, but is not active against other **Gram positive bacteria** or **Gram negative bacteria**. Plantaricin A is heat stable with a molecular weight of >8000 Da. It consists of two **peptides** (molecular weights 2426 and 2497 Da), bacteriocin activity requiring complementary activity of both peptides. Plantaricin B is produced by *L. plantarum* NCDO 1193. It has a narrow inhibitory spectrum against only a few strains of lactic acid bacteria.

Plant diseases Adverse effects in plants caused by infection with **bacteria**, **viruses** or **fungi**, or by infestation with **pests**. Can affect the growth and survival of the whole plant and quality of the **fruits** and other edible parts it produces.

Plant disorders Adverse effects in plants caused by abiotic factors, such as environmental, nitritional and physiological conditions. As with **plant diseases**, the plant and quality of its produce can be affected.

Plant foods Foods derived from plant sources.

Plant growth regulators Chemicals that affect growth of plants. Include endogenous compounds, i.e. **phytohormones**, and exogenously applied chemicals, such as **herbicides** and **antisprouting agents**.

Planting Placing of plants in the ground so that they can take root and grow. Date and density of planting can affect growth of the plants and the yield and quality of produce.

Plant proteins Proteins sourced from plant material as opposed to animal products. Include **vegetable proteins** and **cereal proteins**. Preferred by some consumers due to health benefits. Quality of plant proteins, especially with respect to **amino acids** composition, varies according to source, but many plant breeding programmes have aimed to improve protein quality of individual crops. **Legumes**, particularly **soybeans**, are especially rich in protein.

Plants Multicellular eukaryotic organisms belonging to the Plantae kingdom. Self-supporting plants characteristically exhibit photosynthesis. Source of a wide range of foods, beverages and ingredients.

Plasma Alternative term for **blood**.

Plasmids Autonomously replicating, extrachromosomal, covalently closed circular molecules of **DNA** found in **bacteria**, **fungi**, **algae** and plants. In bacteria, they often carry **genes** conferring antibiotics resistance. Usually non-essential for cell survival under non-selective conditions and may integrate into the host genome. Used widely as expression and **cloning vectors**.

Plasmin EC 3.4.21.7. Serine proteinase that cleaves preferentially after Lys and Arg residues. Derived from **plasminogen** by the proteolytic action of **plasminogen activators**, it is responsible for digestion of fibrin in blood clots. It is also the predominant native proteinase in **milk** where its presence can result in degradation of **casein**, with both desirable and undesirable effects on product quality. Can be used to alter the **functional properties** of **milk proteins**.

Plasminogen The inactive precursor of **plasmin**.

Plasminogen activators EC 3.4.21.68 (t-plasminogen activator) and EC 3.4.21.73 (u-plasminogen activator). The latter is also known as urokinase. Serine proteinases that differ in structure but which both cleave Arg-Val bonds in **plasminogen** to form **plasmin**. Their presence in **milk** can have significant effects on product quality. Have been associated with invasion of cells by certain **pathogens**.

Plasteins Proteins produced by action of **proteinases** on protein hydrolysates (**peptides**). Plastein reactions, i.e. transpeptidation and condensation reactions, have been used to improve nutritional quality, **sensory properties** and/or **functional properties** of proteins, such as **fish proteins**, **soy proteins** and **whey proteins**.

Plasticity Extent to which a substance can be deformed as a result of application of a stress. When stress is applied in excess of a certain value (yield point), deformation is permanent. Below a certain stress, the elastic limit, most substances will recover their original shape when the stress is removed. Such substances are said to be elastic.

Plasticizers Substances, typically organic solvents, which are capable of imparting flexibility to a non-plastic material or improving flexibility of a ceramic mixture. Added during the manufacturing process to decrease brittleness and to promote **plasticity**.

Plastics Synthetic materials made by polymerization, polycondensation, polyaddition or other similar processes from molecules with a lower molecular weight, or by chemical alteration of natural macromolecules. Can be formed into different shapes while soft, generally when heated, and then set into a slightly elastic or rigid form. Synthetic organic polymers which are used as the basis of plastics are referred to as resins. Early plastics were used to make imitations of other materials, but they are now appreciated widely for their own range of useful thermal, electrical, optical and mechanical properties. Major applications of plastics include their use in **containers**, **packaging materials**, construction materials, consumer items, adhesives, **pipes**, textiles and electronic components. Types of plastics used commonly for packaging of foods include **polyethylene, polyvinyl chloride** and **nylon**.

Plastics bags   Bags made from **plastics**. Used widely as **containers** for particulate and solid foods.

Plastics bottles   Bottles made from **plastics**. Used widely as **containers** for liquid foods and **beverages**.

Plastics films   Packaging films, such as **cellulose films** and **polyethylene films**, made from **plastics**. Used to wrap or to make **containers** for foods.

Plate counts   Estimations of the numbers of **microorganisms** in a sample, by means of culturing a solution of the sample on agar plates and counting the number of microbial colonies that grow.

Plesiomonas   Genus of Gram negative, facultatively anaerobic rod-shaped **bacteria** of the family **Enterobacteriaceae**. Occur in surface waters, soil, **fish**, **shellfish**, aquatic animals and mammals. *Plesiomonas shigelloides* is responsible for **gastroenteritis** in humans due to consumption of contaminated food (e.g. fish and shellfish) or water.

Pleurotus   Genus of **edible fungi** including the commercially important **oyster mushrooms** (*Pleurotus ostreatus*). Has a delicate **flavour** and **texture**. Cap colour varies with age. Can be eaten raw or cooked.

Plumcots   **Fruits** that are a hybrid cross between **plums** and **apricots**, with the combined **flavour** of each fruit.

Plum juices   Fruit juices extracted from **plums** (*Prunus domestica*).

Plums   Stone fruits produced by plants of the genus *Prunus*. Vary widely in flesh and skin **colour** and **flavour**, according to variety. Contain about 10% **sugar** and are rich in potassium. Available fresh or canned; dried fruits are known as **prunes**. Eaten out of hand and used in **desserts**, **jams** and **jellies**. Also used to make **liqueurs** and **spirits**, such as **slivovitz**.

Plutonium   Radioactive element, chemical symbol Pu. Isotopes with relative atomic masses of 232-246 have been identified. The most stable isotope is ^{244}Pu.

Pneumatic conveyors   Conveyors containing or operated by air or gas under pressure.

Poffertjes   Small **pancakes** originating from Holland. Usually served hot with **icing sugar** and **butter**.

Pogonias   Genus of **marine fish** containing several species of **croakers** (**drum**). The most commercially important species is *Pogonias cromis* (black drum); distributed along the Atlantic seaboard of North and South America. Marketed fresh and frozen.

Poi   Dish made by **fermentation** of cooked **taro** that has been pounded to a paste.

Polar compounds   Compounds that are ionic or are made up of molecules with a large permanent dipole moment. Commonly used as indicators of oil quality. During repeated heating of **frying oils** in the presence of oxygen, water and foods, **triglycerides** are broken down into polar compounds such as free **fatty acids**, **monoglycerides**, **diglycerides**, **glycerol** and **polymers**. Such decomposition products have a negative effect on the **flavour** and nutritional quality of the **fried foods**. To avoid deterioration in food quality and possible health effects for consumers, there are regulations in force in some countries specifying limits for total polar compound levels in frying oils. Once these values have been reached, the oils are prohibited for use in food processing. Polar compound profiles can also be used in detection of **adulteration** of oils such as virgin **olive oils** with less expensive types.

Polarimetry   Technique in which the identity and quantity of a substance are determined from its effect on the direction of vibration of polarized light.

Polarization   Restriction of the waves of electromagnetic radiation, including light, to one plane or one direction. This property is not directly perceived by the eye but can be detected, in the case of light, by its behaviour after it has interacted with polarizers. Measurement of the degree of polarization of electromagnetic radiation coming from an object reveals valuable information not only about that object but also about any material lying between the object and observer.

Polarography   Electrochemical technique in which current flowing through an electrolysis cell is measured as a function of the potential of the working electrode.

Pole beans   Type of **common beans** (*Phaseolus vulgaris*).

Polenta   Thick **porridge** of Italian origin made with ground **corn** or sometimes **barley** which is boiled in water or stock. Eaten hot with **cheese, gravy, butter** or **oils**, or tomato-based **sauces**. Alternatively, may be cooled, cut into shapes, and baked or fried.

Poliomyelitis   Infectious disease of the central nervous system which may result in muscle paralysis. Caused by a picornavirus, which is excreted in the faeces of an infected person; the disease is most common where sanitation is poor.

Polioviruses   **Viruses** of the genus *Enterovirus* within the family Picornaviridae responsible for poliomyelitis in humans. Transmission may be through the faecal-oral route via contaminated food or water.

Polishing Process in which a surface is made shiny and smooth by rubbing against abrasive materials such as metal, rock or wood. With reference to **rice**, polishing is the final stage in **milling**, in which hulled and pearled rice is spun in cones that are lined with leather or sheepskin. The fully processed form is called polished rice.

Pollack **Marine fish** species (*Pollachius pollachius*) from the cod family (Gadidae); distributed across the northeast Atlantic. Flesh is dry with a delicate, somewhat sweet **flavour**. Marketed fresh (whole, gutted or fillets) or salted.

Pollen Granules produced in the anthers of seed forming plants that contain the male gametes.

Pollination Transfer of pollen from the anther to the stigma of a flower, constituting the first step in the production of **fruits** or **seeds**.

Pollock Alternative term for **coalfish**, or **saithe**.

Polonium Radioactive element, chemical symbol Po. Isotopes of Po have relative atomic mass numbers of between 192 and 218. ^{209}Po is the most stable isotope.

Polyacrylamide gel electrophoresis **Electrophoresis** in which polyacrylamide gel is used as the diffusion medium.

Polyamides Synthetic polymers, including **nylon**, in which the structural units are linked by amide or thioamide groupings. Used as components of **casings** and **packaging materials** for foods.

Polyamines Compounds that contain two or more amine groups. Examples include **putrescine**, **spermidine** and **spermine**.

Polybrominated biphenyls **Organic halogen compounds** which are known **toxins** and suspected **carcinogens**.

Polybrominated diphenyl ethers **Organobromine compounds** used as flame retardants in a variety of commercial products. Persistent and ubiquitous in the environment. There are public health concerns regarding potential adverse effects of polybrominated diphenyl ethers found as contaminants in **fish**.

Polycarbonates Group of synthetic polyesters in which the carboxyl groups are derived from carbonic acid. Used to make reusable **plastics containers** for foods, especially **bottles** for infant feeding.

Polychlorinated biphenyls Toxic **chlorinated hydrocarbons** with many industrial uses, e.g. as pesticide extenders and **plasticizers**. Proven **toxicity** to humans and animals includes adverse clinical effects on the gastrointestinal tract and eyes. Environmental contamination with these compounds and their high stability can allow them to enter the food chain, af-

fecting predominantly animal foods. This has led to decreases in industrial use of these compounds. Preparations include **Arochlor**, Clophen, Fenclor, Kanechlor, Phenoclor, Pyralene and Santotherm. Commonly abbreviated to PCB.

Polychlorinated dibenzodioxins Toxic environmental contaminants produced by municipal waste incinerators and chemical, paper and metallurgical industries. Exposure to these toxic **organochlorine compounds** can occur via the diet, particularly from consumption of animal foods, due to their accumulation in **fats**.

Polychlorinated dibenzofurans Potential toxic contaminants of foods, particularly animal foods, where they accumulate in **fats**. These **organochlorine compounds** are produced as a result of incineration of municipal waste and as wastes from various industrial processes.

Polychlorinated dibenzo-*p*-dioxins One of the **polychlorinated dibenzodioxins**, a group of toxic chemicals which may be contaminants of foods, particularly animal foods.

Polycyclic aromatic hydrocarbons Hydrocarbons comprising two or more ring structures, at least one of which is an aromatic (benzene) ring. Lipophilic pollutants and potential **carcinogens**. Examples that have been found in foods include benzo[*a*]pyrene and **phenanthrene**; foods affected include **cheese**, cooked **meat** and **shellfish**. Commonly abbreviated to PAH and also called **polynuclear aromatic hydrocarbons**.

Polydextrose Low calorie, highly branched polysaccharide composed of randomly linked D-glucopyranose units (average 12-15 units/molecule). Manufactured from **glucose** and **sorbitol** in the presence of **citric acid** or **phosphoric acid**. Used as **sugar substitutes** and **fat substitutes** in **low calorie foods**. Can replace **mouthfeel**, **texture** and humectancy of **sugar**, but does not have a sweet **flavour**. Derivatives of polydextrose with improved flavour are marketed under the **Litesse** brand name.

Polydimethylsiloxane Polymers consisting of dimethyl silicon oxide monomer units. Colourless viscous oil that is insoluble in water but soluble in hydrocarbon solvents. Uses include as **antifoaming agents** or **defoaming agents** in **beverages**, such as **wines** and **fruit juices**, **anticaking agents** in foods, e.g. dried **dessert mixes**, and as a base for manufacture of **chewing gums**. This polymer is also used as an extraction fibre and a separation matrix for analysis of food components. Often abbreviated to PDMS, it is also known as dimethicone.

Polyesters Synthetic resins in which ester groups link the polymer units. They are heated to harden them into a shape which they do not lose when heated subsequently at normal cooking temp. Used to make containers for heating foods in conventional or microwave **ovens**.

Polyethylene Flexible, tough, but lightweight synthetic resin which is a polymer of ethylene and is formed by pressure treatment of ethylene. Used mainly as a packaging material, especially in bags, films and sheets. Density of the polymer varies according to the polymerization process used. **Low density polyethylene** is used for flexible applications, e.g. **polyethylene films**, while **high density polyethylene** is used to make more rigid structures, such as **barrels** and **bottles**. Also known as polythene.

Polyethylene bags Bags made from **polyethylene** which are used for packaging or storage of foods.

Polyethylene films Transparent **packaging films** made from **polyethylene** which are commonly used in packaging of foods. Desirable characteristics include their low cost, resistance to low temperature and tough, moisture-proof and heat sealable nature.

Polyethylene glycol Synthetic polymer which exists as a liquid or waxy solid, depending on the degree of polymerization, and thus molecular weight. Partially soluble in water. A range of applications includes **surfactants, catalysts, flocculants, plasticizers, lubricants, solvents** and **emulsifiers**. Uses in the food industry include modification of **enzymes**, such as **lipases** and **proteinases**, and other **proteins** to alter their properties, and as a plasticizer in edible films and **coatings**.

Polyethylene naphthalate Polyester polymer with characteristics making it suitable for food **packaging** applications. Compared with **polyethyleneterephthalate** (PET) it has improved oxygen barrier properties, heat and chemical resistance, and stiffness, but is more expensive. Its physical and **mechanical properties** make it suitable for manufacturing refillable **containers** and use in hot fill applications. Polyethylene naphthalate is sometimes blended with PET to make **plastics** containers for foods and **beverages**, especially **bottles** for **beer**.

Polyethyleneterephthalate Synthetic resin produced from **ethylene glycol** and terephthalic acid. Used in production of polyester fibres, **plastics bottles** for **beverages**, and food trays for use in conventional and microwave **ovens**. Commonly abbreviated to PET.

Polygalacturonases EC 3.2.1.15. Hydrolyse 1,4-α-D-galactosiduronic linkages in pectate and other galacturonans. Involved in the **ripening** of **fruits**, and are used in the processing of fruits and **vegetables**, and production of **wines**; specifically for improving **cloud** stability in **citrus juices**, mash treatment, **clarification** of **fruit juices**, and **maceration** and **pulping** of plant tissues. Potentially useful for production of oligogalacturonides that can be used as functional food components. Also known as pectinases and pectin depolymerases.

Poly(γ-glutamic acid) High molecular weight polymers composed of **glutamic acid** residues. Produced as bacterial **fermentation products**, mainly by *Bacillus* spp. Uses include as food **thickeners**.

Polyglycerol polyricinoleate Highly viscous, strongly lipophilic liquid which is insoluble in water or ethanol, but soluble in fats and oils. Used in the **chocolate** industry as a **viscosity** reducing agent, and also as an emulsifier in foods. Commonly abbreviated to PGPR.

Polyhydroxyalkanoates **Salts** or **esters** of **poly(hydroxyalkanoic) acids**.

Poly(hydroxyalkanoic) acids **Organic compounds** formed by polymerization of hydroxyalkanoic acids, e.g. **hydroxybutyric acid** and hydroxyvaleric acid. Used in biodegradable **packaging materials**. May be formed as **fermentation** products by **bacteria** grown on food processing wastes such as **whey**.

Polyketides Precursors in the **mycotoxins** biosynthesis pathway in **fungi**.

Poly(β-D-mannuronate) lyases EC 4.2.2.3. Catalyse the eliminative cleavage of **polysaccharides** containing β-D-mannuronate residues to give **oligosaccharides** with 4-deoxy-α-L-erythro-hex-4-enopyranuronosyl groups at their ends. Since they degrade **alginates**, they are also known as alginate lyases.

Polymerase chain reaction Technique usually abbreviated to **PCR**.

Polymerization Chemical combination of simple molecules (monomers) to form long chain molecules (**polymers**) of repeating units. In addition polymerization, the monomers simply add together and no other compound is formed. In condensation polymerization, water, alcohol, or some other small molecule is formed in the reaction.

Polymers Long chain molecules of repeating units formed by the chemical combination of monomers in a process called **polymerization**. Natural organic polymers include proteins, DNA and latexes, such as rubber. Diamond, graphite and quartz are examples of inorganic natural polymers. Synthetic polymers include **inorganic compounds**, such as glass and concrete, but the great majority are **plastics**. Polymers are formed from monomers under the influence of heat, pressure or the action of a catalyst.

Polymorphism   Difference in specific **DNA** sequences among individuals. Useful for genetic linkage studies.

Polymyxins   Group of five **antibiotics** (designated alphabetically A-E) that show specialized activity against **Gram negative bacteria**. Polymyxin B and E are the only examples used in animal husbandry. The former is used primarily as a topical treatment for **mastitis** in cattle, while the latter is used mainly for oral treatment of *Escherichia coli* infections in cattle. Residues in edible tissues after oral administration to animals are usually below detection limits.

Polynuclear aromatic hydrocarbons   Hydrocarbons comprising two or more ring structures, at least one of which is an aromatic (benzene) ring. Lipophilic pollutants and potential **carcinogens**. Examples that have been found in foods include benzo[*a*]pyrene and **phenanthrene**; foods affected include **cheese**, cooked **meat** and **shellfish**. Commonly abbreviated to PAH and also called **polycyclic aromatic hydrocarbons**.

Polyolefins   Polymers, including **polyethylene** and **polypropylene** made from olefin monomers. Used as components of **plastics films** for packaging of foods.

Polyols   Products formed by **hydrogenation** (reduction) of the free aldehyde or ketone groups of **reducing sugars** to produce an alcohol group. Examples include **sorbitol**, **mannitol** and **maltitol**, produced by hydrogenation of **glucose**, **mannose** and **maltose**, respectively. Also known as **sugar alcohols**.

Polypeptides   Unbranched chains of 10 to approximately 100 amino acid residues linked via peptide bonds. In contrast to **proteins**, polypeptides have no secondary or tertiary structure.

Polyphenol oxidases   Alternative term for **catechol oxidases**.

Polyphenols   Compounds containing at least two phenol (hydroxybenzene) groups. Plant polyphenols, including **catechin** and **flavonoids**, are present in **tea**, **coffee**, **fruits**, **fruit juices** and **wines** and have **antioxidative activity**. Polyphenols in **legumes** and **cereals** are regarded as **antinutritional factors**, due mainly to the effects of **tannins**, which reduce protein **digestibility**.

Polyphosphates   **Salts** of polyphosphoric acid, a polymer produced by condensation of two or more **phosphoric acid** molecules. Sodium and potassium salts of **tripolyphosphates** are permitted **food additives** and uses include as **emulsifiers** and **texturizers**.

Polyporus   Genus of **fungi** of the class Homobasidiomycetes. Occur on felled timber and living trees. Some species (e.g. *Polyporus squamosus*) may be used in production of **enzymes**. Several species (e.g. *P. umbellatus* and *P. confluens*) are considered edible.

Polypropylene   Synthetic resin prepared by polymerization of **propylene**. Used as a packaging material for low fat, low sugar foods during microwave heating, but is unsuitable for use in conventional **ovens** because of its limited **heat stability**.

Polysaccharides   **Carbohydrates** that are composed of at least 10 monosaccharide residues linked via glycosidic bonds. **Starch**, **celluloses**, **pectins** and **carrageenans** are all polysaccharides. Polysaccharides have multiple applications in the food industry as **thickeners**, **bulking agents**, **anticaking agents**, **gelling agents**, and substrates for microbial fermentations and manufacture of **sweeteners**.

Polysorbate 60   Produced by reaction of ethylene oxide with monostearic acid esters of sorbitol, with an average of 20 oxyethylene groups per molecule. Used predominantly as **emulsifiers**, e.g. in **cakes**, **coffee whiteners** and non-dairy whipped **toppings**, but also as **foaming agents**, **flavourings** and **dough conditioners**. Also called polyoxyethylene (20)-sorbitan monostearate.

Polystyrene   Synthetic resin made by polymerizing **styrene**. Produced in two forms, i.e. a hard form and a lightweight foam form called expanded polystyrene. There is concern about health hazards associated with migration of styrene monomers, dimers and trimers from **packaging materials** into some types of foods.

Polytetrafluoroethylene   Tough synthetic resin which is used to coat non-stick cooking utensils. Commonly abbreviated to **PTFE**.

Polythene   Alternative term for **polyethylene**.

Polyunsaturated fats   **Fats** and **oils** that contain at least two carbon-carbon double and/or triple bonds due to the presence of **unsaturated fatty acids**. Have lower melting points than saturated fats, and are therefore more likely to be oils at room temperature. Considered more beneficial than saturated fats with respect to their influence on risk of developing **cardiovascular diseases**.

Polyunsaturated fatty acids   **Fatty acids** that contain two or more carbon-carbon double bonds. Have lower melting points than **monounsaturated fatty acids** or **saturated fatty acids** with an identical number of carbon atoms. Hence, **lipids** containing a high proportion of polyunsaturated fatty acids will be more fluid at room temperature. Examples include **linoleic acid**, **linolenic acid** and **arachidonic acid**, with 2, 3 and 4 double bonds, respectively. Commonly abbreviated to PUFA.

Polyurethane Synthetic polymer made from urethane with a wide range of applications. Polyurethane foams have various uses in the food industry, including **immobilization** of cells and **enzymes**, and insulation of brewery tanks and utensils. Polyurethane **adhesives** are used in manufacture of food packaging.

Polyuronides **Pectic substances** present in plant cell walls. Comprise **polysaccharides** composed of uronic acid monomers.

Polyvinyl acetate Synthetic resin which is a polymer of vinyl acetate. Used as a component of gum bases and **flavour** delivery systems in **chewing gums**. Also used in high-gloss **coatings** for foods.

Polyvinyl alcohol Synthetic resin produced by polymerization of vinyl acetate and hydrolysis of the resultant polymer. Biodegradable and suitable for use as food packaging. Commonly abbreviated to PVA.

Polyvinyl chloride Tough, chemically resistant, synthetic resin, which is a polymer of **vinyl chloride**. Low cost material that is moisture-proof but with some oxygen permeability. Used widely in **supermarkets** as a wrapping material for **meat**. Commonly abbreviated to PVC.

Polyvinylidene chloride Transparent, moisture-proof, thermoplastic polymer which has greater **heat stability** than **polyethylene**. Used to make plastics films, e.g. saran, for packaging foods. Commonly abbreviated to PVDC.

Polyvinylpyrrolidinone Substance used in the food industry to control **haze** or **colloidal stability** in **beer** by preventing oxidation and polymerization of **polyphenols**. Similarly used in **clarification** of **wines**.

Polyvinylpyrrolidone Substance used in the food industry to control **haze** or **colloidal stability** in **beer** by preventing oxidation and polymerization of **polyphenols**. Similarly used in **clarification** of **wines**.

Pomaces Wastes or by-products from **fruit juices** manufacture. Solid residue remaining after pressing of **fruits** to extract juices or **musts**.

Pombe Type of **sorghum beer** made in East Africa.

Pome fruits False **fruits**, the flesh of which develops from the receptacle of the flower, enclosing the fused carpels. The carpels form the core (true fruit) after fertilization. Examples include **apples**, **pears** and **quinces**.

Pomegranates **Fruits** produced by *Punica granatum*. The orange to red skin is leathery and encloses a pinkish pulp that contains numerous edible **seeds**. The pulp is scooped out and eaten fresh. Pomegranate

juice is used in **wines** and **cocktails** and is the main ingredient of grenadine syrup.

Pomelos The largest of the **citrus fruits**, produced by *Citrus maxima* or *C. grandis*. Ancestors of the modern **grapefruit**. Closely resemble the grapefruit in appearance, but the flesh is sweeter and less acidic, lacking the **bitterness** of a grapefruit. Rich in **vitamin C** and potassium. Eaten fresh or used to make **jams**, **jellies** and **marmalades**. Also known as **shaddocks**, Chinese grapefruit and **pummelos**.

Pomfret Any of a number of **marine fish** species within the family Bramidae; worldwide distribution. Species valued as food fish include *Brama brama* (pomfret; black sea bream), *B. japonica* (Pacific pomfret), *Taracticthys longipinnis* (bigscale pomfret; long-finned bream) and *Paratromateus niger* (black pomfret). Flesh tends to be tender with a rich, sweet **flavour**. Marketed fresh, frozen and canned; also salted in India.

Pommes frites French term for **potatoes** that have been cut into thick or thin strips, soaked in cold water, dried and deep fried in oil. Also called **chips** (UK), fries or French-fried potatoes or **French fries**.

Ponceau Group of synthetic, mostly red, **azo dyes**, some members of which have uses as food **colorants**.

Ponkans Type of **mandarins** (*Citrus reticulata*).

Pont-l'Eveque cheese French **soft cheese** made from **cow milk**. The edible brown rind is slightly mouldy and ridged as the cheese is cured on straw mats. The interior is soft and yellow. **Flavour** is savoury and piquant.

Poori Puffed deep fried unleavened Indian **bread** made from **wheat flour**. Eaten warm. Plain poori is eaten as an accompaniment to **curries**; can also be flavoured to make a sweet or savoury product.

Popcorn Variety of **corn** with hard kernels that expand on exposure to heat or **microwaves** to form large, fluffy white masses. Also refers to the edible mass formed by this process, which is eaten as a snack food, often flavoured with **salt** or a sweet substance such as toffee.

Popping Process in which **cereals** and grains are expanded by **heating** until the outer skin of the kernels burst with a sudden sharp, explosive sound. Used particularly in the manufacture of **popcorn**.

Poppy seeds Small, kidney shaped, grey-blue seeds produced by *Papaver somniferum*. Have a mild, nutty **flavour** and **aroma** and are used as **toppings** and ingredients for **bread** and other **bakery products**.

Porcine Affecting, resembling or relating to **swine**.

Porins Transmembrane **proteins** present in outer membranes of **Gram negative bacteria**. Porin trimers form channels in the membrane through which transport of small molecules, e.g. **monosaccharides** can occur.

Pork Meat from **swine**, especially when the meat is uncured. Depending on the size of the animal and the part of the swine **carcasses** from which the meat is cut, **colour** of pork varies from pale pink to pinky-red. Raw boar meat and sow meat tend to be a stronger red colour than pork from young swine. On cooking, pork becomes paler and may become almost white in colour. Pork is characterized by clearly noticeable deposits of subcutaneous fat; this fat is white in colour, medium-firm in texture and of a greasy **consistency**. Pork is a particularly rich dietary source of thiamin, containing approximately 10 times as much as **beef**. In some religions, pork is considered as unclean and consumption is forbidden; conversely, in other parts of the world, notably in China and the Pacific, and in other Asian cultures, pork is highly regarded. Pork quality is affected by halothane genotype and Rendement Napole (RN) genotype in swine. Quality is often categorized as being: pale, soft and exudative (PSE defect); reddish-pink, soft and exudative (RSE defect); red, firm and non-exudative (RFN; normal); or dark, firm and dry (DFD defect).

Pork bellies Cuts of swine **carcasses** used in preparation of various foods such as commercial **pork**, **bacon**, bacon bits and specialities including smoked and salted products. Also an important commodity in the futures market.

Pork chops Thick slices of **pork**, usually including an eye' of meat, a rib and a layer of subcutaneous fat.

Pork mince Meat mince prepared from **pork**. Also known as ground pork or minced pork.

Pork sausages **Sausages** prepared from **pork**. Properties of pork make it highly suitable for the preparation of sausages. The majority of sausages include some pork, but pork sausages include a high proportion of pork (lean meat, skin and offal) and pork fat trimmings. Although they may include other types of meat, the proportions of these are lower than the proportion of pork.

Porosity Extent to which a material exhibits **permeability**.

Porphyra Genus of red **seaweeds** found on rocky shorelines around the world. Some species are utilized as foods, including *Porphyra tenera* and *P. yezoensis*. Various names are given to seaweed products formed from members of this genus, including **laver** (England), **nori** (Japan, North America), kim (Korea) and karengo (New Zealand). Cultured on a large scale in some parts of Asia.

Porphyridium Genus of red microalgae of the family Porphyridiaceae. Species include *Porphyridium cru-*

entum and *P. purpureum*. Source of a range of compounds, including **polysaccharides**, the long chain **polyunsaturated fatty acids docosahexaenoic acid** and **eicosapentaenoic acid** and the pigment **phycoerythrin**, which is used in **natural colorants**.

Porphyrins Derivatives of porphin (a cyclic tetrapyrrole) in which the pyrrole β-carbon atoms are variously substituted. Can readily chelate various minerals, the metalloporphyrins being components of several important biological pigments, e.g. **chlorophylls**, cytochromes and **haem**.

Porpoises Marine mammals from the family Phocoenidae; worldwide distribution. Not commercially exploited on a large scale. However, some species are utilized as a source of **meat** and **oils**.

Porridge **Breakfast** food originating from Scotland made by boiling **oatmeal** or other **cereals** in water or **milk**.

Port Sweet **fortified wines** produced from specific local **winemaking** grape cultivars in a delimited area of the upper Douro valley in Portugal. Types include: vintage port; ruby port; and tawny port. Most port is red, but white port is also available.

Portion packs **Packs** which each provide an amount of food suitable for one person.

Port Salut cheese French semi-soft **cheese** made from **cow milk**. Originally made by Trappist monks. Rind is smooth and yellow; interior has an elastic **texture**. Slightly aromatic **flavour** but no pronounced **aroma**. Also known as **Saint-Paulin cheese**.

Possums Tree-dwelling Australian marsupial animals; there are many species, particularly in the family Petauridae. Recently, interest has increased in farming possums for the production of possum **meat**, particularly in New Zealand. Possum **carcasses** (brush-tail possums, *Trichosurus vulpecula*; 1-5 years old) are characterized by a high content of lean meat and a low content of fat. Cooked brush-tail possum meat has acceptable **tenderness**, a high content of protein, a low content of fat and a high content of **unsaturated fats**.

Potable water Water of composition and hygienic and sensory quality permitting its use as **drinking water**.

Potassium Alkali element with the chemical symbol K. A vital element in the human diet.

Potassium bromate Salt that is used primarily in **dough conditioners**. Other food industry uses include in **bleaching agents** and **improvers**. It is also used in **malting** of **barley** for manufacture of **alcoholic beverages**.

Potassium lactate White solid which is produced on a commercial scale by **neutralization** of **lactic acid** with potassium hydroxide. Applications in foods and **beverages** include **flavour enhancers, flavourings, adjuvants, humectants** and **pH** regulators.

Potato chips Thin slices of **potatoes** fried until crisp. Eaten as **snack foods** or served as a garnish or with **dips**. May be flavoured with **salt** or a variety of other **flavourings**. Called **potato crisps** or crisps in the UK.

Potato crisps UK name for thin slices of **potatoes** that are fried until crisp. Eaten as **snack foods** or served as a garnish or with **dips**. May be flavoured with **salt** or a variety of other **flavourings**. Also known as **potato chips** in other countries, including the USA.

Potatoes Edible **tubers** produced by *Solanum tuberosum*, widely cultivated worldwide. Good source of **vitamin C**, a range of **minerals** and **dietary fibre**, with a high water content (approximately 80%). Vary in shape and skin **colour** according to cultivar. Eaten cooked in a number of ways, including boiled, baked, fried and roasted, as well as being used in **soups**, stews and other dishes. Also a source of **starch** and **alcohol**. Green parts of the potato plant, including tubers exposed to light, contain the poisonous glycoalkaloid **solanine**.

Potato flakes Products made by drying thin slices of **potatoes**. Used as an ingredient in doughs or to make instant mashed potato by reconstituting with water or milk.

Potato meal **Flour** prepared from **potatoes**. Can be used as an ingredient in many types of **processed foods**.

Potato peel Outer skin of **potatoes**, sometimes removed before **cooking**, but often retained, as in baked potatoes, boiled potatoes and potato wedges. Consumption of the peel along with the flesh is often recommended as it is rich in **vitamins, minerals** and **fibre**.

Potato purees Made by mashing cooked **potatoes** to a smooth, thick consistency by forcing through **sieves** or blending in food processors. May be combined with other ingredients and served as a side dish or used to thicken **soups** and **sauces**.

Potato salads Salads prepared from boiled or roasted potatoes, cut into chunks. Coated with various dressings, often a mayonnaise-type dressing. Usually served cold, but can be eaten hot.

Potato starch **Starch** isolated from potato tubers.

Potentiometry Technique in which detection is achieved by measuring the change in electric potential between two electrodes placed in the sample solution. One electrode (the indicator electrode) responds to analyte concentration, while the other (the reference electrode) remains at a fixed potential.

Pouches Small, sealed flexible bags which can be used as **containers** for foods. Commonly made from **plastics** or **foils**, and used to store **frozen foods** or **dried foods**.

Pouchong tea Lightly fermented **tea**, intermediate between **green tea** and **oolong tea**.

Poultry The collective term for any domestic or farmed birds including **chickens, chukars, ducks, emus, geese, guinea fowl, ostriches, quails**, rheas, **turkeys** and **waterfowl**. They are reared primarily for **poultry meat** production and production of **eggs**.

Poultry breast A part of poultry carcasses which consists of the breast muscle (**meat**), skin, ribs, sternum and pectoral girdle. It is usually removed from the carcass by cutting through the ribs, near to their attachment to the backbone. Some poultry breast meat is deboned before retail. The breast muscle of larger species of poultry may be processed by deboning and slicing or rolling before retail. Poultry breast meat is lighter in **colour** than meat from the legs and thighs.

Poultry meat **Meat** from poultry. Most of the fat in poultry meat is associated with the skin and can be removed. Skinless poultry meat has low intramuscular and saturated fat contents making it a healthy dietary alternative to red meat. Meat from younger birds tends to be more tender than that from older birds. Developments in poultry husbandry (e.g. intensive production systems), advances in feeds and selective breeding have led to large-scale, rapid production at low prices. Consequently, poultry meat has become an increasingly important part of diets in many countries.

Poultry sausages **Sausages** made from **poultry meat**. They are often prepared from mechanically recovered poultry meat, poultry meat trimmings or poultry thigh meat. Other ingredients may include poultry skin and the less preferred components of poultry **offal**, such as **gizzards** and **hearts**. Poultry fat, pork fat or beef fat may also be included. They may be smoked or unsmoked. They include **chicken sausages** and **turkey frankfurters**.

Poultry science Division of **animal science** dealing with the production, management and distribution of poultry, including those intended for food use or for production of eggs to be used in or as foods.

Powders **Dried foods** in the form of fine particles. Food powders include products, which can be reconstituted (e.g. with **milk** or water) to form liquid foods, and powdered ingredients such as **baking powders** and **spices**.

Pozol **Corn dough** traditionally produced in Mexico and Guatemala by **steeping** corn in lime followed by **cooking** and **fermentation**. The fermented dough is often suspended in water and consumed as a refreshing beverage.

Pralines Cooked mixtures of crushed **nuts** and partly caramelized **sugar**, often used as a centre for **chocolates**. May be ground to a paste for use in pastry or candy **fillings**.

Prato cheese Brazilian semi-hard **cheese** similar to Gouda **cheese**.

Prawn crackers Fried sea food products made from minced **prawns** or **shrimps** mixed with **flour** (usually **tapioca** flour) and **seasonings**. Often consumed as **snack foods** or a meal accompaniment.

Prawns General name for many species of marine and freshwater **crustacea** within the suborder Natantia. The distinction between prawns and **shrimps** is unclear. The term prawns is generally used for larger species within the families Pandlidae, Penaeidae and Palaemonidae. Many species have commercial importance as foods, including *Palaemon serratus* (common prawns), *Penaeus japonicus* (**kuruma prawns**), *P. monodon* (giant tiger prawns) and *P. indicus* (Indian prawns). Marketed fresh, frozen, canned and as pastes.

Prebiotic foods Foods containing nondigestible ingredients with potentially beneficial health effects for the host based on selective simulation of the growth and/or activity of one or a limited number of bacterial species already resident in the colon. Examples of prebiotic components include **inulin** and nondigestible **fructooligosaccharides**. Similarly, **probiotic foods**.

Precipitation Process of forcing a substance in solid form from solution. Achieved through a variety of means, including addition of an agent to the solution and **centrifugation**.

Predictive microbiology Determination of the influence of various chemical, physical and biological factors on microbial growth and survival, typically by means of challenge trials or mathematical models.

Predictive modelling Use of simplified and generalized representations (models) of phenomena to forecast the influence of certain factors on events.

Prepared dishes Types of **convenience foods** similar to **prepared meals**.

Prepared foods Alternative term for **processed foods**.

Prepared meals **Convenience foods** eaten at mealtimes and/or designed to be one of the main **meals** of the day. Similar to **prepared dishes**.

Preservation Process of maintaining a food in its original or existing state by treatment that will prevent its **spoilage** or deterioration. Preservation is achieved by a range of treatments, including **refrigeration**, **freezing**, **canning**, **brining**, **smoking**, **freeze drying**, **drying** and **pickling**.

Preservatives **Additives** that increase **shelf life** of foods and **beverages**. Shelf life is determined by rates of growth of **spoilage** microorganisms and chemical degradation, usually oxidation, of food components. Preservatives are chemicals that inhibit one or both of these processes. Examples of preservatives include **organic acids** (e.g. **lactic acid**, **propionic acid**, **formic acid**), **benzoic acid** derivatives (sodium benzoate, **hydroxybenzoic acid esters**), **sulfur dioxide**, **nitrites** and **antioxidants**.

Preserves Term applied to preserved foods, usually referring to preserved **fruits**. **Fruit preserves** are made by cooking fruits with **sugar** and sometimes also **pectins**. Differ from fruit **jams** in that preserves generally contain larger chunks of fruit, while jams are similar to thick **fruit purees**. Preserves are used in a similar manner to jams. Other types of preserves include **vegetable preserves** and **fish preserves**.

Preserving Alternative term for **preservation**.

Presses Devices used for applying pressure in order to flatten or shape an item, or to extract natural fluids, e.g. **fruit juices** from **fruits** or **oils** from **oilseeds** or **nuts**. For oil extraction, screw presses are commonly used in preference to hydraulic presses because they provide a continuous process, have greater capacity, require less labour and generally remove more oil.

Pressing Process whereby pressure is applied to an item with **presses** in order to flatten or shape it, or to extract natural fluids. Used to produce **fruit juices** from **fruits**, and **vegetable oils** from **oilseeds** and **nuts**.

Pressure The force per unit area applied to a surface. Pressure is usually measured in pascals (Pa), which is defined as 1 Newton per square metre; it can also be measured in millimetres of mercury (mmHg), or millibars. High pressures may be applied in some food manufacturing processes (high pressure processing) or **analytical techniques** as a preservation process or to enhance results, respectively.

Pretzels Small, brittle **biscuits** made from a stiff **dough** typically formed into loose knots which are boiled briefly, glazed with **eggs** and baked. Often topped with salt crystals.

Pricing Determination of the amount of money expected or required in payment for something.

Prickly pears Spiny **fruits** produced by several varieties of **cacti**, especially *Opuntia ficus-indica*. The soft flesh is similar in **texture** to that of **watermelons**. Usually eaten fresh, but also used as an ingredient for **desserts** and **beverages**. Also known as **cactus pears**, Indian figs and barberry figs.

Principal component analysis Statistical technique by which variables in a data matrix are transformed to make them independent of one another. Covariance values are plotted on axes in multidimensional space. The first principal component, describing the majority of the spread of data, corresponds to the first axis in multidimensional space. Higher order axes show less variation, as the data are less correlated.

Printers Equipment, such as computer peripherals, which are used for **printing** text or graphics, e.g. on **labels** for foods. Print quality and printing speed vary greatly between printers. The major types include line printers, matrix printers, letter quality printers and laser printers.

Printing Process of generating printed material, such as **labels** for foods, including text and graphics.

Prion diseases Degenerative, fatal brain **diseases** which are believed to be transmitted by **prions**. Prion diseases are characterized by very long incubation periods. They include **bovine spongiform encephalopathy** (BSE) in **cattle**, **Creutzfeldt-Jakob disease** (CJD) and kuru in man, and **scrapie** in **sheep**. They may also be known as transmissible spongiform encephalopathies.

Prion proteins Alternative term for **prions**.

Prions Submicroscopic particulate **proteins**, which are believed to be the infective agent of prion diseases such as **BSE**, **Creutzfeldt-Jakob disease** and **scrapie**. They resist inactivation by procedures that modify **nucleic acids**. Conversion of soluble prion proteins (PrP) into insoluble, pathogenic, proteinase-resistant isoforms is one of the crucial events in the development of **prion diseases**. To date, the mechanism by which this conversion gives rise to pathogenic events remains unclear; however, increasing evidence suggests that conversion may depend on the role of prions in the prevention of oxidative damage. Also known as prion proteins.

Pristane Member of the branched chain **hydrocarbons** produced by certain zooplankton.

Private labels Products manufactured by one company, but labelled as a store brand.

Proanthocyanidins Condensed **tannins** found in many foods and **beverages**, such as **green tea** and **red wines**, where they contribute to the **flavour**, e.g. with **astringency** and **bitterness**. Possess **antioxidative activity** and **antibacterial activity**. Can form complexes with **proteins** in **beer** which may lead to formation of non-biological **haze**. Also thought to scavenge free radicals. Proanthocyanidins isolated from **cranberries** have been found to inhibit

the **adherence** of *Escherichia coli* to model epithelial surfaces.

Probiotic bacteria **Bacteria** which benefit health by promoting a balanced gastrointestinal **microflora** (e.g. *Bifidobacterium* and *Lactobacillus* species). Used in the preparation of microbial cultures for use in foods and animal feeds.

Probiotic foods Novel foods containing viable **probiotic microorganisms** (particularly **lactic acid bacteria**, but also some bifidobacteria and **yeasts**) that have beneficial effects on the health of the host by improving the microbiological balance of the intestine. Examples include **bifidus milk**, **acidophilus milk** and **yakult**. Similarly, **prebiotic foods**.

Probiotic microorganisms Microorganisms which benefit health by promoting a balanced gastrointestinal **microflora**. Used in the preparation of microbial cultures for use in foods and animal feeds.

Procaryotes Alternative spelling of **prokaryotes**.

Process control Use of computerized systems for automatic control of continuous industrial processes.

Processed cheese Product made from one or more hard or semi-hard **cheese** by milling and heating with water, **emulsifying agents** such as **phosphates** or **citrates**, and other ingredients including **milk** or **whey** powder, **butter**, **cream**, **seasonings** and **flavourings**. The mixture is pasteurized at a high temperature to extend **shelf life** of the product. The heating used during processing stops any further cheese **ripening** or **flavour** development. Soft versions containing 50% water are used as processed **cheese spreads**.

Processed foods Foods which have been subjected to some degree of processing in order to bring about a desired modification, e.g. enhanced **shelf life**, **physicochemical properties**, **sensory properties** or nutritional quality. Examples include **chilled foods**, **frozen foods**, **canned foods**, **ready meals**, **preserves** and **dietetic foods**. Also known as prepared foods.

Processing Treatment of a raw material, such as a food, usually by applying a series of actions or steps, to produce a specific end product.

Processing equipment Machinery used in the processing of foods.

Processing lines Sequences of **processing equipment** units that are integrated in order to manufacture a complete product.

Procyanidins **Polyphenols** found in foods such as **cocoa**, **chocolate**, **green tea** and **red wines** which are thought to exhibit cardioprotective effects due to their ability to scavenge free radicals, inhibit oxidation of lipids and suppress the activation of platelets.

Procymidone Systemic fungicide which inhibits synthesis of **triglycerides** by **fungi**. Used to control a range of fungal **pathogens** (particularly *Botrytis*, *Sclerotinia* and *Monilia* spp.) on **fruits**, **vegetables**, **cereals** and **oilseeds**.

Production The action or process of producing or being produced, or the bulk of a commodity produced in a given country or area.

Product liability A producer's legal responsibility for goods, or the liability of manufacturers and traders for damage or injury caused to purchasers or bystanders by their products.

Product technology Processing procedures employed during the manufacture of foods.

Profitability The monetary difference between the cost of producing and marketing goods or services and the price subsequently received for those goods or services. Profit is an essential competitive feature of buying and selling in the economic system.

Profiteroles Small, **cream puffs** made from baked choux pastry shells, which are filled with **whipped cream** and topped with chocolate **sauces**.

Progesterone One of the steroid **hormones** produced mainly by the corpus luteum which prepares the uterus to receive the fertilized egg and maintains the uterus during pregnancy. Used in cattle breeding to suspend the oestrous cycle and allow the mating of the whole herd to be synchronized. Potentially useful as a marker of **mastitis** in cattle.

Progoitrin **Glucosinolates** found in *Brassica* vegetables that may contribute to their **flavour** and **bitterness**.

Prokaryotes Typically, unicellular organisms, including **bacteria** and **cyanobacteria**, of the kingdom Procaryotae. Characterized by a lack of a defined nucleus, and the possession of a single circular **DNA** molecule and a very small range of organelles. Also spelt procaryotes.

Prolactin Proteinaceous **hormones** secreted by the anterior lobe of the pituitary gland which stimulate secretion of milk in mammals and assist in maintaining the corpus luteum. Also thought to be involved in development of the immune system in neonates.

Prolamins Seed **globulins** that are insoluble in water and soluble in water-ethanol mixtures. Rich in **proline** and **glutamic acid** but contain small amounts of **lysine**, **arginine** and **tryptophan**, resulting in their being of poor nutritional value.

Proline Non-essential amino acid whose structure differs from those of other **amino acids** in that its side chain is bonded to the N of the amino group as

well as the C, making the amino group a secondary amine. Has a strong influence on the secondary structure of **proteins** and is found more abundantly in **collagen** than in other proteins.

Promoters Nucleotide sequences located upstream from **transcription** start sites that are recognized and bound by **DNA-directed DNA polymerases** and other regulatory proteins during the initiation of transcription.

Pronase A commercial preparation of **proteinases** from *Streptomyces griseus* containing at least 4 **enzymes**, including **trypsin** and a neutral metalloproteinase. Used for production of **protein hydrolysates**, and improving the **sensory properties** of dry fermented **sausages** and the **functional properties** of insoluble **gluten**.

Proofers Equipment assisting in the **proofing** (or proving) of **dough**, in which airflow, ambient conditions (e.g. air temperature and **relative humidity**) and handling can all be controlled.

Proofing Stage of **breadmaking** in which **dough** is fermented under controlled conditions. During proofing (or proving), **starch** is converted by **enzymes** into **sugars** that are used as growth substrates by the **yeasts** employed. The breakdown products are **carbon dioxide** and alcohol. As carbon dioxide is produced, it is retained in the tiny cells formed in the protein matrix during mixing, causing them to grow and the dough to expand. Other products of yeast activity, mainly **acids**, are also formed during proofing; they contribute significantly to **flavour** development. Dough expands by a factor of three or four during proofing, and it is important that the skin remains flexible so that it does not tear as it expands. Yeast is at its most active at 35-40°C, so to minimize proofing time, heat transfer to the dough is necessary, to raise its temperature by 10-15°C.

Propanal Aldehyde that exists as a colourless liquid. Can be reduced to **propanol** and oxidized to propanoic acid. Also known as **propionaldehyde**.

Propane Gaseous hydrocarbon of the paraffin series obtained from petroleum. Useful in the food industry for extraction of lipophilic compounds and proteins.

1,2-Propanediol Aliphatic alcohol used primarily in **emulsifiers**. Other uses in foods include in **anticaking agents**, **antioxidants**, **flavourings** and **humectants**. Used in **freezing** media and **solvents** for food processing. Synonym for **propylene glycol**.

Propanil Common name for 3',4'-dichloropropionanilide, a selective contact herbicide with short duration of activity used to control broadleaved and grass weeds among many types of **cereals**

and citrus crops. Also known as DCPA. Classified by WHO as slightly hazardous (WHO III).

Propanol Alcohol containing three carbon atoms which is also known as **propyl alcohol**. Used for extraction of **glutenin** subunits from **wheat** and **phospholipids** from **fish oils**.

Propanone Colourless, flammable, volatile ketone used as a solvent and as a raw material for making **plastics**. Produced commercially by **fermentation** of **corn** or **molasses**, or by controlled oxidation of **hydrocarbons**. Also known as **acetone**.

Propazine Selective systemic herbicide used for pre-emergence control of grasses and broad-leaved weeds around **carrots**, **parsley** and **sorghum** (most other **vegetables** and **cereals** are sensitive). Classified by Environmental Protection Agency as not acutely toxic (Environmental Protection Agency IV).

Propenal Colourless, highly volatile, liquid aldehyde also known as **acrolein**. Can cause formation of undesirable **flavour** in **cider** and **spirits**.

Propham Selective systemic herbicide used to control many annual grasses and broad-leaved weeds around plants including some *Brassica* vegetables, **legumes**, **sugar beets** and **oilseeds**. Also used as a plant growth regulator and **sprouting** inhibitor for stored **potatoes**. Classified by Environmental Protection Agency as not acutely toxic (Environmental Protection Agency IV).

Propineb Dithiocarbamate fungicide used in protection of a wide range of fruit and vegetable plants. Applied to leaves and rapidly degrades, but metabolites may be taken up in small amounts by the plant.

Propionaldehyde Synonym for **propanal**. Aldehyde that exists as a colourless liquid. Can be reduced to **propanol** and oxidized to propanoic acid.

Propionibacteria Bacteria of the genus *Propionibacterium*.

Propionibacteriaceae Family of Gram positive, anaerobic rod-shaped or filamentous **bacteria**. Occur in **dairy products**, and on the skin and in the **gastrointestinal tract** and respiratory tract of humans and animals.

Propionibacterium Genus of Gram positive, anaerobic irregularly-shaped **bacteria** of the family **Propionibacteriaceae**. Occur in soil, **milk** and **dairy products**, and the gastrointestinal tracts of herbivores. Certain species (e.g. *Propionibacterium shermanii*) are used to promote **flavour** development and eye formation during the **ripening** of some types of **hard cheese**. Other species may cause food **spoilage**.

Propionic acid Colourless liquid carboxylic acid. Propionic acid and its derivatives, such as calcium propionate, are used as **preservatives** in foods and **beverages**, where they act as **fungicides**. Can be produced by **fermentation** of **bacteria** such as *Propionibacterium* spp., *Clostridium propionicum* and *Megasphaera elsdenii*.

Propionic acid bacteria Bacteria, usually of the genus *Propionibacterium*, which produce **propionic acid** as a main end product in the **propionic fermentation** of **glucose** or **lactic acid**.

Propionic fermentation Process by which certain **bacteria** (such as *Propionibacterium* spp., *Clostridium propionicum* and *Megasphaera elsdenii*) ferment substrates such as **glucose** and/or **lactic acid** to produce **propionic acid**.

Propionicins Bacteriocins synthesized by *Propionibacterium* spp. Propionicin PLG-1, produced by *P. thoenii* P127, has a bactericidal mode of action. It is heat labile, and effective against a wide range of **Gram positive bacteria** and **Gram negative bacteria**, as well as against some **yeasts** and **fungi**. Different aggregative forms exist: one of 10,000 Da and the other of >150,000 Da.

Propolis Resinous product collected by bees from plant exudates for use in the construction of hives. Has wide applications in medicine, cosmetics and foods, e.g. in **food supplements**. Also reported to have antibiotic or antifungal properties. Other hive products include **honeys**, **beeswax** and **royal jelly**.

Propoxur Non-systemic insecticide used for control of insect **pests** in food storage areas. Also used for control of sucking and chewing **insects** in a range of **fruits**, **vegetables** and **cereals**, **sugar cane** and **cocoa**. Residues tend to be relatively persistent. Classified by WHO as moderately toxic (WHO II).

Propyl alcohol Synonym for propanol. Alcohol containing three carbon atoms. Used for extraction of **glutenin** subunits from **wheat** and **phospholipids** from **fish oils**.

Propylamine Amine containing three carbons atoms.

Propylene Colourless, gaseous hydrocarbon with a **garlic** odour. Also known as propene. Active **ethylene** analogue, which can be used to promote the **ripening** of **fruits**.

Propylene glycol Aliphatic alcohol used primarily in **emulsifiers**. Other uses in foods include as **anticaking agents**, **antioxidants**, **flavourings** and **humectants**. In food processing, propylene glycol is used in freezing media and solvents. Synonym for **1,2-propanediol**.

Propylene oxide Oxide of **propylene** that can be used for **fumigation**. Also used as an intermediate in the synthesis of **propylene glycol**, **glycerol** and propanolamines, and as a solvent for **cellulose acetate**, cellulose nitrates and natural **resins**.

Propyl gallate   Esters of propanol and gallic acid (3,4,5-trihydroxybenzoic acid) with **antioxidative activity**. Soluble in fats and thus used as **antioxidants** for **fats**, including **margarines** and edible **oils**, and **meat** products. Propyl gallate exhibits synergistic antioxidative activity with **BHT** and **BHA**.

Propylparaben   Common name for 4-hydroxybenzoic acid propyl esters. Typical of paraben esters. Uses for propylparaben include as **preservatives** for foods and **beverages**.

Proso millet   Millet belonging to the species *Panicum miliaceum*, which is grown as a cereal food in Asia and Eastern Europe.

***Prosopis africana* seeds**   **Seeds** of a wild tropical plant that grows in Nigeria. Fermented to produce **condiments** (kpaye, okipye or okpehe) that may be used in **soups**, or boiled and made into daddawa cake, used to add **flavour** in cooking. A potential source of **vegetable oils**.

Prostaglandins   Group of compounds formed from **unsaturated fatty acids** with 20 carbon atoms, predominantly **arachidonic acid**, that contain a five membered ring and are mediators of a wide range of physiological processes.

Prostokvasha   **Fermented milk** product popular in the USSR.

Protamine   Antimicrobial peptide with high **arginine** content, usually found in association with **DNA** of spermatozoan nuclei of **fish** (including **salmon**, **carp** and **herring**). Particularly active against **Gram positive bacteria**. Used in food **preservatives** for inhibition of microbial growth, and also shows **emulsifying capacity**. Also known as salmine.

Proteases   Alternative term for **proteinases**.

Proteasomes   Alternative term for **proteinases**.

Proteinases   **Enzymes** that hydrolyse **proteins** by cleavage of peptide bonds. Endoproteinases cleave within protein molecules, while exoproteinases attack the ends of protein chains removing **amino acids** one at a time. They are classified as serine proteinases, thiol proteinases, metalloproteinases or acid proteinases. Some proteinases exhibit a high degree of specificity with respect to the peptide bonds they cleave (e.g. **trypsin**), while others are much less specific (e.g. **papain**). These enzymes are used in all areas of food production, including the **meat**, **brewing**, **cheesemaking** and **breadmaking** industries. Also known by many other names, including proteases, proteasomes and proteolytic enzymes.

Proteinases inhibitors   Substances that have the ability to inhibit the proteolytic activity of certain enzymes. Such inhibitors are found throughout the plant kingdom, particularly among **legumes**. **Trypsin inhibitors** are found in **soybeans**, **lima beans** and **mung beans**. **Chymotrypsin inhibitors** are found in **cereals** and **potatoes**. Proteinase inhibitors are destroyed by heat.

Proteinates   Protein products typically obtained by precipitation from the source material at the **isoelectric points**, followed by a neutralization step (e.g. to form sodium or calcium proteinates). Some of the most widely used proteinates include **caseinates**, total milk proteinates and soy proteinates, which have applications as functional ingredients in **meat** products, **dairy products** and **imitation foods**. Mineral proteinates are also used in animal and human nutrition as a readily absorbed form of mineral complex.

Protein concentrates   Products prepared by extracting **proteins** from animal and plant materials such as **vegetables**, **fish** or **whey**. Protein content varies among preparations. Used to provide protein fortification and enhance **functional properties** in a wide range of foods. Some of the most commonly used concentrates in the food industry are **fish protein concentrates**, **soy protein concentrates** and **whey protein concentrates**.

Protein efficiency ratios   Biological method (commonly abbreviated to PER) for evaluating protein quality in terms of weight gain per amount of protein consumed by a growing animal. PER is used widely in comparing the nutritional values of proteins in individual foods. It assumes that all protein is used for growth and no allowance is made for maintenance.

Protein engineering   Use of **genetic techniques** to modify and enhance the properties of **proteins**.

Protein-glutamine γ-glutamyltransferases   EC 2.3.2.13. Catalyse the formation of amide bonds between side chain glutamine and side chain lysine residues in **proteins** with the elimination of **ammonia**. Used for cross-linking proteins, thus modifying their **functional properties** (e.g. **milk proteins** in production of **yoghurt**, **cereal proteins** in production of **bakery products** and **fish proteins** in production of **surimi** gels). Also used for covalently incorporating individual essential **amino acids** into proteins (e.g. **casein**) and for joining 2 proteins, thus allowing creation of designer proteins. Also known as transglutaminases.

Protein hydrolysates   **Proteins** that have been subjected to **hydrolysis** by treatment with **enzymes**, **acids** or **alkalies**, so that the protein molecule is broken down into **peptides** and free **amino acids**. Easily digestible and used to reduce **antigenicity** of foods. Applications include as ingredients of **medical**

foods, **infant formulas** and hypoallergenic products.

Protein isolates Products prepared by extracting and purifying **proteins** from animal and plant materials. Have similar properties to **protein concentrates**, but typically contain about 90% protein. Examples include soy protein isolates and whey protein isolates.

Proteins Nitrogenous **organic compounds** consisting of linked **amino acids** that are distributed widely in plants and animals. The sequence of amino acids in proteins is determined by the base sequence of their encoding **genes**. They serve many roles, such as **enzymes**, structural elements and **hormones**, and are essential **nutrients**.

Protein values Relative **nutritional values** of **proteins** based on **amino acids** composition, digestibility and availability of the digested products. Also the relative biological value defined in various terms, including the ability of a test protein, fed at various levels of intake, to support nitrogen balance, relative to a standard protein.

Proteoglycans High molecular weight complexes of **proteins** and **polysaccharides** that are major constituents of structural tissues such as bones, cartilage and muscles, and are also found on the surface of cells. Glucosaminoglycans, the polysaccharides in proteoglycans, are polymers of acidic disaccharides containing derivatives of **glucosamine** or **galactosamine**.

Proteolipids Complexes of **proteins** and **lipids** abundant in brains but also found in a wide variety of tissues in animals and plants. In contrast to **lipoproteins**, they are insoluble in water. The proteins in proteolipids have high contents of hydrophobic **amino acids**, while the lipids consist of a mixture of **phosphoglycerides**, **cerebrosides** and sulfatides. In contrast, lipoproteins consist of **phospholipids**, **cholesterol** and **triglycerides**.

Proteolysis Hydrolysis of **proteins** to smaller peptide fractions and their constituent **amino acids**, catalysed by **alkalies**, **acids** or **enzymes** (**proteinases**).

Proteolytic enzymes Alternative term for **proteinases**.

Proteose peptones Small **peptides** in **milk** derived from the breakdown of **casein** by **proteinases**.

Proteus Genus of Gram negative, facultatively anaerobic rod-shaped **bacteria** of the family **Enterobacteriaceae**. Occur in soil, water, **dairy products**, raw **shellfish**, fresh **vegetables** and the gastrointestinal tracts of humans and animals. Some species (e.g. *Proteus vulgaris* and *P. intermedium*) may cause **spoilage** of foods (e.g. **eggs**, **Cottage cheese**, **meat** and shellfish).

Protocatechuic acid Phenolic compound found in many foods and **beverages** which exhibits **antioxidative activity** and is able to scavenge free radicals.

Proton magnetic resonance Spectroscopy technique also known as ^{1}H-NMR (**nuclear magnetic resonance**) in which analysis is based on chemical shifts between non-equivalent protons in the molecule under investigation.

Proton resonance Phenomenon used in **nuclear magnetic resonance** and **proton magnetic resonance** in which protons in a static magnetic field absorb energy from an alternating magnetic field at characteristic frequencies.

Protoplast fusion Fusion of **protoplasts** from different strains, species or genera to form hybrid protoplasts, and ultimately hybrid cells. Protoplasts are mixed together, transferred to appropriate media for cell wall regeneration and the resulting cells are screened for the presence of genetic markers from both parents.

Protoplasts Bacterial and plant cells that lack cell walls. Cell walls can be removed enzymatically or by growth in the presence of antibiotics that block synthesis of cell wall peptidoglycans. Protoplasts can continue to metabolize and can revert to normal cells under appropriate conditions, although they cannot divide. Bacterial protoplasts are prepared more easily from Gram positive cells than from Gram negative cells.

Protozoa Unicellular organisms of the subkingdom Protozoa which lack cell walls. Occur in soil and freshwater, brackish and marine habitats. Some are pathogenic in humans and animals. Transmission is typically via raw **meat** and faecally-contaminated **vegetables**, **salads** and **fruits**.

Providencia Genus of Gram negative, facultatively anaerobic rod-shaped **bacteria** of the family **Enterobacteriaceae**. Occur in soil, **dairy products**, raw **shellfish**, fresh **vegetables** and the **gastrointestinal tract** of humans and animals. *Providencia alcalifaciens* is associated with diarrhoeal illness in humans due to consumption of contaminated foods.

Provitamin A Vitamin precursor for **vitamin A**. Some **carotenoids**, such as α- and β-carotene and β-cryptoxanthin, have provitamin A activity. Provitamins are chemically related to preformed vitamins, but have no vitamin activity unless converted to the biologically active form.

Provolone cheese Italian semi-hard all-purpose **cheese** made from **cow milk**. The rind is yellow and shiny, thin and hard. It may also be waxed. The interior is cream-white or slightly straw coloured with a compact **texture**. Marketed in various types that have been aged for periods of 2-3 months to 2 years, and differ in **flavour** and **aroma**.

Prunasin Cyanogenic glycoside found in a range of plant materials including almond roots and leaves, **vetch seeds** (used as lower priced substitutes for **lentils**), Japanese apricot **seeds**, plum seeds, juice and **peel** of **passion fruits**, and black cherry seeds. Also isolated from fresh **tea leaves**, where it is a precursor for the flavour compound **benzaldehyde**. Present in many immature fruits, but converted to **amygdalin** during maturation.

Prune juices **Fruit juices** prepared from **prunes** (*Prunus domestica*).

Prunes Dried **plums**. Specific varieties, mainly of European plums, are suitable for production of prunes. Plums are dried on the tree where the climate is warm enough, or alternatively dried by artificial means.

Pruning Cutting branches or roots of trees and bushes, usually to a specified length or position. Can have beneficial effects on plant growth as well as on yield and quality of fruits produced.

PSE defect Abbreviation for the **pale soft exudative defect** of **meat**, especially **pork**.

Pseudocereals Plant species that do not belong to the grass family, but produce seeds or fruit that are used in the same way as cereal grain to make **flour** and **bakery products**. Include **buckwheat**, **quinoa** and **amaranth grain**.

Pseudomonadaceae Family of Gram negative, aerobic, curved or straight rod-shaped **bacteria**. Occur in fresh water, salt water and soil. Includes many plant and a few animal **pathogens**.

Pseudomonas Genus of Gram negative, aerobic, curved or straight rod-shaped **bacteria** of the family **Pseudomonadaceae**. Occur in soil, water, **salads** and **meat**. Some species (e.g. *Pseudomonas fluorescens* and *P. fragi*) may cause **spoilage** of meat, **dairy products**, **cream**, **butter**, **eggs** and **fish**. Certain species, e.g. *P. cepacia*, have been reclassified under the new genus ***Burkholderia***.

Pseudoterranova Genus of parasitic **nematodes** of the family Anisakidae. Larvae have been implicated in **anisakiasis**, an infection caused by consumption of contaminated raw or undercooked **sea foods**.

Psicose Sugar with the systematic name D-ribo-2-hexulose of interest for potential use in **sweeteners** or **bulking agents**.

Psocids **Insects**, some species of which e.g. those of the genus ***Liposcelis***, are pests infesting grain stores in hot, humid areas, with consequent adverse effects on grain quality and value. Also infest raw and processed foods in food manufacturing or retail premises as well as in the home. A wide range of commodities is prone to infection, but the insects have a preference for **microorganisms** which can become entangled in their bodies giving a means of disseminating **spoilage**-causing organisms. Also known as booklice.

Psoralens Toxic secondary metabolites found in many **fruits** and **vegetables**. They are potent photosensitizers that can form photoadducts with **nucleic acids** if irradiated with UV light.

PSP Abbreviation for **paralytic shellfish poisoning**.

Psychophysiological effects Effects relating to the relationship between physiological processes and psychological experience.

Psychrobacter Genus of mainly psychrotrophic **Gram negative bacteria** of the family Neisseriaceae. Some species, such as *Psychrobacter immobilis*, can cause **spoilage** of animal foods, including **meat** and **fish** products.

Psychrophiles Organisms, especially **microorganisms**, that grow best at relatively low temperatures. Their optimum growth temperature is generally accepted as being below 20°C.

Psychrotrophs Organisms, especially **microorganisms**, that can grow at relatively low temperatures, but grow optimally within the temperature range of 15 to 20°C.

Psyllium Small, dark red/brown seeds from plants belonging principally to the species *Plantago psyllium*, *P. ovata* or *P. afra*, producing a mucilaginous mass which is often added to foods as a source of soluble **fibre**.

Psyllium gums **Gums** extracted from seeds of **psyllium** (*Plantago ovata*). Used primarily as **stabilizers** for **ice cream** and **sherbet**.

Ptarmigans **Game birds** belonging to the genus *Lagopus* of the Tetraonidae (grouse) family, which are hunted for their **meat**.

PTFE Commonly used abbreviation for **polytetrafluoroethylene**.

Pubs Informal name for public houses, also known as inns. Establishments, found chiefly in the UK, consisting of at least one public room and licensed for the sale and consumption of **alcoholic beverages**. Most pubs now sell meals, often in a separate restaurant area.

Pudding mixes Dried **instant foods** consisting of a mixture of pregelatinized **starch** and other ingredients used to prepare **puddings**, typically by adding **milk**.

Puddings Sweetened, usually cooked, **desserts** made from various ingredients, e.g. flour, fruit, milk and eggs. Include **milk puddings** and steamed sponges. The term may also refer to savoury dishes topped with or surrounded by suet crust or pastry, such as steak and kidney puddings, or to savoury products in a sausage shape enclosed in casings, e.g. **black puddings** or white puddings.

Pu-erh tea Type of China tea which has undergone a microbial **fermentation** process during manufacture.

PUFA Abbreviation for **polyunsaturated fatty acids**.

Puff balls Common name for **edible fungi** of the genus *Lycoperdon*.

Puffed rice Rice grains that are heated under pressure which is then rapidly released, causing the superheated steam in the grain to expand and explode the rice grain. Used in a range of food applications, including **snack foods**, **breakfast cereals** and **confectionery**.

Pufferfish Any of a number of small, predominately **marine fish** in the family Tetraodontidae; widespread in the Indo-Pacific region. Some species are highly esteemed food fish, particularly in Japan. Many species contain potent **neurotoxins**, implicated in severe and often fatal **food poisoning** incidents. Commercially important species include *Takifugu porphyreus* (purple puffer) and *T. vermicularis* (nashi-fugu). Normally marketed fresh and prepared for consumption by specialist chefs able to remove toxic components.

Puffing Method for expanding foods, particularly cereal grains. Grain is subjected to high pressure and/or temperature, before being ejected into a normal atmospheric pressure, causing the samples to expand sharply. Used mainly in the manufacture of **breakfast cereals** such as **puffed rice** and puffed **wheat**, and for making **snack foods** and puffed **rice cakes**.

Puff pastry Light flaky **pastry** formed by alternating layers of fat and **dough** so that, upon **baking**, steam collects between dough layers, causing them to expand and form cavities between the thin pastry layers.

Pullet eggs **Eggs** produced by pullets (young hens usually less than one year old).

Pullulan Extracellular, water-soluble, linear D-glucan produced by *Aureobasidium pullulans*, consisting predominantly of **maltotriose** units linked by (1,6)-α-glucosidic bonds. Useful as a **starch** replacer in **dietetic foods** and as a component of edible films.

Pullulanases Alternative term for **α-dextrin endo-1,6-α-glucosidases**.

Pulpboard Type of **paperboard** in which all plies are usually made from wood pulp, although the centre may sometimes be filled with waste paper.

Pulping Crushing of foods, e.g. **fruits** and **vegetables**, into soft, smooth and moist masses (**pulps**).

Pulps Preparations of a soft, moist consistency, typically obtained by mashing foods, particularly **fruits** or **vegetables**. Used in the manufacture of a wide range of foods and **beverages**, including **fruit juices**, yo-

ghurt and **pie fillings**. Also refers to the solid residue remaining after extraction of juices from fruits and vegetables.

Pulque Mexican **alcoholic beverages** prepared by **fermentation** of sap of the **agave** plant.

Pulsed electric fields Used in food **processing** and **preservation**. A high intensity electric field is delivered as a series of pulses of direct current to the food for a very short period of time while the food is held between two electrodes. This process results in formation of pores in, and breakdown of, cell membranes; the consequences of this can be microbial inactivation and increased yield of **fruit juices** during extraction. The risk of dielectric breakdown of foods limits this type of processing primarily to liquid foods, because uniformity of the applied electrical field would be distorted by air bubbles or suspended solids that usually exist in solid foods.

Pulsed field gel electrophoresis **Gel electrophoresis** technique in which DNA fragments are separated by subjecting the gel to an electric current alternately from two angles at timed intervals. Commonly abbreviated to PFGE.

Pulses Edible seeds of leguminous plants, including various **beans**, **peas** and **lentils**. Mature seeds are dry and can be stored. Also refers to the plants producing these seeds.

Pulverization Reduction into fine particles (powders or dust), usually by crushing, pounding or grinding.

Pummelos Alternative term for **pomelos** or **shaddocks**, the largest of the **citrus fruits**, produced by *Citrus maxima* or *C. grandis* and ancestors of the modern **grapefruit**. Closely resemble the grapefruit in appearance, but the flesh is sweeter and less acidic, lacking the **bitterness** of a grapefruit. Rich in **vitamin C** and potassium. Eaten fresh or used to make **jams**, **jellies** and **marmalades**.

Pumpernickel Dark brown, dense **bread** made with coarsely ground whole **rye flour** and **sourdough**, originating from Germany.

Pumpkins Fruits produced by plants of the genus *Cucurbita*, especially *C. pepo* and *C. maxima*. Contain approximately 90% water, moderate amounts of **vitamin C**, and small amounts of **carotenes**, **starch**, **sugar**, protein, fat and vitamin B complex. Used in **jams** and **pies** and as **vegetables**. Leaves and flowers of the plants can also be eaten. **Pumpkin seeds** are eaten or processed into **pumpkin seed oils**.

Pumpkin seed oils Oils rich in **unsaturated fatty acids**. Frequently used as **salad oils**; also used as an ingredient of **cider vinegar**.

Pumpkin seeds Oilseeds produced by **pumpkins**, *Cucurbita pepo*, which, when roasted and salted, may be consumed as **snack foods**. Rich in **unsaturated fats**, **vitamins** and **minerals**.

Pumps Mechanical devices that use suction or pressure to raise or move liquids or compress gases. Often components of larger pieces of equipment.

Pungency **Sensory properties** relating to the extent to which the **aroma** or **flavour** of a product (usually **onions**, **chillies**, **peppers**, **ginger** and **radishes**) is acrid or pungent.

Pungent principles **Flavour compounds** responsible for **pungency** of foods such as **chillies**, **onions**, **peppers**, **ginger** and **radishes**.

Punnets Small lightweight **containers** or **baskets** for **vegetables** or **fruits**.

Purees Smooth, thick preparations made by mashing from foods, particularly cooked **fruits** and **vegetables**, which have had any coarse fibre removed by **sieving** or similar means.

Purification Removal of contaminants or undesirable components from a substance.

Purines Heterocyclic organic bases that pair with **pyrimidines** in **DNA** and **RNA**, and whose derivatives are important in metabolism. They include **adenine** and guanine, as well as many **alkaloids**, such as **caffeine** and **theophylline**.

Purity Extent to which an item or substance is pure, i.e. free from contaminants and adulterants.

Puroindolines Lipid-binding **cereal proteins** (puroindoline-a and puroindoline-b) found in **wheat** which play a significant role in texture of **bread crumb**. Genetic variation of puroindoline alleles is associated with kernel **hardness** in wheat, a property known to affect **milling** and **baking** qualities.

Purothionin Disulfide-rich protein of the **wheat** endosperm which shows **antimicrobial activity**.

Purslane Common name for plants of the genus *Portulaca*, especially *P. oleracea*. Leafy vegetable that is eaten raw in **salads** or cooked in **soups** or as greens. Rich source of **ω-3 fatty acids**, known to be beneficial in **coronary heart diseases** and some types of **cancer**. **Seeds** are ground into a **meal** that may be used to make **bread**.

Puto Fermented **rice cakes** which are consumed as **breakfast** foods or **snack foods** in the Philippines.

Putrefaction Typically anaerobic, microbial decomposition of substances (especially proteinaceous and fatty products such as **meat** and **fish**) with the production of foul-smelling compounds (e.g. **ammonia**, **hydrogen sulfide**, **cadaverine** and **putrescine**).

Putrescine Foul-smelling biogenic amine formed from the decarboxylation of **ornithine**, usually during **putrefaction**.

PVC Abbreviation commonly used for **polyvinyl chloride**.

PVDC Abbreviation commonly used for **polyvinylidene chloride**.

Pycnometry Technique for determining the **density** of a liquid, using a small bottle of accurately measured volume. Density is determined from the ratio between the weights of a given volume of water and the same volume of sample.

Pyrazines Nitrogen containing, heterocyclic **flavour compounds** found in many foods and **beverages** that can be formed during the **Maillard reaction**.

Pyrene Toxic four ringed polycyclic aromatic hydrocarbon that can contaminate foods and **beverages**.

Pyrethrins Natural insecticidal compounds found in the flower of the pyrethrin daisy, a *Chrysanthemum* sp. native to Kenya. Used in the manufacture of **pyrethroid insecticides**.

Pyrethroid insecticides Class of synthetic **insecticides** based on **pyrethrins**. Widely used for control of insect **pests** on a range of crops. Commonly used examples include **cypermethrin**, **fenvalerate** and **permethrin**.

Pyridine Heterocyclic nitrogenous base that acts as the nucleus of a large number of **organic compounds**, such as **alkaloids**. Used as a solvent, and in the manufacture of various **drugs** and **pesticides**.

Pyridoxal One of the three forms of **vitamin B$_6$**, the aldehyde form, the others being **pyridoxamine** (the amine form) and **pyridoxine** (the alcohol form). The relative proportion of each of the three forms in foods varies considerably. All are equally biologically active.

Pyridoxamine One of the three forms of **vitamin B$_6$**, the amine form, the others being **pyridoxal** (the aldehyde form) and **pyridoxine** (the alcohol form). The relative proportion of each of the three forms in foods varies considerably. All are equally biologically active.

Pyridoxine One of the three forms of **vitamin B$_6$**, the alcohol form, the others being **pyridoxal** (the aldehyde form) and **pyridoxamine** (the amine form). The relative proportion of each of the three forms in foods varies considerably. All are equally biologically active.

Pyrimidines Heterocyclic organic bases that pair with **purines** in **DNA** and **RNA**, and whose derivatives are important in metabolism. Include **cytosine**, **thymine** and **uracil**.

Pyrocarbonic acid diethyl ester Esters with **antimicrobial activity** used as **preservatives** mostly for **beverages**, including **wines**, **alcohol reduced wines**, **fruit juices** and **iced tea**. Also known as dimethyl dicarbonate.

Pyrocatechol Catecholic diphenol that acts as a substrate for **catechol oxidases**.

Pyrococcus Genus of Gram positive, obligately an-aerobic coccoid archaebacteria of the family Thermo-coccaceae. Some species (e.g. *Pyrococcus furiosus*) are used in the production of **enzymes**.

Pyrogallol Phenolic compound also known as pyro-gallic acid that acts as a powerful reducing agent.

Pyroglutamic acid Degradation product of **gluta-mine** found in many types of **cheese**, particularly extensively ripened cheeses produced with thermo-philic **lactic acid bacteria** as **starters**. Pyroglu-tamic acid produced by lactic acid bacteria has been shown to exhibit **antibacterial activity**. Also found in **alcoholic beverages**, **fruit juices**, **meat** and **fruits**, where it can have adverse effects on **sensory properties**.

Pyrolysis Decomposition of chemical substances as a result of high temperatures. Sometimes used in analy-sis of foods by **gas chromatography** and **mass spectroscopy**, and as part of some processing tech-niques to add **flavour** or **colour** to products.

Pyrones Heterocyclic **flavour compounds** found, for example, in roasted **malt** and **chicory**. Can also be produced by microbial **fermentation**. Certain py-rones act as **mycotoxins**, while others have been found to exhibit **antifungal activity**.

Pyrophosphatases Group of **enzymes** within the subclass EC 3.6.1 that catalyse the hydrolysis of di-phosphate bonds, mainly those of nucleoside di- and triphosphates, liberating either a mono- or diphosphate.

Pyrophosphates Compounds containing two phos-phate groups linked together by an ester bond. In-volved in many metabolic reactions in **prokaryotes** and **eukaryotes**.

Pyrroles **Organic nitrogen compounds** that can be formed in foods by the **Maillard reaction** or by other pathways, and contribute to **flavour**. Some pyr-roles exhibit **antimicrobial activity**. The pyrrole ring structure is also found in many important biologi-cal compounds, such as **pigments**, **chlorophylls** and **haem**.

Pyruvaldehyde Organic compound often used as a reagent in organic syntheses and in **flavourings**. Can be formed by the **Maillard reaction** and has been shown to exhibit **antibacterial activity**. Also known as **methylglyoxal**.

Pyruvate carboxylases EC 6.4.1.1. Carboxylating, anaplerotic **enzymes** which produce oxaloacetate from **pyruvic acid**. Important for replenishment of the oxaloacetate consumed during growth of certain **bacteria**. Thought to be required for rapid **coagula-tion of milk** by **lactic acid bacteria**.

Pyruvate decarboxylases EC 4.1.1.1. **Enzymes** which catalyse the decarboxylation of α-keto acids to **aldehydes** and carbon dioxide. Decarboxylate **pyru-vic acid** to **acetaldehyde** irreversibly in lower or-ganisms. Thought to contribute to the formation of im-portant **flavour compounds** such as acyloins and **isoamyl alcohol** in **alcoholic beverages**. Also known as α-carboxylases, pyruvic decarboxylases and α-ketoacid carboxylases.

Pyruvic acid Intermediate in a wide range of aerobic and anaerobic metabolic pathways. Produced as the end product of **glycolysis** and is at the starting point of the Krebs' cycle.

Q

QTL Abbreviation for **quantitative trait loci**.

Quail eggs **Eggs** produced by **quails**. Considered as a delicacy. Consist of approximately 13.05% protein and 11.09% lipids, and have a mean weight of 11 g. **Egg shells** may be a variety of colours, but are often light brown with dark speckles.

Quail meat **Meat** from **quails**, commonly from farmed bobwhite quails (*Colinus virginianus*) or Japanese quails (*Coturnix coturnix*). Farmed quail meat tends to be white in **colour**, delicately flavoured and very tender. In comparison, wild quail meat can be very richly flavoured, but it can also be tough; consequently, it benefits from application of marinades or slow pot-roasting, which soften the meat.

Quails Several species of migratory, short-tailed birds belonging to the Phasianidae family, which are hunted for **quail meat** or farmed for production of quail meat and **quail eggs**.

Quality assurance Planned and systematic actions necessary to provide adequate confidence that goods or services will satisfy given requirements. For the food industry, this is a customer-focused management system, whose aim is to guarantee food safety and consistent product quality by application of production, processing and handling standards. Proactive food safety programmes, in particular those based on **Hazard Analysis Critical Control Point** (HACCP) principles, are the foundation of many food quality assurance systems.

Quality control A system of maintaining standards in manufactured products by testing a sample against the specification.

Quantitative descriptive analysis Comprehensive system used in **sensory analysis** that covers sample collection, assessor screening, vocabulary development, testing and data analysis. Quantitative descriptive analysis (commonly abbreviated to QDA) uses small numbers of highly trained assessors. Once the training sessions have established satisfactory panel performance, and removal of ambiguities and misunderstandings, the test samples can be evaluated. This is carried out in replicated sessions using experimental designs that minimize biases. Three major steps are required: development of standardized vocabulary; quantification of selected sensory characteristics; and analysis of results by parametric statistics.

Quantitative trait loci Location of **genes** that affect traits which can be measured on a quantitative (linear) scale. These traits are usually affected by more than one gene and also by the environment. Examples of quantitative traits are body wt. and plant height. Abbreviated to QTL.

Quarg German **soft cheese** made from **cow milk**. Can be made from whole, skim or semi-skimmed milk or **buttermilk**. **Skim milk powders** are sometimes added, giving a gritty **texture**. Ripens within a few days. The moist, white product has a light taste and is usually sold in pots. Also known as quark.

Quartirolo cheese Italian soft cheese similar to **Taleggio cheese**. Also made widely in Argentina where is is known as Cuartirolo Argentino cheese.

Quassin Triterpenoid produced in the bark of the plant *Quassia amara*. Used as a bittering agent in **foods** and **beverages**.

Quaternary ammonium compounds Cationic surfactant ammonia salts in which the nitrogen atom is bonded to four organic groups. Used as **antiseptics**, **disinfectants** or **preservatives** due to their **antimicrobial activity**. Commonly used for **disinfection** of equipment in **dairies** and **breweries**.

Quercetin Flavonol glycoside distributed widely in **plants**. Found in many **foods** and **beverages**, where it exhibits **antioxidative activity**.

Quercitrin Flavonol glycoside distributed widely in **plants**. Found in many **foods** and **beverages**, where it exhibits **antioxidative activity**.

Queso Blanco cheese Mexican **soft cheese** made from **cow milk**. Traditionally produced from **skim milk** or **whey** coagulated with **lemon juices**. **Flavour** is milky and fresh. Has an elastic **texture** which holds its shape when heated, making it ideal for preparation of dishes such as stuffed chicken breasts, stuffed **peppers**, enchiladas and burritos.

Queso fresco cheese Mexican **soft cheese** made from a mixture of **cow milk** and **goat milk**. Mild, with a fresh acidity and grainy texture; softens but does not melt when heated. Used in cooking and also in **salads**.

Quiches Rich, savoury **tarts** comprising **pastry** cases filled with egg custards containing ingredients such as **vegetables**, **meat**, **cheese** and **sea foods**.

Quillaja saponins Group of **saponins** derived from the tree *Quillaja saponaria*. Non-ionic **surfactants** with good resistance to salt and heat and high stability at acid pH. Used as **foaming agents** in **foods** and **beverages**.

Quinalphos Insecticide and acaricide used for control of a wide range of insect **pests** on **fruits**, **vegetables** and **cereals**. Classified by WHO as moderately toxic (WHO II). Also known as bayrusil and ekalux.

Quince jams **Jams** made using **quinces**.

Quince juices **Fruit juices** extracted from **quinces** (*Cydonia oblonga*).

Quinces **Fruits** produced by *Cydonia vulgaris*. Resemble **pears** in appearance. Good source of potassium and **vitamin C**. Unpalatable when raw, but have a good **flavour** when cooked. Used to make **jams**, **jellies**, **marmalades** and **flavourings**.

Quinic acid Organic acid that, together with **caffeic acid**, is a constituent of **chlorogenic acid**, an antifungal metabolite found in certain higher **plants**. Quinic acid can interact with **proteins**, influencing their function and **digestibility**.

Quinine Bitter alkaloid isolated from cinchona bark, derivatives of which are used in the treatment of malaria. Also used as a bittering agent in **carbonated beverages**, especially **tonic waters**, although high doses are thought to be toxic.

Quinoa A pseudocereal comprising the high protein dried **fruits** and glutinous **seeds** of the plant *Chenopodium quinoa* or *C. album*, which is native to Chile and Peru. Used to make **flour** and **bread**. Rich source of **iron** and **vitamin B_1**.

Quinolones Group of synthetic **antibiotics** used to combat a wide range of diseases in animals and farmed fish. Commonly used examples include **oxolinic acid** and **nalidixic acid**, which show activity against **Gram negative bacteria** only. Second generation quinolones (containing a flourine or piperazino moiety) show broader **antibacterial activity**; examples include **ciprofloxacin** and sarafloxacin.

Quinones Aromatic dioxo compounds that are usually coloured and are constituents of many natural **pigments**; intermediate products of **enzymic browning**. Their derivatives include the K vitamins. They function in aerobic and anaerobic electron transport chains, in photosynthesis, and as carriers of reducing equivalents between dehydrogenases and terminal enzyme complexes.

Quito orange Alternative term for **naranjilla**.

Quorn Trade name for textured **mycoprotein** obtained from the filamentous fungus *Fusarium graminearum* A3/5. Commonly used as **meat substitutes**, e.g. in **sausages** and **ready meals**, or sold as unflavoured chunks or mince for use in home cooked dishes. Originally conceived as a protein-rich food, now usually promoted as a healthy food that is high in **fibre** and low in **calories** and saturated **fats**.

R

Rabadi Traditional fermented food of India, prepared by **fermentation** of a mixture of flour made usually from **pearl millet**, and **buttermilk**. Cereal flour may be partially substituted by that prepared from **soybeans** or other **vegetables**.

Rabbitfish **Marine fish** species (*Chimaera monstrosa*) found in the northeast Atlantic. Of little commercial value, but **livers** are sometimes utilized as a source of **oils**.

Rabbit meat **Meat** from wild or farmed **rabbits**. Rabbit **carcasses** have a high meat to bone ratio; a high proportion of the carcass is edible meat. Meat from young rabbits tends to be more tender and succulent than meat from older rabbits. Rabbits are sold whole or jointed into back legs, forelegs, saddle and fillets. The highest quality meat is found in the rabbit thigh. Farmed rabbit carcasses tend to be larger than those of wild rabbits. Farmed rabbit meat tends to be whiter in **colour**, is covered by a thin layer of fat and is generally more tender, more delicately flavoured and juicier than wild rabbit meat. Wild rabbit meat is very lean and, consequently, can be tough and dry when cooked.

Rabbits Burrowing, plant-eating mammals belonging to the Leporidae family, that are farmed and hunted for **rabbit meat** and fur production.

Rabri Concentrated and sweetened **buffalo milk** product with a flaky/layered **texture**. Popular in India. Traditionally, milk standardized to 6% fat is heated at approximately 90°C with repeated removal of clotted cream (malai), sugar is added to the concentrated milk and finally the clotted cream is added back to the concentrated sweetened milk. In a commercial method, shredded **chhana** or **paneer** is used in place of clotted cream. Rabri has a relatively short **shelf life**.

Racemases Includes members of subclass EC 5.1. These enzymes catalyse the racemization of a centre of chirality and are subdivided according to their substrates; **amino acids** (EC 5.1.1), hydroxy acids (EC 5.1.2) and other compounds (EC 5.1.99).

Racking Process of drawing off **wines** or **beer** from the sediment in the barrel.

Ractopamine β-Adrenergic agonist which increases nitrogen retention and protein synthesis, enhances lipolysis, suppresses lipogenesis and increases rate of weight gain and feed conversion efficiency in farm animals. Rapidly absorbed and eliminated from animal tissues; residues rarely persist in any organs beyond 10 days.

Radiation Energy emitted in the form of electromagnetic waves or subatomic particles.

Radicals Highly reactive molecular species which possess an unpaired electron. Often formed by the splitting of a covalent bond. May react with macromolecules (especially **DNA** and **proteins**), causing them damage.

Radical scavenging activity Ability to trap organic **free radicals** formed by the splitting of molecular bonds. This protects cellular membranes from oxidative destruction and ultimately prevents **DNA** damage caused by the action of the radicals which can lead to **carcinogenesis**. Substances with high radical scavenging activity include antioxidant **vitamins**, such as **α-tocopherol**.

Radioactivity Emission of ionizing radiation or particles caused by the spontaneous disintegration of atomic nuclei.

Radioelements Elements that undergo spontaneous disintegration of their nuclei with the emission of subatomic particles (α-particles and β-particles) or electromagnetic rays (X-rays and γ-rays).

Radiofrequency Electromagnetic wave frequency between audio and infrared. Radiofrequency technology is used in a number of food processing applications, including heating, drying, **tempering**, **defrosting** and **pasteurization**.

Radioimmunoassay Immunological technique in which a substance is measured by its ability to compete with a radioactively labelled form for binding to specific antibodies. Concentration of the substance is determined by comparing inhibition of binding with that caused by a series of standards.

Radioisotopes Isotopic forms of elements that are radioactive and undergo radioactive decay, properties that make them useful in various **analytical techniques** and for studying metabolic pathways.

Radiometry Technique for measurement of incident radiation using radiometers that can be tuned to specific frequencies.

Radionuclides Radioactive species of atoms that decay into products that themselves decay, the sequence of which constitutes a radioactive series.

Radishes Common name for *Raphanus sativus*, the fleshy roots of which are consumed. Roots vary in **colour**, size and shape. Western or small radishes, which contain moderate amounts of **vitamin C**, are generally eaten raw to add **colour**, crispness and **pungency** to **salads** and **sandwiches**. Oriental radishes, such as **Japanese radishes**, produce very large roots which are+ sold in the UK as mouli or rettich. Other types of radish include rat-tailed radishes, which produce edible pods, and leaf radishes, which are grown for fodder.

Radish sprouts Sprouts formed by germination of radish seeds. Rich source of **vitamin C**, **vitamin A**, **calcium** and **folic acid**. Eaten raw in **salads** and **sandwiches**. Have a pungent, peppery **flavour**.

Radium Radioactive element with the chemical symbol Ra.

Radon Radioactive element with the chemical symbol Rn.

Radurization Low-level ionizing radiation treatment designed to enhance the **shelf life** of food by reducing the level of **spoilage microorganisms** present.

Raffinose Oligosaccharide composed of 3 sugar residues, i.e. **fructose**, **glucose** and **galactose**. Considered one of the **antinutritional factors** in **legumes** due to its tendency to cause flatulence.

Raftiline Preparation consisting of **inulin** extracted from **chicory** roots. Used as a fat substitute, sugar substitute, dietary fibre and bulking agent. Applications include low fat **dairy products**, **bakery products** and **processed foods**.

Raftilose Registered trade name of an oligofructose sweetener produced by partial enzymic hydrolysis of **chicory inulin**. Marketed by Orafti Active Food Ingredients, a Belgian food company.

Ragi Cereal plant, *Eleusine coracana*, that is an important food grain in India and Africa. Used in **porridge** and **gruel**, and to make **beer**. Alternative term for **finger millet**. Also the Indonesian name for fermented and dried balls of roasted **rice flour** (other flours may be used as a substitute, e.g. cassava or millet) that contain a mixture of **microorganisms** and are used as **starters** for **fermented foods** such as **tape**.

Ragout Richly seasoned dishes made by stewing **meat**, usually with **vegetables**.

Ragusano cheese Italian **hard cheese** made from raw **cow milk**. **Curd** is heated and stretched until it becomes rubbery before being pressed and left to dry. During **ripening** the **cheese** is rubbed with oil and **vinegar** giving a strong savoury **flavour** to the mature product.

Rahat Alternative term for **Turkish delight**.

Rainbow trout Salmonid **fish** species (*Oncorhynchus mykiss*) predominately found in freshwater; indigenous to geographical areas linked to the East Pacific Ocean, but introduced worldwide. An important food fish with high commercial value; cultured in large numbers around the world. Marketed and consumed in a variety of forms, including fresh, frozen, smoked and canned.

Raising agents **Bakery additives** that are used for chemical leavening of **cakes**. Raising agents, such as baking powders (mixtures of **tartaric acid** and **sodium bicarbonate**), produce CO_2 on addition of liquid, such as water or milk. On **baking**, the gas bubbles expand but are trapped by the protein and **starch** of the flour, and become set as the liquid in the cake mix evaporates.

Raisins Dried **grapes**, usually made from Thompson seedless grapes. Prepared by sun or mechanical **drying**. Rich in **iron** with a high **sugar** content and a range of **vitamins** and **minerals**. Eaten out of hand or used in **bakery products** and various dishes. Golden raisins are amber in **colour** due to treatment with **sulfur dioxide**, and are dried with artificial heat, giving a plumper and moister product that is preferred to common raisins for **cooking**. Muscat raisins are dark and sweet and used in fruitcakes.

Raki Aniseed flavoured **spirits** made in Turkey.

Rakia Spirits made from **grapes** or other **fruits** in Bulgaria and adjacent regions.

Rakkyo Common name for Allium chinense, a plant grown for its bulbs, that resemble small **shallots**. Eaten raw or cooked, but most commonly used for **pickling**.

Raman spectroscopy Technique based on measurement of scattering of incident light from a laser upon striking the sample. Raman scattered light is of a different wavelength from the incident light. The difference in energy between the incident light and the Raman scattered light is the energy required to make a molecule vibrate or rotate. A Raman spectrum is built up of the energy difference at different intensities, with clear bands representing functional groups. From this information, it is possible to determine the structure of compounds present in the sample.

Rambutan **Fruits** produced by *Nephelium lapaceum*. Rich in **vitamin C**. The outer skin is covered with red or yellow spines and encloses the edible white to pink flesh, in the centre of which is a seed. **Flavour** varies from sweet to acid according to **cultivar**; the former are eaten fresh and the later cooked. Fruits are also available canned.

Ram meat   Meat from **rams** (adult male **sheep**), alternatively known as **mutton**. When produced from early maturing breeds, carcass and eating qualities tend to be good. However, meat from older rams tends to be darker in **colour** and may have an undesirable **aroma** and **flavour**.

Ram muscles   Alternative term for **mutton**.

Ramp   Common name for *Allium tricoccum*, a pungent vegetable also known as wild leek. An aroma similar to **onions** is combined with a strong **garlic** flavour. Culinary and medical applications are similar to those of garlic.

Rams   Uncastrated adult male **sheep**. Although often kept solely for breeding, they may be reared for production of **mutton**. They produce lean meat more efficiently than female or castrated male sheep.

Rancidity Sensory properties relating to the extent to which the **flavour** of a product containing fats or oils is perceived to be rancid (sour or stale). Caused by oxidation of **unsaturated fatty acids** in **fats** and **oils**, resulting in the characteristic disagreeable flavour and **aroma**. Occurs slowly and spontaneously, and is accelerated by light, heat and certain minerals. Rancidity in foods may be prevented by proper storage, and/or the addition of **antioxidants**. **Peroxide values** are used as a measure of rancidity of oils and fats.

Randomly amplified polymorphic DNA   Technique usually abbreviated to **RAPD**.

RAPD   Amplification of randomly selected genomic sequences by **PCR** under low stringency conditions using arbitrary primers. Can be used to determine taxonomic identity, study genetic diversity, generate probes and analyse mixed genome samples. Abbreviation for randomly amplified polymorphic DNA.

Rapeseed meal   Residue remaining after **rapeseed oils** have been extracted from **rapeseeds**. Rich in **proteins** and **minerals**, but use in foods is limited due to the presence of **antinutritional factors**, such as **glucosinolates**.

Rapeseed oils   Oils extracted from **rapeseeds**, *Brassica napus*. Rich in **erucic acid**, although varieties producing oils low in erucic acid have been developed. Rich in **monounsaturated fatty acids** and low in **saturated fatty acids**. Often used as cooking oils. Also known as canola oils.

Rapeseeds   Seeds produced by *Brassica napus* and used as a source of **rapeseed oils**. Also known as canola seeds.

Raphia   Genus of **palms**. Stems of some species are the source of **palm wines**.

Ras cheese   Egyptian hard cheese made from **cow milk**, **buffalo milk** or a mixture of both, raw or pasteurized.

Rasogolla   Sweetened dairy product prepared from **chhana**. Chhana is mixed with **flour** and other constituents, divided into balls and cooked in **sugar syrups**.

Raspberries   **Berries** produced by some species of the genus *Rubus*. *R. idaeus* produces red berries, although it has some less common yellow-fruited cultivars. *R. occidentalis* produces black fruit, while purple berries are produced by hybrids. Rich in **vitamin A** and **vitamin C**. Eaten out of hand and used in making **desserts**, **jams**, **jellies** and **beverages**.

Raspberry juices   **Fruit juices** extracted from **raspberries** (*Rubus idaeus*).

Ravioli   Small square parcels of **pasta** which are stuffed with **meat mince** or **cheese** and usually served in tomato-based **sauces**.

Raw milk   **Milk** that has not been heat treated to destroy disease or spoilage causing **microorganisms**. Used to make some products, especially **cheese**, but not usually drunk. Sale of raw milk for drinking is prohibited in many countries. Also called unpasteurized milk.

Ray   General name used for a number of flattened **marine fish** species in the order Rajiformes; worldwide distribution. Generally used synonymously with **skate**. Many species are utilized as food fish, including *Raja clavata* (thornback ray), *R. asterias* (starry ray) and *Leucoraja fullonica* (shagreen ray). Flesh tends to be firm and white with a sweet **flavour**; fins may also be consumed. Marketed fresh, frozen, smoked and salted.

Raya seeds   Seeds extracted from *Brassica juncea* or *B. carinata*. Potential use as **oilseeds**.

Razor shells   Any of a number of marine bivalve **molluscs** with elongated shells. Found in sediments on Atlantic and Pacific shores. Some species are consumed, including *Siliqua patula* (razor clam) and *Ensis ensis* (pod razor).

RDA   Abbreviation for **recommended dietary allowance**.

RDI   Abbreviation for **recommended daily intake**.

Ready meals   **Convenience foods** prepared industrially to a set meals recipe usually by cook freeze or **cook chill processing**, and requiring no further preparation by the consumer other than reheating.

Ready to eat foods   **Convenience foods** that require no further preparation by the consumer, such as **fast foods**, **food bars**, **ready to eat meals** and

ready to eat cereals. Similar to **ready to serve foods**.

Ready to eat meals Convenience foods in the form of **meals** that require no further preparation by the consumer. Similar to **ready meals**.

Ready to serve foods Convenience foods requiring no further preparation by the consumer, other than reheating where appropriate. Examples include ready to serve **dairy desserts**, **gravy**, **salads**, **soups** and **beverages**. Similar to **ready to eat foods**.

Rearing Agricultural term relating to breeding and raising of animals as sources of foods.

Recombinant enzymes Enzymes produced by recombinant DNA techniques. **DNA** encoding the enzyme of interest is manipulated *in vitro* and transformed into an appropriate cell type where it is expressed.

Recombinant microorganisms Microorganisms that contain **DNA** produced as a result of the independent assortment and crossing over of DNA or **genes**, from different sources, during **genetic engineering**.

Recombinant proteins Proteins produced by recombinant DNA techniques. **DNA** encoding the protein of interest is manipulated *in vitro* and transformed into an appropriate cell type where it is expressed.

Recombination Process similar to **reconstitution**, but involving addition of substances other than water which have been removed from the product. Examples include addition of **butterfat** as well as water to **dried skim milk** to make **recombined milk** of the desired fat content.

Recombined foods Products made in a similar way to **reconstituted foods**, but with the addition of substances other than water which have been removed from the product in its original form during processing. Examples include **recombined milk**, made by addition of **butterfat**, as well as water, to **dried skim milk** to achieve the desired fat content in the final product.

Recombined milk Product made by reconstituting **dried milk** with water and other components such as a fat source (e.g. **butter**) to give a composition similar to that of milk.

Recommended daily intake Amounts of nutrients greater than the requirements of almost all members of the population, determined on the basis of the average requirement plus twice the standard deviation, to allow for individual variation in requirements and thus cover the theoretical needs of 97.5% of the population. Commonly abbreviated to RDI.

Recommended dietary allowance Levels of intake of essential **nutrients** that, on the basis of scientific knowledge, are judged to be adequate to meet the nutrient needs on all healthy people. Allowance is used to avoid the implication that these are absolute standards and to emphasize that the levels of nutrient intake recommended are based on a consensus of scientific opinion and should be re-evaluated periodically as new information becomes available. Dietary allowances are higher than physiological requirements in order to allow for a safety factor, which considers **bioavailability** of the nutrients and to allow for individual variations. Commonly abbreviated to RDA.

Reconstituted foods Foods that have undergone **reconstitution** before consumption, often by addition of a liquid. Examples include **soups** and **bakery products** made from **mixes**, and **fruit juices** made from concentrates.

Reconstituted meat products Alternative term for **restructured meat products**.

Reconstitution Restoration of a product to its original state and **consistency**, often achieved by adding a liquid, usually water. Includes addition of water to concentrates and powders.

Rectification One of two general methods, the other being simple **distillation**, used to separate a substance or a mixture of substances from a solution through **vaporization**. Distillation usually involves boiling a liquid and condensing the vapour that forms in a still. In simple distillation, all the distillate is removed from the still after collection. In rectification, part of the distillate flows back into the still. This portion comes into contact with the vapour being condensed and enriches it. Rectification can also be undertaken using large towers (fractionating columns). As the mixture to be separated is heated, its vapours rise through these columns. Substances that boil at the lowest temperatures form the first fractions. Their vapours rise highest and are carried off by pipes near the tops of the fractionating columns. Separate pipes carry off different fractions at various levels. Reflux (return) of some distillate to the columns produces the most efficient conditions for this method of distillation. Rectification can be carried out with a continuous feed of liquid. During manufacture of **vodka**, by-products of distillation, such as methanol, are removed from the distillate by rectification using a continuous still.

Recycling Reuse of renewable resources in an effort to maximize their value, reduce waste, and reduce environmental disturbance. Food packaging wastes such as **paper**, **glass** and **plastics** are often recycled.

Red beans Dark red **beans** used in making chilli con carne and refried beans. Also used in red beans and rice, a dish that is popular in the southern states of the USA.

Red beets Roots of some varieties of *Beta vulgaris*. Eaten cooked as a vegetable and in salads. Also available canned. Leaves are sometimes consumed as a pot herb.

Red cabbages Cabbages containing **anthocyanins** as **pigments**, giving them a red **colour**. Rich source of **vitamin C**. **Flavour** is generally milder and sweeter than that of other types of cabbage. Eaten as a cooked vegetable or raw in **salads** and **coleslaw**. Also popular for **pickling**.

Red crabs Common name used for several species of marine **crabs** occurring along Pacific coasts, principally *Pleuroncodes planipes* and *Chaceon quinquedens*. Marketed in a variety of forms, including fresh cooked whole crab, cooked leg meat, canned meat and pastes.

Redcurrant juices **Fruit juices** extracted from **redcurrants** (*Ribes rubrum*).

Redcurrants Red **berries** produced by *Ribes rubrum*. Rich in **vitamin C**. Eaten out of hand or used as components of **preserves**, **jellies** and **sauces**, especially Cumberland sauce.

Reddish pink soft exudative defect Commonly abbreviated to RSE defect, a condition which affects **pork**. RSE describes one of the four quality conditions into which most pork can be categorized. **Colour** of RSE meat tends to be normal. However, the meat has a poor **water holding capacity**, and drip loss is far greater than in normal red, firm, non-exudative (RFN) pork. Mishandling of swine pre-slaughter increases the incidence of the RSE defect, but the defect is not associated with any particular **halothane** genotype.

Redfish Name given to several different **marine fish** species, but most commonly refers to *Sebastes* spp. Used as a synonym for **rockfish**. In Australia, the name refers to *Centroberyx affinis*. Marketed fresh and frozen.

Red gram Alternative term for **pigeon peas**.

Red hake **Marine fish** species (*Urophycis chuss*) belonging to the family Phycidae. Found in the western North Atlantic Ocean. Marketed fresh, frozen and dried/salted and cooked in various ways, including **steaming** and baking. Small fish are also used in **fish meal** production.

Redox potential Scale of values, measured as electric potential in volts, indicating the ability of a substance or solution to cause reduction or oxidation reactions under non-standard conditions.

Red peppers Term applied to any of several types of red coloured **hot peppers**, such as **chillies**. Also may refer to red **bell peppers**, a milder variety of *Capsicum*.

Red salmon Alternative term for **sockeye salmon**.

Red sea bream **Marine fish** species (*Pagrus major*) distributed around the northwest Pacific. Popular food fish which fetches high prices in Japan; cultured in some coastal regions. Marketed live, fresh, frozen and as a spice-cured product.

Redspot emperor **Marine fish** species (*Lethrinus lentjan*) of the family Lethrinidae which is of high commercial importance. Widely distributed in the Pacific Ocean. Also known as pink ear emperor.

Reducing agents Chemicals capable of the **reduction** of other chemicals, i.e. they donate electrons or hydrogen. During this process, the reducing agents themselves undergo **oxidation**. Also known as **reducing substances**.

Reducing substances Alternative term for **reducing agents**.

Reducing sugars **Sugars** with free aldehyde or ketone groups available for oxidation to form carboxylic acid groups. Reducing sugars are substrates for **Maillard reaction** with **amino acids**. Examples include **glucose**, **maltose**, **lactose** and **mannose**.

Reduction Loss of oxygen from a compound, e.g. removal by **reducing agents**. Also includes reactions in which atoms in the reacting materials gain electrons. Oxidation-reduction reactions always occur simultaneously; if one reactant is oxidized, another must be reduced.

Reductones Chemicals that contain an enediol group, e.g. **ascorbic acid**. Intermediates of the **Maillard reaction** which possess **antioxidative activity**.

Red wines Wines which are red in **colour**, due to the presence of **anthocyanins** extracted from the skins of red **winemaking grapes**. Thought to have beneficial effects on health due to the anthocyanins content.

Reference materials Materials of certified composition that are used as standards in analytical procedures.

Refining Removal of impurities or unwanted elements from a substance. Often used to describe the processing of **sugar** and **oils**.

Reflectance **Optical properties** relating to the measure of the proportion of light or other radiation falling on a surface which is then reflected or scattered.

Reflectivity **Optical properties** relating to the amount of light or other radiation that can be reflected by an item. Rough surfaces reflect in a multitude of directions, and such reflection is said to be diffuse. Smooth, brightly polished or glossy surfaces reflect clearly and sharply at the same angle to the surface as the angle at which the light or heat contacted the surface. **Reflectometers** are instruments used for measuring the luster or sharpness of reflection of a finished surface.

Reflectometers Instruments used to measure the **colour** or **gloss** of **foods** based on their **reflectance** of **light**.

Refractive index Measure of the bending or refraction of a beam of light on entering a denser medium (the ratio between the sine of the angle of incidence of the ray of light and the sine of the angle of refraction). Constant for pure substances under standard conditions. For example, refractive index is used as a measure of **sugar** or **total solids** in solutions, and in determining the **purity** of **oils**.

Refractometry Measurement of **refractive index** using one of the several types of refractometer.

Refrigerants Substances with low vaporization temperatures used to promote the **refrigeration** conditions necessary for **chilling** foods and **beverages**. Examples of refrigerants include **Freons**, ammonia, ice and solid **carbon dioxide**.

Refrigerated foods **Chilled foods** requiring **refrigeration** prior to consumption.

Refrigerated storage Process of keeping objects, usually foods, at a temperature that is significantly lower than that of the surrounding environment in order to extend their **shelf life** by a few days.Refrigeration or cold storage of foods is a gentle method of **preservation**, having minimum adverse effects on **flavour**, **texture** and **nutritional values**. Refrigeration keeps **spoilage** reactions (microbial or enzymic) to a minimum, but does not kill **microorganisms** or inactivate **enzymes**, instead slowing down their deteriorative effects. Household **refrigerators** are usually run at a temperature of 4-7°C. Commercial refrigerators are operated at a slightly lower temperature.

Refrigerated transport Specially designed transport vehicles, such as lorries, rail cars, aeroplanes or cargo ships, with refrigeration systems on board which are designed to protect frozen and perishable foods from high ambient temperatures. The refrigeration systems also cool the hot air mass in the cargo container, and remove the stored heat from the structure of the cargo body. Product integrity is maintained through avoidance of temperature fluctuation.

Refrigeration Process by which heat is removed from an enclosed space or from a substance for the purpose of lowering the temperature. Refrigeration is chiefly used to store foods and **beverages** at low temperatures, thus inhibiting the destructive action of **microorganisms**. **Cooling** caused by the rapid expansion

of gases (**refrigerants**) is the primary means of refrigeration.

Refrigerators Appliances or compartments kept artificially cool by the use of **refrigerants**, and which are used to store foods and **beverages**.Mechanical refrigerators have four basic elements: an evaporator; a compressor; a condenser; and a refrigerant flow control (expansion valve). A refrigerant circulates among the four elements, changing from liquid to gas and back to liquid. In the evaporator, liquid refrigerant evaporates under reduced pressure, so absorbing latent heat of vaporization and cooling the surroundings. The evaporator is at the lowest temperature in the system and heat flows to it. This heat is used to vaporize the refrigerant. The refrigerant vapour is sucked into a compressor, a pump that increases the pressure and then exhausts it at a higher pressure to the condenser. To complete the cycle, the refrigerant must be condensed back to liquid and in doing this it gives up its latent heat of vaporization to a cooling medium such as water or air.

Regenerated cellulose Alternative term for **cellophane**.

Reggianito cheese Argentinean hard cheese made from **cow milk**. Similar to **Parmigiano Reggiano cheese**. Used mostly for grating, in cooking or in toppings on **pasta** dishes.

Reheating Application of heat to a food that has already been thermally processed but then cooled. **Cook chill foods** and **ready meals** often need reheating before consumption.

Rehydrated foods Products made by **reconstitution** of **dried foods**, e.g. **dried vegetables**, with water.

Rehydration Process by which the water or moisture removed in making **dried foods** is replaced, so restoring it to near its original quality.

Reindeer Large migratory ruminant animals (*Rangifer tarandus*) belonging to the Cervidae family. Wild reindeer are hunted and domesticated reindeer are farmed as a source of **reindeer meat**. In some countries, e.g. Sweden, careful management of wild reindeer herds produces a regular crop of animals that can be culled for meat production. Reindeer meat is sometimes referred to as **venison**.

Reindeer meat **Meat** from **reindeer**. It has a low content of fat. In farmed or harvested reindeer, stress during gathering, herding, selection, feeding, road transport and lairage may result in glycogen depletion and hence deterioration of meat quality.

Relative density Ratio of the **density** of a substance to the density of a reference material. For liquids or solids, relative density is the ratio of the density (usually at 20°C) to the density of water (at its temperature of maximum density (4°C). Synomym for **specific gravity** (sp. gr.).

Relative humidity The moisture content of air expressed as the percentage of the maximum possible moisture content of that air at the same temperature and pressure. Commonly abbreviated to RH.

Relishes Pickles or **condiments** with a strong, usually piquant, **flavour** that are served as an accompaniment to foods.

Renaturation Reconstruction of **proteins** or **nucleic acids** that have previously been denatured, such that the molecules resume their original function. Some proteins can be renatured by reversing the conditions that brought about **denaturation**.

Rendering Process applied on a large scale to production of **animal fats** such as **tallow**, **lard**, bone fat and **whale oils**. Consists of cutting or chopping the fatty tissue into small pieces that are boiled in open vats or cooked in steam digesters. The fat gradually liberated from the cells floats to the surface of the water, where it is collected by skimming. Membranous matter is separated from the aqueous phase by pressing in hydraulic or screw presses; in this way, additional fat is obtained. Centrifuges may also be employed in rendering. Cells of the fatty tissues are ruptured in special disintegrators under close temperature control. The protein tissue is separated from the liquid phase in a desludging type of centrifuge, following which a second centrifuge separates the fat from the aqueous protein layer. Compared with conventional rendering, centrifugal methods provide a higher yield of better-quality fat, and the separated protein has potential as an edible meat product.

Rennetability The ease with which **milk** is coagulated using **rennets**.

Rennets Enzymes used to cause **coagulation** of milk during **cheesemaking**. Traditionally extracted from the abomasum of young ruminants, mainly calves (**animal rennets**, **calf rennets**), but alternative forms (e.g. **microbial rennets**, **vegetable rennets**) are now used due to shortages of this type of preparation. The active enzyme is **chymosin**, but pepsin is also present.

Rennet substitutes Enzymes used as alternatives to **animal rennets** for **coagulation** of **milk** during **cheesemaking**. Developed due to shortages of the animal products and in cases where a vegetarian cheese is desired. Substitutes include **microbial rennets** and **vegetable rennets**.

Rennin Alternative term for **chymosin**.

Reporter genes Genes encoding easily assayed products under the control of regulatory elements from other genes. Regulation and localization of expression of the gene of interest can then be studied following transformation into appropriate cells. Examples include the genes encoding luciferases, **β-galactosidases**, chloramphenicol acetyltransferases and green fluorescent proteins.

Resazurin A member of the quinone-imine group of **dyes** that is blue when fully oxidized but is reduced irreversibly to the pink-coloured resorufin when the redox potential is lowered sufficiently. On further reduction, the colourless hydroresorufin is formed. Measurement of resazurin reduction time can be used to determine the **microbiological quality** of **raw milk**.

Reservoir water Water stored in reservoirs. Commonly intended for purification to **drinking water** quality and distribution via the water supply system.

Residence time distribution Distribution of times spent by the various components of a food product through a process vessel. Residence time distribution (RTD) is a critical factor affecting the sizing of holding tubes for **aseptic processing** of particulate foods. Also, in design of continuous **sterilization** equipment for liquid food processing, knowledge of **flow** characteristics, especially residence time distribution, is of prime importance.

Residues Food contaminants derived from a variety of sources, including agricultural chemicals (e.g. **pesticides** and **fertilizers**), veterinary **drugs**, environmental pollution and manufacturing processes.

Resins Group of organic chemicals, usually polymers, which are solid or semi-solid and have high electrical resistance. Used as **chromatography** support materials and for manufacture of **plastics**, including those used as food **packaging materials**, e.g. epoxy resins used for coating of food **containers**.

Resistant starch Starch which is resistant to digestion in the **gastrointestinal tract**. Resistance may be conferred by: protection by a physical barrier, such as plant cell walls, e.g. starch in seeds and legumes; the highly crystalline nature of some **starch granules**, such as those in **bananas**; **retrogradation** of starch in cooked foods; and modification of starch. Regarded as a source of **dietary fibre**.

Resistographs Instruments similar to **farinographs** used to study **rheological properties** of **dough**, and thus evaluate **flour** quality.

Resorcinol Resorcinol and its derivatives are used as **preservatives** in foods, where they exhibit **antioxidative activity** and inhibit **enzymic browning**, and for **stabilization** of **vitamin D** and **vitamin E**. Derivatives of this phenolic compound are also useful in the development of high performance **packaging materials**.

Respiration Metabolic process in animals and plants by which organic substances are broken down into simpler products with the release of energy, which is incorporated into **ATP** and subsequently used for other metabolic processes. In most plants and animals, respiration requires oxygen (aerobic respiration), and carbon dioxide is an end product. Anaerobic respiration is the breakdown of food components such as **glucose** to yield energy in the form of ATP in the absence of oxygen. Anaerobic respiration in **yeasts** produces **ethanol** as a waste product, a process that is the basis of manufacturing **alcoholic beverages**.

Response surface methodology Collection of statistical and mathematical techniques used in developing and optimizing processes, developing new products and improving existing products. Used particularly where several variables affect the process or properties of the product.

Restaurants Any of a wide variety of commercial **catering** establishments where foods and beverages are prepared and served. Types of restaurants include fast food establishments, **cafeterias**, **canteens** and pub restaurants.

Restriction enzymes Endonucleases isolated from **bacteria** which are used to cut **DNA** molecules into smaller pieces. They attach to DNA at specific nucleotide sequences and, according to the enzyme type, either cut randomly somewhere along the length of the molecule or cleave both strands of DNA within the recognition or restriction site. There are 3 types, designated I (EC 3.1.21.3), II (EC 3.1.21.4) and III (EC 3.1.21.5). Type II **enzymes** are used for restriction mapping. In bacteria, the **enzymes** serve as a defence mechanism, preventing or restricting incorporation of foreign DNA from sources such as **viruses** or **plasmids**. Used as tools in **biotechnology**, particularly recombinant DNA technology, and in classification and identification of **microorganisms**, including those found as contaminants in foods or used in food processing. Restriction enzymes are given two-element names, the first three-letter part referring to the bacterial source and the second part being a Roman numeral differentiating enzyme types and enzymes from the same species or strain.

Restriction fragment length polymorphism Commonly abbreviated to **RFLP**. Variation in the length of **DNA** fragments produced by the action of **restriction enzymes**. A result of changes in the DNA code at the site of action of the enzymes, such as by mutation, insertion or deletion. Employed widely in **genetic techniques** for differentiating between organisms.

Restriction modification systems Bacterial enzyme systems comprising restriction **endonucleases** and DNA methyltransferases (e.g. EC 2.1.1.72 and EC 2.1.1.73). In a given bacterium, the modification enzyme methylates the DNA target sequence for the restriction enzyme produced by that strain, thus protecting its genomic DNA from cleavage by the restriction enzyme it produces. Such systems are of potential interest as defence mechanisms against **bacteriophages** in **dairy starters**.

Restructured meat products Small pieces of **meat** reformed into **steaks**, chops and roast-like meat products. They may be difficult to distinguish visually from the real product. Minced, flaked, diced or **mechanically recovered meat** may be used. Often, massaging and **tumbling** are used to extract salt-soluble contractile proteins from the meat pieces. The pieces become coated with these proteins, which subsequently act as an adhesive when the pieces are thermally processed and compressed. Cohesion of the meat pieces also involves gelation of connective tissue proteins. Also known as reconstituted meat products.

Resveratrol Polyphenol found in **grapes** and **wines** that exhibits **antioxidative activity** and is thought to protect against **cardiovascular diseases**.

Retinal Aldehyde derivative of **vitamin A**, originally isolated from animal retina. Formed in the body by cleavage of **β-carotene** in the intestines. Necessary for night vision. Also known as vitamin A aldehyde, retinene or retinaldehyde, the last form being the preferred alternative if the name is liable to be confused with the adjective meaning pertaining to the retina.

Retinoic acid Biologically active acid form of **retinols**; can partially replace retinols in the rat diet. Promotes growth of bone and soft tissue production. However, has no activity in the visual process or the reproductive system and cannot be stored in the body. Retinoic acid is converted by the rat to an unidentified form that is several times as active as the parent compound in conventional **vitamin A** nutritional assays.

Retinoids Compounds consisting of four isoprenoid units joined in a head-to-tail manner. **Vitamin A** is a generic descriptor for retinoids exhibiting qualitatively the biological activity of retinol. While preformed vitamin A occurs only in foods of animal origin, retinoids such as **β-carotene** are found in both **animal foods** and **plant foods**. Retinoids have many activities in the body, including control of cell proliferation, cell differentiation and embryonic development.

Retinols The alcohol form of **vitamin A**. Vitamin A exists in two forms: retinols, which predominate in mammals and **marine fish**; and dehydroretinols, which predominate in **freshwater fish**. Retinols can be reversibly oxidized. Retinols circulate in the blood as a complex with retinol binding protein and transthyretin.

Retinyl palmitate Natural antioxidant which occurs in plant materials such as **vegetable oils**, **celery seeds**, **aniseed** and **allspice**, as well as in **animal fats**. Used in fortification of foods with **vitamin A**. Also known as vitamin A acetate.

Retorting Thermal process that is part of the food **canning** process. Batch retorts, of a still or agitating type, and designed to operate with saturated steam or hot water, are used. By processing under pressure, it is possible to use temperatures of approximately 121°C (250°F), which greatly speeds up the destruction of **microorganisms** and **spores**.

Retort pouches Flexible **containers**, commonly made from **aluminium foils** and plastic laminates. Can withstand in-package **sterilization** of the product that they contain. Some have zipper-type closures and are resealable.

Retrogradation Process in which gelatinized (disordered) **starch** reassociates to form a more ordered structure; under optimal conditions starch may recrystallize. Occurs during cooling of cooked starch.

Retsina **White wines** flavoured with pine resins, produced mainly in Greece.

Reuterin Broad spectrum antimicrobial substance produced by *Lactobacillus* reuteri. Shows potential for use in natural **food preservatives**.

Reverse micelles Aggregates of small molecules such as **surfactants** which assemble in non-aqueous solutions at levels above the critical micellar concentration. In contrast to normal micelles, hydrophilic components associate in the interior of the aggregates. Widely used to manipulate localized solvent polarity, for example in enzyme catalysis, to provide a hydrophilic environment for the **enzymes** used in an otherwise non-aqueous solvent. Also used for selective extraction from mixed solvent systems.

Reverse osmosis Membrane process, driven by a pressure gradient, in which a membrane separates the solvent (generally water) from other components of a solution. With reverse osmosis, the membrane pore size is very small (0.0001-0.001 micrometers) allowing only small amounts of very low molecular weight solutes to pass through. Even small dissolved molecules, such as salts, are retained by the membrane. At this molecular level, high pressures are required of the order of 10-50 bar because osmotic forces come into play. The largest commercial food applications of reverse osmosis are concentration of **whey** produced as a by-product of **cheese** manufacture and **clarification** of **wines** and **beer**. Reverse osmosis systems are additionally used water processing, such as **desalination** of **sea water**. Also known as hyperfiltration.

RFLP Abbreviation for **restriction fragment length polymorphism**.

RH Abbreviation for **relative humidity**.

Rhamnolipids **Glycolipids** which contain **rhamnose** and 3-hydroxy carboxylic acids. Synthesized by **microorganisms**, especially *Pseudomonas* aeruginosa. Used as **biosurfactants** in the food industry.

Rhamnose One of the **methylpentoses**. A deoxysugar composed of 6 carbon atoms which is a component of **pectins**, **mucilage**, **gums** and bacterial **exopolysaccharides**. Also a common glycoside of plant **pigments** and **flavonoids**. Alternatively known as 6-deoxymannose.

α-L-rhamnosidases EC 3.2.1.40. Hydrolyse terminal, non-reducing α-L-rhamnose residues in α-L-rhamnosides. Can be used to reduce the **bitterness** of **citrus juices**, depolymerize **gellan gums** and release **flavour compounds** in **wines**, thus increasing their **aroma**.

Rhea meat **Meat** from **rheas**. The proportion of lean meat from rhea **carcasses** is similar to that obtained from ostrich, broiler chicken, turkey and cattle carcasses. Rhea meat resembles **ostrich meat**, but it has a lower fat content.

Rheas Large flightless **birds** belonging to the order Rheiformes; *Rhea americana* is known as the Greater Rhea and *Pterocnemia pennata* as the Lesser Rhea. Not only are they farmed in their native South America, but also in other parts of the world, e.g. Australia and South Africa. They are used to produce **rhea meat**, rhea eggs, feathers and skins.

Rheological properties **Mechanical properties** relating to the **flow** of materials. In food technology, rheological properties relate to concepts such as **elasticity**, **rigidity**, **shear**, **stretch**, **thixotropy** and **viscosity**.

Rheology Study of the relation between forces exerted on a material and the ensuing deformation as a function of time. In the food industry, rheology provides a scientific basis for subjective measurements such as **mouthfeel**, **spreadability** and pourability.

Rheometers Devices used for measurement of **viscosity**.

Rhizobium Genus of Gram negative, aerobic rod-shaped **bacteria**. Occur in soil. Often symbiotically associated with the root nodules of certain leguminous plants where they carry out nitrogen fixation.

Rhizoctonia Genus of **fungi** of the class Heterobasidiomycetes. Includes some important plant **pathogens**. *Rhizoctonia leguminicola* causes blackpatch disease of clover, while *R. solani* causes damping off and eyespot of **potatoes**.

Rhizomucor Genus of **fungi** of the class Zygomycetes. Occur in soil. *Rhizomucor miehei* and *R. pusillus* produce **proteinases** which are used as **rennet substitutes** in **cheesemaking**.

Rhizopus Genus of **fungi** of the class Zygomycetes. Occur in soil and on **fruits** and **vegetables**. Some species (e.g. *Rhizopus oligosporus*) are used in the preparation of **ontjom**, **ragi**, **bongkrek** and **tempeh**, while others are used in the manufacture of **mycoprotein** for incorporation into foods. *R. stolonifer* may cause **spoilage** of **fruits**, **vegetables** and **bread**, and other species may be responsible for **meat** spoilage.

Rhodobacter Genus of Gram negative, rod-shaped **bacteria** with vesicular-type photosynthetic membranes; transferred from the genus *Rhodopseudomonas*. Occur in freshwater, marine and hypersaline habitats. Some species, especially *Rhodobacter sphaeroides*, produce **polyhydroxyalkanoates** from food processing **effluents**.

Rhodococcus Genus of Gram positive, obligately aerobic coccoid **bacteria** of the family Nocardiaceae. Occur in soil and aquatic habitats. Species may be used in the synthesis of **carotenoids** or **enzymes**. *Rhodococcus erythropolis* is used in the production of lactone hydrolases for the biotransformation (**debittering**) of **terpenes** in **citrus juices**.

Rhodothermus Genus of thermophilic, halophilic **bacteria**, type species *Rhodothermus marinus*. Occur in shallow marine hot springs. May be used in production of **glycosidases**.

Rhodotorula Genus of yeast **fungi** of the class Urediniomycetes. Occur on plants, plant debris, and in seawater and fresh water. Some species (e.g. *Rhodotorula glutinis*) may cause spoilage of **sauerkraut** and **olives**, while others (e.g. *R. mucilaginosa*) may cause **spoilage** of **meat** and **sea foods**.

Rhubarb Common name for *Rheum raphonticum* or *R. rhabarbarum*. The part of the plant that is eaten is the leaf stalk. Although not botanically a fruit, rhubarb is eaten like a fruit with added sugar in **fruit pies**, crumbles, **tarts** and **preserves**. It is also used in **fruit wines**. The leaf blade contains high levels of **oxalic acid**, which can cause poisoning if this part of the plant is consumed.

Riboflavin Synonym for **vitamin B$_2$** and **vitamin G**. A water soluble vitamin which occurs mainly in **yeasts**, **livers**, **milk**, **eggs**, **cheese** and **pulses**; milk and **dairy products** are probably the most important source in the average diet. Occurs in bound form in plant and animal tissues and is not available unless liberated by **cooking**. Resistant to heat, but readily destroyed in the presence of light and alkali. Involved in a wide range of oxidation reactions, of **fats**, **carbohydrates** and **amino acids**. A constituent of the coenzymes flavine adenine dinucleotide (FAD) and flavine mononucleotide (FMN). Deficiency impairs cell oxidation and results clinically in a set of symptoms known as riboflavinosis.

Ribonucleases Hydrolyse ester bonds within **RNA**, acting as either **endonucleases** or exonucleases. Can be used for production of 5'- and 3'-ribonucleotides, which are useful as **seasonings**.

Ribonucleic acids Full form of the abbreviation **RNA**.

Ribonucleosides Compounds that consist of **purines** and **pyrimidines** linked to **ribose**. Ribonucleosides containing the bases **adenine**, guanine, **cytosine**, **uracil**, **thymine** and **hypoxanthine** are called, respectively, **adenosine**, **guanosine**, **cytidine**, **uridine**, thymidine and **inosine**.

Ribonucleotides **Nucleosides** in which the **ribose** sugar contains one or more phosphates.

Ribose Pentose **sugar** that forms, with phosphate, the backbone for ribonucleic acids.

Ribotype **RNA** constitution of organisms or cells, in an analogous manner to **genotype**.

Ribotyping **Genetic techniques** (e.g. **PCR** and detection of **RNA** gene polymorphism) used to determine and compare the RNA constitution of organisms and cells.

Rice Starchy grains produced mainly by *Oryza sativa* that form a staple food, especially in Asia. **Brown rice**, produced by removal of the **hulls**, is regarded as a healthier food than white rice, as vitamin B and **fibre** contents are reduced by removal of the bran and germ. However, **parboiling** of rice before **milling** increases the nutritional quality of white rice. Rice is eaten in many forms, as an accompaniment, or a component of dishes such as paella or risotto. It is also used to make **breakfast cereals** and **infant foods**, and as the starting material in manufacture of **sake**.

Rice beans **Seeds** produced by *Phaseolus calcaratus*. Usually boiled and eaten with or instead of **rice**.

Rice bran Outer layers of **rice** seeds. Used as a source of **rice bran oils** and **protein concentrates** and as a **fibre** ingredient in **bakery products**.

Rice bran oils Oils with high **oxidative stability** which are derived from the outer layers of the rice grain removed during manufacture of white **rice**. Used widely in Japanese cooking as **salad oils** and **frying oils**. Reported to lower serum cholesterol levels due to high contents of **oryzanols**.

Rice bread **Bread** in which **rice flour** is used as a complete or partial substitute for **wheat flour**.

Rice cakes Cakes made with glutinous **rice** that have a soft **texture**. Also known as **arare**.

Rice crackers Crackers made with non-glutinous **rice** that have a hard, rough **texture**.

Rice flour Broken rice grains that are milled and used in **brewing** and **distillation**, as well as making **puddings** and **bakery products** such as **biscuits** and **cakes**.

Rice germ oils Oils extracted from **rice germ**, a by-product of rice **milling**. Rich in **vitamin E**. Major **fatty acids** are **linoleic acid** and **oleic acid**. Benefits for human health include protection against **cardiovascular diseases**, lowering of high **cholesterol** levels and management of menopausal problems.

Rice koji Product prepared by fermenting steamed **rice** with the fungus **Aspergillus** *oryzae*, which converts the **rice starch** into fermentable **sugars**. Used in manufacture of **sake**.

Rice powders Ingredients derived from roasted **rice** grains that are ground into a powder. Often used in **infant foods**.

Rice starch Starch isolated from **rice**.

Rice vinegar Vinegar made from fermented **rice**. Milder and with a gentler **flavour** than other vinegars made from **fruits** and **wines**. Chinese rice vinegars are available in white, red and black varieties, while Japanese rice vinegars tend to be almost colourless. Used in **salad dressings**, a variety of dishes, including **sushi** rice and sweet and sour **meals**, and in **pickles**.

Rice weevils Common name for *Sitophilus oryzae*, serious pests of stored grain and **seeds**. Develop inside whole grain kernels with no external evidence of their presence. May be transported into the domestic environment in infested whole grains or seeds, e.g. **popcorn** or **beans**.

Rice wines Oriental **alcoholic beverages** made from **rice** based **mashes**. **Saccharification** is by **enzymes** of **starters** containing **fungi**, rather than by **malt** enzymes, as in Western alcoholic beverages based on **cereals**.

Ricin Highly toxic lectin occurring in the seeds of **castor beans**. Consists of a toxic A subunit that inactivates ribosomes, and a B subunit that binds to carbohydrates and is specific for galactosyl residues.

Ricinoleic acid Fatty acid found in **castor oils** and other **vegetable oils**. Useful as a precursor for microbial production of **flavour compounds**.

Ricotta cheese Italian **soft cheese** made from cow or ewe milk **whey**. **Citric acid** is used to facilitate separation of proteins from the whey during heating. **Whey proteins** rise and coagulate, after which they are skimmed off and drained for 2 days when the 'cheese' is ready for market. Varieties of Ricotta include Ricotta Salata Moliterna (made from **ewe milk** whey), Ricotta Piemontese (made from **cow milk** whey + 10% milk) and Ricotta Romana (a by-product of Pecorino Romano cheese production).

Rifamycins Antibiotics belonging to the group of naphthalene-ringed ansamycins. Active against **Gram positive bacteria**. Used for treatment and prevention of **mastitis** in cattle during the dry period; also used topically in cattle, sheep, goats and rabbits for treatment of skin bacterial diseases. Absorption through dermal tissues into bloodstream and edible tissues is usually negligible.

Rigidity Rheological properties relating to the extent to which products (such as food gels, plant cells and meat fibres) are rigid, i.e. solid, firm and inflexible.

Rigor Relates to rigidity or stiffness of muscles, as occurs in *rigor mortis*.

Rigor mortis Stiffening of muscles, which accompanies the *post mortem* loss of ATP and glycogen in muscle fibres; it develops gradually after **slaughter** of animals. The physical changes in muscles accompanying development of *rigor mortis* include a loss of **extensibility** and **elasticity**, shortening, and an increase in tension and **firmness**. Stiffening results from the formation of permanent crossbridges between **actins** and **myosin** filaments in the muscles. *Rigor mortis* does not last indefinitely, as after a period of **ageing** or conditioning, the muscles gradually lose their stiffness; resolution of *rigor mortis* results from physical degradation of the muscle structure. In many species, onset and resolution of *rigor mortis* occur more rapidly following **electrical stimulation** of **carcasses**.

Rigorometers Instruments used to measure *rigor mortis* development in **meat** and **fish** on the basis of muscle tension and length.

Rind Alternative term for **peel**.

Rinsing Washing an item with clean water to remove impurities.

Ripened cream Cream that has been ripened naturally or by **fermentation** with **starters**. Used in making **butter**. Also called **sour cream**.

Ripeness Extent to which crops, such as **fruits** or **vegetables**, or **cheese** are ripe (fully developed and mature), and ready for eating.

Ripening Term used in relation to the maturation of **fruits**, **vegetables** or **cheese**. As ripening proceeds, sensory quality of foods improves. Ripening of fruits and vegetables can involve changes in **colour** and **texture**. **Flavour** development is an important stage during the ripening of cheese.

Risk factors Circumstances (e.g. personal habits, environmental exposure) that lead to the enhanced likelihood of an event occurring (e.g. disease development, food contamination).

Risks assessment Estimation of the probability of adverse effects occurring due to exposure to specified **health hazards** or the absence of preventive or beneficial measures.

Risks management Process of minimizing the probability of adverse effects occurring by developing systems to identify, analyse and prevent hazards.

Rissoles Meat products often prepared from cooked **lamb** or **beef**. The **meat** is minced before mixing with the other ingredients, which may include **onions**, **breadcrumbs**, **garlic**, **eggs**, **herbs** and **seasonings**. The mixture is then divided and shaped into round cake shapes; these are coated in **flour** before **cooking**, commonly by **frying**.

River fish Any **freshwater fish** which exist in riverine environments.

RNA **Nucleic acids** consisting of linked **ribonucleotides**, each of which contains the sugar **ribose**, a phosphate group and one of the bases **adenine**, guanine, **cytosine** or **uracil**. Usually single-stranded but can form duplexes with complementary RNA or **DNA** strands. Constitute the **genomes** of many **viruses**. The major RNA species (**mRNA**, **rRNA** and tRNA) are involved in all stages of the synthesis of **proteins** in eukaryotic and prokaryotic cells. Abbreviation for ribonucleic acids.

RNA polymerases Alternative term for **DNA-directed RNA polymerases**.

RNases Alternative name/abbreviation for **ribonucleases**.

Roach **Freshwater fish** species (*Rutilus rutilus*) distributed in lakes and rivers across Europe. Not a popular food fish, but occasionally sold fresh (whole gutted) or as a dried/salted product.

Roasted coffee **Coffee beans** which have been roasted to develop characteristic **flavour** and **aroma**. The degree of **roasting** required is dependent on the intended style of **coffee beverages** to be prepared.

Roasted foods Foods cooked by dry **heating**, usually with added fats, in **ovens**. **Maillard reaction products** contribute to the characteristic roasted **flavour**.

Roasted peanuts **Peanuts** that have been roasted by conventional oven cooking in-shell or shelled, or by microwave or oil cooking out of their shells. Usually seasoned with **salt** or a variety of other **flavourings**, including **garlic**, **paprika** or **chilli**.

Roasting **Cooking** of foods, e.g. tender pieces of meat or vegetables, by prolonged exposure to heat in **ovens** or over a fire. Roasting usually produces foods with a well-browned exterior and, ideally, a moist interior.

Robiola cheese Italian **cream cheese** made from **cow milk**, **goat milk** or a mixture of both. Eaten fresh or aged. During **ageing**, a pink to brown **colour** develops, which intensifies as the process proceeds. **Flavour** is generally tangy. Different varieties of Robiola vary slightly in characteristics due to differences in manufacturing processes. All are typically served as a table cheese, with **olive oils**, **salt**, **pepper** and sometimes a tomato and **anchovy** based sauce.

Robotics The branch of technology concerned with the design, construction and application of robots used for mechanical operation of procedures.

Rock cod General term used for a variety of **marine fish** species in the order Perciformes. The majority of species occur in coastal waters in rocky and reef habitats. Some species are utilized as food fish; normally marketed fresh or frozen.

Rocket Common name for *Eruca sativa*. Generally used in **salads**, although it can be added to **soups** and dishes such as sauteed **vegetables**. **Flavour** is bitter and peppery. Rich source of **iron**, **vitamin A** and **vitamin C**. Also known by several other names, including arugula, roquette, rugula and rucola.

Rockfish General term used for a range of **marine fish** species found in the Atlantic and Pacific oceans, particularly *Sebastodes* spp. Used as a synonym for **redfish**. Several species are utilized as food fish; normally marketed fresh or frozen.

Rodenticides Chemical substances used for control of **rodents** such as mice and rats. Most act as anticoagulants (prevent blood from clotting), causing death from internal bleeding. Some are used for control of rodents in food preparation and storage areas. Examples include bromadiolone, chlorophacinone and cholecalciferol.

Rodents Mammals of the order Rodentia. Occupy a wide range of terrestrial and semiaquatic habitats worldwide. Many species may be **pests** of stored foods (e.g. rats and mice), while others are used as human food (e.g. agoutis).

Roes Eggs from **marine fish** or **freshwater fish**. May also refer to the entire gonads of female fish or to the gonads of male fish (also known as soft roes). Marketed and sold in a variety of forms, including **caviar** (from **sturgeon**), **caviar substitutes** (from a range of fish species), dried/salted and smoked products.

Rohu **Freshwater fish** species (*Labeo rohita*) of high commercial value belonging to the **carp** family

(Cyprinidae). Widely distributed throughout Asia. Marketed fresh.

Rohwurst Raw, fermented, **dry sausages**, conventionally made from frozen raw materials. Rohwurst may be prepared from red meat or **poultry meat**, and may contain pork fat. They may be produced by rapid, moderate or slow fermentation. Addition of starter cultures/micrococci cultures to the sausage emulsion considerably improves reddening of rohwurst. Modern curing methods use starter cultures and additives, such as **glucono-δ-lactone**, to accelerate gel formation; however, these methods tend to cause **flavour** losses in comparison with traditional **curing** methods. Natural and synthetic **flavourings** may be used to enhance the spicy characteristics of the **sausages**.

Roller drying Type of web **drying** in which the material to be dried makes a sinusoidal path around rollers while heat is supplied externally by blowing air.

Roller mills **Mills** that crush or pulverize items by means of rollers that move the material and press it against the sides of a revolving bowl.

Rolling Flattening of an object by passing a roller over it or by passing it between rollers. During **baking**, a rolling pin is used to flatten **dough** into a thin, even layer.

Rollmops Fish products consisting of marinated **herring** fillets wrapped around pickled **vegetables** or slices of **onions** and fastened with small sticks or **cloves**. Packed in **brines** and **vinegar**; may also be packed with **spices**, **mayonnaise** or other **condiments**. Marketed as a semi-preserved product, often with added **preservatives**.

Rolls Small rounded portions of **bread** made from **yeasts**-leavened **dough**. May have a soft or crisp **crust**. Also called **bread rolls**.

Rolmops Alternative spelling for **rollmops**.

Romano cheese Italian **hard cheese** made from **cow milk**, **goat milk**, **ewe milk** (Pecorino Romano cheese), or a combination of cow milk with goat or ewe milk. Traditionally from the area around Rome in Italy. Pale yellow in **colour** and used mostly for **grating**. Alternatively, this cheese can be spray dried and used as a powder. **Flavour** varies according to the type of **cheese milk** used in manufacture, but is usually strong, due to the long **ageing** period.

Roncal cheese Spanish **hard cheese** made from raw **ewe milk**. The surface of the hard rind has a layer of blue grey mould which is sometimes covered in olive oil. The beige interior is firm with small irregular holes and has a sweet herby **flavour** which becomes increasingly tangy as the **cheese** ages.

Rooibos tea Type of **herb tea** from South Africa, produced from leaves of the bush *Aspalathus linearis*. Has strong **antioxidative activity**.

Root crops Produce of plants grown for their edible roots.

Rootstocks Part of the stem of a plant into which a bud or scion is inserted for grafting.

Root vegetables Produce of plants with edible roots, e.g. **carrots**, **turnips**, **salsify** and **celeriac**.

Ropiness Condition responsible for **spoilage** in products including **beer**, **wines** and **bread** due to the presence of certain bacteria (**ropy bacteria**) which form **polysaccharides** and rope-like threads, adversely affecting **viscosity** and **consistency** of the product. In **yoghurt** manufacture, ropy bacteria are sometimes used as **yoghurt starters** to produce a product with the desired consistency.

Ropy bacteria Bacteria which produce **ropiness** in foods. Include *Acetobacter* species causing ropiness in **beer**, *Bacillus* species acting on **bread** and *Leuconostoc* species responsible for **spoilage** of **wines**.

Roquefort cheese French semi-soft **blue cheese** made from **ewe milk**. Traditionally ripened in natural caves under the French village of Roquefort-sur-Soulzon. Interior is creamy and white with blue to green-grey veins. Cheese has a pungent **flavour** with a metallic tang. Frequently used in **dressings** and **salads**.

Roquefortine Mycotoxin produced by various species of *Penicillium* (e.g. *P. roqueforti* and *P. crustosum*).

Rose apples Fruits produced by some species of the genus *Eugenia*, especially *E. jambos*. Usually red and juicy; used in various products, mainly **jellies** and **sauces**. May also be eaten candied.

Rose Bengal A xanthene dye and food colorant also known as Food Red No. 105. Commonly incorporated into nutrient media as a stain for detection of growth of **yeasts** and **fungi**.

Rosehips **Fruits** of the dog rose (*Rosa canina* or *R. rugosa*). Used to make **jellies**, **preserves**, **sauces**, **syrups**, **fruit tea** and other **beverages**. Rosehip syrups are particularly rich in **vitamin C**.

Roselle Common name for *Hibiscus sabdariffa*, and red **colorants** extracted from its **berries**. Roselle is rich in **delphinidin**-based **pigments**.

Rosemary Common name for *Rosmarinus officinalis*, the leaves of which are used as **spices** and commonly used as **flavourings** for meat dishes.

Roses Flowers produced by bushes and shrubs of the genus *Rosa*, the petals of which may be used as the source of **pigments** or **essential oils**. Rose oils are used as **flavourings** in foods. The plants are also the source of **rosehips**, fruits with a number of uses in the food and **beverages** industries.

Rose wines **Wines** which are pink in **colour**, covering a range of shades. The pink colour results from extraction of a small proportion of the **anthocyanins** from the **grape skins** during **winemaking**.

Rosmarinic acid Phenolic compound produced as a secondary metabolite in plants. Of interest in food **preservation** due to its **antioxidative activity** and **antimicrobial activity**. Can be produced by plant suspension cultures.

Rossiiskii cheese Russian **hard cheese**.

Rotaviruses **Viruses** of the family Reoviridae. Occur in the faeces of birds and mammals. Responsible for acute **gastroenteritis** in humans, especially children. Transmitted by the faecal-oral route via foods, such as **salads** and **fruits**, or contaminated **water**.

Roti Flat, unleavened **bread** prepared with **corn flour**.

Rots Fungal or bacterial infections of plant tissues that cause softening, discoloration and disintegration.

Rotting Natural process in which animal or plant tissues decay or decompose due to microbial activity.

Roughage Material derived mainly from plant cell walls which cannot be digested enzymically, but which can be partially broken down by intestinal bacteria to produce **volatile fatty acids** that can then be used as a source of energy. Roughage consists of soluble **fibre** which reduces levels of blood cholesterol and increases the viscosity of the intestinal contents, and insoluble fibre (cellulose and cell walls). Foods with a high fibre content include wholemeal **cereals** and **flour**, **root vegetables**, **nuts** and **fruits**.

Roughness **Physical properties** relating to the extent to which the surface of an item feels rough, i.e. not smooth or glossy.

Roux A base for thickening of **sauces**, prepared by heating together **flour** with **fats**. Sauces produced from this base by addition of liquid (e.g. **milk** or **stocks**) and heating, to thicken the liquid, are known as roux sauces.

Rovral Alternative term for the fungicide **iprodione**.

Rowanberries Scarlet **berries** produced by *Sorbus aucuparia*. Used to make **jams** and **alcoholic beverages**.

Royal jelly Partially digested **honeys** and pollen formed in the stomach of worker honeybees for feeding of bee larvae. Thought to possess beneficial health properties, and thus marketed as a health food.

rRNA Abbreviation for ribosomal **RNA**. The major component of ribosomes, the sites where **mRNA** undergoes **translation** to **proteins**. rRNA is transcribed but not translated.

RSE defect Abbreviation for **reddish pink soft exudative defect**.

Rubber Elastic, tough polymeric substance made synthetically or produced from the latex of *Hevea brasiliensis*, a tropical plant native to the Amazonian rainforest. Used in making various food **contact materials**, including sealing rings for bottle **closures**, teats for infant feeding bottles and **rubber nettings** used to enclose **meat** joints.

Rubber nettings Nettings made with **rubber** thread which are used to enclose joints of **meat** such as **beef** and **ham** to prevent their disintegration during **cooking**. Health concerns are associated with possible formation of **nitrosamines** from vulcanizing agents used in formulating the rubber.

Rubber seeds Seeds produced by *Hevea brasiliensis*, which are rich in **unsaturated fatty acids** and show potential as **oilseeds**.

Rubratoxins **Mycotoxins** produced by certain strains of *Penicillium rubrum*. Cause liver damage, brain lesions and gastrointestinal haemorrhages when ingested by animals.

Rue Common name for *Ruta* spp., the leaves of which are used as **spices**. Added as **natural flavourings** to **bakery products** and **dairy products**.

Rum **Spirits** made by **distillation** of fermented **mashes** based on **sugar cane juices** or **molasses**.

Ruminants Herbivorous even-toed ungulate mammals belonging to the sub-order Ruminantia. They include **cattle**, **sheep**, **goats**, **antelopes** and **deer**. Ruminants regurgitate and re-chew feeds (chew the cud) and have a four-chambered stomach, comprising a rumen, reticulum, omasum and abomasum. Composition of ruminant meat fats and milk fats are greatly affected by microbial activity in the rumen, particularly by the **hydrogenation** of **unsaturated fatty acids** into relatively more **saturated fatty acids**.

Ruminococcus Genus of Gram positive, anaerobic coccoid **bacteria** of the family Clostridiaceae. Occur in the rumen of animals. *Ruminococcus albus* may be used in the production of **glycosidases**, while *R. flavefaciens* is used in the production of **cellulases**.

Runner beans Beans produced by *Phaseolus coccineus*. Popular as a vegetable particularly in the UK. Young pods are eaten, but when older, seeds are removed from the pods before consumption.

Rusks Light, sweet crisp or hard **biscuits** or raised **bread** which are browned in an oven and often used as a food for young children.

Russeting Physiological disorder that affects various plant parts, including **fruits** and **tubers**. Characterized by rough brown areas on the surface of the affected tissue.

Russula Genus of **edible fungi** that contains numerous species varying in **flavour** and quality.

Rust Reddish- or yellowish-brown flaky coating of iron oxide that is formed on iron or steel by oxidation, especially in the presence of moisture.

Rusts Diseases caused by rust fungi in plants, giving them a rusty appearance.

Rutabagas Alternative term for **swedes**.

Rutin Disaccharide derivative of **quercetin**, containing **glucose** and **rhamnose**. Found mainly in **cereals** and at one time known as vitamin P.

Ryazhenka Russian **fermented milk**.

Rye Edible grain from hardy plants belonging to the species *Secale cereale,* used to make **rye bread** and rye **whisky**.

Rye bran Outer layers of the **rye** grain. Used as a source of **fibre**; displays **cholesterol** lowering activity and **anticarcinogenicity**.

Rye bread **Bread** made either entirely from **rye flour** or with a blend of **wheat flour** and rye flour. When made entirely from rye flour, it is often dark grey in **colour** and lacks the **elasticity** of **wheat bread**.

Rye flour **Flour** produced by **milling** of **rye** grains. Available in varying degrees of **purity** and **colour** (light, medium or dark).

Rye malt Fermented **mashes** made from **rye** grain, which are used in the manufacture of rye **whisky**.

S

Sablefish **Marine fish** species (*Anoplopoma fimbria*) distributed across the north Pacific. An important commercial food fish; most of the catch is marketed in Japan. Flesh is soft-textured with a mild **flavour**. Marketed fresh, dried/salted and smoked (known as barbecued Alaska cod). Liver **oils** are a rich source of **vitamin A** and **vitamin D**.

Saccharases Alternative term for **β-fructofuranosidases**.

Saccharides General term for **monosaccharides**, **disaccharides**, **oligosaccharides** and **polysaccharides**.

Saccharification Process by which **oligosaccharides** and **polysaccharides** are degraded to produce smaller **sugar** units. Involves acid, alkali or enzymic (e.g. **cellulases**, **amylases**) hydrolysis of glucosidic bonds. Term is used frequently to describe hydrolysis of wastes, e.g. **sugar cane bagasse** or other lignocellulosic materials to produce substrates for microbial **fermentation**.

Saccharimeters Devices used for measuring degree of rotation produced during transmission of polarized light through a sugar solution. When a standardized saccharimeter is used, this property is a function of the concentration of a sugar solution. In the **sugar** industry the rotation value (Pol) is used as a measure of **sucrose** content due to the low concentrations of other **sugars**.

Saccharin Heterocyclic organic sulfur compound (*o*-benzosulfimide) that has approximately 300-600× the sweetness of **sucrose** and is used in **artificial sweeteners**. Available as the free acid and as sodium or calcium salts. Like **sugar**, saccharin salts are white crystalline solids that are highly soluble in water, but unlike sugar they are non-nutritive and impart a bitter metallic **aftertaste**.

Saccharometers Graduated devices used for determination of the **density** of sugar solutions, based on the level at which the device floats. Also known as hydrometers.

Saccharomyces Genus of yeast **fungi** of the class Saccharomycetes. Occur in foods and **beverages** (e.g. **fruit juices**, **fruits** and **alcoholic beverages**), soil and on human skin. *Saccharomyces cerevisiae* is used in **breadmaking** (**bakers yeasts**) and **brewing** (**brewers yeasts**). *S. cerevisiae* is also used in the manufacture of **spirits**, **wines**, **kefir**, **cider** and **pulque**.

Saccharomycodes Genus of yeast **fungi** of the class Saccharomycetes. *Saccharomycodes ludwigii* may be responsible for **spoilage** of **grape juices** and **wines**, and may also be used in **winemaking**.

Saccharomycopsis Genus of yeast **fungi** of the class Saccharomycetes. Occur on **fruits**, and in soils, foods and the tunnels of wood-boring **beetles**. *Saccharomycopsis fibuligera* may be responsible for the **spoilage** of **bread** and **cereals**.

Saccharose Alternative term for **sucrose**.

Sachalinmint Perennial herb (*Mentha sachalinensis*, syn. *M. canadensis*), the leaves of which are used for flavouring foods.

Sachets Small **packs** or **bags** made of flexible material, that are used to package small quantities of substances, e.g. single servings of foods. Common applications include liquid foods such as **sauces**, **ketchups** and other condiments, and particulate products such as **instant soups**, dried **infant foods** and **coffee granules**.

Sacks Large **bags** usually made of thick **paper**, **plastics** or materials such as hessian. Used for carrying or storing goods, e.g. **potatoes** or **grain**. Less commonly, refers to dry **white wines** formerly imported into the UK from Spain and the Canary Islands.

SADH Alternative term for the plant growth regulator **daminozide**.

Safflower oils Oils extracted from **seeds** of *Carthamus tinctorius* which are rich in **linoleic acid**. Used as **cooking oils**, in **salad dressings** and in the manufacture of **margarines**.

Safflowers Large orange, red or yellow flowers produced by the thistle-like plant, *Carthamus tinctorius*. Used as a source of food **colorants** that may be used as a substitute for **saffron** dye. The plant also has edible leaves and produces seeds from which **safflower oils** may be extracted.

Safflower seeds Oil-rich seeds produced by *Carthamus tinctorius*.

Saffron Dried stigmas from flowers of *Crocus sativus* that are used as yellow **colorants** and **spices**. The principal pigments of saffron are the **carotenoids crocin** and **crocetin**.

Safrole Organic compound found in various **spices** and **essential oils** that has been shown to be carcinogenic in rats. Safrole and its isomer isosafrole are used as **flavourings** in foods.

Sage Common name for *Salvia officinalis*, the leaves of which are used as **spices**. Sage has a warm, **camphor**-like **flavour** and **aroma**, and is often used in **flavourings** for **seasonings**, **soups** and meat dishes.

Sago Starchy substance extracted from the interior of the trunk of sago palms (*Metroxylon sagu*) and other similar plants such as sugar palms (*Arenga pinnuta*). The wet **starch** that is washed out from the bark can be eaten cooked, or dried to produce **flour**. Pearl sago is produced by forcing wet starch through **sieves** and drying; this form is used in **puddings**.

Sailfish Any of a number of large, fast-swimming pelagic **marine fish** from the genus *Istiophorus*; found in tropical and temperate Pacific waters and the Indian Ocean. Commercially important species include *I. albicans* (Atlantic sailfish), *I. Greyi* (Pacific sailfish) and *I. Gladius* (Indo-Pacific sailfish). Marketed fresh, smoked and frozen; also used in preparation of **sashimi** and **sushi**.

Saint-Nectaire cheese French semi-soft **cheese** made from **cow milk**. Rind is pink with a covering of grey mould; the soft interior is ivory to straw coloured. Saint-Nectaire has a fruity **flavour** and characteristic grassy **aroma** due to being cured on a bed of straw for 8 weeks.

Saint-Paulin cheese French semi-soft **cheese** made from **cow milk**. Originally made by Trappist monks. Rind is smooth and leathery, and yellow to orange in colour. Also known as **Port Salut cheese** (licensed name). Saint-Paulin is a mild, creamy dessert or table cheese firm enough for slicing.

Saithe Alternative term for **coalfish** or **pollock**.

Sakacins **Bacteriocins** synthesized by *Lactobacillus* *sake*. Sakacin A, produced by *L. sake* LB706, is heat resistant and bactericidal to sensitive strains. Its inhibitory spectrum includes *Carnobacterium* *piscicola*, *Enterococcus* spp., *L. curvatus*, other *L. sake* strains, *Leuconostoc* sp. and *Listeria* *monocytogenes*. Sakacin A is plasmid encoded.

Sake **Rice wines** made in Japan by **fermentation** of rice **mashes** saccharified with **koji starters**.

Sake yeasts Yeasts (*Saccharomyces* spp.) used for **fermentation** of saccharified **rice mashes** in **sake** manufacture.

Sakuradai **Marine fish** species (*Odontanthias rhodopeplus*) from the sea bass family (Serranidae); occurs in the Indo-West Pacific. Consumed mainly in Japan and Indonesia. Usually marketed fresh.

Salad cream **Salad dressings** similar to **mayonnaise** but of a more fluid **consistency**. Major ingredients include water, **vinegar** and **oils**. Egg yolks and **mustard** provide a characteristic yellow **colour**.

Salad dressings **Condiments** that are served with, and complement the **flavour** of, **salads**. Examples include **mayonnaise**, **French dressing** and **salad cream**.

Salad oils Refined, bleached and deodorized **vegetable oils** used in preparation of **salad dressings**. Oils used in manufacture of commercial salad dressings are also subjected to **winterization** to prevent clouding upon **refrigeration**. Clouding is caused by formation of crystals of high m.p. **triglycerides** and may also be inhibited by addition of anti-clouding agents, namely oxystearin, polyglycerol esters and some **emulsifiers**.

Salads Cold dishes consisting of one or more uncooked **salad vegetables**, such as **tomatoes**, **cucumbers** and **lettuces**, usually sliced or chopped, and often accompanied by a protein source, such as eggs, fish or meat. Also refers to dishes of vegetables served with **dressings**, such as **potato salads** or **coleslaw**, and to cold dishes of cooked rice or pasta mixed with cooked or raw vegetables or fruits. **Fruit salads** usually comprise sliced mixed fruits served in **fruit juices** or **sugar syrups**.

Salad vegetables Vegetables eaten raw in **salads**. Include leafy **green vegetables**, such as **lettuces**, **chicory** and **watercress**, spring onions and **radishes**.

Salami Highly seasoned, raw, dried **sausages**, originally produced in Italy. They are prepared from coarsely comminuted **meat**. There are two major kinds, namely soft salami, which are semi-dry sausages; and dry salami, which are dried slowly to a hard **texture**. Most are made from fresh **pork** and include **garlic**; however, they may be prepared from **beef**, **turkey meat**, **veal**, or from meat mixtures. The majority are cured during preparation, air dried, uncooked and unsmoked, but some smoked versions are produced. Characteristics of salami are affected by: type and amount of meat used; proportion of **lean** to fat; how finely, uniformly or coarsely the fat appears among the lean; choice of **seasonings**; and degree of salting and drying.

Salatrim Fat substitute comprising short and long chain acid triglyceride molecules produced by **interesterification** of short chain **triacylglycerols** with fully hydrogenated **vegetable oils**. Applications include in **confectionery**, **bakery products** and **dairy products**.

Salbutamol β-Adrenergic agonist used to enhance growth rates and improve feed efficiency and lean meat content of animals. Normally depletes rapidly from animal tissues following treatment.

Salchichon Spanish, raw, dry, fermented **pork sausages**, that are very popular in Spain. They are prepared primarily from lean **pork**, but also include **beef** and pork **backfat**. Varieties include Salchichon de Vich (Vich sausage).

Salers cheese French **hard cheese** made from raw **cow milk**. Traditionally, only **milk** from cows grazing mountain pastures in the summer can be used. The hard brown rind becomes rough with **ageing**. The yellow interior has a flowery, grassy **aroma** and a nutty, savoury **flavour**.

Sal fats **Vegetable fats** rich in **stearic acid** and **oleic acid**, derived from the seeds of the sal tree, *Shorea robusta*. **Physical properties** and **melting** behaviour are similar to those of **cocoa butter**, making them useful as **cocoa butter extenders**.

Salicylic acid Aromatic acid with the systematic name 2-hydroxybenzoic acid that is found as the methyl ester in many **essential oils**. In the food industry, it is used in **preservatives**. Used in the pharmaceutical industry in **antiseptics** and aspirin.

Salinity Measure of the total amount of **salt** in foods and **brines**.

Salinomycin Polyether antibiotic and coccidiostat used for prophylaxis of **coccidiosis** in chickens; also used as a growth promoter in swine. Residues present in edible tissues are generally barely detectable after 1 day of withdrawal.

Salmon Any of several medium to large anadromous fish of the family Salmonidae, native to the North Atlantic and North Pacific Oceans and spawning in adjacent streams of Europe, Asia and North America. All are important food **fish** highly prized for the **flavour** of their flesh, which in many species is typically reddish-orange in **colour**. Well-known **Pacific salmon** species include chinook (king) salmon (*Oncorhynchus tshawytscha*), coho (silver) salmon (*O. kisutch*) and sockeye (red) salmon (*O. nerka)*. The **Atlantic salmon**, *Salmo salar*, is the prinipcal salmon species consumed in Europe. Marketed and consumed in a wide variety of forms.

Salmonella Genus of Gram negative, facultatively anaerobic rod-shaped **bacteria** of the family **Enterobacteriaceae**. Occur in soil, water, foods (e.g. raw **meat**, raw **sea foods**, **eggs** and **dairy products**) and the **gastrointestinal tract** of humans and animals (especially **poultry** and **swine**). *Salmonella* Typhi is the causative agent of typhoid fever, while *Salmonella* Typhimurium and *Salmonella* Enteritidis are responsible for **gastroenteritis**. Transmission is via the faecal-oral route by contaminated foods or water.

Salmonellae Bacteria of the genus *Salmonella*.

Salmonellosis Any infection caused by *Salmonella* species. Usually manifests itself as **food poisoning** with severe diarrhoea, nausea, vomiting, fever, headache and abdominal cramps.

Salmon oils **Fish oils** derived from *Salmo salar*. Rich in ω-3 fatty acids, particularly **eicosapentaenoic acid**.

Salsa Literally, the Spanish word for **sauces**. In culinary terms, the term refers to sauces prepared from chopped vegetables, **lemon juices** or **lime juices**, and **spices**. The most common type is tomato-based salsa.

Sal seeds Seeds from the sal tree, *Shorea robusta*, which contain a hard green fat used in **cocoa butter extenders**.

Salsify Name given to two plants of the Compositae family. White salsify is the common name for *Tragopogon porrifolius*. Its white roots are boiled and eaten with melted **butter**, **cream** or **cheese**, or used in soups and stews. Leaves are also eaten, in **salads**. **Black salsify** or scorzonera is *Scorzonera hispanica*, the edible root of which has black skin and white flesh. It is used in the same way as white salsify. Both plants are also known as oyster plant.

Salt Mineral with the chemical formula NaCl, obtained by mining or as residues from evaporation of sea water. Several different forms of this mineral are used as **condiments**; table salt, rock salt and sea salt are all forms marketed for this purpose. Commercial salt often includes other salts, such as **calcium chloride** or magnesium chloride, as **anticaking agents**. Salt has multiple uses in the food industry, primarily in **flavourings**, e.g. salted **butter** and salted **nuts**, and in aqueous solutions (**brines**) as **preservatives**. Other uses include as **dough conditioners** and **curing agents**.

Salted fish Fish products preserved or cured with dry salt or in **brines**, after which they may or may not be dried. In the UK, the term usually refers only to salted white fish species, such as **cod**, **coalfish**, **haddock** and **hake**.

Saltine crackers **Crackers** which are thin and crisp-like and are topped with coarse salt crystals.

Saltiness **Sensory properties** relating to the extent to which a product tastes of **salt**.

Salting Alternative term for **brining**.

Salts Compounds produced from the reaction between **acids** and bases.

Salt substitutes Chemicals used to mimic the flavour and/or applications of **salt**. Concern regarding effects of salt consumption on **blood pressure** has lead to a search for salt substitutes that do not have hypertensive effects. **Potassium**, ammonium and **calcium salts** have been tested as salt substitutes, but these metal ions have been unsuccessful in replacing **sodium**, underlining the importance of sodium ions in perception of **saltiness**. Reductions in salt content of **processed foods** have been possible due to the addition of salt **flavour enhancers** such as **amino acids**, **yeast extracts**, **acetic acid** and **allyl isothiocyanate**.

Samna Egyptian clarified **butter**.

Samphire Herbs (*Crithmum maritimum*) native to Mediterranean and European Atlantic coastlines. Used as **condiments** and incorporated into **pickles** and **salads**. **Essential oils** extracted from the plant display **antimicrobial activity**. Also known as crest marine, rock samphire, marine fennel, **sea fennel** and sampier.

Sampling Collection of samples for analysis. Procedures vary according to type of material and analytical technique to be used.

Samso cheese Danish semi-hard **cheese** made from pasteurized **cow milk**. Swiss-style cheese similar to **Emmental cheese**, with a yellow interior of an elastic **texture** broken up by irregularly sized holes. **Flavour** is mild and nut-like, but a sweet-sour **pungency** develops with **ageing**. Used in a wide variety of ways, from cooked **dishes** to **salads** and **sandwiches**. Alternative spelling is samsoe cheese.

Sandesh Sweetened Indian dairy product made from **chhana**.

Sand lance Marine fish species of the genus *Ammodytes*, some of which are of commercial interest. Similar to sand eels. Widely distributed in the Arctic, Pacific and Atlantic Oceans. Marketed dried, salted or frozen and mainly consumed fried. Also used as a source of **fish meal**.

Sandwiches Snack foods comprising 2 or more slices of **bread** (usually buttered), enclosing sweet or savoury **fillings** (e.g. **meat**, **fish**, **cheese**, **eggs**, **jams**). Variations include open sandwiches and toasted sandwiches. Commercial, pre-packed sandwiches form an important part of the **fast foods** sector in many countries.

Sangak Middle Eastern flat **bread** made from whole wheat **sourdough** and baked in traditional style **ovens**.

Sangria Beverages originating in Spain based on **red wines**, **citrus juices**, **sugar** and water (optionally carbonated water). May be garnished with berries or fruit slices.

Sanitation Establishment and maintenance of environmental conditions conducive to the preservation of public health.

Sanitizers Agents used in **disinfection** or **sterilization**.

Sansa oils Low quality **vegetable oils** that are chemically extracted from press residues of **olives**. May be used as **frying oils**.

San Simon cheese Spanish semi **hard cheese** made from **cow milk**. **Curd** is pressed in pear shaped moulds and smoked to give a woody **flavour**. Rind is glossy and honey to red-brown in **colour**. **Consistency** of the interior is open and supple. Flavour is also buttery with slight acidity.

Santoquin Alternative term for the antioxidant **ethoxyquin**.

Sapodillas Fruits produced by *Manilkara zapota* or *Achras zapota*. Also known as sapota. Brown skinned, with black shiny seeds embedded in the amber to brown pulp. Seeds are removed before consumption of the flesh. Contain moderate amounts of **vitamin C** and approximately 15% sugars. Mainly eaten out of hand, but also used in **fruit salads** and **ice cream**. The plant produces a latex that coagulates into chicle, used in the manufacture of **chewing gums**.

Sapogenins The aglycone components of **saponins** occasionally found free in plants but usually present as **glycosides**. May be triterpenoid or steroid in nature.

Saponification Hydrolysis of **fats** into constituent **glycerol** and **fatty acids** by **boiling** with **alkalies**.

Saponins Glycosides found in many plants, consisting of **sapogenins** and **sugars**. Thought to have a number of beneficial health effects, such as the ability to lower cholesterol levels.

Sapota Alternative term for **sapodillas**.

Saran Class of thermoplastic resins that are polymers of **vinylidene chloride**. Made into transparent films, also called cling films, that are resistant to oils and chemicals and used for wrapping foods. Originally a US trademark. It is also known as saran wrap.

Sarcina Genus of Gram positive, anaerobic coccoid **bacteria** of the family Clostridiaceae. Occur in soil, air, **milk**, **grain** and the **gastrointestinal tract** of humans and animals. Species may cause **spoilage** of milk.

Sarcocystis Genus of parasitic protozoans of the family Sarcocystidae. Occur in reptiles, birds and mammals (especially **sheep**, **cattle** and **swine**). *Sar-*

cocystis hominis and *S. suihominis* may cause infection in humans when contaminated **meat** is consumed.

Sarcodon imbricatum Species of **edible fungi**.

Sarcosine Amino acid derivative (*N*-methylaminoacetic acid) occurring as an intermediate in the metabolism of **choline**.

Sarda Genus of **marine fish** containing several medium-sized **tuna** species; generally known as **bonito**. Commercially important species include *Sarda sarda* (Atlantic bonito), *S. chilliensis* (Pacific bonito) and *S. orientalis* (Oriental bonito). Marketed mainly fresh; also dry-salted, canned and frozen.

Sardine Any of a number of **herring**-like **marine fish** species in the family Clupidae; distribution is worldwide. Many species are also referred to as **pilchards**; the term sardine generally refers to smaller individuals within the species. Commercially important species include *Sardina pilchardus* (European pilchard), *Sardinops sagax* (Pacific sardine) and *S. melanosticta* (Japanese pilchard). Marketed fresh, smoked, salted and dried; particularly popular as a canned product in various **sauces** or **oils**.

Sardinella Genus of **herring**-like **marine fish** in the family Clupidae; worldwide distribution. Commercially important species include *Sardinella aurita* (gilt sardine), *S. longiceps* (oil sardine) and *S. anchovia* (Spanish sardine). Marketed fresh or canned; sometimes processed in the same way as **sardine** and **pilchards**.

Sardine oils **Fish oils** extracted from the body of *Sardina pilchardus*. Contain variable amounts of **eicosapentaenoic acid** and **docosahexaenoic acid**. May be used in the manufacture of **margarines**.

Sargassum Genus of brown **seaweeds** containing a number of free-floating and attached species; distributed across the world. Some species are edible and are consumed directly or used as the basis of **food additives**. Rich source of certain **minerals** and functional **polysaccharides**.

Sarsaparilla **Spices** prepared from the roots of *Smilax* spp. Root extracts from this plant are also used as **flavourings**. Former name for root beer, a beverage containing sarsaparilla extracts.

Sashimi Fish product consisting of thin slices of raw **fish** flesh. Fish commonly used include **tuna**, **halibut**, red snapper, **yellowtail** and **mackerel**. Also known as tsukurimi.

Saskatoon fruits Dark blue to black **berries** produced by *Amelanchier alnifolia*. The mild **flavour** resembles a combination of those of **blueberries** and **cranberries**. Used in pies, preserves and fruit **toppings**. Also known as juneberries, serviceberries and Saskatoon berries.

Satiety State in which the desire or motivation for something no longer exists because the need has been satisfied. In the food sense, satiety relates to the physiological sensation of fullness after consumption of a meal. Satiety can also be sensory-specific, e.g. **texture** and **flavour** specific satiety; this may significantly contribute to overall satiety. Sensory-specific satiety refers to the decrease in the perceived pleasantness of a food after it has been eaten to satiety, and the smaller amount of that food, relative to other foods, that is subsequently eaten.

Satratoxins Trichothecene **mycotoxins** produced by ***Stachybotrys*** *atra*. Cause **mycotoxicosis** (stachybotryotoxicosis) in humans, horses, **cattle** and **poultry**. Responsible for irritation and ulceration of the mucous membranes of the mouth, throat and nose, widespread haemorrhages, leucopaenia and possible death.

Satsuma mandarins Small **citrus fruits** of the **mandarins** (*Citrus reticulata*) family. Almost seedless with a smooth, thin skin. Used in production of canned mandarin oranges. Also called **satsumas**.

Satsumas Alternative term for **satsuma mandarins**.

Saturated fatty acids **Fatty acids** that contain no double bonds. Diets rich in saturated fatty acids are thought to increase the risk of developing **coronary heart diseases**.

Sauce mixes Powders containing all the ingredients required (e.g. fats, flour, seasonings, stabilizers) to produce **sauces** upon reconstitution with water. The reconstituted powders are usually thickened by heating to produce sauces of the required **consistency**.

Sauces **Condiments** of a pourable or spoonable **consistency** that are served as an accompaniment to foods in order to enhance the **flavour** of the food. Sauces may be sweet or savoury, e.g. **apple sauces** and **cheese sauces**, respectively, and may be served as a side dish, poured over the food or used during **cooking**.

Saucisson Raw, dry, **fermented sausages**. Varieties include French and Spanish saucisson. They are prepared from lean meat, generally **pork** and **beef**; other ingredients include pork fat, **spices** and **salt**. **Lactic starters** are often used. The surface of the sausages is often coated with chalk or talc.

Sauerkraut Dish made by fermenting shredded **cabbages**, **salt** and, optionally, **spices**. Rich in **vitamin C** and B vitamins. Sold fresh or in jars or cans. Eaten as a side dish, in **sandwiches** and in **casseroles**.

Saury Any of a number of **marine fish** species in the family Scomberesocidae; distributed worldwide. Commercially important species include *Scomberesox saurus* (Atlantic saury) and *Coloabis saira* (Pacific saury). Flesh of most species has a highly esteemed **flavour**. Marketed fresh, frozen or as a dry-salted product.

Sausage casings Natural, cellulose or collagen casings which are filled with **sausage emulsions** in the preparation of **sausages**. Particular types of sausages are prepared in particular types of casings. For example, sheep intestines are used as casings for chipolatas and **frankfurters**, swine intestines are used as casings for fresh frying sausages, and cellulose casings are used in the preparation of skinless sausages.

Sausage emulsions Fillings for **sausages** prepared from comminuted **meat**, fat, **preservatives**, **spices**, **salt** and sometimes fillers, such as **cereals** or **dried milk** solids. Level of NaCl is controlled in order to improve the binding capacity of sausage emulsions, especially those prepared from non-slaughter-warm meat. **Additives** are often included to help preserve, thicken or colour sausages. Extent of **comminution** of the raw meat materials differs widely, so that sausage emulsions may include small pieces, chunks, chips or slices of meat. **Curing** ingredients may be added during comminution or mixing, either in dry form or as a concentrated solution. Most sausage emulsions are packed into **sausage casings** to produce sausages.

Sausagemeat Fresh **sausages** which are sold in bulk without casings. Often mixed with other meats, formed into patties or balls, or used as an ingredient in **stuffings**.

Sausages Comminuted, seasoned, usually cylindrical, meat products prepared from **sausage emulsions** stuffed into **sausage casings**. Commonly, filled sausage casings are twisted at intervals to form links; these vary in shape and size depending on the type of sausages. Sausage production may also involve **curing**, **smoking**, **fermentation**, shaping and/or **cooking**. Shape or form of particular types of sausages tends to be dictated by tradition. Countries such as France, Italy and Germany have an extensive range of regional speciality sausages. Most sausages are prepared from **pork mince** or **beef mince**, but some are prepared from other meats (e.g. chicken mince or donkey mince) or various types of **offal** (e.g. **livers**). They often include low value meat, such as **mechanically recovered meat** or parts of the carcass that are unattractive to the consumer, e.g. the **intestines** and feet. The six major types of sausages are: fresh (e.g. fresh **pork sausages**); cooked (e.g. **liver sausages**); uncooked smoked (e.g. **mettwurst**); smoked and cooked (e.g. **knackwurst**); semi-dry (e.g. semi-dry **salami**); and dry (e.g. **rohwurst**).

Sauteing **Frying** of foods quickly in a small amount of hot fat or oil in a skillet or special saute pan over direct heat.

Savory Leaves of *Satureia hortensis* (summer savory) or *S. montana* (winter savory) which are used as **spices**. **Essential oils** and extracts of savory leaves are also used as **natural flavourings**.

Savoy cabbages Variety of **cabbages** (*Brassica oleracea*) with wrinkled leaves. Generally have a milder **flavour** than smooth leaved varieties. Used to prepare **coleslaw**.

Saxitoxin Potent neurotoxin produced by **dinoflagellates** (e.g. *Gonyaulax catenella* and *G. tamarensis*). Causes **paralytic shellfish poisoning** in humans who ingest filter-feeding bivalve **molluscs** (e.g. **clams** and **mussels**) which feed on these dinoflagellates.

Scald Necrotic condition in which plant tissues, including **fruits**, appear to have been exposed to high temperature or sunlight, or standing water. Affected fruits, mainly apples, have a dry, brown coloured area on the skin, but quality is not usually otherwise altered.

Scalding Immersion of foods briefly in boiling water. Scalding of **tomatoes** is performed to loosen their skins and facilitate **peeling**.

Scales Alternative term for **weighing machines**.

Scaling Removal of scales from fish skin, generally using blunt knives or special tools called fish scalers.

Scallion Name applied to various types of **onions** which do not develop a bulb at the root. The long, straight green leaves and the white part nearer the bottom are both eaten, raw or cooked. Uses include stir fried dishes, salads, soups and garnishes. Also called green onions and spring onions.

Scallops Common name for marine bivalve **molluscs** in the family Pectinidae; widely distributed in intertidal zones and deeper waters of the Atlantic and Pacific oceans. Most species are valued for the **flavour** and **texture** of flesh, which has a distinct, sweet odour when fresh and is creamy white or slightly orange in **colour**; normally, only the large adductor muscle is eaten. Commercially important species include *Pecten maximum* (great scallops), *P. yessoensis* (Japanese scallops) and *Chlamys opercularis* (queen scallops).

Scampi Italian name for **Norway lobsters** (*Nephrops norvegicus*) or **langoustines**; also refers to the tail meat of this lobster fried in **batters** and other products made from Norway lobsters.

Scanning electron microscopy Electron microscopy technique, usually abbreviated to SEM, in which a focused beam of electrons is used to scan the surfaces of suitably prepared samples. Secondary electrons emitted from the samples are detected and used to create detailed images of the structure of the samples. Advantages over light microscopy include greater magnification (up to 100,000×) and much greater depth of field.

Scenedesmus Genus of green **algae** of the family Scenedesmaceae. Occur in a wide range of freshwater habitats. Some species (e.g. *Scenedesmus quadricauda*) may be used in production of **single cell proteins**.

Schizophyllum **Basidiomycetes** that grow on decayed wood. Some species, mainly *Schizophyllum commune*, are consumed as food, in soups or raw. Can cause the disease basidioneuromycosis in humans. Also used as a source of **enzymes** and **polysaccharides**.

Schizosaccharomyces Genus of **fungi** of the class Schizosaccharomycetes. Occur in **fermented beverages**, **fruit juices**, **dried fruits**, **molasses** and **cereals**. *Schizosaccharomyces pombe* is used in the manufacture of **sorghum beer**.

Schmalzfleisch Comminuted meat products, produced from **pork** and pork fat, seasoned with salt and **spices**; they have a very high content of fat.

Schnapps Strong, dry **spirits**, consumed mainly in Germany, the Netherlands and Scandinavia.

School meals **Meals**, particularly lunches, but sometimes also breakfasts and evening meals, provided for school pupils, usually by a **foods service**. Emphasis is placed on planning healthy menus that appeal to children and adolescents and which provide suitable **nutrients** for these age groups.

Schwanniomyces Genus of yeast **fungi** of the class Saccharomycetes. Occur in soil. Some species (e.g. *Schwanniomyces occidentalis*) are used in the production of certain **enzymes**.

Sclerotinia Genus of **fungi** of the class Leotiomycetes. Some species (e.g. *Sclerotinia fructigna* and *S. trifoliorum*) are responsible for several plant diseases. *S. sclerotiorum* and *S. fructigna* cause **spoilage** of **vegetables** (e.g. **carrots**, **celery**, **cucumbers** and **artichokes**) and **fruits** (e.g. **apples**, **pears** and **peaches**) during storage.

Sclerotium Genus of **fungi** that includes some important plant pathogens that cause rotting. *Sclerotium rolfsii* produces several **glycosidases** as well as the exopolysaccharide scleroglucan, with many potential applications in the food industry, e.g. in **thickeners**.

Scombroid poisoning Poisoning linked to consumption of **fish** containing high levels of **histamine**, which is produced soon after death in fish having naturally high levels of free **histidine**, particularly scombroid or scombroid-like **marine fish** such as **mackerel** and **tuna**. Formation of histamine in fish depends on the temperature at which the fish is kept from time of capture until it is consumed; to minimize risks it is important to refrigerate fish after capture.

Scones Quick breads traditionally prepared with leavened **barley flour** or **oat flour**, **milk** or **buttermilk**, **baking powders**, **sugar**, **salt** and sometimes **cream** and **eggs**, which are then cut into various shapes and baked on a griddle or in **ovens**. Often made with additional ingredients, such as **dried fruits**, **cherries**, **nuts**, **dates** and **cheese**.

Scopoletin 6-Methoxy-7-hydroxycoumarin. Found in a number of higher plants, often as scopolin (scopoletin 7-glucoside). Accumulates in the tissues of certain microbially infected plants and is thought to exhibit antifungal activity. Scopolin may contribute to the **bitterness** of **citrus fruits**.

Scopulariopsis Genus of **fungi** of the class Hyphomycetes. Occur in decaying plant material and foods. *Scopulariopsis brevicaulis* causes **spoilage** of **cereals**, **meat**, **salami**, **cheese** and **eggs**.

Scrambled egg **Eggs** which have been beaten, usually with **milk**, **seasonings** and **butter**, and cooked with stirring to give a lumpy **texture**.

Scrapie One of a group of **prion diseases**, this one affecting **sheep**. Scrapie is characterized by progressive and fatal degeneration of the central nervous system. Deaths occur a few weeks or months after the initial symptoms appear. Mode of transmission of scrapie is not fully understood, but evidence suggests that scrapie has been present in sheep in many parts of Europe for more than 250 years. Experimentally, scrapie has been transmitted to mice, rats, hamsters and goats; however, there is currently no evidence for transmission of scrapie from sheep to man. Scrapie is not currently believed to have a role in the origins of **bovine spongiform encephalopathy** (BSE). Slaughter programmes for scrapie have failed as a means for control, but it may be possible to breed for scrapie resistance.

Sculpin Any of a large number of **marine fish** or **freshwater fish** species in the family Cottidae; most species occur off the Atlantic and Pacific coasts of the USA. Few species have significant commercial importance as food fish.

SDS Abbreviation for sodium dodecyl sulfate.

SDS-PAGE Abbreviation for sodium dodecyl sulfate polyacrylamide gel electrophoresis.

Sea bass Any of a number of **marine fish** in the family Serranidae, many of which are valued food fish; distributed worldwide. Commercially important species include *Dicentrarchus labrax* (European bass), *Centropristis striata* (black sea bass), *Morone saxatilis* (**striped bass**) and *M. chrysops* (white bass). Marketed fresh, frozen and smoked.

Sea bream Any of a number of **marine fish** in the family Sparidae, many of which are valued food fish; distributed in the Atlantic and Mediterranean. Some species are cultured in sea cages. Commercially important species include *Pagrus major* (red sea bream), *Sparus aurata* (gilthead sea bream) and *Pagellus centrodontus* (sea bream). Flesh tends to be lean, with a coarse-grained **texture**. Marketed fresh, frozen, salted, dried and as semi-preserved or canned products.

Sea buckthorn Common name for *Hippophae rhamnoides* and the round yellow-orange **berries** it produces. Fruits are rich in **vitamin C**, **vitamin E**, **carotenoids** and **flavonoids**. Used in **sauces** and jellies, and to make **liqueurs** and **fruit juices**. Also the source of oils with reported healing properties.

Sea buckthorn oils **Oils** extracted from the pulp or seeds of **berries** produced by **sea buckthorn**. Oils vary in composition according to source (pulp or seed), but are consistently rich in **tocopherols**. May be used in foods. Health benefits include potential for protection against **cardiovascular diseases**.

Sea cucumbers Any of the 1100 species of marine invertebrates from class Holothuroidea of the phylum Echinodermata; all have soft cylindrical bodies and are mainly found in shallow tropical waters. Many species are edible, particularly those from the genera *Stichopus* and *Cucumaria*. A popular delicacy in some Asian countries, where they are normally gutted, boiled and dried prior to consumption. Also known as beche de mer and **sea slugs**.

Sea fennel Common name for *Crithmum maritimum*, a herb which grows wild along coastlines, particularly in the Mediterranean and European Atlantic areas. Rich in **vitamin C**. **Sensory properties** are similar to those of **parsley**. Used in **condiments** and in **pickles** and **salads**. **Essential oils** extracted from the plant display **antimicrobial activity**. Also known by a variety of names, including **samphire**, rock samphire, crest marine, sampier and marine fennel.

Sea foods All edible marine and freshwater aquatic organisms; includes **fish** (**finfish**), **shellfish**, aquatic mammals, **plants** and **algae**. Generally regarded as a healthy component of the human diet. Many sea foods are good sources of high quality **proteins**, **unsaturated fatty acids**, **vitamins** and **minerals**, and are low in **fats** and **calories**.

Sea kale Common name for *Crambe maritime*, these **plants** are found on shingle and sandy shores, mainly of the Baltic Sea and Atlantic Ocean. Leaf stalks are blanched by covering when the plants are young, or by a covering of shingle in the wild. Stalks are boiled like **asparagus** and served with mild white **sauces** or **butter**.

Seal blubber Thick, subdermal lipid layer found in **seals**; marine mammals belonging to the family Phocidae. Often forms up to 25% of the animal's total weight and acts as an insulator. May often become contaminated by organochlorine compounds such as **polychlorinated biphenyls** (PCB). Frequently consumed by Arctic inhabitants.

Seal blubber oils **Oils** derived from the subdermal lipid layer (blubber) of **seals**; marine mammals of the family Phocidae. Rich source of ω-3 **polyunsaturated fatty acids**.

Sea lettuces Any of several green **seaweeds** of the genus *Ulva*; distributed on rocky shores worldwide. Consumed raw, cooked, dried, in **soups** or as a deep fried product.

Sealing Process of closing openings in **containers** in such a way as to prevent leakage of the contents or entry of undesirable elements.

Seal meat **Meat** from **seals**. The prime cuts of seal **carcasses** are the flank, flipper and rump sections. Seal meat is dark red in **colour** and has a characteristic **aroma**. Composition of meat is altered when seals are in moulting condition; at this time they shed their hair, reduce feeding substantially and hence lose up to 20% of their **blubber**. Age of seal and type of carcass cut significantly affect sensory quality of the meat.

Seal oils General term for **oils** derived from **seals**, marine mammals belonging to the family Phocidae.

Seals Fish-eating marine mammals belonging to the Phocidae family (eared or true seals) or Otariidae family (the earless or hair seals); there are many species. They are hunted for **seal blubber** and **seal meat**.

Seaming Process of joining together the edges of food cans to form a seal.

Sea mustard Common name for *Undaria pinnatifida*, a member of the brown (Phaeophyta) group of **seaweeds**. Rich source of **dietary fibre**. Extracts possess **antitumour activity** and **antimutagenicity**. Used in production of **soups**, edible **starch gels** (**mook** or muk) and **jams**, and as an ingredient of foods including **cakes** and **kimchies**.

Sea perch General name given to a number of **marine fish** within the family Serranidae (including **grouper** and **sea bass**); particularly refers to *Epinephelus* species.

Sea slugs Any of the 1100 species of marine invertebrates from class Holothuroidea of the phylum Echinodermata; all have soft cylindrical bodies and are found mainly in shallow tropical waters. Many species are edible, particularly those of the genera *Stichopus* and *Cucumaria*. A popular delicacy in some Asian countries, where they are normally gutted, boiled and dried prior to consumption. Also known as **sea cucumbers**.

Seasonings Blends of **spices**, **flavourings** and other **additives**, such as **colorants** and **sweeteners**, that are used to enhance **flavour**, **aroma** and/or overall appearance of foods. Commercial seasonings may also contain **anticaking agents**. Seasonings are often created for use with particular types of food, e.g. barbecue seasonings or chicken seasonings.

Sea squirts Any member of the invertebrate class Ascidiacea (subphylum Urochordata, also called Tunicata), which are found attached to natural and manmade structures in seawater; worldwide distribution. Some species are consumed as a delicacy in some parts of the world; particularly popular in France (often eaten raw with **lemon juices**) and in Japan.

Sea trout Marine form of the **brown trout** (*Salmo trutta*) found in northern Atlantic waters; migrates back into freshwater to spawn. Highly valued as a sport fish and for the **flavour** and **texture** of its flesh. Cultured in some areas of northwest Europe. Marketed fresh, frozen and as a smoked product.

Sea urchin gonads Ovaries and **roes** of **sea urchins** (echinoids); the only part of sea urchins which are consumed. A highly esteemed and valuable delicacy, particularly in Japan (known as uni). Marketed principally as a salted product; also sold fresh and frozen. Used to make **shiokara**.

Sea urchins Any of around 700 species of marine invertebrates in the phylum Echinodermata; worldwide distribution. Generally have rounded hard, calcareous shells and prominent spines. Many species are exploited for their gonads, which are a highly valued delicacy. Also known as echinoids.

Sea water Water from marine environments, characterized by a high salinity and complex physicochemical structure; covers nearly 75% of the earth's surface. In some countries, **desalination** is used to produce **potable water** from sea water.

Seaweeds Multicellular marine **algae** which are fixed to marine substrates by root-like holdfasts; occur in intertidal or subtidal environments worldwide. Subdivided into 4 classes: green (Chlorophyta); brown (Phaeophyta); red (Rhodophyta); and blue-green (Cyanophyta). Many species are edible, providing an excellent source of **vitamins** and **minerals**. **Agar**, **carrageenans** and **alginates** are extracted from some species for use as **food additives**.

Secalins Major **storage proteins** of **rye**.

Sedimentation Settling of matter to the bottom of a liquid by gravitational force so as to separate suspended solids from fluids.

Seedless grapes **Grapes** that contain no seeds. The most commonly eaten varieties include Thomson seedless, flame seedless and ruby seedless. Eaten out of hand, in **salads** and in cooking, e.g. in Veronique dishes.

Seeds Produce of flowering plants; mature fertilized ovules. Contain an embryo and a seed coat, and often an endosperm. Examples include beans, peas, oilseeds and cereals.

Seer fish Group of predominantly **marine fish** of the genus *Scomberomorus* belonging to the family Scombridae (mackerels, tunas, bonitos). Widely distributed in tropical and subtropical waters. Species vary from minor to high commercial value, but all are important game fish. Marketed fresh, dried-salted or smoked, and consumed in a number of ways, including pan-fried, grilled, baked or as spicy fishballs.

Sei-kombu Japanese name for dried seaweed products formed from the kelp species *Laminaria japonica*; used in Japanese cuisine as an ingredient of **stocks** or **seasonings**. Contain significant amounts of **glutamic acid**, the basis of **monosodium glutamate**.

Sekts German **sparkling wines**.

Selenium Essential trace element with the chemical symbol Se. Deficiency can cause keshan disease, a fatal form of cardiomyopathy, and may increase the risk of cancer, while excess can cause balding, garlic breath, intestinal distress and impaired mental functioning. Food sources include **sea foods**, **meat**, and some grains and seeds.

SEM Abbreviation for **scanning electron microscopy**.

Semi skimmed milk **Milk** from which some of the fat has been removed. This low fat product is preferred to **whole milk** by some health conscious consumers, and is used by processors to make low fat **dairy products**. Semi skimmed **cow milk** contains approximately 1.7% fat, compared with approximately 4% in **whole milk**.

Semolina Purified granular middlings from **durum wheat** used principally in the manufacture of **pasta** and **milk puddings**.

Sencor Alternative term for the herbicide **metribuzin**.

Senescence Degeneration of **plants** due to maturation or **ageing**. Stress due to disease or attack by **insects** may induce early senescence.

Sensors Apparatus used in detection by responding to a specific stimulus.

Sensory analysis Analytical techniques used to determine the **sensory properties** of foods. The techniques fall into three main classes: discrimination/difference tests; descriptive tests; and hedonic/affective tests.

Sensory evaluation Alternative term for **sensory analysis**.

Sensory properties Properties that can be detected by the sense organs. For foods, the term relates to the combination of concepts such as **appearance**, **flavour**, **texture**, **astringency** and **aroma**.

Sensory scores Scores given to particular **sensory properties** of foods by panellists during **sensory analysis**.

Sensory thresholds Term used in **sensory analysis** relating to the levels at which perception of increasing concentrations of a stimulus, such as **aroma compounds** or **flavour compounds**, begins. Classical methods for estimating sensory thresholds include probit, graphic, exact, logistic, Spearman-Karber, moving average and up-and-down methods.

Separation Action or state of division into distinct elements, using techniques such as **centrifugation**, **filtration**, **sieving**, **crystallization** and **distillation**. Separation of food components is fundamental for preparation of ingredients to be used in other processes. Some separation methods are used to sort foods into classes based on size, **colour** or shape, to clean them by separating contaminating materials, or to selectively remove water by **evaporation** or **drying**. Centrifugation is used for separation of immiscible liquids and for separation of solids from liquids. Filtration is used for removal of insoluble solids from a suspension by passing it through a filter medium.

Separators Equipment that facilitates the division of items or solutions into distinct elements. Examples include **centrifuges**, **filters** and **sieves**.

Septoria Genus of **fungi** which includes many species responsible for plant diseases. Some species (e.g. *Septoria apiicola*, *S. oxyspora* and *S. nodorum*) cause diseases of plants such as **celery**, **tomatoes** and **cereals** (e.g. **wheat**, **rye** and **barley**).

Sequencing Examination of the sequence of components in a sample to aid in its identification. Components sequenced include bases in **genes**, and **amino acids** in **proteins** or **peptides**.

Sequestrants **Additives** that bind to or form complexes with other chemicals, reducing their reactivity in order to prevent the occurrence of undesirable reactions. Examples of sequestrants include sodium citrate and **EDTA** which are used to chelate calcium ions (e.g. used to modulate the strength of **gellan** gels), and

phosphates that bind to and enhance the stability of **proteins** at low pH.

Serine Non-essential amino acid required for metabolism of fats and fatty acids, muscle growth and a healthy immune system. Abundant in **meat** and **dairy products**, **wheat gluten**, **peanuts** and soy products.

Serological tests Immunological techniques in which **antibodies** in blood serum samples are detected using specific **antigens**.

Serology Study of blood serum with particular reference to components important for **immune response**. Used to detect specific **antigens** or **antibodies**.

Serotonin Hormone derived from **tryptophan** found in humans, animals and plants. Acts as a vasoconstrictor and neurotransmitter. Present in some **tropical fruits** such as **bananas** and **pineapples**. Excessive intake in the diet may lead to myocardial lesions. Also known as **5-hydroxytryptamine**.

Serotype Serologically (antigenically) distinct **variety** or strain of an organism, as defined by antisera against antigenic determinants expressed on the cell surface.

Serotyping Methods for distinguishing between closely related strains of microorganisms based on differences in their surface **antigens**. Strains are exposed to **antibodies** specific for certain antigens and those that interact are detected by **agglutination** or precipitation. Reactions to an appropriate range of antibodies distinguish a strain in terms of its surface antigens.

Serra cheese Portuguese soft, almost spreadable, **cheese** made from **ewe milk** using **vegetable rennets** prepared from cardoon flowers. Traditionally, entirely hand-made, down to breaking of the **curd** by hand. **Flavour** has the slightly burnt toffee character of ewe milk. **Ripening** takes 30-40 days. Also known as Serra da Estrela cheese, after its place of origin.

Serrano ham Cured **ham** produced in Spain using methods similar to those used in Italy to produce Parma ham. Fresh hams are covered with **salt** for approximately 2 weeks to draw off moisture and preserve the **meat**, washed, hung for approximately 6 months and finally air dried. The name derives from the practice of carrying out the air drying phase, which lasts 6 to 18 months, in sheds located at high elevations. Good source of **vitamin B$_1$**, **vitamin B$_2$** and **thiamin**. Served as a snack in thin slices and used to flavour **soups**, vegetable dishes or **pasta** dishes.

Serratia Genus of Gram negative, facultatively anaerobic **bacteria** of the family **Enterobacteriaceae**. Occur in water and soil, and on plants. Some species (e.g. *Serratia liquefaciens* and *S. marcescens*) may be responsible for the **spoilage** of foods (e.g. **meat**, **dairy products**, **shellfish**, **vegetables** and **eggs**).

Serum Blood fraction expressed from clotted blood. Also sometimes used as an alternative term for **whey**, which is produced by the clotting of **milk**.

Sesame Tropical, annual herb, *Sesamum indicum*, which bears small flat seeds, which are used as **toppings** and in **flavourings** for foods as well as being a source of **edible oils**.

Sesame oils Seed oils derived from **sesame seeds**, which are rich in **oleic acid** and **linoleic acid** and have high **oxidative stability** due to the presence of natural **antioxidants**. Contain sesamin and sesamolin. Due to their nut-like **flavour**, the oils are used as **seasonings** as well as cooking oils. Also known as gingelly oils or til oils.

Sesame seed meal Residue remaining when **sesame oils** are extracted from **sesame seeds**. Used as an animal feed, a source of **proteins** and sometimes as a partial substitute for **wheat flour** in **baking**.

Sesame seeds Small flat seeds with a nut-like **flavour** produced by **sesame** (*Sesamum indicum*). Used as toppings for **bakery products**, **flavourings** and as a source of **sesame oils**. Also known as gingelly seeds.

Sesamol Natural phenol **antioxidants** prepared from **sesame oils**.

Sesbania Genus of leguminous plants, the leaves, flowers and **seeds** of which are eaten commonly in India. Seeds and leaves are potential sources of protein. **Gums** extracted from seeds of some species have possible uses in the food industry, e.g. as **thickeners**.

Sesquiterpenoids Volatile compounds produced as secondary metabolites in certain **plants**, **spices** and **essential oils**.

Setting Firming of foods, usually as a result of **cooling**, as with **gelatin**-based dishes, such as **jelly**.

Sevin Alternative term for the insecticide **carbaryl**.

Sevruga Species of **sturgeon** (*Acipenser stellatus*) found in the Caspian Sea; the smallest of the sturgeon exploited commercially. **Roes** are used as **caviar**.

11S Globulins Globulins with a sedimentation coefficient of 11S which constitute one of the main groups of characteristic **storage proteins** in non-cereal grains, such as **beans**, **peas** and **peanuts**.

Shaddocks Alternative term for **pomelos** or **pummelos**, the largest of the **citrus fruits**, produced by *Citrus maxima* or *C. grandis* and ancestors of the modern **grapefruit**. Closely resemble the grapefruit in appearance, but the flesh is sweeter and less acidic, lacking the **bitterness** of a grapefruit. Rich in vita-

min **C** and potassium. Eaten fresh or used to make **jams**, **jellies** and **marmalades**. Also known as Chinese grapefruit.

Shading Complete or partial protection of plants from **sunlight** using cloth or other materials. Prevents **sunburn** and other types of damage and has various effects on the composition and quality of **fruits**.

Shallots Type of **onions** (*Allium ascalonicum*) with many lateral, mild-flavoured bulbs. Eaten raw or cooked. Also used for **pickling**.

Shandy Blend of **beer** with **lemonade**.

Shaping To give a shape or form to a substance, sometimes with the aid of moulds.

Shark fins In culinary terms, can include dorsal, pectoral or tail fins from a few species of **sharks** which are considered a delicacy and are used in Asian, predominantly Chinese, **cooking**. The cartilage in the fin and the **gelatin** which it provides are the components of the fin utilized in cooking. Fins are sold dried, either whole or in shreds. Eaten mainly in shark's fin **soups**, the gelatin imparts a characteristic texture. Also sometimes served after **braising** as a main dish or used in small quantities in **fillings** or **stuffings**.

Sharks Any of numerous cartilaginous, predatory **marine fish**; worldwide distribution. Many species are exploited as a source of food; **shark fins** from several species are used to make **soups**. Marketed fresh, frozen and as dried, salted or smoked products. Liver **oils** are a rich source of **vitamin A**.

Sharon fruit Alternative term for **persimmons**.

Sharpness Sensory properties relating to the extent to which an item tastes sharp, i.e. acid, bitter or astringent.

Shashlik Meat products prepared from **meat**, or meat and **offal**. Ingredients vary between recipes, but may include lean meat, **bacon**, **livers**, **kidneys**, animal fats, **onions**, **peppers** and **gherkins**. The product is cooked on a spit or skewer. Traditional Turkish shashlik is made exclusively from **mutton**, without addition of **offal** or **vegetables**. In Germany, shashlik may contain **pork**, **beef**, bacon, offal, onions and other vegetables.

Shea nut butter Yellowish **vegetable fats** derived from the seed kernels of *Butyrospermum parkii*. Rich in **stearic acid** and **oleic acid**. Resembles **cocoa butter** in its **melting** profile, making it suitable for use in **cocoa butter equivalents**.

Shea nuts Seeds produced by the tree *Butyrospermum parkii*. **Fats** derived from the seeds are used to make **shea nut butter**.

Shear Force that one plane exerts on a neighbouring plane per unit area of contact, and which causes a **deformation** in a direction related to the direction of the applied force. Shear forces are applied during food processing such as **mixing** and **extrusion** and will affect the **texture** of the final product. Shear also occurs during **mastication** of foods.

Shear strength Measure of the resistance of a material, such as a food, to **shear** stress and the associated **deformation** caused by the application of this stress. Peak shear strength is the highest stress sustainable just prior to complete failure of a sample under load; after this, stress cannot be maintained and major strains usually occur by displacement along failure surfaces. For material not previously sheared there is a rapid decline in strength with increasing shear until the residual shear strength is reached. The shear strength of a food will influence the **rheological properties** and **mechanical properties** of the food during processing, and also the **texture** and other **sensory properties** of the food during consumption.

Shear values Measures of the forces experienced by a material, such as a food, undergoing **shear**. Often determined in **meat** after cooking as an indication of **tenderness**.

Sheatfish Freshwater **catfish** species (*Silurus glanis*) found in eastern Europe and Central Asia; occurs mainly in large lakes and rivers. Cultured in some regions on a semi-extensive basis. Marketed fresh, canned and frozen. Also known as wels catfish.

Sheep Ruminants (*Ovis aries*), the majority of which have been domesticated for the production of **lamb**, **mutton**, **ewe milk** and wool. There are many breeds; for example, in the UK there are approximately 50 recognized breeds, various local types and numerous crossbreeds. Current selective breeding programmes aim to produce rapid growth of lean sheep **carcasses**. Different gender and age groups of sheep are known as rams (adult entire males), wethers (adult castrated males), ewes (adult females), tegs (two years of age), hoggets (one year of age), shearlings (15-18 months of age) and lambs (sexually immature animals which are generally less than 6 months of age).

Sheep cheese **Cheese** made from **ewe milk**. Alternative term for ewe, ewe milk or sheep milk cheese.

Sheep meat Alternative term for **mutton**.

Sheep milk Alternative term for **ewe milk**.

Sheep milk cheese **Cheese** made from **ewe milk**. Alternative term for ewe, ewe milk or sheep cheese.

Sheep muscles Alternative term for **mutton**.

Shelf life Time for which a stored item remains usable.

Shellac Solution of resinous exudation from bodies of *Tachardia lacca*, an insect of the same family as the cochineal beetle. Used in **coatings** for foods such as **fruits**, **chocolate** and **sugar confectionery**.

Shellfish General name referring to aquatic invertebrates possessing a shell or exoskeleton, including **crustacea** (**crabs**, **lobsters**, **prawns** and **shrimps**) and **molluscs** (**gastropods**, **bivalves** and **cephalopods**).

Shelling Removal of **husks**, **shells** or pods from foods such as **nuts**, **eggs** and **peas**.

Shells Generally refers to hard and rigid coverings of various invertebrates, mostly calcareous; in other cases chiefly or partially chitinous, horny or siliceous. Shells of some marine **molluscs** and **crustacea** are used by the food industry as a source of calcium carbonate, **chitin** or **glucosamine**. Also used to describe the outer coating of birds' **eggs** (**egg shells**).

Sherbet Artificial fruit-flavoured effervescent powders eaten as **sweets**. When mixed with bicarbonate of soda, **tartaric acid**, **sugar** and **flavourings**, may also be used to make **beverages**. Also a US term for **sorbets**.

Sherry Fortified wines made in a defined region in the vicinity of Jerez de la Frontera in Spain. The main sherry types include Fino, Oloroso, Amontillado, Manzanilla and Palo Cortado. Sherries are aged by the unique solera system of sequential blending of successive vintages. Some sherry types undergo a secondary **fermentation** in which a layer of **yeasts** (flor yeasts) grows on the surface of the **wines** and subsequently dissolves in the wines, imparting a characteristic **flavour** and **aroma**.

Shewanella Genus of Gram negative, facultatively anaerobic, curved or straight rod-shaped **bacteria** of the family Alteromonadaceae. *Shewanella putrefaciens* is responsible for the **spoilage** of **fish** and **meat**.

Shiga like toxins **Cytotoxins** produced by enterohaemorrhagic *Escherichia coli* strains, which are similar to **Shiga toxins**. Also known as **verotoxins** and **Vero cytotoxins**.

Shiga toxins Protein **toxins** produced by some *Shigella* species which have enterotoxic, neurotoxic and cytotoxic activity. Responsible for some of the symptoms of bacillary dysentery caused by *S. dysenteriae*.

Shigella Genus of Gram negative, facultatively anaerobic rod-shaped **bacteria** of the family **Enterobacteriaceae**. Occur in the gastrointestinal tracts of humans and primates, soil, **fruits**, **vegetables** and fresh water. *Shigella sonnei*, *S. boydii*, *S. flexneri* and *S. dysenteriae* are causative agents of bacillary dysentery.

Shigellosis Bacillary dysentery caused by infection with *Shigella* species. Characterized by abdominal cramps, diarrhoea, fever, vomiting, presence of blood, pus or mucus in stools, and tenesmus. Transmission is via the faecal-oral route by contaminated foods (e.g.

salads, **vegetables**, **dairy products** and **poultry meat**) and water.

Shiitake Alternative term for **Lentinus** edodes (renamed *Lentinula edodes*) or Japanese black forest **mushrooms**.

Shiokara Fermented **sea foods**, normally made from **squid**, but also from viscera of **skipjack tuna** or other **marine fish**, or from sea urchin gonads. Fermenting the raw material with **salt** for up to one month produces a brown, salty viscous paste.

Shochu Japanese **spirits** made by **distillation** of fermented **rice mashes**.

Shops Buildings or parts of buildings where goods or services are sold.

Shortbread Sweetened **biscuits** prepared with a high ratio of **butter** or other **shortenings** to **flour**.

Shortening A process that results from changes occurring in numerous connected sarcomeres in the myofibrils of muscles. It occurs during muscle contraction in living animals, but also during *rigor mortis*. Degree of sarcomere shortening is influenced by muscle fibre type (e.g. oxidative vs. glycolytic) and *post mortem* ambient temperature. If ambient temperature decreases rapidly during the onset of *rigor mortis*, muscle fibres contract to a greater extent than at higher ambient temp. This physiological occurrence is referred to as **cold shortening**; severe shortening results in reduced **meat tenderness**. **Electrical stimulation** is used to reduce **toughness** associated with cold shortening in meat.

Shortenings Solid or semi-solid **animal fats** or **vegetable fats** often used in **baking**. By dispersing as a film throughout **batters**, they impart **crispness** or flakiness to **bakery products**.

Shoti Common name for *Curcuma zedoaria*, a plant related to **turmeric**. Young rhizomes are eaten as a vegetable. The dried rhizome is pulverized and used as a spice. Used as a condiment and in manufacture of **flavourings** and bitters. Also known as **zedoary**.

Shoyu Alternative (Japanese) name for **soy sauces**.

Shredding Tearing or cutting of items into strips of material (shreds). This can be achieved either by hand or by using a grater or a food processor fitted with a shredding disk.

Shrikhand Fermented **milk** product usually prepared from **buffalo milk** and popular in India. Also known as srikand and srikhand. Traditionally, the **milk** is fermented with a mixed starter culture (*Streptococcus lactis* and *S. lactis* var. *diacetylactis*) and chakka is prepared by draining off **whey** from the resultant **curd**. Other ingredients, e.g. **sugar**, **colorants** and **flavourings**, are then added to the chakka.

Shrimps General name used for many species of marine and freshwater **crustacea** within the suborder Natantia. The distinction between shrimps and **prawns** is unclear. The term shrimps is generally used for smaller species within the families Pandalidae, Penaeidae and Palaemonidae; particularly refers to the family Crangonidae. Many species have commercial importance as foods, including *Crangon septemspinosus* (sand shrimps), *Pandalus platyceros* (spot shrimps) and *Penaeus monodon* (**tiger shrimps**). Marketed in a variety of forms, including fresh, frozen, smoked, canned and as pastes.

Shrink packaging Transparent, clinging thermoplastic films used to enclose a product or package. When heated, the film shrinks to fit closely to the package.

Shubat Fermented **camel milk** similar to, but thicker than, **koumiss**. Popular in Kazakhstan. Also known as chal.

Shucking Removal of **husks** from corn, or **shells** from **shellfish** such as **oysters** and **clams**.

Shuttle vectors **Cloning vectors** that can replicate in more than one type of organism, thus allowing propagation of **DNA** in either organism.

Sialic acid Organic acid found in animal tissues and fluids, e.g. in **glycolipids**, **mucopolysaccharides** and **gangliosides**. Also found as a component of **milk proteins**. Terminal sialic acid residues in **glycoproteins** or glycolipids in cell membrane components serve as receptor sites. Also known as *N*-acetylneuraminic acid.

Sideritis Genus of plants of the family Lamiaceae, some species of which are used to prepare **herb tea**, especially in Mediterranean countries such as Greece and Turkey.

Siderophores Natural compounds produced by **microorganisms** which chelate ferric ions.

Sides A butchers' term for the two halves of animal **carcasses**, divided along the backbone.

Sieva beans Alternative term for **lima beans**.

Sieves Utensils consisting of a wire or plastic mesh held in a frame used for straining solids from liquids, separating coarser from finer particles or production of **pulps** and **purees**. Also know as **strainers**.

Sieving Process of straining solids from liquids or separating coarser from finer particles using **sieves** or **strainers**. Sieving also incorporates air to make ingredients (such as **flour**) lighter.

Sifters Utensils consisting of a stainless steel or heavy weight plastic mesh used for removal of lumps or large particles from finer particles. Used in **sifting** ingredients such as **flour** or confectioners' sugar.

Sifting Process of passing a dry substance through **sifters** to remove lumps or large particles. Sifting also incorporates air to make ingredients (such as **flour**) lighter.

Sikhe Traditional Korean **beverages** made with **rice** which has been saccharified and fermented.

Silage Fodder which is harvested while green and converted into succulent feed for livestock by **fermentation** in silos. May also be made from **fish** by-catch or wastes. The fish material is chopped or ground prior to addition of acids or of a carbohydrate source for fermentation, the material being preserved by the low pH which develops. Composition of silage fed to animals can affect **milk** or **meat** characteristics.

Silica Silicon dioxide that occurs in crystalline, cryptocrystalline and amorphous hydrated forms. Ubiquitous component of the diet with numerous applications in the food industry, such as **stabilization** of **beer**, **refining** of **vegetable oils**, and **immobilization** of **proteins** and **enzymes**.

Silica gels Gels formed from polymers of silicic acid. When dried, they are termed silica xerogels and are used as desiccators or as **adsorbents**, e.g. for **clarification** of **beer** by **adsorption** of **cloud**-forming **proteins**.

Silicates Salts derived from **silica** or silicic acid, containing silica, oxygen, one or more minerals and possibly hydrogen. Uses include reducing the content of free **fatty acids** in **frying oils**, **immobilization** of **enzymes** and **clarification** of **beverages**.

Silicon Essential, non-metallic element, chemical symbol Si. Always found in a combined state in nature.

Siljo Traditional Ethiopian fermented food made with meal prepared from **horse beans** and an extract of **safflowers**. The cooked slurry made from these components is fermented by **lactic acid bacteria** in mustard powder.

Silos Tall towers or pits which are used for storage. Commonly refers to stores for **grain**, e.g. on a farm or at a mill, but can also be used for storing other commodities including **vegetables** and **milk**. Also applied to airtight structures in which green crops are compressed and stored as **silage** for animal feeding.

Silver Soft, white, metallic element, chemical symbol Ag.

Simazine Selective systemic herbicide used for control of most germinating annual grasses and broad-leaved weeds around a range of crops, but also used as an algicide. Classified by Environmental Protection Agency as not acutely toxic (Environmental Protection Agency IV). Also known as gesatop.

Simmering **Heating** of foods in a liquid, such as water, at a temperature that causes the liquid to bubble gently.

Simmondsin Cyanide-containing glycoside found in defatted jojoba meal, a by-product in the manufacture of **jojoba oils**. Causes food intake inhibition in animals due to satiation. May have potential for use in foods as an appetite suppressant.

Simplesse Fat substitute derived from **whey proteins** or **egg proteins**, and used in foods such as **dairy products**, **condiments**, **margarines** and **sauces**.

Simulated foods **Processed foods** that are modified to simulate another kind of food, e.g. by using **textured vegetable proteins** and **flavourings**, to mimic **texture** and **sensory properties** of the target food. Some of the most popular simulated foods are **meat substitutes** (e.g. for use in **vegetarian foods**), **butter substitutes** and **imitation cream**. Also known as **imitation foods**, **analogues** or **artificial foods**.

Simultaneous saccharification and fermentation Process which involves enzymic **saccharification** of cellulosic **biomass** and simultaneous microbial **fermentation** of the resulting **glucose**, e.g. to **ethanol**. Advantages over the traditional two-stage process include the ability to use lower temperatures, thus reducing operating costs. Although these process can be performed by mixed cultures of an appropriate enzyme-producing microorganism and a fermentative microorganism, recent research has focused on genetic engineering of strains to enable them to ferment cellulosic substrates directly.

Sinapic acid Organic acid which is the major phenolic compound found in **rapeseeds**. Structural changes in sinapic acid have been associated with the darkening of **rapeseed meal** following extraction of **rapeseed oils**, and its presence limits the usefulness of rapeseed meal as a food source.

Sinapine An antinutritional ester with a hot, bitter taste found in the seeds of all *Brassica* spp. Elimination of sinapine from these seeds increases their potential as food sources.

Single cell proteins Protein-rich **biomass** produced by large-scale microbial fermentation using a variety of substrates, such as petroleum fractions or carbohydrates. Used as a source of proteins for use in foods and animal feeds. There is potential for future commercial exploitation of these proteins with advances in **fermentation technology**.

Single cream **Cream** with a fat content of approximately 18%.

Single market An association of countries trading with each other without restrictions or tariffs.

Sinigrin Antinutritional glucosinolate with a bitter taste found in *Brassica* spp.

Sitao Alternative term for **asparagus beans**.

Site directed mutagenesis *In vitro* **mutagenesis** at a specific site in a **DNA** molecule. Various methods can be used, e.g. **oligonucleotides** containing the mutated base sequence are annealed with single-stranded target DNA molecules, usually in **plasmids**, and used as primers for DNA synthesis; the molecules can then be introduced into host cells where subsequent DNA replication segregates the mutant and non-mutant strands.

Sitophilus Species of **weevils** of the family Curculionidae, which are **pests** of crops and stored grains and **cereals**. Include the grain weevil (*Sitophilus granarius*) and the rice weevil (*S. oryzae*).

Sitostanol Phytostanol occurring widely in plants. Can lower levels of total and low density lipoprotein **cholesterol** in blood by inhibiting absorption of cholesterol from the intestine. Used in the form of an ester in production of **functional foods**, such as **spreads**, **food bars** and **yoghurt**, which may have a cholesterol lowering action.

Sitosterol One of the **phytosterols** found commonly in plants and **vegetable oils**, and in certain **algae**. Has been shown to reduce the levels of total cholesterol and low density lipoprotein cholesterol in serum. Exist in α-, β- and γ- isomers.

Skate General name for a number of flattened **marine fish** species in the order Rajiformes; worldwide distribution. Generally used synonymously with **ray**. Commercially important species include *Raja binoculata* (big skate), *R. oxyrhinchus* (longnose skate) and *R. innominita* (smooth skate). Marketed fresh, frozen, smoked and salted; fins are also consumed.

Skatole Biogenic amine which contributes to the development of **taints** in **meat**, particularly **pork**.

Skimmed milk Alternative term for **skim milk**.

Skim milk **Milk** from which virtually all the fat has been removed (fat content is less than 0.5%). Preferred to **whole milk** or **semi skimmed milk** by some health-conscious consumers and used by processors to make low fat **dairy products**. Almost total removal of fat means that skim milk differs greatly from whole milk in **mouthfeel** and also in appearance, having a bluish tinge.

Skim milk powders Products prepared by drying **skim milk** to a low moisture content, giving powders with a long **shelf life**. Also called **dried skim milk** and non-fat dried milk.

Skin The outermost covering of the body, which consists of a thin outer layer, the epidermis, and a thicker inner layer, the dermis. Large quantities of gelatin are produced from **swine skin**. **Chicken skin** also has food uses, particularly in poultry products, such as

sausages. Cattle skins (hides), are by-products of cattle processing, and are generally tanned to produce leather.

Skinning Removal of the **skin** from foods such as **poultry** and **fish** before or after cooking.

Skin spot Disorder in **potatoes** caused by infection with the fungus *Oospora pustulans* (syn. *Polyscytalum pustulans*). Infection occurs in the soil and spreads during storage, particularly at low temperatures. Causes dark spots on the skin that become purple-black when wet.

Skipjack tuna **Marine fish** species (*Katsuwonus pelamis*) which forms the largest part of the world **tuna** catch by volume; widely distributed across the Atlantic and Pacific oceans. Mainly marketed as a canned product, but also sold fresh, frozen, dried, salted and as a semi-preserved product.

Skyr Icelandic **fermented milk** similar to thick **yoghurt**. Served as a dessert with **cream** and **sugar** or **fruits**.

Slaughter The killing of animals and **poultry** for food. In developed and some developing countries, slaughter of animals for **meat** takes place under closely regulated conditions in **slaughterhouses**. Legislation often dictates that food animals should be slaughtered without undue stress and suffering, and that bleeding should be as complete as possible; effective **stunning** is of primary importance in achieving these aims. Both Judaism and Islam prescribe a ritual protocol for the slaughter of animals for human consumption; stunning is not used in kosher or halal slaughter.

Slaughter by-products Alternative term for **offal**.

Slaughterhouses Places where the **slaughter** of animals takes place in a hygienic fashion, and where **carcasses** are prepared for retail to consumers. The term includes **abattoirs** and **butcheries**. Usually, carcasses are examined at slaughterhouses by qualified inspectors and only those carcasses that are free from disease are allowed to leave the premises for retail to consumers.

Slendid **Pectins** based fat substitute used in **low fat foods**.

Slicing Cutting of thin pieces or slices of food from a larger portion using a sharp implement.

Slime bacteria Common term for **bacteria** which produce an **exopolysaccharides** slime. Includes species of **Xanthomonas**, **Leuconostoc**, **Alcaligenes**, **Enterobacter**, **Lactococcus** and **Lactobacillus**. Slime produced may be responsible for characteristic **texture** and **viscosity** of **fermented milk**. May lead to severe quality problems during the storage and processing of certain foods and **beverages**.

Slivovitz **Spirits** made by **distillation** of fermented **mashes** based on **plums**. Manufactured mainly in Serbia, Bosnia and adjacent countries.

Sloes Common name for the wild plum produced by *Prunus spinosa* (also called blackthorn). Too acid for use as a dessert, but commonly used to make **beverages** such as sloe **wines** and sloe **gin**.

Sludges Usually thick, soft, wet mud or similar viscous mixtures, or alternatively any undesirable solids settled out from a treatment process.

Slugs Common name for gastropod **molluscs** of the family Limacidae, which have long fleshy bodies and in which the shell is either vestigial or absent. Occur in damp terrestrial habitats worldwide. Some are plant-eating **pests**.

Small round structured viruses **Viruses** within the Caliciviridae family of **viruses** (e.g. **Norwalk viruses**) which have well-defined surface structures. Responsible for viral **gastroenteritis** transmitted by the faecal-oral route via contaminated foods (e.g. **shellfish**) or water.

Small round viruses **Viruses** with smooth edges and no discernible surface structures. Responsible for viral **gastroenteritis** transmitted by the faecal-oral route via contaminated foods (e.g. **shellfish**) or water.

Smear cheese **Cheese** in which the rind is washed at intervals with water or **brines** to inhibit the growth of unwanted **yeasts** and **fungi**.

Smell Alternative term for **aroma**.

Smen Flavouring made from clarified **butter**, dried **herbs** and **salt**. Aged in earthenware pots until it acquires a **consistency** and **aroma** similar to **Roquefort cheese**. Used in Moroccan dishes.

Smetana **Sour cream** product popular in Eastern Europe. Used as a drink or as **toppings** for many dishes. Also known as smatana.

Smoke Substance obtained by burning certain types of wood, such as hickory, maple, or ash, that is used in the process of preserving or flavouring of foods, including **meat** and **fish**.

Smoke concentrates Concentrated **smoke flavourings** produced by removal of the liquid base in which wood smoke **flavour compounds** are dissolved.

Smoked fish **Fish** which has been processed by **smoking**. Whole gutted or ungutted fish or **fish fillets** are smoked at high or low temperature, sometimes after **brining**, using smoke produced from various types of wood. The kind of wood used affects the **flavour** and **colour** of the product. Cold smoking is usually performed at a temperature not exceeding 30°C.

Hot smoking is carried out at a temperature sufficient to cause thermal **denaturation** of the proteins, usually between 50 and 80°C. Common types of smoked fish include **smoked salmon**, **smoked trout**, **smoked mackerel**, smoked **eels**, smoked **haddock** and **kippers**. Some types are mainly eaten cold, while others, especially smoked haddock and kippers, are usually eaten hot.

Smoked foods Foods preserved and flavoured by treating with **smoke**, e.g. **kippers**, **yellow fish** and smoked meats. In traditional methods, foods are placed directly in the smoke. Other methods use **smoke flavourings** or **liquid smoke**, which are applied to the food. **Sensory properties** of smoked foods are affected by the type of smoke and **smoking** method used.

Smoked mackerel **Mackerel** which has been cooked by **smoking**. Fish can be smoked whole and gutted, or as fillets, by hot smoking at a temperature between 50 and 80°C or cold smoking at up to 30°C. Usually eaten cold in **salads** or made into **pates**.

Smoked salmon **Salmon** which has been smoked by one of two methods, i.e. hot **smoking** or cold smoking. In hot smoking, the process is conducted at a high temperature (50-80°C) and lasts 6-12 hours, the time depending mainly on the size of the **fish** and strength of **flavour** desired. Cold smoking is performed at a much lower temperature (up to 30°C) and may take several weeks to complete. Smoked salmon is usually eaten thinly sliced and cold, often in **salads** and **sandwiches**, although it can be used as an ingredient in many dishes, hot or cold.

Smoked trout **Trout** which has been cooked by **smoking**. The **fish** are most commonly hot smoked, whole or filleted, at a temperature of 50 to 80°C. Can be eaten cold in **salads**, made into **pates** and mousses, or used as an ingredient of **soups** and other hot or cold dishes.

Smoke flavourings **Flavourings** produced by contact of a liquid, usually water or oils, with smoke produced by burning of wood, e.g. hickory, oak or maple.

Smokehouses Sheds or rooms where **smoking** of foods, such as **fish** or **meat**, is performed.

Smoking **Curing** or **preservation**, especially of **meat** or **fish**, by exposure to **smoke** produced by the burning of certain types of wood, such as hickory, maple or ash. Foods can be cold smoked or hot smoked. Hot smoking partially or totally cooks foods.

Smoothies Thick and smooth textured **beverages** made by blending **fruits** with **yoghurt**, **milk**, **ice**, **ice cream** or **frozen yoghurt**.

Smoothness **Sensory properties** relating to the extent to which a product has a smooth **consistency**, i.e. is perceived to be uniform and regular.

Sn Chemical symbol for **tin**.

Snack foods Sweet or savoury foods eaten to provide light sustenance in a quick and convenient format. Eaten between or as an alternative to main **meals**. Popular types include **sandwiches**, **cereal bars** and **potato crisps**. Also known as snacks.

Snacks Alternative term for **snack foods**.

Snail meat Meat from **snails**. The edible portion of snails accounts for <50% of live weight. Snail meat provides a valuable source of dietary protein; for example, snail meat from giant African land snails (*Achatina achatina* and *Archachatina marginata*) typically has a protein content >70%. Proteins from snail meat include all essential amino acids, but the amino acids tend to be present in lower quantities than in mammal meat; snail meat also contains considerable quantities of other N compounds. Biological value of snail proteins is similar to that of **soy proteins**. Snail meat has a high content of **polyunsaturated fatty acids** (PUFA).

Snails A large group of creeping terrestrial and aquatic gastropod **molluscs**; some are herbivorous whilst others are carnivorous. Several species of land snails and marine snails are harvested from the wild or farmed as a source of **snail meat**.

Snake fish **Marine fish** species (*Trachinocephalus myops*) belonging to the family Synodontidae (**lizardfish**) and of minor commercial importance. Distributed worldwide in tropical and warm temperate waters.

Snakehead Name given to a number of **freshwater fish** species from the genus *Channa*; occur in lakes and ponds across South East Asia. Have elongated, cylindrical bodies. Some species are utilized as food fish, including *C. micropeltes* (giant snakehead) and *C. striata* (murrel). Flesh tends to be firm and white with very few bones. Marketed fresh, also used to make **fish pastes** and **soups**.

Snakes Predatory reptiles belonging to the suborder Ophidia; there are many species. Snake meat resembles **chicken meat**, but is chewier and has many small bones. It forms a part of diets in countries including Cameroon, China and Papua New Guinea. In Korea, snakes are used to produce snake **wines**, including salmo-sa, dok-sa and nung-sa.

Snap beans Type of **common beans** (*Phaseolus vulgaris*).

Snapper Any of a number of **marine fish** within the family Lutjanidae; widely distributed across the Atlantic and Pacific oceans. Commercially important species include *Lutjanus campechanus* (red snapper), *L. analis*

(mutton snapper), *Apsilus dentatus* (black snapper) and *Ocyurus chrysurus* (yellowtail snapper). Flesh tends to be lean and firm. Normally marketed fresh.

Snezhok **Fermented milk** beverage flavoured with fruit **syrups**.

SNF Abbreviation for **solids not fat**.

Snoek Alternative term for **pike**, used in the Netherlands.

Snow crabs Marine **crabs** of the genus *Chionoecetes* that occur in cool waters of the north Pacific and northeast Atlantic oceans. The most important species commercially is *C. opilio* (Atlantic snow crabs). Only male crabs have commercial value. Marketed in a variety of forms, including fresh cooked whole crab, cooked leg meat, canned meat and pastes.

Snow peas Type of **peas** (*Pisum sativum*) in which the entire pod is eaten with the seeds inside. The pods are flat and crisp, while the seeds are small and appear immature. Can be eaten raw or cooked (often stir fried). A common ingredient of Chinese dishes. Also known as mangetout peas.

SO₂ Chemical symbol for **sulfur dioxide**.

Soaking Process by which an item is made thoroughly wet by immersion in a liquid.

Soapiness One of the **sensory properties**; relating to the extent to which a product tastes soapy.

Sockeye salmon **Pacific salmon** species (*Oncorhynchus nerka*) found in coastal waters and rivers along the Pacific coasts of North America and Japan. Flesh is highly prized for **flavour** and **texture**. Most of the catch is canned; also marketed as a smoked or salted product. Also known as **red salmon**.

Soda bread Simple type of **bread** leavened with **sodium bicarbonate** and acid instead of **yeasts**. Often enriched with **whey** or **buttermilk**.

Soda water Water carbonated so that it is effervescent when dispensed.

Sodium Soft, silvery, highly reactive alkali metal with the chemical symbol Na, most commonly found in the form of **salt** (NaCl). An essential nutrient in the diet, albeit in moderate quantities; excess intake may result in high blood pressure (**hypertension**).

Sodium acetate Sodium salt of **acetic acid**. Anhydrous and trihydrate forms of this salt are both used as food **additives**. The anhydrous salt is hygroscopic and both forms are highly soluble in water. Uses in foods include as part of pH buffering systems, **flavourings** and **preservatives**.

Sodium ascorbate Sodium salt of **ascorbic acid** (**vitamin C**). In addition to being a source of vitamin C for fortification of foods, this salt has food industry uses in **antioxidants** and **preservatives**. It is also used in **curing** of **meat**.

Sodium bicarbonate Monosodium salt of **carbonic acid** prepared by reaction of sodium carbonate with carbon dioxide and water. In aqueous solution, the bicarbonate tends to decompose, releasing CO_2. Due to this property, sodium bicarbonate is used in **raising agents** for **bakery products**, as an ingredient of **baking powders**, and in the manufacture of **carbonated beverages**. **Baking powders** contain sodium bicarbonate and **tartaric acid**, reaction between which increases CO_2 production. Aqueous bicarbonate solutions are slightly alkaline; solution pH increases with agitation, time and increasing temperature due to loss of CO_2 from the solution. This salt is therefore also used as an alkali and as part of pH buffering systems in foods. Also known as sodium hydrogen carbonate and baking soda.

Sodium caseinate Sodium salt of **casein**. Used in a wide range of foods as a source of protein or to enhance **functional properties** such as **water binding capacity**, **emulsifying capacity**, whitening ability and **whipping capacity**.

Sodium chloride Chemical name for **salt**. Chemical formula is NaCl.

Sodium cyclamate One of the **cyclamates**, **artificial sweeteners** with approximately 30 times the sweetness of **sucrose**. White crystalline solid, highly soluble in water and stable at **baking** temperatures. It has a pleasant **flavour** profile and thus is often used in combination with **saccharin**, a molecule that is sweeter than sodium cyclamate, but which produces a bitter **aftertaste**. Also known as sodium cyclohexylsulfamate and sucaryl sodium.

Sodium hydroxide Highly caustic alkali, chemical formula NaOH.

Sodium lactate Sodium salt of **lactic acid**. Hygroscopic, soluble in water and alcohol and odourless, but with a slight salty **flavour**. Used in **additives** including **preservatives**, **emulsifiers**, **flavour enhancers**, **humectants**, and as part of pH buffering systems. Also known as 2-hydroxypropanoic acid monosodium salt and lacolin.

Sodium metabisulfite Disodium salt of disulfurous acid that forms acidic aqueous solutions. Used predominantly in **preservatives**, but also in **antioxidants**, **flavourings** and **bleaching agents**. Also called (di)sodium pyrosulfite and sodium bisulfite.

Sodium metabisulphite Alternative spelling of **sodium metabisulfite**.

Sodium pyrosulfite Alternative term for **sodium metabisulfite**.

Sodium tripolyphosphate Phosphate often used to improve the **physicochemical properties**, and increase the quality and **shelf life** of meat products.

Soft cheese **Cheese** with a creamy, smooth **texture** made from **milk** with a relatively low **dry matter** content and range of fat contents, **skim milk** soft cheese having a **butterfat** content of <2% and full-fat soft cheese containing at least 20% butterfat. Can be ripened, e.g. **Camembert cheese**, or unripened (fresh), e.g. **Cottage cheese**, **cream cheese**, **fromage frais**, **quarg**.

Soft drinks Non-alcoholic **beverages**, commonly **carbonated beverages**, often with fruit or cola flavours.

Softeners **Additives** that increase **softness** of foods. Examples include **glycerides**, which are added to **bakery products** as crumb softeners and to **chewing gums** to improve **texture**. Also used to describe **chelating agents** that remove ions, e.g. calcium or magnesium ions, from water.

Softening Process whereby products such as **fruits** and **vegetables** lose their **rigidity** and **firmness**, often during **ripening** or **ageing**.

Soft frozen beverages **Frozen beverages** which are served in a partially frozen or slush state.

Softness One of the **sensory properties**; relating to the extent to which a product is firm in **texture**.

Soils Earth in which **plants** are grown. Growth rates of plants, and yield and quality of their **fruits** are influenced by the composition of the soils in which they are grown.

Solanidine Alkaloid present in the sprouts and skin of green **potatoes** (potatoes exposed to light). Formed by hydrolysis of **solanine**. Toxic to humans and not destroyed by **cooking**.

Solanine Alkaloid present in all green parts of the potato plant, including **potatoes** that have been exposed to light. Inhibitor of **cholinesterases**. Poisoning causes gastrointestinal and neurological disorders. Solanine is not destroyed by **cooking**.

Solanum Genus of plants which includes many species that produce commercially important fruits. These include **aubergines**, **potatoes**, **naranjilla** and **pepino**.

Solar driers Equipment used to carry out **solar drying**, a process that depends on the sun as the source of energy. There are two types of solar driers: direct and indirect. In direct solar driers, air is heated in the drying chamber, which acts as both the solar collector and the drier. An indirect drier comprises two parts: a solar collector and a separate drying chamber.

Solar drying **Drying** method that depends on the sun as the source of energy, but which also involves the use of some sort of structure to collect and enhance the solar heat. Solar drying generates higher air temperatures and lower humidities than those produced by sun drying, resulting in faster product drying rates and lower final moisture contents.

Solar energy Radiant energy emitted by the sun that is captured and used during **solar drying** processes or converted into electrical energy.

Solar radiation Radiation emitted by the sun, made up of an extensive range of wavelengths of the spectrum.

Sole Any of a number of marine and estuarine **flatfish** species in the family Soleidae; worldwide distribution. Tend to have lean white or off-white flesh with fine **texture** and mild **flavour**. Commercially important species include *Solea solea* (common sole), *Buglossidium luteum* (yellow sole) and *Microchirus variegatus* (eyed sole). Marketed fresh and frozen.

Solidification Process by which an item becomes hard or solid.

Solid phase extraction Extraction technique for preparation of samples prior to analysis, developed as an alternative to liquid-liquid extraction. Samples are dissolved in solvent and passed through a bed of adsorbent to effect separation of components of interest. Compounds are eluted with small volumes of solvent.

Solid phase microextraction Type of **solid phase extraction** in which samples are adsorbed onto a fused silica fibre coated with a stationary phase. The fibre is then inserted into a GC injector, where the sample is desorbed and analysed.

Solids Particles whose shape and volume are fixed and are not affected by the space available to them, and which have a tendency to resist forces that would alter their shape.

Solids not fat The solids content of **milk** excluding the **fats** content, i.e. the contents of **proteins**, **lactose** and **salts**. Used as an index of milk quality. Commonly abbreviated to SNF. Milk contains on average 8.6% SNF.

Solid state fermentation Fermentation of **microorganisms** on a solid support of low moisture content under non-septic conditions. Energy requirements are low but the process can yield high product concentrations. In addition, **downstream processing** is facilitated. A variety of agricultural residues (e.g. wheat straw, rice hulls and corn cobs) have been used as supports for production of **enzymes** and secondary metabolites.

Solubility Extent to which one substance dissolves in another. Normal solubility records the maximum mass of a solid that can be dissolved in a specified mass of water to form a saturated solution, and is measured in kilograms per metre cubed. When solubility is exceeded, excess solid appears as a precipitate. Solubility is temperature-dependent. Generally, for a solid in a liquid, solubility increases with temperature; for a gas, solubility decreases with temperature.

Solubilization Process by which a substance is made soluble or more soluble, especially in water.

Soluble solids Particles that can be dissolved in fluids, especially water.

Solvents Liquids that dissolve other substances to form solutions. Polar solvents are compounds such as water and liquid ammonia, which have dipole moments and consequently high dielectric constants. These solvents are capable of dissolving ionic compounds or covalent compounds that ionize. Nonpolar solvents are compounds such as ethoxyethane and benzene, which do not have permanent dipole moments. These do not dissolve ionic compounds but will dissolve nonpolar covalent compounds. Solvents can be further categorized according to their proton-donating and accepting properties. Amphiprotic solvents self-ionize and can therefore act both as proton donators and acceptors. A typical example is water. Aprotic solvents neither accept nor donate protons; tetrachloromethane (carbon tetrachloride) is an example.

Somatic cells Animal cells that are not involved in reproduction. In **milk**, most of the somatic cells are white blood cells (**leukocytes**) that cross into milk from the bloodstream to destroy **bacteria**. The level of somatic cells in **milk** is an indicator of udder health, increasing in cases of infection, e.g. **mastitis**.

Somatotrophin Alternative term for **somatotropin**.

Somatotropin Alternative term for growth hormone, a substance produced by the anterior lobe of the pituitary gland which stimulates the synthesis of proteins, mobilizes reserves of fats and increases blood glucose levels. Recombinant bovine somatotropin may be administered to cattle to modify **milk** production, growth rate, or composition of cattle carcasses or **beef**. This application is permitted in some countries but prohibited in others due to concerns about the safety of food products.

Somen Thin, white Japanese **noodles** made from **wheat flour** and sometimes **egg yolks**. Often served chilled, as well as in **soups**.

Sonication Process of disrupting biological materials such as **bacteria**, **plants** or foods using high-frequency sound waves. Used widely in preparation and extraction of samples prior to analysis.

Sorbates Salts of **sorbic acid**. Sorbates, including sodium, potassium and calcium sorbates, are used as **preservatives** for foods, particularly **cheese**, and **beverages**, including **wines**.

Sorbestrin Thermally stable preparation formed from **fatty acid esters** of **sorbitol** and sorbitol anhydrides. Used as a substitute for **vegetable oils** in applications such as **frying oils**, **salad dressings** and **mayonnaise**.

Sorbets Water ices made from water, **sugar** and sometimes **eggs**, flavoured with **fruit purees** or **fruit juices** and sometimes **alcoholic beverages** (e.g. **champagne**). Frequently served between courses of **meals** to act as a refresher.

Sorbic acid Organic acid, solutions of which exhibit **antimicrobial activity**. The free acid and it salts (**sorbates**) are used as food **preservatives**. Sorbic acid also has uses in **acidulants** and **flavourings**. Systematic name is 2,4-hexadienoic acid.

Sorbitan Emulsifier formed via cyclization of **sorbitol**. Many sorbitan-based food **emulsifiers** are available commercially. Most are sorbitan esters of **fatty acids**, such as sorbitan oleate and sorbitan stearate.

Sorbitol Sugar alcohol (polyol) produced by reduction of **glucose** or **fructose**. Occurs naturally and has approximately 0.5× the **sweetness** of **sucrose**. Digestion of sorbitol yields fructose, making it suitable for use as a sweetener for diabetic foods. Also known as glucitol.

Sorbitol dehydrogenases Alternative term for **L-iditol 2-dehydrogenases**.

Sorbose Monosaccharide of 6 C atoms (**hexoses**) that is also one of the **ketoses**. L-Sorbose is an intermediate in the synthesis of **ascorbic acid**.

Sorghum Grain produced by cereal plants belonging to the species *Sorghum vulgare* and *S. bicolor*. Seeds are dark brown/red or white/yellow. Dark seeds have a high content of **tannins**, which decrease **palatability** and protein digestibility. White/yellow seeds are thus preferred for food applications, which include porridge, bread and beer. Also a source of **starch** and **syrups**.

Sorghum beer Beer brewed with **sorghum** as the main source of fermentable **carbohydrates**. Most sorghum beers are traditional African beer types, but conventional Western-style beer may be brewed with sorghum.

Sorghum malt Malt prepared from red **sorghum** varieties, which has high **diastatic activity**. Usually rich in **α-amylases**. Used in **brewing**, **weaning foods** and **breakfast cereals**.

Sorgo Synonym for the sugar crop, **sweet sorghum**.

Sorption A term that encompasses the various processes by which one substance binds to another, especially the processes of **absorption** and **adsorption**.

Sorrel Common name for *Rumex acetosa*, the leaves of which are used as vegetables and **spices** in **soups**, **salads** and **sauces**. Sorrel has a sharp, astringent **flavour**, similar to that of **rhubarb**, due to a high content of **oxalic acid**.

Sorting Systematic arrangement of items in groups or grades.

Soudjouk Spicy **fermented sausages** popular in Turkey. Ingredients include **beef mince**, **tallow**, **seasonings** and **spices**. Eaten cooked or uncooked. Many alternative spellings, including soujouk, soudjut and soudjouck.

Souffles Light spongy egg products which may be flavoured with sweet or savoury ingredients such as **jams** or **cheese**. Usually made by incorporating beaten **egg whites** into **sauces** containing **egg yolks**, **flavourings**, **flour** and **butter**, and cooking the mixture.

Soup mixes **Mixes**, usually powdered, that are recombined, typically with water, to form **soups**.

Soups Liquid foods typically made from **stocks** to which are added various vegetables and sometimes cereals, pasta, meat or fish. The term covers many types of product, including: clear, e.g. consommes; creamy, with all ingredients liquidized and often with cream added, e.g. cream of chicken; or thick, with chunks of ingredients floating in the clear liquid base, e.g. **broths**. Soups are generally eaten hot, but some types, e.g. vichyssoise, are usually consumed chilled. Other popular types include **borshch**, bouillabaisse and minestrone. Some types are also available as **soup mixes** or **instant soups**.

Sour cherries **Cherries** produced by *Prunus cerasus*. Suitable for **cooking**, they are commonly incorporated into **pies** and **jams**. Also used in manufacture of **fruit juices** and **liqueurs** such as **kirsch**. Available canned, frozen and dried. Include **morello cherries** and amarelle cherries.

Sour cream Commercial product made by **fermentation** of homogenized pasteurized **cream** with **lactic acid bacteria**. Used in **cooking** and as a component of **dips**. Also known as soured cream and ripened cream.

Sourdough **Dough** which has either been fermented by **microorganisms** naturally present in **flour** and/or other ingredients, or by added microbial cultures, e.g. **lactic acid bacteria**. **Fermentation** of the dough produces **organic acids**, which impart a desirable sour **flavour** to the dough. Used to make sourdough bread.

Sour milk **Milk** that has become rancid due to breakdown of **fats** or a **fermented milk**. The latter is produced by **fermentation** of milk (of various species) by **lactic acid bacteria** (**starters**). During fermentation, **lactose** is converted into **lactic acid**, **aroma compounds** are formed and **milk proteins** are partly decomposed to **peptides** and free **amino acids**, improving **digestibility** of the milk.

Sourness One of the **sensory properties**; relating to the extent to which a product tastes sour, i.e. tart, bitter or sharp.

Soursop **Fruits** produced by *Annona muricata*. Closely related to **sugar apples** and **custard apples**. The white flesh is embedded with black seeds and has a **flavour** reminiscent of **pineapples** and **mangoes**. Pulp is used in making **beverages**, such as **fruit juices**, and products such as **sherbet** and **custards**. Also known as guanabana.

Sous vide Food processing and packaging technique in which fresh ingredients are combined into specific dishes or **meals**, vacuum packaged in individual portion pouches, cooked under vacuum and then chilled.

Sous vide foods Vacuum-sealed pouches of **chilled foods** preserved by the **sous vide** process. Foods preserved in this manner undergo minimal heat processing and thus have improved **shelf life** compared with non-vacuum cook chill methods. Improved eating quality benefits have also been reported. Foods commonly processed in this way include **fruits** in **syrups** and some **meals** used in **catering**.

Sous vide meals Individual meal portions preserved in vacuum-sealed pouches by the **sous vide** process.

Southern blotting Method for detecting specific **DNA** fragments. DNA is digested with restriction endonucleases, separated by gel electrophoresis, denatured and transferred to a chemically reactive matrix (e.g. nitrocellulose or nylon), on which the DNA fragments bind covalently in a pattern identical to that on the original gel. After blotting, target molecules are detected through the use of labelled complementary single-stranded DNA or **RNA** molecules.

Southern peas Type of **cowpeas** (*Vigna unguiculata*).

Sovetskii cheese Russian **hard cheese**.

Sowing Scattering of plant seeds on or in the earth. Sowing date and rate can affect subsequent plant growth, as well as yield and quality of produce.

Sows Adult female **swine** that have produced their first litter of piglets.

Soyabeans Alternative term for **soybeans**.

Soybean lecithins Alternative term for **soy lecithins**.

Soybean oils **Oils** extracted from seeds of *Glycine max* (**soybeans**). Contain **palmitic acid**, **oleic acid**, **linoleic acid** and **linolenic acid**. Used as **salad oils** or **cooking oils**, as well as in **margarines** and **shortenings**. By-products obtained during processing include **lecithins**, **tocopherols** and **phytosterols**. Also known as soy oils.

Soybeans Seeds produced by *Glycine max*. Rich in high quality **soy proteins**, unsaturated **soybean oils**, B vitamins and **minerals**. Eaten whole or split, or germinated to produce **bean sprouts**. Numerous soy products are made from the seeds, including **soymilk**, **cheese**-like products (**tofu**, **tempeh**, **miso**) and **meat substitutes** (**soy meal**, **soy protein concentrates**, soy protein isolates). Soybean plants tolerant of specific herbicides were the first genetically modified crops to be produced on a large scale.

Soybean sprouts **Legume sprouts** produced by **germination** of **soybeans**. Rich in protein, **vitamins** and minerals. Widely used in Asian dishes such as egg rolls and stir fried meals. Also used in **soups**, **casseroles**, **sauces**, **bakery products**, and raw in **salads**. Dried sprouts can be eaten as **snack foods** or used as a substitute for **nuts** in **bakery products** or dishes.

Soy beverages **Beverages** derived predominantly from **soybeans** or their products. Include **soymilk**.

Soy cheese Creamy product made from **soymilk**. Used as a replacement for **cheese** or **sour cream**.

Soy curd Product made from **soymilk**, by coagulation, draining and **pressing** in a manner similar to that used in **cheesemaking**. Rich in protein. Available packaged in water, vacuum packaged or frozen. Used in a variety of dishes, including **soups**, **casseroles** and **sauces**. Also called **tofu**.

Soy flour Alternative term for **soy meal**.

Soy globulins Storage proteins of **soybeans**. Main **soy proteins** are **glycinin** and **β-conglycinin**.

Soy glycinin One of the main **soy proteins**. An 11S storage protein that, along with **β-conglycinin** (7S globulin), makes up approximately 70% of **storage proteins** in **soybeans**.

Soy ice cream Frozen dessert made from **soymilk** and used as a substitute for conventional **ice cream**.

Soy infant formulas Products made by mixing soy protein isolates with **fats** and **carbohydrates** to give a composition similar to that of **human milk**. Used mainly to feed infants who are allergic to **cow milk** or suffer from **lactose intolerance**.

Soy lecithins **Lecithins** extracted from **soybeans** and used as **emulsifiers** in foods. Also called soy lecithins.

Soy meal **Flour** made by grinding roasted, dehulled **soybeans**. Full fat soy meal is made from soybeans that still contain oil. Defatted soy meal is made using soybeans from which **soybean oils** have been extracted. Good source of **soy proteins**, iron, calcium, B vitamins and **fibre**. Used in **baking** and in **thickeners** for **sauces**.

Soymilk Product prepared by cooking dehulled, ground **soybeans** in water and filtering of the solid matter (**okara**). Rich in B vitamins, protein and iron. Used similarly to **milk** as a beverage, as the basis of soy products such as **soy yoghurt**, **soy ice cream** and **soy cheese**, and in **cooking** and **baking**. Available in regular, low-fat and flavoured forms or as a powder.

Soy oils Alternative term for **soybean oils**.

Soy pastes Fermented products prepared from cooked **soybeans**. Include Japanese **miso** and Korean **doenjang**. Used mainly as **seasonings**.

Soy protein concentrates Protein concentrates made by extracting **sugars** from defatted soy flakes, leaving **proteins** and **fibre**. Used to make **meat substitutes** and in a variety of products, such as **cereal products**, **bakery products**, **beverages** and **gravy**.

Soy proteins Storage proteins found in **soybeans**. Nutritional and health-promoting properties, combined with **functional properties** make them useful and widely-used ingredients in food processing.

Soy purees Preparations made by **mashing** or blending cooked **soybeans**. Used in **infant foods** and **beverages**.

Soy sauces Sauces produced by **fermentation** of a soybean mash prepared by grinding soybeans with water. A fungus, often *Aspergillus oryzae*, is added to the soybean mash to initiate fermentation. Duration of the fermentation process and addition of other ingredients influences **sensory properties** of the sauces. Soy sauces fermented for a shorter period have a less rich **flavour** than those fermented for a longer period. Addition of **molasses** produces richer, darker soy sauces, while inclusion of **wheat** in the fermentation produces lighter products.

Soy 11S globulins One of the two major types of **soy proteins** (the other group are the 7S **globulins**) that together make up 70% of the total **storage proteins** in **soybeans**.

Soy 7S globulins One of the two major types of **soy proteins** (the other group are the 11S **globulins**) that together make up 70% of the total **storage proteins** in **soybeans**. Trimeric **glycoproteins** comprising α, α′ and β subunits, which together form **conglycinin**. Responsible for **softness** and **adhesion** properties of soy products.

Soy yoghurt Creamy product made from **soymilk**. Used as a substitute for **cream cheese** or **sour cream**.

Space flight foods Meals designed for consumption in the confined microgravity environment encountered on space flight programmes. Originally bite-sized cubes or squeezed from a tube, space flight foods have now evolved into more appetizing meals that can in-

corporate frozen, refrigerated and ambient foods. A typical meal tray could include a foil beverage pouch, and individual servings of lightweight easily rehydrated foods, **intermediate moisture foods** and thermostabilized, aseptic fill, natural form foods. Early research into providing assurance against microbial contamination in space led to development of the **HACCP** concept.

Spaghetti Pasta in the form of long strands approximately 2 mm in diameter.

Spanish mackerel Alternative term for **chub mackerel**.

Sparkling winemaking Processes involved in manufacture of **sparkling wines**.

Sparkling wines Wines which contain sufficient dissolved **carbon dioxide** to result in effervescence when the bottle is opened. The high carbon dioxide content may be achieved by secondary **fermentation** (in the bottle or in a tank) or by **carbonation**.

Spearmint Common name for *Mentha spicata*, the leaves of which are used as **spices**. Has a sweet, minty (fresh and cool) **flavour** due predominantly to the flavour compound L-carvone. **Essential oils** distilled from spearmint are also used as flavourings, particularly for **chewing gums**.

Species identification Recognition of the animal source of products containing **meat**, **fish** or **milk**. Used to detect **adulteration** or establish **authenticity**. Methods used to identify the species of origin include **electrophoresis**, **isoelectric focusing** and **genetic techniques**. Can also refer to determination of microbial species.

Specific conductivity Measure of the ability of a substance to transport an electric charge. Inversely proportional to the electrical resistance of a substance and dependent upon its dimensions (length/cross-section) if the substance is a solid. For a solution, specific conductivity is measured between electrodes spaced 1 cm apart with a cross section of 1 cm^2. For dilute solutions, specific conductivity is proportional to electrolyte concentration.

Specific gravity Ratio of the **density** of a substance to the density of a reference material. For a liquid or solid, specific gravity is the ratio of its density (usually at 20°C) to the density of water (at its temperature of maximum density (4°C)). Synonymous with **relative density**. Abbreviated to sp. gr.

Specific heat Heat capacity of a substance per unit mass. The amount of energy required to raise the temperature of unit mass of an object by a unit increment in temperature (measured in Joules per Kelvin per kilogram).

Specific rotation Optical properties relating to the rotation that a beam of light of a given wavelength undergoes, relative to its plane of polarization, as it passes through a solution of a given density, path length, concentration and temperature.

Speck Cured, smoked meat products, made primarily of **pork** and traditionally produced in the Alpine region of Italy. Production includes **trimming**, dry curing and massaging, **smoking** and drying of pork. Periodic surface treatment of the **meat** with a mixture of **spices** results in formation of an outer crust, which helps to prevent bacterial contamination.

Spectinomycin Broad-spectrum aminocyclitol antibiotic produced by *Streptomyces spectabilis*. Used in the treatment of a variety of enteric, respiratory and other infections in farm animals. Exhibits low **toxicity** and is normally excreted rapidly from animal tissues. Also known as actinospectacin.

Spectra Pattern of properties arranged in order of increasing or decreasing magnitude. In analytical applications, the property measured varies according to the analytical technique being employed. In mass spectroscopy, a mass spectrum with a range of masses is produced. An emission spectrum represents the range of radiations emitted when a substance is heated, bombarded by electrons or ions, or absorbs photons. An absorption spectrum shows the energies absorbed from a continuous spectrum of radiation by an absorbing medium. Spectra produced by an unknown substance can be compared with those of a standard to give information about the composition of the sample.

Spectrofluorometry Spectroscopy technique in which the intensity of fluorescence of a sample is measured as a function of wavelength. A pair of monochromators is used, one of which selects the excitation wavelength and the other the emission wavelength.

Spectrometry Alternative term for **spectroscopy**.

Spectrophotofluorometry Alternative term for **spectrofluorometry**.

Spectrophotometry Alternative term for **spectroscopy**.

Spectroscopy Series of techniques in which absorption or emission of radiant energy of various wavelengths is used to measure chemical concentrations or structures. Includes atomic emission, atomic absorption, IR and mass spectroscopy.

Spelt Coarse, hardy type of **wheat** (*Triticum spelta*) cultivated predominantly in Europe. Unripe grains are used in **soups**. Also known as **spelt wheat**.

Spelt wheat Alternative term for **spelt**.

Spermidine Polybasic amine present almost universally in **prokaryotes** and **eukaryotes**. Affects the structure of **nucleic acids** and the activity of **enzymes**, and may play a role in the synthesis of **proteins** in **bacteria**. Also required for the replication of at least some **bacteriophages**. This biogenic amine is found in many foods, where it may contribute to **toxicity** and **spoilage**.

Spermine Polybasic amine present almost universally in **prokaryotes** and **eukaryotes**. This biogenic amine is found in many foods, where it may contribute to **toxicity** and **spoilage**. Has been shown to act synergistically with **tocopherols** in inhibiting the oxidation of polyunsaturated oils.

sp. gr. Abbreviation for **specific gravity**.

Sphingolipids Lipids derived from the amino alcohol sphingosine. Include **sphingomyelin**, **cerebrosides** and GM **gangliosides**. Thought to have a number of beneficial effects on health, including **anti-tumour activity**, and are therefore potentially useful as components of **functional foods**. Good sources include **dairy products**, **meat**, **eggs** and **soybeans**.

Sphingomonas Genus of Gram negative, aerobic rod-shaped **bacteria** of the family Sphingomonadaceae. *Sphingomonas paucimobilis* is used in the production of **gellan gums**.

Sphingomyelin Sphingolipid in which the amino group at C-2 of sphingosine forms an amide bond with a long chain fatty acid, while the terminal (C-1) hydroxyl is esterified with phosphorylcholine. Exhibits **antitumour activity** in **animal models** and may play a role in cholesterol metabolism.

Spices Aromatic plants or parts of plants, e.g. roots, leaves or seeds, in various forms (native, dried, ground, whole) used primarily for their **flavour** rather than for any nutritional benefit.

Spina bifida Specific type of neural tube defect (also known as rachischisis), present at birth, in which the backbone fails to fuse properly, leaving the spinal cord and its coverings exposed. The defect commonly occurs in the lower spine. Recent evidence suggests that the risk of spina bifida is reduced if extra **folic acid** is included in the diet of women in the first three months of pregnancy.

Spinach Common name for *Spinacea oleracea*. Leaves are eaten raw or boiled, or as ingredients of **soups**, **pasta** and dishes such as **souffles** and **quiches**. Rich in nutrients, including protein, **fibre**, β-carotene, iron, **vitamin C**, **vitamin E** and B vitamins.

Spinal cord tissues Tissues associated with the part of the central nervous system in vertebrates which is lodged in the vertebral canal and from which spinal nerves emerge. Due to concerns about a possible link between variant **Creutzfeldt-Jakob disease** (CJD) in humans and **bovine spongiform encephalopathy** (BSE) in cattle, controls are in place in **abattoirs** and **slaughterhouses** to exclude BSE risk materials, such as spinal cord tissues and other **central nervous system tissues**, from the human food chain. Measures include processing of **carcasses** without splitting the spine or removal of spinal cord tissues prior to splitting. BSE risk materials are considered a source of BSE **prions**, consumption of which could potentially result in the development of CJD. In addition, techniques have been developed to screen **meat** and meat products for the presence of spinal cord tissues.

Spinning **Texturization** process usually applied to **protein isolates**. For example, biodegradable films can be prepared by spinning soy protein isolates in a coagulating buffer, and wet spinning methods can be used to produce edible protein fibres from a variety of materials, such as **soy proteins**, **casein** and blood plasma proteins. The term can also be used to describe the process used in the manufacture of **cotton candy**. **Chocolate** can be spun moulded.

Spinosad Insecticide used to control a variety of **insects**, including **fruit flies**, **thrips**, leaf miners and certain **beetles** in fruit and vegetable crops. Effective at low usage rates and of short residual activity. Has low toxicity to humans and wildlife.

Spiny lobsters Alternative term for **crawfish**; marine lobster species within the genera *Palinurus* and *Panulirus*.

Spiramycin Macrolide antibiotic produced by *Streptomyces* *ambofaciens*. Used for treatment and control of a number of bacterial and mycoplasmal infections in animals. Distributes widely in tissues following absorption from the gut, but is normally excreted rapidly.

Spirits **Alcoholic beverages** with high **ethanol** contents, made by **distillation** of fermented **mashes** derived from **fruits**, **cereals**, **root crops**, **sugar cane** or other sources of fermentable **sugars**.

Spirulina Genus of cyanobacteria. Occur in warm saline environments. Some species (e.g. *Spirulina platensis*) are used in the production of **single cell proteins**.

Spleens A part of edible **offal**. The spleen is the largest lymphatic organ in the body, and has a sponge-like structure. In animal **carcasses**, it lies in the upper left abdomen, between the stomach and the diaphragm. Spleens of cattle, sheep and goats have been identified as risk materials in relation to the transmission of **prion diseases**; consequently, in many countries they are banned from the food chain.

Splitting Breaking forcibly into parts. For example, the cutting of animal **carcasses** into left and right

sides using a saw during processing. Also relates to undesirable processes, such as the damage that can occur to **fruits** (such as **tomatoes**, **cherries** and **grapes**) when their peel splits upon absorption of excess water, and fruit splitting, a physiological disorder of peel development in **citrus fruits**. Water stage fruit split is an erratic and complex problem often causing major crop losses to susceptible cultivars of **pecan nuts**. In the beverage industry, **corks** placed in wine bottles can be susceptible to splitting. Problems are also associated with premalting (splitting) of **malting barley**, which is thought to be caused by alternating periods of sunny and rainy weather during ripening of the grain. **Egg shells** can split during boiling; this results from excess internal pressure in the egg, due to the egg contents having a higher coeff. of thermal expansion than the shell. Canned **kidney beans** are liable to split during storage, and **sausage casings** can split during **cooking**.

Spoilage Deterioration of a food by chemical, physical or microbial means.

Spoilage bacteria **Bacteria** typically involved in the **spoilage** of foods.

Spoilage yeasts **Yeasts** typically involved in the **spoilage** of foods.

Spondias Genus of tropical plants, some species of which produce good quality **fruits**, including **caja**, jocote, ciruela fruit, ambarella and African plums. Fruits are eaten fresh, cooked or dried, and made into jellies or **beverages**, such as **fruit juices**. Flower clusters from *Spondias mangifera* are consumed as a vegetable or in **salads**. Some species of *Spondias* are a source of **gums** suitable for food applications.

Sponge **Dough** used in **breadmaking** which contains a proportion of the flour, all of the **yeasts**, yeast foods, **malt** and sufficient water to make a stiff dough. Fats may also be added, together with a proportion of salt; this controls **fermentation** which takes place over 3-5 hours.

Sponge cakes Light, porous **cakes** made using self-raising **flour**, **sugar**, beaten **eggs** and **flavourings**. **Butter** or **oils** may be added, although many sponge cakes contain no **shortenings**.

Spores Usually unicellular, dormant reproductive or resting bodies produced by **microorganisms** under conditions of environmental stress (e.g. extremes of temperature and dehydration). Resistant to unfavourable environmental conditions, and capable of germinating and developing into vegetative cells when environmental conditions are favourable, without fusion with another cell.

Sporobolomyces Genus of yeast **fungi** of the class Urediniomycetes. Occur on decaying plant material.

Sporobolomyces roseus is used as a **biocontrol** agent in the control of postharvest diseases of **fruits** and **vegetables**.

Sporotrichum   Genus of **fungi**. *Sporotrichum carnis* is commonly involved in the **spoilage** of fresh and refrigerated **meat**.

Sports drinks **Soft drinks** formulated to enhance or maintain the performance of sports people, or to improve their recovery after a sporting event or training session. Generally contain ingredients such as sugars and electrolytes.

Sports foods   Products formulated to contain precise levels of **nutrients** and other ingredients intended to enhance sports performance in athletes.

Sporulation   Process by which **spores** develop in **microorganisms**.

Sprat   Small **herring**-like **marine fish** species (*Sprattus sprattus*) distributed across the northeastern Atlantic. Marketed fresh and frozen (whole, ungutted), smoked, canned (headed, tailed, gutted and packed in **oils** or **tomato sauces**) and as a component of **fish pastes** (dyed red to distinguish them from sardine pastes). Also known as brisling.

Spray dried foods **Dried foods** prepared by **spray drying** slurries or liquids. Foods dried in this manner include **milk** and **eggs**.

Spray driers   Equipment for manufacture of **dried foods** from liquids, such as production of **dried milk** from liquid milk, by **spray drying**. Liquids are sprayed as a fine mist into a hot-air chamber, where they dehydrate; solids fall to the bottom of the chamber as dry powders.

Spray drying   Process for manufacture of **dried foods** from liquids. The liquid food is generally preconcentrated by **evaporation** to reduce the water content. The concentrate is then introduced as a fine spray or mist into a tower or chamber with heated air. As the small droplets make intimate contact with the heated air, they flash off their moisture, become small particles, drop to the bottom of the tower and are removed. The advantages of spray drying over other types of **drying** include the need for only a low heat and short time, which leads to better quality product.

Spreadability **Texture** term relating to the ease with which a product can be spread.

Spreads   General term for preparations spread onto products such as **bread** or **crackers**, sometimes in place of **butter**. May be low in fat, and either sweet or savoury.

Springbok meat **Meat** from **springboks**. Springbok **carcasses** are commonly used to produce fresh venison-type meat and **biltong**.

Springboks   African gazelles (*Antidorcas marsupialis*) which are hunted, often as part of controlled culling programmes, for **springbok meat** production.

Springiness   One of the **sensory properties**; relating to the extent to which a product springs back quickly when squeezed, bent, pressed or stretched.

Spring waters **Mineral waters** derived from springs or similar sources.

Sprouting   Term synonymous with **germination**, meaning the process whereby seeds or spores begin to grow. Also describes the production of sprouts in **potatoes** and other **tubers** during **storage**. Sprouting can be controlled by storing susceptible vegetables in the dark and at low temperatures, or by the use of sprouting inhibitors.

Squalene   Phenolic compound with **antioxidative activity** that is found in **olive oils** and **fish oils**. Has also been found to exhibit **antitumour activity** *in vitro* and in **animal models**.

Squash **Fruit juice beverages** (mainly based on **citrus fruits**) containing comminuted whole **fruits** (including **peel**). Commonly retailed as concentrates for dilution with water by the consumer.

Squashes   Fruits produced by plants of the genus *Cucurbita*, including *C. pepo* and *C. maxima*. Produce of this genus also include **marrows** and **pumpkins**. Summer squashes are immature fruits with a soft skin that are mainly used as a table vegetable. Winter squashes are mature fruits used in a variety of ways, such as in **pies** and **jams**, as well as being eaten as vegetables. Contain mainly water (usually at least 90%), with small amounts of **starch**, **sugar**, **fats**, **proteins**, **carotenes** and B vitamins, and moderate amounts of **vitamin C**.

Squash seeds   Seeds contained in fruits (squashes) produced by plants of the genus *Cucurbita*. Kernels are eaten raw or cooked, and used as a source of **oils**.

Squid   Marine cephalopod **molluscs** within the family Loliginidae; worldwide distribution. Flesh is firm and chewy, with a somewhat sweet **flavour**. Commercially important species include *Loligo vulgaris* (European squid), *L. pealei* (longfin inshore squid), *Todarodes pacificus* (Japanese flying squid) and *Ommastrephes bartrami* (flying squid). Marketed fresh and frozen (whole, ungutted; split, gutted) and as salted, semi-preserved, dried and canned products.

Squid ika shiokara   Fermented **sea foods** made from flesh and viscera of **squid**. Brown, salty viscous pastes made by fermenting the raw material with **salt** for up to one month; often flavoured with **sake** during **fermentation**.

Squid oils Oils derived from **squid** viscera. Generally rich in **docosahexaenoic acid** and **eicosapentaenoic acid**.

Srikand **Fermented milk** product usually prepared from **buffalo milk** and popular in India. Also known as **shrikhand** or srikhand. Traditionally, milk is fermented with a mixed starter culture (*Streptococcus lactis* and *S. lactis* var. *diacetylactis*) and chakka is prepared by draining off **whey** from the resultant **curd**. Other ingredients, e.g. **sugar, colorants, flavourings**, are then added to the chakka.

Srikhand Alternative term for **shrikhand** or **srikand**.

Stabilization Process of making or becoming stable. **Stabilizers** such as **agar, alginates, carrageenans** and **gums** are used for the stabilization of foods.

Stabilizers **Additives** included in food formulations to prevent separation of ingredients and thus improve appearance and **shelf life**. Common uses include stabilization of oil and water components in **emulsions**, e.g. in **salad dressings**, of air incorporation into foams, e.g. in **whipped cream**, and of **proteins** in **beer**, precipitation of proteins producing **cloudiness**. Examples of stabilizers include **gums** and **hydrocolloids**.

Stachybotrys Genus of **fungi** of the class Hyphomycetes. Occur in soil, hay and other plant products. *Stachybotrys alternans* may be responsible for food **spoilage**. Ingestion or inhalation of **satratoxins** produced by *S. atra* on foods may cause stachybotryotoxicosis (a **mycotoxicosis**) in humans, horses, **cattle** and **poultry**.

Stachyose Non-reducing tetrasaccharide found in legumes and other plants, hydrolysis of which gives two molecules of **galactose**, and one each of **glucose** and **fructose**.

Staining Marking or **discoloration** with something that is not easily removed, such as penetrative **dyes**, **pigments** or chemicals.

Stainless steel Type of **steel** which contains chromium. Resistant to tarnishing and rusting. Widely used in equipment and utensils for the food industry.

Staling Process by which foods cease to be fresh or pleasant to eat. For example, **bread** becomes dry and hardened when stale, due to changes in the structure of **starch**.

Standardization Process by which substances and procedures are made uniform. In the dairy industry, the term refers to adjustment of the fat content of **milk** to a given level. Milk from different batches is blended to the desired fat content. Used especially to ensure the uniform quality of **cheese milk**.

Standards Something used as a measure, norm or model in comparative evaluations; a benchmark or specification.

Stanol esters **Fatty acid esters** of plant **stanols** (**phytostanols**). Commonly used in enrichment of foods such as **spreads, yoghurt** and **food bars** to produce products which may have a **cholesterol** lowering action. Reduce levels of total and low density lipoprotein cholesterol in blood by inhibiting absorption of cholesterol in the intestine.

Stanols Hydrogenation products of **sterols** which occur naturally in plants (**phytosterols**). Less abundant than the corresponding plant sterols. Like plant sterols, stanols reduce levels of total and low density lipoprotein **cholesterol** in blood by inhibiting absorption of cholesterol in the intestine. **Stanol esters** are commonly used in enrichment of foods such as **spreads, yoghurt** and **food bars** to produce products which may have a cholesterol lowering action.

Staphylococcus Genus of Gram positive, facultatively anaerobic coccoid **bacteria** of the family Micrococcaceae. Occur on the skin and mucous membranes of humans and animals. *Staphylococcus aureus* may be responsible for **food poisoning** due to consumption of contaminated foods (e.g. **meat** and meat products, **eggs, salads, bakery products** and **dairy products**). *S. carnosus* is used as a starter culture in the manufacture of **fermented sausages**.

Star anise Common name for *Illicium verum*, fruits of which are used as **spices**. The main aromatic compound present is **anethole**. Used to flavour **bakery products, beverages, meat** products and **sugar confectionery**.

Star apples Apple-sized **fruits** produced by trees of the genus *Chrysophyllum*, predominantly *C. cainito* and *C. africanum*. Round, with a white or purple rind which is green at the calyx. A soft, white, sweet pulp surrounds a centre containing seed cells. Flesh is scooped out of the bitter tasting rind and eaten raw, often mixed with other fruits, or mixed with **orange juices**. May also be made into **jams**. **Oils** extracted from the **seeds** are sometimes used as **cooking oils**. Also known as caimito.

Starch Polysaccharide that is the main energy store of plants. Composed of molecules of **amyloses** and **amylopectins**. Amount of each polymer, which varies between plant species, influences the **functional properties** of starch, such as gel forming ability of starch pastes. In addition to its role in cereal **flour** or **meal** used as a base for **breadmaking**, and manufacture of other **bakery products** and **pasta**, starch has many applications in foods, including as **thickeners, anticaking agents, coatings** and **binding agents**. Starch is often chemically or physically modified in order to improve its applicability for food processing, e.g. to increase **thermal stability** or alter the **texture**.

Starch granules Native structure of starch, comprising discrete aggregates of amylose and amylopectin. Arrangement of the starch polymers is highly organized and some crystalline regions are present due to strong interactions between amylopectin chains. Granules also contain minor amounts of protein, lipid, ash and moisture. Starch granule size and composition vary between plant species and varieties.

Starch hydrolysates **Sugar syrups** produced by hydrolysis of **starch** slurries. Starch hydrolysis is commonly achieved by the action of acids, e.g. hydrochloric acid, or **amylases**; degree of hydrolysis determines the saccharide composition of the syrups. Dextrose equivalent of a hydrolysate is a measure of the degree of hydrolysis relative to the dextrose (D-**glucose**) content, i.e. 100% dextrose equivalent denotes full hydrolysis. **Glucose syrups**, **maltose syrups** and **maltodextrins** are starch hydrolysates and are substrates for other starch-based **sweeteners**, such as **fructose high corn syrups** and crystalline **sugars**.

Starch synthases EC 2.4.1.21. Transfer the **glucose** moiety from ADPglucose to glucose-containing **polysaccharides** by means of 1,4-α-linkages. The entry also covers glycogen synthases that utilize ADPglucose. Several isoforms are found in plant tissues where they are responsible for synthesis of **starch**.

Starch syrups **Sugar syrups** produced from **starch** by hydrolysis with acids or **amylases**. Sugar composition of the syrups is dependent on the degree of hydrolysis, which is measured in terms of the dextrose content of the syrup (the dextrose equivalent value).

Star fruit **Fruits** produced by *Averrhoa carambola*. Waxy in appearance, the juicy, yellow fruits are star-shaped in cross-section. Contain relatively high amounts of **vitamin C** and approximately 7% total sugar. Used in **beverages**, **fruit salads**, **tarts** and **preserves**. Also known as **carambolas** and **five fingers**.

Starters Microbial cultures used to initiate **fermentation**. Mixtures of specific strains are used to produce the desired properties in the product. Types include **cheese starters**, **yoghurt starters** and **butter starters**.

Statistical analysis Group of mathematical techniques by which analytical results can be examined on the basis of probability theory.

Steaks Thick slices of high-quality **meat** taken from the hindquarters of animal carcasses. They are usually cooked by **grilling** or **frying**.

Steam Hot vapour into which water is converted when heated. Condenses in the air into a mist of miniature water droplets. Used as a source of energy or in **cooking** of foods.

Steamed bread **Bread** prepared by **baking dough** in **ovens** which are heated to a constant temperature using closed pipes through which steam is passed.

Steaming **Cooking** of foods by heating in steam produced from boiling water. The food to be steamed can be placed in steaming apparatus over boiling or simmering water in a covered pan. Steaming has advantages over **boiling** in terms of retention of **flavour**, **colour**, shape, **texture** and **nutrients** content of foods.

Stearic acid A saturated fatty acid which contains 18 carbon atoms. Found abundantly in animals and plants. Even though consumption of **saturated fatty acids** has been linked with an increased the risk of **coronary heart diseases**, data suggest that stearic acid may be neutral with respect to effects on serum **cholesterol** levels.

Stearin **Triglycerides** present in both animal fats and vegetable fats; found particularly in solid fats, such as tallow and cocoa butter. May also be synthesized by **esterification** of **stearic acid** with **glycerol**. Uses include as **emulsifiers** and surface-finishing agents for **chocolate** and **sugar confectionery**. Also known as tristearin, glyceryl tristearate and octadecanoic acid 1,2,3-propanetriyl ester.

Stearoyl lactylates Salts of the stearoyl lactylate anion prepared by reaction of **stearic acid** with **lactic acid**. The nature of the cation in the salt influences the **functional properties** of the lactylate, e.g. the sodium salt is soluble in water whereas calcium stearoyl lactylate is not. Uses include as **emulsifiers**, **dough conditioners** and **stabilizers**.

Steel Strong, hard grey or bluish-grey alloy made from iron with carbon and usually other elements. Used widely as a structural and fabricating material. Also refers to a rod of roughened steel which is used for sharpening **knives**.

Steeping Soaking of ingredients such as **tea leaves**, **herbs** and **spices** in water or other liquid until the **flavour** is infused into the liquid. The liquid used is usually hot. Also refers to soaking of **barley** or other **cereals** as part of the **malting** process, and during which **imbibition** occurs prior to **germination**.

Steers Castrated, adult male **cattle**, which are widely used for **beef** production. Compared with bulls, steers are easier to handle and their **carcasses** are less affected by stress related conditions, such as the **DFD defect**. However, steers grow more slowly, convert feed less efficiently and achieve lower carcass weights than bulls. Steer meat tends to be lighter in **colour** than bull beef.

Stellar Fat substitute based on starches such as **potato starch** or **tapioca**. Used as a bulking agent or texture modifier in **salad dressings**, **condiments**, **sauces**, **bakery products** and **dairy products**.

Stenotrophomonas Genus of aerobic **Gram negative bacteria**. *Stentrophomonas maltophilia* is a multidrug resistant opportunistic pathogen found in moist environments, including **water** and foods. Some strains found in **fish** can produce the biogenic amine **cadaverine**.

Sterculic acid One of the **fatty acids**; has a branched, odd-numbered, unsaturated C chain structure. A potent inhibitor of **desaturases**.

Stereoisomers Molecules with the same molecular formula and the same functional groups, but with different spatial arrangements, e.g. optical **isomers**.

Sterigmatocystins Carcinogenic and hepatotoxic **mycotoxins** produced by certain *Aspergillus* species (e.g. *A. nidulans* and *A. versicolor*) growing on foods (e.g. **cereals**, **fruits**, **coffee beans** and **cheese**).

Sterilization Destruction of all **microorganisms** and **spores** in or on a material, such as food, by various means, including the application of chemicals, heat, **radiation** or **filtration**. Conventional sterilization involves in-container sterilization, usually at temperatures between 115 and 120°C for 20-30 minutes. Commercial sterilization does not always meet this definition, because some harmless, heat resistant **bacteria** may still be present. The criterion for food sterility is a process which will ensure no surviving botulism bacteria or their spores. The common guideline is to use a multiple of 12 for the D-value (121°C) of ***Clostridium*** *botulinum*, or its equivalent.

Sterilized milk **Milk** that has been heated at a high temperature (e.g. 110°C for 30-40minutes, 130°C for 30 seconds or 150°C for less than a second) to kill all **bacteria** and increase **shelf life**. Similar to UHT (ultra high temperature) milk. Has a distinctive **flavour**.

Steroids Complex polycyclic lipids with a hydrocarbon nucleus. Include **sterols**, **bile acids**, various **hormones** and **saponins**.

Sterols Steroid **alcohols** found widely in animals and plants which have an aliphatic hydrocarbon side chain of 8-10 C atoms at the 17-β position and a hydroxyl group at the 3-β position.

Stevia rebaudiana Plants native to South America, the leaves of which have a sweet **flavour**. Analyses have revealed the presence of at least eight sweet compounds in the leaves, the most widely used of which is **stevioside**.

Stevioside Terpenoid obtained from the leaves of *Stevia rebaudiana* that has 200-300 times the sweetness of **sucrose** and is stable at baking temperatures. However, it can also have an undesirable bitter and liquorice-like **aftertaste**. Stevioside **sweeteners** are commercially available, although they have not been approved as **food additives** in all countries, e.g. the USA. Also known as steviosin.

Stewing **Cooking** foods slowly and for a long period of time in a small amount of liquid in a closed dish or pan to make a stew. Stews usually contain **meat**, **vegetables** and a thick soup-like broth. Stewing not only tenderizes tough pieces of meat but also allows the **flavour** of the ingredient components to blend.

Stickiness One of the **rheological properties**; relating to the extent to which an item is cohesive or adhesive. This term also relates to the extent to which a food adheres to the palate during **mastication**.

Sticking Process of **adhesion**.

Stiffness One of the **rheological properties**; relating to the extent to which an item is stiff, i.e. firm and rigid. When stress is applied to a material, strain is produced in the direction of the stress; stiffness is the ratio of the stress divided by the strain.

Stigmastadienes Dehydration products of **sitosterol**, formed in **vegetable oils** during high temperature processing steps of the **refining** procedure such as **bleaching** and **deodorization**. The main product is 3,5-stigmastadiene. Stigmastadienes are not usually formed in high levels in virgin **olive oils**, where production involves nonthermal processes such as **centrifugation** and **pressing**. Stigmastadienes can be used as indicators of the presence of refined vegetable oils in virgin olive oils or to differentiate thermally treated oils from those that have been cold pressed.

Stigmasterol Plant sterol, also found in **milk**, deficiency of which can cause muscular atrophy and calcium phosphate deposits in muscles and joints. Oxidation of stigmasterol (e.g. during the **heating** of **vegetable oils**) can result in formation of carcinogenic products.

Stillage Liquid wastes or by-products from **distilleries**, **breweries** or **wineries**. May be used as animal feeds or in culture media for **microorganisms**.

Stilton cheese English semi-hard cheese made from **cow milk**. Available in blue and white varieties. Stilton has a rich and mellow flavour and a piquant aftertaste, but is milder than **Roquefort cheese** or **Gor-**

gonzola cheese. The wrinkled rind is not edible. Maturation takes 6-8 months.

Stir frying Cooking method in which food is cut into small pieces and fried over a very high heat in a pan with a large surface area, e.g. a wok, with constant stirring. Very small amounts of oil or fat are used. Associated particularly with Asian dishes.

Stirring Manual or automated processing action involving circular movements of a utensil (e.g. a spoon) within a food mixture. Allows ingredients to become well mixed together and where required, distributes heat throughout the mixture.

Stocks Juices obtained by simmering meat, bones, vegetables or fish, usually with **seasonings**, in water or other liquid. Used as a base for **soups** and accompaniments such as **gravy** and **sauces**. Available commercially as liquid products or in dried form.

Stollen Rich **bread** originating from Germany which is prepared by fermentation with **yeasts**. Usually contains **dried fruits** and **nuts**, and is topped with **icing sugar**. Traditionally eaten at Christmas. Sometimes called **christstollen**.

Stomachs A part of edible **offal**. **Tripe**, usually obtained from the rumen and reticulum of cattle, is used as an ingredient in some **sausages**. Swine stomachs are also used as an ingredient in some sausages. Cleaned animal stomachs are used as containers for various traditional meat dishes, including **haggis**.

Stone fruits Fruits with a thin skin, middle fleshy region and a single, central stone, containing the seed. Include **plums**, **peaches**, **apricots** and **cherries**.

Stones Alternative term for **seeds** found in the middle of some fruits, such as **stone fruits**.

Stoppers Plugs for sealing holes, particularly for sealing the necks of **bottles**. Also known as stopples in the USA.

Storage Maintenance of commodities, for example fresh or processed foods, under controlled conditions for extended durations while maintaining quality. Undesirable quality changes that may occur during storage include changes in nutrient levels or **colour**, development of **off flavour** or loss of **texture**. Most foods benefit from storage at a constant, low temperature (**cold storage** or **frozen storage**) where the rates of most degradative reactions decrease and quality losses are minimized. However, some products, e.g. **canned foods** or **dried foods**, are processed in such a way that they may be kept at ambient temperature with no loss in quality. Careful control of atmospheric gases, such as **oxygen**, **carbon dioxide** and ethylene (**controlled atmosphere storage**), is important in extending the storage life of many products, such as **fruits** and **vegetables**.

Storage life The time for which a stored item remains usable.

Storage proteins Proteins that accumulate within commodities such as **cereals**, **seeds** and **legumes**, and serve as nitrogen sources essential for germination. They usually occur in an aggregated state within membrane surrounded vesicles (e.g. protein bodies and aleurone grains), and are often built from a number of different polypeptide chains. Storage proteins have no enzymic activity. In cereals, storage proteins are deposited in the endosperm; in legumes, they are deposited in the cotyledon. Seed storage proteins are synthesized in large quantities over a limited period of time. In dicots, these proteins are deposited in the embryo as well as in the endosperm of the developing seed. Storage proteins are deficient in several essential **amino acids**, and because they account for the majority of the protein, seed proteins generally have limited nutritional value. Storage proteins of different cereals have distinct structural characteristics that are responsible for their unique **functional properties**. For example, it is the composition of seed storage proteins (**gliadins** and **glutenin**) that dictates flour and **dough** properties.

Stores Places, such as rooms or warehouses, where items such as foods are kept under controlled conditions for extended durations, for future use or sale.

Stout Strong, dark, top-fermented **beer** brewed mainly in the UK and Ireland.

Stoves Devices for **cooking** or **heating** of foods. Operated by burning fuel or using electricity.

Stracchino cheese Alternative term for **Taleggio cheese**.

Strainers Devices for straining liquids, semi liquids or dry ingredients to separate out any undesirable solid matter. These utensils have a perforated or mesh bottom, and are usually made from stainless steel, plastic or aluminium. Available in a variety of sizes, shapes and mesh densities.

Strawberries Juicy **fruits** produced by plants of the genus *Fragaria*, particularly *Fragaria × ananassa*. Consist of swollen flower receptacles with the pips or seeds (true fruits) embedded on the surface. Rich in **vitamin C**. Eaten out of hand and in **fruit salads**, or used to make **desserts**, **jams**, **jellies** and **toppings**. Available virtually all year round.

Strawberry jams Jams made using **strawberries**.

Strawberry juices Fruit juices extracted from **strawberries** (*Fragraria* x *ananassa*).

Strawberry tree fruits Red **fruits** produced by the strawberry tree (*Arbutus unedo*). Similar in appearance to **strawberries**, but with a very bitter **flavour**. Used in **jams**, **wines** and **liqueurs**, and as a source of sugar.

Straw mushrooms Common name for the **edible fungi** *Volvariella diplasia.*

Street foods **Fast foods** sold by street vendors, particularly in developing countries. Often associated with high microbiological risk due to lack of hygienic food preparation and holding areas.

Streptococcus Genus of Gram positive, anaerobic coccoid **lactic acid bacteria** of the family Streptococcaceae. Occur on the skin, mucous membranes and in the **gastrointestinal tract** of humans and animals. *Streptococcus salivarius* subsp. *thermophilus* is used in **starters** for **yoghurt** and **cheese** (e.g. **Emmental cheese** and **Parmesan cheese**) manufacture. *S. agalactiae* and *S. uberis* may be responsible for **mastitis** in cattle. *S. pyogenes* is the causative agent of strep throat and scarlet fever, which can be transmitted via contaminated foods (e.g. **milk**, **dairy products**, **eggs** and **salads**). Other species (*S. faecalis*, *S. faecium*, *S. durans*, *S. avium* and *S. bovis*) are responsible for diarrhoeal disease via ingestion of contaminated foods (e.g. **meat** products, milk and cheese).

Streptomyces Genus of Gram positive, aerobic filamentous **bacteria** of the family Streptomycetaceae. Occur in soil, water, and as parasites of humans, animals and plants. Some species may cause **taints** in **wines**, water and **shellfish**. Other species may cause diseases of crops (e.g. **potatoes** and **sugar beets**).

Streptomycin Aminoglycoside antibiotic produced by certain strains of **Streptomyces** *griseus*. Active against many **Gram negative bacteria**. Used to treat systemic and enteric infections in animals and also for growth promoting purposes. Residues may persist for long periods in **kidneys** but normally deplete rapidly in other commonly consumed tissues.

Streptoverticillium Obsolete genus, species of which have been transferred to the genus **Streptomyces**.

Stresnil Alternative term for the sedative **azaperone**.

Stress Any unusual events or conditions which bring about physiological or behavioural changes in animals. In addition to fear and physical trauma, it includes environmental factors such as cold, heat, humidity, light, sound and wind. The term stress also describes the results of such events or conditions. Stress often occurs when animals are faced with unfamiliar, threatening or harmful situations. Transport to markets or **abattoirs** and poor pre-slaughter management of animals are widely recognized as causes of animal stress. Animal stress is not only an **animal welfare** issue, but is also associated with various defects in meat including the **DFD defect** and the **PSE defect**. Susceptibility to

stress differs greatly between species, breeds, genders and individual animals.

Stress proteins Proteins which are synthesized by an organism in response to environmental stress, e.g. heat shock, exposure to toxic substances, exposure to **UV radiation** or viral infection. Examples include the **heat shock proteins**. Produced to protect the organism from destructive consequences of the stress conditions encountered, but also play a role in normal cell physiology. Appear to act as molecular **chaperones**, assisting in the folding/refolding of other proteins. Prevent stress-induced protein aggregation by binding to surfaces exposed as a result of destabilization of protein structure. May also be involved in repair of damaged proteins.

Stress relaxation One of the **rheological properties**; relating to the process of stress decay, i.e. the stress response that is apparent after subjecting a material to a certain strain.

Stress resistance Ability of an organism to withstand environmental stress.

Stretch One of the **rheological properties**; relating to the ability of an item to be drawn out in length (extended).

Stretching Making something that is soft or elastic longer or wider without tearing or breaking. An integral part of the manufacture of some cheeses, e.g. **Mozzarella cheese**, where the **curd** is stretched during processing.

String beans Type of **common beans** (*Phaseolus vulgaris*).

Striped bass **Marine fish** species (*Morone saxatilis*) belonging to the family Moronidae. Distributed in the western Atlantic Ocean and northern Gulf of Mexico. Produced commercially by **aquaculture**. Marketed fresh and consumed mainly broiled or baked.

Stroke Sudden attack of weakness often affecting just one side of the body. Brain tissue is damaged due to blockage of a blood vessel as a result of thrombosis, atherosclerosis or haemorrhage. Severity of the stroke depends on the region of the brain affected and the extent of damage. **Hypertension** and hypercholesterolaemia are major risk factors.

Strontium Metallic element with the chemical symbol Sr.

Structural genes Genes that encode substances such as **enzymes**, structural **proteins** and **RNA** molecules, rather than genes that serve regulatory purposes.

Structured lipids **Lipids** that have been modified to change the position and/or the composition of their constituent **fatty acids**. Typically **triacylglycerols**

containing mixtures of medium and long chain fatty acids.

Strudels Sweet or savoury **pastries** made from a **dough** of high-**gluten flour**, **eggs** and a high proportion of liquid, causing the dough to become highly malleable. The dough is then stretched out to paper thinness, and used to enclose **fruits**, e.g. sliced **apples**, or **cheese fillings**.

Stuffings Savoury mixtures of chopped and seasoned ingredients which are either used to stuff poultry or other meat joints prior to **roasting**, or served as a meat accompaniment.

Stunning Methods used to immobilize animals and **birds** before **slaughter**. Includes **electrical stunning**, captive bolt (projectile) stunning and CO_2 immobilization. Stunning is carried out immediately before bleeding; it aims to render the animal unconscious without stopping the action of the heart, which aids the bleeding procedure. Although stunning procedures involve some **stress**, they decrease stress responses when compared with bleeding without immobilization; consequently, stunning influences the properties and composition of **meat**. Overall effectiveness of stunning depends on the design and careful operation of the equipment used.

Sturgeon Any of a number of **marine fish** or **freshwater fish** from the family Acipenseridae (subclass Chondrostei); found in temperate waters of the Northern Hemisphere. Most species live in the sea and migrate into rivers (possibly once in several years) to spawn in spring or summer; a few others are confined to fresh water. Flesh tends to be fatty with firm **texture**. Marketed fresh, frozen, smoked, salted and canned. **Roes** from some species are highly valued as **caviar**.

Styrene Unsaturated liquid hydrocarbon, which is a by-product of petroleum manufacture. Polymerized to make resins and plastics that are used as **packaging materials** for foods. There is concern about **health hazards** associated with **migration** of styrene monomers, dimers and trimers from packaging materials into some types of foods.

Suberin Aromatic polymer similar to **lignin**, to which are aliphatic components such as ω-hydroxy acids, dicarboxylic acids and long chain alcohols are attached. Found in **waxes** and in the cell walls of plants, and also deposited at wound sites.

Subtilins **Bacteriocins** produced by *Bacillus subtilis*.

Subtilisins EC 3.4.21.62. A group of **proteinases** (serine endoproteinases) produced by *Bacillus* spp. that exhibit broad specificity, but with a preference for large uncharged residues in the P1 position. Variants include subtilisin BPN' and subtilisin Carlsberg. Used for production of **protein hydrolysates** and **casein phosphopeptides**, and for **meat tenderization**.

Succinic acid Dicarboxylic acid with a number of applications in the food industry. Can be produced industrially by microbial **fermentation** and is the main flavour component produced by **yeasts** during **sake** manufacture.

Succinoglucans Microbial **polysaccharides** produced by, for example, *Agrobacterium* radiobacter. Have properties similar to those of **xanthan gums**.

Succinylation Introduction of succinyl groups into a compound or substance. Usually achieved by reaction with succinic anhydride. Such modification is used to alter the **physicochemical properties**, **functional properties** or nutritional quality of substances such as **proteins** and **starch**. Succinylation has also been used to modify the properties of **enzymes** such as **papain**.

Succulence One of the **sensory properties**; relating to the extent to which a product (e.g. **meat** or **fish**) is succulent or juicy. Degree of succulence of a food can be measured using a succulometer, in which samples are compressed to squeeze out juices, the volume of which can then be recorded.

Sucralose Non-nutritive, **artificial sweeteners** produced by chlorination of sucrose; three hydroxyl groups on the sugar are substituted by chlorine atoms. Sucralose has 400-800 times the sweetness of **sucrose**, a flavour profile similar to that of sucrose and no **aftertaste**. Sucralose is stable at **baking**, **pasteurization** and **extrusion** temperatures. It is less reactive than sucrose and thus interacts less with components of foods or **beverages** to which it has been added.

Sucrases Alternative term for **sucrose α-glucosidases**.

Sucrose Disaccharide comprising a molecule of **glucose** and a molecule of **fructose**. Sucrose occurs naturally and is extracted commercially from **sugar cane** and **sugar beets** to yield the crystalline sweetener marketed as **sugar**. **Sweetness** of sucrose is the milestone by which sweetness of all other **sugars** and/or **sweeteners** is compared.

Sucrose acetate isobutyrate Mixture of **esters** of **sucrose** esterified with **acetic acid** and isobutyric acid. Produced by reaction of food grade sucrose with acetic anhydride and isobutyric anhydride in the presence of a catalyst. Used as a stabilizer or weighting agent to increase the **specific gravity** of flavouring oils in citrus based **beverages**. Commonly abbreviated to SAIB.

Sucrose α-glucosidases EC 3.2.1.48. Hydrolyse **sucrose** and **maltose** by an α-D-glucosidase-type action. Also known as sucrases.

Sucrose-phosphate synthases EC 2.4.1.14. Catalyse the conversion of UDPglucose and D-fructose 6-phosphate to UDP and sucrose 6-phosphate. Involved in **sucrose** and **starch** biosynthesis in plants, and in the **ripening** of **fruits**.

Sucrose polyesters Esters of **sucrose** and **fatty acids** (C12 to >C20) that are not absorbed on their way through the gastrointestinal tract and may act as **fat substitutes** in **shortenings**, **spreads** and other foods.

Sucrose synthases EC 2.4.1.13. Catalyse the conversion of UDPglucose and D-fructose to UDP and **sucrose**. Involved in sucrose and **starch** biosynthesis in plants.

Sucuk Turkish, raw, cured **sausages**, prepared mainly from **beef** with added **mutton** and sheep fat. Sucuk are dry cured for 7-10 days and retailed unsmoked. They are usually eaten warm.

Suet Hard, white fatty tissue surrounding the kidneys of cattle and sheep. Used in **baking**, **frying** and in the manufacture of **tallow**.

Sufu Cream cheese-type **fermented foods** made from **tofu** and eaten widely in China. **Fermentation** of tofu is carried out commercially using *Actinomucor elegans*, followed by **brining** and ageing. Sufu is eaten as an appetizer, as a relish, cooked with **vegetables** or **meat**, or in the same manner as **cheese**.

Sugar Commercial name for crystalline **sucrose** extracted from either **sugar cane** or **sugar beets**, purified to at least 98% purity.

Sugar alcohols Products formed when aldehyde or ketone groups of **sugars** are hydrogenated (reduced) to alcohol groups. Examples include **sorbitol**, **mannitol** and **lactitol**, produced by **hydrogenation** of **glucose**, **mannose** and **lactose**, respectively. Also known as **polyols**.

Sugar almonds Shelled **almonds** covered with a hard **sugar** coating, which is often coloured. Often given as symbols of good luck at religious occasions such as weddings and christenings.

Sugar apples **Fruits** produced by *Annona squamosa* and related to **cherimoya**, for which they are sometimes mistaken. The egg-shaped fruits have a thick, yellowish-green skin and sweet yellow custard-like flesh containing dark seeds. Rich source of **vitamin C**. The flesh is eaten with a spoon as a fresh fruit or used in **fruit salads**, **milkshakes**, **yoghurt** and **desserts**. Also known as sweet sop.

Sugar beet cossettes Thin slices cut from **sugar beets** in the initial stage of sugar processing, from which **sugar** is extracted. Cossettes are used to increase the surface area available for, and efficiency of, sugar extraction.

Sugar beet juices Alternative term for **beet sugar juices**.

Sugar beet molasses Alternative term for **beet sugar molasses**.

Sugar beets Roots produced by *Beta vulgaris*, plants from which **sugar** is extracted commercially. Sugar is present in specific cells of the tap root of the plant. Major sugar beet producing regions in the world include Europe and the USA.

Sugar cane Tropical grass of the genus *Saccharum*, stalks of which are a rich source of **sugar**. *S. officinarum* is the main species cultivated for commercial sugar production.

Sugar cane bagasse **Cane sugar** processing waste that is composed of unextracted **sugar** and the remains of the **sugar cane** after milling. Used as a fuel source, in feeds, as a substrate for microbial **fermentation** and for paper and board manufacture. Also called bagasse or megass.

Sugar cane juices Alternative term for **cane sugar juices**.

Sugar cane molasses Alternative term for **cane molasses**.

Sugar cones **Ice cream cones** that have been formulated to have a particularly crisp **texture**.

Sugar confectionery Collective term for foods which have **sugar** as a principal component, e.g. **chocolate**, **candy**, **fudges**, **jelly confectionery**, **sweets** and **toffees**.

Sugar crops Plants that are cultivated for **sugar** production, including **sugar beets**, **sugar cane** and **sweet sorghum**.

Sugar cubes Cubes produced by moulding or compression of moistened **granulated sugar**.

Sugar juices **Sugar** containing solutions obtained by crushing **sugar cane**, or by hot water extraction of **sugar beet cossettes**. Sugar is crystallized from the juices following removal of impurities.

Sugar manufacture Alternative term for **sugar processes**.

Sugar pans Vessels, usually made of metal, e.g. steel plate, in which **evaporation** of **sugar juices** and **crystallization** of sugar are performed.

Sugar processes Processes involved in the manufacture of **sugar**, such as **carbonatation**, **liming**, **evaporation** and **crystallization**.

Sugar refineries Factories where raw **cane sugar** is purified to produce **granulated sugar**.

Sugars General term for saccharides or their derivatives that have a sweet **flavour**.

Sugar substitutes Chemicals used to mimic the **flavour** and applications of **sucrose**, e.g. **sweeteners**.

Sugar syrups Concentrated aqueous solutions of **sugars**. Include syrups of individual sugars, such as **glucose syrups** and **fructose syrups**, and syrups extracted from specified sources, e.g. **corn syrups** and **maple syrups**.

Sulfadiazine Sulfonamide antibiotic active against a range of **microorganisms** and used to treat diseases such as **toxoplasmosis**, meningitis and pneumonia.

Sulfadimidine Alternative term for **sulfamethazine**.

Sulfamates Alternative term for **cyclamates**.

Sulfamethazine Sulfonamide drug used primarily for control of atrophic rhinitis and other infections in swine and cattle. Also used as a growth promoter. Normally absorbed and excreted rapidly; elimination is generally more rapid when the drug is injected.

Sulfanilamide Sulfonamide drug used to treat a range of bacterial and protozoal infections in animals. Often used in combination with other **sulfonamides**.

Sulfates Inorganic salts of **sulfuric acid**.

Sulfathiazole Sulfonamide drug used as a coccidiostat in animals. Often used in combination with other **sulfonamides**. Also used as a growth promoter and for treatment of foul brood in bees. Normally absorbed and excreted rapidly by animals.

Sulfhydryl groups Reactive SH groups that are effective at chelating aluminium and other toxic minerals. Mediate the formation of disulfide bonds in **proteins** and other compounds.

Sulfides Sulfur compounds in which the S atom can be bound to inorganic or organic moieties. In inorganic sulfides, the S atom may be linked to metals or nonmetals, while in organic sulfides, the S atom is linked to two hydrocarbon groups.

Sulfitation Use of salts of **sulfurous acid**, mainly **sulfites**, for applications including inhibition of bacterial growth, prevention of **spoilage** or oxidation, and control of **browning** in foods. Sulfites, which may be added as **preservatives** to packaged and processed foods, can cause a severe allergic response in certain individuals.

Sulfites Inorganic salts of sulfurous acid that are used as food **preservatives** since they exhibit **antimicrobial activity** and **antioxidative activity**, and prevent **enzymic browning**. However, they are potentially cytotoxic and mutagenic, and may be allergenic to hypersensitive individuals. Hence, their use is regulated strictly.

Sulfolobus Genus of Gram negative, aerobic or facultatively anaerobic, coccoid or irregularly-shaped **bacteria** of the family Sulfolobaceae. Occur in sulfur-rich hot springs. *Sulfolobus solfataricus* is used in the production of several thermostable **enzymes** (e.g. thermostable **β-glucosidases**).

Sulfonamides Group of synthetic **organic compounds** with a broad spectrum of activity. Widely used for control of bacterial and protozoal infections in animals, particularly infectious diseases of the digestive and respiratory tracts. Administered to animals by all known routes, often at dosages noticeably higher than those for **antibiotics**. Residues are normally eliminated much earlier from **livers**, **kidneys** and **milk** than from muscle or **adipose tissues**. Examples include **sulfamethazine**, **sulfanilamide** and **sulfathiazole**.

Sulfur Non-metallic element with the chemical symbol S. Essential in that it is a component of **cysteine**, **methionine**, **vitamin B₁** and **biotin**. However, there appears to be no requirement for S in any other form.

Sulfur dioxide Gas that is used in **preservatives** and **bleaching agents**, e.g. for **beet sugar**, and in **stabilizers** for **vitamin C**. Degrades **vitamin B₁**, and thus it is not recommended for use in foods rich in this vitamin.

Sulfuric acid Inorganic acid with the chemical formula H_2SO_4.

Sulfurous acid Aqueous solutions of **sulfur dioxide** used in **preservatives**.

Sulphadimidine Alternative spelling for **sulfadimidine**.

Sulphamates Alternative spelling for **sulfamates**.

Sulphamethazine Alternative spelling for **sulfamethazine**.

Sulphanilamide Alternative spelling for **sulfanilamide**.

Sulphates Alternative spelling for **sulfates**.

Sulphathiazole Alternative spelling for **sulfathiazole**.

Sulphides Alternative spelling for **sulfides**.

Sulphitation Alternative spelling for **sulfitation**.

Sulphites Alternative spelling for **sulfites**.

Sulphonamides Alternative spelling for **sulfonamides**.

Sulphur Alternative spelling for **sulfur**.

Sulphur dioxide Alternative spelling for **sulfur dioxide**.

Sulphuric acid Alternative spelling for **sulfuric acid**.

Sulphurous acid Alternative spelling for **sulfurous acid**.

Sulphydryl groups Alternative spelling for **sulfhydryl groups**.

Sultanas **Dried fruits** prepared from golden sultana **grapes** by **drying** in the sun or mechanically. Rich in iron with a high **sugar** content, a range of **vitamins** and **minerals**, and a moderate level of dietary fibre. Eaten out of hand or used in **bakery products** and various dishes. Also called seedless raisins.

Suluguni cheese Georgian mild semi-soft **cheese** made from **ewe milk**. Often eaten fried or grilled, or in a variety of dishes.

Sumac Common name for the plant *Rhus coriaria* and its dark purple-red **berries**, which are dried and used whole or ground as **spices**, giving a fruity, sour and astringent **flavour** to foods. Widely used in cooking in the Middle East, especially in Lebanese cuisine. Alternative names include sumaq, sumach, shumac and Sicilian sumac.

Summer sausages Spicy, semi-dry fermented **sausages**, which are cooked and dried after **fermentation**. Commonly prepared from **pork** and/or **beef**, but may also be prepared from meat mixtures including **chicken meat** or **turkey meat**. Natural **pigments**, e.g. **betalaines**, may be used to simulate a cured meat **colour** in summer sausages. High quality is achieved by use of frozen concentrated **lactic acid bacteria starters** and control of **lean** to fat ratios in the meat. Varieties include landjaeger and thuringer.

Sunburn Damage to plants and **fruits** (particularly **apples**) caused by exposure to intense **sunlight**. Causes necrotic lesions on the fruit and browning of the flesh underneath. May be controlled by shading or cooling. Also called sunscald.

Sunett Registered trade name for the artificial sweetener **acesulfame K**.

Sunfish A variety of **freshwater fish** and **marine fish**; particularly refers to North American freshwater fish of the genus *Lepomis*, e.g. *L. macrochirus* (bluegill sunfish), some of which are popular food fish. Also refers to the large marine fish species *Mola mola*.

Sunflower meal Cake remaining after extraction of **sunflower oils** from **sunflower seeds**. Contains high levels of **polyphenols**, which must be removed before the meal is used in foods. Source of **proteins**, which may be isolated and used in foods.

Sunflower oils Oils extracted from **sunflower seeds** (*Helianthus annus*). Rich in **linoleic acid** and **oleic acid** and low in **saturated fatty acids**. Used as **salad oils** and **cooking oils** as well as in the manufacture of **margarines** and **shortenings**.

Sunflowers Plants belonging to the species *Helianthus annus*. Characterized by a long stem and large, composite yellow flower heads, which produce **sunflower seeds** rich in **sunflower oils**.

Sunflower seeds Seeds produced by **sunflowers** (*Helianthus annus*). Rich in **vitamin B$_1$**, **proteins**, **iron** and **niacin**. May be eaten dried and roasted as **snack foods**, incorporated into **bakery products** or used as a source of **sunflower oils**.

Sunlight Light emitted from the sun. Used in sun drying and **solar drying** of foods.

Sunset Yellow Orange monoazo dye used in **artificial colorants**. Soluble in water or glycerol but only slightly soluble in ethanol. Has a reddish-yellow hue in concentrated solution that becomes yellow on dilution; the dye is colour stable at **extrusion** temperatures, pH 3-8 and in the presence of **organic acids** and alkalis commonly used in food processing, such as **citric acid** and **sodium bicarbonate**. Sunset Yellow is often blended with **tartrazine** and used in **colorants** for **low fat spreads**. Also used, in combination with other colorants, to colour a range of products, including **bakery products**, **beverages** (e.g. **cola beverages**), **sugar confectionery** and **ice cream**. Also known as FD&C Yellow No. 6 and Food Yellow 3 and CI 15985.

Supercritical CO$_2$ extraction Extraction process that uses supercritical **carbon dioxide** (CO$_2$) as the selective solvent. The polarity of CO$_2$ limits its use to extractions of relatively apolar or moderately polar solutes. Thus, a small amount of a polar organic solvent (e.g. methanol, acetonitrile, water), called a modifier or entrainer, is usually added to the supercritical fluid for extraction of more polar compounds. CO$_2$ is frequently used as the extraction solvent in **supercritical fluid extraction** because it is in a supercritical state at a relatively low temperature (31°C) and pressure (73 atmospheres), making it a suitable choice from an instrumental point of view. Extraction using supercritical CO$_2$ also avoids the use of dangerous or toxic organic solvents and the gas is easily removed by reducing the pressure.

Supercritical fluid chromatography **Chromatography** technique that uses a supercritical fluid as the mobile phase. Developed for analysis of substances not separated effectively using **liquid chromatography** or **gas chromatography**, including **triglycerides** and **fatty acids**.

Supercritical fluid extraction Extraction process that uses a supercritical fluid as the selective solvent. A supercritical fluid is the mobile phase of a substance intermediate between a liquid and a vapour, maintained at a temperature greater than its critical point. **Carbon dioxide** (CO$_2$) becomes a supercritical fluid when held above its critical temperature and pressure. Instrumentation for supercritical fluid extraction consists of a solvent supply, a pump, a cooler to cool the pump head, an extraction cell that is mounted in a ceramic heater tube, a heater controller to monitor the temperature of the extraction cell, a restrictor connected to the outlet of the cell, a restrictor heater, and a collection vial. The possibility of varying the solvent strength of the supercritical fluid by alteration of pressure makes supercritical fluid extraction extremely versatile in its applications.

Supercritical HPLC   HPLC technique in which a supercritical fluid is used as the mobile phase.

Supermarkets   Large self-service **shops** selling foods and household goods.

Superoxide dismutases   EC 1.15.1.1. Metalloenzymes that are important in protecting cells against oxidative stress via their ability to scavenge toxic superoxide radicals, catalysing their dismutation to molecular oxygen and H_2O_2. The enzymes from most **eukaryotes** contain both Cu and Zn, while those from mitochondria and most **prokaryotes** contain Mn or Fe.

Superoxides   Inorganic compounds that contain the superoxide radical or ion. Formed when very reactive metallic elements (e.g. sodium, potassium, rubidium and caesium) react with oxygen. Powerful **oxidizing agents** and strong bases. Generated in prokaryotic and eukaryotic cells, where they are potentially harmful.

Supracide   Alternative term for the insecticide **methidathion**.

Surface active agents   Substances such as **surfactants** that reduce **surface tension** by interaction with non-mixing substances at phase boundaries.

Surface active properties Functional properties relating to the ability of a compound to reduce the **surface tension** of a liquid, thereby increasing **wettability** or blending ability. **Surfactants** used as additives in the food industry have surface active properties.

Surface tension   Force on the surface of a liquid that makes it behave as if the surface has an elastic membrane. Caused by forces between the molecules of the liquid: molecules at the surface experience forces from below, whereas those in the interior are acted on by intermolecular forces from all sides. The surface tension of water is very strong, due to intermolecular hydrogen bonding. Surface tension causes a meniscus to form, liquids to rise up capillary tubes, paper to absorb water, and droplets and bubbles to form. It is measured in Newtons per metre.

Surfactants   Substances that concentrate at phase boundaries and reduce **surface tension**. Contain hydrophilic and hydrophobic regions which align at inter-

faces to promote mixing of phases. Above a particular concentration, the critical micellar concentration, surfactants form micelles which encapsulate one phase within the other. Used to produce oil/water **emulsions** and for encapsulation of lipid soluble **flavourings** in processed foods. **Emulsifiers**, such as **fatty acid esters**, are surfactants as are sodium dodecyl sulfate (SDS) and Tween.

Surfactin   Cyclic lipopeptide antibiotic produced by certain strains of **Bacillus** *licheniformis* and *B. subtilis*. Strong biosurfactant having a number of potential food processing applications, including in **emulsifying agents**, **foaming agents** and **stabilizers**. Shows moderate **antimicrobial activity**.

Surimi   Fish products comprising refined, stabilized, frozen **fish mince**. Refining and stabilization are achieved by washing repeatedly in fresh water to remove soluble protein, straining, pressing to restore water content to natural levels (approximately 80%), followed by incorporation of **sugar**, **sorbitol** and **polyphosphates**. Used to make products such as **kamaboko**, **fish sausages** and sea food analogues such as **imitation crabmeat**.

Susceptors   Alternative term for **microwave susceptors**.

Sushi   Japanese sea food dishes which normally consist of thin slices of fresh raw **fish** flesh or seaweed wrapped around a cake of boiled **rice**. The term is also used for dishes consisting of fresh raw fish flesh placed on boiled rice flavoured with **vinegar**.

Suspension cultures   Cell cultures maintained in liquid media, which grow in suspension rather than attached to a surface within the culture vessel. Can include cultures of **plants**, animals and **microorganisms**. May be used in the manufacture of **fermented foods** and **beverages** (e.g. **dairy products**, **beer** and **wines**), but also for more specialized **fermentation products**, including some food ingredients.

Swedes   Root vegetables of the cabbage family with a shape similar to that of **turnips**, with which they are often confused. Rich in **vitamin C** and **potassium**, they also contain small amounts of **vitamin A**. The sweet yellow flesh is eaten in similar ways to turnips, i.e. boiled, mashed, eaten raw, or used in soups and stews. Also known as rutabagas and Swedish turnips.

Sweet almonds Nuts produced by varieties of **almonds** (*Amygdalus communis*). Eaten raw, dried, or roasted and salted, and used in **bakery products** such as **bread** and **pastry**. Almond **meal** is also used in bakery products and **confectionery**. Kernels are a source of sweet **almond oils**.

Sweet basil   Common name for *Ocimum basilicum*, the leaves and flowers of which are used as **spices**.

Sweet basil has a sweet, spicy **flavour** reminiscent of **mint**. Predominant **flavour compounds** include methylchavicol, **eucalyptol** and estragol. Sweet basil **essential oils** and **oleoresins** are also commercially available and are used as **flavourings**.

Sweetbreads Butchers' term for **pancreas** glands (gut sweetbreads) and thymus glands (chest sweetbreads) from animal **carcasses**. They form a part of edible **offal**.

Sweet cherries **Cherries** produced by *Prunus avium*. Eaten out of hand or in **pies** and other **desserts**, and used in **beverages**. Available dried and canned as well as fresh. Also known as geans and mazzards.

Sweet chestnuts Nuts produced by *Castanea sativa*. Contain more **starch** and fewer **oils** than other nuts, and are eaten as vegetables. Eaten whole or used as ingredients in dishes such as **soups**, stews and **stuffings**. Also shelled and preserved whole in **sugar syrups**, when they are known as **marrons glaces**.

Sweetcorn Variety of **corn**, kernels of which are sweet when young.

Sweet cream Cream in which no **acidity** has developed. Used to make sweet cream butter, the cream being ripened by warming only, with no addition of **butter starters**.

Sweeteners Additives with a sweet **flavour** that are added to foods as **sugar substitutes**. Grouped according to the nutritional value of the sweetener into: nutritive sweeteners that may be metabolized and/or incorporated into the glycolytic pathway in cells to produce energy, e.g. **starch**-derived sweeteners; fruit-derived sweeteners, e.g. **honeys**, **lactose** and **maple syrups**; and non-nutritive or non-carbohydrate based sweeteners. Sweeteners may also be classified as natural (existing in nature), e.g. carbohydrate-derived sweeteners, **stevioside**, **thaumatin**, **glycyrrhizin**, or artificial (produced by organic synthesis and not present in nature), e.g. **sucralose**, **aspartame**, **acesulfame K**, **cyclamates**, **saccharin**.

Sweet limes Citrus fruits produced by *Citrus limettioides* or *C. lumia*. **Peel** is greenish to orange-yellow when the fruits are ripe. The juicy pulp is pale yellow in **colour**, with a non-acid, faintly bitter **flavour**. Eaten out of hand, cooked or preserved. Peel is a source of essential oils with a strong **aroma** of **lemons**. Sometimes confused with the sweet lemon (*C. limetta*).

Sweetmeats Any sweetened delicacy, especially **sweets** or, less commonly, **cakes**.

Sweetness One of the **sensory properties**; relating to the extent to which a product tastes sweet. Sweetness of **artificial sweeteners** is often expressed in relation to that of **sugar** (**sucrose**).

Sweet peppers **Fruits** produced by *Capsicum annuum*. Vary in size, shape and **colour**, but all are mild in **flavour**. Rich in **carotenes** and **vitamin C**. Although yellow, purple, red and orange types are available, sweet peppers are usually picked when green. Eaten raw in **salads** or as crudites, or cooked, sometimes stuffed with **rice**, **meat** or **vegetables**. Sweet peppers include **bell peppers**.

Sweet potatoes Common name for edible **tubers** of *Ipomoea batatas*. Vary in shape and **colour** of skin and flesh. Rich in **vitamin C**; orange- and yellow-fleshed cultivars contain high levels of **carotenes**. Eaten cooked in the same way as **potatoes**. Also a good source of **starch** and **alcohol**.

Sweet potato starch Starch isolated from tubers of **sweet potatoes**.

Sweets Small shaped pieces of **confectionery**, which are usually made with **sugar** or **chocolate**.

Sweet sorghum Tropical grass that has a sweet flavoured stalk from which **sugar** is manufactured.

Swelling An increase in the volume of a gel or solid associated with uptake of a liquid or gas.

Swine Wild or domesticated omnivorous mammals belonging to the Suidae family; they include pigs and wild boars. Swine are farmed or hunted for the production of **bacon**, **ham**, **pork**, edible **offal** and other products. Different gender and age groups of swine are known as boars (adult entire males), barrows, hogs or bars (adult castrated males), sows (adult females after producing their first litter of piglets), gilts, hilts, yelts or yilts (young sexually mature females to the end of their first pregnancy) and piglets or piglings (sexually immature animals, usually <10 weeks old).

Swine fever An infectious, notifiable viral disease of **swine**. Although it is caused by an RNA virus, *Salmonella* Cholerae suis and *Pasteurella* multicida are commonly involved in the aetiology of the disease. Swine fever is characterized by a refusal to eat, fever, foul-smelling diarrhoea, distressed breathing, discharge from the eyes and general weakness. The disease may take an acute or chronic form. If swine are slaughtered in the incubative stage and **carcasses** are chilled immediately, the **viruses** can persist in bone marrow, frozen **pork** and **bacon**. Consequently, the disease may be transmitted to healthy animals if they are fed on **offal**, slaughterhouse wastes or waste food prepared from infected animals. Outbreaks of such **animal diseases** are controlled by animal slaughter, burning or burial of infected carcasses, and restriction of transportation and export of swine and swine products.

Swine kidneys **Kidneys** from **swine**; they are a part of edible **offal**. Swine kidneys have a strong **flavour**, and are commonly used to add richness to **pates** and **terrines**.

Swine livers **Livers** from **swine**; they are a part of edible **offal**. Swine livers are strongly flavoured, dark in **colour** and may have a mealy **texture**. Commonly, they are cooked by braising or are minced for use in **liver sausages**, **pates** and **terrines**.

Swine muscles Alternative term for **pork**.

Swine skin **Skin** from **swine**. The skin has a high content of soluble **collagen**. A large proportion of swine skin is used to prepare **gelatin** and **aspic** products. **Pork** rinds (skin with adhering fat) and swine skin connective tissue are used widely as ingredients in **sausages**. The crisp, fatty skin of roast pork is known as crackling.

Swiss chard Common name for a type of *Beta vulgaris*. Member of the beet family that is grown for its large leaves, which are eaten as **leafy vegetables** in a similar way to **spinach**. Can also be used raw in salads, or incorporated savoury dishes and **stuffings**. Rich in **vitamin A**, **vitamin C**, **potassium** and **iron**. Also known as **leaf beet**, white beet, silver beet and spinach beet.

Swiss cheese A pale yellow **cheese** with large holes and a slightly nutty **flavour** that is made in Switzerland, e.g. **Emmental cheese** and **Gruyere cheese**. Also a US term for any **hard cheese** that contains relatively large bubbles of air.

Swiss rolls Thin **sponge cakes** which are covered on one side with **jams** and rolled into cylinders. Called jelly rolls in the USA.

Sword beans Seeds produced by *Canavalia gladiata*. Used in a similar way to **jack beans**.

Swordfish Large **marine fish** species (*Xiphias gladius*) with a long, flat, sword-like bill; found in tropical and temperate waters around the world. A commercially important food fish. Red flesh tends to be firm-textured with a mild **flavour**. Marketed fresh (whole, gutted or steaks) and frozen. Liver **oils** are used as a source of **vitamins**.

Syagrus Genus of **palms**. Fruits of some species are used as food; seeds are the source of **palm oils**.

Synechococcus Genus of unicellular **cyanobacteria**, generally found in marine habitats, but can also survive hypersaline environments and hot springs. Produce **phycocyanin**, and sometimes grown in **bioreactors** for the production of this and other commercially useful **pigments**.

Syneresis Contraction of a substance, usually a gel, when allowed to stand, and the resulting exudation of liquid from the gel. Control of syneresis is a key step for increasing **curd** yield and improving **cheese** quality. Also important for **yoghurt** quality. Syneresis depends on a combination of specific and nonspecific interactions at the protein level, many of which also occur during curd formation.

Syringic acid Phenolic isoflavone with radical scavenging activity and **antioxidative activity**. Found in various foods and **beverages**, including soy products, **alcoholic beverages** and **olive oils**. Has also been shown to possess **antibacterial activity**.

Syrups Aqueous solutions of **sugars** or **starch hydrolysates**.

T

2,4,5-T Herbicide used to control weeds among a range of fruits, vegetables and cereals. Also known as 2,4,5-trichlorophenoxyacetic acid.

Table grapes Species of **grapes** grown for eating as opposed to **winemaking** or **drying**. They are seeded or seedless **fruits** of the genus *Vitis*, the most important species of which is *V. vinifera*. While most grapes are grown as **winemaking grapes**, significant amounts are produced as table grapes. Table grapes have a firmer flesh and lower level of **acidity** than winemaking grapes. All grapes are rich in **sugar**, but contain little **vitamin C**. **Organic acids** include **tartaric acid** and **malic acid** in approximately equal amounts. Black grapes contain **anthocyanins**. Table grapes are eaten out of hand, or used in **salads**, **pies** and other **desserts**.

Table jellies Fruit flavoured sweetened **desserts** set with **gelatin** or similar **gelling agents**. Known as jello in the USA.

Tacos **Pancakes** made from **corn flour** which are filled with **meat mince**, **cheese** or **beans**, together with piquant **sauces**, before being fried.

Taco shells Crisp food products made from **corn masa dough** which are shaped into thin discs and formed into a U-shape before being fried. Often filled with cooked **beef mince sauces** and topped with shredded lettuce and grated **cheese**.

Taenia Genus of parasitic tapeworms of the class Cestoda. *Taenia solium* is associated with **pork**, while *T. saginata* is associated with **beef**. Infection in humans is usually transmitted by eating raw or undercooked beef or pork.

Tagatose Ketose monosaccharide comprising six carbon atoms (**hexoses**); an isomer of **galactose**. Has **sweetness** similar to that of **sucrose** but no calorific value, making it suitable as a low-calorie sweetener and bulking agent. Formed by bacterial **fermentation** using **galactitol** as substrate or produced from **lactose** via **isomerization** of **galactose**.

Tagliatelli **Pasta** formed into narrow flat ribbons.

Tahini Paste made from ground **sesame seeds**. Used as an ingredient of **humous** and also as the base for **sauces**.

Taints **Sensory properties** relating to the perception of **off flavour** or **off odour** in a product. Taints in foods can be related to, for example, **warmed over flavour** in **ready meals** or **boar taint** in **pork** products.

Take away foods Cooked dishes, often **fast foods**, which are sold at **restaurants** or other **catering** outlets for consumption off the premises.

Takju **Rice wines** manufactured in Korea.

Talaromyces Genus of **fungi** of the class Eurotiomycetes whose species are anamorphs of ***Penicillium*** species. Occur in soil. Some species (e.g. *Talaromyces macrosporus* and *T. flavus*) may cause **spoilage** of **fruit juices**. *T. flavus* is also used as a **biocontrol** agent against certain plant **pathogens**.

Taleggio cheese Italian semi-soft **cheese** made from **cow milk**. Also known as Stracchino. **Flavour** is buttery and fruity. Rind is pinkish-grey and the interior is white. **Ripening** lasts 25-50 days. Also produced as a cooked **curd** variety that is firmer and similar to **Mozzarella cheese**. Taleggio is an excellent dessert cheese.

Tallow Solid **animal fats** normally derived from cattle or sheep tissue, containing high levels of **saturated fatty acids** and **monounsaturated fatty acids** (**triglycerides** of **stearic acid**, **palmitic acid** and **oleic acid**). White, flavourless, odourless and solid at room temperature. Usually prepared by heating **suet** under pressure in closed vessels. Used for **frying** and in shortenings.

Tamales Concentric layered **corn** products, traditionally produced in Mexico. Some tamales include seasoned **meat**, for example beef tamales, but others are prepared without meat, for example green tamales.

Tamarillos **Fruits** produced by *Cyphomandra betacea*. Skin is yellow to deep red in **colour**, while the flesh varies from yellow-orange to purple. Contains numerous seeds. Rich in potassium and **carotenes**, with moderate amounts of **vitamin E** and **vitamin C**. Best eaten cooked, with the bitter tasting skin removed. Used in a range of products, including juices, **sauces**, **chutneys** and **relishes**. Also known as tree tomatoes.

Tamarinds Common name for **fruits** of *Tamarindus indica*. The brown, curved pods contain a sticky pulp studded with up to 10 starchy seeds that can be eaten as a pulse. The sweet-sour flavoured pulp is extracted and used in a variety of foods, including **sweetmeats**, **curries**, **preserves** and **chutneys**. Leaves and flowers of the plant are also eaten in India.

Tamper evident closures Closures designed to ensure that any unauthorized interference is evident.

Tamper evident packaging Packaging designed to ensure that any unauthorized interference is evident.

Tangelo Citrus fruits that are a cross between **tangerines** and **pummelos**. The most popular varieties are the minneola, with few seeds and a tart, sweet **flavour**, and the Orlando, a many seeded fruit with a mild, sweet flavour.

Tangerine juices Fruit juices extracted from **tangerines** (*Citrus reticulata*).

Tangerines Small, loose skinned **citrus fruits** (*Citrus reticulata*). Relatively good source of vitamin C. This species also includes **mandarins** and **satsumas**, names tending to be used indiscriminately. Tangerines tend to be darker in **colour** than mandarins. Consumed as fresh or as a dessert, often as canned segments. Used in several citrus hybrids.

Tangle Alternative term for brown **seaweeds** of the genus *Laminaria*.

Tangor Citrus fruits that are a cross between **tangerines** and **oranges**. Similar in **flavour** to oranges, but contain many seeds.

Tania Common name for *Xanthosoma sagittifolium*, the corm of which is processed in a similar way to **taro**. Nutritionally similar to taro also, although the **starch** is more difficult to digest. Sometimes used as the base for preparation of **fufu**. Also known as tannia, taniers, yautia or new **cocoyams**.

Taniers Alternative term for **tania**.

Tanks Large storage chambers or containers, particularly for gases or liquids.

Tannases EC 3.1.1.20. Catalyse the breakdown of hydrolysable **tannins** and gallic acid esters. Used in the manufacture of **instant tea**, **clarification** of coffee-flavoured **soft drinks**, production of **gallic acid**, removal of unwanted tannins from foods, and processing of **wines** and **fruit juices**.

Tannic acid Polyphenol which displays **antimutagenicity**, **anticarcinogenicity** and **antioxidative activity**. Used as a food additive, a clarifying agent and a refining agent, but may inhibit the **absorption** of dietary iron.

Tannins Complex polyhydroxybenzoic acid derivatives found in many foods. **Antinutritional factors** inhibiting the **bioavailability** of **vitamins** and **minerals**, and may be carcinogenic. However, also possess **antimicrobial activity**, **antioxidative activity** and **antitumour activity**.

Tanoor Thin Middle Eastern leavened flat **bread** made from high-extraction **wheat flour**.

Tanshen Common name for *Salina miltiorrhiza*, the roots of which are used widely in Chinese herbal medicine. Extracts display **antitumour activity**, **antimutagenicity** and **antioxidative activity**. Also known as dan shen.

Tansy Common name for *Tanacetum vulgare*, the leaves and tops of which are used as **herbs** with a bitter **flavour**. Leaves are used in preparation of **herb tea**, **salads** and herb **stuffings**. Tansy **essential oils** and extracts contain significant amounts of the toxin α-thujone. Only α-thujone free tansy oils are permitted as food additives and their use is limited to **alcoholic beverages**.

Tape Indonesian sweet and sour fermented alcoholic **rice** product made by inoculating steamed glutinous rice with traditional rice flour **starters** containing **fungi**, **yeasts** and **bacteria**, and incubating in an airtight container. Also known as **tape ketan**. Eaten as a snack food. Can also be made from **cassava** (**tape ketela**).

Tape ketan Indonesian alcoholic snack food made by inoculating steamed glutinous **rice** with a combination of **starters** and incubating in an airtight container. Also known as **tape**.

Tape ketela Product made by **fermentation** of **cassava** to improve the protein content and overall nutritive value of the vegetable. Microorganisms used are usually *Amylomyces rouxii* and a yeast such as *Endomycopsis burtonii*. Sun dried and used as an ingredient in **soups**.

Tapioca Starch extracted from tubers of **cassava** (*Manihot esculenta*). Also called **cassava starch**.

Tapioca starch Alternative term for **tapioca**.

Tap water Water supplied to consumers via the water mains system; usually suitable for use as **drinking water**.

Tarag Asian **fermented milk** of a variety of species.

Tarama Fermented fish product containing fish **roes** (usually from **carp**) mixed with **salt**, **breadcrumbs**, Feta cheese, **olive oils** and **lemon juices**.

Tarhana Traditional Turkish version of **kishk**, a fermented **wheat flour-yoghurt** mixture used in **soups**. The Greek version of kishk is known as **trahanas**.

Taro Common name for *Colocasia esculenta* or *C. antiquorum*. The corm is eaten cooked; if not well enough cooked, irritation of the mouth results due to oxalate crystals. Used as a vegetable, in **soups** and stews, processed to make **fufu** or fermented to produce **poi**. Subsidiary corms (cormels), known as eddo in China and Japan, and leaves are also eaten. Taro is a good source of **potassium** and **fibre**. Leaves contain **carotenes** and are rich in **vitamin C**.

Tarragon Common name for *Artemisia dracunculus*, the leaves of which have a sweet, mild anise-like **flavour** and are used as **spices**. The predominant flavour compound is estragole, also known as *p*-allylanisole and methyl chavicol. Tarragon is used in **flavourings** for foods such as **meat** and meat products, flavoured **vinegar** and **pickles**. Leaf **essential oils** are extracted and also used as flavourings. Also known as estragon.

Tartaric acid Organic acid present in **fruits** and isolated from potassium tartrate films produced as a by-product in **winemaking**. Tartaric acid, as well as sodium and calcium **tartrates**, have many uses as **food additives**, including as **flavourings** (**acidulants**) imparting a fruity flavour, **humectants**, **antioxidants**, **sequestrants** and as part of a pH buffering system. Tartaric acid is also a substrate for production of the raising agent, cream of tartar (potassium hydrogen tartrate) which is an ingredient of **baking powders**. Systematic name is 2,3-dihydroxybutanedioic acid.

Tartrates Salts of **tartaric acid**. Crystallization of tartrates in **wines** is a problem, since the **wines** are then generally considered unacceptable by consumers.

Tartrazine Synthetic bright yellow pyrozole dye used in **artificial colorants** for foods and **beverages**. In aqueous solution, tartrazine shows high stability when exposed to acids and alkalis, moderate stability to light and heat (stable at **extrusion** and **baking** temperatures) and poor stability in the presence of **ascorbic acid**. Synonymous with FD&C Yellow 5 and CI 19140.

Tarts Open **pastry** cases made with shortcrust pastry, which are frequently baked blind (or empty) and then filled with sweet **fillings** such as **fruits**, **jams** or **custards**, or sometimes savoury mixtures, e.g. **cheese** or **vegetables**.

Taste Sensation produced by stimulation of the taste buds on the tongue. The tongue can distinguish five separate tastes (sweet, salt, sour, bitter and savoury/**umami**). Often used as an alternative term for **flavour**.

Taste panels Groups of individuals, untrained or trained, used to sample products and assess their **flavour**, with a view to providing an insight into consumer preferences. Taste panels are used in research, product development and for purposes of evaluating new and competitive products, and are not restricted to evaluating flavour. **Texture**, **colour** and many other quality factors can be measured meaningfully.

Taste thresholds Alternative term for **flavour thresholds**.

Taurine Aminosulfonic acid synthesized from **cysteine** and **methionine**. Abundant in **animal proteins** but is not found in **vegetable proteins**. Hence, vegetarians with insufficient cysteine and methionine intakes may have difficulty producing taurine.

Taxonomy Study of the theory, practice and rules of classification and nomenclature of living and extinct organisms. The principles of taxonomy were established in the 18th century by the work of Linnaeus. As far as possible, organisms are arranged into a hierarchy of groups (called taxa) based on degrees of relationship (phylogeny). When knowledge of the evolution of a group is lacking, taxonomy is based on structural and other similarities. An organism is classified according to a hierarchical system as follows: kingdom, phylum, class, order, family, genus, species.

TBA reactive substances Abbreviation for thiobarbituric acid reactive substances (TBARS). Name applied jointly to **malonaldehyde** and the other substances formed during lipid oxidation, as measured in terms of **thiobarbituric acid values** (TBA values) determined from reaction with thiobarbituric acid (TBA). TBARS values are expressed as mg malonaldehyde equivalents per kg of sample. Care must be taken when comparing TBARS values between different studies because of the many variations that have been developed for performing the TBA test.

TBARS Abbreviation for **TBA reactive substances**.

TBA values Abbreviation for **thiobarbituric acid values**.

TBHQ Abbreviation for *tert*-**butylhydroquinone**.

TDE Persistent non-systemic organochlorine insecticide used to control a wide range of **insects**. Use on crops has generally been displaced by less persistent **insecticides**. Can occur as a degradation product of **DDT**. Classified by WHO as moderately toxic (WHO II). Also known as DDD.

Tea Hot or cold **beverages** made by infusion of dry, prepared leaves of *Camellia sinensis* in water. The main types are **black tea**, in manufacture of which the fresh **tea leaves** have undergone **fermentation** before **drying**, and **green tea**, in which the fresh tea leaves have not undergone this fermentation. **Oolong tea** and **pouchong tea** have undergone partial **fermentation**, and are intermediate in character between green and black teas.

Tea bags **Tea** packaged in small portion-size permeable bags for easy preparation of **tea beverages**.

Tea beverages Hot or cold **beverages** prepared from **tea leaves** or infusions.

Tea granules **Instant tea** products comprising **granules** of dry tea extracts which are reconstituted into **tea beverages** on addition of water.

Tea leaves Fresh or processed leaves of the tea plant, *Camellia sinensis*.

Tea powders **Instant tea** products comprising powdered dry tea extracts which are reconstituted into **tea beverages** on addition of water.

Teas Hot or cold **beverages** prepared by infusion of dry plant leaves, flowers or other plant parts. The type usually referred to as tea is made from the leaves of *Camellia sinensis*; other types include **mate**, **rooibos tea**, **honeybush tea** and a wide range of types of **herb tea** and **fruit tea**.

Tea seed oils **Vegetable oils** extracted from the seeds of tea species such as *Thea sasangua* or *Camellia oleifera*. Used as **salad oils** and **cooking oils**.

Tea tree oils **Essential oils** distilled from leaves of *Melaleuca alternifolia*, a tree native to Australia and certain parts of Asia. Major constituents of the oils are terpinen-4-ol, 1,8-cineole and γ-terpinene. The oils have a warm, spicy **flavour**. Tea tree oils exhibit **antimicrobial activity** and are used as an antiseptic. Although more commonly used for their therapeutic properties, tea tree oils are also used as food **flavourings**, including as a substitute for **nutmeg**.

Technetium Metallic element with the chemical symbol Tc.

Tecto Alternative term for the anthelmintic **thiabendazole**.

Teff Tropical **millet**, *Eragrostis abyssinica* or *E. tef*, which is native to northeastern Africa and southeastern Arabia and is used as a cereal crop and livestock feed. Used to make the traditional flat **bread**, **injera**.

Tehineh **Pastes** made from ground, dehulled, dry roasted **sesame seeds**.

Teleme cheese Greek **soft cheese** prepared from **ewe milk** or **cow milk**. Now also made in California, USA. Similar to **Brie cheese**, with a tangy flavour that develops as the cheese ages.

Telemetry Process of transmitting readings from instruments or measurements by radio or a telecommunications link.

TEM Abbreviation for **transmission electron microscopy**.

Temephos Non-systemic insecticide used primarily for control of mosquito and midge larvae and certain aquatic **insects** in urban and agricultural environments. Also used for controlling lice on animals (including humans). Residues may contaminate **water supplies**. Classified by Environmental Protection Agency as slightly toxic.

Temik Alternative term for the insecticide **aldicarb**.

Temp. abuse indicators Devices used to give an indication of whether products have been exposed to inappropriate temperatures that could cause damage during transport, distribution or storage. For example, indicators can be used to show whether **frozen foods** have been thawed during handling or storage; **thawing** during distribution can potentially affect quality and safety. Indicator devices often produce a visible, irreversible **colour** change to show when temperature abuse has occurred. Microbial indicators may also be used to detect exposure to temperature abuse, especially in animal **carcasses**. For example, poultry products that have been maintained at the correct temperature will have fairly constant counts of **coliforms**, while those that have been warmed will have higher counts.

Tempe Alternative term for **tempeh**.

Tempeh Product generally made by **fermentation** of **soybeans**, sometimes mixed with **cereals**. Used as **meat extenders** or **meat substitutes**. Cooked in a variety of ways or added to dishes such as **sauces**, **soups** and **casseroles**. Some types of tempeh are made from other materials, e.g. **bongkrek** is made by fermentation of presscake of **coconuts** or **coconut milk** residue.

Temper Measure of the degree of crystallization of **cocoa butter** in **chocolate** and the type of crystals present.

Temperature Degree or intensity of heat present in a substance or object or its surroundings, usually measured using **thermometers**.

Tempering Stabilization of **chocolate** by application of a **melting** and **cooling** process. Chocolate is tempered to stabilize the **cocoa butter**, a fat that can form crystals and cause **bloom** in the finished product. The classic tempering method includes the following stages: melting of the chocolate; working two-thirds of the melted chocolate on a marble slab with a metal spatula until it becomes thick; transferring the thickened chocolate back into the remaining melted chocolate; and reheating the product.

Tempura Japanese dish prepared from **vegetables**, **fish** or **shellfish**, fried in **batters**.

Tench **Freshwater fish** species (*Tinca tinca*) from the carp family (Cyprinidae); distributed across Europe and western Asia. Marketed fresh and frozen and as a canned product. Also known as **lin**.

Tenderization Mechanical or chemical processes by which **meat** can be made easier to cut or chew, so improving its **tenderness**. Mechanical methods break down tough fibres in the meat, usually through pounding. Pounders can be made of metal or wood, and can be a variety of shapes and sizes. Chemical methods that can also be applied to soften meat fibres include application of long, slow **cooking, marination** in acidic **marinades** and use of commercial meat tenderizers. Most meat tenderizers are composed primarily of **papain**, an enzyme extracted from **papayas**; they can also contain **salt, sugar** (usually **glucose**) and **anticaking agents** (usually calcium stearate).

Tenderness **Sensory properties** related to the extent to which a product, such as **meat**, is tender, i.e. soft, palatable and chewable. Tenderness can be measured using **tenderometers**.

Tenderometers Instruments used to measure **tenderness** or the stage of maturity of produce, particularly **peas**, on the basis of the force required to cause shearing.

Tenjan Alternative term for **doenjang**.

Tenox Registered trade name for a series of natural and synthetic **antioxidants** manufactured by Eastman Chemical.

Tensile strength Measure of the resistance that a material produces to a pulling stress (tensile stress); measured in Newtons per square metre.

Tensiometry Measurement of surface tension.

Tenuazonic acid Mycotoxin produced mainly by *Alternaria alternata* growing on foods (e.g. **fruits, vegetables** and **cereals**).

Tepary beans Seeds produced by *Phaseolus acutifolius*, a plant that grows well under drought conditions. Vary greatly in shape and **colour**. Dried seeds are soaked before **cooking** or are ground into **meal**. **Pinto beans** may be substituted for tepary beans in recipes.

Tequila Mexican **spirits** made by **distillation** of fermented sap of the **agave** plant.

Teratogenesis Process leading to developmental abnormalities in the foetus.

Teratogenicity Capacity of a substance to produce teratogenic effects, i.e. to cause developmental abnormalities in the foetus.

Terbuthylazine Broad-spectrum herbicide used for pre- or post-emergence control of weeds around a wide range of food plants such as **fruits, vegetables, cocoa, coffee** and **sugar cane**. Classified by WHO as slightly toxic (WHO III).

Terfezia Genus of **edible fungi** including the desert truffles and the poor man's truffle.

Termitomyces Genus of **edible fungi**.

Terpenes Unsaturated **hydrocarbons** consisting of isoprene units found in many higher plants and **essential oils**. Typically, **volatile compounds** with pleasant odours used as **flavourings**. Terpenes are major components of **citrus essential oils** but, since they are not responsible for the characteristic **flavour** and readily oxidize and polymerize to produce unpleasant flavours, they are generally removed by distillation or solvent extraction.

Terpenoids **Volatile compounds** found in plants and **essential oils** which are important for **flavour**. Certain terpenoids exhibit **antioxidative activity, anticarcinogenicity** and **antimutagenicity**.

Terpinene **Flavour compounds** found in plants and **essential oils** that have been found to inhibit food **spoilage yeasts**.

Terpineol Monocyclic monoterpene alcohol used in flavourings. Found naturally in **essential oils, citrus juices** and **wines**, and can be produced by microbial transformation of **limonene**.

Terpinyl acetate Flavour compound with **antifungal activity** that is found in **essential oils**.

Terrines Foods, particularly **pates**, which are cooked and served in earthenware tureens (or terrines). A pate made in this way is also referred to as pate en terrine.

Terroir Total environment in which a grapevine is grown for the purpose of producing **winemaking grapes**. Includes a great many factors, including soil, climate, location and cultivation conditions.

tert-Butylhydroquinone Commonly abbreviated to **TBHQ**. An antioxidant used in foods, including meat products, **vegetable oils, potato crisps** and **cereal products**.

Testosterone Male sex hormone produced by the interstitial cells of the testis of mammals. Used to promote muscular development in certain animals.

Tetilla cheese Spanish semi-soft **cheese** made from **cow milk**. Rind is pale yellow and ridged. The cheese has a fresh lemony **flavour** and a creamy **consistency**; fat content is 25%. **Ripening** is completed in 2-3 weeks.

Tetrachlorodibenzo-p-dioxins Potent **toxins** released into the environment from, for example, industrial sources that can then find their way as contaminants into the food chain.

Tetrachloroisophthalonitrile Alternative term for the fungicide **chlorothalonil**.

Tetrachloromethane Synonym for **carbon tetrachloride**. Organic halogen compound and versatile organic solvent whose use has diminished since the discovery that it is carcinogenic. May be used in **fu-**

migants. Can occur as a contaminant of treated **drinking water**.

Tetracyclines Broad-spectrum **antibiotics** widely used in animals, both for prevention and treatment of disease and as feed additives to promote growth. Tend to accumulate in **livers** and may persist in the bloodstream (and **milk** in cattle) for considerable time periods following administration. May also accumulate and persist in bone tissue and **eggs**. Commonly used examples include **chlortetracycline**, **doxycycline** and **oxytetracycline**.

Tetradifon Non-systemic contact acaricide used to control plant eating **mites** on a wide range of **fruits** and **vegetables**, **hops** and **tea**. Classified by Environmental Protection Agency as slightly toxic (Environmental Protection Agency III).

Tetrahydrofolate Biochemically active form of **folic acid**. Coenzyme of various reactions involved in the metabolism of **amino acids**, **purines** and **pyrimidines**. Many foods are rich in folates, including green **leafy vegetables**, **livers**, **fruits** and **yeast extracts**.

Tetrahydrophthalimide Primary degradation product of the fungicide **captan**.

Tetrazoles Group of **organic nitrogen compounds** derived from tetrazole, a heterocyclic organic compound comprising four nitrogen atoms and a single carbon atom.

Tetrodotoxin Highly toxic and potentially lethal neurotoxin found in many species of **pufferfish**. Produced by **bacteria** which colonize the fish. Responsible for poisoning caused by consumption of contaminated pufferfish.

Texture **Sensory properties** relating to the feel of a surface or product, or the impression created by a surface structure or the general physical appearance of a surface. A major factor affecting the **mouthfeel** and quality of a food.

Textured vegetable proteins Plant protein products that are shaped and textured to form particles, or shaped pieces, such as chunks and strips, usually by spinning or **extrusion** technology. Typically formulated with added **colorants** and **flavourings**, and used as **meat substitutes**. **Soy proteins** are most commonly used, although other proteins, such as **wheat gluten**, can also be used. Commonly abbreviated to TVP.

Texture profile analysis Analysis of the texture of a food in terms of mechanical properties, geometrical characteristics, and fat and moisture contents, at specific points during the **mastication** process.

Texturization Process by which **sensory properties** of a substance are altered, e.g. to produce a particular feel, appearance or **consistency**.

Texturizers **Additives** that improve the **texture** of foods. Examples include **gums**, **hydrocolloids** and **polydextrose**, used as **fat substitutes** to add **body** to **low fat foods** and calcium chloride, which is added to canned **fruits** and **vegetables** to maintain **firmness** of the product.

Texturizing agents Substances which act as **texturizers**, improving the **texture** of foods.

Texturometers Devices used to measure **texture** properties of foods, by analysis of physical attributes such as hardness, cohesiveness and crush resistance.

Thaumatin Non-nutritive natural **sweeteners** isolated from **fruits** of *Thaumatococcus danielli*, a plant native to West Africa. The sweet **flavour** of *T. danielli* fruits is attributed to two proteins of approximately 22 kDa, designated thaumatin I and II. Both thaumatin proteins are approximately 1000-2000 times as sweet as **sucrose** (weight for weight). Commercial thaumatin preparations are complexed with aluminium to improve their stability. Thaumatin is soluble in water and alcohols and is synergistic with acesulfame K and saccharin. Aqueous solutions of the sweetener have high thermal stability and are stable over the pH range 2-10. However, factors which influence thaumatin structure, e.g. reducing agents, affect its sweetness. Although used as a sweetener, thaumatin has a liquorice-like **aftertaste**. It is commonly used in **flavour enhancers**, e.g. in **chewing gums**. Synonymous with **katemfe** and sold under the trade name Talin.

Thawing Transition of an item from a frozen to an unfrozen state.

Theaflavins **Flavonoids** which contribute significantly to the **colour** and **flavour** of **black tea**, and are used as markers of quality. Possess **antitumour activity** and **antioxidative activity**.

Theanine Amino acid found in **tea**. As well as improving the **flavour** of tea, theanine has a relaxing effect, improves learning ability and lowers blood pressure. Has also been found to help prevent D-galactosamine-induced liver injury in rats.

Thearubigins Flavonoid **pigments** found in **tea** which contribute to the **flavour**, depth of **colour** and **body**.

Theobromine Purine alkaloid similar to **caffeine** that is found in **cocoa**, **chocolate**, **soft drinks** and **tea**. Acts as a stimulant and may be toxic.

Theophylline Purine alkaloid that contributes to the **flavour** of and is used as a marker of quality in **tea**, **coffee**, **soft drinks** and **chocolate**. Acts as a stimulant.

Thermal capacity **Thermophysical properties** relating to the extent to which a material can retain heat.

Thermal conductivity **Thermophysical properties** relating to the rate of conduction of heat through a material, measured in Joules per second per metre per Kelvin.

Thermal diffusivity **Thermophysical properties** relating to the extent to which an item diffuses or spreads heat throughout its mass.

Thermal expansion Increase in size (e.g. length, volume, surface area) of a body in response to **heating**. For liquids, expansivity observed directly is called the apparent expansivity, as the container holding the liquid will have expanded also with the rise in temperature. Absolute expansivity is the apparent expansivity plus the volume expansivity of the container.

Thermal processes Processes involving **heating** that are used to produce desirable changes in products, such as protein coagulation, starch swelling, textural softening and formation of **aroma compounds**. Undesirable changes can also occur with application of thermal processes, such as losses of **vitamins** and **minerals**, and loss of fresh appearance, **flavour** and **texture**. Examples of thermal processes used in the food industry are: HTST processing; LTLT processing; electric heating; **ohmic heating**; microwave heating; and **blanching**.

Thermal processing Application of **heating** methods to the processing of foods. Techniques in the category include: HTST processing; LTLT processing; electric heating; **ohmic heating**; microwave heating; and **blanching**.

Thermal properties Properties that influence the **heating** rate and response to heating of a material.

Thermal stability **Thermophysical properties** relating to the ability of materials to maintain stability when subjected to various temperatures of applied heat. If food ingredients or **additives** are heat stable, it is possible for them to be used successfully in products which have to be thermally processed. Synonymous with **heat stability**.

Thermistors Semiconductors used for measuring temperature on the basis that their electrical resistance decreases with increasing temperature.

Thermization Heat treatment of foods at a temp. lower than that used for **pasteurization**, with an upper limit of about 65°C for 20 s. Thermization is less severe for the product and associated **microorganisms** than pasteurization.

Thermoanaerobacter Genus of Gram positive, anaerobic rod-shaped thermophilic **bacteria**. Some species are used in the production of thermostable **proteinases**.

Thermoanaerobacterium Genus of Gram positive, anaerobic rod-shaped thermophilic **bacteria**. Some species (e.g. *Thermoanaerobacterium thermosaccharolyticum)* are used in the production of thermostable **proteinases**.

Thermoascus Genus of thermophilic filamentous **fungi**. The species *Thermoascus aurantiacus* shows strong cellulose degrading activity and is a source of a number of **glycosidases**, including **xylan endo-1,3-β-xylosidases** and **cellulolytic enzymes**.

Thermocouples Devices for measuring or sensing a temperature difference, consisting of two wires of different metals connected at two points, between which a voltage is developed in proportion to any temperature difference.

Thermoluminescence Luminescence produced by heating a solid substance. Caused by emission of photons of light by free electrons and holes trapped in the solid.

Thermolysins Neutral, heat-stable, zinc-containing metalloproteinases (EC 3,4.24.4.) produced by ***Bacillus*** *thermoproteolyticus*. Most heat-stable **proteinases** available commercially; remain active up to 80°C.

Thermometers Instruments for measuring and indicating temperature, typically consisting of a graduated glass tube containing mercury or alcohol which expands when heated and contracts when the temperature falls. Thermometers are tailored for different purposes. For example, specific instruments are available for use during the manufacture of **sugar confectionery** or **cooking** of **meat** (to ascertain that the meat has reached the desired degree of doneness), and also for temperature monitoring in **freezers**, **refrigerators** and **ovens**.

Thermomonospora Genus of Gram positive, aerobic, thermophilic, filamentous **bacteria** of the family Thermomonosporaceae. Occur in soil and compost. Some species may be used in the production of thermostable **proteinases**.

Thermomyces Genus of **fungi** whose species (e.g. *Thermomyces lanuginosus*) are used in the production of thermostable **proteinases**.

Thermophiles Organisms, especially **microorganisms**, that grow best at relatively high temperatures. Their optimum growth temperature is generally accepted as being above 50°C.

Thermophilic bacteria **Bacteria** that are **thermophiles**.

Thermophysical properties Properties that influence the **heating** rate and response to heating of a material. Examples of thermophysical properties are **thermal conductivity** (the ability of a material to conduct heat) and **specific heat** (the ability of a material to store heat).

Thermostats Devices that automatically regulate temperature to a specified value or range, or activate devices at a set temperature.

Thermotoga Genus of rod shaped hyperthermophilic **bacteria** belonging to the order Thermotogales. The species *Thermotoga maritima* and *T. neapolitana* metabolize many simple and complex **carbohydrates** and are a source of a number of **enzymes** including **glycosidases**.

Thermus Genus of Gram negative, aerobic, rod-shaped or filamentous, thermophilic **bacteria**. Occur in hot springs, hot water tanks and thermally polluted rivers. *Thermus thermophilus* is used in the production of thermostable **proteinases**.

Thiabendazole Broad-spectrum anthelmintic used to treat a range of roundworm and cestode infections in cattle, sheep, swine and poultry. Also used in food **preservatives** and agricultural **fungicides**. Normally metabolized and excreted rapidly by animals, residues rarely persisting in any tissues beyond 7 days post-treatment.

Thiamin Synonym for **vitamin B_1** and **vitamin F**. Member of the water soluble **vitamin B group**. Active in the form thiamin pyrophosphate, a coenzyme for decarboxylation reactions in carbohydrate metabolism. Helps to maintain normal nervous system activity and regulates muscle tone of the gastrointestinal tract. Severe deficiency is clinically recognized as beriberi. Thiamin is found in unrefined **cereals**, **beans**, **meat** (especially **livers**, **kidneys**, **hearts** and **pork**), **yeasts**, **potatoes**, **peas** and **nuts**. Cooking losses can be as much as 50%.

Thiamine Alternative spelling for **thiamin**.

Thiamphenicol Recently introduced synthetic antibiotic (**chloramphenicol** analogue) intended for treatment and control of respiratory and intestinal diseases in cattle and poultry. Also intended for intermammary administration in cattle and as a replacement for **antibiotics** used in aquaculture that present long depletion times. Post-treatment depletion in animal tissues is relatively rapid.

Thiazoles Volatile **flavour compounds** found, for example, in cooked **meat** and **beer**. May also cause **off flavour**.

Thickeners **Additives** that increase the **viscosity** of foods. Unlike **gelling agents**, do not promote the formation of gels. **Gums** and **starch** are important thickeners in the food industry.

Thickening Process of making or becoming thicker and usually more viscous. For example, **sauces** are thickened using **corn starch**.

Thickness As well as relating to **consistency** and **viscosity**, this term relates to measurement of the depth of a substance such as **backfat** on animal carcasses.

Thidiazuron Plant growth regulator with defoliation activity; used to stimulate fruit growth in a range of **fruits**, including **apples**, **grapes** and **kiwifruit**.

Thielaviopsis Genus of **fungi** of the class Hyphomycetes. Species may cause **spoilage** of **fruits** and **vegetables**. *Thielaviopsis paradoxa* causes black rot of **pineapples**, and *T. basicola* causes black root rot of **carrots**.

Thin layer chromatography **Chromatography** technique in which sample components are separated as the sample travels, under the influence of a solvent, up an inert plate coated with a sorbent. Commonly abbreviated to TLC.

Thinning In plant cultivation, removal of young plants to allow remaining plants more room to grow, or removal of selected fruits from a plant so that the other fruits can increase in size.

Thiobarbituric acid values Values (commonly abbreviated to TBA values) used for assessing lipid oxidation in foods and other biological systems, using thiobarbituric acid (TBA). Two molecules of TBA react with one molecule of malonaldehyde to produce a red pigment; the amount of pigment produced is measured spectrophotometrically. Extent of lipid oxidation, reported as the TBA value, is expressed as milligrams of malonaldehyde equivalents per kilogram of sample, or as micromoles of malonaldehyde equivalents per gram of sample. The TBA test may be performed directly on the sample, its extracts or distillate.

Thiocyanates Alternative term for **isothiocyanates**.

Thiodan Alternative term for the insecticide **endosulfan**.

Thioesters **Esters** containing sulfur instead of oxygen. Important **aroma compounds** often added to **processed foods**. Can be prepared by lipase-catalysed **esterification** of **fatty acids** with short- and long-chain **thiols**.

Thioglucosidases EC 3.2.3.1. **Glycosidases** that hydrolyse *S*-glycosyl compounds. Have a wide specificity for **thioglycosides**, forming a thiol and a sugar. Responsible for hydrolysis of **glucosinolates** in cruciferous plants, producing compounds with an undesirable **flavour** that may also be toxic. Also known as myrosinases, sinigrinases and sinigrases.

Thioglycolic acid Toxic organic acid also known as 2-mercaptoacetic acid, α-mercaptoacetic acid and thiovanic acid.

Thioglycosides Sulfur-containing **glycosides** found in cruciferous vegetables that show **anticarcinogenicity**. They are useful as glycosyl donors in the synthesis of complex **carbohydrates**.

Thiols Compounds containing **sulfhydryl groups**, i.e. in which the oxygen of an alcohol is replaced with sulfur. These compounds have extremely unpleasant odours.

Thionins Low molecular weight **proteins** which occur in seeds of several plant species and show **antimicrobial activity**.

Thiophanate-methyl Systemic fungicide used for control of a wide range of fungal diseases on **fruits**, **vegetables** and **cereals**. Classified by Environmental Protection Agency as not acutely toxic (Environmental Protection Agency IV). Also known as methylthiophanate and Pelt 44.

Thiophenes Sulfur-containing **volatile compounds** that contribute to the **flavour** of many foods and **beverages**.

Thiouracil Drug which inhibits production of thyroid hormones and results in increased water retention in muscle tissue. Sometimes used illegally to increase meat yield in animals.

Thiourea Primary degradation product of the **fungicides chlorothalonil** and **mancozeb**, which are used on a wide range of **fruits**, **vegetables** and **cereals**. Also used as a photographic fixative, textile-treating agent and as a starting material for various dyes and drugs. Residues may contaminate **water supplies**.

Thiram Protective fungicide applied to foliage to control fungal diseases on **pome fruits**, **stone fruits**, **grapes** and *Brassica* vegetables. Also used to control **fungi** in stored **fruits** and **cereals**. Classified by WHO as slightly toxic (WHO III).

Thistles Group of **plants** including many species used as **vegetables**. Such types include **globe artichokes** (*Cynara scolymus*), **cardoons** (*C. cardunculus*) and milk thistles (*Silybum marianum*). Parts which are eaten include flowers, leaves, stems and seeds. Extracts of dried cardoon flowers are used as **vegetable rennets** in **cheesemaking**.

Thixotropy Property of a material that enables it to stiffen in a relatively short time on standing, while, upon agitation or manipulation, it can change to a very soft consistency or to a fluid of high **viscosity**, the process being completely reversible.

Threadfin bream Any of several **marine fish** species in the genus *Nemipterus*; distributed across the Indo-Pacific. Commercially important species include *Nemipterus japonicus* (Japanese threadfin bream) and *N. virgatus* (golde threadfin bream). Marketed mainly fresh, but also frozen, steamed, dried-salted, dry-smoked, fermented or made into **fish balls** and **fish meal**.

Threonine Aminohydroxybutyric acid. An essential amino acid.

Threshers Machines that separate grain from other debris.

Thrips Common name for members of the insect order Thysanoptera. Pests of a wide variety of crops (e.g. **citrus fruits**, **vegetables** and cereal grains). Some species are important vectors of **fungi** and **viruses** responsible for **plant diseases**.

Thromboelastographs Instruments used in the food industry to monitor **gelation**, e.g. milk **coagulation**, by measuring gel **firmness**.

Thujone Toxic ketone present in **absinthe**, and certain herbal plants, **essential oils** and natural **flavourings**.

Thyme Common name for plants native to Mediterranean countries of the genus *Thymus*, leaves and flowering tops of which are used as **spices**. The most commonly used variety is *T. vulgaris*; other spice varieties include *T. citriodorus* (lemon thyme), *T. zygis* and *T. serpyllum* (wild thyme). The predominant **flavour compounds** of thyme are **thymol** and **carvacrol**. Thyme extracts and **essential oils** are used as **flavourings** in the food industry.

Thymine Pyrimidine base that pairs with **adenine** in **DNA**. In **RNA**, it is replaced by **uracil**.

Thymol Phenolic derivative of **cymene** that is isomeric with **carvacrol**. Present in **essential oils**, and exhibits **antioxidative activity** and **antimicrobial activity**.

Thyristors Process control charging units used to convert three-phase power to direct current.

Thyroxine Iodine-containing hormone derived from **tyrosine** that is produced by the thyroid gland.

Tigernuts Stem tubers of *Cyperus esculentus*, cultivated in West Africa. Eaten raw or roasted, and used to make alcoholic and non-alcoholic **beverages**. Also a source of **oils** of potential food use. Alternatively known as **chufa nuts**; also spelt tiger nuts.

Tiger shrimps Species of **shellfish** (*Penaeus monodon*) which is the largest of the commercially available types of **shrimps**. As well as being widely distributed in the seas around Asia, Australia and the eastern coast of Africa, tiger shrimps are major **aquaculture** products of Australia and south east Asia. Characterized by grey/blue shells with black stripes and also stripes on the peeled meat. Shell turns red when cooked. White flesh is tinged orange or red depending on whether it is cooked in or out of the shell. Also known as black tiger shrimps and giant tiger shrimps.

Tilapia Any of a number of **freshwater fish** in the family Cichlidae, particularly those within the genus *Oreochromis*. Occur in lakes and rivers across Africa; introduced for aquacultural purposes in many other parts of the world. Commercially important species include *Oreochromis niloticus* (**Nile tilapia**) and *O. mossambicus* (Mozambique tilapia). Flesh tends to be white or light pink in **colour** and firm, with a sweet and mild **flavour**. Marketed fresh and frozen.

Tilmicosin Macrolide antibiotic used as a veterinary antibacterial agent in food-producing animals. Major residues in treated animals are of the parent compound, and are most persistent in the **kidneys** and **livers**. In muscle, residues persist at the injection site. Due to persistence in **milk**, tilmicosin is not recommended for treatment of lactating **cattle**.

Til oils Alternative term for **sesame oils**.

Tilsit cheese German semi-hard cheese made from **cow milk**. Buttery and fruity **flavour** with a spicy tinge, and mildly pungent **aroma**. Rind is crusty and yellow-beige in **colour**. Interior is supple with small irregular holes. Tilsit is considered an excellent sandwich cheese.

Time intensity **Sensory analysis** techniques used to measure the intensity of a specific food attribute as a function of time. Usually used to investigate the temporal behaviour of **flavour compounds**, such as sweet and bitter molecules, and the release of **volatile compounds** from foods. Such techniques are important in the reformulation of foods that results in structural modification.

Time temp. indicators Devices designed to monitor and register accumulated temperature exposure of foods over time. Used to alert the distributor or consumer to conditions which may render a particular food hazardous. Usually fixed to the product at the point of distribution and read by the receiving establishment. Time temp. indicators have been used on food rations employed in the armed services, as such rations may be subjected to high temperatures during transit and may also be stored and used in high-heat locations. On rations, each time temp. indicator consists of an outer reference ring and an inner circle. The inner circle darkens with time, and darkens more quickly as the temperature increases; therefore, the darker the circle, the less fresh the food.

Time temp. integrators Simple **quality control** devices and process evaluation tools that monitor food temperature exposure history and relate it to **shelf life** behaviour. Time temp. integrators should give accurate information and be easy to use, should be incorporated into food without disturbing heat transfer and should quantify the impact of the process on a target attribute that results in a specific kinetic requirement. Time temp. integrators are classified according to working principle, type of response, origin, and application and location in food, and can be biological (microbiological and enzymic), chemical or physical systems.

Tin Silvery-white metal, with the chemical symbol Sn. Also refers to various metal **containers** used for food storage or preparation. Examples include lidded airtight storage containers made of **tin plate** or **aluminium**, open-topped metal containers used for **baking** food, e.g. **cakes**, and sealed containers made from tin plate or aluminium used for preserving foods. In the UK, the term is often used as being synonymous with the term **cans**.

Tin plate Iron or sheet **steel** which is coated with the chemical element **tin**. Used to make **containers** and **cans** for food **storage** and **preservation**.

Tipburn Necrosis of plant apical or marginal tissues, affecting only a small part of the leaf. Possibly caused by internal **water stress** induced by salt or wind desiccation.

Titin Family of very large **proteins** found in the sarcomere of striated muscle. Degradation of titin improves the **tenderness** of **meat** during *post mortem* storage.

Titratable acidity Measure of the total **acidity** in a sample, both as free hydrogen ions and as hydrogen ions still bound to undissociated acids. Determined by addition of a standardized base to the sample until a predetermined endpoint is reached. The endpoint may be assessed by a change in the colour of an indicator at a particular pH. This test can be used to determine milk quality and to monitor the progress of **fermentation** in **cheese** and **fermented milk**.

Titration Technique in which reagent solution is added to the analyte until the reaction is complete. Commonly based on oxidation-reduction or acid-base reactions, complex formation or precipitation. The end point of the reaction may be measured by a range of methods, including **spectroscopy**, change in colour of an indicator or changes in voltage or current passing between a pair of electrodes in the reaction solution.

Titrimetry Alternative term for **titration**. A method for determining the amount of an analyte by reacting it with a reagent solution until the supply of analyte is exhausted.

TLC Abbreviation for **thin layer chromatography**.

TMTD Alternative term for the fungicide **thiram**.

Toast Sliced **bread** which has been placed near a fire or grill so that it becomes brown and crisp.

Toast bread **Bread** suitable for making **toast**.

Toasting **Cooking** or **browning** of a food, e.g. **bread**, **almonds** or other **nuts**, by exposure to radiant heat.

Tochu tea Aqueous extract of *Eucommia ulmoides* leaves which is drunk as a **herb tea** in Japan. Displays **antimutagenicity**.

Tocols Complex alcohols of the chromanol type. Tocols are generically termed **tocopherols**. Several tocopherols have been isolated, but only four have **vitamin E** activity.

α-Tocopherol The major contributor to **vitamin E** activity in foods. Rich sources of this fat-soluble vitamin include vegetable oils, margarines, wheat germ, nuts, seeds, sea foods, beef, eggs, fruits and vegetables. α-Tocopherol is a powerful antioxidant that protects polyunsaturated fats and vitamin A from oxidation in the gastrointestinal tract. α-Tocopherol also prolongs the life of red blood cells and protects lung tissue from the adverse effects of pollution. α-Tocopherol is included among **GRAS substances** and is one of the **antioxidants** used in the food industry to retard rancidity in foods containing polyunsaturated fats.

α-Tocopherol acetate Alternative term for **vitamin E acetate**.

Tocopherols Members of the **vitamin E** group that are fat soluble and have **antioxidative activity**. In chemical terms, tocopherols are **terpenoids**. Four isomers exist that have vitamin E activity - α-, β-, γ- and δ-tocopherols, the most important of which is α-tocopherol. Tocopherols are found in **wheat germ oils**, **butter**, **egg yolks** and **leafy vegetables**, and are important in the stabilization of cell membranes by protecting them from the damaging effects of oxygen free radicals, which are produced by various disease processes and toxic substances.

α-Tocopheryl acetate Alternative term for **α-tocopherol acetate/vitamin E acetate**.

Tocotrienol One of the main groups of compounds wth **vitamin E** activity (the other being **tocopherols**). Four isomers exist - α-, β-, γ- and δ-tocotrienols. Tocotrienols are found in **vegetable oils**, seeds and **leafy vegetables**. These compounds function primarily as antioxidants in cell membranes, protecting **unsaturated fatty acids** from oxidative damage.

Toddy Type of palm wine made in Southeast Asia by **fermentation** of sap of coconut **palms** (*Cocos nucifera*) or other palm species.

Toffees Hard **sugar confectionery** products made from boiling together **butter** or **vegetable oils**, **milk** and **sugar**. Similar to **caramels**, although the temperature used to boil the ingredients is higher than that used for caramels.

Tofu **Soy curd** product with a **texture** similar to that of compressed **Cottage cheese**. Made like **cheese** by **coagulation** of **soymilk** and draining of the curd. A good source of **proteins** and B vitamins. Available in firm, soft and silken forms that have different uses. Firm **tofu** is cubed and cooked or added to a variety of dishes. Other forms are used as substitutes for **sour cream** or **yoghurt**.

Tolerance Maximum level of a given, potentially harmful, substance (e.g. **mycotoxins**, **heavy metals**, **pesticides**) permitted in foods or **beverages**.

Tomatillos **Fruits** produced by *Physalis ixocarpa*. Related to, but larger than the **cape gooseberries**. Rich in **vitamin A**, **vitamin C** and **niacin**. Although classified as fruits, they are eaten as **vegetables**, almost always cooked, as this enhances their **flavour**. Used in **sauces**, such as **salsa**, stews, **casseroles** and **preserves**. Also known as jamberries.

Tomatine Glycoalkaloid saponin present in high concentrations in green **tomatoes**. Toxic to many **fungi** and **bacteria**.

Tomato catsups Catsups in which the main ingredient is **tomatoes**. Popular as an accompaniment for **French fries**, **burgers** and many other foods. Also known as **tomato ketchups**.

Tomato concentrates Products made by **concentration** of **tomato pulps** by processes such as **reverse osmosis**, **evaporation** and **ultrafiltration**. Uses include as **flavour enhancers** or in the manufacture of **tomato juices**.

Tomatoes **Fruits** produced by *Lycopersicon esculentum*. Vary in **colour** (red or yellow), size and shape, according to variety. Contain more than 90% water, the **carotenes lycopene** and **β-carotene**, vitamin B complex, **vitamin E** and moderate amounts of **vitamin C**, as well as a range of minerals. Tomatoes also contain the non-toxic alkaloid **tomatine**, amounts of which decrease as the fruits ripen. Consumed raw, cooked, as ingredients of a wide range of products, including **soups**, **sauces**, **casseroles**, pastes and purees, or in juices. Available canned and dried. Green tomatoes are used in **pickles** and **chutneys**. **Genetic engineering** has been used to produce tomatoes (e.g. Flavr Savr tomatoes) with improved **shelf life**, **flavour** and processing properties.

Tomato juices **Fruit juices** prepared from **tomatoes** (*Lycopersicon esculentum*). Drunk on their own (sometimes with **Worcestershire sauces** added) or mixed with other **beverages**. May also be used as the base of **sauces** and in various dishes.

Tomato ketchups Ketchups in which the main ingredient is **tomatoes**. Popular as an accompaniment for **French fries**, **burgers** and many other foods. Also known as **tomato catsups**.

Tomato pastes Rich concentrates produced from **tomatoes** by cooking, straining and reducing. Used as the base for **sauces** and **soups**. Available commercially in cans, jars and tubes.

Tomato pulps The soft, succulent parts of **tomatoes** or preparations made from them by mashing and **concentration**. Used in the preparation of many cooked dishes.

Tomato purees Smooth, thick liquids produced from **tomatoes** by cooking and straining. Used as the base for **soups** and **sauces**. Available commercially in jars, cans and tubes.

Tomato sauces Condiments produced from **tomatoes**, **seasonings** and other additives. Tomato based **sauces** are used as **toppings** for **pizzas** and **pasta** dishes and in many other dishes, such as stews and **casseroles**.

Tomato seed oils **Vegetable oils** extracted from **tomato seeds** produced as a by-product in **canning** of **tomatoes**. High in **unsaturated fatty acids**. Used as cooking oils.

Tomato seeds Seeds contained in **tomatoes** (*Lycopersicon esculentum*) and produced as a by-product of tomato **canning**. Contain **oils** and **proteins** rich in **unsaturated fatty acids** and **lysine**, respectively. **Tomato seed oils** may be used as **cooking oils**.

Tomato skins Outer surface of **tomatoes**. Removed during manufacture of many tomato products and thus is a waste product of the tomato processing industry. Rich source of **pigments**, including the carotenoid **lycopene**, which is valued for its health benefits.

Tongues A part of edible **offal**, often sourced from calves, lambs, oxen and pigs. **Tenderness**, **flavour** and **texture** vary with species and age of the source animal. Tongues may be sold fresh or brined; brining produces a pink **colour** and intensifies **flavour**. They are eaten hot or cold after boiling, skinning and slicing, or are used to produce meat products, such as **brawn**.

Tonic waters Carbonated **soft drinks** containing bitter compounds such as **quinine**.

Top fermenting yeasts Brewers yeasts which are non-flocculent and remain at the top of the **beer** during **fermentation**. Commonly used for **ale** and other British style types of beer.

Topinambour Alternative term for **Jerusalem artichokes**.

Toppings Sweet or savoury food items such as **sauces**, **pizza fillings** or **icings**, used to garnish/top other foods.

Top shells Any of a number of marine gastropod **molluscs** within the family Trochidae; found in intertidal and deeper waters around the world. A few species are consumed, including members of the genus *Omphalius*. Marketed fresh (shelled or unshelled) and frozen (unshelled).

Tordon Alternative term for the herbicide **picloram**.

Torte Rich **cakes** comprising either cake mixture baked in a **pastry** case or several thin layers of **sponge cakes**, filled with various ingredients such as **fruits**, **nuts**, **chocolate** and **cream**.

Tortellini Pasta shaped into small rings, stuffed with **meat** or **cheese** and often served with **sauces**.

Torten Rich **cakes** which often have **pastry** or sponge bases and are enriched with **cream**, **fruits** or **nuts**.

Tortilla chips Popular salted **snack foods**. Typically prepared by cutting extruded **corn masa** into **chips**, **baking** and **frying**. Eaten in the same way as **potato crisps** or as an accompaniment to **dips**. Also available flavoured with a variety of **flavourings**.

Tortillas Round, thin unleavened **pancakes** originating from Mexico which are traditionally made with **corn flour** and baked on a hot surface. Also known in Colombia as **arepas**.

Torulaspora Genus of yeast **fungi** of the class Saccharomycetes. Occur in soil, faeces, **wines**, fermenting cucumber **brines** and **fruit juices**. *Torulaspora delbrueckii* is used in **winemaking**, and is responsible for the **spoilage** of **fruit juice concentrates**, **cheese** and **wines**.

Torula yeast Highly nutritious **yeasts** (*Candida utilis*) grown on media such as ethanol and sulfite liquor wastes. Rich source of **proteins** and **vitamins** (especially B vitamins). Used as an animal feed supplement and a food additive.

Torulopsis Obsolete name for a genus of **yeasts** whose species have been reclassified into the genus ***Candida***.

Total quality management Management philosophy geared towards continuous improvement of product quality to meet, exceed and anticipate customer requirements.

Total solids Total amount of **solids** in a product. Commonly abbreviated to TS.

Total soluble solids Total amount of **soluble solids** in a product. Commonly abbreviated to TSS.

Toughness **Sensory properties** relating to the extent to which a product such as **meat** is hard to chew or cut due to its innate resistance, hardness and leathery texture. In a physical sense, toughness is defined as the energy required to propagate a fracture by a given crack area, generally derived from the area under a force-extension curve.

Toxaphene Alternative term for the insecticide **camphechlor**.

Toxicity Quality or degree of being poisonous.

Toxicology Scientific study of the nature, effects and detection of **toxins**, and the treatment of conditions caused by them.

Toxic substances Alternative term for **toxins**.

Toxins Poisonous substances, especially those that are produced by one living organism, and are poisonous to other living organisms.

Toxoplasma Genus of parasitic protozoans of the class Coccidia. Species are intracellular **parasites** of birds and mammals, including domestic cats and humans. *Toxoplasma gondii* is the causative agent of **toxoplasmosis**.

Toxoplasmosis Acute or chronic disease of humans and animals caused by *Toxoplasma* gondii. Transmission in humans is usually via ingestion of contaminated raw or undercooked **meat** (especially **pork** or **mutton**), or by contact with cat faeces. Symptoms range from an asymptomatic, or mild influenza-like disease, to an extensive fulminating disease that may cause damage to the brain, eyes, skeletal and cardiac muscles, liver and lungs. Can be transmitted transplacentally to cause congenital disease.

Traceability The ease with which origin or developmental history of something can be found by investigation.

Trace elements Elements that are essential **nutrients** but are required only in minute amounts (mg or micrograms/day) by humans. Examples are chromium, copper, manganese and zinc.

Trace metals Alternative term for **trace elements**.

Trade agreements Treaties designed to facilitate trade between two nations or a group of nations. In the absence of trade agreements, many nations impose special taxes (tariffs) and take other actions to discourage importation of foreign goods. Trade agreements usually seek to reduce or eliminate such barriers.

Trademarks Words or symbols established by use or legally registered as representing a product or company. The term 'trade name' may sometimes be used to refer to a name that has the status of a trademark.

Trahanas Greek name for **kishk**, a fermented **wheat flour**-**yoghurt** mixture used in **soups**. Known as **tarhana** in Turkey.

Trametes Genus of **fungi** of the class Homobasidiomycetes. Occur on wood. *Trametes versicolor* is used in the production of several **enzymes** used in **bioremediation** processes, e.g. **laccases** and **catechol oxidases**.

Tranquilizers General term for **drugs** that act on the central nervous system and are used primarily in the treatment of anxiety and psychiatric disorders that have an anxiety-related component. Major use in farm animals is for sedation prior to and during handling or transportation, usually in the form of barbiturates such as **azaperone**, nembutal and propiopromazine.

Transaminases EC 2.6.1. Also known as aminotransferases, these enzymes transfer amino groups from a donor, usually **amino acids**, to an acceptor, usually 2-oxo-acids, in a cyclic process. Most are pyridoxal phosphate proteins. The reaction also involves oxidoreduction; donors are oxidized to ketones, while acceptors are reduced. However, since the transfer of the amino group is the most prominent feature of the reaction, these enzymes are classified as aminotransferases rather than oxidoreductases.

Transcription Process by which **RNA** copies of template **DNA** strands are synthesized, catalysed by **DNA-directed RNA polymerases**. The initial products of transcription are typically processed and/or modified to give the mature RNA products, e.g. **mRNA**, **rRNA** and tRNA. In RNA **viruses**, RNA acts as the template for transcription; in this case the process is catalysed by RNA-directed RNA polymerases.

Transducers Devices that transform one type of energy to another.

Transesterification Process by which fatty acyl residues are transferred to **triglycerides** in a mixture of triglycerides and **fatty acids**. Can be catalysed by **lipases**, and may be used to modify the composition and properties of **fats** and **oils**.

trans **Fatty acids** **Fatty acids** produced during the **hydrogenation** of **fats** and **oils**, which are found in foods such as vegetable **shortenings**, **margarines** and partially hydrogenated **vegetable oils**. Thought to have several adverse effects on health, such as increased risk of **coronary heart diseases**, increased levels of cholesterol and low density lipoproteins, and reduced levels of high density lipoproteins.

Transferases EC 2. **Enzymes** that transfer a group, e.g. a methyl, acyl or glycosyl group, from one compound (the donor) to another (the acceptor). In many cases, **coenzymes** carrying the group to be transferred act as the donor.

Transferrins **Proteins** that transport Fe into cells. Found in the plasma of vertebrates and used as indicators of Fe status.

Transformation Process by which exogenous **DNA** is taken up by recipient cells, sphaeroplasts or protoplasts. The DNA may be in the form of **plasmids** that can replicate autonomously, or may be a fragment that can integrate into the host **chromosomes**. Transformation can occur naturally in some **bacteria**, but in other bacteria and eukaryotic microorganisms, it can only occur after cells have been permeabilized by artificial methods. Also refers to conversion of cultured cells to a malignant phenotype.

Transgenes Foreign **genes** introduced into the **genomes** of transgenic organisms early in development. Transgenes are present in both somatic and germ cells, and are inherited by offspring in a Mendelian fashion.

Transgenic animals Genetically engineered animals or their offspring that contain genetic material from at least one unrelated organism inserted into their **genomes**.

Transgenic plants Genetically engineered plants or their offspring that contain genetic material from at least one unrelated organism inserted into their **genomes**.

Transglucosylases Members of sub-class EC 2.4; synonymous with **glucosyltransferases**. **Enzymes** that transfer a glucosyl group from a donor to an acceptor.

Transglutaminases Alternative term for protein-glutamine γ-glutamyltransferases.

Transglycosylation Transfer of glycosyl groups, or saccharides, from a donor to an acceptor, with **enzymes** of the group **glycosyltransferases** as **catalysts**. This type of modification is performed to alter the **physicochemical properties** or **functional properties** of a natural compound, e.g. to improve the **solubility** of **neohesperidin dihydrochalcone** or to decrease the **bitterness** of **naringin**.

Translation Process by which **polypeptides** are assembled at ribosomes using **mRNA** molecules as templates. **Amino acids** are carried to the ribosome by specific tRNA molecules where they are incorporated into the growing chain in a sequence specified by the nucleotide sequence of the mRNA template.

Translucency **Optical properties** relating to the extent to which an object diffuses light passing through it, so that objects cannot be seen clearly.

Transmissible spongiform encephalopathies Alternative term for **prion diseases**.

Transmission electron microscopy **Electron microscopy** technique in which the image forming rays are passed through or transmitted by the sample. Commonly abbreviated to TEM.

Transparency **Optical properties** relating to the extent to which an item allows light to pass through it so that bodies can be clearly seen.

Transpeptidases **Enzymes** that catalyse the formation of an amide linkage between a free amino group and a carbonyl group within an existing peptide linkage.

Transposable elements DNA segments that can translocate from one site to another, either in the same replicon or in a different replicon in the same cell. Extensive sequence homology between transposable elements and their target sites is not required. Transposable elements are normal components of elements such as **chromosomes**, **plasmids** and phage **genomes**, and occur in both prokaryotes and eukaryotes. Some transposable elements are highly specific with respect to their target sites, whereas others appear to insert randomly.

Transposition Process by which **transposable elements** translocate from one site to another. Different elements use different methods for transposition, which is normally a rare event, and insertion leads to duplication of a short sequence of the target **DNA**, resulting in the formation of direct repeats flanking the inserted element. Transposition can result in gene **mutations** and/or may have significant effects on **gene expression**. Occasionally, transposable elements can excise from their insertion sites.

Transposons **Transposable elements** that can move from one site to another within **chromosomes**. They contain inverted repeats at either end and, in addition to encoding functions necessary for **transposition** (including the enzyme (transposase) that catalyses their insertion), also carry **genes** with unrelated functions, e.g. **antibiotics resistance**, production of **toxins** or **lactose** metabolism.

Trappist cheese Cheeses made by Trappist monks worldwide. Include **Port Salut cheese**.

Travnik cheese **Cheese** originating from Travnik, in Bosnia.

Treacle Low purity, thick, brown syrup produced as a by-product of **sugar refining**. Called **molasses** in the USA and Canada.

Tree tomatoes Alternative term for **tamarillos**.

Treflan Alternative term for the herbicide **trifluralin**.

α,α-Trehalases EC 3.2.1.28. Hydrolyse the disaccharide **trehalose** into 2 units of its monomer, D-glucose. Can be used for analytical determination of trehalose concentrations.

Trehalose Disaccharide composed of two molecules of **glucose** linked via an α-1,1-glucosidic bond. Isolated from **fungi**, including **yeasts**.

Trematodes **Liver flukes** which belong to the class Trematoda, e.g. *Fasciola* hepatica.

Tremorgens Neurotoxic **mycotoxins** (e.g. penitrem and alfatrem) produced by various fungi (e.g. *Penicillium*, *Aspergillus* and *Claviceps* species). Ingestion of contaminated foods and feeds by humans and animals can lead to weakness, tremors, convulsions and death.

Trenbolone acetate Synthetic anabolic steroid with similar hormonal activity to testosterone but with greater anabolic activity. Used legally for growth-promoting purposes in animals, mainly in young cattle. Following administration, rapidly hydrolyses to two major metabolites; residues of these metabolites may persist in tissues for considerable periods.

Triacylglycerol lipases EC 3.1.1.3. Hydrolyse **triacylglycerols** to **diacylglycerols** and free **fatty acids**. Usually referred to as **lipases**.

Triacylglycerols **Lipids** composed of **glycerol** esterifed at all three of its constituent carbon atoms with one or more **fatty acids**. Triglycerides are components of natural **fats** and **oils** and have multiple uses in the food industry, including as **emulsifiers**, **coatings** and encapsulating agents. Synonymous with **triglycerides**.

Triadimefon Systemic fungicide used for control of a variety of fungal diseases in many different **fruits**, **vegetables** and **cereals**. Classified by WHO as slightly toxic (WHO III).

Triazophos Non-systemic broad-spectrum insecticide and acaricide used for control of a wide range of **insects** and **mites** in **fruits**, **vegetables** and **cereals**. Also used for control of some free-living **nematodes** (particularly in **strawberries**) and as a bulb dip for **garlic**. Classified by WHO as highly hazardous (WHO Ib).

Tribolium Genus of small **beetles** of the family Tenebrionidae. *Tribolium castaneum* (red flour beetle) and *T. confusum* (confused flour beetle) are pests of **flour**, as well as stored **cereals** (e.g. **rice** and **wheat**).

Tributyltin Component of anti-fouling paints which are used on the hulls of ships. Can be released into the water and accumulate as **contaminants** in **sea foods**.

Tricaprylin Triglyceride of **glycerol** esterified with three molecules of **caprylic acid** (octanoic acid). Used in **transesterification** reactions to synthesize **structured lipids** incorporating desirable **fatty acids** such as **eicosapentaenoic acid** or **conjugated linoleic acid**. Also called glyceryl tricaprylate and caprylic acid triglyceride.

Trichinae Parasitic **nematodes** of the genus *Trichinella*.

Trichinella Genus of parasitic **nematodes** of the class Enoplea. *Trichinella spiralis* is the causative agent of **trichinosis**.

Trichinosis Infection caused by *Trichinella* spiralis. Transmission is via ingestion of larvae in undercooked

meat (especially **pork**). Larvae, which hatch from eggs laid by female worms in the small intestine, bore through the intestinal wall and migrate around the body causing disease. Characterized by diarrhoea, nausea, delirium, fever, abdominal pain, muscle pain and swelling of the eyes. The lungs, nervous system and heart may be affected in more advanced cases. Sometimes fatal.

Trichlorfon Non-systemic organophosphorus insecticide used for control of a wide range of insect **pests** in crops, stored **fruits**, **vegetables** and **cereals**. Also used in animal husbandry. Rapidly hydrolyses in plants and degrades rapidly in soil. Classified by WHO as moderately toxic (WHO II). Also known as chlorophos.

Trichloroanisole Chlorinated hydrocarbon with a very low sensory threshold which is most often associated with cork **taints** in **wines**.

Trichloroethylene Industrial solvent, prolonged exposure to which can cause cardiotoxicity and neurological impairment. Industrial pollution can cause **contamination** of **drinking water** sources with this compound. Irrigation of garden vegetables with contaminated water can result in uptake of trichloroethylene into plant tissues.

Trichloromethane Volatile compound often found in foods. Can occur as a contaminant of treated **drinking water**.

Trichoderma Genus of **fungi** that occurs in soil and on wood. May cause **spoilage** of **citrus fruits**, **cereals** and **peanuts**. *Trichoderma hazianum* is responsible for the spoilage of citrus fruits and cereals (e.g. **corn**, **rice** and **wheat**). *T. viride* causes rots of citrus fruits and spoilage of stored grains (e.g. wheat, rice and **barley**) and peanuts. Some species (e.g. *T. virens*) parasitize disease-causing fungi, making them useful **biocontrol** agents.

Tricholoma Genus of **edible fungi** that contains a number of species varying in **flavour** and quality. *Tricholoma caligatum* is commonly known as matsutake.

Trichosporon Genus of yeast **fungi** of the class Heterobasidiomycetes. Occur in water, soil and faeces, and on plants, wood pulp and human skin. Some species (e.g. *Trichosporon pullulans* and *T. variabile*) may cause **spoilage** of fresh **fish** and **shellfish**, **poultry meat**, **beef mince**, cooked **sausages**, **cheese** and **bread**. *T. pullulans* is used in production of **idli**. Some species are used in manufacture of surface-ripened cheeses such as **Limburg cheese** and **Gruyere cheese**.

Trichothecenes Group of **mycotoxins** produced by various fungi, such as *Fusarium*, *Myrothecium* and

Trichothecium. Include **deoxynivalenol**, **T2 toxin**, **diacetoxyscirpenol**, **trichothecin**, **nivalenol** and **fusarenon X**. Mainly infect cereal grains (e.g. **wheat**, **barley** and **corn**). Ingestion of contaminated foods and feeds can lead to haemorrhagic **gastroenteritis**, lung and brain haemorrhages, and bone marrow damage, accompanied by vomiting, headache, fever and nausea.

Trichothecin Trichothecene produced by *Trichothecium roseum*.

Trichothecium Genus of **fungi** of the class Hyphomycetes. *Trichothecium roseum* causes pink rot of **fruits** and **vegetables** (e.g. **gherkins**, **tomatoes**, **melons**, **apples** and **grapes**), and may also cause **spoilage** of **bread**. *T. roseum* is also responsible for producing **trichothecenes** (including **trichothecin**) on foods.

Trifluralin Selective herbicide used for pre-emergence control of many annual grasses and broad-leaved weeds around plants producing **vegetables**, **fruits**, **oilseeds**, **sugar beets** and **sugar cane**. Also used in combination with **linuron** for control of weeds in winter cereals. Classified by Environmental Protection Agency as not acutely toxic (Environmental Protection Agency IV).

Triglycerides **Lipids** composed of **glycerol** esterified at all three of its constituent carbon atoms with one or more **fatty acids**. Triglycerides are components of natural **fats** and **oils** and have multiple uses in the food industry, including as **emulsifiers**, **coatings** and encapsulating agents. Synonymous with **triacylglycerols**.

Trigonelline Alkaloid found in green **coffee beans** that has been implicated in mutagenic activity of **roasted coffee**.

Trihalomethanes Volatile compounds that may be formed during **chlorination** of **drinking water** and which are thought to be carcinogenic.

Triiodobenzoic acid Plant growth regulator that can increase the oil content of oilseeds following spraying of the plants but that can also increase the **spoilage** rate of **fruits**.

Triiodothyronine One of the iodine-containing hormones produced by the thyroid gland; also produced by conversion of **thyroxine**.

Trilinolein Polyunsaturated triglyceride formed from **linoleic acid**. Found in **frying oils** such as **sunflower oils** and **linseed oils**. Thermal decomposition of trilinolein during **deep frying** can lead to formation of **volatile compounds** giving rise to undesirable odours.

Trimethoprim Sulfonamide drug used for treatment of respiratory and intestinal infections in cattle, swine, sheep, goats, poultry and farmed fish. Often used in combination with other **sulfonamides**. Rapidly and widely distributes around tissues following administration. Normally depletes rapidly in farm animals; rate of depletion in farmed fish is greatly dependent on water temperature.

Trimethylamine Volatile compound found in **sea foods** that has a characteristic herring-like **aroma**. Associated with the onset of microbial **spoilage** in ice-stored **fish**. Hence, analysis of trimethylamine content is used to evaluate fish quality and **freshness**.

Trimming Making an item neat by cutting away irregular or unwanted parts. In the food industry, usually applied to removal of **fats** from **meat**.

Triolein Triglyceride of **glycerol** esterified with three molecules of **oleic acid** (9-octadecenoic acid). A natural component of **fats** and **oils**, triolein is used in the food industry in **stabilizers** and in solvents for **flavourings** and fat-soluble vitamins. Also called glyceryl trioleate.

Tripalmitin Triglyceride of **glycerol** esterified with three molecules of **palmitic acid** (hexadecanoic acid). A natural component of **fats** and **oils**, tripalmitin is used in the food industry in **additives** for the manufacture of compressed **sweets**. Also known by other names, including glyceryl tripalmitate.

Tripe A part of edible **offal**, generally comprising the lining of the four-chambered stomach of ruminants, particularly of calves and oxen. Although, tripe is usually produced from **cattle**, **sheep** tripe is used to make **haggis**, and **lamb** tripe, thinner than that of oxen or calves, may be used as a wrapping for savoury **stuffings**. Different parts of the cattle stomachs are used to make different kinds of tripe: the rumen is used to produce blanket tripe, which has a rough texture and varies in thickness; the reticulum is used to produce characteristically patterned honeycomb tripe; book tripe, also known as bible tripe, comes from the omasum; and reed tripe, also known as black tripe, is prepared from the abomasum. Usually, tripe is cleaned, trimmed of fat, parboiled and bleached before sale as dressed tripe. Tripe may also be used to make **sausage casings**. It has a high connective tissue content; on boiling, much of this is converted into **gelatin**. Cooked tripe has a mild **flavour** and slippery **texture**.

Tripolyphosphates Phosphates used to enhance the **tenderness**, **juiciness** and **flavour** of **meat**, and to inhibit oxidation of **lipids**. Include **sodium tripolyphosphate**.

Trisodium phosphate Phosphate that can be used in the food industry to sanitize **meat**, particularly chicken carcasses, and to prevent **discoloration** of ground **garlic**.

Tristearin Triglyceride of **glycerol** esterified with three molecules of **stearic acid** (octadecanoic acid). A natural component of **fats** and **oils**, tristearin is used in the food industry in **food additives** such as surface finishing agents, **lubricants**, **emulsifiers**, encapsulating agents and **crystallization** accelerator agents. Also known as glyceryl tristearate.

Triticale High-yielding hybrid of **wheat** (*Triticum* spp.) and **rye** (*Secale* spp.) which combines the resilience of rye with the particular elastic **baking properties** of wheat. Often used in multigrain **bread**.

Tritium Long-lived, radioactive isotope of **hydrogen**. Suitable for use in autoradiography and easy to incorporate into complex molecules for use in experimental studies.

Tritordeum Hybrid of **barley** and **wheat**.

Tropical fruits **Fruits** grown in countries of the tropics (on either side of the equator), or in hot and humid conditions. Include a great many species, such as **mangoes**, **pineapples**, **pomegranates**, **bananas**, **papayas**, **lychees**, **guavas** and **tamarinds**.

Tropomyosin **Myofibrillar proteins** with high contents of acidic and basic **amino acids**. Tropomyosin represents approximately 8-10% of myofibrillar protein. Molecules of tropomyosin consist of two coiled peptide chains, attached to each other end to end. The molecules form long, thin, filamentous strands.

Troponin Complex of three **proteins** found in striated muscle, where it is associated with **tropomyosin** and **actins** on the thin filaments, conferring calcium sensitivity.

Trout Any of several anadromous fish of the family Salmonidae, native to rivers and streams of Europe, Asia and North America; usually restricted to freshwater, though some types migrate to the sea between spawnings. The most important species commercially is *Onchorhynchus mykiss* (**rainbow trout**), which is cultured around the world. Other important species include *Salmo trutta* (**brown trout**/**sea trout**) and *O. clarki* (cutthroat trout). Flesh is usually pale orange-pink, sometimes a deeper red-pink (young trout are often white-fleshed), with a firm yet creamy texture and moderate to high fat content. Marketed fresh, frozen and as a smoked or canned product.

Trub Precipitates, comprising coagulated **proteins**, **polyphenols** and **carbohydrates**, which form during boiling of **worts** in the **beer brewing** process. Also termed break; may be divided into hot break, formed during boiling, and cold break, formed during subsequent cooling.

Trucks Alternative name, used especially in Canada and the USA, for **lorries**. Large motor vehicles designed to transport heavy loads. Used in a wide range of applications, including transport of livestock to **slaughterhouses**, carriage of grain and other raw materials to processing facilities and transfer of **processed foods** from factories to retail premises. The term also describes vehicles used for carrying freight on a railway. Forklift trucks are vehicles with power operated horizontal prongs that can be raised and lowered and are used for transporting goods, especially those stacked on pallets, in **warehouses** and factories.

Truffles Alternative term for **edible fungi** of the genus *Tuber*.

Trumpet shells Any of a number of marine gastropod **molluscs** within the family Cymatiidae; occur in intertidal regions and deeper waters in tropical and southern temperate areas. Flesh of some species is consumed; occasionally used to make **preserves**.

Trussing Process of tying up the wings and legs of poultry **carcasses** in preparation for **cooking**. Skewers, thread, string or pins may be used. Helps the food to maintain a compact shape during cooking.

Trypsin EC 3.4.21.4. Highly specific serine **proteinases** that hydrolyse peptide bonds in which **arginine** or **lysine** provides the carbonyl group.

Trypsin inhibitors **Proteins** found in a range of foods, including **soybeans**, **peanuts**, **peas**, **lentils**, and raw **egg whites**, which inhibit the activity of **trypsin**. Denatured, and hence inactivated, by heating.

Tryptamine Biogenic amine formed by microbial decarboxylation of **tryptophan**. May be formed in foods such as ripened **cheese**, **chocolate**, **wines** and **fermented foods**. Consumption of contaminated foods can cause increased blood pressure and **migraine**.

Tryptophan Essential amino acid important in the synthesis of **haemoglobin**, plasma **proteins** and **nicotinic acid**.

Tryptophol Phenolic compound found in **beer** and **wines**, the levels of which can be used to distinguish beer types.

TS Abbreviation for **total solids**.

TSS Abbreviation for **total soluble solids**.

Tsukemono Japanese **vegetable pickles**. Popular types include pickled **turnips**, **carrots**, Chinese **cabbages**, **aubergines**, **burdock** and giant **radishes**. Ingredients can also include **miso** and **sake**.

T2 toxin Acutely toxic **trichothecenes** produced by *Fusarium* species (e.g. *Fusarium tricinctum* and *F. sporotrichiodes*).

Tuba **Alcoholic beverages** made by **fermentation** of the sap of coconut **palms**.

Tuber Genus of **edible fungi** including the British truffle, *Tuber aestivum*, and French Perigord **truffles**, *T. melanosporum*. Grow underground in woods, and are irregularly shaped. The solid flesh is light brown with white veins. Perigord truffles are used to make pate de foie gras.

Tuberculosis Infectious disease most commonly caused by the bacillus **Mycobacterium** *tuberculosis* which is characterized by the formation of nodular lesions (tubercules) in the tissues. Tuberculosis is associated with poor living conditions, such as nutritional deficiency and inadequate housing. Transmission of tuberculosis is by inhalation of infected droplets. Treatment is by long-term administration of **antibiotics**.

Tubers Swollen and fleshy underground stems of plants, usually high in starch. Include potatoes.

Tulum cheese Turkish **cheese** made from **goat milk** or **cow milk**. Crumbly **texture**. Used in dishes or as an appetizer.

Tumbling As well as being a process by which surface irregularities are removed from an item by rotating it in a tumbling barrel, this term also refers to a process by which the quality of **meat** can be improved. The mechanical action of tumbling alters the structure of muscle proteins. Tumbling can also be used to increase the rate of uptake of **marinades** by meat pieces.

Tumours Growths in the body caused by the abnormal proliferation of cells. Some food components are thought to possess **antitumour activity**. Tumours may be benign (i.e. grow at one site only) or malignant (i.e. they destroy the tissue in which they arise and spread to other parts of the body). Benign tumours, which are covered by a capsule, are usually harmless but may become very large, exerting pressure on neighbouring tissues and producing severe effects. In malignant tumours, which are not enclosed by a capsule, cell division is rapid; cells show partial or complete loss of function and bear little resemblance to the tissue cells from which they originated. Malignant tumours cause extensive damage.

Tuna Any of several species of large pelagic **marine fish** in the family Scombridae; worldwide distribution. Most species have high commercial importance, particularly *Thunnus alalunga* (albacore), *T. obesus* (**bigeye tuna**), *T. albacares* (**yellowfin tuna**) and *Katsuwonus pelamis* (**skipjack tuna**). Marketed in a variety of forms, including fresh and frozen (whole, gutted or fillets), canned, salted, dried and semi-preserved. Also used in a variety of prepared dishes, such as tuna sausages, tuna roll and tuna pastes. Also known as tunny.

Tuna oils **Fish oils** which are one of the richest sources of **docosahexaenoic acid**.

Tunny Alternative term for **tuna**.

Turban shells Any of a number of marine gastropod **molluscs** within the family Turbinidae; distributed in intertidal zones and deeper waters across the Indo-Pacific. Flesh of several species is consumed; typically served grilled with **soy sauces**.

Turbidimetry Measurement of turbidity of a solution, usually using a turbidimeter, an instrument that records the loss of intensity of a light beam passed through a solution containing suspended particles.

Turbidity **Optical properties** relating to the extent to which a solution is turbid, i.e. cloudy or hazy. Turbidity in solutions is caused by the presence of finely suspended matter.

Turbot Name given to a number of marine **flatfish** species within the family Pleuronectidae; most occur in the northern Atlantic. Commercially important species include *Scopthalmus maximus* (European turbot) and *Reinhardtius hippoglossoides* (Greenland turbot). Flesh of most species is highly esteemed and tends to be white, firm with low fat content and delicate **flavour**. Marketed fresh and frozen.

Turkey frankfurters **Frankfurters** prepared from **turkey meat**. They are often prepared from turkey thigh meat and/or turkey meat trimmings or mechanically recovered turkey meat. Other ingredients may include turkey fat, pork fat or beef fat.

Turkey ham Cured turkey products prepared from boneless thigh meat after removal of the skin and surface fat. They may contain other ingredients, such as salt, dextrose, sodium nitrate and sodium. Turkey mince, prepared from trimmings removed from the turkey thigh during **boning** and **trimming**, may be added as a binder.

Turkey livers Relatively large abdominal organs consisting of several lobes of a continuous parenchymal mass covered by a capsule which form part of the edible **offal** in turkey **carcasses**. Used to make **stocks** and **gravy** or eaten in a variety of other ways, including fried or as ingredients in **stuffings** and **pates**. May contain high levels of vitamin A, particularly if poultry are given retinol-supplemented feeds. Also rich sources of iron and B vitamins.

Turkey meat Meat from **turkeys**. Many turkeys are sold whole, sometimes they are injected with **butter** or **vegetable oils** and are marketed as self-basting. Turkey breast meat contains less **myoglobin** than turkey drumstick or thigh meat. As a result of genetic selection based on the economic traits of turkey **carcasses**, the turkey industry suffers from the occurrence of several metabolic and musculoskeletal disorders. Poor **water holding capacity** in turkey breast meat is thought to be caused by similar factors to those underlying the **PSE defect** in **pork**.

Turkey mince **Meat mince** prepared from **turkey meat**. It may be prepared specifically from light or dark turkey meat. Mince prepared from light coloured turkey meat has a lower content of saturated fats than mince prepared from dark turkey meat. Also known as ground turkey.

Turkey patties **Meat patties** prepared from **turkey mince**.

Turkeys Large **birds** (*Meleagris gallopavo*) which belong to the pheasant family. Turkeys are reared throughout the world for **turkey meat** production. Different gender and age groups of turkeys are known as toms, stags or cocks (adult entire males; >26 weeks of age), hens (adult females; >26 weeks of age), turkey growers (sexually immature young birds; 8-26 week of age) and poults (sexually immature birds which have down rather than feathers).

Turkey sausages **Sausages**, both fresh and cured, made from **turkey meat**. Varieties include **turkey frankfurters**, **bratwurst**, **hot dogs**, kielbasa, **salami** and **wieners**. The majority are prepared from coarsely comminuted dark turkey meat or **mechanically recovered meat**. Products may contain binders and extenders, such as calcium lactate, **carrageenans**, **cereals**, **soy meal**, **soy proteins**, vegetable starch and **whey**.

Turkish delight Soft **jelly confectionery** originally of Turkish origin, made by cooking flavoured **syrups** and **corn starch** together slowly, leaving the mixture to set, cutting into cubes and rolling in **icing sugar**. Flavours are usually based on **orange juices** or **lemon juices**, with rose water or orange flower water. Alternatively, a mint **flavour** is produced by adding **peppermint essential oils** or creme de menthe **liqueurs**. Colour varies according to the ingredients used, but is usually white, pink or green. Also known as lokum, lukum or rahat.

Turmeric Common name for a plant native to Asia, *Curcuma longa*, the dried ground rhizomes of which are used as **spices**. Turmeric is deep yellow in colour due to the presence of **curcumin**, desmethoxycurcumin and bisdesmethoxycurcumin. Used in **natural colorants**, particularly in **mustard**, **pickles** and other spicy **condiments**, curry **seasonings**, and **fats** and **oils**. The predominant flavour compound of turmeric is turmerone. The majority of commercially available turmeric is cultivated in India, leading to the

alternative name, Indian saffron. Also known as CI natural yellow 3 and CI 75300. Extracts and **essential oils** of *C. longa* rhizomes are also used as **colorants** and **flavourings**.

Turnip rooted celery Alternative term for **celeriac**.

Turnips Common name for the root form of *Brassica campestris* or *B. rapa*. Roots are used in soups and stews or as a separate vegetable dish, while the leaves, or spring greens, are eaten as a vegetable. The root contains moderate amounts of **sugar** and **vitamin C**; leaves contain large amounts of **vitamin C** and also reasonable amounts of **carotenes**.

Turron **Nougat** originating from Spain which is made with **almonds**, **sugar**, **honeys** and **egg whites**.

Turtles Several species of freshwater or marine, shelled reptiles belonging to the order Chelonia, that are hunted for their **meat** and **shells**. Turtle eggs may also be eaten. Turtle meat has good **flavour**, but because of its chewiness, it tends to be used to prepare **soups**. Most turtle meat is produced from sea turtles; however, meat from freshwater terrapins is often considered to have the best sensory properties amongst turtle meats.

Tutane Alternative term for **butylamine**.

Tvaroh Czech **soft cheese** similar to **quarg**.

Tvorog Russian **soft cheese** similar to **quarg**. Served as a dessert with various degrees of sweetness, sometimes with **sour cream** or **jams**.

Twarog Polish **soft cheese** similar to **quarg**.

Tykmaelk Danish **fermented milk**.

Tylosin Macrolide antibiotic produced by *Streptomyces fradiae*. Used primarily to treat the chronic respiratory disease complex in chickens and infectious sinusitis in turkeys; also effective against cattle respiratory diseases and swine dysentery and is sometimes used as a growth promoter in swine. Excreted relatively slowly from tissues; withdrawal periods range from 5 days (turkeys) to 21 days (swine). Can pass into **milk** and **eggs**. Not permitted for use in laying hens, and **milk** from treated cows may not be used until 3 days after final treatment.

Typhoid Infectious disease of the digestive tract caused by *Salmonella* Typhii. Transmission is by drinking infected water, usually where there is no clean water supply. Symptoms, which begin 10-14 days after ingestion of the bacterium, include fever, headache, cough, loss of appetite, and constipation; a characteristic red rash may appear. If left untreated, increasing production of **toxins** causes delirium, coma and death. Treatment is by administration of fluids and the antibiotic chloramphenicol.

Tyramine Biogenic amine formed by microbial decarboxylation of **tyrosine**. May be formed in foods such as ripened **cheese**, **chocolate**, **wines** and **fermented foods**. Consumption of contaminated foods can cause increased blood pressure and **migraine**.

Tyrophagus   Genus of **mites** of the class Arachnida. *Tyrophagus putrescentiae* and *T. longior* are common pests of stored foods (e.g. **corn**, **wheat**, **barley**, **bran** and **wheat flour**).

Tyrosinases   Catalyse the oxidation of L-tyrosine. Exhibit activity of both **catechol oxidases** and **monophenol monooxygenases**.

Tyrosine   Non-essential amino acid which can be synthesized from **phenylalanine** in humans. Important precursor of adrenaline, noradrenaline, thyroxine and **melanins**. Tyrosine isomers can also be formed by γ-irradiation of phenylalanine and their detection can therefore be used as an indicator of **irradiation** of foods.

Tyrosol   Main phenolic compound found in **olive oils**, where it is thought to protect against oxidative stress.

Tzatziki   A Greek speciality **yoghurt** dip containing **cucumbers**, **mint** and **garlic**.

U

UASB bioreactors Abbreviation for **upflow anaerobic sludge blanket bioreactors**.

Udon Thick Japanese **noodles** prepared from **wheat flour**, often used in **soups** or **broths**.

Ugba Protein-rich product produced by **solid state fermentation** of **African oil beans**. Used as **snack foods** or **condiments**.

UHT cream **Cream** heated by **UHT treatment** to prolong its **shelf life**. Also known as long life cream.

UHT milk **Milk** heated by **UHT treatment** to prolong shelf life. Also known as long life milk.

UHT treatment Abbreviation for ultra-high temperature treatment, a brief, intense heat treatment (direct or indirect) used to sterilize foods prior to packaging. Kills all **microorganisms** that would otherwise spoil the product. Following UHT treatment, foods are filled into pre-sterilized containers in a sterile atmosphere. Food products processed by UHT treatment include liquid products (e.g. **milk**, some **fruit juices**, **cream**, **yoghurt**, **wines**, **salad dressings**), foods with discrete particles (e.g. **infant foods**, tomato products, some **fruit juices** and **vegetable juices**, **soups**), and foods containing larger particles (e.g. stews).

Uji Thin, fermented **porridge** made from **corn flour**, **sorghum** flour or **cassava** meal, either singly or in mixtures. Often used in Ghana and Kenya in **infant foods**. Also known as **koko**.

Ulluco Common name for *Ullucus tuberosus*, an important tuber crop of the Andean region. **Tubers** are produced in a wide range of shapes and bright colours. Their flesh is white to yellow in **colour** with a smooth **texture** and nutty **flavour**. Leaves, which are similar in texture to **spinach**, are also eaten as a vegetable, representing a good source of protein, **calcium** and **carotenes**.

Ultracentrifugation **Centrifugation** in **centrifuges** which have the ability to develop centrifugal fields of up to 100,000 times that of the gravitational field. Ultracentrifugation is generally used for analytical purposes, such as the determination of physicochemical properties of food **polysaccharides** using sedimentation analysis.

Ultrafiltration Selective membrane separation process, driven by a pressure gradient, in which suspended solids, colloids, emulsified solids such as fat-protein complexes, and dissolved macromolecules with molecular weight in the range 10,000-100,000 Da are retained by the **membranes**. Molecules that do not pass through the membranes constitute the retentate. Lower molecular weight dissolved materials that pass through the membrane under a driving force of relatively low hydrostatic pressure (1-10 bar) are the permeate. Ultrafiltration is generally used in the **concentration** and **fractionation** of large molecules from materials such as **cheese whey** and **milk**.

Ultrapasteurization Process of **heating** foods, especially **milk** and **liquid egg** products, at a high temperature for a short time, sufficient to kill any **pathogens** present. Used to extend the **shelf life** of the product without greatly affecting its nutritional properties. A typical process for ultrapasteurization of milk would involve heating at 280°F for at least 2 seconds. Ultrapasteurized products are aseptically packaged and stored under refrigeration.

Ultrasonics The science and application of ultrasonic waves that have a frequency above those that are audible, generally defined as above 20,000 hertz.

Ultrasound Sound or other vibrations having an ultrasonic frequency. Generally, ultrasound is classified as any acoustic wave above the normal range of human hearing, i.e. above 20,000 hertz, but, in practice, the term usually refers to a much higher frequency used for a specific application.

Ultraviolet Relating to electromagnetic radiation having a wavelength just shorter than that of violet light but longer than that of **X-rays**. Abbreviated to UV.

Ultraviolet radiation Electromagnetic radiation having a wavelength just shorter than that of violet light but longer than that of **X-rays**. Abbreviated to UV radiation.

Ultraviolet spectrophotometry Alternative term for **UV spectroscopy**.

Ulva lactuca Species of green **seaweeds** distributed on rocky shores worldwide. Consumed raw, cooked, dried, in **soups** or as a deep fried product. Rich source of **vitamins** and **minerals**, particularly **vitamin B$_1$**, **vitamin C**, iron and iodine. Also known as **sea lettuces**.

Umami **Sensory properties** relating to the perception of savoury **flavour**, particularly that of **monosodium glutamate**, **proteins**, certain **amino acids**, and the **ribonucleotides** inosinate and guanylate. Derived from the Japanese word for savoury taste.

Ume Alternative term for **Japanese apricots**.

UMP Abbreviation for the nucleotide **uridine monophosphate**, also known as uridylic acid.

Undaria Genus of brown **seaweeds** occurring on natural and man-made substrates along coasts of many parts of the world. The most important species in commercial terms is *Undaria pinnatifida*, which is cultured on a large scale in parts of Asia, particularly Japan. Used in **soups**; also consumed as a toasted, sugar-coated and canned product. Also known as **wakame** and wakami.

Undecanone Aroma compound found in foods such as **milk**, **cheese** and **spices**, which can also be produced by microbial **biotransformations**.

Unsaponifiable matter Substances present in **fats** and **oils** which are not **glycerides** and which are resistant to **saponification** with strong **alkalies**. Content varies among different types of oils and fats, and can thus be used as a source of information for their characterization and authentication.

Unsaturated fats **Fats**, found at high levels in **vegetable oils**, that contain one or more carbon-carbon double or triple bonds. Thought to lower plasma cholesterol levels and reduce the risk of **cardiovascular diseases** when used to replace saturated fats in the diet.

Unsaturated fatty acids **Fatty acids** containing one or more carbon-carbon double bond. Those that contain one double bond are termed **monounsaturated fatty acids** and include **oleic acid**, while those that contain two or more double bonds are termed **polyunsaturated fatty acids** and include **linoleic acid**. Found at high levels in **vegetable oils** and **fish oils**, and thought to lower plasma cholesterol levels and reduce the risk of **coronary heart diseases**.

Unsaturation State in which an organic compound contains double or triple bonds and thus shows increased capacity for reaction relative to saturated compounds. Used especially with respect to **fats** and **oils**. The degree of unsaturation refers to the number of double and triple bonds within the compound. This is expressed in terms of **iodine values**, determined by the weight of iodine absorbed by the substance under investigation. With respect to fats and oils, degree of unsaturation is important for their characteristics and health considerations, unsaturated forms having benefits with respect to blood **cholesterol** levels and risk of **cardiovascular diseases** development.

Upflow anaerobic sludge blanket bioreactors **Bioreactors** in which anaerobic digestion is performed by **microorganisms** that form thick flocculations maintained in a suspended state near the bottom of the reactor. Used for **bioremediation** of **wastes** and **waste water** from the food industry.

Uracil Pyrimidine base that replaces **thymine** in **RNA**, where it pairs with **adenine**. Also a constituent of **uridine**.

Uranium Radioactive metallic element with the chemical symbol U.

Urd beans Alternative term for **black gram**.

Urea Synonym for **carbamide**. The excretory product of nitrogen metabolism produced in the livers of mammals following the breakdown of **amino acids**. Formation during the **fermentation** of **wines** is a cause for concern, since it is a precursor of **ethyl carbamate**, a carcinogen. As well as being used as a fertilizer, it is also utilized as a feed supplement for ruminants leading to its presence in **milk**.

Ureases EC 3.5.1.5. Convert **urea** to CO_2 and NH_3. Used in the food industry for removal of urea from foods and **beverages**, and for preventing formation of the carcinogen **ethyl carbamate**. Also used to measure urea concentrations and have been used to control **pH** during **lactic fermentation**, thus enhancing **lactic acid** production. These **enzymes** are important **virulence factors** in certain bacterial **pathogens**.

Urethane Synonym for **ethyl carbamate**. Organic nitrogen compound derived from **urea**, which in pure form is a white or colourless, crystalline solid. Soluble in water, alcohol and ether, and slightly soluble in oils. A possible carcinogen that is used in **pesticides** and **fungicides**. Formed in **wines**, other **alcoholic beverages** and **fermented foods** during processing or storage.

Uric acid End product of purine metabolism in certain mammals, and the main nitrogenous excretory product in birds, reptiles and some invertebrates. Responsible for gout in humans. It is thought that consumption of **caffeine**-rich beverages such as tea and coffee may reduce serum levels of uric acid. May be useful as an indicator of **insects** infestation of **cereals** and extruded products.

Uridine Nucleoside in which **uracil** is bound covalently to **ribose**.

Uridine monophosphate Nucleotide usually abbreviated to UMP and also known as uridylic acid.

Uronic acids Carboxylic acids, e.g. **glucuronic acid** and **galacturonic acid**, formed by oxidation of

hexoses. Found in certain **polysaccharides**, such as **pectins** and **alginates**.

Urticaria Itchy skin rash of raised spots (weals) on a reddened background, resulting from release of **histamine** by mast cells. Acute urticaria represents an immediate response to such **allergens** as **sea foods** or **strawberries**. Also known as nettle rash or hives.

UV Abbreviation for **ultraviolet**.

UV radiation Abbreviation for **ultraviolet radiation**.

UV spectroscopy Spectroscopy in which samples are identified on the basis of absorption of light of **ultraviolet** wavelength.

V

Vaccenic acid One of the *trans*-18:1 **fatty acids** present at significant levels in **milk fats** as well as in other foods.

Vacuum A space entirely devoid of matter or from which the air has been completely removed. In practical terms, a vacuum is an enclosed region of space in which the pressure has been reduced (below normal atmospheric pressure) sufficiently so that processes occurring within the region are unaffected by the residual matter.

Vacuum cooling Technique based on liquid **evaporation** which produces a rapid **cooling** effect in products containing free water. Suitable only where removal of the free water will not cause structural damage and where there is no barrier, e.g. a thick wax cuticle, to water loss. Subjecting suitable products to **vacuum** pressure allows part of the water contained in them to boil out at relatively low temperatures. Used successfully in reducing postharvest deterioration in **fruits** and **vegetables**, thus prolonging **shelf life**, during processing of some products, including liquid foods and **bakery products**, and rapid cooling of cooked **meat**, fish products and **ready meals**.

Vacuum drying Removal of liquid from a solid material while in a vacuum system, to lower the temperature at which **evaporation** takes place and thus prevent heat damage to the material.

Vacuum evaporation **Concentration** technique in which the use of high temperatures is avoided by subjecting the substance to a **vacuum**, causing it to boil at a lower temperature. The process is performed in a chamber surrounded by a water jacket through which water is circulated to control temperature. Particularly useful for products where heat-induced protein **denaturation** should be avoided, e.g. **liquid egg whites** and **skim milk**.

Vacuum packaging Packaging process in which some or all of the air is removed from flexible or rigid containers before sealing. This form of packaging is used to preserve **flavour**, inhibit bacterial growth and prolong the **shelf life** of food.

Vacuum pans Sealed devices that control the **crystallization** of solids from liquids by lowering the pressure within the sealed container. Vacuum pans are widely used for crystallization during the manufacture of **sugar**.

Valeraldehyde Synonym for **pentanal**. Organic compound present in many foods that has an unpleasant odour and a low odour threshold value. One of the main compounds that can cause **off odour** in **sake**.

Valeric acid Synonym for **pentanoic acid**. Volatile fatty acid comprising 5 carbon atoms and a single carboxylic acid group. Contributes to the **aroma** of mature **cheese**. Uses include as a reactant in production of **aroma compounds** and **flavourings**. Also one of the main malodorous pollutants from livestock houses.

Valine Essential amino acid important for growth. Good sources include **soy meal**, **brown rice**, **Cottage cheese**, **fish**, **meat**, **nuts** and **legumes**.

Valtellina Casera cheese Italian semi **hard cheese** made on an artisanal or semi industrial scale from semi skimmed **cow milk**. Granted controlled Denomination of Origin status. Rind has a characteristic straw-yellow **colour** which intensifies with **ripening**. **Flavour** is sweet with a note of **dried fruits**. Eaten on its own or as an ingredient of a range of local cooked dishes and **salads**.

Valves Mechanical devices, either manual or automatic, for controlling the passage of fluids through pipes or ducts.

Vanadium Element with the chemical symbol V that is intermediate between the metals and non-metals.

Vanaspati Grainy hydrogenated **vegetable oils** used as an alternative to **ghee** in India and Pakistan. Similar to **margarines** and often fortified with **vitamin A** and **vitamin D**.

Vanilla **Natural flavourings** produced by curing of fully grown but unripe beans (pods) of *Vanilla planifolia* or *V. tahitensis*. Curing causes hydrolysis of glucovanilla to produce glucose and the flavour compound, **vanillin**. Glucose is then involved in **nonenzymic browning** via the **Maillard reaction** with bean proteins. Major vanilla producing countries are Mexico, Madagascar and Tahiti, each country producing vanilla with a distinctive flavour profile. Although vanillin is the main flavour component of vanilla it comprises only about 3% of the total **flavour com-**

pounds and **aroma compounds**. Thus composition of minor flavour and aroma compounds is an important determinant of **flavour**.

Vanillic acid Phenolic compound produced as an intermediate in **bioconversions** of **ferulic acid** to **vanillin**. Also found as a pollutant in **olive oil mills effluents**.

Vanillin Substituted phenol that is the main flavour compound of **vanilla**. Synthetic vanillin is also manufactured for use in **flavourings**. Used as a cheaper alternative to vanilla in a wide range of foods, such as **ice cream**, **bakery products**, **sugar confectionery** and **beverages**.

Vapona Alternative term for the insecticide **dichlorvos**.

Vaporization Process by which moisture or another substance is diffused or suspended in the air, becoming converted into vapour. Examples include the rapid change of water into **steam**, especially in **boilers**.

var Abbreviation generally applied to **variety**.

Variety Taxonomic rank below subspecies, usually abbreviated to var. Varieties are usually the result of selective breeding and diverge from the parent in relatively minor ways. Varieties may be distinguished within a given subspecies by, for example, metabolic and/or physiological properties (biovar. or biotype), morphology (morphovar. or morphotype), **pathogenicity** for specific hosts (pathovar. (pv.) or pathotype), susceptibility to lysis by specific **bacteriophages** (phagovar. or phagotype) or serological characteristics (serovar. or **serotype**). However, these terms are often used loosely, in a non-taxonomic sense.

Varnishes Resins dissolved in liquids which are used to coat wood or metals. Form a transparent, shiny, hard surface when dry. Varnishes based on epoxy resins are often used for coating the interior of food **cans**.

Vats Large tubs or tanks used to hold or store liquids. Examples include fermentation vats used in **winemaking** and vats used during **cheesemaking**.

Veal **Meat** from young calves, usually **cattle** which are slaughtered at <20 weeks of age. Commonly, veal is produced under semi-intensive systems in which calves are fed on milk-based concentrated feeds to produce very light-coloured (white or pink) meats. Veal calves are prevented from feeding on fibrous feeds in order to prevent development of darker coloured, stronger flavoured meat. Typically, veal is very lean and tender, and has a delicate **flavour**. The highest quality veal tends to be produced from calves slaughtered at 12-16 weeks of age at body weight of 70-90 kg; these calves are often of French lineage, being from breeds such as the Belgian blue or Charolais. Veal is expensive to produce and, sometimes, calves are treated with growth promoters (e.g. anabolic steroids) to increase the weight of veal **carcasses**.

Vectors Autonomously replicating **DNA** molecules (e.g. **plasmids**, cosmids, **viruses** and yeast artificial **chromosomes**) into which foreign DNA fragments can be inserted. They can then be transformed into suitable host cells and propagated. In addition to origins of replication, vectors usually contain selectable markers that allow selection of recombinant cells. They may also contain sequences that direct expression of cloned genes in host cells.

Vegan diet Strict **vegetarian diet** which contains no **animal foods** of any kind.

Vegan foods **Vegetarian foods** suitable for a **vegan diet**, i.e. excluding meat, eggs, milk, butter, cheese and all other **animal foods**.

Vegetable burgers Patties made from mashed or chopped **vegetables**, sometimes also containing cereal or nut ingredients, eaten as an alternative to meat-based **burgers** such as **beefburgers**. Commonly used ingredients include **beans**, **mushrooms**, **onions** and **carrots**. **Spices** and **condiments** are added to produce the desired **flavour**. Health benefits compared with meat-based burgers include low fat and **sodium** contents, little or no **cholesterol** content and increased **dietary fibre** levels. Also known as veggie burgers.

Vegetable fats Lipid-rich vegetable products that are solid at room temperature. May be produced by **hydrogenation** of **vegetable oils**. Used in **cooking** and as food ingredients. Include **cocoa butter**, **sal fats**, **shea nut butter** and **vanaspati**.

Vegetable juice beverages **Beverages** prepared from **vegetable juices** with addition of other ingredients.

Vegetable juices Juices extracted from **vegetables**. Drunk as **beverages** in a similar way to **fruit juices**. Include **carrot juices** and **cabbage juices**.

Vegetable nectars **Vegetable juice beverages** made by addition of water and/or **sugar**, and optionally other ingredients, to **vegetable juices**.

Vegetable oils Lipid-rich vegetable products that are liquid at room temperature. Extracted from plant material including seeds, fruit or nuts. Often contain **phytosterols**. Used widely as **cooking oils** and **salad oils** and as **flavourings**. Include **cottonseed oils**, **olive oils**, **sunflower oils**, **soybean oils** and **essential oils**.

Vegetable pickles **Vegetables** preserved in liquids such as **brines** or **vinegar** and eaten as an accompaniment to a meal. Examples include **pickled onions** and **cucumber pickles**.

Vegetable preserves **Vegetables** that have been preserved by immersing in brines, vinegar or oils.

Vegetable proteins **Proteins** sourced from vegetable tissue. Preferred by some consumers due to health benefits. Quality of vegetable proteins, especially with respect to **amino acids** composition, varies according to source, but many plant breeding programmes have aimed to improve protein quality of individual crops. **Legumes**, particularly **soybeans**, are especially rich in protein. **Textured vegetable proteins**, usually derived from soybeans, are used as **meat substitutes** and **meat extenders**.

Vegetable pulps Preparations made from **vegetables** by mashing the cooked flesh. Used as ingredients in various dishes, such as **soups**, **sauces** and **casseroles**.

Vegetable purees **Vegetables** that have been mashed, usually after cooking, to a smooth, thick consistency by various means, such as forcing through sieves or blending in food processors. Used as garnishes, side dishes or ingredients in dishes such as **sauces** and **soups**, or **beverages**.

Vegetable rennets **Enzymes** sourced from plant materials that are used as substitutes for **animal rennets** in **coagulation** of milk for **cheesemaking**. Include enzymes extracted from flowers of **cardoons** or curdle thistle (*Cynara cardunculus*).

Vegetables Plants cultivated for an edible part, e.g. root, tuber, leaf or flower buds (as in **broccoli** and **cauliflowers**), or the edible parts of such plants.

Vegetable salads Dishes prepared from a mixture of **vegetables**, raw or cooked, sometimes served in **sauces** or **dressings**.

Vegetarian diet Diet based on **plant foods**, and which excludes **meat** and **fish**, and, in some cases, other **animal foods**. Lacto-ovo vegetarians consume **dairy products** and **eggs**, while those following a **vegan diet** consume no animal products at all. Vegetarianism is adopted for a variety of reasons, including ethical and religious beliefs as well as for nutritional/health benefits. The positive health effects reported for the diet have been attributed to relatively low contents of **fats** and **cholesterol** and the high contents of some **vitamins** and minerals. Inclusion of supplements in the diet may be necessary to prevent the risk of deficiency in **vitamin B$_{12}$** and some minerals, such as iron, zinc and iodine.

Vegetarian foods Meat-free foods suitable for inclusion in a **vegetarian diet**. Include **pasta**, soy products, vegetable burgers and simulated **meat substitutes**. Much of the recent growth in the vegetarian food market has been fuelled by non-vegetarians who are keen to cut down on meat consumption and who perceive vegetarian foods as a healthy option.

Veillonella Genus of Gram negative, anaerobic coccoid **bacteria**. Occur as **parasites** in the mouth, and gastrointestinal and respiratory tracts of humans and animals. Species may be included in competitive exclusion cultures, which are fed to animals (e.g. **poultry**) to prevent intestinal colonization by **pathogens** (e.g. *Salmonella* species).

Velvet beans Seeds produced by *Mucuna pruriens* rich in **proteins** and **fibre** but containing **antinutritional factors** that must be destroyed by **cooking** prior to consumption.

Vendace **Freshwater fish** species (*Coregonus albula*) from the family Salmonidae; distributed across northwest Europe. Normally marketed fresh; in Sweden, **roes** are used as **caviar substitutes**.

Vending machines Machines that dispense articles such as packaged foods or beverages, usually when a coin or token is inserted.

Venison **Meat** from **deer**. It is very lean and has a strong gamey **flavour** and **aroma**, which may be decreased by **marination** before cooking. The prime cuts are from the loin areas of deer **carcasses**. Pre-slaughter **stress**, particularly the holding of farmed or harvested wild deer in unfamiliar surroundings before slaughter, is associated with high ultimate pH values in deer carcasses and venison with a dark cutting appearance. In broader use, the term is used to describe meat from antelopes, caribou, elks, moose and reindeer. Also known as deer meat.

Veratryl alcohol Aryl alcohol (3,4-dimethoxybenzyl alcohol) synthesized by white rot **fungi** and involved in activation of their ligninolytic enzyme systems. Enzymes act on plant material and can be used for various functions, including removal of **phenols** from **fruit juices**, treatment of **olive oil mills effluents** and detoxification of lignocellulosic hydrolysates.

Verbascose Oligosaccharide composed of **fructose**, **galactose** and **glucose** residues.

Vermicelli **Pasta** formed into very long, thin strands.

Vermouths **Aperitifs** based on **wines** flavoured with **herbs** and **spices**, including **wormwood** flowers (*Artemisia absinthium*).

Vernonia Genus of plants producing a seed oil rich in vernolic acid and containing **triacylglycerols** with epoxidized fatty acid moieties.

Vero cytotoxins Alternative term for **verotoxins** and **Shiga like toxins**. So called because of their cytotoxic activity in African Green Monkey Kidney (Vero) cells.

Verotoxins **Cytotoxins** produced by enterohaemorrhagic *Escherichia coli* strains, which are similar to

Shiga toxins. Alternative term for **Vero cytotoxins** and **Shiga like toxins**.

Verrucosidin Potent neurotoxin produced by **Penicillium** species, such as *P. polonicum* and *P. aurantiogriseum*, particularly on meat products including **sausages** and **dry cured ham**.

Verruculogen Tremorgenic mycotoxin produced by species of **Penicillium**, **Neosartorya** *fischeri* and **Aspergillus** *fumigatus*, **fungi** responsible for **spoilage** of foods.

Versicolorin Precursors in the **aflatoxin B$_1$** biosynthesis pathway in **fungi**. Occur as versicolorin A and versicolorin B.

Verticillium Genus of **fungi** of the class Sordariomycetes. May be responsible for **plant diseases** and food **spoilage**. *Verticillium psalliotae* causes brown spots on cultured **mushrooms**, *V. dahliae* causes wilt in **potatoes**, and *V. theobromae* causes crown rot of **bananas**. *V. lecanii* is used as a **biocontrol** agent against various insect **pests**.

Vetch seeds **Seeds** produced by plants of the genus *Vicia*, especially *V. sativa*, common vetch. High in protein, making them a popular feed for ruminants. Resemble **lentils** when split, making them a potential low cost substitute for lentils in human nutrition. However, there is concern over **toxicity** to monogastric species due to the presence of **neurotoxins** such as γ-glutamyl-β-cyanoalanine and other precursors of cyanide formation. **Toxins** may be removed by appropriate **steeping** and **cooking** procedures.

Veterinary inspection Governmental surveillance of food producing animals to ensure a clean, wholesome, disease-free **meat** supply that is without adulteration. There are approximately 70 diseases that animals can transmit to man; for this reason, inspections are made by veterinarians at places of animal slaughter and at meat processing facilities.

Vibrio Genus of Gram negative, facultatively anaerobic, straight or curved rod-shaped **bacteria** of the family Vibrionaceae. Occur in freshwater and marine habitats. *Vibrio cholerae* is the causative agent of **cholera**, which is often transmitted via contaminated foods (e.g. **shellfish**) and water. *V. parahaemolyticus* and *V. vulnificus* are responsible for **gastroenteritis**, and are often transmitted via contaminated shellfish.

Vicilin One of the main **storage proteins** of legumes.

Vicine Antinutritional glycoside present in **faba beans** that can cause favism (haemolytic anaemia), thus limiting the nutritional value of these **beans**.

Video image analysis Computer-aided technique in which photographic images of a sample are analysed to give information about particle structure and dispersion.

Vienna sausages Small, cooked, smoked **sausages** often served as an hors d'oeuvre; they take their name from the city of Vienna, Austria. Traditional, Vienna sausages are twisted into a chain of links. More commonly, however, they are open-ended sausages, which are canned in brine.

Vilia Finnish **fermented milk**.

Vinasse Liquid wastes remaining in the still after **fermentation** of **beverages** such as **wines** in the manufacture of **spirits**.

Vinegar Fermented condiment that is essentially a solution of ≥4% **acetic acid**. The word is derived from the French, meaning sour wine, as vinegar was originally produced as an unwanted by-product of **winemaking**. Several types of vinegar with characteristic flavour profiles are produced by **fermentation** of various substrates, including **apples**, **cider**, **grape musts**, **wines** and **malt**. Vinegar fermentation is a 2-stage process. The initial **alcoholic fermentation** of sugars in the chosen substrate is carried out by **Saccharomyces** spp., while the **acetic fermentation** of the alcohol produced to acetic acid is carried out by **acetic acid bacteria** in the presence of O_2. Due to the acidic nature of vinegar, it is also used in acidifying agents and preservatives.

Vine leaves Leaves of grape **vines** used to wrap foods prior to cooking, as in dolmades. Also used in **salads** and garnishes. Available fresh or canned in **brines**.

Vines Plants of the genus *Vitis*, generally *V. vinifera*, which produce **grapes**. The leaves of the plants are also eaten, being used to wrap foods prior to cooking, as in dolmades, and also eaten in **salads** and garnishes.

Viniculture Alternative term for **viticulture**.

Vinification Alternative term for **winemaking**.

Vinyl chloride Flammable, possibly carcinogenic, gas which is polymerized to make **polyvinyl chloride**. Also used as a propellant in **aerosols**. Synonym chloroethene.

Vinylidene chloride Colourless liquid which is polymerized to make the thermoplastic material polyvinylidene chloride (PVDC). Synonym 1,1-dichloroethene.

Violaxanthin Xanthophyll carotenoid pigment found in **algae** and certain **fruits**, e.g. **kiwifruit, olives, grapes** and **mangoes**.

Viomellein Mycotoxin produced by species of **Aspergillus** and **Penicillium**. May be synthesized in stored **cereals** contaminated with these **fungi**.

Vioxan A preparation of the insecticide **carbaryl**.

Virginiamycin Mixture of peptolide **antibiotics** produced by *Streptomyces* *virginiae*. Primarily effective against **Gram positive bacteria**. Often used as a growth promoter in non-ruminant animals and to increase production of **eggs** in hens. Also effective against necrotic enteritis in broilers and against dysentry in swine. Not significantly absorbed by treated animals. Residues in edible tissues are normally undetected, so no withdrawal period has been established.

Viridicatin Mycotoxin produced by **fungi** of the genus *Penicillium*, including *P. cyclopium* and *P. discolor*. Strains producing the toxin have been isolated from a wide range of food types.

Viridicatol Mycotoxin produced by **fungi** of the genus *Penicillium*, including *P. cyclopium* and *P. discolor*. Strains producing the toxin have been isolated from a wide range of food types.

Viriditoxin Teratogenic mycotoxin produced by some species of *Aspergillus*, including *A. fumigatus* and *A. viridinutans*, and also by *Paecilomyces variotii*. Strains producing the toxin have been isolated from a range of agricultural commodities.

Virulence Capacity of **pathogens** to cause disease. Generally indicated by the severity of infection in the host.

Virulence factors Properties of or substances produced by **pathogens** which confer **virulence** on them. Include cell motility, and production of **cytotoxins** and **adhesins**.

Viruses Non-cellular **microorganisms** that consist of a core of **RNA** or **DNA** enclosed in a protein coat and, in some forms, a protective outer membrane. Can live and reproduce only in susceptible living microbial, plant, human and animal host cells. Causative agents of many important diseases of humans, animals and plants.

Viscera Soft internal organs of the body, usually those contained in the abdominal cavity. In animals, **fish** and **birds** processed for food, the viscera (removed by **evisceration** or **gutting**) are often discarded as waste products. However, **fish** processing wastes have shown potential for recovery of **lipids** and **proteins**.

Viscoelasticity Rheological properties relating to the reaction of a product to a stress or strain, consisting partly of a viscous element and partly of an elastic one.

Viscometers Instruments for measuring the **viscosity** of liquids. Also called viscosimeters.

Viscometry Measurement of **viscosity** of a liquid, usually performed with **viscometers**.

Viscosity Measure of the ease with which a fluid can flow when subjected to shear stress, measured in Newton seconds per square metre or Pascal seconds. Low viscosity, e.g. that of a gas, allows **flow** through a fine tube to be quite rapid, whereas high viscosity (as with thick **oils**) makes motion sluggish. Viscosity arises from the intermolecular forces in a fluid (internal friction); the stronger these forces, the greater the viscosity. With a rise in temperature, attraction between the molecules is reduced, enabling them to move more freely.

Vision systems Systems of visual feedback based on various devices, such as video cameras, photo cells, or other apparatus, allowing a robot to recognize objects or measure their characteristics. Vision systems are widely employed in **quality control** processes in the food industry.

Vital gluten Wheat protein complex separated from **starch** in a **wheat flour dough** and dried. Used to improve strength of **bread dough**.

Vitamers Group of compounds varying in structure but displaying qualitatively similar biological activities with respect to specific vitamins. Collectively referred to by the name of the vitamin involved.

Vitamin A Group of fat soluble compounds (**retinoids**) which exist in several isomeric forms and occur preformed only in foods of animal origin. The two vitamin A forms are: **retinols**, which predominate in mammals and **marine fish**; and dehydroretinols, which predominate in **freshwater fish**. Vitamin A is present in yellow and green leafy plants as **provitamin A**, of which there are several forms. The most important ones in human nutrition are the **carotenoids**, α- and β-carotene and **cryptoxanthin**. These are converted to the active vitamin in the intestinal wall and liver. Richest sources of preformed retinols are **fish liver oils**, **egg yolks** and fortified **milk**. Biologically active carotenoids are found in dark green **leafy vegetables** and yellow **fruits** and **vegetables**, such as **squashes** and **carrots**. In humans, common signs of vitamin A deficiency are poor growth, lowered resistance to infection, night blindness and rough scaly skin. Severe deficiency leads to keratomalacia and xerophthalmia.

Vitamin antagonists **Antinutritional factors** which are present in some natural foods and do not function as **vitamins**, even though they are chemically related to them. As a result, they cause vitamin deficiencies where the body is unable to distinguish them from true vitamins, and incorporates them into essential body compounds.

Vitamin B₁ Former name for **thiamin**.

Vitamin B₁₂ Synonym for **cyanocobalamin**. Member of the **vitamin B group**, found in foods of animal origin such as **livers**, **fish** and **eggs**. Vitamin B_{12} is the coenzyme for methionine synthase (EC 2.1.1.13), an enzyme important for the metabolism of **folic acid**, and methylmalonyl coenzyme A mutase (EC 5.4.99.2). **Absorption** of this vitamin requires the presence of an intrinsic factor. Failure of absorption, rather than dietary deficiency, is the major cause of pernicious anaemia.

Vitamin B$_{13}$ Synonym for **orotic acid**. An intermediate in the biosynthesis of **pyrimidines**, and growth factor for some **microorganisms**.

Vitamin B$_2$ Former name for **riboflavin**.

Vitamin B$_6$ Vitamin which exists in three forms - **pyridoxine** (the alcohol form), **pyridoxal** (the aldehyde form) and **pyridoxamine** (the amine form). The relative proportion of each of the three forms in foods varies considerably. All are equally biologically active.

Vitamin B complex Alternative term for **vitamin B group**.

Vitamin B group Group of water soluble **vitamins** generally found together in nature and basically related in function, although unrelated chemically. These include **vitamin B$_1$** (**thiamin**), **vitamin B$_2$** (**riboflavin**) the **vitamin B$_6$** group (**pyridoxine, pyridoxal** and **pyridoxamine**), the **vitamin B$_{12}$** group (the **cobalamins**), **nicotinic acid** (**niacin**), **folic acid** (pteroylglutamic acid), **pantothenic acid** and **biotin**.

Vitamin C Synonym for **ascorbic acid**, an antioxidant nutrient present in a wide range of foods. Necessary for growth of bones and teeth, for maintenance of blood vessel walls and subcutaneous tissues, and for wound healing; dietary deficiency results in scurvy. Used in **food additives**, with applications in **food antioxidants** and **bakery additives**.

Vitamin D Group of several related **sterols**. The most important members are **vitamin D$_2$** (**ergocalciferol** or **calciferol**) and **vitamin D$_3$** (**cholecalciferol**). The former is synthesized by **irradiation** of the plant provitamin **ergosterol**, and the latter is produced from the provitamin 7-dehydrocholesterol (found underneath the skin) on exposure to UV light from the sun. Vitamin D is also considered to be a prohormone. **Fish liver oils** and foods fortified with vitamin D are the major dietary sources; smaller amounts are found in **livers, egg yolks, sardine** and **salmon**. Severe deficiency in children results in rickets; deficiency in adults leads to osteomalacia.

Vitamin D$_2$ Synonym for **calciferol** and **ergocalciferol**; one of the group of **sterols** which constitute **vitamin D**. Synthesized by **irradiation** of the plant provitamin **ergosterol**.

Vitamin D$_3$ Synonym for **cholecalciferol**; one of the group of **sterols** which constitute **vitamin D**. Fat soluble vitamin necessary for formation of the skeleton and for mineral homeostasis. Produced on exposure to UV light from the sun from the provitamin 7-dehydrocholesterol, which is found in human skin.

Vitamin E Two main groups of compounds have **vitamin E** activity - **tocopherols** and **tocotrienol**. There are 4 isomers of each: α-, β-, γ- and δ-tocopherols; and α-, β-, γ- and δ-tocotrienols. Each has differing vitamin potency. Vitamin E functions primarily as an antioxidant in cell membranes, protecting **unsaturated fatty acids** from oxidative damage. Vitamin E contents of foods are expressed as mg α-tocopherol equivalent; **leafy vegetables**, seeds and most **vegetable oils** are good sources.

Vitamin E acetate Esterified form of **vitamin E** which has no **antioxidative activity** until the acetate is removed in the intestine as it is absorbed. The acetate form is more stable with respect to storage time and temperature than unesterified forms.

Vitamin F Obsolete name for **thiamin**.

Vitamin G Obsolete name for **riboflavin**.

Vitamin H Obsolete name for **biotin**.

Vitamin K Group of fat-soluble **vitamins** essential for production of prothrombin and several other **proteins** involved in the blood clotting system, and the bone protein osteocalcin. Deficiency causes impaired blood coagulation and haemorrhage; vitamin K is sometimes called the antihaemorrhagic vitamin. Two groups of compounds have vitamin K activity: **phylloquinone**, found in all green plants; and a variety of **menaquinones** synthesized by intestinal **bacteria**. Dietary deficiency is unknown, except when associated with general malabsorption diseases.

Vitamin K$_1$ Synonym for **phylloquinone**. Fat-soluble vitamin found in all green plants. Especially abundant in **alfalfa** and green **leafy vegetables**. Essential for production of prothrombin, and several other proteins involved in the blood clotting system, and the bone protein osteocalcin. Deficiency causes impaired blood coagulation and haemorrhage. Two groups of compounds have **vitamin K** activity: phylloquinones; and a variety of **menaquinones** synthesized by intestinal **bacteria**.

Vitamin K$_3$ Synonym for **menadione**. Synthetic compound with **vitamin K** activity, used in prevention and treatment of hypoprothrombinaemia, secondary to factors that limit absorption or synthesis of vitamin K. Two to three times more potent than naturally occurring vitamin K.

Vitamin K$_2$ series Synonym for **menaquinones**. Variety of metabolites with **vitamin K** activity synthesized mainly by intestinal **bacteria**. Also found in **meat, livers, eggs** and **cheese**. Formerly called farnoquinone.

Vitamin P Group of plant **bioflavonoids**, including **rutin**, **naringin**, **hesperidin**, eriodictin and citrin, which affect the strength of capillaries in the body. Bioflavonoids are found as natural **pigments** in **vegetables**, **fruits** and **cereals**. In addition to their effect on capillary fragility, it is claimed that bioflavonoids function as follows: they are active antioxidative compounds in foods; they possess a metal-chelating capacity; they have a synergistic effect on **ascorbic acid**; they possess bacteriostatic and/or antibiotic activity; and they possess anticarcinogenic activity.

Vitamin PP Obsolete name for **niacin**.

Vitamins Groups of **nutrients** which are essential in small amounts for most living organisms to maintain normal health and development.

Vitamin U Synonym for *S*-methylmethionine. A compound found in raw **cabbages**, other **green vegetables**, **beer** and **citrus juices**. Thought to assist in healing of skin ulcers and ulcers in the digestive tract; also has an effect on secretory, acid-forming and enzymic functions of the intestinal tract.

Viticulture Cultivation of **vines** for production of **winemaking grapes** or **table grapes**.

Vitreosity Extent to which a substance resembles glass with respect to properties such as hardness, brittleness, transparency and structure.

Vitrification Phenomenon whereby a substance is cooled rapidly to a low temperature such that the water it contains forms a glass-like solid without undergoing **crystallization**. The temperature at which the transition into a glassy solid occurs is the **glass transition temp.** Glass formation can result in stabilization of non-equilibrium systems, including most foods. In the glassy state, physicochemical deterioration is inhibited, effectively preserving the system. Vitrification temperature can be used as an indicator of food safety and storage stability.

Vla Dutch custard-type viscous dairy dessert made with **milk**, **carrageenans**, **modified starches** and **flavourings**.

Vodka **Spirits**, originating in Russia and northeast Europe, made from **grain** or **potatoes**. Generally rectified to have neutral **flavour** and **aroma**, but some types contain added **flavourings**.

Volatile compounds Compounds that are readily vaporized. Often have a characteristic **aroma** and are therefore often **flavour compounds** and **aroma compounds**.

Volatile fatty acids **Fatty acids** that, apart from being present in some foods, are produced by **bacteria** in the human intestine and the rumen of cattle from undigested **starch** and **dietary fibre**. To some extent, they can be absorbed and used as a source of energy. Volatile fatty acids formed in the colon may show **anticarcinogenicity**.

Volatile organic compounds Non-methane **hydrocarbons** produced as industrial pollutants.

Voltammetry Electrochemical technique in which the relationship between voltage and current flowing between electrodes in a reaction solution is measured. Utilizes a working electrode, where the reaction occurs, an auxiliary electrode for current flow and a reference electrode that is used to measure the potential of the working electrode.

Volumetric analysis **Titration** technique based on measurement of the volume of reagent required to react completely with the analyte.

Volvariella Genus of **edible fungi** that include **padi straw mushrooms** (*Volvariella volvacea*) and **straw mushrooms** (*V. diplasia*). Another widely consumed species is *V. speciosa*, easily confused with some poisonous *Amanita* species.

Volvatoxins Cardiotoxic proteins produced by ***Volvariella*** *volvacea* (straw mushrooms). Exist as volvatoxin A1 and volvatoxin A2.

Vomitoxin Synonym for **deoxynivalenol**. A trichothecene produced by ***Fusarium*** species.

W

Wafers Light, thin, crisp **biscuits** served as an accompaniment to **desserts** or **ice cream**, or eaten sandwiched together with sweet or savoury **fillings** or coated with **chocolate**.

Waffles Light, crisp, indented raised **cakes** leavened with baking **powders** or **yeasts** and typically baked in a special waffle iron, which cooks both sides simultaneously. Often consumed as a **breakfast** food, accompanied by **maple syrups**. May also be eaten as **desserts**, topped with **cream** or **ice cream**.

Wakame Common name for *Undaria* pinatifida, one of the the brown **seaweeds**. Used in **soups** and also consumed as a toasted, sugar-coated and canned product. Alternative spelling is **wakami**.

Wakami Alternative spelling of **wakame**; one of the brown **seaweeds** in the genus *Undaria*.

Walleye Freshwater fish species (*Stizostedion vitreum*) belonging to the family Percidae; distributed across North America. Flesh is highly esteemed for its **flavour** and **texture**. Cultured in some parts of North America. Marketed fresh and frozen.

Walleye pollack Alternative term for **Alaska pollack**.

Walnut oils Relatively expensive **oils** extracted from **walnuts**. The distinctive nutty **flavour** and **aroma** make them popular for use in **salad dressings**, drizzling on to cooked foods and in **cooking**. Sometimes used as an alternative to **olive oils**. To prevent development of **rancidity**, walnut oils are best stored in a cool, dry location, out of direct sunlight.

Walnuts **Nuts** produced by trees of the genus *Juglans*, the most economically important species being *J. regia* (common or Persian walnuts), *J. nigra* (black walnuts) and *J. cinerea* (**butternuts** or white walnuts). Ripe nuts are rich in **vitamin E** and B group vitamins, while younger fruits also contain **vitamin C**. Used as dessert nuts, and as ingredients in **confectionery**, **bakery products** and **ice cream**. **Oils** extracted from the nuts contain a high proportion of **unsaturated fatty acids** and have a range of food uses.

Walruses Large, carnivorous marine mammals (*Odobenus rosmarus*) belonging to the family Odobenidae in the order Pinnipedia. They are hunted for their **meat**, particularly by the northern Inuit and Indian communities in the Canadian Arctic and northern coastal British Columbia regions. Characteristics of walrus meat include: a high content of protein, with a biological value similar to that of **beef**; a darker **colour** than beef; and a distinctive **flavour**. Walrus **blubber** forms a part of traditional diets in some areas, but may be associated with health risks due to bioaccumulation of **organochlorine pesticides** and other **contaminants**. In Arctic regions, **trichinosis** is commonly associated with consumption of raw or inadequately cooked walrus meat.

Warehouses Large buildings in which raw materials or manufactured goods are stored.

Warmed over flavour Characteristic **off flavour** primarily associated with cooked **meat** and **poultry meat** in chilled **ready meals** and other **cook chill foods**. In cooked meat and poultry held at chilled storage temperatures, this stale, oxidized **flavour** becomes apparent within a short time (48 hours), particularly if the product is stored under air. **Modified atmosphere packaging** under low oxygen levels helps to delay the onset of oxidative warmed over flavour.

Warming The process by which an item is heated slightly to the point of being warm.

Wasabi Pungent spices produced from the roots of *Wasabia japonica*. Used most commonly in Japanese cuisine and for flavouring of **condiments**. Also known as Japanese horseradish.

Wastes Unusable, unwanted or discarded materials. In the food industry, wastes can result from application of processing procedures, and consist of solids such as **pomaces**, **feathers** and **sludges**. By recycling, some materials in wastes can be reclaimed for further use.

Waste water Unusable, discarded water (**effluents**) resulting from processing procedures. In the food industry, waste water is commonly produced by **breweries**, **dairies**, **distilleries**, olive oil mills and palm oil mills. Must be disposed of safely, often after treatment, to minimize pollution.

Water Colourless, odourless and tasteless liquid with the chemical formula H_2O, which is essential for plant and animal survival. Widely drunk as a beverage, usually after some form of **disinfection**. Used in the food and beverage industries in many ways, including as an ingredient, in the form of process water, and in cooling and heating systems.

Water activity Measure of the **water vapour** generated by the moisture present in a hygroscopic product. Defined as the ratio of the partial pressure of water vapour to the partial pressure of water vapour above pure water at the same temperature. In foods, it represents water not bound to food molecules; the level of unbound water has marked effects on the chemical, microbiological and enzymic stability of foods. Commonly abbreviated to a_w.

Water binding capacity Extent to which a substance can bind water.

Water convolvulus Common name for *Ipomoea aquatica*, a plant grown in China, Taiwan and Vietnam, also known as **water spinach**. Stems and leaves are eaten as vegetables, either boiled or stir fried; stems are also used as ingredients in **pickles**.

Watercore Internal defect that affects mainly **apples**, but also **pears** and sometimes other **fruits**. Characterized by water-soaked appearance of some or all of the flesh.

Watercress Dark green leafy plant (*Nasturtium officinale* or *N. microphyllum* x *officinale*). Rich in proteins, **iron**, **carotenes** and **vitamin C**; also contains **vitamin E**, group B vitamins and other minerals. Used in **salads**, garnishes, **soups** and cooked as a vegetable.

Water dropwort Common name for *Oenanthe stolonifera*. Young shoots and leaves from the plant are used in China as **flavourings** for **fish soups** and **poultry** dishes.

Waterfowl Wetland **birds** such as ducks, geese and swans, which belong to the order Anseriformes. The term is most commonly used for wetland **game birds**, some of which are hunted for their **meat**.

Water holding capacity Extent to which a substance can hold and retain water. Related to the solubility of the sample.

Water ices Frozen **sugar confectionery** made from water and sugar and flavoured with **fruit juices**, **fruit purees** or other fruit **flavourings**. Used to make some types of **ice lollies**.

Wateriness One of the **sensory properties**; relating to the extent to which a product is watery, i.e. runny and wet.

Watermelons Large globose or oblong **fruits** produced by *Citrullus lanatus* or *C. vulgaris*. Good source of **vitamin A** and **vitamin C**. **Colour** of rind and flesh varies according to variety. Flesh contains numerous **seeds** that are rich in **proteins** and **oils**, and can be eaten dry or roasted.

Watermelon seeds Seeds from **watermelons** of the genus *Citrullus*. Mature seeds are roasted and salted for consumption as **snack foods** and have potential use as **oilseeds**.

Water pollution Contamination of water resources with substances (usually toxic chemicals or waste matter) which can be harmful to organisms living in the water, or to those that drink it or are otherwise exposed to it.

Water sorption Attachment of water to a substance.

Water spinach Common name for *Ipomoea aquatica*, a plant native to India and South East Asia but grown widely in other regions. Due to its invasive and aggressive nature, the plant poses a serious threat to waterways in the southern USA and is considered a noxious weed. Stems and leaves are eaten as a vegetable, often stir fried. Rich in **proteins** and **minerals**, especially **iron**. Alternative names include swamp cabbage and **water convolvulus**.

Water stress Condition caused in plants by lack of sufficient water for growth, as in drought. Can have adverse effects on growth and quality of edible plant parts, e.g. **fruits** and leaves.

Water supplies **Drinking water** supplied to the public and industry by a water supply company or authority.

Water vapour Water that is in its gaseous state, especially when below its boiling point.

Wax beans Type of **common beans** (*Phaseolus vulgaris*).

Wax coatings Wax-based materials used to coat and preserve the quality of **fruits** and some types of **cheese**.

Waxes White translucent materials including **beeswax**, but also a wide variety of similar viscous substances, such as **carnauba wax**. Used as **coatings** for foods or to make candles and polishes.

Wax esters Long-chain **fatty acid esters** present in **vegetable oils** which can also be synthesized by **lipases**, either from free **fatty acids** or through degradation of **triacylglycerols**.

Wax gourds Juicy-textured **fruits** of *Benincasa hispida* that are used as vegetables. Can be stir-fried, used in preparation of sweet **pickles**, added to **soups**, or stuffed with meat or vegetables and steamed. Also known as ash **pumpkins**, ash **gourds**, Chinese fuzzy **gourds** and Chinese preserving **melons**.

Weaning Process of gradually replacing mother's milk or milk substitute with other types of food in the diet of an infant or other young mammal. For infants, **weaning foods** are initially of a puree-like **consistency** and are often based on **cereals**, but other textures and types of food are introduced as the process proceeds.

Weaning foods **Infant foods** used during the transition from consuming solely **human milk** or **infant formulas** to introduction of a mixed diet. Types of weaning food differ widely between cultures, but initial weaning foods are frequently based on **cereals**, of a puree-like **consistency**, and are introduced individually in order to detect **allergies** to particular foods.

Weevils Common name for various **insects** of the family Curculionidae. Also known as snout **beetles**. Often highly destructive pests of crops and stored cereal grains, e.g. the **alfalfa** weevil (*Hypera postica*), the **grain** weevil (*Sitophilus granarius*) and the **rice** weevil (*S. oryzae*). Larvae of some species can be destructive to **fruits**, **nuts** and **grain**.

Weighing Process of determining the weight of an object.

Weighing machines Devices, also called scales, used to determine the weight of an object. The simplest weighing mechanism is the equal-arm balance, which consists of a bar with a pan hanging from each end and a support (fulcrum) at the centre of the bar. Precision balances used in scientific laboratories can measure the weight of small amounts of material down to the nearest 1 millionth of a gram. Such weighing machines are enclosed in glass or plastic to prevent wind drafts and temperature variations from affecting the measurements. Electronic scales, which use electricity to measure loads, are faster and generally more accurate than their mechanical counterparts; in addition, they can be incorporated into computer systems, which makes them more useful and efficient than mechanical scales.

Weissella Relatively newly identified genus of **Leuconostoc**-like **lactic acid bacteria**. Includes *Weissella paramesenteroides*, which was formerly know as *L. paramesenteroides* and also other closely related species. The type species is *Weissella viridescens*. Found in a range of foods, including **fermented foods**.

Weisswurst White German **sausages** made with **veal**, **cream** and **eggs**. Eaten fried or poached and traditionally served in Germany during the Oktoberfest, accompanied by sweet mustard, **rye bread** and **beer**.

Well water Water derived from wells. May be used as **drinking water**.

Welsh onions Common name for *Allium fistulosum*. Rich in **vitamin C**; also contains a range of other **vitamins**, **carotenes** and group B vitamins. Very small bulbs, but hollow, cylindrical leaves that are used is **salads** and **soups**. The whole plant may be cooked. Also known as Japanese leeks, Japanese bunching onions, ciboule and cibol.

Western blotting Method for detecting specific proteins. Proteins are separated by gel electrophoresis and transferred to a suitable matrix (e.g. nitrocellulose or PVDF), on which the proteins bind in a pattern identical to that on the original gel. After blotting, target molecules are detected through the use of labelled **antibodies** specific for the proteins of interest. Alternatively, proteins can be detected through the use of specific, unlabelled primary antibodies followed by addition of labelled secondary anti-antibodies.

Wet milling Process for separation of a substance into its constituent parts by a combination of chemical and mechanical means. Used mainly in processing of **corn**, but can also be applied to other **cereals** such as **sorghum**, **wheat** and **rice**. Cereals are steeped in water with or without sulfur dioxide to soften the kernels before removal of the **germ** and separation of the other components. The main product is **starch**, which can be further processed in the case of corn to manufacture **sweeteners** or **ethanol**. Other products include **fibre**, **gluten** and **oils**, such as **corn fibre oils**.

Wettability One of the **physical properties**; relating to the ability of a solid to absorb a liquid, such as water, as it spreads over the surface of the solid.

Whale meat **Meat** from **whales**, which is eaten in Japan, Norway, Iceland, Greenland, the Faroe Islands and other Arctic regions. For example, Eskimos living in whaling villages consume raw, frozen, boiled and fried whale meat; they also eat mekiqag, a whale meat product, prepared by very slow cooking of the meat in its own juices. There is growing recognition that whale products may contain high concentrations of toxic chemicals, such as **heavy metals** and **organochlorine compounds**.

Whale oils **Oils** derived from the blubber of **whales** of the order Cetacea. Contain **wax esters** and **triacylglycerols**. Uses include the manufacture of **margarines**. Also known as spermaceti.

Whales Large, air-breathing marine mammals belonging to seven families, namely: Delphinidae, Physeteridae, Monodontidae, Ziphiidae, Eschrichtidae, Balaenopteridae and Balaenidae. Many species of whales have been killed in large numbers by commercial whalers and are now rare. They are hunted to provide **whale meat**, **blubber**, **whale oils** and edible **offal**. Whale products are traditional foods to some ethnic groups, e.g. the Eskimos, for whom the most important whale parts are whale meat and muktuk (a layer of blubber with skin attached).

Wheat Grain of cereal grasses belonging to the genus *Triticum* (particularly *T. aestivum*, and *T. durum*) which contains **gluten**, a protein complex important for the **breadmaking** properties of this grain. Used to make many food products, including **pasta** and **breakfast cereals**; wheat flour is used widely to make bakery products such as **biscuits**, **cakes** and **bread**.

Wheat beer **Beer** made from **mashes** derived wholly or partially from **wheat malt**, rather than the more common **barley malt**.

Wheat bran Protective outer layer of the wheat grain which is removed from commercial **flour** by bolting or **sifting**. Added to foods such as **breakfast cereals** or **bread** as a source of **fibre**.

Wheat bread Bread made from **wheat flour**. White wheat breads are made from finely sifted wheat flour, while whole wheat bread is prepared by incorporating the fibre-rich outer layers of the wheat grain.

Wheat breadmaking Process by which **bread** is made from **wheat**.

Wheat dough Unbaked, thick, plastic mixture of **wheat flour** and a liquid, such as water or **milk**. May contain **yeasts** or **baking powders** as leavening agents. Used predominantly to make **bread**; dough used to make other products, e.g. **pizzas**, **biscuits**, **noodles**, may vary in composition from **bread dough**.

Wheat fibre Fibre extracted from **wheat**.

Wheat flour Product resulting from grinding **wheat** grains. **Wholemeal** flours are obtained by grinding whole wheat grains, while white flour is produced by separating **wheat germ** and **wheat bran** from the endosperm. Used to prepare a range of **bakery products** such as **bread**, **cakes** and **biscuits**.

Wheat germ Vitamin- and lipid-rich embryo (sprouting portion) of the wheat grain. **Milling** of grain to produce white **wheat flour** results in separation of the germ, which may then be used to enrich **bread** and **breakfast cereals**. Also used in **dietary supplements**.

Wheat germ oils Oils extracted from seeds of wheat (*Triticum aestivum*). Rich in **linoleic acid** and **tocopherols**; also contain α-linolenic acid.

Wheat gluten Complex formed when wheat **proteins** are mixed with water. Consists of **glutenin** and **gliadins**. **Gluten** forms an elastic network during **kneading** of **dough**, which is important for the **texture** of the **bread**. Gluten content of **wheat** varies among varieties.

Wheat malt Germinated wheat grains used in **brewing** and **distillation**.

Wheat starch Starch isolated from **wheat**.

Whelks Shellfish, including several species of marine gastropod **molluscs** of the family Buccinidae; worldwide distribution. Flesh of many species is tenderized by pounding prior to consumption. Commercially important species include *Buccinum undatum* (common whelks) and *Neptunea antiqua* (red whelks). Marketed fresh (in shell; cooked or uncooked), semi-preserved (in **vinegar** and **salt**) and canned.

Whey Liquid formed by **coagulation** of **milk** during **cheesemaking**. The solid portion (**curd**) is processed further to make **cheese**. Whey is sometimes used in making **whey cheese**, but is produced in large amounts as a waste, disposal of which poses problems for the dairy industry. Although mainly used in animal feeds, whey can be utilized as an ingredient in some foods and as a fermentation substrate. Also called serum or lactoserum.

Whey beverages Drinks, sometimes **sports drinks** or nutritional beverages for specific population groups, based on **whey**. Can be alcoholic or non-alcoholic.

Whey cheese Cheese prepared by concentrating **whey** and coagulating the **proteins** with heat and **acids**. The resulting curd is strained and possibly pressed. **Milk** or **cream** may be added to increase fat content or improve cheese **flavour**. **Ricotta cheese** is a well-known whey cheese.

Whey concentrates Concentrates prepared from **whey**. Used in a variety of foods to supplement their nutritional value. Uses include preparation of **sports foods** and **sports drinks**, and dietetic products.

Whey protein concentrates Products prepared from **whey** by separation of **whey proteins** using **precipitation** or **ultrafiltration**. Precipitation at a high temperature and low pH followed by **centrifugation** produces a concentrate of denatured, insoluble whey proteins. Ultrafiltration followed by vacuum evaporation and spray drying produces a concentrate of non-denatured, soluble proteins. Concentrates varying in composition can be made by controlling manufacturing conditions. Uses include adjustment of protein contents of various products, including **infant formulas**, dietetic products and protein-enriched foods for specific groups of people, e.g. athletes. **Foaming properties** of whey protein concentrates make them suitable for use in aerated foods and as replacements for **egg whites**.

Whey proteins Milk proteins that remain in **whey** after manufacture of **cheese**. Sometimes called serum proteins. Consist of albumins (**α-lactalbumin** and serum albumin) and globulins (mainly **β-lactoglobulin**).

Whipped cream Cream in which the volume has been increased (overrun) by 90-100% by whipping in air. Available commercially in aerosol cans, the product containing **sugar** in addition to cream.

Whipping Beating of ingredients, particularly **cream** and **egg whites**, during which air is incorporated into them, increasing their volume and creating a froth.

Whipping capacity The extent to which a food can be whipped.

Whipping cream **Cream** with a fat content of approximately 34% that can be whipped to approximately double its volume.

Whipping properties **Functional properties** relating to the ability of a food to be whipped, increasing the volume by incorporation of air.

Whiskey Alternative spelling of **whisky**. This spelling is generally used for Irish and American whiskies. **Spirits** made by **distillation** of fermented **mashes** made from saccharified **cereals**, using raw materials, distillation conditions and ageing periods as specified by national regulations for the specific whiskey type.

Whisky Alternative spelling of **whiskey**. This spelling is commonly used for Scotch and Canadian whiskies. **Spirits** made by **distillation** of fermented **mashes** made from saccharified **cereals**, using raw materials, distillation conditions and ageing periods as specified by national regulations for the specific whisky type.

White amur Alternative term for **grass carp**.

Whitebait General name used for young **marine fish** of various **herring**-like species, including *Clupea harengus* (Atlantic herring) and *Sprattus sprattus* (European sprat). Often consumed as a fried product, sometimes in **batters** (whitebait **fritters**).

White beans Type of **common beans** (*Phaseolus vulgaris*).

White cabbages Variety of *Brassica oleracea*. **Cabbages** with white heads that mature in winter.

White cheese **Fresh cheese** that is either uncured or only slightly cured. High moisture content and perishable.

Whitecurrants White **berries** produced by *Ribes sativum*. Rich in **vitamin C**. Eaten out of hand or as components of **preserves, jellies** and **sauces**.

White fish General name referring to white-fleshed **marine fish** in which the main fat reserves are in the **livers**, particularly gadoid species such as **cod, haddock, whiting** and **coalfish**.

Whitefish Any of several marine and **freshwater fish** within the genera *Coregonus* and *Prosopium*; distributed in the North Atlantic or in lakes across northern Europe and North America. Commercially important species include *Coregonus clupeaformis* (lake whitefish) and *C. albula* (**vendace**). Marketed fresh and frozen.

White lupins Common name for the white-flowered plant *Lupinus albus* or *L. termis*. Pods contain large, off-white seeds that are rich in **proteins** and **oils**. Seeds are sometimes used as **coffee substitutes** and their flour as a replacement for **soy meal**. Potentially toxic **alkaloids** in **lupin seeds** are removed by washing in water.

White mould cheese Creamy and smooth **cheese** with white *Penicillium* mould grown on the outside.

White mustard Common name for *Sinapis alba*, seeds of which are ground to produce **spices**. When reconstituted with water, the spice develops a pungent **aroma** due to formation of **allyl isothiocyanate**. **Turmeric** is often added to the mustard powders to produce a bright yellow coloration, leading to the alternative name, **yellow mustard**.

Whiteners Substances used to whiten or bleach foods such as **flour** or **fish**. May be used as substitutes for fresh **milk** in beverages including **coffee** (**coffee whiteners**), **tea** or **cocoa**, or in **sauces**. Available as liquids or powders. These are prepared from **milk proteins** or non-dairy proteins (e.g. **soy proteins**) and **fats**, blended with other ingredients such as **sugar, emulsifiers, stabilizers**, buffers, **flavourings** and **colorants**.

Whiteness One of the **optical properties**; relating to the extent to which an item is white, i.e. snowy and milky in appearance.

White pepper Common name for *Piper nigrum*, fruit of which are ground to produce **spices**. Compared with **black pepper**, which is produced from fully grown, but unripe, fruit of *P. nigrum*, white pepper has a more delicate **flavour**. The major flavour compound of white pepper is **piperine**.

White pickled cheese **White cheese** pickled in **brines**. Alternative term for brine ripened **cheese**.

White sugar Purified crystalline **sugar** containing approximately 1% moisture. Dried to produce **granulated sugar**.

White truffles Common name for *Tuber* magnatum, a truffle growing in the Piedmont region of Italy. **Colour** ranges from white, sometimes veined with pink, to grey-brown, depending on the type of tree on whose roots it originates. Eaten raw, shaved over egg dishes, **pasta**, risotto, **salads** and other light warm dishes just before serving to intensify the **wild garlic**-like **aroma**. Due to their superior aroma and scarcity, white truffles are very expensive.

White tuna Generally refers to flesh from the **albacore** (*Thunnus alalunga*), which is lighter-coloured than flesh from other **tuna**.

White whales Alternative term for **beluga whales**.

White wines Wines with a white to golden yellow **colour**. May be made from white **winemaking grapes** or alternatively from red winemaking grapes by a technique which avoids extraction of **anthocyanins** from the **grape skins**.

Whiting Name given to a variety of **marine fish** species, the majority being in the cod and hake families (Gadidae and Merlucciidae). Particularly refers to *Merlangius merlangus*, a commercially important species found in the north Atlantic Ocean. Marketed fresh and frozen (whole, or single and block fillets) and as smoked or canned products.

WHO Abbreviation for **World Health Organization**.

Wholegrain foods Foods made from whole, unrefined grains or wholegrain ingredients. Wholegrains contain the entire edible parts of a grain kernel, i.e. the **germ**, endosperm and **bran**, and are rich in many nutrients which are generally lost during refining. In addition, wholegrains are low in fat and **cholesterol**. Wholegrain foods include **wholemeal bakery products** and **pasta**, some **breakfast cereals** and **brown rice**. Consumption of wholegrain foods has been associated with a number of health benefits including reduced risks of developing certain cancers and heart disease.

Wholemeal **Flour** or **bread** made from the entire cereal grain with none of the **bran** or **germ** removed.

Whole milk Milk from which none of the fat has been removed. Fat content of milk varies according to species, being approximately 4% in **cow milk**. Milk is also available in other forms from which some (**semi skimmed milk**) or almost all (**skim milk**) of the fat has been removed. These other forms are preferred by some consumers wishing to limit their intake of **fats**.

Wieners Cooked, smoked **frankfurters**, which take their name from the city of Vienna (Wien), Austria. Some wieners are prepared in edible natural casings; these **sausages** are often considered more traditional, and tend to cost more than skinless varieties. Traditionally, wieners are braided in groups of links.

Wild boar meat Meat from **wild boars**. It is similar to **pork**, but has a redder **colour**, a lower content of fat and a stronger **flavour**. It may be infested with larvae of *Trichinella spiralis* and therefore must be cooked thoroughly before eating to prevent **trichinosis**.

Wild boars Wild **swine** (*Sus scrofa*) of the family Suidae from which most domestic swine have been bred. They are hunted for **wild boar meat**.

Wild cabbage Type of *Brassica oleracea* that grows wild on coastal cliffs. Evolved into many varieties grown for their edible stem, leaves, buds or flowers.

Wildebeests Large African **antelopes** belonging to the genus *Connochaetes*; they are also known as gnus. There are two species, namely the white-tailed gnu (*C. gnou*), which is now a protected species, and the blue wildebeest or brindled gnu (*C. taurinus*). They are hunted for their meat, particularly in East Africa where controlled culling is carried out to harvest wildebeest meat.

Wild garlic Wild plants of the genus *Allium* used in **flavourings** or as a vegetable, and having beneficial effects on health. Commonly consumed species include *A. ursinum* and *A. victorialis*.

Wild mushrooms **Mushrooms** that grow in the wild and are prized for their exotic **flavour**. Since many wild species are poisonous, great care must be taken to identify the edible species when picking them.

Wild rice Long grain aquatic grass with a nutty **flavour**. Chinese wild rice is *Oryza latifolia*, while North American wild rice is produced by plants of the genus *Zizania*. Due to the high costs of this cereal, it is often eaten mixed with other **rice** varieties or **bulgur** wheat.

Wild vegetables **Plants** that are harvested from the wild rather than being cultivated and are eaten as **vegetables**.

Wild yeasts Naturally-occurring strains of **yeasts**.

Wine coolers **Beverages** made by blending **wines** with other ingredients, including water, **fruit juices**, **sugar**, **flavourings** and ice.

Wine distillates Intermediate products or finished **spirits** made by **distillation** of **wines**.

Wine gums **Sugar confectionery** products with a chewy texture made with **sucrose**, **glucose** and either **gum arabic** or **gelatin**. Often fruit-flavoured. Similar to **fruit gums** and to fruit **jellies**, although the latter are softer due to a higher moisture content.

Winemaking Process of manufacture of **wines**. The basic process comprises crushing **grapes**, **alcoholic fermentation** of the **grape juices** and ageing of the **wines**. Many additional processes may be applied, including **maceration**, **clarification**, **chaptalization**, **filtration**, **fining** and, in the case of **sparkling winemaking**, secondary **fermentation**.

Winemaking grapes Grape cultivars used primarily for **winemaking**, and having characteristics making them especially suitable for this application. Mainly *Vitis vinifera*, but other *Vitis* spp. or their hybrids with *V. vinifera* are also used for winemaking.

Wineries Industrial establishments where **wines** are manufactured.

Wines Alcoholic beverages manufactured by **alcoholic fermentation** of fruit **musts** or **fruit juices**. Generally refers to beverages produced from **grapes** (*Vitis* spp., mainly *V. vinifera*). **Fruit wines**

are made from other fruit musts or juices. The term wines may also be used to refer to **rice wines** (made from saccharified **rice mashes**), and **palm wines** (made from palm sap).

Wines manufacture   Alternative term for **wine-making**.

Wine vinegar   **Vinegar** produced by **acetic fermentation** of **wines**, e.g. **red wines**, **white wines** or **sherry**. Wine vinegar has a wine-like **flavour** and is used more as a flavouring than as a condiment, e.g. as an ingredient of **salad dressings**.

Wine yeasts   **Yeasts** used for **fermentation** of **grape musts** to produce **wines**. May be spontaneously occurring yeasts, or pure yeasts cultures. Mainly *Saccharomyces* spp., although other genera of yeasts may play a role in the early stages of **fermentation**.

Winged beans   **Beans** produced by *Psophocarpus tetragonolobus*. Rich in protein. As well as the **seeds**, immature green pods, leaves and root **tubers** of the plant are eaten. Also known as **goa beans** and **asparagus peas**.

Winnowers   Devices for blowing air through grain in order to remove the chaff. Winnowing is also used to separate the shell and some of the germ from **cocoa beans** during manufacture of **chocolate**.

Winterization   Removal of traces of **waxes** and higher melting **glycerides**, or **stearin**, from **fats**. Waxes are generally removed by rapid **chilling** and **filtration**. Separation of **stearin** usually requires very slow **cooling** in order to form crystals that are large enough to be removed by filtration or **centrifugation**. **Cottonseed oils** and **groundnut oils** are winterized to produce **salad oils** that remain liquid at low temperatures. **Tallow** and other animal fats are winterized for simultaneous production of hard **fats** and oleo oil. Also known as destearination.

Withering   Process whereby plant material or foods become dry and shrivelled. Controlled withering can be undertaken either chemically or physically (including techniques such as freeze withering, solar withering and warm air withering). Withering is commonly the first stage in the processing of **teas**. In some regions, **wines** are made from **grapes** which have been partially dried by withering in the sun before pressing.

Witloof   Type of **chicory**.

Wood   Hard fibrous material which forms the main substance of the branches and trunk of trees. Used as a packaging material, particularly for making wooden **barrels**, **baskets**, **crates** and some **fibreboard**. Physicochemical properties of wood have major effects on the **aroma** and **flavour** of alcoholic beverages stored and/or aged in wooden barrels, or foods exposed to **wood smoke** during processing.

Woodcock   Long-billed **game birds** of the sandpiper family. Valued highly as a food. Includes the American woodcock (*Philohela minor*) and the European woodcock (*Scolopax rusticula*).

Wood pigeons   Eurasian **pigeons** that may be hunted as **game birds** for their **meat**. Synonymous with the ring dove (*Columba palumbus*).

Wood smoke   Smoke produced from the burning of wood. The type of wood used (e.g. oak, hickory, mesquite) influences the properties of the smoke and governs its application. Used in **flavourings** and/or **preservatives**. Foods which are commonly processed using smoke include **fish** and **meat**. **Smoke flavourings** may be added to barbecue **sauces** or **marinades**.

Woolliness   Extent to which products, usually **fruits**, have a woolly texture, i.e. are dry and spongy. Woolliness is an adverse sensory property and physiological disorder, involving lack of **juiciness**, internal **browning** and inability to ripen, without variation in tissue moisture. It is associated with an imbalance in pectolytic enzymic activity during storage. Onset of woolliness can be quantified instrumentally and is characterized as a lack of **crispness**, low hardness values and low juiciness.

Worcestershire sauces   **Condiments** produced by **fermentation** with **yeasts** of a mixture of **fruit juices**, **vegetable juices**, **syrups** and **amino acids**.

World Health Organization   The World Health Organization (WHO) is a specialized agency of the United Nations (UN) that helps countries to improve their health services and coordinates international action against diseases.

World Trade Organization   The World Trade Organization (WTO) is an international body based in Geneva, Switzerland, that promotes and enforces the provisions of trade laws and regulations. The WTO has the authority to administer and police new and existing free trade agreements, to oversee world trade practices, and to settle trade disputes among member states. The WTO was established in 1994 when the members of the **General Agreement on Trade and Tariffs** (GATT), a treaty and international trade organization, signed a new trade pact. The WTO was created to replace GATT, and began operation on 1 January 1995. The WTO has a significantly broader scope than GATT, expanding the GATT agreement to include trade in services and protections for intellectual property. 128 nations were contracting parties to the new GATT pact at the end of 1994, and became members of the WTO. By early 2000, the WTO had 136 members, and about 30 other countries had applied for membership. The WTO is controlled by a general council made up of member states' ambassadors who also serve on various subsidiary and specialist committees. The ministerial conference, which meets every two years and appoints the WTO's director-general, oversees the General Council.

Wormwood Common name for *Artemesia absinthium*, leaves and flowering tops of which are used to produce spices. Wormwood has a bitter **flavour**. It is used in **natural flavourings** for **vermouths**. *A. absinthium* extracts and **essential oils** are also used as **flavourings**. The plant also contains α-thujone, which is a convulsant at high concentrations; hence, in some countries such as the USA, foods and **beverages** containing wormwood are permitted only if **thujone** is not present.

Worts Clarified extracts prepared from **mashes** based on **malt**, sometimes with addition of **brewing adjuncts**, and subsequently fermented to form **beer**. Worts are generally boiled with **hops** to extract hop **bitter compounds**.

Wrapping Packaging, e.g. paper or soft material, used to cover or protect a food, particularly during retail and after selection by the consumer.

Wreckfish **Marine fish** species (*Polyprion americanus*) belonging to the family Polyprionidae and of minor commercial importance. Distributed in the Atlantic Ocean, western Indian Ocean and southwest Pacific Ocean. Marketed fresh, frozen or cooked in a variety of ways.

Wuerstel Small sized, frankfurter style **sausages**, traditionally made in Italy. They have high fats content.

X

Xanthan Extracellular heteropolysaccharide produced by **Xanthomonas** *campestris*. Uses in the food industry include in **gelling agents**, gel **stabilizers**, **thickeners** and crystallization inhibitors.

Xanthan gums **Gums** produced by the bacterium **Xanthomonas** *campestris*. These gums are **exopolysaccharides** composed of repeating pentasaccharide units comprising a cellulose backbone and trisaccharide side chains of D-mannose and D-glucuronic acid residues. The gums also contain variable quantities of **pyruvic acid**. Used widely in the food industry as **thickeners** due to their ability to produce highly viscous, highly stable aqueous solutions. Other uses include as **emulsifiers**, **stabilizers** and **binding agents**, and to provide **body**, e.g. in **low fat foods**.

Xanthene dyes **Pigments** derived from xanthene. Examples of those used as food **colorants** include **rose bengal, erythrosine** and **phloxine**.

Xanthine dehydrogenases EC 1.1.1.204. **Enzymes** that catalyse the conversion of **hypoxanthine** to xanthine and the further oxidation of xanthine to **uric acid**. Also act on a variety of **purines** and **aldehydes**. Major proteins of bovine **milk fat globule membranes**. Animal enzymes can be converted to **xanthine oxidases** by storage at -20°C or by treatment with **proteinases**, organic solvents or thiol reagents. In animal **livers**, the enzyme exists mainly as the dehydrogenase, but in other tissues it is found almost entirely in the form of xanthine oxidase.

Xanthine oxidases EC 1.1.3.22. **Enzymes** that convert xanthine to **uric acid** and H_2O_2, but also oxidize **hypoxanthine**, and certain **purines**, pterins and **aldehydes**. Under certain conditions, toxic superoxides are generated rather than peroxides; **green tea** and **seaweeds** extracts, together with certain **flavonoids**, have been found to inhibit this process. In **milk**, these enzymes are thought to constitute a natural bacterial defence mechanism, since they can produce nitric oxide radicals with **antibacterial activity**. Xanthine oxidases have also been used in **biosensors** for evaluating the **freshness** of **meat**.

Xanthohumol Prenylated chalcone present in **hops** and **beer**. Possesses a range of properties beneficial for health, including **antioxidative activity**, anti-carcinogenicity, **antimutagenicity** and protection against **osteoporosis** and **atherosclerosis**.

Xanthomegnin Hepatotoxic mycotoxin produced by certain species of **Aspergillus**, **Penicillium** and *Trichophyton*.

Xanthomonas Genus of Gram negative, aerobic rod-shaped **bacteria**. Several species are plant pathogens (e.g. *Xanthomonas campestris, X. fragariae, X. ampelina* and *X. abilineans*). *X. campestris* causes black rot of **cabbages** and **cauliflowers**, common blight of **beans**, and bacterial spot of **tomatoes** and **peppers**. *X. campestris* is also used in the production of **xanthan gums**. Several species may cause **spoilage** of raw chilled **meat**, **fish** and egg products.

Xanthophylls Group of neutral yellow or brown carotenoid **pigments** that are oxygenated derivatives of **carotenes** and distributed widely in plants. Useful as food **colorants**.

Xanthotoxin Furanocoumarin toxin produced by **celery** in response to infection by certain **fungi** and **bacteria** and after various stress treatments. Consumption or contact with affected celery can cause phototoxic skin reactions or bullous dermatitis.

Xenobiotics Substances that are foreign to living organisms. Can be synthetic or naturally occurring compounds. Examples include **drugs**, **pesticides** and **carcinogens**.

Xerocomus Genus of **edible fungi**, commonly consumed species including *Xerocomus badius* and *X. subtomentosus*.

X-ray fluorescence spectroscopy **Spectroscopy** technique in which the sample is irradiated with X-rays, causing emission of a characteristic X-ray photon and fluorescence, which is measured using a spectrophotometer.

X-rays Penetrating electromagnetic radiation of very short wavelength, able to pass through many materials. X-rays are produced by bombarding a target, usually made of tungsten, with high-speed electrons. The shorter the wavelength of the X-ray, the greater is its energy and its penetrating power. Longer wavelengths, near the UV-ray band of the electromagnetic spectrum, are known as soft X-rays. The shorter wavelengths, closer to and overlapping the gamma-ray range, are called hard X-rays. A mixture of many different wavelengths is known as white X-rays, as opposed to monochromatic X-rays, which represent only a single wavelength. X-rays are used in the food industry for a wide range of analytical purposes, including detection of foreign bodies and contaminants in manufactured foods.

Xylan Polysaccharide found in the cell walls of plants, where it forms the bulk of the **hemicelluloses** component. Consists of (1→4)-β-linked D-xylose residues with side chains of other sugars, such as (4-*O*-methyl)-α-D-glucopyranosyluronic acid and α-L-arabinofuranosyl residues. (1→3)-linkages may also be present and the molecule may be acetylated.

Xylanases Alternative term for **xylan degrading enzymes** and **xylan endo-1,3-β-xylosidases**.

Xylan degrading enzymes General term for **glycosidases** that hydrolyse and degrade **xylan**.

Xylan endo-1,3-β-xylosidases EC 3.2.1.32. **Xylan degrading enzymes** that catalyse the random hydrolysis of 1,3-β-D-xylosidic linkages in 1,3-β-D-xylans. Useful as **dough** improvers and volume-increasing agents in **bread** and **bakery products**, and for **wheat starch** separation. Also known as xylanases and endo-1,3-β-xylanases.

Xylan 1,4-β-xylosidases EC 3.2.1.37. **Xylan degrading enzymes** that hydrolyse 1,4-β-D-xylans, removing successive D-xylose residues from the non-reducing termini. Also known as xylobiases, β-xylosidases and exo-1,4-β-xylosidases, these enzymes are useful for utilization of **xylan**-containing substrates. Also hydrolyse xylobiose.

Xylitol Naturally occurring polyol comprising 5 carbon atoms which has equivalent **sweetness** to **sucrose**. Manufactured by **hydrogenation** of **xylose**. Used in **sweeteners**, especially for **low sugar confectionery**, since it is non-cariogenic.

Xylitol dehydrogenases Alternative term for **D-xylulose reductases**.

Xylobiases Alternative term for **xylan 1,4-β-xylosidases**.

Xyloglucans Polysaccharides found in the **hemicelluloses** component of plant cell walls. Consist of (1→4)-linked **glucose** residues, most of which have a **xylose** residue side chain attached. **Galactose, arabinose** and **fucose** may also be present.

Xylooligosaccharides Oligosaccharides that contain **xylose** residues. Useful as **sweeteners** and as prebiotics. Thought to be indigestible, and **animal models** have suggested that they may reduce serum cholesterol levels and repress peroxidation of lipids induced by a high cholesterol diet.

Xylose Aldose monosaccharide comprising 5 carbon atoms which may be produced by hydrolysis of **xylan**. Substrate for manufacture of **xylitol** and **xylulose**. Has approximately 0.7 times the **sweetness** of **sucrose** and is used as a sweetener for **diabetic foods**.

Xylose isomerases EC 5.3.1.5. **Enzymes** that catalyse the **isomerization** of D-xylose and D-xylulose. Also isomerize D-ribose and D-glucose, and are useful for isomerization of **glucose** to **fructose** in the production of **fructose high corn syrups**. The name **glucose isomerases** is still widely used for these enzymes.

Xylose reductases Alternative term for **aldehyde reductases**.

β-Xylosidases Alternative term for **xylan 1,4-β-xylosidases**.

Xylulose Ketose monosaccharide comprising 5 carbon atoms (pentose) that is an isomer of **xylose**. May be formed by aldose-ketose **isomerization** of xylose using bacterial **xylose isomerases**.

D-Xylulose reductases EC 1.1.1.9. **Enzymes** involved in **fermentation** of **xylose** and production of **xylitol** from **xylulose**. These enzymes have been expressed in *Saccharomyces cerevisiae*, together with **aldehyde reductases** and xylulokinases, and recombinant cells have been used for fermentation of xylose to **ethanol**. They have also been used in xylitol **biosensors** for on-line control of xylitol production by **yeasts**.

Y

Yacon Edible tubers of *Polymnia sonchifolia* that are usually eaten raw. Contain high contents of **inulin**. The sweetish water chestnut-like **flavour** develops after exposure to the sun for a few days. Used for production of **alcohol** and **sweeteners**.

Yakifu Japanese bakery product made by mixing **gluten** with **starch** or **wheat flour** and **baking**.

Yakju Alcoholic beverages of the **rice wines** type, produced in Korea.

Yak meat Meat from **yaks**. It has higher protein, **thiamin**, iron, potassium and sodium contents, and lower fat and **riboflavin** contents than **beef**. In sensory terms, yak meat is described as very juicy, but sweetish, with a metallic off-flavour, due to its high iron content.

Yak milk Milk obtained from yaks, and drunk predominantly in Tibet, but also in Mongolia and India. Pink in **colour**. In Tibet, it is processed into 3 products: crispy oil (butter oil produced by separation of milk and used in cooking and making butter tea); sour milk made from **whole milk** or the **skim milk** remaining after removal of crispy oil; and milk solids residue resulting from boiling **skim milk** (made into yak milk **cheese**).

Yaks Large stocky ruminants belonging to the Bovidae family. Wild yaks are a protected species, but domesticated yaks (*Bos grunniens*) are reared to provide **yak meat**, **yak milk**, hair and hides. The domesticated yak is the dominant dairy animal in the pastoral areas of the Qinghai-Tibet plateau in China.

Yakult Brand of fermented **skim milk** drink containing live *Lactobacillus* casei strain Shirota. It is one of the **probiotic foods** that is drunk to help maintain the health of the **gastrointestinal tract**.

Yam beans Common name for tubers of *Pachyrrhizus erosus*. Thinly sliced tubers are eaten raw in **salads** or cooked in dishes such as **soups** and stews. Used as a substitute for water chestnuts. Young pods of the plants may also be eaten.

Yams Starchy underground tubers of the genus *Dioscorea*. Good source of potassium and zinc, but contain only small amounts of **vitamin C**; yellow-fleshed varieties contain **carotenes**. Eaten cooked in the same way as **potatoes**, and sometimes processed into fufu. **Flavour** resembles that of potatoes. In the USA the name is used for **sweet potatoes** with orange flesh.

Yam starch Starch isolated from **yams**.

Yard-long beans Alternative term for **asparagus beans**.

Yarrow Pungent, aromatic herb of the genus *Achillea*, especially *A. millefolium*. Used sparingly in **salads** and **soups**, and to make **herb tea**.

Yarrowia Genus of **fungi** of the class Saccharomycetes. *Yarrowia lipolytica* is responsible for **spoilage** of certain foods, e.g. **yoghurt**, **butter**, **margarines**, **meat mince** and **cheese**.

Yeast biomass Quantitative estimate of the total population of **yeasts** present in a given habitat, in terms of mass, volume or energy.

Yeast extracts Water soluble fraction of autolysed **yeasts**. During autolysis, yeast **enzymes** hydrolyse cytoplasmic proteins and carbohydrates. Insoluble cell wall material (cellulose) is removed, e.g. by centrifugation, to leave a clear extract of water soluble cellular material that is rich in **amino acids** and other **nutrients**. Yeast extracts are used as **flavourings**, as a source of nutrients for microbial **fermentation**, and as a source of B group **vitamins** for **fortification** of foods.

Yeast proteins Proteins produced by **yeasts**.

Yeasts Unicellular **fungi** of the phylum Ascomycota that reproduce by fission or budding, and are capable of fermenting **carbohydrates** into **alcohol** and carbon dioxide. Some are responsible for food **spoilage**, while others are economically important as agents in **breadmaking**, **brewing** and **winemaking**, and in the production of **single cell proteins**, B vitamins and other **fermentation products**.

Yellowfin Alternative term for **yellowfin tuna**.

Yellowfin tuna Marine fish species (*Thunnus albacares*), which forms the second largest part of the world **tuna** catch after **skipjack tuna**; widely distributed across the Atlantic and Pacific oceans. Marketed mainly as a canned product, but also sold fresh, frozen, dried, salted and as a semi-preserved product.

Yellow fish General name used for salted cold smoked white **fish fillets**, which usually develop a yellow colour after smoking; particularly refers to smoked **haddock**.

Yellow mustard Synonym for **white mustard** (produced from seeds of *Sinapis alba*), to which **turmeric** has been added to produce a bright yellow coloration.

Yellow perch **Freshwater fish** species (*Perca flavescens*) of commercial importance belonging to the family Percidae. Widely distributed in rivers and lakes of America and Canada. Marketed fresh or frozen and cooked by pan **frying**, **broiling** or **baking**.

Yellowtail Any of several **marine fish** species of the genus *Seriola* (family Carangidae); distributed across warmer regions of the Atlantic and Pacific oceans. The most important food fish species in commercial terms is *S. quinqueradiata*, which is cultured on a large scale in Japan. Marketed fresh, salted and dried; also canned (smoked flesh packed in **oils**). Also known as amberjack.

Yerba mate Tree (*Ilex paraguariensis*) which grows in South America, the leaves and twigs of which are dried, seasoned and made into a popular local infusion beverage, called **mate** or sometimes yerba mate.

Yersinia Genus of Gram negative, facultatively anaerobic rod-shaped **bacteria** of the family **Enterobacteriaceae**. Occur in soil and water, and in the gastrointestinal tracts of animals (e.g. **swine**, beavers and squirrels). *Yersinia enterocolitica* and *Y. pseudotuberculosis* are the causative agents of **yersiniosis**.

Yersiniosis Disease of humans or animals caused by *Yersinia enterocolitica* and *Y. pseudotuberculosis*. Frequently characterized by **gastroenteritis** with diarrhoea and/or vomiting, and accompanying fever and abdominal pain. Transmission in humans is usually via ingestion of contaminated water and foods (e.g. **meat**, **fish**, **shellfish**, **milk**, **dairy products**, **fruits** and **vegetables**).

Yessotoxins Class of **shellfish toxins** produced by **dinoflagellates**. Can produce enterotoxic effects in humans following ingestion of molluscan shellfish (e.g. **clams**, **mussels**, **oysters** and **scallops**) which filter feed on these dinoflagellates.

Yield stress Stress at which the yield strength of a material is exceeded and elastic behaviour gives way to viscous behaviour. If continued, the stress may lead to failure stress, beyond which failure occurs. Measured in Newtons.

Ymer Danish **fermented milk**.

Yoghurt **Fermented milk** of creamy texture that can be prepared from milk of many species, but most often is made from **cow milk**. Can be made from **whole milk**, **semi skimmed milk** or **skim milk**, in a range of thicknesses, stirred or set, and in plain or flavoured varieties. **Flavoured yoghurt** is mixed with **sugar** and **flavourings** or **fruits**. Also made into **frozen yoghurt**, a product resembling soft serve ice cream. Commercially, yoghurt is made using **yoghurt starters** (generally *Lactobacillus* bulgaricus and *Streptococcus* thermophilus). Other bacteria beneficial to gastrointestinal health, e.g. *L. acidophilus* and *Bifidobacterium bifidum*, may also be added. **Pasteurization** destroys the bacteria in yoghurt; unpasteurized product is known as live yoghurt. Yoghurt is rich in **calcium** and **iodine** and a source of protein and B vitamins. Many spelling variants for yoghurt are used in various parts of the world, including yogurt, yoghourt and yogourt.

Yoghurt beverages Drinks based on yoghurt. Include many **health beverages** as well as fruit containing beverages such as smoothies.

Yoghurt starters Microbial cultures inoculated into **milk** to produce **acidity** by **fermentation** during manufacture of **yoghurt**. Commercial starter preparations generally contain *Lactobacillus* bulgaricus and *Streptococcus* thermophilus.

Yokan Japanese **confectionery** products made with **agar** (gelling agent), **sugar** and **adzuki beans** paste, together with **persimmons** and **chestnuts**, which are used as **flavourings**.

Youngberries Dark red **berries** produced by *Rubus ursinus*, a hybrid between **dewberries** and **blackberries**.

Yuba Product made from the skin that forms on the surface of **soymilk** during heating. The skin is hung up to dry in sheets or sticks. Used in **meat substitutes**, wrapped round other foods or eaten alone after **deep frying**.

Yucca Trees belonging to the genus *Yucca* which grow mainly in the USA and Mexico. Extracts of some species, especially *Y. brevifolia* and *Y. schidigera*, are used as **foaming agents** in foods and **beverages**, including root beer, cocktail mixes and whipped drinks. Yucca extracts are also used as feed additives.

Yukwa Traditional Korean snack food made by **deep frying** gelatinized waxy **rice dough**, which has previously been steamed, punched and moulded.

Yusho Disease caused by ingestion of **edible oils** which became contaminated with **polychlorinated biphenyls** (PCB) on the Japanese island of Kyushu in 1968.

Yuzu **Citrus fruits** (*Citrus junos*) cultivated mainly for the rind which has a characteristic **aroma** and is used as a garnish or **flavour** enhancer in a variety of dishes. Source of **essential oils**.

Z

Zabadi **Fermented milk** resembling **yoghurt** that is popular in the Middle East. Sometimes served as a dessert with thick **syrups**. Alternative term for **zabady**.

Zabady Alternative term for **zabadi**.

Zearalenol Alcohol derivative of **zearalenone** with oestrogenic activity, which may be used as an anabolic growth promoter in food-producing animals. Use is banned in some countries. Animals may carry out *in vivo* metabolic conversion of zearalenone to zearalenol. Also known as **F2 toxin**.

Zearalenone Synonym for **F2 toxin**. A mycotoxin produced by *Fusarium graminearum*, *F. culmorum* and other ***Fusarium*** species. May be formed when the fungus grows on damp cereal **grain** (e.g. **wheat**, **barley** and **corn**) used as animal feeds. Has oestrogenic activity and can cause hyperoestrogenism in **swine**, **cattle** and **poultry**.

Zeatin Naturally occurring cytokinin derived from **adenine** which plays a role in the growth and development of plants.

Zeaxanthin Member the **xanthophylls** group of carotenoid **pigments** and an isomer of **lutein**. May contribute to visual health. Found in many plants, certain **algae** and **egg yolks**, and used in food **colorants**.

Zedoary Common name for *Curcuma zedoaria*, a plant related to **turmeric**. Young rhizomes are eaten as a vegetable. The dried rhizome is pulverized and used as a spice. Used as a condiment and in manufacture of **flavourings** and bitters. Also known as **shoti**.

Zefir Traditional Russian foamed **confectionery** products, similar to **meringues**.

Zein Prolamin which accounts for approximately half of the total **storage proteins** in corn. Contains minimal concentrations of **lysine** and **tryptophan**, but is rich in **leucine**.

Zeleny values Indicators of **wheat** protein quality for **breadmaking**, providing estimates based on sedimentation of swollen **gluten** and **starch** suspended in a solution of **lactic acid**.

Zeolites Crystalline, hydrated alkali-aluminium silicates. Useful as **catalysts** for production of **invert** sugar from **sucrose**, **downstream processing** of **flavour compounds**, detoxification of contaminated foods and feeds, and as molecular sieves.

Zeranol Anabolic growth promoter with oestrogenic activity which may be used in food-producing animals. Use has been banned in the EU since 1988. May be formed in animals by *in vivo* metabolism of ***Fusarium*** species **mycotoxins** (e.g. **zearalenone**) present in feeds.

Zinc Essential trace element, chemical symbol Zn. Important for growth and is part of the active site of many **enzymes**, where it is usually required for activity.

Zineb Foliar fungicide used for control of downy mildew, **blights** and other fungal diseases in **leafy vegetables, potatoes, tomatoes, berries, stone fruits** and **pome fruits**. Classified by Environmental Protection Agency as not acutely toxic (Environmental Protection Agency IV).

Zingerone One of the primary **pungent principles** of **ginger**, displaying **antioxidative activity**.

Ziram Foliar fungicide used for control of many fungal diseases in a wide range of **fruits** and **vegetables**. Also applied to plants as an animal repellent. Classified by WHO as slightly toxic (WHO III).

Zireh Name used in some parts of the world for **black cumin** (*Nigella sativa*). The dark brown crescent-shaped fruits or seeds are used as a spice and as the source of **essential oils** rich in monoterpene **aldehydes** and terpene **hydrocarbons** such as **cuminaldehyde** and γ-**terpinene**.

Zn Chemical symbol for **zinc**.

Zolone Alternative term for the insecticide **phosalone**.

Zoonoses A group of infectious and parasitic **diseases** which are transmissible from animals to man, e.g. **brucellosis, salmonellosis** and **trichinosis**. Many disease organisms affect only humans or particular animals; however, zoonotic organisms can adapt themselves to many different species.

Z-Trim Thermally stable fat substitute derived from insoluble **fibre** from **cereals** or legume **hulls** used in **bakery products** and **dairy products**.

Zucchini   Alternative (US) name for **courgettes**.

Zucchini squashes   Alternative term for **courgettes**.

Zwieback   Sweetened **bread** originating from Germany. The **dough** contains **eggs** and **butter** and is baked, sliced and baked a second time to form a type of **rusks**.

Zygosaccharomyces   Genus of **fungi** of the class Saccharomycetes. *Zygosaccharomyces rouxii* is responsible for **spoilage** of certain foods (e.g. **musts, fruit juice concentrates, confectionery** and **honeys**), and is important in the manufacture of **miso, soy sauces** and **ogi**. *Z. bailii* causes spoilage of **mayonnaise, salad dressings, pickles, mustard, ketchups, carbonated beverages** and some **wines**.

Zymomonas   Genus of Gram negative, facultatively anaerobic rod-shaped **bacteria** of the family Sphingomonadaceae. Occur in fermenting **beverages** and plants. Some species (e.g. *Zymomonas anaerobia*) may cause **spoilage** of **alcoholic beverages** (e.g. **cider** and **beer**). *Z. mobilis* is used in the production of **pulque** and **palm wines**. **Levansucrases** produced by *Z. mobilis* are used in hydrolysis of **sucrose** to **levans** and **ethanol**.

APPENDIX A: THE GREEK ALPHABET

Letter		
Upper case	Lower case	Name
A	α	alpha
B	β	beta
Γ	γ	gamma
Δ	δ	delta
E	ε	epsilon
Z	ζ	zeta
H	η	eta
Θ	θ	theta
I	ι	iota
K	κ	kappa
Λ	λ	lambda
M	μ	mu
N	ν	nu
Ξ	ξ	xi
O	o	omicron
Π	π	pi
P	ρ	rho
Σ	σ	sigma
T	τ	tau
Υ	υ	upsilon
Φ	φ	phi
X	χ	chi
Ψ	ψ	psi
Ω	ω	omega

APPENDIX B: SCIENTIFIC SOCIETIES AND ORGANISATIONS IN THE FOOD SCIENCES

International Union of Food Science and Technology (IUFoST)

The International Union of Food Science and Technology is the world organisation for food science and technology and is a full scientific member of the International Council for Sciences (ICSU). The chief aims of IUFoST are to promote international cooperation, support international progress, advance technology, stimulate education and teaching, and to foster professionalism and professional organisation.

Contact information

IUFoST Secretariat
PO Box 61021
No. 19, 511 Maplegrove Road
Oakville, Ontario
Canada L6J 6X0
Phone: +1 905 815 1926
Fax: +1 905 815 1574
Email: secretariat@iufost.org
Web site: www.iufost.org

IUFoST has four regional bodies:

European Federation of Food Science and Technology (EFFoST)

The European Federation of Food Science and Technology is a regional grouping of IUFoST which acts as an international umbrella organisation for European food science and technology. Its primary purposes are to provide a framework for cooperation among national, regional and global professional societies and their members; enhance food science and technology competencies in Europe; and improve public understanding of food science and technology.

Contact information

EFFoST Secretariat,
c/o A&F
Bornsesteeg 59
PO Box 17
6700 AA Wageningen
The Netherlands
Phone: +31 317 475 000
Fax: +31 317 475 347
Email: info@effost.org
Web site: www.effost.org

The Working Party on IUFoST Activity in Africa (ECSAAFoST)

This body works with other national institutes in food science and technology in Africa to carry out at least one IUFoST sponsored activity in the region each year. An activity may be a conference or symposium or participation in an education programme. The objective is to increase awareness of the role of the Union and it is also a means of identifying how IUFoST can be most effective in the region.

Federation of Institutes of Food Science and Technology in ASEAN (FIFSTA)

Activities of this body include joint efforts between seven Asian countries to develop a vital interest in and help set standards for the food industry, through conferences, committees addressing specific regional issues, and workshops.

FIFSTA aims to promote cooperation and exchange of scientific and technical information among scientists, food technologists and specialists; support progress in both theoretical and applied areas of food science; advance technology in the processing, manufacturing, preservation, storage and distribution of food products; stimulate appropriate education and training in food science and technology; and foster professionalism and professional organization among food scientists and technologists.

Asociación Latinoamericano y del Caribe en Ciencia y Tecnología de Alimentos (ALACCTA)

ALACCTA unites Latin-American and Caribbean Associations of Food Science and Technology. A regional seminar is organized every other year as well as international courses with experts from all over the world. Sixteen countries have joined the Association and on going efforts are being made to pool research efforts and find scholarship funds. Details of member organizations can be found on the ALACCTA web site, www.publitec.com/alaccta.htm.

Selected national scientific societies for food science and technology

Australia
Australian Institute of Food Science and Technology Inc.
Suite 2, Level 2, 191 Botany Road
Waterloo, NSW 2017, Australia
Phone: +61 2 8399 3996
Fax: +61 2 8399 3997
Web site: www.aifst.asn.au

Canada
Canadian Institute of Food Science And Technology
3-1750 The Queensway Suite 1311
Toronto, Ontario M9C 5H5, Canada
Phone: +1 905 271 8338
Fax: +1 905 271 8344
Email: cifst@cifst.ca
Web site: www.cifst.ca

China
Chinese Institute of Food Science and Technology
Room 201 Zhongke Mansion, No.75 Deng shikou Street
Dongcheng District, Beijing, P. R. China 100006
Phone: +86 10 652 65374
Fax: +86 10 652 64731
Email: cifst@public.bta.net.cn
Web site: www.cifst.org.cn

Italy
Associazone Italiana di Technologia Alimentare
Strada Farini, 31
43100 Parma, Italy
Phone & Fax: + 39 521 230 507
Email: aitaer@tin.it
Web site: www.aitaer.com

Japan
Japanese Society for Food Science and Technology
Shokuhin Sogo Kenkyujo
2-1-12 Kannondai
Tsukuba-shi
Ibaraki 305-8642, Japan
Phone: +81 298 38 8116
Fax: +81 298-38-7153
Email: info@jsfst.or.jp
Web site: www.jsfst.or.jp

Singapore
Singapore Institute of Food Science and Technology
Singapore Professional Centre (SPC) Peoples Association West Blk, Room 4, Stadium Link
Singapore 397750
Phone: +65 62 568 890
Email: info@sifst.org.sg
Web site: www.sifst.org.sg

South Africa
The South African Association for Food Science and Technology
van der Walt & Co.
Attn.: Mr Jean Venter
P O Box 868
2160 Ferndale, South Africa
Phone: +27 11 789 1384
Fax: +27 11 789 1385
Email: saafost@vdw.co.za
Web site: www.saafost.org.za

UK
Institute of Food Science & Technology
5 Cambridge Court, 210 Shepherd's Bush Road
London W6 7NJ, UK
Phone +44 20 7603 6316
Fax: +44 20 7602 9936
Email: info@ifst.org
Web site: www.ifst.org

USA
Institute of Food Technologists
525 W. Van Buren, Ste. 1000
Chicago, IL 60607, USA
Phone: +1 312.782.8424
Fax: +1 312.782.8348
Email: info@ift.org
Web site: www.ift.org

APPENDIX C: WEB RESOURCES IN THE FOOD SCIENCES

Shown below is a collection of web resources of relevance to the food science, food technology and human nutrition communities. For details of food-related societies, please see Appendix B.

Food Science Central
http://www.foodsciencecentral.com
A gateway to free and subscription based information relating to the world of food science, food technology and food-related human nutrition. The site includes feature articles, reports on important papers published in leading food science journals, and details of products and services offered by IFIS Publishing.

FSTA Direct
http://www.fstadirect.com
Offers web access to *FSTA – Food Science and Technology Abstracts®*, a database composed of an extensive collection of abstracts prepared from the world's food science, food technology and food-related human nutrition literature.

Arbor Nutrition Guide
http://arborcom.com
A gateway providing a wealth of nutrition-related resources on the web classified under the headings Applied Nutrition, Food Science, Clinical Nutrition and Food. Topics arranged below these headings may be browsed, while specific sites may be located using the site search engine.

CAB International (CABI)
http://www.cabi.org
This site serves to further CABI's aims of generating, disseminating and using knowledge in the applied biosciences to enhance development, human welfare and the environment.

Campden and Chorleywood Food Research Association (CCFRA)
http://www.campden.co.uk
CCFRA Group is the UK's largest independent membership-based organization carrying out research and development for the food and drinks industry worldwide. Its website includes details of current research, member services, legislation information and training timetables.

Codex Alimentarius
http://www.codexalimentarius.net
Website of the Codex Alimentarius Commission which aims to develop food standards, guidelines and related texts such as codes of practice under the Joint FAO/WHO Food Standards Programme.

UK Department for Environmental, Food and Rural Affairs (DEFRA)
http://www.defra.gov.uk
DEFRA's remit is the pursuit of sustainable development, weaving together economic, social and environmental concerns. Information on the DEFRA website aims to further this outlook.

Deutsche Landwirtschafts Gesellschaft eV (DLG)
http://www.dlg.org
DLG is one of the key organizations in the German agricultural and food sector and aims to translate scientific findings into practice. The website details current research programmes and events.

Food and Agriculture Organization (FAO)

http://www.fao.org

This site acts to further the FAO's goals of leading international efforts to defeat hunger, with particular reference to: putting information within reach; sharing policy expertise; providing a meeting place for nations; and bringing knowledge to the field.

US Food and Drug Administration (FDA)

http://www.fda.gov

The FDA is responsible for protecting the public health by assuring the safety, efficacy and security of human and veterinary drugs, biological products, medical devices, food, cosmetics, and products that emit radiation. The website includes information on hot topics, reference materials and FDA-regulated products.

Food Law

http://www.foodlaw.rdg.ac.uk/index.htm

Provides resources on UK, European and international legislation including food additives, labelling and hygiene.

Food and Nutrition Information Center (FNIC)

http://www.nal.usda.gov/fnic/

The FNIC web site provides a directory to credible, accurate, and practical resources for a wide audience. Visitors can find material such as printable format educational materials, government reports and research papers.

Food Navigator

http://www.foodnavigator.com

A specialized news service, broadcast as a free access website, as well as e-newsletters to registered subscribers, which is built around a proactive news agenda that adds value to product announcements.

UK Food Standards Agency (FSA)

http://www.food.gov.uk

The FSA provides advice and information to the public and Government on food safety, nutrition and diet. Its website includes information on a variety of topics including food labelling, genetically modified foods and BSE.

Institute of Food Research (IFR)

http://www.ifr.ac.uk

The IFR is concerned with the safety and quality of food, and improving diet and health in people. Its website provides resources on food science topics, information sheets, IFR publications and news releases.

International Food Information Council (IFIC)

http://www.ific.org

This site aims to provide a resource on food safety and nutrition and communicate science-based information to health and nutrition professionals, educators, journalists, government officials and consumers.

International Portal on Food Safety, Animal and Plant Health

http://www.ipfsaph.org

Developed by FAO, this portal provides a single access point for authorized official international and national information across the sectors of food safety, animal and plant health.

Just Food

http://www.just-food.com

A rapidly growing food trade website providing instant access to over 1500 reports, books and research products from leading market information providers, as well as news, industry announcements, feature articles and discussion forums.

Leatherhead Food International

http://www.lfra.co.uk

Leatherhead Food International is a global and independent provider of food information, market intelligence and technical and food research services. Its website details the different products and services on offer.

US National Agricultural Library

http://warp.nal.usda.gov/

The National Agricultural Library is one of the world's largest and most accessible agricultural research libraries and plays a vital role in supporting research, education, and applied agriculture. The website provides online access to its library catalogue, AGRICOLA, as well as to details of publications and services.

US Department of Agriculture (USDA)

http://www.usda.gov

The USDA aims to provide leadership on food, agriculture, natural resources and related issues based on sound public policy, the best available science and efficient management. This website offers information about the USDA's agencies and offices and allows users to browse the site either by type of audience or subject.

World Food Net

http://www.worldfoodnet.com

Described as an online gathering place for the international food processing and supply industry, the site hosts an online suppliers directory and buyers guides, as well as the latest product launches and links to upcoming events.

World Health Organization (WHO)

http://www.who.int

Published by the WHO, the United Nations specialized agency for health, this website provides health-related details on member countries, together with information on specific health topics, WHO publications and research tools.